# Biology: Exploring Life

# Biology: Exploring Life

**GIL D. BRUM**
Professor of Biology and Botany
California State Polytechnic University, Pomona

**LARRY K. McKANE**
Professor of Biology and Microbiology
California State Polytechnic University, Pomona

John Wiley & Sons

New York        Chichester        Brisbane        Toronto        Singapore

PRODUCTION, DESIGN AND ART DIRECTION
Hudson River Studio
    Dominique Fitch-Febvrel
    Susan Posmentier
    John Sovjani

ILLUSTRATIONS
David Mascaro

Blaise Zito Associates, Inc.
Viga Jesionowski, Art Coordinator
Barbara Niemczyc
Danielle Dubray
Alfred Corring
Simon Chan

ILLUSTRATORS
Robert Villani
George Kelvin
Howard Friedman
Briar Lee Mitchell
Manabu Saito
Henry Iken

Copyright © 1989, by John Wiley & Sons, Inc.

All rights reserved. Published simultaneously in Canada.

Reproduction or translation of any part of
this work beyond that permitted by Sections
107 and 108 of the 1976 United States Copyright
Act without the permission of the copyright
owner is unlawful. Requests for permission
or further information should be addressed to
the Permissions Department, John Wiley & Sons.

*Library of Congress Cataloging in Publication Data:*

Brum, Gilbert D.
    Biology : exploring life.

    Bibliography: p.
    1. Biology.   I. McKane, Larry.   II. Title.

QH308.2.B78   1989          574          88-20877
ISBN 0-471-81339-7

Printed in the United States of America

10  9  8  7  6  5  4  3  2  1

# *Dedication*

To The Student
We hope this text helps you grow and discover the thrill of exploring life.

To Our Families
Your love and understanding helped shape this book. We're home again.

# Preface

**A Process Oriented Book *for College and University Introductory Biology***

***Biology: Exploring Life*** guides the student on a captivating journey of exploration and discovery into the world of life. We have reversed the traditional accent on static structures and terminology in favor of highlighting the *processes* of biology; bringing biology to life by emphasizing the ongoing quest for scientific understanding and the dynamic processes of life.

Our goal is to become partners with the instructor in teaching biology to college students. The biology teacher's greatest pedagogic asset is the basic desire of students to understand themselves as well as the natural world around them. We seek to arouse this fascination, building knowledge and insight that will enable them to make "real life" judgements as modern biology takes on greater significance in everyday life.

## A Balanced Treatment

Biology is more than a compilation of facts and structures to be memorized. At the foundation of biology are basic concepts that tie facts and terms into a meaningful and integrated whole. By balancing concepts and facts, we help students better visualize the integrated processes of life so they come away not only knowing "what", but also "how".

Biology is not only about humans; it is about all life. We use the student's natural interest in human biology (and other animals) as a backdrop to describe many biological concepts, but we balance this with an equal emphasis on plants and microorganisms, areas often slighted in introductory texts. This balanced approach enables the student to fully appreciate and understand the diversity of life and the critical interactions necessary for perpetuating life on earth.

## A Student-Centered Book

This book is for the student. It is a teaching-learning facilitator, designed to first pique interest and then reveal new facts and concepts. This is a book that sparks curiosity and makes students think. By arousing curiosity, building a foundation, and creating a desire to know more, students are better able to synthesize information, see natural relationships, and gain a sense of accomplishment and fulfillment as they acquire new insights and knowledge about the natural world.

Engaging the student in active learning is best achieved by biologists, not by professional writers. People who *do* biology know how science works and have the insight and broad knowledge needed to make connections, inspire true understanding, invent analogies with sharper clarity, inject more relevant examples, and challenge the student's intellect.

## Learning Strategies

Our goal has been to write visually, painting verbal images that help students grasp the process and essence of biology. In addition to proven techniques, we have introduced many new pedagogic strategies that make this book an effective teaching tool.

- **Connections Series**—revealing the interconnectedness of all areas of biology. This innovation goes beyond cross-referencing and accomplishes far more than identifying biology's "unifying themes." The Connections Series is devoted to the major concepts that resurface throughout biology. The series includes:

  — ATP
  — Homeostasis
  — Totipotency
  — Multicellular
      Organization
  — AIDS and other
      Virus Infections
  — Cancer and the Cell Cycle
  — Cancer and Gene Regulation
  — Evidence for Evolution
  — Plasmids and Gene Transfer
  — The advantages and
      drawbacks of sexual
      reproduction

  An "articulation icon" visually connects related topics that may be separated by hundreds of pages. By assembling connected facts and concepts into an integrated whole, the student forges a deeper understanding of biology.

- **Species Concept Series**—clarifies the "definition" of a species. Beginning with an overview, the sequence allows students to "grow into" a more complete understanding of what a species is, while acquiring insight into why definitions of a species differ from kingdom to kingdom, and sometimes from biologist to biologist.

- **Discovery Series**—explores specific eposides in the quest for knowledge. Students acquire insight into the *process of science* and the growth of scientific ideas and explanations. The reader experiences the dramas of scientific endeavor, while gaining an appreciation for the way science is performed in the real world.

- **Animal Behavior Series.** Each aspect of animal behavior is integrated into the chapters where they naturally fit, next to the biological principles that form the foundations of the behavior. The neurological basis of behavior, for example, is reinforced by discussing it with the nervous system rather than in a far removed chapter devoted to behavior alone.

- **Biolines** — highlight biology's fascinating facts, applications, and "real life" lessons. Each bioline enhances interest while providing additional information that enlivens the "mainstream" of biological information. Many are remarkable stories with unexpected twists that reveal nature to be as inventive and interesting as any imaginative novelist.

- **Synopsis — beyond Summaries and Key Terms.** The end-of-chapter material continues to teach by integrating and clustering material from the chapter into two effective innovations.

   *Main Concepts* — reemphasizes the major principles, underlying trends and emerging themes in the chapter. Such understanding facilitates all levels of learning.

   *Key Term Integrator* — clusters related terms in statements that reinforce the relationships between terms by joining them into meaningful units. It is not a glossary, but an organizer that conveys meaning beyond mere definitions.

- **From Overview to Understanding.** Each chapter begins with an outline that visually depicts the chapters organization, and ends with **Review and Synthesis** questions and activities that reinforce, challenge, allow for self-assessment, encourage problem solving, and promote deeper understanding.

- **New Images with Improved Meaning.** Biology is visual and dynamic. In addition to realistic three-dimensional renderings, our illustrations add motion that reveals the process of biology, as well as making it more inviting and enjoyable to learn. You will not find the same old diagrams rehashed in this book. The use of full color throughout the text, combined with recent breakthroughs in understanding and how students learn has enabled us to open new doors to improve the teaching value of each illustration. Virtually every color in an illustration has a purpose. For example, we use color to help the student follow complex chemical pathways, to trace the flow of energy, electrons, and nutrients, to recognize recurring objects and processes, and to reinforce the location of the important molecules of life. This is the first book to standardize all objects and colors throughout, making it easier for students to mentally cross-reference related material presented many chapters apart. (The cell membrane in Chapter 4 looks exactly the same as the cell membrane in Chapters 6, 7, and 10.)

- **Standardized Representations.** Many structures and processes recur in biology, remainders of the unity of life. We have invented a recognizable icon to identify each of these repeating images. Our standard icons are presented on the following page for easy reference:

| **Icon** | **Description** |
|---|---|
|  | ATP |
|  | Energized state |
|  | Energized electrons |
|  | Hydrogen bonds |
|  | Energy-rich phosphate bond |
|  | Phospholipids |
|  | Electron transport, gradual loss of energy as energy-rich electrons pass from cytochrome to cytochrome |
|  | High-energy arrow |
|  | carbon |
|  | hydrogen |
|  | oxygen |
|  | nitrogen |
|  | sulfur |
|  | phosphate |

# Acknowledgments

The evolution of this text has spread over 5 years and involved many people. Although it is impossible to mention each by name, we wish to acknowledge the talent and dedication of each of them here. This text evolved through two distinct phases—a development phase and a production phase—both of which were critical to getting a book of this ambition, vision, and quality in your hands. We extend our sincerest gratitude and thanks to the following people:

## The Development Phase

Cliff Mills, who shared our vision and enthusiasm, and helped gather the resources and generate the creative spirit that laid the foundation for this project;

Fred Corey, who saw the potential and launched this incredible journey;

Katie Vignery, who skillfully guided the project through the "rough period" and helped us realize our potential;

Serje Seminoff and John Wiley and Sons, Inc. for having confidence in us and providing the support that enabled this project to reach new levels.

## The Production Phase

Rick Leyh, who kept the project alive and helped it grow;

Linda Larrabee, who, despite being handed the editorial reins late in the project, orchestrated the many details needed to carry this complex project to completion;

Stella Kupferberg, who knows how vital photographs are to understanding biology and helped us uncover photographs with style, grace, and meaning;

Hudson River Studio, David Mascaro and Blaise Zito Studio for skillfully helping us turn our manuscript and sketches into a perfect blend of science and art; and

Ed Burke, our designer, production manager, art consultant, and friend, thank you for a work of beauty and magic.

Finally, the expertise and diligence of the following reviewers greatly enriched this book; your efforts helped make this book as clear and as accurate as possible:

Alan Atherly, Iowa State University
Edmund Bedecarrax, City College of San Francisco
Gerald Bergtrom, University of Wisconsin at Milwaukee

Richard Bliss, Yuba College
Richard Boohar, University of Nebraska at Lincoln
Clyde Bottrell, Tarrant County Junior College
J. D. Brammer, North Dakota State University
Allyn Bregman, SUNY at New Paltz
Gary Brusca, Humboldt State University
Jack Burk, California State University, Fullerton
Marvin Cantor, California State University,
    Northridge
David Cotter, Georgia College
Loren Denny, Southwest Missouri State University
Thomas Emmel, University of Florida
Alan Feduccia, University of North Carolina
    at Chapel Hill
Michael Gains, University of Kansas
S. K. Gangwere, Wayne State University
Paul Goldstein, University of North Carolina
    at Charlotte
Frederick Hagerman, Ohio State University
W. R. Hawkins, Mt. San Antonio College
Vernon Hendricks, Brevard Community College
Paul Hertz, Barnard College
Howard Hetzel, Illinois State University
Duane Jeffery, Brigham Young University
Claudia Jones, University of Pittsburgh
John Kmetz, Kean College of New Jersey
Lawrence Levine, Wayne State University
Jon Maki, Eastern Kentucky University
William McEowen, Mesa Community College
Roger Milkman, University of Iowa
Helen Miller, Oklahoma State University
John Mutchmor, Iowa State University
Mary Ann Phillippi, Southern Illinois
    University at Carbondale
R. Douglas Powers, Boston College
Charles Ralph, Colorado State University
Aryan Roest, California State Polytechnic
    University, San Luis Obispo
Robert Romans, Bowling Green State University
Richard G. Rose, West Valley College
A. G. Scarbrough, Towson State University
John Schmitt, Ohio State University
Marilyn Shopper, Johnson County Community College
Ralph Sulerud, Augsburg College
Tom Terry, University of Connecticut
James Thorp, Cornell University
Michael Torelli, University of California, Davis
Terry Trobec, Oakton Community College
Len Troncale, California State Polytechnic
    University, Pomona
Richard Van Norman, University of Utah
David Whitenberg, Southwest Texas State University
P. Kelly Williams, University of Dayton

# Brief Contents

# Contents

# 1

## *Unifying Themes*

To understand life you must explore the obvious and the subtle, as well as, all levels in between.

Chapter 1

# Biology: Exploring Life

You became a biologist long before you opened this book. It happens to everyone sometime during childhood, a time that marks an awakening of the fascination we all have for living things, for **organisms**. We explore ant colonies and inspect sprouting seeds. We watch hypnotically as a spider spins its web. We later surround ourselves with houseplants and enrich our homes with pets. For a few of us, the fascination grows and becomes more focused. Casual curiosity gives way to the irresistible lure of exploration, and an interesting pastime ripens into an exciting search for knowledge and discovery. These are biology's "lifers," scientists who are engaged in a lifelong adventure, a quest to make rational sense of living phenomena.

Biology is the study of life. This definition may sound dishearteningly flat, a tedious memorization exercise stocked with countless disconnected facts. But modern biology is anything but flat. It is a multidimensional, dynamic, creative activity that does something quite remarkable—it replaces mystery with understanding. No list of terms or facts can produce understanding, any more than a pile of unassembled gears and springs explains how a clock works. Individual biological facts reveal little about how life works until they are unified by concepts or principles that explain how these parts interact and what process is created by the combination. Biologists are detectives who use bits of information as clues to solve the complex mysteries of life. The work of biologists also yields countless practical bonuses, such as controlling deadly diseases, increasing crop yields, and proposing measures to preserve our environment.

The result of these pursuits is a developing body of information and concepts that is so immense and broad in scope that not even a professional biologist can thoroughly master all its areas. In distilling this information for use as a general introduction, we have struggled to retain the challenge and excitement of biology. This intellectual adventure takes you from the microscopic landscape of molecules and cells to the global community of organisms. The discoveries you make in this book can rekindle your fascination with living phenomena and reaffirm your connectedness with all living things.

## Distinguishing the Living from the Inanimate

Most of us have little trouble identifying something as being alive or inanimate. The diverse organisms shown in Figure 1-1 are easily distinguished from the nonliving environmental components. But try to define what distinguishes the two, and you may find yourself at a loss. You'll discover

**Figure 1-1**
As simple as it is to identify each object in the photograph as either alive or inanimate, justifying your choices would be much more difficult. Most characteristics popularly associated with living things (growth, movement, reproduction, consumption of food) may also be properties of nonliving entities. Clouds and mineral crystals grow; rivers, air, and clouds move; fire consumes "food" (fuel) as it "grows" and "reproduces" giving rise to "progeny" fires. Yet none of these is alive. Defining "life" requires a combination of many properties, no one of which can, by itself, be considered *the* criterion of life.

that many of the properties that flag an object as being alive cannot be found in all organisms, or may be exhibited by some nonliving things. You may have selected the ability to move as a "characteristic of life." But is a redwood tree not alive because it cannot move? Is a river alive because it can? Rather than defining life, biologists describe it, usually as a list of activities and properties that characterize all living things.

- Organisms maintain a high degree of organization.

- They acquire and use energy.

- They reproduce, transmitting their characteristics to offspring by heredity.

- They grow and develop.

- They respond to stimuli.

- They possess adaptations to their environment.

- They maintain a state of homeostasis, the stabilizing of conditions inside the organism to resist fluctuations in the environment.

As you examine these properties in more detail on the next few pages, bear in mind that these not only "define" life, they are responsible for its success, its nonstop presence on this planet for 3.5 billion years. Mountains crumble, continents collide, and life on earth persists in spite of the changes (Figure 1-2).

## High Degree of Organization

Even the simplest organisms are almost unfathomably complex. Atoms and molecules, themselves very complex, are assembled into exquisitely ordered structures that comprise an organism. This building process takes form in stages of complexity, each stage creating the "building blocks" for assembly at the next level (see "Levels of Biological Organization" later in this chapter). The unique way in which these constituents are organized generates the characteristics of each particular organism. This high degree of order reveals itself in many ways, some of which are quite striking (Fig. 1-3a).

a

c

b

**Figure 1-2**
The Enduring Nature of Life. *(a)* This moss-covered skull testifies that life goes on in spite of the death of an individual. *(b)* This Bristle cone pine has endured cold, lack of soil, and slashing winds in this same spot for thousands of years. *(c)* The plants in this river scene still retain characteristics of their ancestors that lived millions of years ago. The river sediment contains bacteria that resemble organisms living 3 billion years ago.

## Acquisition and Use of Energy

Manufacturing and maintaining such complexity requires constant input of energy. Molecules must be extracted from the organism's surroundings and assembled into forms that are distinct from those in the environment. Some organisms fuel these processes by directly transforming sunlight into usable energy, a process known as **photosynthesis**. Photosynthesizers are **autotrophs**, organisms that produce their own food (auto = self, troph = feeding). Other types of organisms are **heterotrophs** (hetero = other). These organisms acquire energy from the tissues of other organisms. The snake in Figure 1-3b, for example, is about to harvest the energy stored in the body of a frog, which obtained its fuel by eating photosynthetic plants.

## Reproduction and Heredity

The perseverance of a particular type of organism requires individuals to *reproduce,* that is, make offspring that can carry on the parent's traits. **Heredity** is the passage of genetic traits to offspring, which consequently are similar or identical to the parent (or parents) — see Figure 1-3c. Similar individuals are produced generation after generation, so the characteristics and functions of a particular kind of organism persist long after the parents have died.

## Growth and Development

Organisms cannot survive and perpetuate their kind without growing at some time during their life. *Growth,* an increase in size, is usually accompanied by *development,* a change in appearance and abilities. Although some very simple organisms, notably bacteria, show few developmental changes as they increase in size, virtually all other organisms change their form and capabilities dramatically as they grow (Figure 1-3d). An acorn develops into an oak tree. A caterpillar changes into a butterfly. A fertilized human egg becomes a fully developed person. Different developmental stages of a particular organism may inhabit very different environments. A tadpole, for example, cannot survive out of water, whereas the same organism as an adult toad spends its life mostly on dry land.

Developmental changes prepare an organism for its major role at each stage of life. The notorious appetite of caterpillars keeps them eating voraciously, building a huge nutrient surplus stored as fat. Suddenly the gluttony stops, and the caterpillar begins its transformation into a moth or butterfly. Lacking mouthparts, some types of moths must live their entire adult lives on these stored nutrient supplies. The moth pursues its major "business" of locating a mate and reproducing without having to stop for dinner.

## Response to Stimuli

The ability to respond to **stimuli** (changes in the internal or external environment) is literally a matter of life and death. Responses may help an organism escape predators, capture prey, optimize its exposure to sunlight, move away from detrimental environmental conditions, move toward a source of water or other resources, locate mates, change growth patterns according to season, and perform many other activities necessary for survival. Some responses must be very rapid to succeed. The angler fish responds to the presence of a smaller fish so rapidly that its strike cannot be seen by the human eye; the dinner simply disappears. Plants generally respond to stimuli more slowly than do animals, but they respond nonetheless (Figure 1-3e).

## Adaptation to Environment

Biological success for a particular type of organism depends on how well suited it is to its environment. It can survive only in environments that supply all its essential needs, and then only if it possesses the equipment needed for acquiring those environmental resources. Even then, it can succeed only if it is able to survive the adverse conditions to which it is exposed. In other words, it must be adapted to its particular set of environmental conditions. **Adaptations** are traits that improve the suitability of one type of organism to its environment. (This includes not only its external surroundings, but the environment *inside* the organism as well.)

Individual organisms, however, do not adapt. They respond to changes, but the development of adaptations is a multigenerational process. Adaptations are genetically determined. They are, like all inheritable traits, determined by stable units of genetic information called **genes.** An individual organism cannot simply change its genes and develop a new trait. In passing on copies of its genetic endowment to offspring, however, a small number of the genes do change, the result of occasional mistakes in gene duplication, for example. These and other types of genetic variation assure that at least slight (and sometimes major) differences appear with each new generation. Any of these new or altered traits that increase an organism's suitability to the conditions it encounters improves the ability of that individual to survive and produce more offspring. Because genetic traits are stable, the improved characteristic is passed on to each successive generation. The new or altered trait has become a stable adaptation.

Examples of adaptations are available on every organism you see. The subdued color of some moths better enables them to avoid detection by birds or other predators. The trunk of an elephant provides it access to a supply of edible leaves on high branches that are unavailable to most of the competition (other leaf-eating animals). Elephants need a huge amount of food, more than a short-nosed elephant could get from "its share" of the lower foliage, considering the much greater competition for these leaves.

Pick out a noticeable trait on the nearest living thing. Although you may have selected a neutral trait (one that provides no particular biological advantage), you've probably selected an adaptive trait. The adaptation you have chosen is probably some trait of *anatomy* (structural characteristics). A few of you may have selected a property that reveals a more subtle adaptation in *physiology* (biological processes and function), such as eyesight or taste. The ability of microorganisms to thrive in heat that is too extreme for other organisms is another physiological adaptation (Figure 1-3f). But many adaptations in structure and function would be useless without an accompanying adaptation in *behavior.* Spiders, for example, not only possess silk-producing glands, but also inherit the automatic behavioral adaptations ("instincts") that compel them to weave a characteristic web and instruct them in its use for capturing prey. The close relationship between behavior and anatomy/physiology forms one of the central themes in this book, one that frequently reappears in the text as a series of reports entitled "Animal Behavior."

**Figure 1-3**
**Life's seven properties illustrated.** *(a)* The delicate intricacy of this soft coral, one of the simplest animals, reveals just a hint of its elaborate degree of organization. *(b)* This snake is about to harvest energy from a frog. *(c)* The resemblance within this family of swallow-tailed bee eaters is the dual product of reproduction and heredity. *(d)* All organisms develop from a single cell. This week old salmon embryo is acquiring the form of an adult fish, but still obtains all its nutrients from the yolk sac protruding from its midsection. *(e)* Sculptured tree roots whose final form reflects its ability to respond to the presence of a boulder. *(f)* Adaptations to temperature differences are revealed by the concentric bands of color in this aerial photograph of a hot sulfur pool at Yellowstone National Park. The pond, which is hottest in the center and coolest at its rim, is colored by enormous populations of microorganisms concentrated in each ring-shaped thermal zone. Each distinctly pigmented type of microorganism is specifically adapted to (and limited to) one narrow temperature range. *(g)* Even when exposed to subzero air temperatures, thermal homeostasis is maintained in these snow monkeys, whose body temperatures remain steady in spite of the cold.

c

g

## Homeostasis

Conditions in the external environment often fluctuate to such extremes that an organism could not survive if its internal conditions were to similarly fluctuate. Maintaining fairly constant internal conditions, a "steady state" called **homeostasis**, is essential to maintaining life. Constant physical and chemical adjustments inside the organism stabilize each internal condition individually. Any excess or deficiency activates homeostatic mechanisms that reverse the impending imbalance. The heat in your body, for example, is maintained at a steady 98.6° F (37° C), even in very cold or hot surroundings. Homeostatic control centers detect deviations from this "set point" and instruct the body to react with heat-conserving or heat-dissapating measures (Figure 1-3g).

Although thermal homeostasis is a property of only a few types of organisms, all living things must regulate and stabilize their internal chemical environment. An excess of a particular chemical inside an organism can fatally disrupt life processes. Regulatory mechanisms govern the production or destruction of each type of chemical, which is usually maintained at a steady concentration. The synthesis of the chemical is "switched off" when there is enough and "turned on" again when it is needed. Virtually all the chemical changes *(reactions)* which an organism undergoes are subject to homeostatic regulation to prevent unbridled reactions from generating lethal imbalances. This is no small task, given the millions of reactions that each organism performs.

Together, these chemical reactions constitute an organism's **metabolism** (the sum of all the chemical activity in an organism). Regulating metabolism is controlling the very essence of life. Metabolism accounts for every aspect of living, homeostasis being but one of these aspects. It is a universal activity essential to all seven of life's properties.

# The Hierarchies of Life

## Levels of Biological Organization

The construction of a living organism from a supply of inanimate molecules is not a direct transition. Rather, the high degree of biological organization found in organisms emerges in steps, or levels. The simpler structures in the lower levels combine to form the more complex structures of the next level, which then can interact to form even more complex units. This hierarchy of organization gains in complexity one step at a time until it forms the largest, most complex, and all-encompassing biological level (Figure 1-4). As highly ordered as organisms are, they become subunits that form even more complex levels of organization.

It all begins with protons, electrons, and neutrons. These three kinds of subatomic (smaller than atoms) particles combine to form atoms. An infinite variety of molecules can be created from different combinations of atoms. Molecules join together to form the *subcellular components* (cellular structures called *organelles* and a fluid-type substance called *cytoplasm*). These components provide the materials for constructing a **cell**, the smallest unit that is itself considered alive (Chapter 6). For some **unicellular** organisms (like the *Paramecium* in Figure 1-8 on page 14), the cell *is* the organism. But most forms of life are **multicellular**, consisting of hundreds to trillions of cells, depending on the organism's size and complexity. In complex organisms, the "jump" from cells to complete multicellular organism is not as direct as the simplified single step in Figure 1-4 suggests. Rather, it develops in the following sequence, each structure composed from the units on the previous level.

- Cells.

- **Tissues**, functionally active "fabrics" composed of specialized cells (examples—muscle tissue in animals, bark tissue in plants).

- **Organs**, body parts that perform specialized functions (examples—stomach in animals, roots in plants).

- **Organ Systems**, groups of functionally related organs (examples—digestive system in animals, root system in plants).

- The complete multicellular organism.

Groups of organisms combine with each other and with their inanimate environment to form even higher levels of order. Individuals of the same species inhabiting the same area constitute a **population**. Different populations interact with each other to form **communities**, the organisms that inhabit a particular **ecosystem** (the organizational level that consists of the community in an area plus the nonliving environment). All the world's ecosystems combine to form the **biosphere**, which is at the top of the hierarchy of biological organization on earth.

But there is something else to be learned from this hierarchy, something much more profound and fascinating. Increasing complexity not only generates a higher order of structural organization, but may also create an extra property, something more than the sum of the parts used to form the structure. These novel phenomena, these "extras," emerge at specific levels, depending on the complexity of the new property. Consider, for example, the ability to direct the occurrence of simple metabolic reactions, a property that emerges from the structures at the molecular level of complexity. This is the lowest level that allows structures to be complex enough to acquire the ability to perform these reactions. Yet this new property is elementary compared to the remarkable extra phenomenon that appears at the fifth degree of complexity, life itself.

Life could not emerge at lower levels of organization any more than a story could be created using single alphabetical letters without joining them into progressive levels of complexity. Individual letters must first combine to form a more complex structure, a *word,* from which emerges an additional property, the word's meaning. Words are ordered into sentences, and another novel property emerges, a statement. These levels of organization increase in steps until the final structure, a manuscript, is formed at a level complex enough for the final emerging phenomenon, the story itself. A word is more than a group of letters in a particular order, just as a cell is more than a highly ordered collection of subcellular components. Ultimately, it is these extra properties that epitomize the essence and significance of the hierarchy of biological organization.

## The Taxonomic Hierarchy

The sheer complexity of organisms seems so bewildering that some people believe life to be unfathomable by the rational mind. Biologists constantly chip away at this notion by exploring and discovering real (and fully understandable) explanations for phenomena. Our ability to solve these natural mysteries would be severely handicapped, however, without another hierarchy of organization that creates meaningful order out of the overwhelming variety of organisms to be studied. This orderly system enables us to distinguish clearly between different types of organisms, allowing us to classify and assign names to the almost two million known types of living organisms as well as their extinct ancestors. This formal system of naming, cataloguing, and describing organisms is the science of **taxonomy**. In the absence of this system, we would be unable to identify and study organisms without being hopelessly misled by deceptive appearances. For example, two organisms with little common

---

**Figure 1-4**
**Levels of biological organization.** Each single-step jump increases structural complexity and may also generate a unique property distinct from the structure.

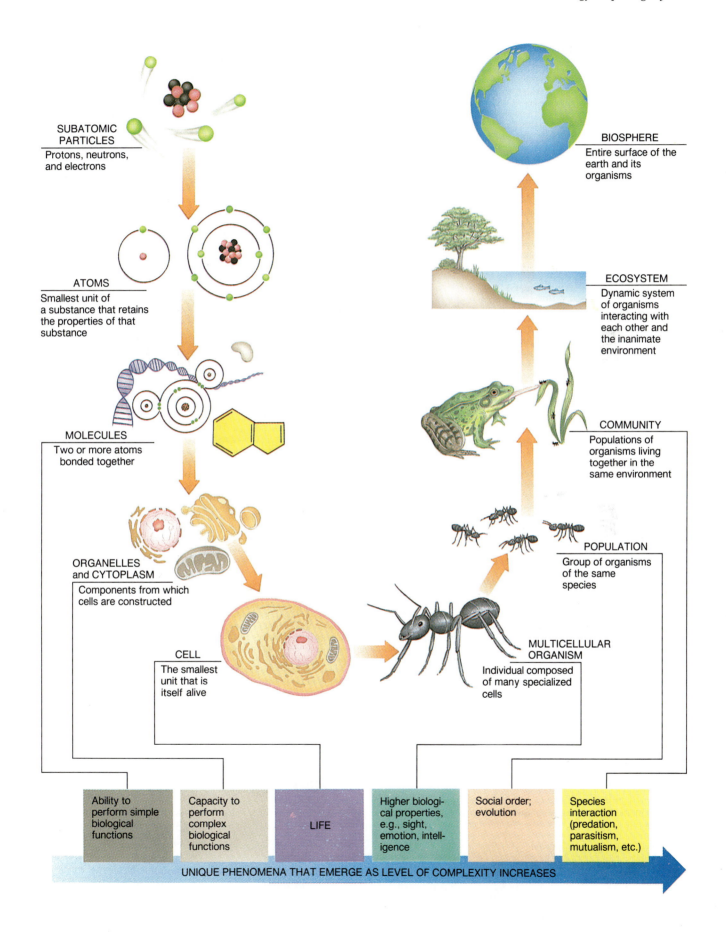

SUBATOMIC PARTICLES
Protons, neutrons, and electrons

ATOMS
Smallest unit of a substance that retains the properties of that substance

MOLECULES
Two or more atoms bonded together

ORGANELLES and CYTOPLASM
Components from which cells are constructed

CELL
The smallest unit that is itself alive

BIOSPHERE
Entire surface of the earth and its organisms

ECOSYSTEM
Dynamic system of organisms interacting with each other and the inanimate environment

COMMUNITY
Populations of organisms living together in the same environment

POPULATION
Group of organisms of the same species

MULTICELLULAR ORGANISM
Individual composed of many specialized cells

Ability to perform simple biological functions

Capacity to perform complex biological functions

LIFE

Higher biological properties, e.g., sight, emotion, intelligence

Social order; evolution

Species interaction (predation, parasitism, mutualism, etc.)

UNIQUE PHENOMENA THAT EMERGE AS LEVEL OF COMPLEXITY INCREASES

kinship may mimic one another in appearance. Equally deceiving, two organisms with distinctly different looks may in fact be the same type (Figure 1-5).

Taxonomists generate a "library" of information that forms a foundation on which all biological investigation depends. This formally organized library enables biologists to know the identity and fundamental properties of any catalogued organism, as well as its relationship with other varieties of organisms. Without a formal taxonomic scheme, biologists could not even identify an organism to the mutual agreement of others studying (or believing they were studying) the same organism. The results of such "unintegrated" research would be confusion, making the complexity of life seem even more bewildering. Thanks to taxonomists, however, scientific order prevails.

Taxonomy perpetuates order by examining each type of organism and describing its properties, defining the set of features that distinguishes that particular type of organism from all others. (Taxonomists have described more than 1.8 million distinct species, which represents only a fraction of the estimated total of 5 to 30 million species.) From this body of descriptive information, taxonomists create the two orderly systems, that are so necessary to biology.

- *A system of nomenclature* that assigns a specific name to each type of organism, a label that doesn't vary from biologist to biologist.

- *A system of classification* that groups organisms into categories according to similarities in major properties. This formal system has a dual purpose. First, it provides a standard set of criteria that can be used to ascertain the identity of an organism from its observable properties. Second, the classification scheme reveals the degree of kinship, and thus ancestral relationships, among different groups of organisms.

## ASSIGNING SCIENTIFIC NAMES

Using standard names is no less important for organisms than for people.

a

b

**Figure 1-5**
**Deceptive similarities, misleading differences.** *(a)* These two "palms" could easily be mistaken for closely related plants based on their similar appearances, but the cycad (left) is not a palm, or even a flowering plant. It is no more related to the true palm (right) than a pine tree is to a rose. *(b)* The close relationship between these two animals is disguised by their distinct appearances. In fact, both are amphibians, but the legless cecilian looks more like a snake.

Imagine the chaos if every person were given a different name by each acquaintance, who in turn used that very name for many other people. Communicating by mail would be impossible, reports on job performance would be useless, as would college transcripts (even if one could be assembled). Fortunately, your name evokes a whole order of personal characteristics and actions, a unique set of properties that describes you and only you. To a biologist, the name of a particular organism launches the same type of association, the distinct set of properties that not only identifies but also characterizes that type of organism.

Unlike the names of people, however, individual organisms are not assigned their own scientific names. Instead, we use scientific names to identify a single *type* of organism, what we call a **species**. Your scientific name is the same one given to all other people, for we are all the same type of organism, all members of one species, *Homo sapiens.* Each species possesses consistent, recognizable features that distinguish it from every other species (see Connections: The Species Concept).

Like all species, humans are assigned a name consisting of two latinized words according to our uniform **binomial system of nomenclature.** (Binomial loosely means "two names.") Latinizing these names may strike you as being overly formal and difficult, but it would be infinitely more confusing if each scientific name reflected the native language of the biologist who provided the label. Some scientific names would be in English, others would be written with Chinese characters, and a few others would contain only Arabic letters. The use of Latin standardizes the language of nomenclature, so that all species receive names expressed in the same language, using the same alphabet.

Scientific names are always written in a standard order to prevent accidentally switching the categories of the two words in the binomial. The first word in the pair always identifies the organism's *genus,* a group that contains closely related species. The second word, called the *specific epithet,* singles out from within that genus one kind of organism, the only group on earth identified by that particular pair of terms. Another way of stating this idea is that, while closely related species may have the same genus name, *no two species have the same binomial name.* Accidental duplications are avoided (or corrected) by using strict *rules of nomenclature* to assure that all organisms are named according to the same set of international rules.

Another convention establishes a standard format that helps distinguish scientific names from other types of latinized words that may appear in pairs. By this convention, the genus name is always capitalized, followed by the specific epithet written as a lower case word. Both terms are underlined or italicized. *Crotalus ruben,* for example, is instantly recognizable as a scientific name of an organism (in this case, a rattlesnake).

Ideally, a scientific name should provide information about an organism in addition to merely identifying it. The word *Homo* (Latin for human) that identifies our genus is modified by the Latin term *sapiens,* which means "wise," creating a scientific name meaning "wise human." Your pencil was probably made from the wood of *Pinus strobus,* a white pine tree. The somewhat intimidating name *Saccharomyces cerevisiae* is loaded with information about this uniquely popular organism. *Saccharo* refers to sugar, *myces* tells us the organism is a fungus, and *cerevisiae* is derived from the Latin term for brewer or beer. This organism is common brewer's yeast, a fungus that ferments sugar to alcohol.

---

| CONNECTIONS | Animal versus Plant Species (page 380) |
| | Defining Moneran Species (page 559) |
| | Ecological Species (page 694) |

### THE SPECIES CONCEPT: MORPHOLOGICAL SPECIES

Although the *species* level is the most fundamental in the taxonomic hierarchy, biologists do not all agree on what criteria to use to define a species. Biologists do concur that species are "real" (that is, the distinction between two species is a natural one), and that each species represents a particular kind of organism. All the individuals of a particular species have similar features that distinguish them from the individuals of other species. But what features should be chosen, and on how many unique features does it take to make a species? Should only visible traits (morphology) be used, or could differences in internal anatomy, breeding behavior, environmental habitat requirements, or even biochemistry be used? Some biologists argue that a species should be recognized solely on whether individuals are successful in producing fertile offspring; if they can, or if they have the *potential* to successfully breed, then they are members of the same species. Each of these criteria is valid as a basis for defining a species, but unfortunately, they often do not produce the same results.

Because biologists use different approaches for defining a species, we will explore the concept of a species throughout this book in a series of Connections Boxes entitled "The Species Concept". Each box discusses a different criterion for answering the question "what is a species?"

We begin our exploration of the species concept by looking at the *traditional approach* of defining species, on the basis of differences in visible traits that can be passed on to offspring. Using this approach, there is no doubt that a horse is a different species than a donkey, and a yellow pine is different species than a Jeffrey pine, even though the horse and donkey can successfully interbreed, and so can the yellow and Jeffrey pine.

To incorporate large numbers of traits, and to lessen the subjectivity associated with identifying distinguishing features, some traditional taxonomists have recently enlisted computers. These taxonomists use computers to calculate degrees of similarities between organisms based on combinations of many traits. The results are plotted on a graph, and the taxonomist then defines a species (or any other taxonomic level) on the basis of where breaks occur on the graph, each break separating clusters of similar groups.

Whether the taxonomist relies on one or a few distinct traits, or uses calculations of similarities based on large numbers of traits, the traditional approach works, for it produces clear demarcations between different species. Since this approach is based on visible characteristics, it also enables taxonomists to produce an easy-to-use system for identification, one that can be used to determine the identity of any organism, except newly discovered ones that have not yet been assigned to a species.

---

### CLASSIFICATION OF ORGANISMS

Taxonomists classify organisms by sorting them into groups according to **phylogenetic** relationships. This means that the organisms in the same group have a similar ancestral history and are therefore more closely related than are organisms in different groups. ("Phylogenetic" refers to *evolutionary* relationships, revealing how recently two types of organisms shared a common ancestor. This concept will be discussed later in this chapter.) Such a "natural system" of classification is preferable to an "artifical system" which relies on arbitrary criteria (such as eye color) to group organisms. An artificial system of classification might combine all animals with brown eyes in the same category, thus lumping together such unrelated and genetically different organisms as pelicans, walruses, bears, all apes, and only some humans. On the other hand, a "natural system" of classification groups organisms according to evolutionary characteristics, such as an animal's skeletal type or a plant's ability to produce flowers. A natural system of classification is rich in information because it reflects the evolutionary history of organisms, as well as their relationships to organisms in other groups.

Systematic classification also provides a practical tool for identifying the species of an organism by examining its properties and then using taxonomic guides that lead us, property by property, to the name of the only species that "fits" the description.

Organisms are classified into categorical levels that form a taxonomic hierarchy, starting with millions of categories (the individual species) at the lowest level. The next level consists of larger groups, each one (called a *genus*) containing one or more closely related species. The third level is created by separating the members of the previous level into *families,* each family consisting of several similar genera (plural of genus). The sequence continues in this manner in steps, each step composed of groups formed by clustering the categories of the previous level. Related families form an *order,* related orders form a *class,* related classes form a *division* (or *phylum* in animals), and related divisions (or phyla) form the highest level of biological classification, a *kingdom.* The system arranges organisms in successive categorical levels, each of which is called a **taxon** (plural, "taxa"). We call this system **the Hierarchy of Biological Classification** (Figure 1-6).

The currently accepted scheme of classification maintains that all organisms on earth belong to one of five kingdoms (Figures 1-7 and 1-8). The simplest organisms, bacteria and their relatives, belong to the Monera kingdom. All these single-celled organisms are **prokaryotic,** a word that identifies a group of cells with very simple cellular anatomy, free of organelles and the

complicated membrane networks that characterize the more complex **eukaryotic** cells. (Formal definitions of these two cell types require the use of terms that are not introduced until Chapter 6.). The four other kingdoms contain organisms composed of eukaryotic cells. The somewhat larger and more structurally complex organisms of the next kingdom (the Protists) are the simplest species with eu-

karyotic structure, most being unicellular with some simple multicellular members as well. Members of the Fungus kingdom are nonphotosynthetic organisms, such as mushrooms, molds, and yeasts, that generally acquire nutrients and energy by decomposing dead organisms and feces. Members of the Plant kingdom are also multicellular, but obtain their energy through photosynthesis.

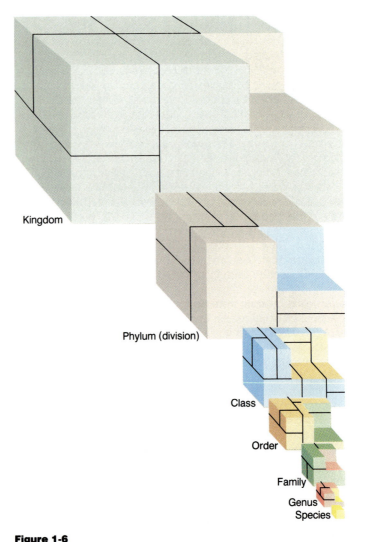

**Figure 1-6**
**Taxonomic levels used in the classification of life.** Members of the same group at a particular level (taxon) are more related to each other than they are to members of the other groups in the same taxon. For example, all the organisms in the same genus resemble each other more than they do members of the other genera in that family. The smallest group, the species, consists of the most closely related organisms. At the other end of the spectrum are the kingdoms, the largest, most encompassing groups that contain the greatest number of organisms. Every organism on earth is grouped into one of five kingdoms.

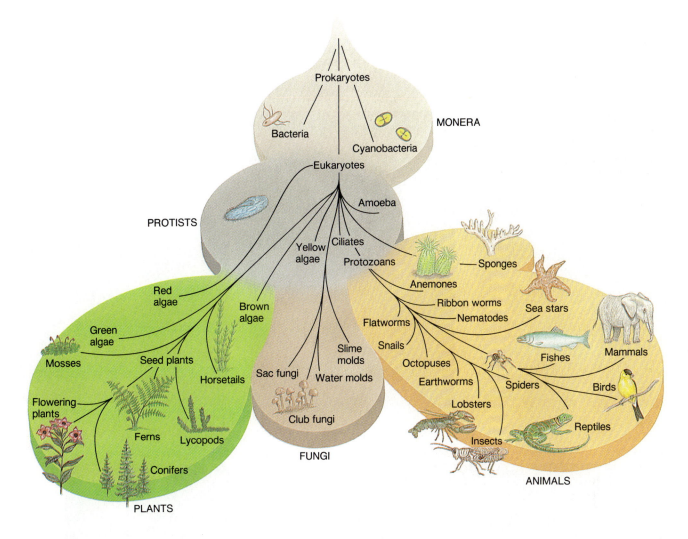

**Figure 1-7**
**The modern five-kingdom scheme of biological classification.** This diagram shows the relationship between the five kingdoms and the relative numbers of species that each contains. The animal kingdom, with its 1.3 million species, is the largest kingdom; in other words, it contains the greatest diversity.

The Animal kingdom contains multicellular organisms that obtain energy and nutrients by consuming other organisms (or their wastes). To suceed at this, most of them need elaborate systems for movement, for capturing prey, for digestion, and for circulating the nutrients to all parts of the organism. These are just a few of these essentials that require most animals to possess a structural complexity that exceeds that of the organisms in the other kingdoms.

## Unity Embedded in Diversity

The diversity of life is astounding. We share this planet with as many as 30 million other species, each of which consitutes a distinct form of life. With such bewildering variety, one might identify "diversity" as the central theme in biology. But despite this diversity, there is a remarkable amount of sameness among all organisms. Hidden beneath the obvious differ-

ences lie striking similarities, even among such radically different appearing organisms as people, spiders, ferns, and bacteria. A look at a few of these similarities clearly reveals the *unity of life* as another of biology's principal themes, a theme that is just as prevalent as diversity.

The unity of life becomes apparent when we examine the fundamental levels of biological organization. A person and a mushroom, for example, seem to share few similarities on the

**Figure 1-8**
**Representative organisms from each kingdom.** *(a)* rod-shaped bacteria—Monera Kingdom; *(b) Parmecium,* a unicellular animal-like organism—Protista (protist) Kingdom; *(c)* poisonous *Amanita* mushroom—Fungi (fungus) Kingdom *(d)* ferns—Plantae (plant) Kingdom; *(e)* a blue snake—Animalia (animal) Kingdom.

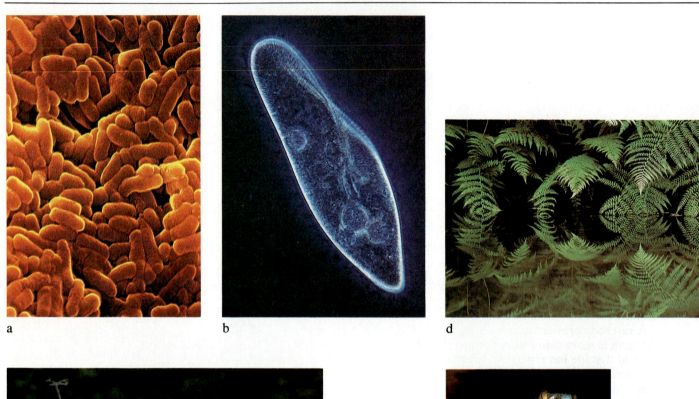

a

b

d

c

e

level of the whole organism, but are quite similar on the cellular and molecular levels, resembling all other species as well. Every organism is composed of cells. We are all built from the same four classes of biological molecules (Chapter 4). All living things extract molecules from their environment and rearrange them into other molecules needed for their own growth and reproduction. During these metabolic processes, organisms assemble and disassemble chemicals in small steps, each step encouraged by a particular **enzyme**. (An enzyme is a biological molecule that promotes a specific chemical reaction.) The universal use of enzymes to build organisms and carry out metabolic functions is another clear example of unity (Chapter 5).

Humans and mushrooms, like all other organisms, use the same molecule (DNA) to store genetic information. In both organisms, these instructions are encoded as discrete units of genetic information, the *genes.* Furthermore, all organisms use the same genetic "language" to encode these genetic instructions. (A particular gene orders the production of the same protein whether in human cells or in bacterial cells.) The universal nature of this genetic code is a particularly striking example of biological unity, a testimony to the kinship of virtually all species.

## Evolution—Unity and Diversity Explained

One theme underlies all biology. It explains why millions of species exist rather than just one type of organism. At the same time, this principle accounts for the unity that exists among even diverse organisms. This principle, **evolution**, states simply that species change through time. New species of organisms emerge and older forms of life disappear. In other words, life evolves.

A vast body of evidence reveals that the earth is more than 4 billion years old and that life has resided on earth for about 3.5 billion of those years. The evidence also shows that humans, and every other living species, evolved from preexisting species. A new species may be better adapted to its environment than its predecessor, or better able to succeed after some environmental change, or able to exploit a new source of nutrients for which there is less competition. Some of these new species replace their ancestral species. Other emerging species coexist with their predecessors or inhabit a different environment, increasing the number and variety of species. Over billions of years, this diversifying process has created millions of new species, each one a unique collection of adaptations that suit it for its particular environment.

Yet all of these diverse species are descendants of the simple cells that were the earliest life forms. It is this common ancestry that accounts for the unity of life. You and that mushroom are so similar because you shared the same ancestor far back in time, and both of you retain many of the same molecular, genetic, and cellular endowments provided by that ancestor.

Evolution, like all natural phenomena, occurs by mechanisms that are amenable to rational explanation. It is best understood when examined in steps. First, evolutionary change occurs over generations; it does not change the individual organism during the course of its lifetime. Second, at any given time there is great genetic variation among the members of a particular species. (For example, notice the variation in the population of *Homo sapiens*—no two of the earth's 5 billion people are identical, including "identical" twins.) This variation is due to genetic changes that inevitably occur during reproduction, as explained in the earlier section, "Adaptation to Environment." Although the variant traits are inheritable (passed on to offspring), they are *not* produced in response to any *need* for a particular adaptation. In fact, most of them have no advantage whatsoever, and many are even detrimental.

The fate of a new trait is determined by the environmental conditions to which the organism possessing the trait is exposed. The trait is perpetuated if it improves an organism's ability to survive adverse conditons, to compete for nutrients, to avoid predators; in short any improvement that ultimately increases the number of offspring it produces, offspring that inherit the adaptive trait. As a result, the number of organisms with the advantageous trait increases at a faster rate than similar organisms that lack it. Generation by generation, the more adapted variant increases in numbers until finally becoming an established member of that community. A new species of organism has evolved. The opposite fate occurs when genetic variation creates detrimental traits that lessen an organism's ability to survive (and thus reproduce) in its particular environment. These organisms leave fewer (or no) offspring to carry on the detrimental trait. Both the detrimental trait and the species that bears it disappear. The same result occurs when the environment changes, altering the effect of a particular trait. A well-camouflaged animal, for example, would find itself glaringly visible to predators if the color of the background foliage changed from green to orange.

This process illustrates how new species can evolve to suit their specific environment. The underlying mechanism that creates evolutionary change is a two-stage process, genetic variation followed by natural selection.

- **Genetic variation.** This stage provides a "smorgasbord" of different varieties of organisms, individuals with new properties, and combinations of properties different from each other. The greater the variety of organisms, the greater the likelihood that at least one organism will possess an improved combination of traits compared to those of the parent's (see Connections—"Asexual versus Sexual Reproduction," Chapter 12).

- **Natural selection.** The environment then determines which of these variants will prosper reproductively (perhaps forming a new species) and which will die without leaving offspring. This selection process does not *cause* individuals to change; it merely determines which individuals (and ultimately which species) succeed and which do not. In this way, new species emerge with adaptations well suited to their environment (Figure 1-9).

## Making Connections

The evolution of species embodies much more detail than is introduced in this brief overview. (These details are the subject of Chapter 29.) A "bare bones" understanding of natural selection, however, prepares you for many of the fundamental biological concepts you will soon encounter. It also acquaints you with biology's fundamental theme, one that unifies all areas of the science. It is a core concept that ties together seemingly unrelated pieces of biological information to form an interconnected body of knowledge. These associations reveal new information and an increased depth of understanding that simply

**Figure 1-9**
**Evolution and adaptations — two cases of mistaken identity.** Both of these insects are successful products of natural selection. Well-camouflaged individuals are not as noticeable to predators, who eat individuals with more conspicuous bodies, eliminating them before they can leave offspring. Well-disguised individuals live long enough to reproduce, so their numbers increase whereas more conspicuous versions disappear (unless some other adaptation, such as noxious-tasting flesh, discourages predators). Although the strategy of both insects shown here is the same (camouflage), their different habitats have selected for strikingly different results.

doesn't exist in the same collection of facts considered separately as isolated bits of information. Biology abounds with such "unifiers" that forge individual facts together into more meaningful, richly satisfying sources of knowledge.

To help make these connections, many topics in this book are presented in "Connections Boxes," each of which discusses a concept that applies to several areas of biology. The Connections Box is accompanied by a linkage icon that identifies the differ-

ent areas unified by that topic. This approach helps you see biology as more than a collection of isolated facts. It is a multilayered puzzle that becomes more understandable — and more enjoyable — with each new fit between facts.

# Synopsis

## MAIN CONCEPTS

- Seven properties characterize life.
    Complex organization.
    Ability to obtain usable energy to maintain this complex and unique identity.
    Production of offspring that acquire the parental traits by heredity.
    Growth and development.
    Response to stimuli.
    Adaptation that improves an organism's fitness to its environment.
    Homeostasis, maintaining internal stability.

- The hierarchy of biological organization is as follows.

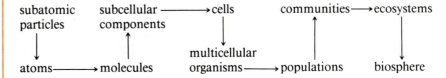

New properties emerge at different levels, novel properties that exceed the sum of the parts. "Life" is such a property.

- Biologists classify organisms in the following hierarchy of categories:

| Kingdom | most organisms per group | [LOW] |
| Division (or Phylum) | | |
| Class | | similarity among organisms in the same group |
| Order | | |
| Family | | |
| Genus | | |
| Species | fewest organisms per group | [HIGH] |

The genus and specific epithet to which a particular type of organism belongs provides the binomial name that identifies that species.

- Evolution explains the diversity and unity of life.
    Natural selection generates new species from preexisting species. The following sequence illustrates how this often happens:

**Genetic variability**                                             **Natural selection**

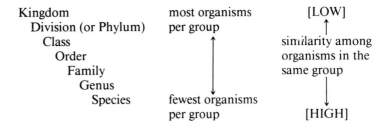

Although genes are stable, some genetic change occurs during reproduction. This creates great variability in the types of organisms produced, so even the individuals of the same species differ from each other in a few traits.

Organisms with detrimental traits die or leave fewer offspring; those with favorable traits survive longer and thus leave more offspring that carry on the favorable traits. Organisms with favorable traits proliferate in that environment; natural selection has produced a new species.

The evolutionary origins of all species go back to the earth's original cells. Common ancestry has created the abundant similarities shared by even the most diverse organisms.

## KEY TERM INTEGRATOR

### Animate Versus Inanimate

**organism**    A living entity, capable of maintaining exquisite organization, obtaining and using energy, reproducing offspring that show **heredity**, growing in size (usually with accompanying developmental changes), responding to **stimuli**, and **homeostasis**. **Metabolism** also characterizes organisms, as do suitable **adaptations** to habitat.

### The Hierarchies of Life

**cells**    The smallest living units. A particular cell is either **prokaryotic** or **eukaryotic** in structure. A **unicellular organism** consists of a single cell; a **multicellular organism** contains many cells, usually arranged into **tissues**, which form **organs**, then **organ systems** that are assembled into complete organisms. **Populations, communities, ecosystems**, and the **biosphere** are the higher levels of organization.

**taxon**    A level of organization in the classification of organisms. The level that contains the most organisms is the **kingdom**; **species** contain the fewest. Classification and nomenclature constitute **taxonomy**. Organisms are named using a **binomial system of nomenclature**, which assigns each species a unique two-word latinized name.

### Unity and Diversity

**evolution**    Process by which species change over generations. **Natural selection** produces new and more diverse species specifically tailored to their environment. Evolution also accounts for life's unity, such as the universal reliance on **genes** and **enzymes**.

## Review and Synthesis

1. For each of the seven properties that characterize life, try to identify an inanimate object that shows that characteristic. Why would each of these objects still be considered inanimate in spite of their one "life-like" feature?

2. Rearrange each list from the most fundamental to the most encompassing.
   — [ecosystems, subatomic particles, cells, organs, the biosphere, populations, molecules]
   — [order, class, genus, family, phylum (or division), kingdom, species]

3. Consider this statement: "In all _____ there are one or more cells." Which of the following words could be used in the blank without creating a false statement? organism; populations; organelles; molecules.

4. Using familiar processes, create an analogy that reveals how novel properties emerge as levels of order increase. Begin by drawing an organizational hierarchy for one of the following: composing a song; building a house; writing a novel. Identify the intangible "extra" entities that emerge as the level of complexity increases.

5. Considering the process of natural selection, how does greater variation within a species strengthen the chances that the species will avoid extinction in the event some new deadly environmental condition appears?

6. The trunk of elephants was used as an example of adaptation. Explain how the combination of genetic variation, competition for food, and natural selection account for the development of this adaptation, and why are there no short-nosed elephants?

7. An examination of bone structure reveals a close similarity between the anatomy of a bat's wing and that of a human hand. In fact, they are two variations on the same theme. How does evolution account for the coexistence of diversity and unity between these two structures?

8. Evolution produces species that contain, in addition to their adaptations, traits that are neutral — they provide neither advantage nor disadvantage. Considering the existence of such useless "passenger" traits, explain why natural selection is better described as "survival of the adequate" rather than "survival of the fittest."

## Additional Readings

Ayala, F., and T. Dobzhansky (eds.). 1974. *Studies in the Philosophy of Biology.* University of California Press, Berkeley. (Intermediate.)

Gingerich, O. 1982. "The Galileo Affair." *Scientific American* 247:132–143. (Introductory.)

Lewin, R. 1982. *Thread of Life: The Smithsonian Looks at Evolution.* Smithsonian Books. (Introductory.)

Thomas, L. 1974. *Lives of a Cell: Notes of a Biology Watcher.* Viking Press, New York. (Introductory.)

# Chapter 2

# *The Process of Science*

Science is the ongoing process of exploring and discovering the way nature works. We may never know everything there is to know about life and the universe, but we can continue to expand our understanding by making observations, asking questions, and seeking their answers. A brief look at how biologists are tackling one problem—identifying the mechanism by which we smell—will give us some insight into how science works.

Scientists have uncovered some interesting facts about the roles smells play in our daily lives.

- **Smell and Taste:** At least 75 percent of the "flavors" in food and drinks are not really tastes, but aromas.

- **Smell and Sex:** Over 25 percent of people with a loss in smelling ability lose some sexual drive.

- **Smell and Habits:** Days before we are even born we have the ability to smell. Scientists are now investigating whether the smell of amniotic fluid, altered because a mother smoked or had a spicy meal, gave her child a taste for cigarettes, curry, or garlic.

- **The Odor Detectors:** Each person has some 20 million olfactory nerve endings dangling from the roof of each nasal cavity (Figure 2-1). Suspended in a thin film of mucus, the nerve endings detect many thousands of odorants from the outside world, from the pungent smell of air pollution to the sweet smells of perfumes.

Although we know a great deal about the cells and tissues involved in olfaction (smelling), identifying the exact mechanism by which olfactory receptors work is still one of the challenges facing biologists.

Scientists have advanced a number of proposals in an attempt to explain this mechanism. In the language of science, each proposal is called a **hypothesis**, a tentative explanation for an observation or a phenomenon.

A scientific hypothesis has to be testable and consistent with observed facts, although not all observations contribute to the formulation of a hy-

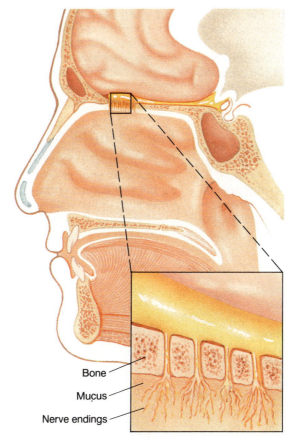

**Figure 2-1**
"Being led by the nose," we are able to smell our way to the dinner table or away from someone who forgot to shower. People are capable of detecting a tremendous range of odors. As we breathe in, airborne aromas travel up through the nose, swirl around in the back of the mouth, and flow up to the nasal cavity where millions of tiny olfactory nerve endings dangle in a film of mucus. Odorants (aromatic molecules) stimulate the nerve endings, sending a signal to the brain which identifies the odor. But how the nerve endings are triggered by the odorants still remains a mystery. Scientists are busy proposing hypotheses (explanations) and designing experiments to solve what many of us would consider a very basic question, "how do we smell?"

Bone
Mucus
Nerve endings

pothesis. For example, only the final observation in the above list was used as the foundation for hypothesis on how olfactory receptors work.

One hypothesis, proposed in the 1950s, suggested that different nerve endings responded to different "smelly" substances; in other words, each nerve ending had a specific receptor site for each odorant. The second part of the hypothesis proposed that when an appropriate odorant passed by that nerve, the nerve cooled slightly and the lowered temperature triggered a specific smell. To test whether this hypothesis was correct, experiments were designed to measure differences

in temperatures between the surrounding air and nerve endings. After conducting these tests, however, investigators were unable to measure even tiny temperature changes in nerves exposed to smelly air. Furthermore, people continued to smell when the air and nerve endings were at the same temperature. Clearly, the results of these tests did not support this hypothesis, at least not the part about temperature.

Today, some scientists continue to hypothesize that nerve endings and odors do indeed fit like a lock and key, as the first part of the original hypothesis proposed. Still other scientists argue against this notion, pointing out that if we evolve receptors for each smell, then how can we smell newly invented odors such as the smell of a new car or one of the 10,500 chemical compounds that are invented each week? There is no way to evolve that many chemical receptors that rapidly. In their quest to find an answer, the scientists who disagree with the lock-and-key hypothesis have proposed their own hypotheses for how olfactory receptors work. Investigators on both sides continue to design experiments to test which explanation is cor-

rect. Through experimentation, as well as debates and exchange of information between researchers, eventually we will know the answer to the question "how do olfactory receptors work."

## The Scientific Method

This introduction helps to illustrate how scientists often go about discovering the way nature works. As in this example, scientific investigations frequently include the following steps.

1. *Making careful observations.* Example: People are able to discriminate between many thousands of specific smells. Each person has a total of 40 million olfactory nerves.

2. *Formulating one or more explanations, hypotheses, to account for the observations.* Olfactory nerves have a specific recognition site for each odorant. When an odorant passes the appropriate nerve, it lowers nerve temperature, which in turn triggers the specific smell.

3. *Designing and conducting experi-*

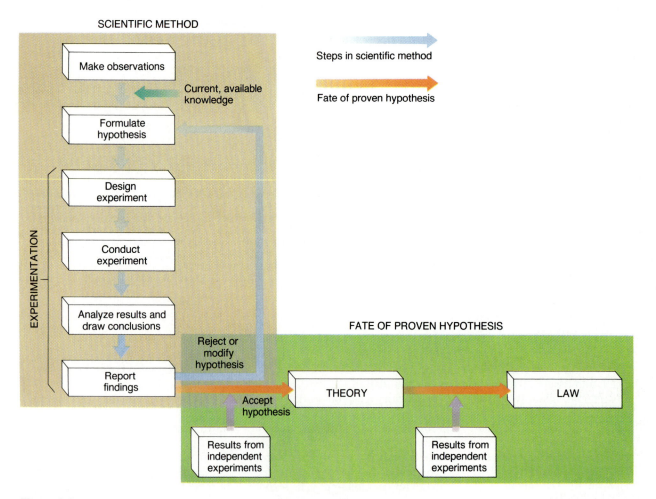

**Figure 2-2**
**The Scientific Method.** Many scientific investigations start with an observation of a phenomenon. Based on current knowledge from a number of sources (other researchers, scientific journals, scientific meetings, for example), a plausible explanation for the observation is formulated. This explanation is called a hypothesis. The validity of the hypothesis is then tested through experimentation. The results enable the investigator to conclude whether the hypothesis accurately accounts for the observed phenomenon. If it does, the hypothesis is accepted. If not, the hypothesis is rejected or modified. Hypotheses may eventually become theories, and theories may become principles or laws if consistently supported by further experimentation.

*ments to determine if a hypothesis is correct.* Determine if people can smell when the temperature of the air and nerves is equal. Measure the difference between the temperature of the air in the nasal cavity and specific nerves when exposed to different odorants.

4. *Analyzing results and drawing conclusions.* People continue to smell when the air and all olfactory nerve endings are at the same temperature. No difference in temperature was found between that of the nerves and the surrounding smelly air.

5. *Accepting, rejecting, or modifying a hypothesis to be consistent with the results of experiments.* Reject the "temperature difference" part of the hypothesis.

6. *Sharing results with other scientists.* The results of this study have sparked entirely new hypotheses, as well as hypotheses that maintain, as the original temperature hypothesis, that there are specific receptor sites for specific odorants. However, none of the subsequent hypotheses includes temperature difference as the trigger for smell.

This step-by-step approach to scientific inquiry is often referred to as the **scientific method**. (The entire process is illustrated in Figure 2-2.) Although many discoveries have followed just such a series of steps, many others have not. Because there are many approaches to studying nature, making a list of steps and calling it "the scientific method" may be misleading. This may give the impression of a cookbook approach that minimizes the importance of creativity which is so vital to scientific advancement (see Discovery: Creativity and the Scientific Method). In addition, a set of "rules" (*the* scientific method) suggests that science is very formal and ritualistic and is the exclusive domain of those who follow the prescribed path. If all science can be practiced by merely following a sequence of steps, then why does the process of scientific inquiry abound with so many false starts and even more deadends?

Yet, even though a great deal of scientific inquiry does not follow these steps, "the scientific method" forms the general ground rules that scientists use to formulate explanations (hypotheses) for how nature works, to test the accuracy of those explanations, and to report the findings so that others can benefit from the new information.

# Science as a Process

Scientists believe that all natural phenomena in the universe have rational, verifiable explanations. Such explanations are called hypotheses, theories, or laws, depending on how many times the explanation has survived the scrutiny of scientific testing without being disproven.

## Formulating Hypotheses

Hypotheses are the foundation on which science grows. An hypothesis is a plausible explanation for a particular observation or phenomenon, worded in such a way that it can be tested by experimentation. (For example, "the devil made it happen" is *not* a testable hypothesis.) The results of experiments tell the investigator whether the hypothesis should be accepted as is, whether it should be rejected because it is not supported by the results, or whether it can be modified to be consistent with the new findings. Repeated refinement of hypotheses is one of the ways science corrects itself and grows. In other words, science profits from its mistakes, as well as its successes.

For example, until about 50 years ago, the long-held explanation for the slipperiness of snow and ice was that the pressure from skis and ice skates lowered the freezing point of water, causing snow and ice to melt immediately beneath them. This film of water was said to allow skiers and skaters to slide freely over frozen surfaces (Figure 2-3). But more recent experiments have revealed that even a heavy skater produces only enough pressure to lower the melting point of water by one-tenth of one degree. This is certainly not enough pressure to melt ice when temperatures fall as low as −20° C (−4° F) and some hardy people continue to ice skate. (Even an elephant could not produce enough pressure to melt ice at that temperature.)

In a series of elegant experiments performed in the 1940s, a physicist at Cambridge University discovered that the friction of skis on very cold snow (about −60° C or −85° F) equaled that on dry sand. The results of these experiments led to a new explanation for how ice and snow melts beneath skates and skis; the friction of skates and skis produces sufficient heat to melt ice and snow.

## From Hypothesis to Theory

When an hypothesis has been repeatedly verified by observation and experimentation, it becomes a **theory**. Because scientific theories are backed by repeated tests, they are not mere speculations or guesses, as the term is sometimes popularly used.

But many theories are not just the product of one hypothesis. A theory may incorporate a number of verified hypotheses. For example, the atomic theory (all substances are formed from combinations of atoms) combines a number of hypotheses from both chemistry and physics. As with other scientific theories, the accuracy of the atomic theory has been verified time after time.

For a theory to continue being accepted as correct, it must consistently predict the results of new experiments or related phenomena. For instance, one of the most important theories of biology is the theory of evolution — the explanation for how species change over time. Evolutionary theory explains why there are fossilized organisms, long since extinct. It accounts for the observed changes in the characteristics of organisms over repeated generations, changes that are not only revealed in the fossil record, but take place today as well. Even similarities in the anatomy of related organisms and their nearly indistinguishable appearance during embryological development are explained by evolutionary theory. No one has yet discovered any scientific evidence that contradicts the theory of evolution or its role in the origin and ancestry of organisms on earth.

Yet, in science even the most trusted theories must withstand the challenge of further experimentation. Like hypotheses, theories either continue to

**Figure 2-3**
It's friction, not pressure as was originally thought that melts the snow under skis. Through continued experiments scientists make the world understandable.

be accepted, or are modified or discarded if new experiments reveal contradictory results.

Sometimes the invention of a new technology reveals information that sparks new hypotheses and theories. The invention of the microscope in the 1600s, for example, revolutionized biological thinking. By revealing that all living things, even giant redwood trees and whales, were constructed of tiny compartments called cells, the microscope was directly responsible for the development of the cell theory, which is as fundamental to biology as the atomic theory is to chemistry and physics.

Sometimes the invention of a new technology discloses a flaw in an existing hypothesis or theory. Again, it was the invention of the microscope that forced biologists to reconsider their practice of classifying all organisms as simply plants or animals. By examining microscopic organisms, biologists discovered many organisms that possessed both plant and animal characteristics, necessitating the formation of a third kingdom, the Protista. When the electron microscope was invented in the 1930s, internal cell characteristics were revealed that could not be

seen with even the best light microscope. Investigtors soon discovered that some cells had small internal structures (organelles), whereas other cells did not. This discovery led to a new theory that there were two fundamental cell types: those with organelles and those without. This new insight convinced biologists to establish yet another kingdom—the Monera kingdom—for all organisms whose cells lacked organelles. (The five kingdom classification system is discussed in Chapter 1.)

Most hypotheses and theories are the product of **inductive reasoning**, the assimilation of specific observations into a generalized explanation. The theory of evolution typifies inductive reasoning because it is a comprehensive explanation synthesized from many observations and hypotheses about how organisms change over generations (Chapters 29 and 30).

Once formulated, an hypothesis or theory is tested by **deductive reasoning**, using a generalization to deduce or predict a specific event or phenomenon. If a hypothesis or theory cannot predict successfully, it is modified or rejected. One historical example of how this works is when Galileo tested Aristotle's theory that the velocity of a falling object is directly proportional to its weight. Although Aristotle's theory seems "self-evident," it proved to be incorrect—a 10-pound ball and a 1-pound ball hit the ground at the same instant when dropped simultaneously from the same height. This example emphasizes the importance of repeated testing in science, for in this case testing resulted in the rejection of a long-held theory.

## From Theory to Law

If experimentation and observation consistently support a theory, it becomes a **principle** or **law**. But like all phases of scientific inquiry, even principles and laws must hold up to the unrelenting challenges of new discoveries, improved technology, and ongoing experimentation.

## The Central Role of Experimentation

Since hypotheses, theories, and principles or laws are subject to continued

experimental testing to verify or reconfirm their accuracy, experimentation is the very cornerstone of the process of science (see Discovery: The Explorers). To help us understand this central process, we will divide experimentation into four basic elements (Figure 2-2).

1. Designing experiments.

2. Conducting tests.

3. Analyzing results and drawing conclusions.

4. Reporting findings.

## DESIGNING EXPERIMENTS
Experiments are designed to test the validity of a hypothesis, theory, or law. Each experiment requires precise planning so as to eliminate bias and avoid reaching false conclusions. To do so, all factors that might affect the outcome of the experiment must be kept constant except for the one that is being tested. Such experiments are called **controlled experiments**.

Controlled experiments reveal a cause and effect relationship between an intentional change and the observed results. In other words, the experiment is designed so that the investigator is sure the results are caused by the experimental condition, and not just by coincidence.

To prove cause and effect requires a **control**, that is, a duplicate of the experiment identical in every way except for the one factor being tested. Similar results from the experimental test and the control reveal occurrences that will have happened anyway, changes that cannot be attributed to the presence of the experimental condition. In this way, controlled experiments reduce the number of variables to one, so that any change that occurred can be interpreted as the direct result of the one variable being tested. Cause and effect is proven.

To conduct controlled experiments on living organisms, biologists divide a *test population* of organisms into two subgroups: an *experimental group* and a *control group.* For example, to test the hypothesis that the food additive red dye #2 causes cancer (Figure 2-4), a test population of 100 rats was divided into two groups: an experimen-

# D I S C O V E R Y

## CREATIVITY AND THE SCIENTIFIC METHOD

Knowledge builds over time, with each new discovery blossoming from the seeds planted by previous researchers. Isaac Newton, for example, remarked, "if I have seen farther than other men, it is by standing on the shoulders of giants." In other words, it was the work of Copernicus, Brahe, Kepler, Galileo, and the ancient Greeks that enabled Newton to formulate many laws on the nature of the physical universe.

Most great scientific achievements are the result of creative genius, diligence, and many long hours of dedicated research. However, some revolutionary discoveries are simply the result of accident or flashes of insight, as in the following examples.

**Accidents and Scientific Discovery**. Although many scientific studies follow well-defined experimental procedures, occasionally an accident produces some unexpected results that lead an alert scientist to a great scientific discovery.

- *Microbe against microbe — The discovery of penicillin.*
  While trying to find a cure for influenza, Alexander Fleming accidentally discovered a substance that changed medicine forever — penicillin.

  In 1928, Fleming left some petri dishes containing bacterial cultures by an open window in his research laboratory. After a few days, he noticed that some of the dishes had become contaminated with a green mold. As he was discarding his ruined experiment, he alertly observed that the mold was surrounded by a ring in which there were no bacteria. But what, Fleming asked, was preventing the bacteria from growing near the mold?

  Fleming then conducted a series of experiments to demonstrate that the mold produced a substance that killed the bacteria. Fleming named this substance "penicillin" after the name of the contaminating mold, *Penicilliun notatum*. Penicillin was the first antibiotic (anti = against, bio = life), a microbe-killing substance produced by a fungus or a bacterium.

  Fleming's discovery soon led to the use of penicillin to treat a wide range of bacterial diseases and ushered in the "era" of antibiotics in medicine. For this discovery, Fleming shared the Nobel Prize for Medicine in 1945.

- *Pitting microbe against microbe to combat disease.*
  Louis Pasteur believed in the germ theory of disease, that is, that many diseases are caused by microorganisms growing inside one's body. (This theory may seem obvious to us now, but in the mid-1800s it was hotly debated.) In 1879, Pasteur was studying chicken cholera, a disease that quickly kills fowl. Managing to isolate the disease-causing microorganism from the blood of infected chickens, Pasteur cultured the microorganisms in chicken broth and demonstrated through experiments that microorganisms grown from these cultures maintained the ability to cause cholera when injected into healthy chickens.

  During his investigations, Pasteur took a summer vacation, leaving cultures with chicken cholera microbes in his laboratory. When he returned four months later, he resumed his studies. He injected chickens with these old cultures, expecting the chickens to come down with cholera as they always did in the past. But this time they didn't. Somehow the old bacterial cultures had lost the ability to cause the disease.

  Pasteur performed a new set of experiments, this time injecting two groups of chickens with freshly transferred microbes: (1) the healthy chickens that had not received any injections and (2) the chickens he had injected with the old culture. As expected, the first group of chickens got cholera. But the second group of chickens remained healthy. Pasteur had accidentally discovered the basic principle of vaccination — injecting a "weakened" version of a disease-causing microbe enables an animal to withstand exposures to the dangerous version of the microorganism. More specifically, weakened microbes can be used as *vaccines* to stimulate *immunity* against the corresponding disease.

  Pasteur went on to develop a vaccine against rabies. His accidental discovery enabled other scientists to develop vaccines against polio, rubella, mumps, diphtheria, tetanus, and others.

**Flashes of Insight**. Sometimes even the way our brain works affects the process of scientific discovery. While conscious, the brain is often so preoccupied with processing events and experiences that it is unable to make new connections between seemingly unrelated facts. Yet, while sleeping or when relaxing, a flash of insight connects previously unrelated phenomena, and a new level of understanding is achieved, as in our case of "The Apple That Changed the View of the Universe — Newton's Law of Universal Gravitation."

Does an apple falling from a tree have anything to do with the movement of the moon and planets? To Isaac Newton, watching just such an event sparked a connection that had never been made before — that the force which pulled the apple to the ground is also responsible for helping to hold the moon and planets in their places.

In 1665, there was nothing new about the notion that the earth's gravity pulls objects to its surface. But most people believed that the forces affecting earthly things were different from the forces controlling heavenly bodies. Newton's great contribution was in mathematically proving that gravity affected all objects in the universe. He showed that gravitational force was proportional to the difference in the mass of the attracted bodies (the earth is much more massive than an apple) and that the force diminished according to the square of the distance between the bodies. With this discovery, he could account for the elliptical orbits of moons and planets, as well as an apple falling from a tree.

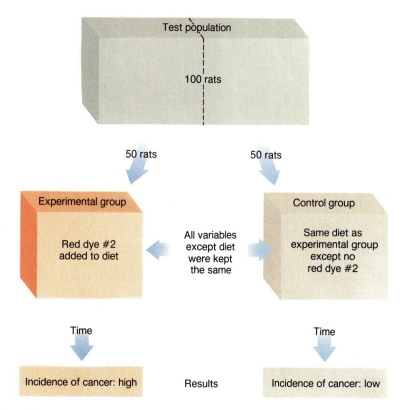

**Figure 2-4**

Does red dye #2 increase the incidence of cancer? Around the turn of the century, it was not unusual to test the health effects of food preservatives (and other chemicals) on human subjects, even children. But today not only is the human "guinea pig" approach considered inhumane, but also the length of time for ill effects or diseases to develop (often between 20 and 40 years) makes the approach dangerous particularly if a disease spreads rapidly. Because of many physical and physiological similarities to humans, shorter lived rodents have now replaced human subjects.

Using such experiments, scientists tested the hypothesis that red dye #2 increases cancer in laboratory rats. All factors were identical in both the test group and control group except for the addition of red dye #2 to the food of the test group. The results of this experiment supported the hypothesis, as did a number of similar studies. Because there is a strong correlation between some causes of cancers in humans and in rats, the FDA banned red dye #2 as a food additive.

tal group which received food with red dye #2, and a control group which received exactly the same food but without red dye #2. All other variables (such as light, temperature, quantity of food, and amount of water) were identical for the two groups. The results from a number of such studies demonstrated that red dye #2 did indeed increase the incidence of cancer in rats. Because rats and humans are biologically similar, the Food and Drug Administration (FDA) banned red dye #2 as a food additive.

Controlled experiments on people often include "blind" and "double-blind" tests to reduce the chances that psychological effects will influence the results. For example, in a blind test to

analyze the effects of a particular tranquilizer, the experimental group received a pill containing the tranquilizer and the control group received a *placebo,* the same pill but without the tranquilizer. Because neither group knows if it is part of the experimental group (receiving the tranquilizer) or the control group (receiving the placebo), this kind of experiment is called a "blind test." When the researchers themselves are unaware of which subject is receiving the tranquilizer or the placebo, the test is called a "double-blind test." Neither the researchers nor the subjects know which individuals are part of the experimental group and which are part of the control group. Double-blind tests reduce the chances

of bias that may be introduced when researchers try to anticipate an outcome.

Two other critical elements of experimental design are (1) the size of the test population, and (2) the manner in which individuals are selected for the test population. For example, we would not expect the FDA to ban red dye #2 if the results of the experiment were obtained by examining only a few, specially chosen rats. In general, the larger the size of the test population and the more randomly the individuals are chosen for the test population, the less bias and subjectivity are introduced into the experiment. Bias and subjectivity can be responsible for invalidating the results of experiments.

## CONDUCTING TESTS

There is no definitive way to conduct a test, for most experiments differ in design, equipment, surroundings, number of researchers, and several other factors. But if there is one rule that should invariably apply when conducting experimental tests, it is to take painstakingly detailed notes at every step of the way. In this way, any factor that may possibly affect the outcome of the experiment is documented and can be reviewed during analysis.

Sometimes the significance of a factor is not apparent to the original investigators but becomes clear to later researchers. For example, biologists have frequently described a rather gruesome "love story" between male and female praying mantises. This version of the sex life of the praying mantis is based on a number of investigations that have been done over a 200-year period. However, all these studies were conducted in the same way and so ended up with similar results. These results are now being challenged by new studies.

The original "love story" goes like this. During mating, the smaller male praying mantis approaches and mounts a female. As he begins to copulate, she turns and decapitates him with her powerful jaws. The headless male continues mating, and the female is inseminated. In attempts to explain this grisly behavior, scientists have speculated that perhaps copula-

# DISCOVERY

## THE EXPLORERS: BIOLOGISTS DISCOVERING THE SECRETS OF LIFE

Although biologists rarely become politicians or economic leaders, they nonetheless help shape the future of the world. How different life would be, for example, if biologists had not provided us with our understanding of (and our weapons against) the microorganisms that cause smallpox, polio, diphtheria, typhoid fever, and dozens of other fatal diseases. Modern biologists (see photos) are not only continuing their century-old war against disease, but are also tackling many other tough problems that face the modern world—air pollution, waste and toxic chemical disposal, malnutrition and human starvation, acid rain, and energy shortages, to mention a few. Some biologists are creating and testing vaccines against the deadly disease AIDS (*a*cquired *i*mmuno*d*eficiency *s*yndrome), while others genetically manipulate corn and other crop plants to engineer strains that will enable farmers to produce more food on less land and at lower costs.

There are also biologists seeking the answers to scientific questions simply for the sake of knowledge and increased understanding. The question of what killed the dinosaurs, for example, may have no practical application, but it is something most of us would like to know nonetheless.

Sometimes, however, without any practical motivation for the research, a single discovery can lead to hundreds of practical applications. Consider one example: the discovery of the molecular structure of DNA, the genetic material of life. Our understanding of the structure of DNA and how it works has profoundly improved medicine, agriculture, horticulture, animal husbandry, and other applied fields, in addition to spawning a whole new field of applied biology, called **genetic engineering**. With the knowledge of DNA structure, genetic engineers are able to manipulate DNA in order to construct organisms with a desirable combination of traits. For example, they have developed bacteria with the ability to produce the human hormone insulin and plants with the ability to manufacture their own fertilizer. (Some of the current techniques and advances in genetic engineering are discussed in the final chapter of this book, "The Biotechnology Revolution.")

Knowledge with no obvious application may sometimes seem like useless information. But as the discovery of the structure of DNA illustrates, pure knowledge may lead to many new applications. In fact, our increased understanding of fundamental biological principles will likely spawn cures for cancer, mental illness, AIDS, and even the common cold. So the question "What good is this information?" is easy to answer. It prepares the mind for the unexpected, in other words, for the future. No one can predict what the future holds, but the better equipped the mind, the better the ability to make connections between seemingly unrelated phenomena, connections that lead to insight and new discoveries.

**Figure 2-5**
New evidence now casts doubt on what was once believed to be a very gruesome mating ritual between male and female praying mantises. The new results disprove the conclusion that males are always decapitated by a female during mating. Furthermore, recent studies show that "beheading" is not necessary for copulation and insemination, as was originally suggested. Researchers conducting these new tests on the mating behavior of praying mantises suggest that the earlier observations of "beheadings" by females may have been caused by the experimental conditions themselves, specifically bright laboratory lights and inadequate supplies of food for the larger females.

tion and ejaculation are inhibited by the male's own brain. If this were so, insemination of a female would be possible only after the male's brain was removed, a reflex response similar to a chicken running around with its head cut off. When we analyze the behavior objectively, as scientists do, we see that the benefits of this behavior are twofold; not only does the female receive a supply of sperm for reproduction, but she gains nourishment as she eats her mate, nourishment that might contribute to the development of the young themselves.

But recent investigations at the University of California at Santa Cruz reveal that male praying mantises do not lose their heads over females (Figure 2-5). In videotapes of mating praying mantises, the females did not decapitate a single male, either before or during copulation. The tapes show that both the male and female mantises perform a courtship dance before copulation, and although the female dance does begin with what appears to be a threatening stance, it ends with a receptive posture for copulation. The male mounts the female, begins copulation, and, with his head in place, inseminates the female.

So, why did the male mantises lose their heads in all the earlier experiments, but not in the Santa Cruz study? The Santa Cruz researchers explain that in previous studies, mantises were always observed under very bright lights, a condition that perhaps distracted the mantises and prevented the natural courtship dance. The dimly lit, videotaped matings observed at Santa Cruz also suggest that the original "headless lover" hypoth-

esis (that is, the male's brain had to be removed before copulation and insemination) is incorrect. But why would the females in the earlier studies eat the heads off the males? The Santa Cruz researchers believe that the cannibalism was more the result of hunger than of some horrid sexual instinct. Since a female mantis eats about 15 full-grown crickets each day, previous investigators may simply not have given the females enough food. If new studies continue to support these findings, biologists will be telling a new "love story" for the praying mantis, one that is definitely less repulsive, especially to males.

The praying mantis "love story" illustrates why scientists repeat experiments to verify results and are very skeptical of observations that cannot be duplicated by others or are reported by only one individual. If something is factual, the results should be consistent from test to test, and from scientist to scientist.

## ANALYZING RESULTS AND DRAWING CONCLUSIONS

All variables that might affect the outcome of an experiment must be accounted for during the analysis of the results. A number of statistical techniques helps biologists eliminate bias, identify variables, and reveal relationships that may not be obvious from observation alone. Some statistics also help determine whether the observed results were caused by the variable being tested, or simply were due to chance alone.

Sometimes it is impossible to run controlled experiments in which only one variable is tested. In many ecological studies, for instance, a number of variables (temperature, wind, exposure to light, for example) cannot be controlled and yet might affect the outcome. In these studies, biologists frequently rely on statistical procedures that help identify the variables that have the greatest impact on the outcome.

Experimental results are of little value if they don't suggest conclusions based on interpretation of the data. Results are usually much easier to interpret, and patterns and trends are more obvious if the results are arranged in tables or graphs, as Figure 2-6 illustrates.

After all variables are considered, all results synthesized and tabulated, and all statistics completed and interpreted, what conclusions can be drawn? Did the results support the hypothesis? Should the hypothesis be rejected, or can it be modified to agree with the findings? What new insights do the results reveal, or what trends are suggested? And, inevitably, what is the next step? The results of one experiment can inspire the formulation of many new hypotheses and may even open the door to a whole new arena of scientific interest. Finding the cure for cancer is a classic example. When research indicated that the body's own immune defenses might play a role in combating cancer, a whole new series of studies were initiated on "immune boosters," such as natural hormones that stimulate the body's immune system.

## REPORTING FINDINGS

Unless the results and conclusions gained from an experiment are shared with other scientists, the potential value of the experiment remains hidden. Sharing what one has discovered also helps to speed up the process of discovery. It sparks the imagination of other scientists and redirects other experiments. For example, when James Watson and Francis Crick reported their model of the structure of DNA, the genetic material of life, a revolution was unleashed in biochemical and genetic research. With this knowledge, investigators soon learned how cells and genes work. We are already enjoying the triumphs of these discoveries, and as molecular biology and biotechnology continue to advance based on the Watson and Crick model, not even the researchers themselves can predict what the future accomplishments will be.

Reporting findings has yet another effect: sometimes it generates controversy among scientists themselves. But controversy, too, helps science to grow, as we will see throughout the text.

(a) WORDS
Your chances of being killed in a passenger car or taxi are about 1 in 4 million in any single trip, compared to 1 in 40 million on a train or on an airline, or 1 in 106 million on a bus.

(b) TABLES

| Transportation mode | Number of deaths | Billions of passenger miles |
|---|---|---|
| • Auto or taxi | 22,859 | 2,154.4 |
| • Airline | 218 | 213.6 |
| • Train | 9 | 10.9 |
| • Bus | 32 | 88.8 |

(c) GRAPHS

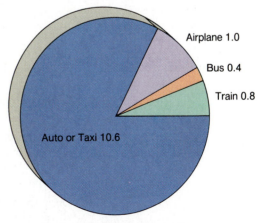

Airplane 1.0
Bus 0.4
Train 0.8
Auto or Taxi 10.6

Deaths per billion passenger miles

**Figure 2-6**
**Words, tables, or graphs?** Often, the way the results are presented helps an investigator (and reader) see trends more clearly or better understand the relationship between variables. In this figure, the number of deaths for different modes of transportation are compared (deaths per billion of passenger miles). Which presentation do you feel shows the relationship best?

## Synopsis

### MAIN CONCEPTS

- Scientists believe that the world is understandable and that rational explanations exist for all phenomena.

- Scientific investigations commonly follow an investigative strategy known as the scientific method. The process includes
  —making careful observations.
  —formulating hypotheses, theories, and laws.
  —designing and conducting experiments to test the validity of hypotheses, theories, and laws.
  —analyzing results and drawing conclusions.
  —accepting, rejecting, or modifying hypotheses, theories, and laws.

- One or more hypotheses that have been repeatedly supported by experimentation become a theory, and if further experimentation continues to support the theory, it eventually becomes a law or principle.

- A controlled experiment produces results that are due to the effect of a single experimental variable. Controlled experiments help distinguish cause and effect from coincidence, and reduce the influence of bias. The use of a control allows the investigator to determine the effect of one factor at a time.

### KEY TERM INTEGRATOR

The Scientific Method: The Process of Science

| | |
|---|---|
| **hypothesis** | A tentative explanation for an observed phenomenon. Hypotheses are tested through **experimentation.** |
| **theory** | A general explanation derived from one or more hypotheses that have been repeatedly verified by experimentation. |
| **law** or **principle** | A theory that has been repeatedly verified. |
| **inductive reasoning** | A logical process in which a generalization is developed to explain several specific facts. Hypotheses and theories are formulated by inductive reasoning. |
| **deductive reasoning** | A logical process in which hypotheses and theories are tested, using a generalization to deduce or predict the outcome of a specific event or set of circumstances. |
| **controlled experiments** | Experiments with **controls** that eliminate the influence of all variables except for the one being tested. |

## Review and Synthesis

1. Explain why some scientists argue against the existence of a single, all-purpose "scientific method."

2. Formulate a hypothesis based on the following set of observations.
   a. Some organisms are composed of only one cell.
   b. Large organisms are composed of many small cells.
   c. Every organism, small or large, begins life as a single cell.

3. Design a *controlled* experiment to test the hypothesis that high doses of vitamin C reduce the incidence of colds. Can you use a test population of human subjects? What are the normal limitations, and how will they affect the results?

4. Why do scientists have so much confidence in a law or principle?

**5.** How would you apply the scientific method to test the accuracy of the following statements? (Get together with some of your classmates and compare the strengths and weaknesses of one another's approach.)

a. An ostrich sticks its head in the ground to avoid impending danger.

b. Disasters occur in groups of three.

c. Large doses of vitamin E improve one's energy.

d. A black cat crossing your path will bring bad luck.

## Additional Readings

Asimov, I. 1984. *Asimov's New Guide to Science.* Basic Books, New York. (Introductory.)

Branscomb, L. 1986. "Science in 2006." *American Scientist* 74: 650–658. (Introductory.)

Lederman, L. M. 1984. "The value of fundamental science." *Scientific American* 251:40–47. (Introductory.)

Lenihan, J., and J. Fleming 1979. *Science in Action.* Institute of Physics, Bristol and London, England. (Introductory.)

*Science 84.* November 1984. "20 Discoveries that shaped our lives." American Association for the Advancement of Science. 5th Anniversary issue. (Introductory.)

# 2

# Chemical Foundations of Life

At the most fundamental level, all organisms are chemicals. Understanding life requires understanding chemistry.

Chapter 3

# The Atomic Basis of Life

To explore life adequately, one must venture into an invisible realm. This is the realm of chemicals and chemical reactions, where miniscule particles combine with each other to forge everything in the universe, including that which is alive. Chemistry represents biology on its most fundamental and essential level, that which ultimately accounts for what we are and what we do as living entities. Probing the chemical basis of life is a first step in understanding biology. It reveals how different particles can be arranged and rearranged to form an unlimited variety of organisms.

## Atoms and Elements

As remarkable as it may seem, all substances are composed of three types of particles—protons, neutrons, and electrons. These three kinds of minute particles combine to form **atoms**, the fundamental units of matter that can enter into chemical reactions. From atoms, all more complex forms of matter are constructed.

Every atom has a central nucleus that contains one or more **protons**, particles with a positive electrical charge. In addition, the nucleus usually contains electrically neutral (uncharged) particles called **neutrons**. The atomic nucleus therefore carries a positive charge, one unit of charge for each proton. Smaller, negatively charged particles, called **electrons**, orbit around the nucleus. Particles with opposite charges of equal magnitude neutralize each other, one charge canceling out the other. Thus, atoms with equal numbers of protons and electrons have no net charge. They are electrically neutral.

### Elements

The nucleus of each type of atom possesses a characteristic number of protons called the **atomic number.** The number of protons in the atom determines its identity and its chemical properties. Atomic numbers of naturally occurring atoms range from *1* (for hydrogen, the smallest atom) to *92* (for uranium, the largest). These 92 elements do not include the 14 elements (93 through 106) that have been synthesized by human technology.

The simplest substances that retain the structure and properties of an atom are **elements.** All atoms with the same atomic number constitute the same element. For example, any atom with eight protons is an atom of the element oxygen. A six-proton nucleus distinguishes every atom of the element carbon. Carbon and oxygen are two elements especially important to life. Fewer than one-fourth of the 92 natural elements participate in life processes (Table 3-1). Of these, six elements (the first six shown in the table) are the most biologically prevalent, comprising about 99 percent of all living matter (Figure 3-1). The other elements shown in the table, even though present in smaller amounts, are nonetheless essential for most organisms.

**Figure 3-1**
**Different but similar.** Each of these strikingly different organisms is composed almost entirely of the same six chemical elements, plus small or trace amounts of various other elements.

## TABLE 3-1

| Elements That Constitute the Chemistry of Life | | | |
|---|---|---|---|
| **Element** | **Symbol**[a] | **Atomic Number** | **Atomic Weight** |
| Hydrogen | H | 1 | 1 |
| Carbon | C | 6 | 12 |
| Nitrogen | N | 7 | 14 |
| Oxygen | O | 8 | 16 |
| Phosphorus | P | 15 | 31 |
| Sulfur | S | 16 | 35 |
| Fluorine | F | 9 | 19 |
| Sodium | Na | 11 | 23 |
| Magnesium | Mg | 12 | 24 |
| Silicon | Si | 14 | 28 |
| Chlorine | Cl | 17 | 35 |
| Potassium | K | 19 | 39 |
| Calcium | Ca | 20 | 40 |
| Vanadium | V | 23 | 51 |
| Manganese | Mn | 25 | 55 |
| Iron | Fe | 26 | 56 |
| Cobalt | Co | 27 | 59 |
| Zinc | Zn | 30 | 65 |
| Selenium | Se | 34 | 79 |
| Iodine | I | 53 | 127 |

[a] *These symbols are used as chemical shorthand for describing elements and molecules.*

Each element also has a characteristic **atomic weight** (sometimes called "atomic mass") that is roughly equivalent to the total number of protons and neutrons in the nucleus. Protons and neutrons are virtually identical in weight, each arbitrarily assigned a mass value of 1. Thus, oxygen, with its eight protons and eight neutrons, has an atomic weight of 16. Electrons have little effect on the total weight of an atom. It would require 1840 electrons to add one unit of mass to an atom, and even the largest natural element has fewer than one-twentieth that many electrons.

## Isotopes

Although an element's atomic number is constant, its atoms may vary in the number of neutrons they possess. Each element may therefore have atoms with different atomic weights. Carbon, for example, usually has six protons and six neutrons, making its atomic weight 12. Yet some carbon atoms contain one, two, or three additional neutrons, giving them atomic weights of 13, 14, and 15, respectively. In other words, carbon has four **isotopes**, each a different form of the same element. Most elements have several isotopes, which are identified by a superscript number in front of the atomic symbol. For example, $^{12}C$ denotes the isotope carbon-12.

Some isotopes are radioactive because of their instability. When these "radioisotopes" decay to a more stable atomic configuration, they release high-energy radiation. This discharged energy can be detected by instruments such as a Geiger counter or by photographic film. Radioisotopes play an indispensable role in biological and medical research (see Bioline: Putting Radioisotopes to Work).

## The Behavior of Electrons

The number of protons in the nucleus dictates the chemical behavior of an element by determining the number of electrons in the atom. There is one electron per proton in an electrically neutral atom. Electrons tend to orbit around the nucleus in energy levels called **shells** (Figure 3-2). Electrons don't follow a flat circular pathway but sweep around the nucleus in spherical or dumbbell-shaped orbits. The electrons move so rapidly in their orbits that they are envisioned as "electron clouds."

Each shell has a maximum number of electrons it can hold. When a shell becomes filled, any additional electrons must go into the next shell further away from the nucleus. The first shell can contain only two electrons; the next two energy levels can accept

# B I O L I N E

## PUTTING RADIOISOTOPES TO WORK

Radioactive isotopes have helped reveal biological information that may have otherwise escaped detection. Even with the most powerful microscopes, chemical processes remain invisible. Radioisotopes, however, can be followed throughout a chemical reaction with a radiation-sensitive machine. Radioisotopes used this way are called *tracers* because they can easily be tracked and located. Tracers incorporated into specific biochemicals are introduced into an organism, and the fate of that chemical can be tracked by periodically determining what compounds contain the isotope. In this way scientists have charted the complex chemical odysseys of molecules as they are processed by living organisms. Radioisotopes revealed how plants capture simple molecules and, using solar energy, transform them into the complex molecules of which leaves, stems, roots, and new plants are constructed. Growing plants in an atmosphere saturated with radioactive carbon dioxide ($^{14}CO_2$) allowed scientists to determine the chemical mechanism by which plants build the bulk of their tissues from $CO_2$ extracted from air. Other radioisotopes helped map the routes used for transporting substances throughout plants.

Tracers have provided answers to some of our most fundamental questions, for example how genes dictate all the physical characteristics of heredity, and how they duplicate themselves precisely so that offspring have characteristics similar to those of their parents. Radioactive isotopes were also the key to cracking the genetic code, the "language" used by cells that enables them to follow the genetic instructions encoded in their genes. (The details of the genetic code are discussed in Chapter 27.)

Radioisotopes are also used for diagnosing and treating disease. Radioactive iodine ($^{131}I$), for example, is used to detect abnormal thyroid function. Isotope tracers provide a means of observing blood flow, often revealing dangerous circulatory obstructions caused by blood clots or narrowing of arteries. Many highly sensitive laboratory tests for diagnosing allergies, cancers, and other disorders also utilize radioactive isotopes.

Although many radioisotopes are relatively harmless, some emit enough energy to injure or kill living tissue and cells. The lethal effect of these isotopes has been harnessed and used for several purposes. Many types of medical and laboratory supplies are sterilized by exposure to radiation emitted from these isotopes, thus reducing the danger of contaminating patients with disease-causing microorganisms. Radiation may also be used to kill cancer cells that grow rampantly inside people. Many cancer patients are alive today because their malignant tumors were showered with lethal radiation.

Radioactive isotopes also provide accurate geological "clocks" that reveal the approximate age of very old objects, a fossil, for example. Each radioactive element has a characteristic **half-life**, the time required for half of it to decay into its stable, nonradioactive form. To illustrate, carbon-14 has a half-life of 5730 years. After 5730 years, only half of the original amount of $^{14}C$ remains. During the next 5730 years, 50 percent of that amount decays, and so on.

Once an object's ratio of radioactive $^{14}C$ to $^{12}C$ (the most common isotope of carbon) has been determined, its age can be calculated. In all living organisms, this ratio is the same as the ratio of $^{14}C$ to $^{12}C$ in atmospheric carbon dioxide. When an organism dies, however, it stops accumulating carbon. As its $^{14}C$ decays, the relative proportion of radioactive carbon to nonradioactive carbon declines steadily according to the isotope's half-life. Because the rate of decay is constant, it dependably identifies an object's age. Radioisotopes ignore all environmental fluctuations and continue decaying with relentless precision. Every fossil younger than 70,000 years contains a built-in "carbon clock" that reveals its age. (The sabre-tooth tiger, whose skull is in the accompanying photo, is 1 million years old.) Although specimens older than 70,000 years have too little $^{14}C$ left for accurate determination, their age can still be determined by using radioisotopes with much longer half-lives than carbon-14. Radioactive potassium ($^{40}K$), for example, decays (forming argon-40) at a half-life rate of 1.3 billion years, helping biologists reconstruct evolutionary events that occurred even billions of years ago.

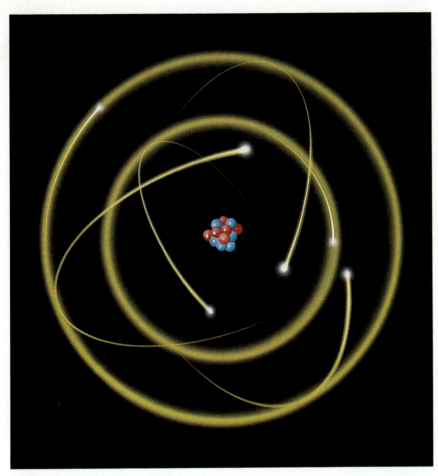

**Figure 3-2**
**Simplified version of a typical atom.** The nucleus of this carbon atom contains 6 protons and 6 neutrons. Six electrons orbit the nucleus in two shells. Although useful in understanding atomic structure, this simplified model (called a Bohr's atom) fails to show the enormous distance between nucleus and electrons, and inaccurately represents the shape of the shells. The shells are more like "clouds" created by electrons traveling around the nucleus at about the speed of light.

eight each. Thus, the six electrons of the carbon atom depicted in Figure 3-2 are distributed in two shells in a 2:4 arrangement, that is, a pair in the innermost shell and the other four electrons in the second shell. The element phosphorus has its 15 electrons distributed as 2:8:5. Sulfur's 16 electrons form a 2:8:6 arrangement, differing from chlorine (17 electrons) by one less electron in its third shell (Table 3-2).

Electrons therefore tend to be oriented in a hierarchy around the nucleus, so that every neutral atom of a particular element has the same number of electrons in its outermost shell. These outer-shell electrons determine some fundamental chemical properties of the atom, namely, its tendency to lose electrons, to take electrons from other atoms, or to share electrons with other atoms. Such atomic tendencies dictate the chemical reactivity of an element and determine its role in the chemistry of life.

## Chemical Reactions and Molecular Forces

Most atoms tend to establish stable partnerships with other atoms, forming larger, multi-atomic complexes

called **molecules.** These reactions do not occur haphazardly, however, but are governed by a fundamental principle—that is, an atom is most stable when its outermost electron shell is filled. Most electrically neutral atoms fall short of this ideal.

Two or more atoms, however, can team up to form a molecule that is more stable (less reactive) than its separate individual atoms. For example, when an atom whose outer shell is only two electrons shy of the eight needed to fill the shell encounters an atom with only two electrons in its outer shell, they may join together. By transferring or sharing electrons, each atom fills its outer shell while maintaining the overall electric neutrality of the team. Such a partnership between atoms links them together, forming a **chemical bond.** Two types of chemical bonds are fundamental to life—ionic bonds and covalent bonds.

### Ionic Bonds

Some atoms are so electron-attractive that they can capture electrons from other atoms, those that have a weaker hold on their one or two outer-shell electrons. The "electron donor" atoms are more stable without these electrons which readily jump to electron acceptor atoms.

For example, when the elements sodium (a metal) and chlorine (a gas) are mixed, one electron migrates from one element to the other, transforming these two materials into sodium chloride—neither a gas nor a metal, but ordinary table salt. The single electron in sodium's outer shell is readily transferred to an atom of chlorine, filling the one-electron deficit in chlorine's outer shell. This transaction produces two electron-satisfied atoms, each with a saturated outer shell. The electron transfer, however, creates an electrical imbalance in each atom because the electron number no longer corresponds to the proton number. In other words, both atoms have acquired an electrical charge. The atom that captured the extra electron acquired a negative charge, whereas the electron donor became positively charged. Such electrically charged atoms are called **ions.** The oppositely charged ions attract one another in much the same way as do protons and electrons, forming a chemical linkage called an **ionic bond** (Figure 3-3). The resulting molecule is electrically neutral.

Organisms capitalize on the mutual attraction of oppositely charged ions to accomplish a number of biological tasks, many of which are discussed in later chapters.

## Covalent Bonds

Some atoms *share* their electrons with other atoms, forming a linkage quite different from ionic bonds. In these linkages, called **covalent bonds,** the participating atoms lose no electrons. Instead, the outermost shells of two atoms share the same electrons. A few examples of covalent bonding are presented in Figure 3-4. Hydrogen, with its single electron, can satisfy its requirements for two outer shell electrons by combining with an identical atom, forming a hydrogen molecule, linking the two atoms together by a *single covalent bond* (one pair of shared electrons). Oxygen's requirement for two electrons can be satisfied by combining with two hydrogen atoms, forming water ($H_2O$). Oxygen can also satisfy its two-electron deficit by bonding with an identical atom, forming an oxygen molecule ($O_2$). The

TABLE 3-2

| | | Number of Electrons in | | | | |
|---|---|---|---|---|---|---|
| **Element** | **Atomic Number** | **Shell 1** | **Shell 2** | **Shell 3** | **Electrons in Outermost Shell** | **Number of Electrons that would Produce a "Full" Outermost Shell** |
| Hydrogen | 1 | 1 | — | — | 1 | (gain or lose) 1 |
| Carbon | 6 | 2 | 4 | — | 4 | (gain) 4 |
| Nitrogen | 7 | 2 | 5 | — | 5 | (gain) 3 |
| Oxygen | 8 | 2 | 6 | — | 6 | (gain) 2 |
| Sodium | 11 | 2 | 8 | 1 | 1 | (gain) 1 |
| Magnesium | 12 | 2 | 8 | 2 | 2 | (lose) 2 |
| Phosphorus | 15 | 2 | 8 | 5 | 5 | (gain) 3 |
| Sulfur | 16 | 2 | 8 | 6 | 6 | (gain) 2 |
| Chlorine | 17 | 2 | 8 | 7 | 7 | (gain) 1 |

**Electron Distribution in Several Biologically Important Elements**

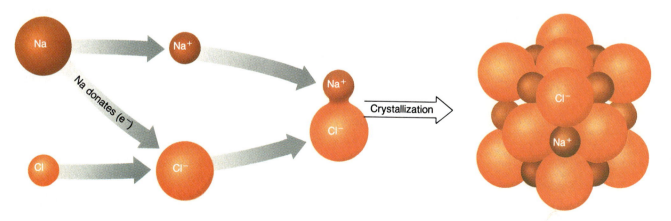

**Figure 3-3**
**Formation of an ionic bond.** Sodium chloride (NaCl), common table salt, typifies the product of ionic bonding. Sodium readily donates its one outer-shell electron to chlorine, which has one unfilled electron space. Because of their opposite charges, the resulting ions ($Na^+$ and $Cl^-$) strongly attract each other and, in the absence of water, form tight ionic bonds that hold adjacent ions together in a cube-shaped crystal of salt.

Hydrogen

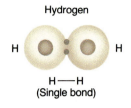

H——H
(Single bond)

Water

Nucleus

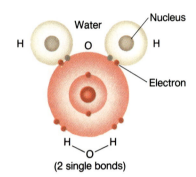

Electron

(2 single bonds)

Oxygen

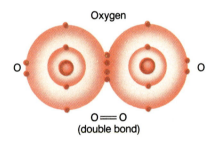

O═══O
(double bond)

Nitrogen

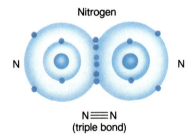

N≡≡≡N
(triple bond)

Methane

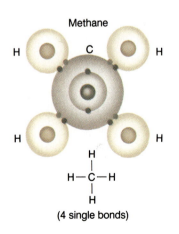

(4 single bonds)

**Figure 3-4**
**Examples of covalent bonding.** The shared electron pairs (in color) orbit around every nucleus in the molecule.

two atoms of oxygen are bound together by a *double bond,* one bond for each of the two electron pairs shared. Some elements can even form *triple bonds,* sharing three pairs of electrons, as illustrated by molecular nitrogen ($N_2$). Some atoms can form four covalent bonds, although not with just one other atom. For example, carbon shares its electrons with four molecules of hydrogen to form methane ($CH_4$).

## POLARITY IN COVALENTLY BONDED MOLECULES

In some electrically neutral molecules, the electrons are not shared equally by all the atoms because one atom is more electron-attractive than its partners. This shift of electron distribution concentrates negative charges in the area of this atom at the expense of the molecule's less attractive atoms, where a slight positive charge accumulates. The result is a **polar** molecule, one with an unequal charge distribution that creates distinct positive and negative regions, or *poles.* (In contrast, **nonpolar** molecules have an equal charge distribution throughout their structure and thus lack charged poles.)

Water is one of the most important polar molecules in the biosphere. Water's single oxygen atom attracts electrons more forcefully than do the two hydrogens. In fact, the electrons spend about 95 percent of their time at the oxygen end of the molecule, making water extremely polar.

As a result, water has a negative pole opposite two positive poles. The polarity of water provides it with important properties on which the very existence of life on this planet depends. One of these properties is the ability to form hydrogen bonds.

## Hydrogen Bonds

Another kind of chemical coupler pulls polar molecules close to one another. This linkage, the **hydrogen bond,** is formed when two molecules share an atom of hydrogen. Positively charged hydrogens of one polar molecule align next to negative portions of another molecule. Water clearly illustrates how this attractive force works. Adjacent water molecules are drawn together at their oppositely charged regions, generating a weak attraction similar to that between opposite poles of a magnet (Figure 3-5). In effect, the oxygen atoms of two neighboring water molecules share a common hydrogen atom. The sharing is not equal, of course, for the covalent bond that links the hydrogen to one of the oxygens is about 20 times stronger than the hydrogen bond that links it to the other. Although weak, hydrogen bonds are so plentiful in water that, at room temperature, most of the water molecules remain closely associated in a liquid state. Even so, water molecules occasionally break away from the intramolecular attraction and escape the liquid as water vapor. This process is called evaporation.

Hydrogen bonding is not limited to water but is common among polar molecules in general. Hydrogen bonds stabilize giant protein molecules and nucleic acids (see Chapter 4). Other examples throughout this book illustrate life's dependence on hydrogen bonding.

## Hydrophobic Interactions

Polar molecules are attracted to water molecules, that is, they are **hydrophilic** (hydro = water; philia = loving). Hydrophilic substances readily dissolve in water. *Nonpolar* substances, on the other hand, are insoluble in water. These molecules lack charged regions that would be attracted to the poles of water molecules. When nonpolar compounds are mixed with water, the attraction between the polar water molecules forces these nonpolar, **hydrophobic** (phobia=fearing) substances into aggregates that minimize exposure to their polar surroundings. This

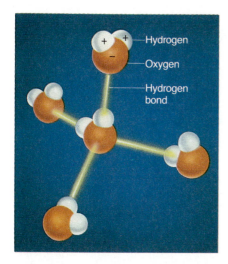

**Figure 3-5**
**Hydrogen bond formation between adjacent water molecules.** These weak "bonds" are really attractions between oppositely charged portions of adjacent molecules. Each $H_2O$ molecule can bond to four other water molecules.

explains why fats and oils, both of which are hydrophobic substances, form globules in water. Hydrophobic forces hold together cell membranes and many other biologically important structures (see Chapters 4 and 6).

## Molecular Collisions

Chemical reactions occur only when the reactants physically contact one another. (**Reactants** are the molecules or atoms that are changed to **products** during a reaction.) Atoms and molecules are in constant random motion and inevitably collide with other atoms and molecules. Many of these collisions simply change the direction in which the particles are moving. Sometimes, however, the force of collision is great enough to disrupt chemical bonds or form new ones, that is, to trigger a chemical reaction.

Heat increases the rate of chemical reactions by increasing molecular velocity. Faster moving particles collide more often than do those that travel more slowly. They also strike with greater force, increasing the likelihood that a molecular collision will exceed the energy necessary to break existing bonds and form new ones. The mini-

mum collision energy needed to produce a chemical reaction is the **activation energy** for that reaction.

## Energy Conversions

The formation of chemical bonds often requires an overall input of energy, much of which is retained in the bond. This **chemical energy** stored in the molecule is a measurable amount; each specific compound contains a particular amount of energy. In general, the larger and more organized a molecule is, the more chemical energy it contains. Because chemical energy can be released as heat, the standard units for reporting the amount of energy contained in compounds is the **calorie**, a measure of heat. One calorie is defined as the amount of heat energy necessary to elevate the temperature of one gram of water—about half a thimbleful—by one degree (from $14.5°$ C to $15.5°$ C).

The chemical energy trapped in these bonds maintains the stable organization of the molecule. This stored energy resists the tendency of the individual atoms in the molecule to travel their separate ways. Maintaining order is hard work, and work requires energy. Matter tends to resist order, to become more chaotic and less organized. In other words, it seeks its lowest energy state, like water flowing downstream. Becoming more organized, or even maintaining the present state of organization, requires a constant input of energy. This is a fundamental law of nature.

## Entropy

By "law," nature is wasteful. Energy transfers are far from being 100 percent efficient. Consequently, much of an organism's chemical energy escapes when it is consumed by another organism. In fact, any time energy is converted from one form to another, some of it inevitably is lost as useless energy, usually in the form of highly dissipated heat. When an automobile converts the chemical energy of gasoline into the mechanical energy of motion,

about 85 percent of it is wasted as heat. (This explains why cars need a cooling system.) Although living cells are almost three times more efficient than cars, much useful energy still escapes during the chemical processes of life. All organisms must comply with this inescapable law of thermodynamics*: during any process that consumes or releases energy, some useful energy is lost. This tendency toward energy escape even has a special name—**entropy**—to distinguish it from the release of usable energy. Entropy is therefore a measure of disorganization.

Entropy is the ultimate winner in the energy game. Every time energy "changes hands," the universe's entropy increases. Eventually, entropy will envelop everything, locking the entire universe into a state of perpetual randomness. Fortunately, we have billions of years left before this state of entropic doom arrives.

## Struggling Against Randomization

Each organism represents a temporary departure from matter's relentless march toward disorganization. An organism is one of nature's most intricately organized entities, and it remains alive only as long as it continues to defy entropy—to resist the dictates of a randomizing universe. Living requires a constant source of energy for an organism to maintain itself as an organized entity, to build more of itself (growth), to replace molecules as they wear out, and to produce offspring. Much of this energy resides in the chemical bonds of an organism's complex molecules. This chemical energy maintains molecular order.

Chemical energy, indeed, stored energy of any kind, is called **potential energy**. When potential energy is changed into **kinetic energy** (energy in

---

* *Thermodynamics is the study of energy transformations. The first law of thermodynamics states that energy can neither be created nor destroyed. The second law of thermodynamics, as stated above, deals with the continuous increase of entropy in the universe.*

motion), it may be expressed as movement, heat, light, sound, electricity, or one of several other forms. Each form of energy can be converted to another, as the following diagram illustrates:

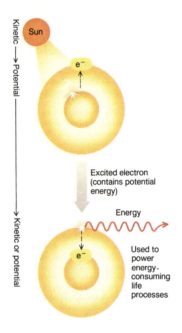

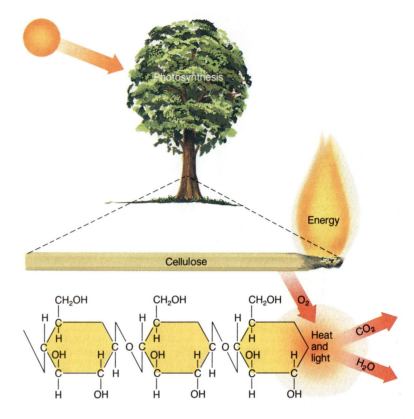

**Figure 3-6**
**Unleashing trapped "sunlight."** The heat and light released from burning wood was originally provided by the sun, trapped by photosynthesis and stored in cellulose and other chemicals of which the plant is composed. This chemical energy becomes kinetic energy as the wood burns. Oxygen combines with the carbons and hydrogens in cellulose, forming carbon dioxide ($CO_2$) and water ($H_2O$). The energy that held the large molecule together is released as heat, light, and sound.

Because it takes energy to move electrons against the pull of the positively charged nucleus of an atom, an electron that has jumped to a shell further away is at a higher energy level. Such electron excitability is fundamental to life on earth.

Some organisms, such as plants, rely on electron excitability to transform light energy into chemical energy during *photosynthesis* (Chapter 9). This energy ends up trapped in the chemical bonds of the plant's complex molecules. Cellulose, for example, is a highly organized, complex molecule that helps form the structural matrix of wood. The molecule contains enormous amounts of chemical energy. Applying a flame to cellulose provides the activation energy needed to start a chain reaction that randomizes the distribution of atoms, lowering the overall energy state of these atoms. Chemical bonds rapidly break as the atoms combine with oxygen from the atmosphere, and the wood burns (Figure 3-6). The atoms in cellulose leave the organized molecule and are randomly dispersed as small, low-energy

molecules ($CO_2$ and $H_2O$). The chemical energy is released as kinetic energy —the heat, light, and sound of fire. A cellulose-digesting organism can also break these chemical bonds and "burn" the cellulose, though more slowly. It harvests the released energy to fuel its own biological processes.

In short, the living organized state requires energy and is itself a reservoir of stored energy, which is released when the organized state deteriorates. Much of this released energy is captured by opportunistic organisms, such as scavengers and microbes that decompose dead organisms, or by organisms that kill their energy sources (Figure 3-7). In all cases, the deterioration of one individual's organized state generates increased organization in

the energy recipient. Such transfers of chemical energy from one organism to another fuel the biological machinery of virtually all the nonphotosynthetic organisms on earth, including people. We harvest this potential by "burning" our food in a controlled process that releases its chemical energy in a form that is safe and usable (see Connections: ATP). Much of this energy is used for the formation of new chemical compounds that constitute our tissues and perform our life processes.

## The Life-Supporting Properties of Water

In addition to energy, life on earth depends on water. In fact, most biologists

a

b

**Figure 3-7**
**Biological transfers of energy.** *(a)* Energy is introduced into the biosphere by photosynthesis. Plant leaves transform sunlight into chemical energy stored in the plant's tissues. Some animals, such as this snow hare, obtain their energy directly from plants. *(b)* These animals serve as an energy source for other animals, this linx for example.

believe water to be essential to the existence of life anywhere in the universe. Planets without water are believed to be devoid of life. About 70 to 80 percent of a typical organism is composed of water. The life-supporting properties of water are largely due to the molecule's polar nature and its hydrogen-bonding ability.

## Water as a Solvent

One of water's most important life-sustaining properties is its effectiveness as a solvent. A **solvent** is a substance in which another material, called the **solute**, dissolves by separating into individual molecules or ions (Figure 3-8). The resulting mixture is a **solution.** Most of the chemicals essential to life are dissolved in water.

Water can dissolve more types of substances in greater quantities than can any other medium. Water is such an efficient solvent that by the time a raindrop completes its fall, it has become a solution containing many dissolved gases, mainly nitrogen, oxygen, and carbon dioxide. The molecular oxygen ($O_2$) that dissolves in lakes, streams, and oceans is sufficient to supply huge communities of fish and other oxygen-dependent underwater dwellers (Figure 3-9).

Some substances do not dissolve when mixed with water or some other fluid. These substances tend to settle out of the mixture and fall in the direction of gravitational pull. However, if the particles are quite small, less than 0.0001 centimeters in diameter, they may remain suspended in the fluid, forming a **colloidal suspension** (or sim-

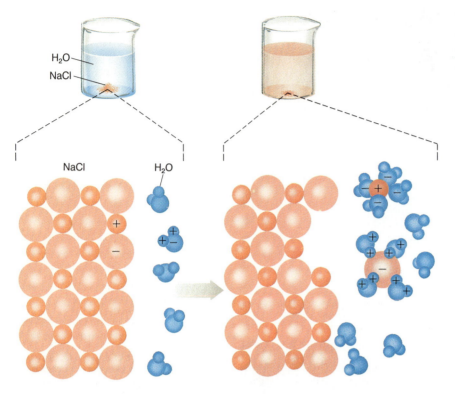

**Figure 3-8**
**Example of solution formation.** Ions in the salt (NaCl) crystal attract the part of the water molecule that bears the opposite charge. The strength of this attraction exceeds the ions' attraction to their oppositely charged neighbors in the crystal, and water wins the molecular "tug-of-war." The ions go into solution, and the salt crystal (the solute) dissolves in the water (the solvent).

**Figure 3-9**
**External gills atop these striking nudibranchs (sea slugs)** extract enough dissolved oxygen from water to allow them to live permanently in the ocean without having to surface for air.

ply a colloid). Common examples of colloids include milk, gelatin, and fog (water droplets suspended in air). Another colloidal suspension is *cytoplasm,* the semifluid substance in cells. Most of life's intracellular processes occur in this water-based colloid.

## Surface Tension and Capillary Action

The attraction between water molecules creates a high *surface tension,* that is, the liquid's surface resists being broken. A leaf, for example, floats on top of a pond, as though supported by a film along the water's surface. This "film" is the product of the cohesive force generated by hydrogen bonding between adjacent water molecules. It takes force to overcome this molecular attraction and break the water's surface. Some small animals actually walk on water by distributing their weight over enough surface area to prevent disruption of the hydrogen bonding (Figure 3-10).

Water molecules are attracted not only to each other, but also to any hydrophilic substance. Glass, for example, has a hydrophilic surface that attracts water molecules with enough force to overcome the pull of gravity. The results of this type of attraction are apparent when one dips a narrow glass tube into water. The water seems to "crawl" up the tube, pulled along by its attraction to the hydrophilic surface. As these water molecules rise, they pull along adjacent molecules to which they are hydrogen bonded, so that the water in the center of the tube also rises. This tendency of water to be pulled up a small-diameter tube is called *capillary action.* Water also enters the minute hollow spaces of porous substances made of cellulose, which attracts water much like glass does (this explains why a towel or a sponge soaks up water.) When dry wood is immersed in water, capillary action pulls the liquid into the pores with enough force to cause the wood to expand. Seed germination is assisted by the expansion of the inner tissues as they soak up water. Eventually, enough pressure is generated to break

**CONNECTIONS**

## ATP—"SPENDABLE" ENERGY

Organisms perform a multitude of biological chores that require expenditures of energy. For instance, they must manufacture new molecules, a process called **biosynthesis.** Somehow the cell must transfer energy released from its energy-generating reactions to those processes that require an input of energy. The problem is compounded by the fact that the energy generated by one reaction may not be needed immediately; the cell must carry it around until it is finally spent on transactions that require energy. All forms of life solve this problem in the same way: by temporarily storing energy in the chemical bonds of ATP (once again spotlighting the unity of life).

ATP (*A*denosine *Tri*Phosphate) is composed of a large molecule *(adenine),* linked by a sugar (ribose) to three small phosphate units (see figure 1). Attaching the third phosphate to the rest of the molecule absorbs a large amount of energy that is captured in this high-energy phosphate bond. This reservoir of energy is immediately available to an organism, which breaks the phosphate linkage, converting ATP to ADP (adenosine *di*phosphate). The released chemical energy is applied to the work the cell is doing at that moment. In this way, ATP carries energy from **exergonic** (energy-releasing) reactions to **endergonic** (energy-absorbing) reactions (see figure 2). To survive, an organism must have enough ATP to cover the cost of constantly maintaining or increasing order. Thus, ATP formation, often simply referred to as **phosphorylation**, must be performed with unceasing vigilance if life is to be maintained.

Some biologists speak of energy as the currency on which biological economics is based and ATP as "money in the cell's pocket." Extending this analogy, we may say that phosphorylation is the process by which cells become rich.

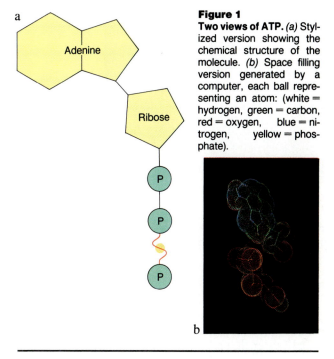

**Figure 1**
**Two views of ATP.** *(a)* Stylized version showing the chemical structure of the molecule. *(b)* Space filling version generated by a computer, each ball representing an atom: (white = hydrogen, green = carbon, red = oxygen, blue = nitrogen, yellow = phosphate).

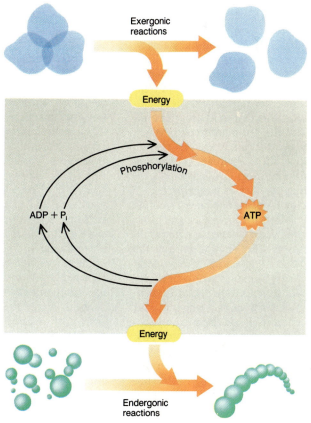

**Figure 2**
**Producing and "spending" the cell's energy currency.** Endergonic and exergonic reactions are coupled by ATP, the common denominator in energy production and utilization. Exergonic reactions generate ATP; endergonic reactions utilize ATP.

**Figure 3-10**
**Surface tension** of water provides enough support to allow a water strider to walk on the liquid's surface as though it were a flexible membrane.

open the seed coat and allow the developing root and shoot to protrude.

## Temperature Regulation

Life is temperature sensitive; too much or too little heat can be fatal. Because organisms contain large amounts of water, they are remarkably flexible in their ability to withstand changes in the amount of heat to which they are exposed. This flexibility is due to water's thermal stability. Water can absorb or lose large amounts of heat without undergoing a correspondingly large change in temperature. This resistance to temperature change is due to hydrogen bonding between water molecules, which break apart when water is heated. Breaking these hydrogen bonds absorbs the heat, so less is available for elevating the water's temperature.

Water is also an efficient cooling agent. When it evaporates, the water vapor molecules that escape the liquid carry away heat energy, cooling the liquid that is left behind. This property is exploited by many organisms to prevent fatal heat accumulation in hot environments. Sweating is one way your body cools itself. As the sweat evaporates, it cools the surface of your body, an especially effective strategy in dry environments that encourage rapid evaporation. The strategy works less well in humid air because the rate of evaporation decreases as the amount of moisture in the air increases. (The dryer the air, the faster water evaporates.) In humid conditions sweat accumulates rather than evaporates, a fact that accounts for the common complaint, "It's not the heat, it's the humidity."

## Ice Formation

As the temperature of water decreases, the movement of the molecules slows down and they crowd in closer to one another. In other words, as water gets colder, it becomes denser, until it reaches 4° C. At this temperature, water is at its densest state. As the temperature approaches 0° C, however, this trend reverses itself. The molecules form more hydrogen bonds as they lose kinetic energy. At 0° C or below, water freezes to a solid, forming a rigid molecular lattice in which each molecule is hydrogen bonded firmly to four other water molecules (Figure 3-11). The molecules are held firmly apart, so water expands as it freezes, the same amount of water occupying more volume as a solid than a liquid. This explains why ice floats.

Freezing is fatal to many organisms because the expansion of water damages cell membranes and tissues, especially if large, sharp-edged crystals form. Yet freezing sometimes helps maintain life. The blanket of ice that covers the surface of frozen lakes and oceans provides insulation against additional heat loss. The water beneath the ice stays warm enough to remain a liquid, allowing life to continue during winter months when the temperature at the surface is well below freezing.

## Acids and Bases

The intricate processes of life can be profoundly affected by some ionic compounds, notably those that release **hydrogen ions** (H$^+$) or **hydroxide ions** (OH$^-$) when they dissolve in water. Substances that release H$^+$ (but not OH$^-$) when they dissolve are **acids.** Hydrochloric acid (HCl), for example, ionizes in water to form H$^+$ and Cl$^-$. This increases the number of H$^+$ ions, forming an acidic solution. A substance that releases OH$^-$ (but not H$^+$) when dissolved in water is called a **base.** Sodium hydroxide (NaOH) is a base that dissociates into Na$^+$ and OH$^-$ in water, forming an alkaline (basic) solution. Water itself is neither acidic nor alkaline. Although it has a slight tendency to dissociate into hydrogen and hydroxide ions, the numbers of each are identical because an H$^+$ is released with each OH$^-$ produced.

$$H_2O \xrightarrow{\text{dissociates}} H^+ + OH^-$$

Water is, therefore, a neutral substance.

The ease with which a molecule releases its H$^+$ or its OH$^-$ determines its relative strength as an acid or base. Hydrochloric acid, for example, readily dissolves in water and is a strong acid. Weak acids, such as the acetic acid of vinegar, and weak bases, the Mg(OH$_2$) in antacids and Milk of Magnesia, dissociate less readily and release fewer H$^+$ or OH$^-$ ions. The strength of an acid or base is numerically expressed by using the **pH scale** (Figure 3-12). The scale ranges from 0 to 14, 0 being the strongest acid, 14 the strongest base, and 7.0 being exactly neutral. A solution can be *neutralized*—that is, it can be brought to a pH of 7.0—by mixing it with another solution that has a pH of the same strength, but on the opposite side of 7, thereby equalizing the number of hydrogen and hydroxide ions. In other words, acids and bases of equal strength neutralize one another when mixed.

Most biological processes are acutely sensitive to pH and occur only within a narrow range on either side of neutrality. Even slight changes in pH can impede the chemical reactions on which life depends. Organisms protect themselves from pH fluctuations with

**buffers,** chemicals that couple with free hydrogen and hydroxide ions, thereby resisting changes in pH. Without these buffers to protect internal fluids from becoming too acidic or alkaline, an organism could not survive. Were it not for one of our own buffer systems, such common activities as eating high protein food or rapid breathing would change the pH of our blood from its normal of 7.4, a potentially fatal fluctuation.

**Figure 3-11**
**Why water expands as it freezes.** In liquid water, the kinetic energy causes the hydrogen bonds between adjacent molecules to break and reform continuously, allowing the water molecules to come into close proximity to one another. In ice, the kinetic energy of the water molecules has decreased to the point that hydrogen bonds do not break, so each molecule is held firmly at an equal distance from all four of its neighbors. The rigidity of these four hydrogen bonds keeps the molecules of ice further apart than in the liquid, so water expands as it freezes.

**Figure 3-12**
**The pH scale.** The acidity or alkalinity of a solution is indicated by a number between 0 and 14. The number reveals the relative proportion of $H^+$ and $OH^-$ ions, which are equal in number at a pH of 7 (neutrality). Acidic solutions have pH values below 7, alkaline (basic) solutions are greater than 7. Each unit on the scale represents a 10-fold change in the hydrogen ion concentration. For example, lemon juice (pH = 2) is 10 times more acidic than vinegar (pH = 3).

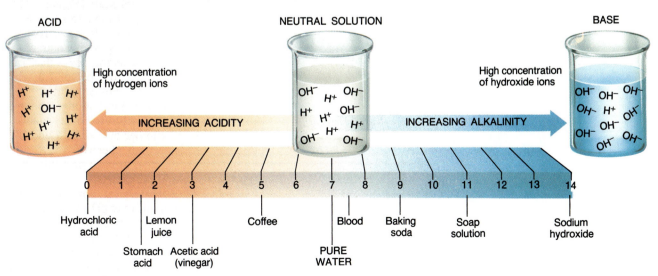

## Synopsis

### MAIN CONCEPTS

- **Organisms are built from protons, neutrons, and electrons, the three components of atoms.** An atom contains a positively charged central nucleus of protons and neutrons encircled by orbiting electrons. The number of protons in the nucleus dictates the atom's elemental identity. In neutral atoms, the number of electrons and protons is identical. In ions, however, the atom has acquired an electric charge by having either more or fewer electrons than protons. Distinct forms of the same element that differ from each other only in their number of neutrons are isotopes.

- **Chemical bonds hold atoms together to form molecules.** Ionic bonds unite oppositely charged ions formed when one atom transfers electrons to another type of atom, leaving both with a filled outer electron shell. Covalent bonds are stable partnerships formed when atoms share their outer-shell electrons, each participant gaining a filled shell.

- **Energy is required to hold together the atoms of highly organized molecules.** This energy, chemical energy, can be harvested by breaking the chemical bonds. ATP traps energy in an accessible form; this energy is available to an organism by simply breaking a high-energy phosphate bond. Much of the energy, however, is lost to entropy.

- **Molecules can influence one another without reacting.** Concentrations of electric charges in polar compounds create poles that form weak attractions with oppositely charged poles of other molecules. Polar compounds are hydrophilic. Two polar molecules are often held together by the attraction between a positively charged hydrogen atom on one molecule and the negative pole on another molecule. Nonpolar compounds are hydrophobic molecules that, in water, interact with other nonpolar molecules in configurations that minimize exposure to water.

- **Water is the solvent on which life depends.** Water's polarity and hydrogen bonding allow it to dissolve most of life's chemicals or form colloidal suspensions. They also account for water's surface tension and capillary action, and help protect organisms against temperature fluctuations. When dissolved in water, acids increase the relative proportion of hydrogen ions (which lowers pH) and bases increase the relative proportions of hydroxide ions (which elevates pH). Neutral solutions have a pH of 7.0, the same as pure water. Some substances resist changes in pH; these buffers help protect cytoplasm and tissue fluids from pH fluctuations.

### KEY TERM INTEGRATOR

Atomic Structure

| | |
|---|---|
| **atoms** | Entities composed of **electrons** orbiting a nucleus of **protons** and **neutrons**. The **atomic number** of an atom determines what **element** it is. **Isotopes** of an element have different **atomic weights**, and some may be radioactive. |

Chemical Reactions, Molecular Forces, and Energy

| | |
|---|---|
| **chemical bonds** | Linkages between atoms. Some **molecules** are held together by **covalent bonds** or by **ionic bonds**. **Hydrogen bonds** are weaker associations formed between **polar** molecules attracted to the same hydrogen. Unlike these **hydrophilic** polar compounds, **nonpolar** substances are **hydrophobic** and form aggregates in water. |

| | |
|---|---|
| chemical reactions | Chemical changes that convert **reactants** to **products**. The **activation energy** needed to initiate a reaction is measured in **calories**, as are the **potential energy** stored in chemical bonds and **kinetic energy** released when chemical bonds are broken. **Biosynthesis** and other **endergonic** biological reactions acquire their energy from **exergonic** reactions. The energy transfer often requires an intermediate step, **phosphorylation**, which safely captures energy in ATP. |

Water and Solutions

| | |
|---|---|
| **surface tension** | Hydrogen bonding causes water molecules to "stick together," forming a surface that resists breaking. |
| **solutions** | Fluids that contain a **solute** dissolved in a **solvent**. Polar molecules readily dissolve in water. Larger particles form **colloidal suspensions** in water. Some solutions have a neutral **pH**, which is often kept that way by **buffers**. Other solutions are **acids**, and still others are **bases**. |

## Review and Synthesis

1. How does the number of protons in an atom determine the chemical reactiveness of the element? What would be the effect of small changes in the number of neutrons only? In the number of electrons?

2. Explain what is *wrong* (if anything) with the description of each key term

   | | |
   |---|---|
   | element | all atoms with the same atomic weight |
   | radioisotope | the most stable form of an element |
   | shell | always filled to their electron capacity |
   | ionic bond | chemical partnership in which atoms share electrons |
   | covalent bonds | may be single, double, triple, or quadruple |
   | hydrophobic compounds | polar molecules that form hydrogen bonds with each other and dissolve in water. |
   | reactants | converted to products in chemical reactions |
   | entropy | useful energy released from endergonic reactions and stored in ATP |
   | pH | measurement of acidity or alkalinity. The neutral pH is 0. |

3. Compare ionic and covalent bonding with respect to the number of bonds that can exist between two atoms. What happens to each type of chemical bond when the molecule dissolves in water?

4. Why does an excited electron represent potential energy? What happens to this energy when the electron returns to a shell closer to the atomic nucleus?

5. Discuss how hydrogen bonding provides water its four life-supporting properties.

6. When carbon dioxide dissolves in water, some of it combines with water to form $H_2CO_3$, which then ionizes to $H^+$ and $HCO_3^-$. What effect would $CO_2$ have on the pH of water-based fluids? What protects organisms from this effect?

## Additional Readings

Breed, A., T. Rodella, and R. Basmajian. 1975. *Through the Molecular Maze.* William Kaufman Co. (Intermediate.)

Frieden, E. 1972. "The chemical elements of life." *Scientific American* 227: 52–64. (Intermediate.)

Summerlin, L. 1981. *Chemistry for the Life Sciences.* Random House, New York (Introductory.)

Moeller, T. 1982. *Inorganic Chemistry: A Modern Introduction.* John Wiley and Sons, New York (Introductory.)

Chapter 4

# Biochemicals— The Molecules of Life

**A**n organism is an entity created by chemical reactions; it requires constant chemical changes to preserve its unique molecular assemblage. Some of these chemical transformations harvest life-sustaining energy from the sun or from other chemicals. But an organism needs more than just energy. It must also extract chemicals from the environment and transform them into the molecules that comprise its own tissues. All these chemical changes that occur under an organism's direction constitute its *metabolism* (Chapter 1). Metabolism builds all the molecules that collectively comprise the organism. Many of these molecules are thousands of times larger than a water molecule. Some of them are also the most complex chemicals on earth, perhaps in the universe. Yet these molecular giants share one common characteristic with even the smallest compounds of life—they all have chemical "skeletons" composed of carbon.

## The Importance of Carbon

Nearly every molecule that contains carbon atoms belongs to a special category of compounds, the *organic chemicals*. Originally, the term "organic" referred to molecules that could be manufactured only by living organisms (in contrast to inorganic chemicals). Chemists eventually discovered that all compounds called "organic" contained carbon, a discovery that led to the redefining of **organic** as simply "carbon-containing."* Our ability to manufacture plastics and other synthetic organic compounds from inorganic materials has made the original definition of organic obsolete. To distinguish synthetic and natural organic compounds, we use the term **biochem**

**icals** to refer to the molecules of life, those manufactured and used by organisms.

Carbon's chemical properties make it ideal as the central element on which life is based. Perhaps the most important of these properties is its ability to form long chains with itself. Without these chains, the complex molecules of life could not exist. Carbon chains provide the chemical backbone of biochemicals, a backbone that may be linear, branched, or cyclic:

$$C—C—C—C—C—C$$
linear

cyclic

branched

These carbon skeletons would be biologically worthless, however, if not for the ability of each carbon atom to form four covalent bonds. This means that even after forming one or two bonds with other carbons to form a chain, each carbon still has at least two free bonding sites to which additional chemical groups and elements (espe-

cially hydrogen, oxygen, nitrogen, phosphorus, or still more carbons) can attach. For example, each of the six carbons in the sugar glucose is attached to four elements or chemical groups. All the elements that collaborate to form glucose are chemically satisfied by this arrangement as shown in Figure 4-1a. Carbon can also form double and triple bonds, either with itself ($—C{=}C—$) ($—C{\equiv}C—$) or with another element such as oxygen ($—C{=}O$).

A carbon atom may also be bonded to a chemical group composed of several atoms rather than to a single element. These accessory chemical entities, called **functional groups**, help determine the identity and chemical properties of the compound. The properties of various functional groups are discussed later in this chapter with the compounds in which they are characteristically found.

## The Four Biochemical Families

The chemicals that form the structure and perform the functions of life are highly organized molecules. Compared to small molecules like water, many of these organic compounds are enormous, consisting of hundreds to millions of carbon atoms and their accessory elements. These enormous chemicals are called **macromolecules** (macro = large). The size of these mo-

---

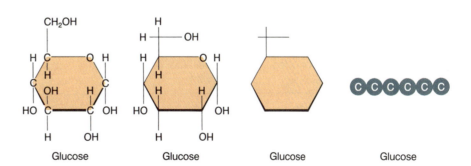

**Figure 4-1**
**Three structural conventions for representing organic compounds.** All three representations are of glucose. *(a)* Fully detailed molecule, all atoms shown. *(b)* Carbons omitted, all other atoms shown. *(c)* Skeletal representation, with no atomic detail shown. *(d)* Convention used throughout this textbook; each ball represents a carbon.

lecular giants contributes to their ability to perform specific tasks with rigorous precision.

Macromolecules are constructed by assembling together small molecular subunits, often in a linear process that resembles coupling railroad cars onto a train. Each subunit is called a **monomer** (mono = one; mer = part), and the macromolecule is referred to as a **polymer** (poly = many). There are 40 different types of monomers from which all polymers, and therefore all organisms on earth, are constructed.

Every macromolecule falls into one of four fundamental families of organic compounds.

- Carbohydrates.
- Lipids.
- Nucleic acids.
- Proteins.

The macromolecules in each of these families perform functions that are essential to life (Table 4-1).

## Carbohydrates

**Carbohydrates** are a group of compounds that include sugars and all larger molecules constructed of sugar subunits. Carbohydrates contain three elements—carbon, hydrogen, and oxygen—in a ratio that associates each carbon with two hydrogens and one atom of oxygen (the elements of water). The name "carbohydrate" reflects this ratio (carbo = carbon, hydrate = addition of water). Single carbohydrate subunits, called *simple sugars,* are small compounds, the most important of which contain six carbons (*glucose* and *fructose*) or five carbons (*ribose).* Sugars supply energy to organisms that cannot harness it from sunlight. They also provide plants a means of storing the sun's energy by trapping it in the chemical bonds of these sugars. The bonds holding the sugar together are broken, and the energy is released whenever needed either by the plant or by another organism that has consumed the plant's tissues. In other words, carbohydrates fuel the life processes of all organisms.

Although small, sugars can be coupled together to form giant carbohydrates called **polysaccharides** (saccharide = sugar). These macromolecules are energy storage molecules that bank surplus energy in a readily accessible form. When the organism eventually needs the chemical energy stored in the polysaccharide, it disassembles the polymer and releases free simple sugars for use as fuel.

The sugars of polysaccharides are covalently bonded together by a **glycosidic bond** (Figure 4-2). This bond is formed by a process called **dehydration synthesis,** which literally means "assembling by removing water." The synthesis process removes an atom of hydrogen (H) from one sugar and a hydroxyl group (OH) from another. The removed H and OH react with each other to form water, $H_2O$, a byproduct of dehydration synthesis. Removing the H and OH creates free reactive sites that join together, chemi-

TABLE 4-1

| Major Macromolecules Found in Living Systems | | | |
| --- | --- | --- | --- |
| **Macromolecule** | **Class of Monomer** | **Some Major Functions** | **Examples** |
| Polysaccharide (large carbohydrates) | Sugars | Energy storage; physical structure | Starch, glycogen, cellulose |
| Lipid: | | | |
| Triglycerides | Fatty acids and glycerol | Energy storage; thermal insulation; shock absorption | Fat; oil |
| Phospholipids | Fatty acids, glycerol, phosphate, and an R group[a] | Foundation for membranes | Plasma membrane |
| Waxes | Fatty acids and long-chain alcohols | Waterproofing; protection against infection | Cutin; suberin; ear wax; beeswax |
| Steroids | 4-ringed structure[b] | Membrane stability; hormones | Cholesterol; testosterone; estrogen |
| Nucleic acid | Ribonucleotides; deoxyribonucleotides | Inheritance; ultimate director of metabolism | DNA; RNA |
| Protein | Amino acids | Catalysts for metabolic reactions; hormones; oxygen transport; physical structure | Hormones (oxytocin, vasopressin); hemoglobin; keratin; collagen; and a class of proteins called enzymes |

[a] R group = a variable portion of a molecule.
[b] steroids are neither polymers nor macromolecules.

cally linking the two sugars to form a *disaccharide* (di = two). The best known (and sweetest) disaccharide is sucrose, common table sugar. Additional sugars can be added to the chain by the same process, generating a polysaccharide.

Sugars can be released from polysaccharides by reversing dehydration synthesis. Reinserting a water molecule between the monomers severs the bond holding the sugars together, and their free reactive ends are "capped" with the H and OH. Depolymerization thus occurs by **hydrolysis** (hydro = water, lysis = split or dissolve), as shown at the right.

Mammals (such as yourself) bank much of their surplus sugar by forming **glycogen**, a highly branched polysaccharide composed entirely of glucose monomers. Glycogen is stored chiefly in the liver and muscles. Whenever the body demands additional energy, hydrolysis of glycogen releases glucose from the liver into the bloodstream. During muscular activity, hydrolysis of glycogen in the active muscles concentrates a supply of glucose in those tissues.

Another source of glucose is **starch** (amylose and amylopectin), the polysaccharides most commonly used by plants for energy storage. Like glycogen, starch consists entirely of glucose monomers, but its lack of branching imparts physical characteristics that are very different from glycogen. Even though animals can't produce starch, they can readily digest it by hydrolysis of its glycosidic bonds.

Some polysaccharides play no role in energy storage but engage in other activities. *Structural polysaccharides* provide support and protection for the organism (Figure 4-3). Many animals, for example, secrete a layer of **chitin**, a polysaccharide that hardens to form a strong external skeleton on their surface. The earth's most abundant polysaccharide is a structural polymer called **cellulose**, which is the fibrous matrix of wood, cotton, and all plants. Cellulose cannot be disassembled by the digestive enzymes in humans, so this vast reserve remains unavailable to us as a direct source of energy. Ironically, cellulose is composed entirely of glucose, the same monomer found in

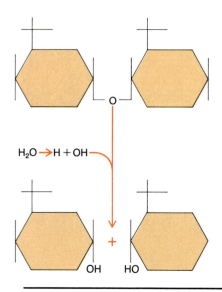

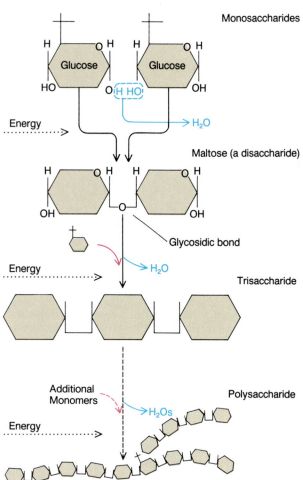

**Figure 4-2**
**Formation of a glycosidic bond between sugars by dehydration synthesis.** Joining just two sugars produces a disaccharide; adding another generates a trisaccharide. Each end of the growing carbohydrate chain has a reaction group available for the addition of another monomer, regardless of chain length. Branching of the chain occurs when the OH of the uppermost carbon in a glucose molecule participates in dehydration synthesis.

**Figure 4-3**
**A showcase of structural polysaccharides.** The organisms in this picture reveal a few of the natural roles of polysaccharides. The fibrous polysaccharide that supports and shapes the pitcher plant is cellulose. The internal organs and skin of these frogs are secured in place by structural polysaccharides that strengthen connective tissue. Although not shown, insects that fall prey to the carnivorous pitcher plant possess external skeletons consisting of the rigid polysaccharide chitin.

microscopic cellulose decomposers, the world would be permanently littered with dead bodies of plants. Some cellulose-digesting microorganisms reside in a special digestive chamber (the rumen) of cattle. Although cows cannot digest the grass and hay that make up much of their diet, their microscopic passengers do the job for them. Even termites depend on cellulose-digesting microbes living in their stomachs to digest their wooden meals.

## Lipids

One group of organic compounds consists of a variety of molecules whose only common property is their *water insolubility;* that is, they are nonpolar, hydrophobic molecules that do not dissolve in water (Chapter 3). These compounds are **lipids.** They include fats, oils, waxes, phospholipids, and steroids.

Long-term energy storage is one of the principal functions of lipids. Recall that ATP is "spendable" energy, "money in the cell's pocket" (review Connections: ATP, in Chapter 3). If we extended this analogy, lipids would represent the organism's energy savings, or "money in the biological bank." But organisms must carry the "bank" around with them. This is why lipids excel as energy assets; they provide the most weight-efficient means for an organism to store and save surplus energy. More than twice as much energy can be banked in lipids as in an equivalent weight of proteins or carbohydrates, making fat an especially "high-calorie" substance. The flammability of candle wax, cooking oil, and other lipids testifies to their high-energy content. Energy stored in lipids, however, is not as quickly retrieved for immediate use in living cells as is the energy in glycogen and starch (polysaccharides).

When an animal's energy intake exceeds its needs, the surplus is stored by building reserves of fats, a frustrating reality for many persons struggling to conform to society's current ideal of slenderness (see Bioline: Lipids). Plants don't make fat but store their energy in oils. Both fats and oils consist of three long-chain molecules of **fatty acid** coupled to a single molecule

starch, one of the most easily digested polymers. The two polysaccharides differ, however, in the orientation of the glucose molecules and the way they are linked together (Figure 4-4). The bonds of starch are readily hydrolyzed by an enzyme in our digestive tracts. But we have no enzymes that can break the bonds of cellulose and release its glucose molecules. Thus, the

polysaccharide passes intact through our digestive tracts, providing fiber (roughage) that aids in the formation and elimination of feces, but supplying us with no energy. For starving people, this is an especially tragic twist in our biochemical development.

Other organisms, principally microorganisms, do have the necessary enzymes for digesting cellulose. If not for

of **glycerol** (Figure 4-5). These lipids are therefore called "triglycerides" (tri = three). Assembling a triglyceride by dehydration synthesis requires an input of energy, which is used to fasten the *carboxyl* ("organic acid") group (—COOH) of each fatty acid chain to one of the three hydroxyl (OH) groups of glycerol.

If the resulting triglyceride is a liquid at "room temperature" (25° C), the lipid is considered an *oil;* if it is solid, it is called a *fat.* In general, the greater the number of *unsaturated* (double or triple) bonds that exist between the carbons of the fatty acids, the lower will be the melting temperature of the lipid. The profusion of unsaturated bonds in the oils accounts for their liquid state. Almost all the linkages in the carbon chain of fats, on the other hand, are single bonds, that is, they are *saturated* with hydrogens. Unsaturated fatty acids are often intentionally saturated to convert an oil to a solid, a chemical process called **hydrogenation.**

$$H - \overset{\displaystyle H}{\underset{\displaystyle H}{C}} - \overset{\displaystyle }{\underset{\displaystyle H}{C}} = \overset{\displaystyle }{\underset{\displaystyle H}{C}} - \overset{\displaystyle H}{\underset{\displaystyle H}{C}} - H + 2H$$

(one unsaturated
bond)

(hydrogenation) ↓

$$H - \overset{\displaystyle H}{\underset{\displaystyle H}{C}} - \overset{\displaystyle H}{\underset{\displaystyle H}{C}} - \overset{\displaystyle H}{\underset{\displaystyle H}{C}} - \overset{\displaystyle H}{\underset{\displaystyle H}{C}} - H$$

(no unsaturated
bonds)

This transforms double or triple bonds to single bonds. (Even partial hydrogenation converts some vegetable oils to a solid for use in shortening or margarine.) The "polyunsaturated" products promoted in popular advertisements are made with vegetable oils rich in double bonds, even after they have been partially hydrogenated. Unsaturated triglycerides in the human diet may be less likely to promote circulatory and heart disease than animal fat, such as that found in butter.

Some lipids, namely, the **phospholipids**, act as the major structural component of the membranes that surround cells and separate them from their environments. The glycerol of these structural lipids is attached to only two fatty acid chains instead of three. The third glycerol carbon carries a phosphate group ($PO_3^-$) linked to a "variable group," a charged chemical entity that differs from one phospholipid to another. (Such variable groups are represented by the letter "R," in general formulas.) Because of its electric charge, this end of the molecule is polar, so it is soluble in water. Thus, phospholipids have hydrophilic ends that seek water and hydrophobic ends that don't. The phospholipids of cells align with one another, so that their hydrophobic "tails" are buried inside the two hydrophilic layers formed by the phosphate groups (Figure 4-6). This produces a sheet composed of two molecular layers—a phospholipid bilayer—that form the fundamental structure of membranes. The dependency of life on membranes is discussed in Chapters 6 and 7.

The category "lipids" also includes a group of compounds called **steroids** that qualify as lipids because of their hydrophobic nature. All steroids have the basic four-ringed molecular skeleton represented by cholesterol (Figure 4-7). In humans and other mammals,

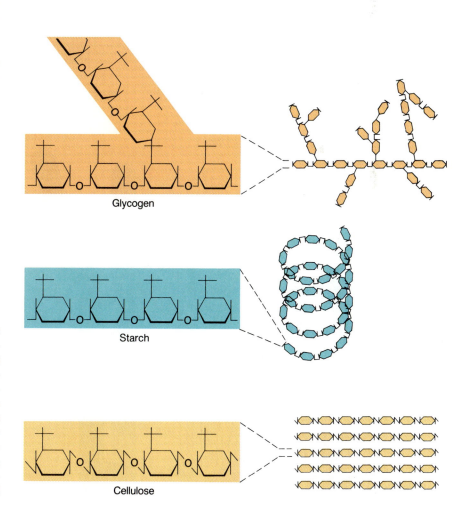

Glycogen

Starch

Cellulose

**Figure 4-4**
**Three polysaccharides—identical sugar monomers, dramatically different properties.** Glycogen, starch, and cellulose are each composed entirely of glucose subunits, yet their chemical and physical properties are very dissimilar. These differences are due to the distinct ways that the monomers are linked together and arranged.

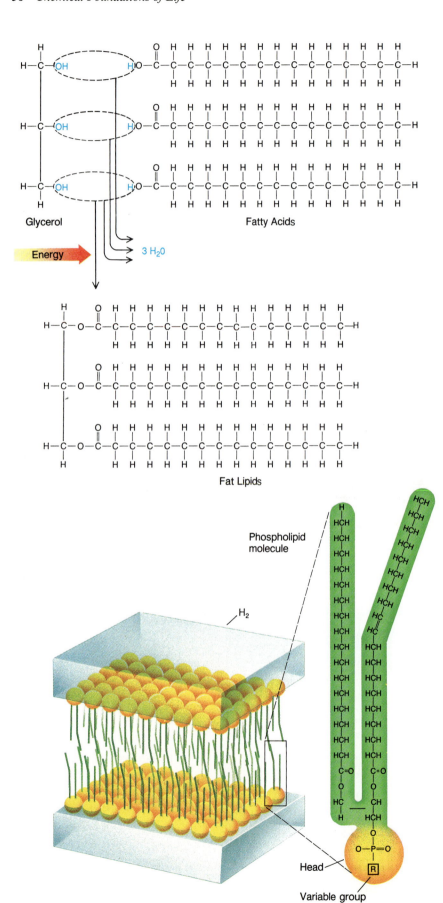

Glycerol

Fatty Acids

Energy

3 H₂O

Fat Lipids

Phospholipid molecule

H₂

Head

Variable group

**Figure 4-5**
**Formation of a triglyceride molecule by dehydration synthesis.** One glycerol and three long-chain fatty acids combine to form, in this case, a fat. The carboxyl group (—COOH) of each fatty acid reacts with one of the three hydroxyl groups (—OH) of the glycerol. The reaction is accompanied by the removal of water. Hydrolysis reverses this process, freeing the fatty acids for disassembly and the release of large amounts of usable energy.

**Figure 4-6**
**Phospholipids—molecules that are both soluble and insoluble in water.** The polar phosphate "head" is attracted to water, while the nonpolar fatty acid "tails" extend in the opposite direction, away from water. This "molecular schizophrenia" results in molecules that, in cells, align so that they develop double-layered sheets. All life depends on cell membranes formed from these phospholipid sheets. (In some cell membranes, the fatty acid tails overlap at their tips, eliminating the gap between layers.)

# B I O L I N E

## LIPIDS: A BULGING NUISANCE OR ESSENTIAL ADAPTATION?

Many people consider the tendency to accumulate fat a nuisance, detracting from our physical health and appeal. Fat deposits, however, contribute to our well-being. They form a natural buffer, a sort of packing material, that absorbs shock and helps cushion fragile internal organs. It also retards loss of body heat by acting as thermal insulation surrounding much of the body. This role is particularly evident in the thick layer of blubber characteristic of whales, walruses, and many other mammals that live in cold conditions. Fat helps maintain healthy skin and hair, carries fat-soluble vitamins (A, D, E, and K), and supplies us with essential fatty acids (such as linoleic acid) needed for proper growth and development. Fat may also accent our sex appeal. The distribution of fatty deposits in women differs considerably from that of men, a difference controlled by sex hormones. Many biologists believe that these differences function as visual sex attractants between men and women.

Our bodies are efficient at forming fat to store surplus energy. Excess fat accumulation, however, can be detrimental. It puts constant strain on the heart, forcing it to pump blood further. (Each additional pound of fat requires an extra 8 miles of vessels.) It contributes to circulatory diseases and shortens our lives. Exercise discourages fat accumulation by consuming energy; if our body's energy demands exceed our caloric intake, we usually lose weight.

The mechanisms for regulating fat formation were inherited from our prehistoric ancestors who rarely enjoyed the luxury of regular meals. Fat reserves were critical for surviving those hungry intervals between successful hunts. Furthermore, the level of physical activity was probably greater in prehistoric times than it is today. Making a day-to-day living meant constant physical work for men and women alike. There were no Neanderthal executives, no cave men and women sitting at computers. Although circumstances have changed, we have inherited our ancestors' fat storage regulators, which were set long ago, before human endeavor made sedentary lifestyles possible.

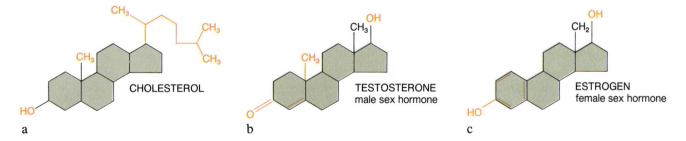

a          b          c

**Figure 4-7**
The structure of steroids. *(a)* All steroids share the basic four-ring skeleton *(light brown)*. (Although not shown, a carbon atom occupies the point of each angle.) The orange parts of the molecule are unique to this steroid, cholesterol. *(b)* In humans, the seemingly minor differences in chemical structure between testosterone and estrogen generate profound biological differences. Testosterone induces male characteristics, for example, a deep voice and facial hair; estrogen stimulates development of female characteristics such as breast enlargement.

cholesterol is a component of cell membranes and a precursor for building vitamin D and several types of *hormones,* chemical messengers sent from one part of the body to other parts, orchestrating many of the body's processes. For example, sexual maturation is coordinated by *testosterone* (in males) and *estrogen* (in females); both of these hormones are steroids.

## Nucleic Acids (DNA and RNA)

One of the properties distinguishing the living from the inanimate is an ability to reproduce offspring that have characteristics similar or identical to those of their parent(s). In other words, living organisms preserve their individual characteristics by reproduction and heredity (Chapter 1). To do so, an organism must have some mechanism for recording and housing precise instructions that direct the manufacture of inherited characteristics, "written" in a form that can be passed on to (and "read" by) the offspring. These instructions are encoded in the polymer **deoxyribonucleic acid (DNA)**, the material from which genes are made. DNA (and therefore genes) are found in the cell's chromosomes. Another type of nucleic acid, however, is not part of the chromosome. It is **ribonucleic acid (RNA)**, which is needed to translate DNA's genetic information into expressed cellular characteristics. Genetic messages encoded in nucleic acid molecules are fundamental to all organisms.

Nucleic acids are constructed from monomers called **nucleotides.** (See page 456, for the chemical structure of these monomers.) DNA is composed entirely of four types of nucleotides, the molecular "letters" that spell out the genetic message. (RNA also consists of four nucleotide types.) The four letters in the genetic alphabet can be arranged in a nearly infinite number of sequences. Thus, the number of genetic messages that can be encoded in DNA is virtually unlimited, making possible countless numbers of unique organisms. Discovery of the molecular mechanism by which genetic informa-

a

b

c

**Figure 4-8**
**The protein gallery** shows just a few of the thousands of biological structures composed predominantly of protein. *(a)* The fabric of feathers used for thermal insulation, flying, and sex recognition among birds. *(b)* The silk spun by spiders to form webs. *(c)* The lenses of eyes, as in this net casting (Dinopsis) spider.

tion is stored in chromosomes and translated into expressed properties has answered fundamental questions that have puzzled scientists for thousands of years. This fascinating mechanism is described in Chapter 27.

The energetics of nucleic acid construction differ from other macromolecules because of the structural similarity of the monomers to ATP and other compounds used for energy currency. Adenosine (one of the four monomers in DNA and RNA) is also the core of the ATP molecule. By removing two phosphates from ATP, adenosine can be added directly to the growing nucleic acid chain, the adenine attached by its sugar and one remaining phosphate. The removal of the two phosphates also releases the energy needed for the addition of the nucleotide. All nucleotides can form high-energy compounds similar to ATP. Therefore, an entire nucleic acid polymer is constructed from the potential energy stored in the free subunits from which it is assembled.

## Proteins

If you were to drive off all the water from an organism's body, more than half the dried remains would consist of protein. Protein constitutes the major component of muscle and provides the structure of hair, feathers, nails, skin, bone, cartilage, ligaments, tendons, and thousands of other structures (Figure 4-8). Protein also forms much of the substructure of individual cells of every organism.

The importance of proteins to the structure of life, however, does not overshadow their role as performers of complex biological tasks. Proteins determine much of what gets into and out of a cell, regulate the expression of genes, and provide muscles and some internal cell structures their ability to move. Hemoglobin, the red protein in blood cells, carries oxygen for distribution throughout the body of every vertebrate (animals with backbones). Other important proteins include hormones that regulate the delicate balance of sugar metabolism (insulin), the maintenance of blood pressure (vasopressin), and the timing of uterine contractions during childbirth (oxytocin). Our ability to survive constant

contact with bacteria, viruses, and other microorganisms is largely due to protective proteins called *antibodies* that patrol our tissues and body fluids, attacking microbial intruders that manage to enter the body.

Another essential group of proteins are *enzymes,* the mediators of metabolism. Enzymes supervise the chemical events in an organism. Thousands of different types of enzymes collaborate to direct the development and maintenance of an organism. They determine what chemical reactions occur in an organism. The influence of enzymes is so pervasive that the next chapter is devoted exclusively to discussing these proteins and their role in directing the life processes.

## THE MOLECULAR ANATOMY OF PROTEINS

Proteins are polymers made of amino acid monomers. There are 20 naturally occurring amino acids in proteins, each characterized by an *amino group* ($-N-H$) and a *carboxyl group* ($-COOH$) attached to the same carbon atom. Another chemical group, also attached to this carbon, varies in structure and distinguishes the 20 amino acids from each other. This R group determines each amino acid's characteristic properties, such as electric charge and water solubility (Table 4-2).

The amino group of one amino acid can react with the carboxyl group of another, releasing a molecule of water and forming a **peptide bond** that chemically links the two monomers (Figure 4-9). The bonding process can continue indefinitely because free reactive groups remain available at each end of the chain regardless of the number of amino acids in the sequence. Most proteins contain at least 100 amino acids, and some contain more than

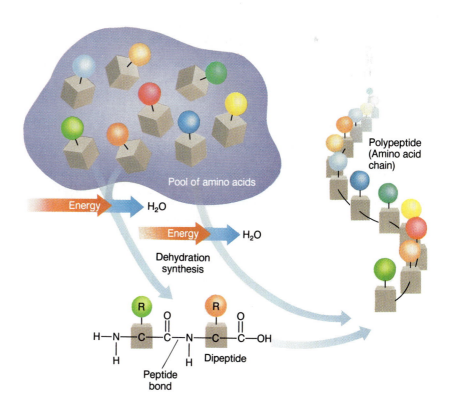

**Figure 4-9**
**Formation of a peptide bond** between two amino acids by dehydration synthesis. The chain can continue to grow by attaching more amino acids.

1000. Even one of the smallest proteins, insulin, contains more than 50 amino acids.

The great variety among proteins is due to differences in the types of amino acids in the polymer and the sequence in which they are arranged.

Amino acid content and sequence determine the properties of the protein by dictating the shape the molecule assumes when it folds into its final compacted form. Proteins first fold into a shape that buries hydrophobic amino acids in the center of the protein, placing the hydrophilic amino acid on the polymer's surface in contact with the surrounding water. Attraction between oppositely charged amino acids bends the protein chain into a shape that brings these charges close together. The protein is also held by covalent bonds between the sulfide groups of two sulfur-containing amino acids. These linkages are called disulfide bonds (—S—S—). These and other forces combine to shape each protein into a precise three-dimensional configuration.

Protein configuration has several levels of organization. The most complicated proteins acquire their final shapes in four distinct levels of complexity (Figure 4-10). Even the simplest proteins have a *primary structure*—the sequence of amino acids in the linear chain. In many proteins, hydrogen bonding twists some regions of the chain into a helix, and the protein assumes a *secondary structure.* Hydrogen bonding may also hold several amino acid chains parallel to one another, creating an accordion-shaped molecular sheet. This pleated sheet is another form of secondary structure. Some proteins progress no further in their complexity. Fibrous proteins, such as keratin of hair, are long, coiled strands of proteins with helical-type secondary structure.

The secondary helices tend to resemble tightly coiled springs forming a hollow tube. In many proteins, these long "tubes" may fold into an even more complex configuration, the *tertiary structure.* This folding creates a globular protein with a definite shape held stable by hydrogen bonds, disulfide bonds, electrical charge, and hydrophobic-hydrophilic forces.

Some proteins are complexes of two or more polypeptide chains. These complexes constitute the *quaternary structure.* Hemoglobin is a protein with quaternary complexity. This oxygen-carrying protein consists of four polypeptide chains—two identical pairs of subunits.

The shape of *structural proteins* is important to their roles as building blocks of biological structures. Shape also dictates what biological activities *functional proteins* can perform. The importance of amino acid sequence in

TABLE 4-2

| The 20 Amino Acids Found in Proteins | | | |
|---|---|---|---|
| **Amino Acid** | **Abbreviation** | **Water Solubility** | **Charge** |
| Glycine | (gly) | Hydrophobic | Neutral |
| Alanine | (ala) | Hydrophobic | Neutral |
| Valine | (val) | Hydrophobic | Neutral |
| Leucine | (leu) | Hydrophobic | Neutral |
| Isoleucine | (ile) | Hydrophobic | Neutral |
| Methionine | (met) | Hydrophobic | Neutral |
| Phenylalanine | (phe) | Hydrophobic | Neutral |
| Tryptophan | (trp) | Hydrophobic | Neutral |
| Proline | (pro) | Hydrophobic | Neutral |
| Serine | (ser) | Hydrophilic | Neutral |
| Threonine | (thr) | Hydrophilic | Neutral |
| Cystine | (cys) | Hydrophilic | Neutral |
| Tyrosine | (tyr) | Hydrophilic | Neutral |
| Asparagine | (asn) | Hydrophilic | Neutral |
| Glutamine | (gln) | Hydrophilic | Neutral |
| Aspartic acid | (asp) | Hydrophilic | Negative |
| Glutamic acid | (glu) | Hydrophilic | Negative |
| Lysine | (lys) | Hydrophilic | Positive |
| Arginine | (arg) | Hydrophilic | Positive |
| Histidine | (his) | Hydrophilic | Positive |

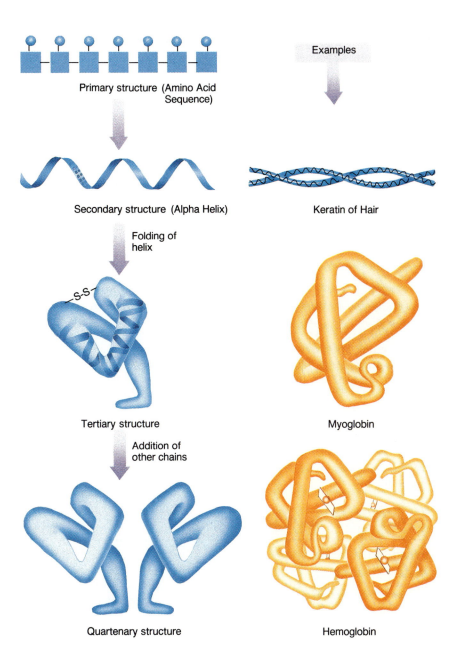

Primary structure (Amino Acid Sequence)

Examples

Secondary structure (Alpha Helix)

Keratin of Hair

Folding of helix

Tertiary structure

Myoglobin

Addition of other chains

Quartenary structure

Hemoglobin

**Figure 4-10**
**A protein acquires its active form in a series of steps.** The sequence of amino acids (the primary structure) determines its final shape, which in turn determines the activity of the protein. The final configuration of some proteins is the secondary structure. Others, the globular proteins, acquire additional complexity by folding into a tertiary conformation. Quaternary proteins contain several peptide chains that join to form a single protein.

determining the shape (and therefore the function) of a protein is dramatically illustrated by the consequences of changing just one amino acid in hemoglobin. Substituting a hydrophobic, uncharged amino acid (valine) for a hydrophilic, negatively charged amino acid (glutamic acid) distorts the protein and seriously impairs its ability to carry oxygen. The distorted hemoglobin elongates red blood cells into a sickle shape that clogs vessels.

This often fatal disease is called sickle cell anemia.

## Molecular Hybrids

Many cell functions are performed by macromolecules that join forces, forming complexes of two kinds of polymers. These "mixed marriages" between molecules are quite common,

the organism profiting from the best of both types of compounds. For example, the lipid portion of some **lipoproteins** helps integrate the protein into the hydrophobic interior of the cell's membranes. Other molecular hybrids include combinations of carbohydrate and lipid (**glycolipids**), carbohydrate and protein (**glycoprotein**), and nucleic acid and protein (**nucleoproteins**). Their various functions are discussed throughout this book.

## Synopsis

### MAIN CONCEPTS

- **Carbon is the central element in all organic compounds.** Its versatile properties include its ability to form four covalent bonds, link up with other carbons to form straight, branching, or ring-shaped chains, attach to other functional chemical groups, and form single, double, or triple bonds. Its versatility makes possible the extraordinarily complex molecules on which life depends.

- **Most biochemicals are macromolecules.** Macromolecules constitute the fabric of an organism and conduct its needed functions. Most macromolecules are polymers constructed by linking together the same class of subunits—monomers—into long chains. Macromolecules are constructed by the energy-consuming process of dehydration synthesis, the formation of a chemical bond between two chemicals by removing an H and OH from their adjacent ends. Hydrolysis reverses this process, splitting chemical bonds between adjacent monomers by adding an H and OH. Disassembling macromolecules releases energy.

- **Four families of biochemicals.** Carbohydrates are simple sugars and larger molecules created by linking together many sugars. Polysaccharides are the largest of these carbohydrates. Some polysaccharides store energy (glycogen and starch are examples); others (such as cellulose) are structural polysaccharides. Lipids are hydrophobic organic compounds that contain fatty acids or the four-ringed steroid structure. Triglycerides (fats and oils) are the most weight-efficient molecules for storing energy. Phospholipids form double-layered molecular sheets essential to the structure of cell membranes. Steroids are smaller lipids that function as hormones or join to and modify phospholipids. Nucleic acids (DNA and RNA) are linear polymers responsible for heredity and expression of genetic information. Proteins are long chains of amino acids. The sequence of amino acids in a protein determines the shape of the protein, which in turn determines the protein's biological role. Some are structural proteins that make up the physical components of an organism. Others, such as hormones, antibodies, hemoglobin, and enzymes, are functional proteins.

### KEY TERM INTEGRATOR

#### The Importance of Carbon

| | |
|---|---|
| **organic** | Refers to compounds that contain carbon. All **biochemicals** are organic, and many of these are **macromolecules**. These are usually **polymers**, the family of which depends on the kind of **monomers** it contains. The **functional group** on the monomers dictates the category to which the subunit belongs. Most macromolecules are assembled by **dehydration synthesis** and disassembled by **hydrolysis** of the bonds between monomers. |

#### Families of Biochemicals

| | |
|---|---|
| **carbohydrates** | Sugars and complexes of sugars. **Polysaccharides** are polymers composed of sugars linked together by **glycosidic bonds.** Energy storage polysaccharides include **starch** and **glycogen;** others, such as **cellulose** and **chitin,** provide structure and protection. |
| **lipids** | Class of hydrophobic biochemicals that includes fats, oils, waxes, and **steroids.** Fats and oils are complexes of three **fatty acids** attached to **glycerol.** The presence of a polar phosphate group accounts for the ability of **phospholipids** to form the **bilayers** fundamental to cell membranes. |

| | |
|---|---|
| **nucleic acids** | Polymers composed of **nucleotides.** They are **deoxyribonucleic acid (DNA)** and **ribonucleic acid (RNA).** |
| **proteins** | Polymers of **amino acids** linked together by **peptide bonds** into linear chains that are folded into specific shapes essential to their functions. |
| **molecular hybrids** | Molecules that contain two or more biochemical types. **Lipoproteins, glycolipids, glycoproteins,** and **nucleoproteins** are all composites of different families of organic molecules. |

## Review and Synthesis

1. Explain why the term "dehydration synthesis" is used to describe the formation of chemical bonds between monomers. Why is the reverse process called hydrolysis?

2. Discuss how carbon's ability to form four chemical bonds is critical to life.

3. Some people equate "organic" with natural and wholesome. Why is it inaccurate to do so? What are some examples of unnatural or unwholesome organic compounds?

4. Another myth states that camels store water in their humps, which is actually for storing fat. From what you know about hydrolysis, would utilizing this stored fat produce more water or use it up?

5. Describe how the calories in the carbohydrates you eat can end up as calories in fat. Why is that better than using carbohydrates to store the energy?

6. Protein deprivation is the most common form of human malnutrition. Why is protein such an important component of our diets?

7. Complete the following table.

| Compound | Class of Compound | Monomer(s) | Function |
|---|---|---|---|
| Starch | _____ | _____ | _____ |
| | Dipeptide | _____ | |
| _____ | _____ | Unsaturated fatty acids and glycerol | _____ |
| DNA | _____ | _____ | _____ |

## Additional Readings

Asimov, I. 1966. *The World of Carbon.* Rev. ed. Collier Book Co. (Introductory.)

Bloomfield, M. 1980. *Chemistry and the Living Organism.* 2nd ed. John Wiley and Sons, New York (Introductory.)

Dickerson, R. E. 1978. "Chemical evolution and the origin of life." *Scientific American* 239:70–86 (Intermediate.)

Hill, J. W. 1979. *Chemistry for Changing Times.* 3rd ed. Burgess Publishing Co. (Introductory.)

Stryer, L. 1981. *Biochemistry.* 2nd ed. W. H. Freeman Co., San Francisco (Intermediate to advanced.)

Chapter 5

# Enzymes—
# Directors
# of Metabolism

Every organism is a precisely co-ordinated collection of chemicals, each built by biochemical reactions. Every detail of your body, from the hair atop your head down to your toenails, is the product of such reactions. Your eyes, for example, owe their color to pigments manufactured by metabolic activity. The chemical changes that construct these pigments don't happen automatically or haphazardly, however. They require the tight supervision provided by enzymes, which may be thought of as "metabolic traffic directors." Enzymes act on pre-pigment chemicals, changing them step by step until they become the molecules that ultimately produce eye color. Without this set of enzymes, the chemicals would not be directed along this metabolic pathway, and your eyes would be another color.

This is the most universal job of proteins, to direct and organize biochemical processes, so that molecular chaos doesn't reign in place of life. **Enzymes** are proteins that act as biological **catalysts**, substances that accelerate a particular chemical reaction that would not occur at significant rates in the absence of the enzyme. In this way, enzymes determine which biochemical reactions occur in an organism. Most of your inherited characteristics developed because your cells produced enzymes that encouraged the biochemical construction of each particular trait. In other words, an organism is the product of its unique combination of enzymes (Figure 5-1).

Most biochemical reactions occur spontaneously but at extremely slow rates. Chance encounters between reactants usually fail to bring them together in an orientation that allows them to react, or with enough energy to overcome their resistance to change. Enzymes align molecules in a configuration that promotes the chemical change. They may strain a particular chemical bond until it breaks, or they may position molecules in an orientation that favors their joining together. In both instances, the enzyme lowers the reaction's *activation energy,* the amount of energy required to trigger the reaction (Chapter

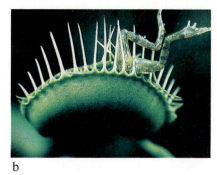

a     b

**Figure 5-1**
**A sampler of enzyme performance.** The properties and activities of every organism depend on enzymes. Here are two examples. *(a)* The unusual color of this blue frog is due to a pigment manufactured by enzymes, a different set than that of his green counterparts of the same species. *(b)* Digestive enzymes secreted by the Venus fly trap will disassemble the body of this imprisoned frog. Other enzymes reassemble the chemicals released from the frog's flesh into plant tissue.

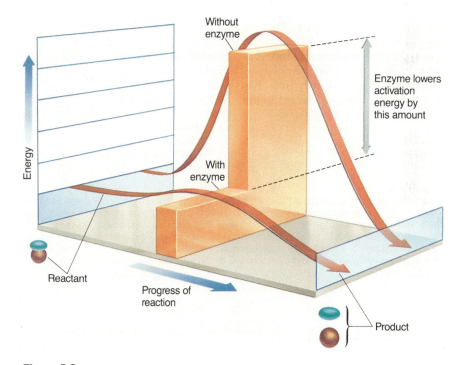

**Figure 5-2**
**Activation energy and enzymes.** Activation energy is the energy required to initiate a chemical reaction, to overcome the reactants' chemical inertia (their tendency to resist change). This energy barrier must be hurdled before a chemical reaction occurs. Enzymes reduce activation energy, so a reaction proceeds at a much faster rate in the presence of the appropriate enzyme than without it.

3). With activation energy lowered by the appropriate enzyme, a reaction proceeds much more rapidly—sometimes several thousand times faster (Figure 5-2).

## Properties of Enzymes

Enzymes are proteins. These macro-molecules create the complex shapes that make enzymes highly specific in their catalytic action. In general, an enzyme recognizes only one set of **substrates** (the reactants) and converts them to one set of **products**. This specificity is largely due to complementary configurations between the enzyme and its substrates. The area on the enzyme that binds a substrate is the **active site.** The substrate fits into the enzyme's active site with a high degree of precision (Figure 5-3). Although the fit is initially loose, the presence of the substrate in the active site distorts the enzyme and induces a tighter fit. If the shapes are not complementary, nothing happens. This high degree of specificity enables cells to control their metabolic reactions by regulating the types and amounts of enzymes produced.

Enzymes are remarkably efficient in the reactions they catalyze. Even in very low concentrations, enzymes cause chemical reactions to proceed at very rapid rates. This efficiency is largely the result of recycling. Enzymes escape being consumed in the reactions they catalyze, so the catalyst always reappears intact after the substrate has been converted to the product. An enzyme is reused until it eventually wears out, at which time it must be replaced.

Some enzymes require the assistance of small "helper" molecules, without which they cannot perform their catalytic function. These helpers temporarily associate with the enzyme (or substrates) and are released when the reaction is completed. If the molecular helper is a metallic ion, it is called a **cofactor**; if it is an organic molecule, it is referred to as a **coenzyme** (Figure 5-4). Plants and many types of microorganisms manufacture their own coenzymes. Humans and other animals, however, require external sources of several coenzymes (or closely related compounds that are converted to coenzymes). The nutritional role of many **vitamins** is to provide the precursors for essential coenzymes that the organism is unable to synthesize (Table 5-1).

# Conditions Affecting Enzyme Activity

## Heat

Increases in temperature boost enzyme activity for the same reason that heat accelerates chemical reactions in general—it increases molecular velocity. The result is a greater frequency and force of molecular collisions, so that reactions occur more often.

An organism's growth rate is the net product of its enzyme activity; in general, the faster the enzymes catalyze, the more rapidly an organism grows. The reaction-enhancing effects of heat explain why most organisms grow more rapidly at higher temperatures. Because these organisms have internal temperatures that cannot be held constant, their growth is accelerated by higher environmental temperatures and slowed by lower temperatures. People and other mammals, as well as birds, maintain steady internal temperatures, even when external condi-

**Figure 5-3**
**A fictitious enzyme in action.** *(a)* Substrates "A" and "B" are enzymatically altered to form product "C." The presence of the substrates in the active sites induces a better fit, in this case, bringing the molecules together in an orientation that encourages reaction. The enzyme is recycled, so it continues to catalyze the reaction. *(b)* The enzyme–substrate fit is more realistically revealed by this computer-generated version of a substrate (the nucleic acid RNA, shown here in green) being attacked and disassembled by the enzyme ribonuclease (in purple).

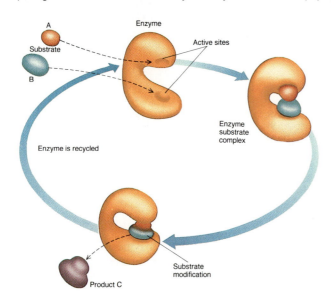

a

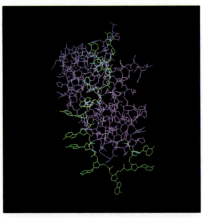

b

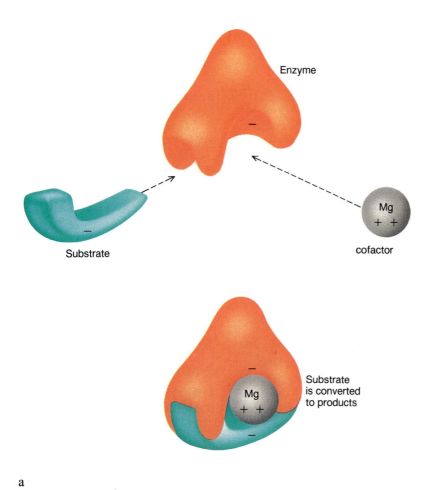

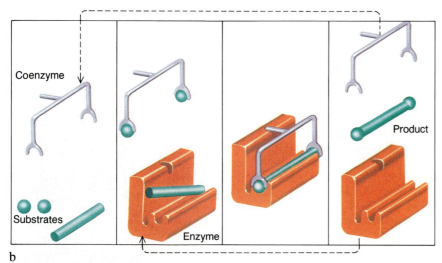

**Figure 5-4**
**Cofactors and coenzymes.** Schematic representation of how they assist enzyme-mediated reactions. *(a)* The positive magnesium ion neutralizes the negative charges on the enzyme and substrate that would otherwise repel each other and prevent interaction. *(b)* Some coenzymes act as vehicles that carry molecules to enzymes and assist in their proper alignment. Like enzymes, coenzymes are recycled.

tions are less than optimal. You would therefore grow about as fast in Hawaii as in Alaska.

Above a critical temperature, however, growth stops. This is the temperature at which enzymes (and proteins in general) suffer heat damage. Excessive heat disrupts hydrogen bonding and other intramolecular forces that stabilize secondary and tertiary protein configuration. Enzymes change shape, may lose solubility, and *coagulate,* in much the same way as does the protein in a cooked egg. In other words, the protein **denatures**—it unfolds and loses its capacity to function properly. Once an enzyme has been physically damaged, it usually cannot be repaired. This is one of the reasons why it is safer to eat cooked food, especially meats and eggs, than raw foods. Proper cooking denatures the proteins of any disease-causing microbes residing in the food. Pasteurization (controlled heating) of milk and other dairy foods has dramatically reduced the spread of serious milk-borne diseases such as tuberculosis. Without functional enzymes, even the most dangerous microorganisms pose no threat of infection.

Unlike excessive heat, cold does no permanent damage to enzymes. Nonetheless, cold can be used to combat food-borne disease and spoilage by reducing the rates of enzyme-mediated reactions, thus retarding the growth of microorganisms. Refrigeration has made the modern world a much safer place for people to live.

**pH**

Enzymes are also denatured by changes in pH. An enzyme that functions optimally at neutral pH (7.0) will usually be inactivated when its environment becomes too alkaline or acidic. The excess hydrogen ions ($H^+$) or hydroxyl ions ($OH^-$) destabilize the molecular forces, the protein unravels, and the enzyme ceases to function. To prevent this outcome, the pH of cytoplasm in cells must be maintained at a pH close to 7.0, the optimum for metabolism. Buffers in cytoplasm resist potentially fatal pH fluctuations.

As with temperature, pH is instrumental in preserving some foods. The enzymes of most spoilage microbes

TABLE 5-1

## Coenzyme activity of B-complex vitamins

| Vitamin | Coenzyme | Types of Reactions Assisted |
|---|---|---|
| Thiamin ($B_1$) | TTP | Removal of $CO_2$ from molecule (decarboxylation) |
| Riboflavin ($B_2$) | FAD | Hydrogen carrier in energy-generating reactions |
| Pyridoxine ($B_6$) | Pyridoxyl phosphate | $NH_2$ transfer (transamination); decarboxylation of amino acids; removal of $H_2O$ (dehydration) from amino acids |
| Cobalamin ($B_{12}$) | Cobalamine | Protein and nucleic acid metabolism |
| Niacin (nicotinic acid) | NAD;NADP | Hydrogen carrier in energy-generating reactions and biosynthesis |
| Pantothenic acid ($B_5$) | Coenzyme A | Transfer of small organic molecular fragments in respiration and fatty acid metabolism |
| Folicin | Folic acid | Single-carbon transfer in nucleic acid and amino acid metabolism |
| Vitamin H | Biotin | $CO_2$ transfer (transcarboxylation) |

cannot tolerate the acidic conditions of sauerkraut, pickles, and many cheeses.

## Concentration Effects

Relative concentrations of substrates and products affect the rate of enzymatically catalyzed reactions. Increased concentrations of substrates relative to the product speed up the reaction, whereas a low substrate-to-product concentration ratio slows the reaction or halts it altogether. The reason for this effect is the *reversibility* of chemical reactions, including those that are enzyme-mediated. Consider the following hypothetical reaction:

$$A + B \longrightarrow C$$

The reaction occurs in the indicated direction because, under existing conditions, the products are more stable than the reactants. Although a few of the products will react in the opposite direction, regenerating the less stable reactants, the overall reaction favors the products, and the reaction proceeds to the right.

Eventually, so much product accumulates and so little of the reactants remain that the rate of conversion in both directions stops changing. At this point, the equation is in **dynamic equilibrium**.

$$A + B \rightleftharpoons C$$

Although equilibrium favors a higher concentration of products (indicated by the longer arrow in that direction), *net* change stops when the product/reactant (P/R) ratio increases to the equilibrium point. This balance can be shifted to the right by reducing the P/R ratio, as would occur if the product concentration was decreased or if reactants were added to the system. The opposite situation (an increase in product concentration or reduction in amount of reactants) drives the reaction back to the left.

The concentration of the enzyme itself plays a role in determining the reaction rate, with higher concentrations increasing the speed of chemical conversion. In fact, cells regulate many of their metabolic reactions by either turning on or shutting off synthesis of

the corresponding enzyme. For example, in some organisms, the enzyme that digests lactose (the sugar in milk) is produced only when the sugar is available. In this way, organisms can avoid wasting energy and resources by selectively stopping the production of an enzyme when it is not needed.

Enzyme concentrations are also reduced by chemicals that are detrimental to an organism's metabolic well-being. If the concentration of even one essential enzyme drops below a critical level, the corresponding life-sustaining reaction stops, as does the organism's growth. Our chemical arsenal against undesirable microorganisms is well stocked with growth-retarding enzyme inhibitors (see Bioline: Saving Lives).

## Recognizing Enzyme Names

Most enzymes have been named by placing an "ase" at the end of a root word that describes either the substrate or the type of reaction catalyzed. For example, an enzyme that accelerates the hydrolysis of a protein to free amino acids is called a *protease* (root word = protein, the substrate). Removal of a carboxyl group (-COOH), a common event in cellular metabolism, is mediated by a *decarboxylase* (root word = decarboxylation, the type of reaction). Alternatively, enzymes that break down their substrates are often named by adding the suffix "lytic" to the name of the corresponding substrate. A protease, for example, may also be called a *proteolytic enzyme,* literally one that splits apart protein. Finally, a few enzymes that were named before nomenclature was standardized retain their traditional names. Two of the best known are *trypsin* and *pepsin,* both of which are proteolytic enzymes that initiate protein digestion in mammals (Chapter 18).

Because of their role in virtually all biological activities, enzymes are discussed frequently throughout this textbook. Familiarizing yourself now with their nomenclature will help you recognize these important proteins and make connections that might otherwise escape your appreciation.

# B I O L I N E

## SAVING LIVES BY INHIBITING ENZYMES

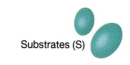

Substrates (S)

Inhibitor (I)

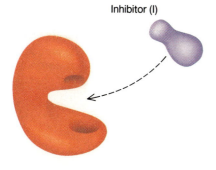

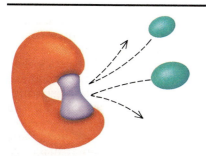

Enzyme-inhibitor complex
(active site blocked)

**Figure 1.**
**Competitive inhibition of enzyme activity** is due to the structural similarity between the inhibitor and the substrate(s). The enzyme's active site recognizes both the inhibitor (I) and the substrates (S), but the inhibitor cannot be converted to products.

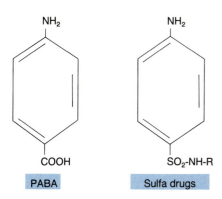

**Figure 2.**

You may be alive today because of our ability to chemically inhibit enzyme activity. Many of the diseases that ravaged our ancestors, killing about one-fourth of their children before they reached puberty, are treatable today with chemicals that block the action of enzymes essential to disease-causing bacteria but not needed by our cells. Such inhibitors may be safely introduced into an infected person's body, where they selectively inhibit bacterial metabolism without impeding that person's own metabolism.

These inhibitory compounds work by competing with the substrate for the active site on the corresponding enzyme. The shape of the inhibitor molecule mimics that of the normal substrate and readily fits into the enzyme's active site. Unlike the substrate, however, the inhibitor cannot be converted to products and, while in the active site, prevents normal substrate binding (Figure 1). In this state, the enzyme ceases to function. Such **competitive inhibition** of enzyme activity is highly specific for a particular metabolic reaction. Only reactions with substrates closely resembling the inhibitor molecules are affected.

*Sulfa drugs* provide an example of competitive inhibition that has saved countless human lives. These agents are given to people to fight bacteria that cause many types of diseases, from urinary bladder infections to pneumonia. The drugs block the ability of bacteria to transform *para-aminobenzoic acid (PABA)* to the essential coenzyme *folic acid.* The structural similarities between sulfa drugs and PABA, Figure 2, create metabolic confusion in bacteria.

If enough of the drug is present to out-compete PABA for a seat in the enzyme's active site, folic acid synthesis abruptly halts. Unable to produce folic acid, or even absorb any that may be in the surrounding tissue fluids of the infected person, the invading bacteria stop growing. All the enzymatic reactions that depend on this coenzyme, most of which are metabolically essential, grind to a standstill. Within hours or days the patient begins to recover. Although many critical metabolic reactions in our bodies also depend on folic acid, we cannot manufacture it. All of our folic acid is obtained from the vitamin *folicin* in our diet. For this reason, we are not affected by the sulfa drugs in the same way as bacteria, which must make their own folic acid. As a result, we survive the treatment unimpaired.

Enzymes may also be inactivated by heavy metals—lead, silver, mercury, and arsenic. In contrast to competitive inhibitors, these poisonous substances bear no resemblance to the enzyme's substrates, but nonspecifically bind to protein. Some of these inhibitors inactivate the enzyme by altering its shape. Others bind in place of cofactors the enzyme needs for activity. In either case, critical metabolic functions are lost. These inhibitors affect enzymes in all cells rather than a specific reaction in one particular type of cell. Our cells are as susceptible to their lethal effects as are bacteria. Introducing heavy metal inhibitors into an infected person's body to cure a disease would eliminate the patient along with the microbes.

Sometimes heavy metals are accidentally introduced into a person, accounting for the tragic consequences of lead poisoning in people who inhale too much leaded gasoline or in children who eat lead-containing paint that has peeled from walls of older dwellings. Laws that restrict the use of lead in paint and gasoline have reduced the incidence of lead poisoning in many countries. Similar poisoning follows ingestion of food that contains high concentrations of mercury. In Japan mercury poisoning has been associated with consumption of fish caught in mercury-contaminated water.

Yet even heavy metals can be used to fight disease. Mercury and silver can be safely applied to the body's surface as mercuric chloride and silver nitrate, preventing or treating minor infections before they become life-threatening diseases.

## Synopsis

### MAIN CONCEPTS

• **All living things are manufactured by enzymes.** Enzymes determine which reactions occur. Each enzyme (or set of enzymes) is responsible for the development of a specific product. Enzymes direct the metabolic traffic into a precise pattern that generates a unique organism.

• **The properties of enzymes account for their catalytic efficiency.** They are specific, recognizing one and only one set of substrates and converting them to a particular set of products. They are composed of protein. The fit between the protein's configuration and substrate dictates the specificity of the active site. Enzymes are never consumed in the reactions they catalyze. Some enzymes require molecular helpers—coenzymes or cofactors that carry substrates to the enzyme or help create a proper fit between enzyme and substrate(s).

• **Enzyme activity is sensitive to several factors.** Higher temperatures accelerate activity, whereas cold retards activity. Excessive heat denatures protein, destroying enzyme activity. Enzymes are also pH sensitive. Neutral pH promotes enzyme activity; excessive acidity or alkalinity impairs enzymes. The product/reactant concentration affect reaction rate, which slows as the product accumulates. Inhibitor compounds compete with substrates and retard reaction rates or damage enzymes in general.

### KEY TERM INTEGRATOR

Properties of Enzymes

| | |
|---|---|
| **enzymes** | Biological **catalysts**, substances that accelerate the rate of a specific chemical reaction. An enzyme's **active site** binds a particular set of **substrates**, which are then modified to **products**. |
| coenzymes | Relatively small organic "helper" molecules (supplied by **vitamins** in the diet of some organisms) needed by some enzymes for activity. Still other enzymes require the participation of inorganic **cofactors**. |

Conditions Affecting Enzyme Activity

| | |
|---|---|
| **denature** | To alter a protein's physical structure with subsequent loss of enzyme activity, often due to heat, acidity, or alkalinity. |
| **dynamic equilibrium** | A stable state in which there is no net change in a chemical reaction. **Competitive inhibition** shifts the equilibrium in the direction of substrates. |

# Review and Synthesis

1. Which of the following are properties of enzymes?
   __composed of protein
   __activity determined by sequence of nucleotides in the primary structure
   __high degree of specificity for substrate(s)
   __unchanged by the reactions they catalyze
   __irreversibly damaged by exposure to cold
   __may require cofactors or coenzymes for activity
   __heat and pH-sensitive
   __work faster in higher concentrations of product
   __shift the dynamic equilibrium of a reaction to the right

2. Discuss the relationship between enzyme specificity and protein configuration as described in Chapter 4.

3. How might a particular enzyme lower the activation energy of a reaction that joins two substrates into a single-molecule product?

4. Inadequate dietary protein is the most common nutritional deficiency among people, followed by vitamin deficiencies. In some ways, both of these types of deficiencies would have the same molecular effect. Describe the similarity.

5. What is the relationship between enzyme activity and each of the following?
   —increasing temperature (two answers)
   —cold
   —growth rate
   —the presence of a structural analog of the substrate (an analog is a molecule with a similar shape)
   —lead poisoning
   —extreme acidity

6. Why do competitive inhibitors make better drugs for fighting infection than do heavy metals and other nonspecific enzyme inhibitors?

# Additional Readings

Becker, W. 1986. *The World of the Cell* (Chapter 6). Benjamin–Cummings. Menlo Park. (Introductory.)

Doolittle, R. 1984. "Proteins." *Scientific American* (an issue devoted entirely to the "Molecules of Life"), October. (Intermediate.)

Fersht, A. 1985. *Enzyme Structure and Function.* 2nd ed. W. H. Freeman, San Francisco. (Advanced.)

Koshland, D. 1973. "Protein shape and biological control." *Scientific American* 229 (Intermediate.)

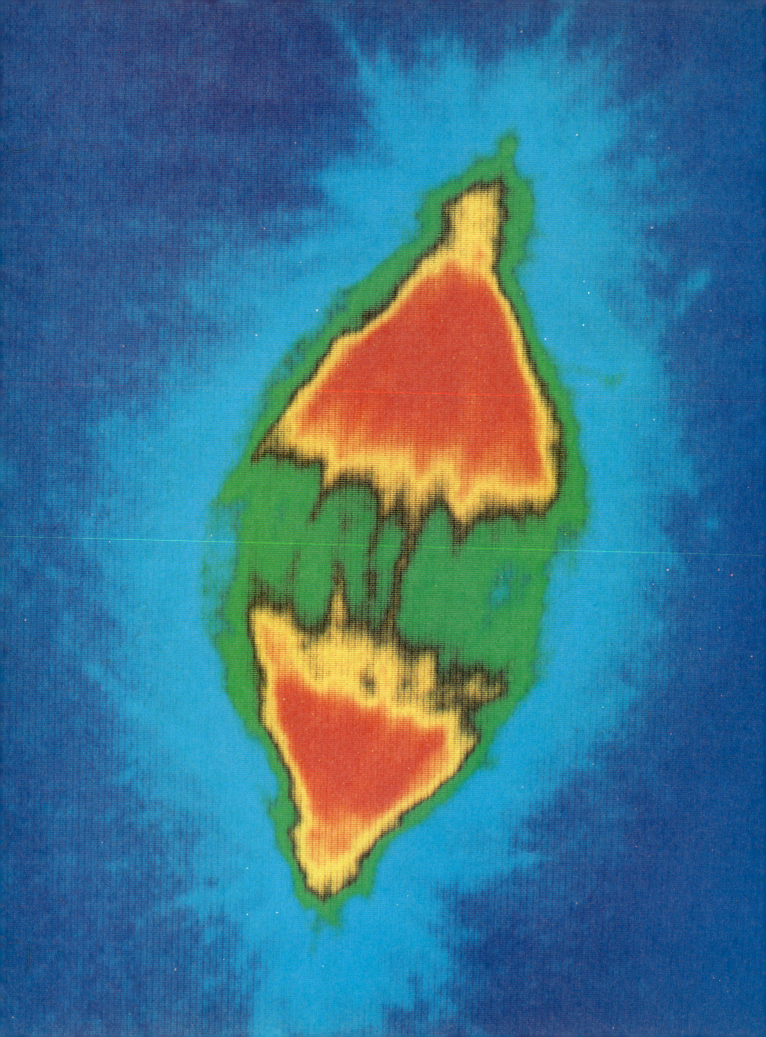

# 3

## *The Cell*

All organisms start life as a single cell. Through cell division, shown here, organisms grow and reproduce.

Chapter 6

# Cell Structures and Functions

*I*n these times of burgeoning computer technology, we have grown accustomed to hearing how smaller and smaller microchips are able to store and process greater and greater amounts of information. A 75-volume encyclopedia, for instance, can now be stored on a single chip and be "read" by a computer in about one second. As remarkable as microchips are, there is another microscopic package that can do this and more—the **cell**. In addition to storing an enormous amount of information (a human cell houses some 40,000 pieces of information, from the contours of a thumb print to a person's rate of hair growth), the cell is the center of life itself, screening, sheltering, organizing, and coordinating a multitude of life-sustaining chemical reactions. The cell is the *fundamental unit of life,* and all living organisms are composed of cells.

- *Nucleus or nucleoid.* All instructions for growth, development, and reproduction are stored in a specific region of the cell. Bacteria and cyanobacteria (monerans) localize such instructions in an area called the **nucleoid**, whereas the cells of all other organisms house hereditary information within a membrane-enclosed **nucleus**.

- *Cytoplasm.* All cells contain **cytoplasm**, a viscous substance containing chemical compounds, particles, and membranes that work together to sustain life. Cytoplasm is a general term that includes all parts of the cell except the plasma membrane and nucleus or nucleoid.

- *Protoplasm.* Together, the plasma membrane, nucleus (or nucleoid), and cytoplasm form the **protoplasm**, a very general term that refers to the overall living parts of a cell.

## Unicellular and Multicellular Organisms

Some of the smallest organisms are **unicellular**, that is, they are composed of just a single cell that contains all the parts essential for life. However, most of the 2 million or so living species are **multicellular**. Large multicellular organisms, like yourself, are complexes of billions or even trillions of cells, many of which perform only one or a few specialized functions for the whole organism. The community of cells that make up a multicellular organism integrate and coordinate life processes so that all perform as a well-orchestrated unit.

## Prokaryotic and Eukaryotic Cells

Cells come in two basic types; prokaryotic and eukaryotic (Figure 6-1). **Prokaryotic cells** (pro = before, kar-

## Cell Characteristics

Cells come in a variety of shapes and sizes. Some are square, round, rectangular, spiraled, star-shaped, and a few can even change shape depending on their activity or location. Cells range in size from the relatively enormous yolk cell of an ostrich egg (a cell that is about the size of an orange), to extremely small bacteria (more than a thousand bacteria would be needed to fill the dot in the *i* in the word "bacteria").

As diverse as cells are in appearance and function, living cells all share the following basic features.

- *Plasma membrane.* All living cells are surrounded by a **plasma membrane** (also called cell membrane). The plasma membrane is a molecular boundary that separates one cell from adjacent cells, and helps safeguard a cell's delicate machinery and chemicals from the disruptive external environment. Because of its unique structure, the plasma membrane also controls the shuttling of substances in and out of the cell (Chapter 7), as well as provides a surface on which many life-sustaining chemical reactions take place.

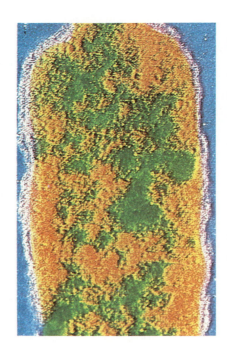

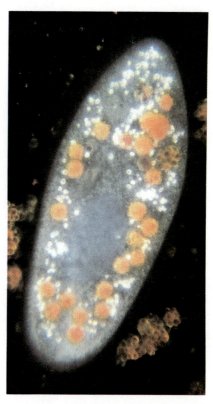

**Figure 6-1**
**Prokaryotic versus eukaryotic cells.** A comparison of the two fundamental cell types.

yotic = nucleus) are the simpler cells, possessing neither a nucleus nor the organelles that are found in the cytoplasm of **eukaryotic cells** (eu = true). **Organelles** are internal cell structures, each of which plays a specific functional role in the cell. With a nucleus and organelles, eukaryotic cells have a more advanced form of cellular organization than prokaryotic cells.

## Limitations to Cell Size

Nearly all cells are microscopic (Figure 6-2), that is, they are too small to be seen without a microscope. Prokaryotic cells seldom reach diameters greater than just a few micrometers (each micrometer—$\mu$m— is one-mil-

lionth of a meter, or 0.000039 inches), and most eukaryotic cells range in diameter from only 10 $\mu$m to 100 $\mu$m (see Appendix I, The Metric System). In general, any organism larger than 100 $\mu$m in diameter is composed of more than one cell.

Why are cells so small? For a cell to be able to perform life functions, all of its living interior parts must be supplied with nutrients, as well as be purged of wastes that might accumulate to toxic levels. This requires a constant exchange of materials between the cell and its surrounding environment. The rate of exchange depends on the cell's size, or more specifically, on the relationship between the sur-

face area of the plasma membrane (through which materials enter and exit) and the volume of the cell. This relationship is called the **surface area-to-volume ratio**.

As a cell grows, its internal volume enlarges faster than the surface area of the plasma membrane, as illustrated in Figure 6-3. This creates two problems. First, because all nutrients and wastes must enter and exit the cell by crossing the plasma membrane, there is a minimum membrane surface area for a particular volume that allows exchanges to take place fast enough for normal metabolism. Second, inside the cell, nutrients and waste molecules travel through the cytoplasm by **diffu-**

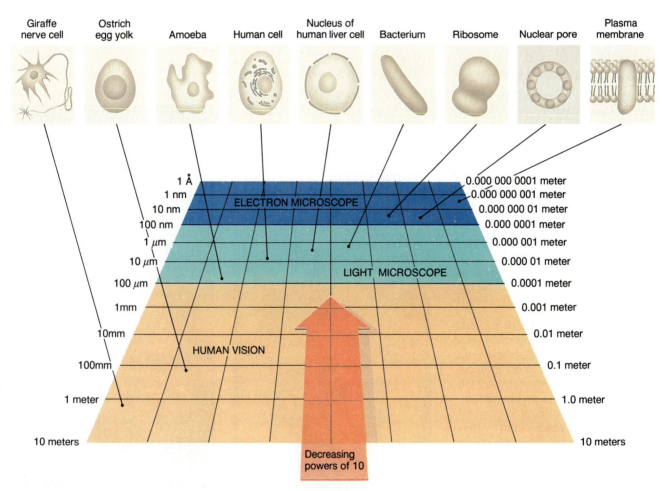

**Figure 6-2**
**Relative sizes of cells and cell components.** Each unit of measurement is one-tenth as large as the preceding unit.

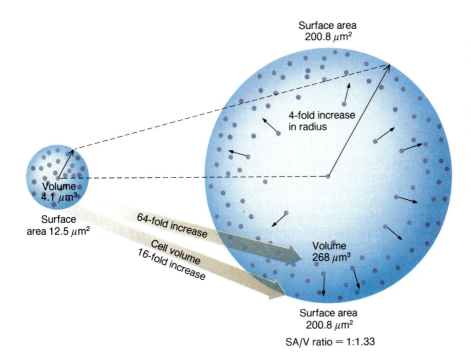

Surface area
200.8 μm²

4-fold increase
in radius

Surface area
200.8 μm²

Volume
4.1 μm³

Surface
area 12.5 μm²

64-fold increase

Cell volume
16-fold increase

Volume
268 μm³

SA/V ratio = 1:1.33

**Figure 6-3**
**Relationship between surface area and volume.** Quadrupling the radius of a cell increases its surface area 16 times and its volume 64 times. The plasma membrane of the larger cell would have to take in and release 64 times as many nutrients and wastes as the smaller cell, but through a surface only 16 times as large. Even if this could be done, nutrients (shown as blue dots) would move very slowly to the cell's center, drastically decreasing metabolism there. Metabolism would be restricted to areas close to the surface of the large cell. On the other hand, metabolism could occur throughout a small cell.

sion (that is, by flowing from areas of greater concentration to areas of lesser concentration—this process is described further in Chapter 7). In small cells, nutrients diffuse to all parts of the cytoplasm, and wastes do not accumulate in the cell's center. As cell volume increases, however, the inner recesses become too distant from the plasma membrane surface for nutrients and wastes to diffuse in or out quickly enough for normal metabolic rates. The cell's interior eventually starves from lack of nutrients, or the cell poisons itself with its own accumulated wastes. Thus, most cells are small, thereby maintaining a surface area-to-volume ratio that allows for optimum exchange rates.

A few cells, however, do attain somewhat large sizes (a nerve cell, for instance) by being long and narrow, or by having convoluted cell margins. These characteristics greatly increase a cell's surface area-to-volume ratio, bringing parts of the cytoplasm close enough to the plasma membrane to permit normal exchange and circulation. In addition, some larger cells have special fibers that pull small bags of cell material to and from the plasma membrane, thus bypassing the limits set by diffusion.

## Discovering the Cell

The invention of the microscope in the mid-1600s sparked one of the greatest revolutions in the history of biology. Before then, biological observations were restricted to the *macroscopic* level, that is, that which is seen by the naked eye. But the invention of the microscope opened up an entirely new world, enabling biologists to explore life at its most fundamental level, the cell (see Bioline: Microscopes).

Robert Hooke was one of the first scientists to see cells. In fact, it was Hooke who gave cells their name. While viewing thin slices of cork under one of the first microscopes, Hooke thought the rows of tiny, rectangular compartments making up the cork resembled the "little rooms" in a monastery, or cells as they were called, so he gave them the same name.

Anton van Leeuwenhoek was one of the first persons to see and describe living bacteria, the smallest of all cells. Leeuwenhoek worked with a simple (single-lensed) microscope, which was little more than a powerful magnifying glass. Because he was an expert lensmaker, Leeuwenhoek's simple microscope magnified objects about 300 times, almost 10 times greater than the

primitive double-lensed microscope used by other scientists at that time.

As microscopes gradually improved, biologists began to see a consistent pattern—all living things were made up of cells. In 1838, Matthias Schleiden proposed that all plants were constructed of cells. The following year, a zoologist named Theodor Schwann extended this generalization to include animals as well. Summarizing the new observations gained from observing the microscopic structure of life, Schwann proposed the first two parts of the **cell theory**.

1. *All organisms are composed of one or more cells.*

2. *The cell is the basic organizational unit of life.*

After studying cell reproduction, 10 years later Rudolf Virchow proposed the third part of the cell theory

3. *All cells arise from preexisting cells.*

Thus, no matter how different organisms appear, all are similar in that they are made up of one or more cells. The cell is the most fundamental structure to harbor and sustain life, as well as to give rise to new life. Consequently, understanding the cell is the very cornerstone of all biology.

## Cell Structures and Functions

With the great diversity in cell size and shape, there is no "typical cell." In this chapter, we describe "generalized" plant and animal cells, cellular "models" that contain a combination of features found in eukaryotic plant and animal cells (Figure 6-4).

### Cell Walls

Some cells have a rigid outer casing called a **cell wall** (Figure 6-5). Cell walls give support, slow dehydration, and prevent a cell from bursting when internal pressure builds due to an influx of water. Bacteria, cyanobacteria, some protists (the algae), fungi, and plants have cells with walls, whereas all animal and nonalgae protist cells do not have cell walls.

Cell walls are constructed of materials that produce a strong, yet porous, coat. Cell wall materials are synthesized inside the cell and then are exported across the plasma membrane to the outside where the cell wall is assembled. Because it is outside the plasma membrane, the cell wall is not considered to be living.

In multicellular plants, the walls of adjacent cells are bonded together, forming an intermediate layer known as the **middle lamella** (Figure 6-5a and b). Cells joined by a middle lamella exchange materials through connecting passages called **plasmodesmata** (Figure 6-5a).

Two types of walls are found in plant cells: primary and secondary cell walls (Figure 6-5b and c). Every plant cell has at least a primary cell wall. However, some specialized cells also manufacture a layered secondary cell wall that improves strength and resiliency. Plant cells with secondary cell walls are found clustered in stems, as well as in the stalks that support leaves, flowers, and fruits. This placement helps support and maintain the orientation of these structures.

Both primary and secondary plant walls are constructed primarily of cellulose, long chains of glucose molecules (Chapter 3). The cellulose molecules intertwine to form bundles called **microfibrils** (Figure 6-5d), and microfibrils wrap around each other to form a stronger unit called a **macrofibril**. Micro- and macrofibrils are aligned in a random overlapping pattern in the

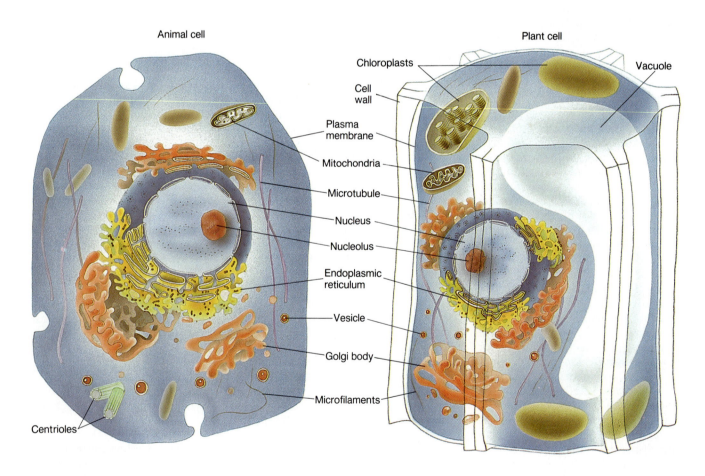

**Figure 6-4**
**Generalized structure of eukaryotic cells.** Plant and animal cells both have a plasma membrane, nucleus, cytoplasm, and cellular organelles. They differ in the types of organelles they contain and by the presence of cell walls in plant cells.

# B I O L I N E

## MICROSCOPES: EXPLORING THE DETAILS OF LIFE

a

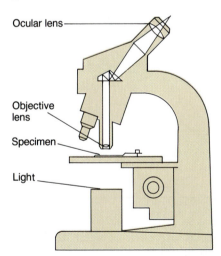

Ocular lens

Objective lens

Specimen

Light

b

c

**M**icroscopes give biologists the power to magnify and distinguish objects that are beyond our normal range of sight. They provide opportunities to probe the details of the structure of life and to survey the fundamental unit of life—the cell.

### The Light Microscope

Light microscopes illuminate a specimen by reflecting light off an object or by transmitting light through it. A *simple light microscope* is much like a magnifying glass *(a);* it increases the size of an object equal to the magnification of the single lens. But a *compound microscope* has two lenses *(b),* so the magnification is compounded (multiplied), that is, the enlarged image produced by the first lens (the objective lens) is even further magnified by the second lens (the ocular lens). For example, if the ocular lens of a compound microscope magnified 10× (times) and the objective lens 40×, the total magnification will be 10 multiplied by 40, or 400×. Thus, compound light microscopes can magnify objects many times larger than a simple light microscope. The maximum effective magnification of the compound light microscope is about 2000×.

Although many biological breakthroughs have been achieved by viewing life structures through light microscopes, there is an important limitation to them, a limitation caused by the very nature of light itself. Light microscopes have limited **resolving power**—the ability to distinguish two very close objects as being separate from each other. The resolving power of the light microscope is about 250 nm (nanometers). This means that no matter what the magnification, objects closer to each other than 250 nm will be seen as a single, blurred object through the light microscope. Since many cell structures are smaller than 200 nm, they remained invisible to biologists until the invention of the electron microscope—an instrument that uses a beam of electrons instead of light. Electron microscopes can generate resolving powers 1000 times greater than the light microscope. Because electrons travel in waves with lengths much shorter than visible light, greater resolving power is possible than with light.

### The Transmission Electron Microscope

The **transmission electron microscope** (abbreviated TEM) works by shooting electrons through very thinly sliced specimens *(c).* Magnetic fields in the electron microscope act as lenses that bend the beam of electrons in the same way that glass lenses bend light. The result is an enormously magnified image than can be viewed by projecting the electron beam onto either a phosphorescent screen (similar to a TV screen) or high-contrast photographic film. The two-dimensional image has remarkable detail (Figure 6-8, for example). The resolving power of the electron microscope, between 0.2 and 1.0 nm, combined with its magnification potential of up to 1 million times, has provided biologists with a new understanding of cellular structure and has helped scientists unravel the details of how cells work.

### The Scanning Electron Microscope

Between 1940 and 1960, a different type of electron microscope was developed. Instead of shooting electrons through a specimen, the **scanning electron microscope** (SEM) showers electrons back and forth across the surface of a specimen prepared with a thin metal coating. The electrons are deflected by the metallic surface, producing a startlingly clear three-dimensional image of the surface characteristics of the specimen (Figure 6-20a). Although the scanning electron microscope has a lower resolution than the TEM and can only provide views of an object's surface, the depth and clarity of the SEM image has given biologists an exciting view of the structure of life.

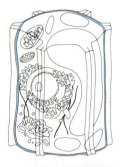

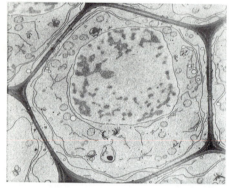

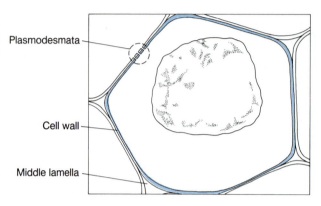

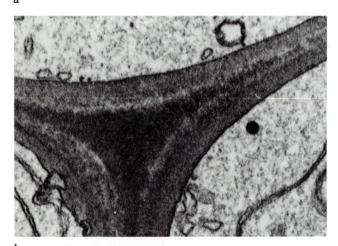

Plasmodesmata

Cell wall

Middle lamella

a

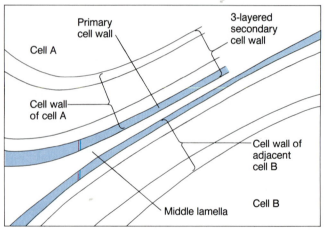

Primary
cell wall

3-layered
secondary
cell wall

Cell A

Cell wall
of cell A

Cell wall of
adjacent
cell B

Cell B

Middle lamella

b

primary cell wall, and, in cells with secondary walls, the fibrils align at different angles in each layer.

Secondary walls are often impregnated with a hardening substance called *lignin.* Lignified secondary wall layers are responsible for the strength of wood and natural fibers. Some secondary cell walls also contain *suberin* or *cutin,* waxes that restrict water movement. Cells on the surfaces of leaves and stems are waterproofed by a layer of cutin on the side of the cell that faces the external environment. This layer helps prevent water loss from internal tissues.

Other plant cells have secondary cell walls that are completely impregnated with waxes. Unless these cells have pores to allow water to move through the water-repellent secondary wall, the cell will eventually die of dehydration, leaving only the cell wall and a hollow area where the protoplasm used to be. In most plants, it is through the hollow passages of dead cells that water and minerals are moved from the roots to stems, leaves, and flowers (Chapter 13).

Even with a surrounding wall, living cells are capable of dividing to produce new cells. During division, a new cell

wall is deposited between daughter cells, as shown in Figure 6-6.

## The Plasma Membrane

A living cell must not only remain separated from its environment, but it must continually interact with its surroundings in order to acquire nutrients and rid itself of wastes. The plasma membrane is a dynamic barrier that accomplishes both of these life-supporting tasks.

The plasma membrane is a fluid *bilayer* (bi = two) of phospholipids in which proteins and a few other types of macromolecules are suspended (Fig-

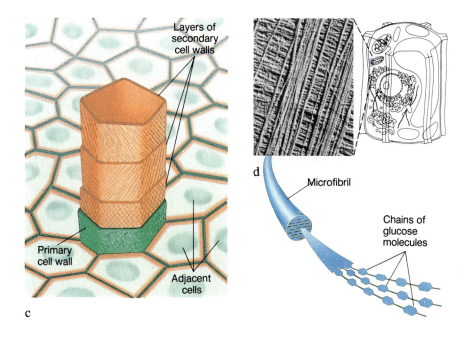

c

d

Microfibril

Chains of glucose molecules

Layers of secondary cell walls

Primary cell wall

Adjacent cells

**Figure 6-5**
*(a)* Electron micrograph of a plant cell wall, middle lamella, and interconnecting plasmodesmata. *(b)* Middle lamella and cell walls of adjacent cells, showing the primary cell wall and layers of the secondary wall. *(c)* A diagrammatic plant cell wall, telescoped to reveal the primary cell wall, layers of the secondary cell wall, and the arrangement of cellulose fibrils. *(d)* A surface view of the layers of parallel fibrils in a secondary cell wall. Each microfibril is a complex of cellulose chains, held together by hydrogen bonds between adjacent glucose molecules to form flat, sheetlike strips.

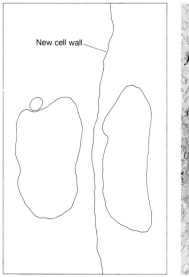

New cell wall

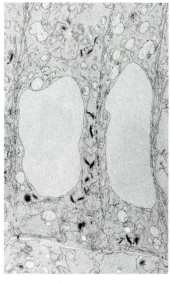

**Figure 6-6**
**The formation of a new cell wall following plant cell division.**

ure 6-7). Proteins that span the two phospholipid layers either form small channels that allow certain substances to move through the membrane, or they ferry specific substances across the membrane, transporting them into or out of the cell. (The manner in which materials are transported through the plasma membrane is described in the next chapter.) Other membrane proteins are attached or embedded to one phospholipid layer or between the two layers. In addition to proteins, the plasma membrane also contains *glycolipids* (carbohydrate-lipid complexes) and *glycoproteins* (carbohydrate-polypeptide complexes). In humans, glycoproteins act as receptors for hormones and antibodies.

Even with the most powerful electron microscope, researchers are still unable to see individual molecules of the plasma membrane. How, then, do scientists know what the plasma membrane looks like? Chemical analysis reveals that 50 to 60 percent of the dry weight of the plasma membrane is protein, 30 to 40 percent is phospholipid, and between 1 and 10 percent is carbohydrate. These findings, together with the results of experiments on how the plasma membrane works, have given investigators enough information to propose visual models to describe the arrangement of chemicals that make up the plasma membrane.

During the late 1960s, a new procedure was discovered that allowed researchers to actually see some structural details of the membrane. By freezing the plasma membrane, investigators could split apart (fracture) the two phospholipid layers and then observe the inner membrane surfaces under an electron microscope. Using this procedure, appropriately named freeze-fracturing, investigators saw protein-size particles dotting the membrane. This suggested the presence of round, *globular* proteins in the membrane, not uniform protein and lipid *layers* as was previously hypothesized. These discoveries led to the revised model of the plasma membrane, the **fluid mosaic model**. According to the fluid mosaic model, the phospholipid bilayer is liquid, having a viscosity similar to that of light household oil. The globular proteins float in the fluid phospholipid bilayer, forming patterns resembling an inlaid mosaic —hence, the name *fluid mosaic*. The fluid mosaic model is now the favored explanation for the architecture of the plasma membrane, not only because it is consistent with the way the membrane looks following freeze-fracturing, but also because this arrangement

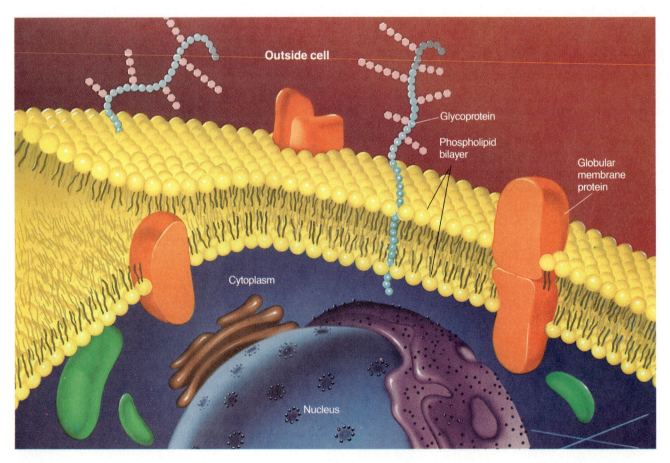

**Figure 6-7**
**Structure of the plasma membrane.**

helps explain how the plasma membrane is able to govern the entrance and exit of materials into and out of the cell as described in the next chapter.

## Cell Organelles

All eukaryotic cells contain specialized organelles that carry out specific metabolic activities. Many organelles have a surrounding membrane, shaping a discrete membrane-bound structure. Other organelles form a system of folded membranes, providing surfaces that help organize complicated chemical reactions. These surfaces also partition the inside of the cell into smaller compartments so that metabolic activities are separated and do not interfere with each other. These internal membranes are similar to the plasma membrane in both composition and construction. Indeed, some are infoldings of the plasma membrane itself.

The principal organelles in eukaryotic cells and their primary functions are overviewed in Table 6-1.

### THE NUCLEUS

The most prominent structure in a eukaryotic cell, the nucleus, also has the most prominent function (Figure 6-8). The nucleus houses the hereditary information of an organism. This information includes all of the genetic instructions for (1) governing the synthesis of chemicals for normal cell functioning, (2) controlling the production of a new cell or an entire multicellular organism, and (3) directing the development of genetically acquired characteristics. Because of these supervisory functions, the nucleus is often called "the control center" of a cell.

The genetic information in the nucleus governs the synthesis of specific enzymes that, in turn, catalyze particular metabolic processes or contribute to the formation of a certain trait. All of this information is stored in molecules of DNA found in the chromosomes. In addition to containing chromosomes, the nucleus is surrounded by a nuclear envelope that encloses the

TABLE 6-1

| The Primary Functions of Cell Organelles | |
|---|---|
| **Organelle** | **Primary Functions** |
| Nucleus | Storage of hereditary information, control of cell activities |
| Endoplasmic reticulum (ER) | Protein synthesis, distribution of material throughout cell |
| Ribosomes | Sites of protein synthesis |
| Mitochondria | Chemical energy conversions for cell metabolism |
| Plastids (Plant cells only) | Conversion of light energy into chemical energy, storage of food and pigments |
| Golgi complex | Synthesis, packaging, and distribution of materials |
| Lysosomes | Digestion, waste removal, discharge |
| Microbodies | Chemical conversions, discharge |
| Vacuoles | Digestion, storage |
| Microfilaments and microtubules | Cellular structure, movement of internal cell parts |
| Cilia and flagella | Locomotion, production of currents that draw in food |

nucleoplasm and one or more small bodies called nucleoli.

### Chromosomes

An organism's genes are located in its **chromosomes**, dark-staining structures composed of roughly half DNA and half protein (Figure 6-9). The number of chromosomes in the nucleus varies according to species; for example, one type of pine tree has 24, the common sunflower has 34, and humans have 46. Actually, chromosomes are visible only when a cell divides. The remainder of the time, chromosomal material is unraveled into a loosely organized collection of fibers called **chromatin**.

### Nuclear Envelope

The **nuclear envelope** is a double membrane that partitions the nuclear contents from the rest of the cell (Figures 6-8 and 6-10). Both the inner and outer membrane layers are similar in structure to the plasma membrane (that is, a fluid mosaic of phospholipids and proteins). However, the nuclear envelope is studded with complicated pores that control the passage of chemicals to and from the nucleus (Figures 6-10). For instance, proteins needed for making new DNA in the nucleus are synthesized in the cytoplasm and then transported through nuclear pores to DNA assembly sites. Conversely, chemical instructions for

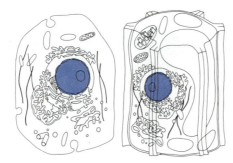

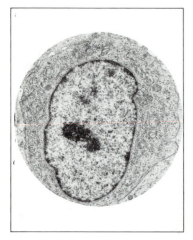 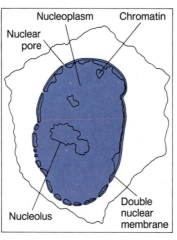

Nucleoplasm    Chromatin

Nuclear
pore

Nucleolus

Double
nuclear
membrane

**Figure 6-8**
**The nucleus.**

synthesizing proteins are formed in the nucleus and then transported through nuclear pores to the cytoplasm where protein synthesis occurs. Thus, the nuclear envelope is an important gateway between the control center of the cell (the nucleus) and the cytoplasm, the region of greater metabolic activity.

Nucleoli

The nucleus often contains one or more darker regions called **nucleoli** (Figure 6-8). Each nucleolus is a site for constructing the two subunits that will combine to form **ribosomes**, the sites of protein synthesis in the cell's cytoplasm. Because cells constantly manufacture proteins, a continual supply of ribosome subunits are needed for normal cell functioning.

Nucleoplasm

The **nucleoplasm** is the semifluid substance in the nucleus containing pro-

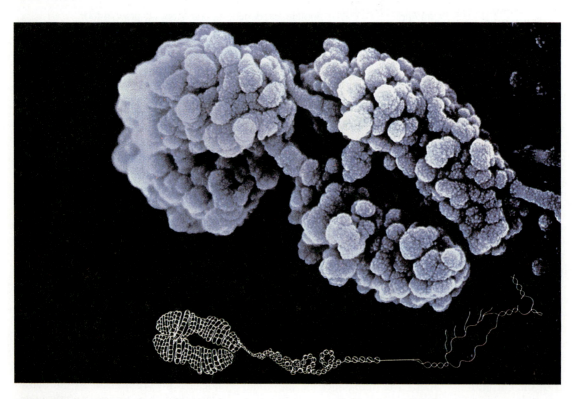

**Figure 6-9**
**Chromosomes.** During cell division in a eukaryotic cell, the genetic material takes on the form of tightly coiled chromosomes.

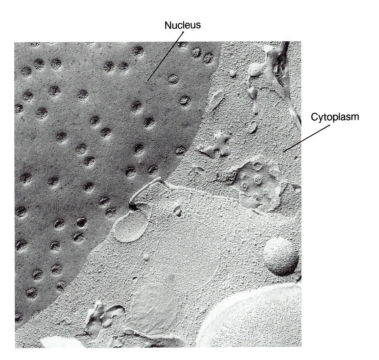

**Figure 6-10**
**Nuclear pores.** Labeled freeze-fracture electron micrograph showing complex nuclear pores.

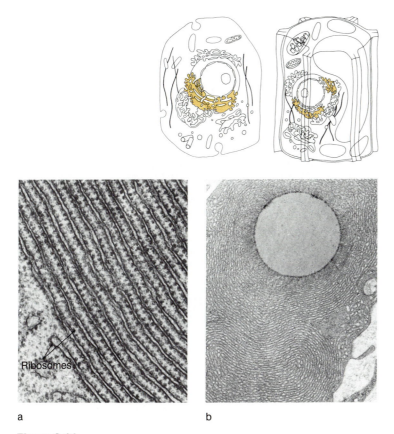

a                                    b

**Figure 6-11**
**Endoplasmic reticulum.** *(a)* Electron micrograph labeled rough endoplasmic reticulum (RER). *(b)* Labeled electron micrograph of smooth endoplasmic reticulum (SER).

tein, granules, and fibrils (Figure 6-8). Some scientists hypothesize that the nucleoplasm also includes a structural frame that supports nuclear structures. This molecular "skeleton" has been called the *nuclear matrix.*

## ENDOPLASMIC RETICULUM

Under the light microscope, a cell's cytoplasm appears watery, without any structure whatsoever. However, high-resolution magnification with an electron microscope reveals that the cytoplasm contains not only tiny subcellular organelles, but also a complicated membrane system that extends to virtually all parts of the cell's interior. This elaborate system of folded, stacked, and tubular membranes is called the **endoplasmic reticulum**, or **ER** for short (Figure 6-11).

The continuous membranes of the endoplasmic reticulum form a circulation system within the cell. This system creates passageways that channel materials to different locations or provide corridors for exporting materials produced by the cell during **secretion.** Materials move more quickly through these membrane passages than would be possible by diffusing through the cell's viscous cytoplasm. In fact, without ER many nutrients would not be able to travel fast enough to nourish the inner recesses of larger cells. The same is true for the movement of some toxic waste products produced during metabolism. Without ER, wastes might accumulate in remote areas to concentrates high enough to kill the cell.

ER occurs in one of two forms, depending on whether or not ribosomes are attached to the membrane surfaces. **Rough ER (RER)** appears bumpy (rough) in electron micrographs because of many attached ribosomes (Figure 6-11*a*). ER without ribosomes is called **smooth ER (SER)** (Figure 6-11*b*). SER is generally more tubular, whereas RER is usually composed of flattened sacs of membranes.

Proteins destined for export from a cell are synthesized on the rough ER. Because RER membranes are continually undergoing changes in structure, rough endoplasmic reticulum even synthesizes the proteins that eventually become part of its own membrane. The assembling of amino acids into proteins takes place at the attached ribosomes.

Like RER, SER forms transport channels for circulating materials, but since there are no attached ribosomes, protein synthesis does not take place on SER. However, SER is the site for synthesis of some other chemicals, such as some fats and a few hormones.

The total amount of ER in a cell, as well as the relative proportions of RER to SER, varies between cells with different functions. Many of the cells that line the digestive system of animals, for example, manufacture large amounts of digestive enzymes and, therefore, contain extensive RER. On the other hand, cells that secrete chemicals other than proteins—steroid hormones, for instance—contain more SER. But whether a cell has mostly RER or SER, the endoplasmic

reticulum greatly increases the surface area for chemical reactions, establishes internal cell compartments, and provides channels for the transport of materials within a cell.

## RIBOSOMES

Ribosomes are not only found attached to the rough ER membranes, but also occur "free floating" in the cytoplasm of prokaryotic and eukaryotic cells alike. In addition, they are found in mitochondria and chloroplasts, two important energy conversion organelles of eukaryotic cells.

Although there are several types of ribosomes, all have the same function —protein synthesis. Ribosomes act as chemical "matchmakers," matching amino acids with the molecular instructions from the nucleus to construct a specific protein. (How protein synthesis actually works is described in

Chapter 27.)

## MITOCHONDRIA

Life demands continuous metabolism; an average cell executes some 60 types of metabolic reactions every *second*. Many of these reactions require the energy in ATP, the form of chemical energy used by the cells of all organisms. Virtually all the ATP manufactured by a eukaryotic cell is produced in **mitochondria** (plural) through a series of chemical reactions called *cellular respiration* (Chapter 10). Because they supply chemical power to run metabolism, mitochondria are frequently called the "powerhouses" of a cell (Figure 6-12).

As with other organelles, the number of mitochondria in a cell depends on the cell's function. A few very specialized cells have no mitochondria in them (in which case, they get ATP

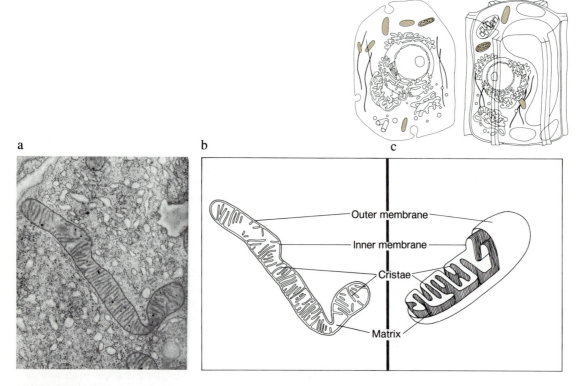

a          b          c

**Figure 6-12**
**A mitochondrion.** *(a)* An electron micrograph, *(b)* a labeled sketch, and *(c)* three-dimensional detail of a mitochondrion.

from the mitochondria in adjacent cells), others have only a few, and still others contain hundreds or thousands of mitochondria.

Generally, mitochondria are sausage-shaped, although their contours often change as they flow through the cytoplasm of a cell (Figure 6-12a). But regardless of shape, each mitochondrion is constructed of two membranes—an outer membrane surrounding an elaborately folded inner membrane. The labyrinth of convolutions created by the inner membrane, which is similar in appearance to ribbon candy, forms the *cristae* of the mitochondrion (Figure 6-12b and c). The double membrane system of a mitochondrion produces an *intermembrane space* (between the outer and inner membranes) and a *matrix* (between the cristae folds) (Figure 6-12c). Both the intermembrane space and the matrix contain enzymes necessary for producing ATP during cellular respiration.

## PLASTIDS

**Plastids** belong to a category of similar organelles found only in plant cells. All plastids have two outer membranes that enclose a fluid called the **stroma**. Three types of plastids store plant materials.

- *Leucoplasts* store proteins, oils, and starch

- *Amyloplasts* store only starch

- *Chromoplasts* store colored pigments

But the most important plastids in all biology are **chloroplasts** (Figure 6-13), the organelles that carry out photosynthesis by converting light energy into chemical energy (Chapter 9). Virtually all life on earth depends on photosynthesis. Each chloroplast contains an elaborate system of internal **thylakoid membranes**, flattened membrane sacs in which light-capturing pigments (mainly **chlorophyll**) are embedded. Some thylakoids are arranged in stacks called **grana**, whereas other membranes extend from granum to granum to form **intergranal thylakoids** (Figure 6-13b and c).

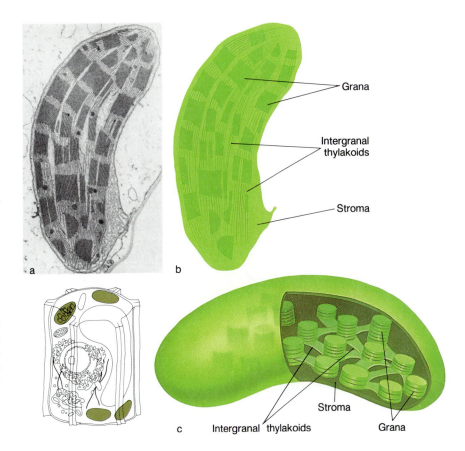

**Figure 6-13**
**Chloroplast.** *(a)* An electron micrograph, *(b)* a labeled sketch, *(c)* a three-dimensional diagram of a chloroplast.

Because chlorophyll reflects green light, chlorophyll-rich thylakoid membranes give plants their green color. The spectacular changes in leaf color of some plants in the fall occurs because the plant breaks down its own chlorophyll pigments, thereby unmasking the remaining red, orange, and yellow photosynthetic pigments (Figure 6-13d).

Both plastids and mitochondria contain their own DNA. Mitochondrial and plastid DNA is similar to that of a prokaryotic cell. This observation, among others, has sparked a number of fascinating hypotheses on the origin of organelles in eukaryotic cells (Chapter 29).

## Golgi Complex

Materials continually enter and exit cells. One of the functions of the Golgi complex is to package substances for secretion from the cell.

In some cells, the Golgi complex is a system of loosely arranged membranes, whereas in others it is highly organized, with stacks of flattened membrane sacs called **cisternae** (Figure 6-14). A single stack of cisternae forms a **Golgi body** (sometimes called a **dictyosome** in plant cells), and all the Golgi bodies in a cell collectively comprise the **Golgi complex**.

Each disk-shaped cisterna has a lacey edge formed by dividing and reconnecting tubules. The edges swell as

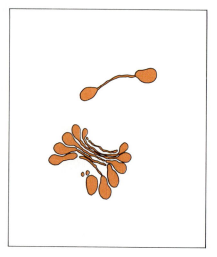

a

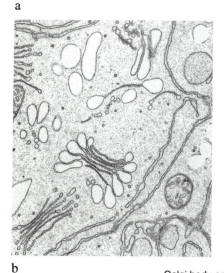

b

Golgi body or dictyosome

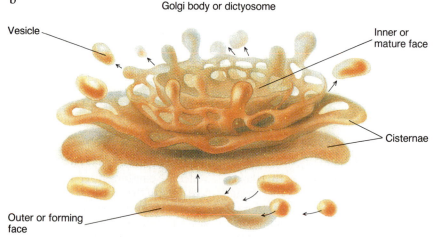

Vesicle

Inner or mature face

Cisternae

Outer or forming face

**Figure 6-14**
**The Golgi complex.** *(a)* An electron micrograph and *(b)* a labeled sketch of a Golgi body. All of the Golgi bodies in a cell constitute the cell's Golgi complex.

chemicals accumulate. When they reach a certain size, the swollen areas pinch off to form membrane-enclosed **vesicles**. Some vesicles remain in the cell, storing important chemicals, whereas others move to the cell surface and fuse with the plasma membrane to secrete their contents to the outside. During secretion, the Golgi complex forms a crucial link in a *secretory chain* (or *pathway*) that carries materials from their site of synthesis to the cell's outside (Figure 6-15).

As with most other organelles, the number of Golgi bodies varies according to the type of cell, its function, and its stage of growth. Dividing plant cells, for example, contain many Golgi bodies (dictyosomes) clustered near the area where a new cell wall is being manufactured. These Golgi bodies produce vesicles that contain material needed to construct cell walls between the newly formed cells (Figure 6-16). Animal secretory cells, such as those that manufacture hormones, each contain thousands of Golgi bodies.

## LYSOSOMES

**Lysosomes** are one type of storage vesicle produced by the Golgi complex. Lysosomes contain powerful hydrolytic enzymes that are capable of digesting every macromolecule in the cell. The membrane surrounding the lysosomes keeps these lethal enzymes safely sequestered. But to remain an effective life-saving barrier, the lysosome membrane must be continually renewed.

Why do cells house such potentially injurious chemicals? The chemicals in lysosomes dispose of impaired or worn-out mitochondria and other cell structures. In some organisms, lysosomes digest trapped food particles. In a few cases, an organism's own cells are deliberately digested by lysosomes. The lysosomes inside the cells that make up a tadpole's tail gradually break down the tail, thereby providing an additional source of nutrients to support its development into a frog.

## MICROBODIES

**Microbodies**, like lysosomes, are a type of membrane-enclosed organelle that contains enzymes (Figure 6-17).

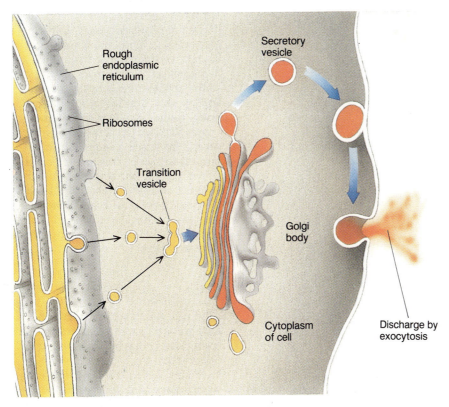

**Figure 6-15**
**The secretory pathway.** Proteins destined for secretion are synthesized on the RER, packaged in a transition vesicle, and then transferred to a Golgi body that concentrates and stores them in secretory vesicles. When a vesicle's membrane fuses with the plasma membrane, the proteins are discharged (secreted) to the outside.

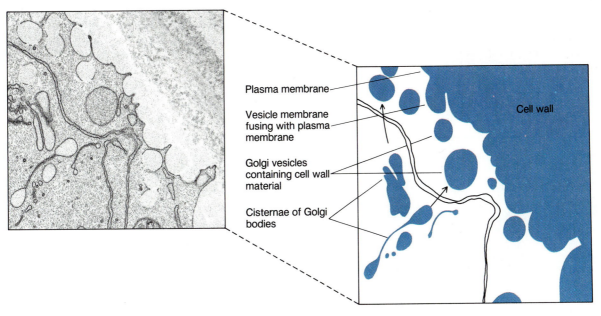

**Figure 6-16**
**Cell wall synthesis.** Cell wall precursors are packaged by the Golgi bodies into vesicles that fuse with the plasma membrane to release material to the outside where they are added to the growing plant cell wall.

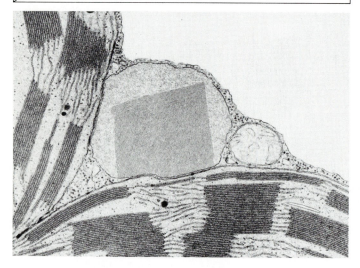

**Figure 6-17**
**A tobacco leaf cell microbody containing a large crystallized enzyme.** Microbodies contain a variety of enzymes that break down specific substances.

However, unlike lysosomes, microbodies do not contain hazardous hydrolytic enzymes. One kind of microbody contains enzymes that help degrade hydrogen peroxide (a toxic product of some metabolic reactions) into water and oxygen. Another type of microbody, found in plants, contains enzymes that convert fats and oils into sugars. Such conversions are important during seed germination so that stored lipids in the seed or plant embryo can be digested. This provides the developing plant with an immediate source of chemical energy until it can begin manufacturing its own sugars by photosynthesis.

## VACUOLES

Although plant and animal cells both have **vacuoles**, these structures are not identical in both types of cells. Animal cells contain many types of vacuoles, including those used for eliminating water, entrapping food particles, digestion, and discharging wastes, among others. In plants, the term "vacuole" is generally used to describe a large organelle found in mature cells. The plant vacuole often occupies most of the cell's volume, sometimes more than 90 percent (Figure 6-18). It is surrounded by a **tonoplast**, a single membrane that governs which materials are exchanged between the cytoplasm and the fluid inside the vacuole. This fluid, the *cell sap,* is a solution that varies in

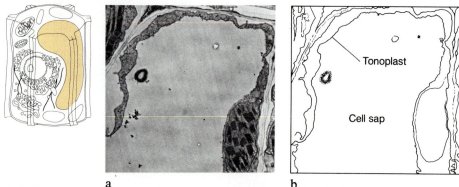

a                                          b

**Figure 6-18**
**Plant vacuole.** *(a)* An electron micrograph. *(b)* A labeled sketch of a plant vacuole.

content depending on the age and specialization of the cell. Cell sap contains water, and sometimes gases (like oxygen, nitrogen, and carbon dioxide), acids, salts, sugars, crystals, and pigments. The color of beets and the petals of some flowers are due to water-soluble pigments stored in the vacuole of each cells. Plant vacuoles function to store materials and to maintain high internal water pressure which aids in the support of the plant.

Botanists have recently learned that toxic substances may accumulate in the vacuole to concentrations that would be lethal if in the cytoplasm. This finding suggests that the vacuole may also function as a receptacle for isolating metabolic wastes and toxic chemicals. Unlike animals which have elaborate excretory systems for expelling toxic wastes, plants apparently "excrete" such substances into their own vacuoles. The vacuole's tonoplast safely partitions the toxins from the plant's own protoplasm.

## MICROFILAMENTS AND MICROTUBULES

A living cell bustles with metabolic activity. Materials constantly enter and exit the cell; genetic instructions flow from the nucleus to the cytoplasm; proteins and other substances are synthesized or degraded by cellular organelles; and many manufactured products are steadily packaged for storage or secretion. To maintain such a high level of activity, the cell's contents must not only be in constant operation, but in constant motion as well. The one-way streaming of cell contents is called **cyclosis** or **cytoplasmic streaming**.

Cyclosis is the result of an internal current produced by contracting **microfilaments**, solid protein fibers that use energy of ATP to shorten and exert a physical force. The rate of cyclosis is affected by the presence of a cell skeleton called the **cytoplasmic lattice** or **cytoskeleton** (Figure 6-19). The cytoskeleton extends throughout the cytoplasm and is constructed of cross-linked microfilaments, intermediate-sized filaments, and **microtubules**. Microtubules, hollow tubes built from repeating protein units of *tubulin* are an important component of the cytoplasmic lattice. In addition, they move chromosomes during cell division, and are the key constituents of cilia and flagella, the organelles used for locomotion.

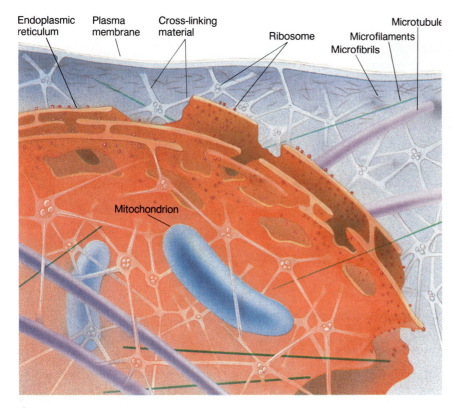

**Figure 6-19**
**Cytoskeleton.** Microtubules, microfilaments, and interconnecting strands make up a cell's cytoskeleton. The cytoskeleton helps direct the flow of cyclosis and supports the cell's organelles.

## CILIA AND FLAGELLA

Some cells have the ability to propel themselves through a watery environment or to create currents that move food and other materials over their surfaces. The structures responsible for these movements are cilia and flagella.

**Cilia** are short, hairlike structures projecting from the surfaces of some cells that beat in coordinated waves (Figure 6-20). Generally, cilia are found in large numbers and are densely packed.

Whiplike **flagella** are longer than cilia and are fewer in number. Flagella whip back and forth, causing a cell to move much as a tadpole does as it swims.

Cilia and flagella are powered by microtubules. The microtubules are arranged in a specific pattern called a *9 + 2 arrangement* — nine pairs of microtubules encircling a pair of central microtubules (Figure 6-21). Tiny armlike projections called *dynein arms* extend along one side of each pair of microtubules. When the dynein arms attach to a neighboring microtubule, the breakdown of ATP is triggered. The released energy causes the dynein arm to reattach at a lower point on its neighboring microtubule. The progressive attachment and reattachment slides adjacent microtubules in opposite directions, forcing the cilium or flagellum to bend to one side. When the dynein arms rapidly attach and reattach at successively higher points, the paired microtubules slide in the other direction and the flagellum or cilium whips to the other side.

Each cilium and flagellum has a *basal body* where it enters the cell. In the basal body, the microtubules lose their "9 + 2 arrangement," clustering into groups of three that form a circle of microtubules. The basal body is also illustrated in Figure 6-21.

## The Fundamental Unit of Life?

Is all life cellular? **Viruses** are minute structures composed only of hereditary information (DNA or RNA) surrounded by a protein coat. Since viruses lack a plasma membrane, nucleus, and cytoplasm, they cannot metabolize on their own. Yet, when a virus invades a cell, the viral nucleic acids enslave the host cell. Under the

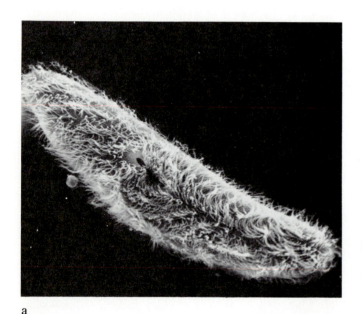
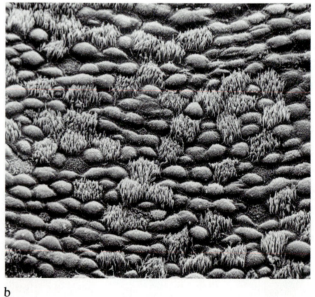

a                                                    b

**Figure 6-20**
**Cilia.** *(a)* A scanning electron micrograph of a *Paramecium*. The numerous cilia enable the cell to propel itself through water while creating currents that channel food into its "mouth." *(b)* A scanning electron micrograph of the human trachea (windpipe). Mucus secreted by goblet cells (G) traps dust and microbes. The coordinated beating of the cilia lining the trachea moves the mucus blanket and its captives away from the lungs and into the throat where it can be swallowed.

direction of the virus nucleic acids, the host cell relinquishes control of its energy and cellular machinery and begins producing new viruses. Sometimes the host cell is filled with so many viruses that it explodes, broadcasting a flood of new viruses that can then invade more host cells.

Are viruses living because they can commandeer a host cell to build new copies of themselves (recall from Chapter 1 that reproduction and heredity is one of the basic characteristics of life), or are viruses not living because they cannot metabolize or grow on their own? Biologists still debate this point. (We discuss viruses further in Chapter 31.)

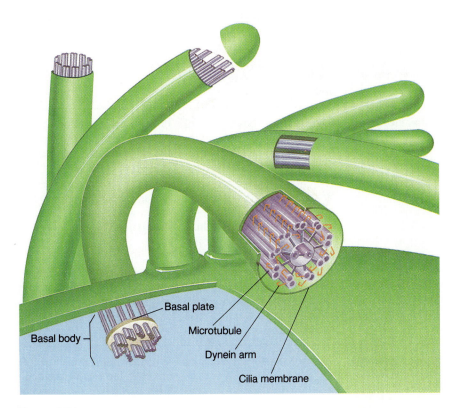

**Figure 6-21**
**Cilia structure and movement.** The 9 + 2 microtubule arrangement of cilia is shown in this sketch. A cilium moves when the dynein arms move along neighboring microtubules, sliding the two microtubules of each pair in opposite directions. The sketch also shows the microtubule arrangement in the basal body.

## Synopsis

### MAIN CONCEPTS

- **The cell theory has three parts.**
  All organisms are composed of one or more cells.
  The cell is the basic organizational unit of life.
  All cells arise from preexisting cells.

- **Living cells continually carry out metabolism and are in constant interaction with their surroundings.**

- **Living cells are surrounded by a plasma membrane,** store hereditary information in a nucleus or nucleoid, and have cytoplasm, the site of most metabolic reactions. The living cell's protoplasm includes the plasma membrane, nucleus or nucleoid, and cytoplasm.

- **Cells have a minimum plasma membrane surface area-to-cell volume ratio** that allows nutrient and metabolic waste exchanges to occur at rates adequate to maintain normal metabolism.

- **Cells are either prokaryotic or eukaryotic.** Eukaryotic cells contain a true nucleus and membrane-bound organelles, whereas prokaryotic cells do not. All multicellular organisms are constructed of eukaryotic cells. Unicellular organisms are prokaryotic or eukaryotic.

### KEY TERM INTEGRATOR

| Cell Structure | Characteristics | Functions |
| --- | --- | --- |
| **cell walls** | Deposited outside cell; most constructed of carbohydrates; plant cells have **primary** and sometimes **secondary** wall layers | Support; resistance to internal and external pressure; water relations |
| **plasma membrane** | Surrounding boundary of cell, 2 layers of phospholipids with embedded proteins and carbohydrates | Protection; regulation of uptake and release of materials; site of chemical reactions |
| **nucleus** | Prominent, often round; double membrane layer with pores, **chromosomes (chromatin)**, **nucleolus**, and **nucleoplasm** | Storage site of hereditary information; control of cell metabolism |
| **endoplasmic reticulum (ER)** | System of membranes extending throughout the cell; forms channels, tubes, and membrane sacs | Channels for transporting materials and **secretion** |
| **rough endoplasmic reticulum (RER)** | ER with ribosomes attached | Synthesis and transport of proteins for export or new RER |

| | | |
|---|---|---|
| **smooth endo-plasmic reticulum (SER)** | ER lacking ribosomes | Synthesis of lipids and hormones; transport of substances in cell |
| **ribosomes** | RNA + protein; free floating or attached to ER | protein synthesis |
| **mitochondria** | Organelle with two membranes; folded inner membrane forms **cristae** | Location of most cellular respiratory reactions that produce ATP |
| **plastids** | Membrane-bound organelles; chloroplasts have inner **thylakoid membranes** containing photosynthetic pigments | Storage of lipids, fats, and carbohydrates; chloroplasts trap and convert light energy into chemical energy during photosynthesis |
| **Golgi bodies** | Stacks of flattened membranes **(cisternae)**; small **vesicles** pinched off from ends | Synthesis or modification of chemicals; packaging materials for export or storage; formation of lysosomes |
| **lysosomes** | Membrane-enclosed vesicles containing digestive enzymes | Breakdown of worn-out cell parts or unwanted materials; digestion |
| **microbodies** | Membrane-enclosed sacs containing various enzymes | Site of various metabolic reactions |
| **vacuoles** | Membrane-bound sacs; **tonoplast** membrane surrounds large central vacuole in plant cells | Digestion, transport, and storage of materials (excretion of wastes in plant cells) |
| **microfilaments** | Solid rods of contractible proteins | Structural, movement of cell structures during **cyclosis** or **cytoplasmic streaming** |
| **microtubules** | Hollow rods of protein | Support of cell structures, cell locomotion, (components of cilia and flagella), movement of chromosomes during cell division |
| **cilia** | Short extensions from cell surface consisting of microtubules in a **9 + 2 arrangement** | Cell locomotion or movement of materials over cell surface |
| **flagella** | Long extensions from cell surface consisting of microtubules in a 9 + 2 arrangement | Cell locomotion |

# Review and Synthesis

1. Complete the diagram by filling in the following terms

   | | |
   |---|---|
   | chromoplasts | leucoplasts |
   | nucleolus | thylakoids |
   | chromosomes | rough |
   | grana | smooth |
   | plastids | matrix |
   | cristae | cisternae |
   | vesicles | nucleoplasm |

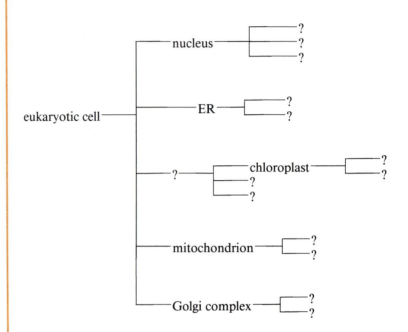

2. Give five generalizations about how cells work. (For example: The number and kinds of cellular organelles vary with the type of cell, its function, and its age.)

3. Match the cell structure with its function.

   | Structure | Function |
   |---|---|
   | a. microfilament | 1. support |
   | b. ribosome | 2. protein synthesis |
   | c. vacuole | 3. cyclosis |
   | d. cytoskeleton | 4. prevents cell from bursting |
   | e. cell wall | 5. storage, excretion |

4. Rewrite the following so that they are true statements.
   a. Animal cells produce primary and sometimes secondary cell walls.
   b. Chromatin unravels into chromosomes.
   c. A collection of cristae form the Golgi complex of a cell.
   d. Flagella are shorter and more numerous than cilia.

5. Cross out the terms that don't belong in each of the indicated categories.

   | *Prokaryotic cell* | *Protoplasm* |
   |---|---|
   | cell wall | plasma membrane |
   | plasma membrane | nucleus |
   | nucleus | cell wall |
   | mitochondrion | cytoplasm |
   | cytoplasm | middle lamella |

| *Plasma membrane* | *Plant cell walls* |
|---|---|
| phospholipid bilayer | cellulose |
| chromosomes | microfibrils |
| carbohydrates | lignin |
| microtubules | protein |
| globular proteins | phospholipids |

6. Using diagrams, illustrate the relationship between changes in the surface of the plasma membrane and cell volume as the diameter of a circular cell increases. Why is surface area-to-cell volume ratio so important?

7. List the similarities and differences between prokaryotic and eukaryotic cells.

8. Prepare a chart comparing the similarities and differences between plant and animal cells.

## Additional Readings

Allen, R. 1987. "The microtubule as an intracellular engine." *Scientific American* 256:42–49.

Brachet, J. 1961. "The living cell." *Scientific American* 205:39. (General introduction.)

Geise, A. 1979. *Cell Physiology.* Saunders, Philadelphia. (Advanced.)

Scharf, D. 1977. *Magnifications. Photography with the Scanning Electron Microscope.* Schocken Books, New York. (Intermediate.)

Thorpe, N. 1984. *Cell Biology.* John Wiley and Sons, New York. (Intermediate.)

Wolfe, S. 1985. *Cell Ultrastructure.* Wadsworth, Belmont, Calif. (Intermediate.)

Chapter 7

# Membrane Transport

Although it is easy to see how organelles such as the nucleus, mitochondria, and chloroplasts enable cells to perform life processes, we might dismiss the plasma membrane as a mere barrier between the cell's contents and its environment. But the plasma membrane also does much more for a cell.

Delicate life-sustaining metabolic reactions require chemicals that are very different in type and proportions from those found in the cell's surrounding environment. The plasma membrane is the boundary that isolates essential substances and maintains conditions suitable for life by regulating the passage of materials into and out of the cell. In addition to this vital transport function, the plasma membrane detects and responds to external changes, as well as receives chemical messengers from other cells, such as hormones or chemicals that transmit impulses from nerve cells. In some cells, the plasma membrane participates in cell movement, secretion, absorption, recognition of other cells, and the relay of impulses from one cell to another.

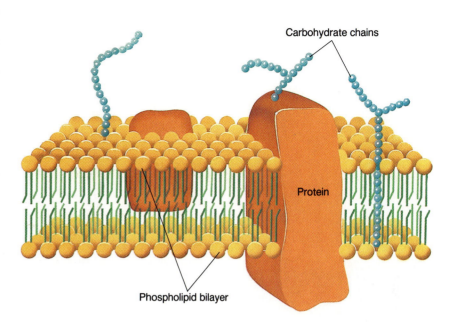

**Figure 7-1**
**The plasma membrane: a fluid mosaic.** The plasma membrane is composed of two layers of phospholipids. Globular proteins float in the phospholipid bilayer, and carbohydrate groups project out from the proteins and phospholipids.

## Properties of the Plasma Membrane

How does the plasma membrane perform these many life-supporting tasks? The answer is found in the chemical structure of the membrane. You learned in Chapter 6 that all cell membranes are basically similar—a fluid mosaic of globular-shaped proteins floating in two layers of phospholipids. The plasma membrane may also have various carbohydrates protruding from its outer surface (Figure 7-1). These chains of sugars are attached either to the membrane proteins, forming glycoproteins, or to phospholipids, forming glycolipids. The projecting carbohydrate groups act as receptor sites that recognize and bind with specific substances, enabling cells to receive chemical communications from other cells (hormones), detect chemical changes in their surroundings, or recognize other cells. When chemicals bind with their complementary carbo-

hydrate receptor sites, they often initiate changes in the cell's metabolism, some of which can have profound effects on our health (see Discovery: Glycoproteins and Heart Disease).

Both major components of the plasma membrane, the proteins and phospholipid bilayer, affect whether substances enter or leave a cell. The phospholipid portion of the plasma membrane blocks the passage of most molecules, allowing water and a few ions to move through freely. Membrane proteins, on the other hand, actually *control* the entrance and exit of "selected" materials, substances that specifically bind to the protein. These "selected" molecules either pass through channels formed within the proteins or between clusters of proteins, or they are transported through the membrane when the protein rotates or changes its shape (Figure 7-2). The proteins also control how fast these molecules are transported across the membrane.

Together, then, the phospholipid bilayer and globular proteins make the plasma membrane both **differentially**

**permeable** and **selectively permeable**. The plasma membrane is differentially permeable because it allows some materials to pass through freely, yet it completely restricts the passage of others. The plasma membrane is also selectively permeable because membrane proteins control which molecules are to be transported. Selective permeability enables a cell to import and accumulate essential molecules to concentrations high enough for normal metabolism, as well as to export wastes and other substances that might stall metabolism.

## How Materials Pass Through the Plasma Membrane

Materials move through the plasma membrane by **passive** or **active transport**. During passive transport, molecules spontaneously move from the side of the membrane where they are more abundant (a region of higher concentration) to the side of the mem-

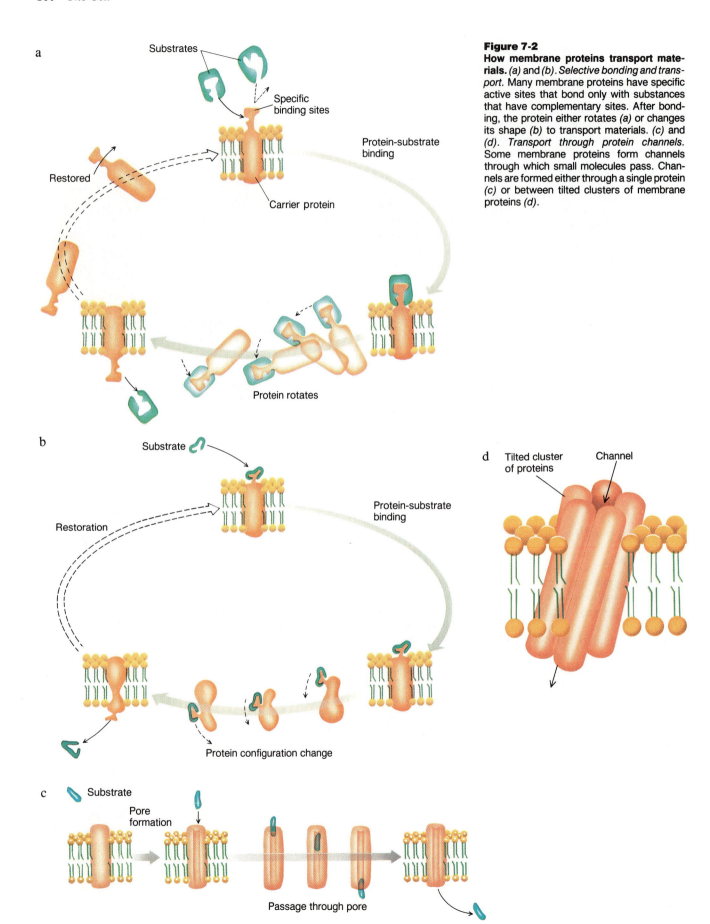

a

Substrates

Specific
binding sites

Protein-substrate
binding

Restored

Carrier protein

Protein rotates

**Figure 7-2**
**How membrane proteins transport materials.** *(a)* and *(b). Selective bonding and transport.* Many membrane proteins have specific active sites that bond only with substances that have complementary sites. After bonding, the protein either rotates *(a)* or changes its shape *(b)* to transport materials. *(c)* and *(d). Transport through protein channels.* Some membrane proteins form channels through which small molecules pass. Channels are formed either through a single protein *(c)* or between tilted clusters of membrane proteins *(d).*

b

Substrate

Protein-substrate
binding

Restoration

Protein configuration change

d    Tilted cluster
of proteins                Channel

c    Substrate

Pore
formation

Passage through pore

## D I S C O V E R Y

## GLYCOPROTEINS AND HEART DISEASE

Joseph Goldstein and Michael Brown won the 1973 Nobel Prize in Medicine for their discovery that glycoproteins control the level of cholesterol in human blood. Their investigations proved that the glycoproteins projecting from the surface of cells are receptors for cholesterol-containing particles called low-density lipoproteins (LDLs) that circulate in the blood. When LDLs bind with receptor glycoproteins, they are taken into the cell and used to make cell membranes, hormones, and bile. Liver cells normally remove 70 percent of the LDLs from blood, excess LDLs that could otherwise build up in blood vessels and eventually cause atherosclerosis (narrowing of arteries) or a heart attack.

The greater the number of glycoprotein receptors, the more LDLs are removed from the blood. A normal cell has some 20,000 LDL receptor glycoproteins on its plasma membrane. However, one in about 500 people has abnormally small numbers of receptors, and these individuals have an increased risk for heart attacks because fewer LDLs are removed from the blood.

Goldstein and Brown found that people with few LDL receptors have inherited a defective gene that represses the forma-tion of glycoproteins. One in every million people inherits two of these genes, and such individuals produce no LDL receptors at all. Children with both defective genes often have heart attacks before they reach age 6.

Then why do people with a normal number of LDL receptors develop atherosclerosis and have heart attacks? When a cell's cholesterol needs are met, glycoprotein receptors shut off. People with a high-fat diet quickly saturate their need for cho-lesterol; so as a result, cells turn off receptor glycoproteins and blood cholesterol levels skyrocket. A diet high in fats sup-presses the manufacture of LDL glycoprotein receptors by as much as 90 percent. With more LDLs in the bloodstream, the risk of atherosclerosis and heart attacks greatly increases.

The breakthroughs achieved by Goldstein and Brown have paved the way to understanding and treating other receptor-related diseases, such as diabetes, pernicious anemia, and myasthenia gravis. Don't be surprised if you soon hear that the medical treatment for a disease is aimed directly at the plasma membrane's glycoproteins, a new medical frontier that, like so many others, was explored first by biologists.

---

brane where there are fewer numbers of that molecule (a region of lower concentration). When substances move from regions of higher to lower concentrations, they are said to move *along a concentration gradient*. Mole-cules that pass through the plasma membrane by passive transport spon-taneously move along their own par-ticular concentration gradient.

Conversely, active transport is not spontaneous. The cell must use energy stored in the bonds of ATP for active transport of substances across the plasma membrane. Materials may be actively transported along a concen-tration gradient, or *against a concen-tration gradient* (from regions of lower concentration to regions of higher concentration), depending on the needs of the cell.

Passive and active transport mecha-nisms are summarized in Table 7-1.

TABLE 7-1

### How Materials are Transported Through the Plasma Membrane

Types of Passive Transport (Spontaneous)

Diffusion: Substances move from areas of high concentration to areas of low concentration.
Osmosis: Diffusion of *water* through a differentially permeable membrane.
Facilitated Diffusion: Diffusion is assisted by carrier proteins in the plasma membrane.

Types of Active Transport (Require ATP)

Active Transport: Movement of substances by membrane proteins.
Exocytosis: Discharge of materials housed in vesicles.
Endocytosis: Engulfment of small cells, particles, or compounds into a membrane sac formed from the plasma membrane.

# Passive Transport
## DIFFUSION

Substances that cross the plasma membrane by passive transport always move by **diffusion**, the net movement of identical molecules from areas of high concentration to areas of low concentration. Because all molecules are in constant, random motion, they spontaneously diffuse along their own concentration gradient. Diffusion eventually leads to the equal concentration of identical molecules throughout. Equal concentration, however, does not mean that molecular motion stops. Rather, it signifies only that molecules are uniformly dispersed and that there is no *net* change in concentration between any two regions.

The rate of diffusion varies depending on the following conditions.

- *State of matter.* Diffusion is slowest in solids, faster in liquids, and fastest in gases.

- *Temperature.* Heat accelerates molecular velocity, thereby increasing the rate of diffusion; cold decreases velocity and lowers the diffusion rate.

- *Size of molecule.* Smaller molecules travel faster than larger ones at the same temperature.

- *Difference in concentration of identical molecules between areas.* The greater the difference in concentration between two areas, the faster the rate of diffusion.

Only a few kinds of small molecules are able to diffuse through the plasma membrane. They include water, oxygen, and carbon dioxide, molecules that enter and exit cells in very large quantities.

Molecules that diffuse through the plasma membrane either temporarily dissolve in the phospholipid bilayer as they cross, or travel through small transient pores in the phospholipid bilayer or through channels formed by the membrane proteins.

## OSMOSIS

The diffusion of *water* through a differentially permeable membrane has a special name, **osmosis**. Only "free"

water molecules, that is, water molecules that are not bonded to other types of molecules (a sugar, a protein, or another solute), are available for osmosis. Therefore, the concentration of solutes in a solution is a primary factor in determining the direction of osmosis (Figure 7-3).

The direction of osmosis can have dramatic consequences for a cell. If a cell has a lower solute concentration' than its surroundings (and therefore more free water molecules), water will move out of the cell by osmosis, shrinking the cell and causing it to collapse. Put another way, cells shrink when surrounded by **hypertonic** solutions, solutions with higher solute concentrations (hyper = more) than found inside the cell (Figure 7-4a). Salt water and sugar solutions are hypertonic to most cells.

On the other hand, cells suspended in **hypotonic** solutions accumulate water because the concentration of solutes outside the cell is less (hypo = less) than it is inside (Figure 7-4b). Since hypotonic solutions have more free water molecules than the cell,

water diffuses into the cell by osmosis. Unless the cell has a nonexpandable cell wall, or unless it expends energy to remove excess water, the cell will swell in hypotonic solutions.

There are limits to the amount of shrinkage and swelling that a cell can withstand before its ability to function is compromised. Red blood cells, for instance, are particularly vulnerable to changes in solute concentration because they have no mechanism for removing excess water. To prevent red blood cells from bursting or shrinking, blood plasma must be neither hypotonic nor hypertonic, but **isotonic**. The concentration of free water molecules is the same inside a cell as it is in the surrounding isotonic solution (iso = same). In isotonic solutions, the diffusion of water into and out of the cell is equal, so there is no net change in the concentration of free water inside the cell (Figure 7-4c).

Osmosis is affected not only by solute concentration, but also by pressure (Figure 7-5). For example, when a plant cell is placed in a hypotonic solution, water diffuses into the cell by os-

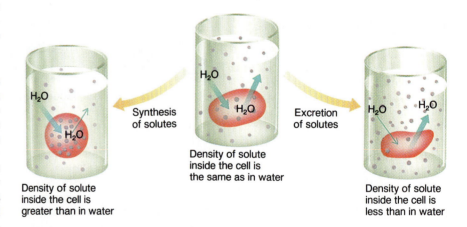

**Figure 7-3**
**Solute concentration and osmosis.** Many metabolic reactions produce solutes inside a cell. These solutes bond with available water molecules, thereby reducing the number of water molecules that are free to diffuse. When the number of free water molecules inside a cell falls below that of an adjacent cell or the surrounding fluid, water diffuses into the cell by osmosis. Conversely, as solutes are used up or secreted, the number of free water molecules inside a cell may increase above that of its surroundings, causing water to diffuse out by osmosis. When solute concentrations are equal between a cell and its surroundings, the rate of osmosis is equal in both directions and there is no net uptake or release of water by the cell.

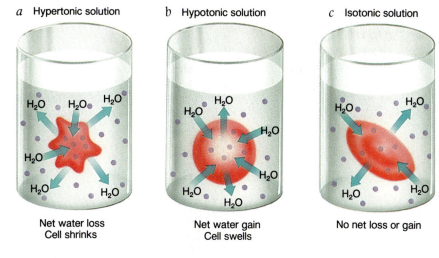

*a* Hypertonic solution     *b* Hypotonic solution     *c* Isotonic solution

Net water loss
Cell shrinks

Net water gain
Cell swells

No net loss or gain

**Figure 7-4**
**The effects of hypertonic, hypotonic, and isotonic solutions on osmosis.** *(a)* A cell placed in a hypertonic solution (one containing a higher solute concentration than the cell) soon shrinks because of a net loss of water by osmosis. *(b)* A cell placed in a hypotonic solution (one having a lower solute concentration than the cell) swells because of a net gain of water by osmosis. *(c)* A cell in an isotonic solution (equal solute concentrations) neither swells nor shrinks because it gains and loses equal amounts of water.

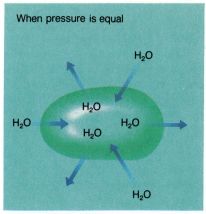

When pressure is equal

$H_2O$

When pressure is greater inside cell

Pressure

**Figure 7-5**
**Pressure and osmosis.** *(a)* When a cell is suspended in a solution with equal solute concentration and pressure, there is no net gain or loss of free water molecules. *(b)* Increasing the pressure in a cell, and not solutes, forces free water out of the cell, increasing the rate of osmosis.

mosis. However, because the cell has a rigid cell wall that resists expanding, the plant cell will not swell to the point of bursting as does an animal cell. Instead, internal water pressure builds as water continues to diffuse into the plant cell. This internal water pressure is called **turgor pressure**. Eventually, turgor pressure builds high enough to stop the influx of more water, even though the solute concentration inside the cell remains higher than the surrounding solution or adjacent cell; internal pressure stops water movement. In multicellular plants, water will diffuse from a cell with higher turgor pressure into an adjacent cell with lower turgor pressure, even when the cells have the same solute concentration.

In hypertonic solutions, the protoplast of a plant cell shrinks away from its cell wall. This process is called **plasmolysis** (Figure 7-6). Plasmolysis causes land plants to wilt because much of the support of nonwoody plants or of the nonwoody parts of plants (for example, leaves) results from the turgor pressure of water-engorged cytoplasm pushing against rigid cell walls.

## FACILITATED DIFFUSION

Like diffusion and osmosis, the transport of materials by facilitated diffusion is always along a concentration gradient—from a region of high concentration to a region of low concentration. During **facilitated diffusion**, "carrier" proteins aid in the transport of molecules through the membrane, increasing the molecule's rate of diffusion. (Facilitated diffusion is also called *carrier-assisted transport*.)

Carrier proteins form channels through the membrane as they change their shape. Different carrier proteins have different hydrophilic groups lining the channels. Some researchers suggest that these hydrophilic groups bind with specific molecules or ions. Once these substances enter the channel, the carrier protein changes shape, increasing the rate of diffusion through the channel, as illustrated in Figure 7-7. Lipids, glucose, glycerol, and urea are molecules commonly transported through the plasma membrane by facilitated diffusion.

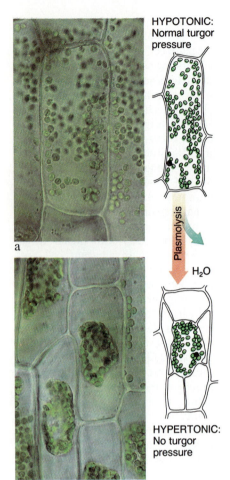

HYPOTONIC:
Normal turgor
pressure

Plasmolysis

H₂O

HYPERTONIC:
No turgor
pressure

a

b

**Figure 7-6**
**Plasmolysis.** Most of the fresh water that surrounds aquatic plants is hypotonic to plant cells. Water therefore tends to flow into the cells, creating high turgor pressure *(a)*. If the plant is placed in a hypertonic solution, the cell loses water and the plasma membrane pulls away from the cell wall, a process known as plasmolysis *(b)*. If the cells of land plants evaporate large amounts of water to dry air, turgor pressure plunges, plasmolysis results, and the plant wilts.

## Active Transport

**Active transport** is an energy-dependent process that relies on the energy stored in the bonds of ATP to transport materials across the plasma membrane. Energy is required because substances are being moved against their concentration gradient, from more dilute regions to more concentrated regions.

Because active transport requires energy, it is either directly or indirectly coupled with the hydrolysis of ATP. When transport is directly coupled with ATP breakdown, it is called *primary active transport*; when transport is indirectly coupled with ATP hydrolysis, it is called *secondary active transport*, or *cotransport*. During secondary active transport, the passive transport of one substance is combined with the active transport of another. Primary and secondary active transport mechanisms are illustrated in Figure 7-8.

During both types of active transport, membrane proteins bind with specific passenger molecules. The energy in ATP is used to change the orientation or configuration of the protein so that substrates are transported through the membrane (Figure 7-2*a* and *b*). Exchanges of sodium and potassium ions in nerve cells, as well as the import and export of amino acids and large sugars, require active transport. Active transport occurs in all organisms and, in some cells, consumes more than half of their energy expenditures.

## EXOCYTOSIS AND ENDOCYTOSIS

**Exocytosis** and **endocytosis** are types of active transport that move large molecules, particles, or even other cells across the plasma membrane. Exocytosis (exo = outside, cyto = cell) exports materials out of a cell, whereas endocytosis (endo = inside) imports particles or small cells into a cell.

Before substances can be exported by exocytosis, they must first be surrounded by a membrane, forming a secretion vesicle. (This process was illustrated in Figure 6-18.) The vesicle membrane then fuses with the plasma membrane to expel the material (Figure 7-9). Examples of exocytosis in-

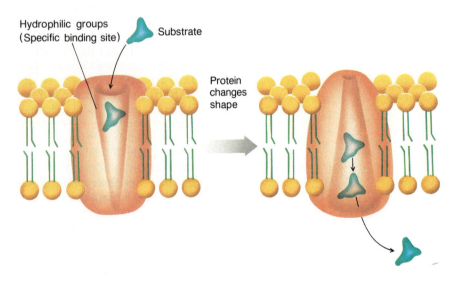

Hydrophilic groups
(Specific binding site)

Substrate

Protein
changes
shape

**Figure 7-7**
**Facilitated diffusion.** Carrier proteins have specific binding sites that match particular substrates. Proteins may change shape as they move substances through the channel.

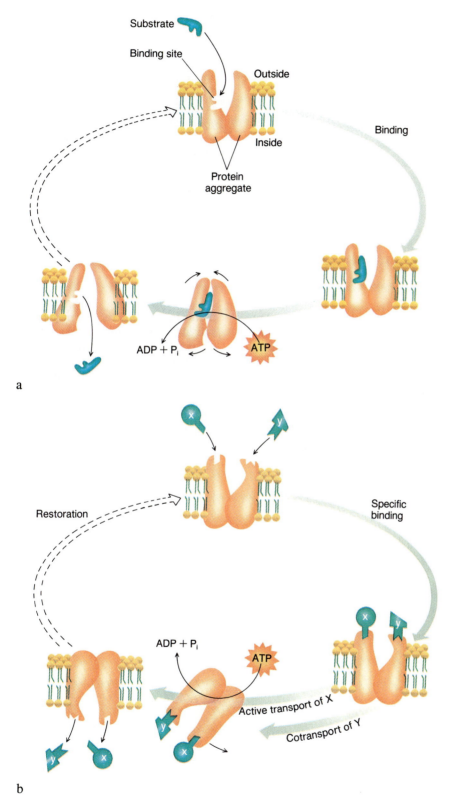

a

b

**Figure 7-9**
**Exocytosis.** A substance to be exported from a cell is housed within a secretion vesicle. The surrounding vesicle membrane fuses with the plasma membrane, releasing the substance to the outside.

**Figure 7-8**
**Active transport.** *(a) Primary active transport.* Single membrane proteins or clusters of proteins form channels through the plasma membrane. The energy of ATP is used to flip-flop or change protein configuration, thereby transporting materials through the membrane. *(b) Secondary active transport.* The active transport of one substrate (x) results in the passive transport of another (y). Both mechanisms rely on the hydrolysis of ATP.

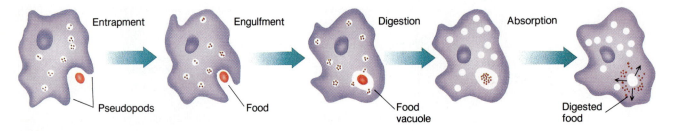

Entrapment

Pseudopods

Engulfment

Food

Digestion

Food
vacuole

Absorption

Digested
food

**Figure 7-10**
**Phagocytosis by an amoeba.** A food particle engulfed by pseudopodia becomes enclosed
within phagocytic food vacuole. The food particle is digested by enzymes discharged from
lysosomes that fuse with the food vacuole membrane. The digested food is then absorbed into
the amoeba's cytoplasm.

clude the release of materials for constructing plant cell walls, and the secretion of hormones by certain animal cells.

There are two forms of endocytosis, depending on whether the cell is ingesting a solid particle or a liquid. Endocytosis of a solid particle or cell is called **phagocytosis** (phago = eating), whereas endocytosis of a liquid is called **pinocytosis** (pino = drinking).

An example of phagocytosis is presented in Figure 7-10. Here, a unicellular amoeba engulfs its microscopic meal, a bacterium, by producing lobelike extensions called *pseudopodia*. The pseudopodia eventually enclose the bacterium in a *phagocytic vacuole*. Once the bacterium is enclosed, the amoeba releases digestive enzymes into the vacuole. The bacterium is digested, and the nutrients are transported through the vacuole membrane and absorbed by the amoeba. Unabsorbed wastes are discharged by exocytosis. In much the same way as an amoeba eats a bacterium, the white blood cells in your circulatory system defend you against invading microorganisms; they engulf the microorganism in a phagocytic vacuole and then destroy them.

# Synopsis

## MAIN CONCEPTS

- **The cell's plasma membrane is a dynamic boundary that isolates and maintains critical levels of essential chemicals.**

- **The plasma membrane's functions include**
  - Regulating the passage of essential materials into and out of the cell.
  - Receiving outside information so that the cell can respond to changes in its surrounding environment and chemical messages from other cells
  - Recognizing other cells and specific chemicals.
  - Aiding in the movement of some cells, or substances past the cell.
  - Secreting substances from the cell.
  - Absorbing materials into the cell.
  - Transmitting chemicals and impulses.

- **The plasma membrane is a double layer of fluid phospholipids.** Globular-shaped proteins float in the phospholipid bilayer, and various carbohydrate groups, attached to the proteins and phospholipids, extend out from the membrane surface.

- **The plasma membrane is both differentially and selectively permeable.**

- **Substances move through the plasma membrane by passive or active transport.**

- **Since water is critical for normal cell functioning, the direction of osmosis can alter cell metabolism.** Pressure and concentration of solutes inside and outside the cell are the two primary factors that affect the direction and rate of osmosis.

## KEY TERM INTEGRATOR

### Membrane Structure: A Fluid Mosaic

| | |
|---|---|
| **plasma membrane** | Composed of many globular-shaped proteins embedded in a fluid **phospholipid bilayer**. **Glycoproteins** and **glycolipids** project from the membrane surface. |

### Membrane Transport

| | |
|---|---|
| **differentially permeable** | Passage of some substances and not others through the membrane. In addition, membrane proteins transport specific substances, making the plasma membrane **selectively permeable** as well. |
| **transport** | Movement across the plasma membrane either by **passive transport** or **active transport**. |
| **diffusion** | Movement of substances from areas of high to low concentrations along a **concentration gradient**. Membrane proteins assist transport during **facilitated diffusion**. |
| **osmosis** | The diffusion of water through a differentially permeable membrane. When surrounded by a **hypotonic solution**, a cell gains water. When surrounded by a **hypertonic solution**, the cell loses water, causing a decrease in internal **turgor** pressure in plant cells and eventual **plasmolysis**. When surrounded by an **isotonic solution**, water enters and exits at the same rate. |

| active transport | Requires the energy in the bonds of ATP move substances through the plasma membrane against a concentration gradient. During **exocytosis**, substances are encased in membranes and actively transported out of a cell. **Endocytosis** is the engulfment of materials into a cell. When substances are liquid, the process is **pinocytosis**. Solid substances are encased within a **phagocytic vacuole** during **phagocytosis** |
|---|---|

# Review and Synthesis

1. Complete the diagram using the following terms.

| | |
|---|---|
| endocytosis | glycolipids |
| differential | passive transport |
| exocytosis | glycoproteins |
| globular proteins | selective |
| diffusion | facilitated diffusion |
| active transport | phospholipid bilayer |
| osmosis | |

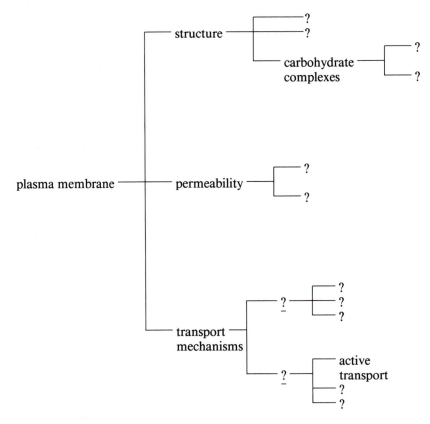

2. Describe five ways in which membrane proteins work to select and transport materials through the plasma membrane.

3. Match the structures with their functions.

| Structure | Function |
|---|---|
| glycoproteins | active transport |
| phospholipids | hormone recognition |
| globular proteins | digestion |
| phagocytic vacuole | diffusion of lipids |

**4.** Circle the chemicals that cross the plasma membrane by diffusion alone. (Do not circle molecules assisted by membrane proteins during facilitated diffusion.)

| | |
|---|---|
| oxygen | glucose |
| amino acids | water |
| proteins | carbon dioxide |
| sodium | potassium |
| glycerol | urea |

5. Rewrite the following so that they are true statements.
   a. Exocytosis includes phagocytosis of liquids and pinocytosis of solid particles.
   b. Some cells rely on active transport to move substances along their particular concentration gradient.
   c. The "carriers" in carrier-assisted transport, also called osmosis, are the phospholipids in the plasma membrane.
   d. A cell suspended in a solution that is hypertonic will spontaneously take up both water and solutes by osmosis.

## Additional Readings

Bretscher, M. 1985. "The molecules of the cell membrane." *Scientific American* 253:100–108.

Capaldi, R. 1974. "A dynamic model of cell membranes." *Scientific American* 230:26.

Finean J., B. Coleman, and R. Michell. 1978. *Membranes and Their Cellular Functions.* Oxford University Press, Oxford, England. (Intermediate.)

Thorpe, N. 1984. *Cell Biology.* John Wiley and Sons, New York. (Intermediate.)

Unwin, N., and R. Henderson. 1984. "The structure of proteins in biological membranes." *Scientific American* 250:78–95.

Chapter 8

# *Metabolic Universals*

*T*he struggle by all organisms to obtain a constant supply of energy and nutrients is a central theme of life. These resources are needed for **biosynthesis**, the construction of new compounds by a cell. Several strategies for obtaining these resources are available to organisms. Photosynthetic organisms capture solar energy and use it to assemble carbon dioxide and other simple nutrients into the complex organic molecules that ultimately constitute their tissues. Consumers and decomposers harvest nutrients and chemical energy from organic molecules produced by other living things, reorganizing these compounds to form their own cellular materials. Some strategies, however, are universal—they are employed by every living organism on earth. Although a few tactics, such as using ATP as energy currency, have already been discussed, several important metabolic approaches common to all organisms will be introduced in this chapter. These universal strategies provide more keys to understanding life and its metabolic dynamics. They also illustrate the unity of life, since all cells, from the most primitive to the most sophisticated, depend on similar metabolic mechanisms.

## Metabolic Pathways

Houses and proteins have much in common. Both are complex structures that must be assembled in sequential steps, each building on the previous steps. Construction is therefore a process that consumes time and energy. Reversing this process, however, can be a rapid, dramatic event—houses and complex molecules can be disassembled in a single explosive step. Proteins, for example, break apart into their simplest components when they burn, releasing their chemical energy

as heat. This sudden deterioration of order is much like dynamiting a building. Little remains of the protein except molecular rubble. If its components are to be reused, the protein must be disassembled in coordinated steps, taken apart bit by bit to generate usable materials for building new compounds and to release energy in small, manageable amounts. Every metabolic task, whether it builds complex molecules or dismantles them, proceeds in an orderly sequence of chemical changes rather than in one single-step reaction. Each sequence of reactions is called a **metabolic pathway**.

A complete metabolic pathway consumes substrates and converts them to end products. The process also temporarily generates several intermediate compounds revealing hypothetical pathway, as shown in the illustration at the bottom of the page, in which A is the *substrate* and E is the final *product*; B, C, and D are called **metabolic intermediates**. In contrast to the above linear sequence of reactions, some metabolic pathways form cycles. In these **cyclic pathways**, metabolic intermediates are regenerated, while assisting the conversion of substrate (S) to product (P).

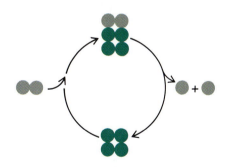

Whether linear or cyclic, different pathways may share a common intermediate that forms a regulatory link between two pathways. The hundreds of different metabolic pathways in a cell are coordinated in this way. For

example, one of the intermediates of the 19-step pathway for dismantling glucose to carbon dioxide and water is also a substrate for the fat-producing pathways in animals (Figure 8-1). An abundance of glucose generates an excess supply of the common intermediate, which is diverted from the energy-releasing pathway to the biosynthetic sequence for storage as fat. Linking pathways in this fashion provides a cell with the metabolic "switchtracks" needed to redirect intermediate compounds into those pathways that satisfy the organism's changing needs at any given moment.

## Catabolism and Anabolism

All metabolic reactions fall into one of two categories depending on whether the products are more or less complex than the substrates. **Catabolism** refers to reactions that degrade complex compounds into simpler molecules, usually with the release of the chemical energy that held the atoms of the larger molecule together. Most catabolic reactions are therefore exergonic (energy releasing). **Anabolism** is the biosynthesis of complex molecules from simpler components. Since molecular construction requires energy, anabolic reactions are endergonic (energy absorbing). In most organisms, the energy needed to drive anabolic reactions is provided by the sun or by catabolic degradation of chemicals. In both cases, energy is transferred in ATP from energy-generating processes to anabolic reactions (see Connections: ATP, Chapter 3).

## Oxidation and Reduction

Anabolism and catabolism cannot be understood in terms of molecular complexity and energy transfers alone because, in all these reactions, electrons are transferred as well. Reactions in which electrons are lost by one compound and gained by another are called oxidations and reductions, or *redox reactions*. **Oxidation** is the loss of electrons by a molecule; **reduction** is the gain of electrons. Electrons lost during oxidation are not simply re-

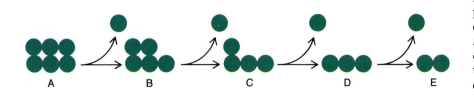

A      B      C      D      E

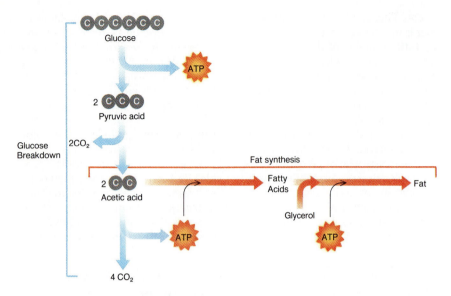

**Figure 8-1**
**Metabolic pathways linked by a shared intermediate.** Glucose degradation and fat assembly are integrated by acetic acid, a metabolic intermediate in both pathways. When glucose is plentiful, excess energy can be stored by diverting surplus acetic acids into the fat synthesis pathway. When the organism's external energy resources are scarce, it can disassemble fat into acetic acids, which are further broken down to $CO_2$ for a large yield of ATP, the cell's usable energy.

leased free into the cytoplasm, however, but are immediately captured by another molecule, which is reduced in the process. Therefore, whenever a substance is oxidized, something else must be reduced.

Catabolic reactions are oxidative, not only providing energy but also releasing electrons. Biosynthesis, on the other hand, is reductive, absorbing both energy and electrons. Therefore, the products of biosynthesis not only contain more energy than the oxidized molecules from which they were assembled, but are also a potential source of electrons. That is, they are in a more reduced state. As a general rule, highly reduced compounds are more energy rich than highly oxidized compounds.

The oxidation of the six carbons in glucose illustrates both principles discussed in the previous two paragraphs.

In this reaction, one compound (glucose) is oxidized and one substance (oxygen) is reduced. Glucose is "ripe" for oxidation because its carbons are in a highly reduced state. The molecule therefore contains substantial chemical energy, which is released when oxygen strips away electrons from the sugar. This breaks the covalent bonds that hold the glucose molecule together, forming carbon dioxide ($CO_2$), carbon in its most oxidized state. The electrons transferred to oxygen are accompanied by the hydrogens from glucose, so each oxygen atom combines with two hydrogen atoms to form water. The reaction releases 3811 calories per gram of glucose, the difference in energy content between the reduced and oxidized states.

Let's return to the transfer of hydrogens from carbon to oxygen in this oxidation–reduction reaction. With a

few exceptions, all such reactions transfer hydrogen ions ($H^+$, a solitary proton) and electrons ($e^-$) together. Consequently, we can determine which molecules are highly reduced simply by counting the number of hydrogen atoms per carbon atom. The lack of hydrogen atoms in $CO_2$ tells us that it is a highly oxidized molecule with virtually no energy remaining that can be tapped by a cell.

## Electron Carriers

The complete oxidation of glucose by organisms is much more complex than simply burning the sugar in the presence of oxygen (although the net reaction is the same as that shown in the previous equation). Rather than releasing energy, $CO_2$, and water in a single burst, the biological process requires a 19-step metabolic pathway. In addition, electrons released during glucose oxidation are not transferred directly to oxygen, but are temporarily picked up by one of two *coenzymes* (Chapter 5) that function as electron-carrier molecules. These two coenzymes are **NAD$^+$** (nicotinamide adenine dinucleotide) and **FAD** (flavin adenine dinucleotide). Each of these compounds can be reduced by a pair of electrons released during the oxidation of an organic compound, forming the reduced coenzymes NADH and FADH$_2$ (Figure 8-2). Thus, in their reduced forms, these coenzymes are sources of electrons (and protons).

These reduced carrier molecules also represent a source of potential energy that can be used to manufacture ATP. The electrons released during the oxidation of glucose still contain much of the energy originally stored in the chemical bonds of glucose. Reduced coenzymes carry these energy-rich electrons to the **electron transport system**, a specialized molecular apparatus that helps extract the energy from the electrons so it can be used to form ATP (Figure 8-3).

Another option of electron-carrying coenzymes is to transfer their electron cargo from oxidation reactions to biosynthetic reactions. Biosynthesis requires a source of electrons (and protons) as well as energy to build highly reduced (electron-rich) complex molecules. The coenzyme **NADP$^+$** (nico-

$$6\,O_2 + C_6H_{12}O_6 \xrightarrow[\text{of glucose}]{\text{oxidation}} 6\,CO_2 + 6H_2O + ENERGY$$

GLUCOSE
(highly reduced,
energy rich)

CARBON DIOXIDE
(highly oxidized,
energy poor)

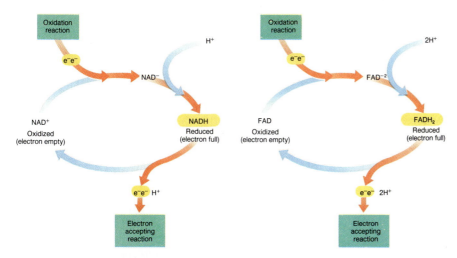

**Figure 8-2**
**Electron carriers in action.** NAD⁺ and FAD work in a similar fashion: each picks up a pair of electrons released from an oxidation reaction and carries them to a reductive reaction site. The orange color follows the path of the electrons. Although not shown, NADH may transfer its electrons to FAD, which is reduced in the process.

tinamide adenine dinucleotide phosphate) transfers electrons and protons from oxidative reactions to sites where biosynthesis is occurring. A cell's reservoir of NADPH (reduced NADP⁺) represents its **reducing power**, its capacity to continue synthesizing the molecules of life (assuming supplies of ATP are adequate).

## Energized Electron Cycle

The electrons carried by NADH, NADPH, and FADH₂ are energy-rich electrons. The electrons have absorbed large amounts of energy either from the sun or from oxidation of chemical bonds during catabolism. (This is analogous to winding a spring, thereby storing potential energy that can be used when the spring's tension is released.) A cell has options as to what to do with these high-energy electrons. Some will be used for their reducing power, as discussed in the previous paragraph. Others are cashed in for their energy, forming molecules of ATP. These options are depicted in Figure 8-4.

Energy for biological work travels in electrons. Electrons not only transfer energy from one compound to another during metabolism, but they also transfer energy from one organism to another. Let's follow the fate

of a bit of solar energy captured by a photosynthetic cell. (The journey is illustrated in Figure 8-5.)

Electrons in the cell absorb energy from sunlight and become **photoexcited**, jumping to a higher energy level (Chapter 3). Some of these photoexcited electrons help reduce several molecules of carbon dioxide to form a sugar molecule during photosynthesis. Once that has happened, the energized electrons reside in the chemical bonds of the reduced sugar. All the complex molecules that constitute the plant's tissues contain these energy-rich electrons. When an animal (or a microorganism) consumes the plant, it oxidizes these compounds, removing their energized electrons. Some of the electron's energy is immediately used to form ATP. Most of the energy, however, remains with the electrons that are stripped from glucose and transferred to electron-carrying coenzymes, such as NAD⁺.

Some of these coenzyme-bound electrons are used as reducing power for biosynthesis of new molecules that constitute the organism's tissues. Here they retain their energy until the molecules are oxidized, either when needed by the owner or when consumed by another organism. The electrons that were not used for biosynthesis are stripped of their energy during *respira-*

*tion*, an ATP-generating process that oxidizes organic compounds and returns the released electrons to their initial low-energy states by running them through the electron transport system. The de-energized electrons and protons join with oxygen to form water. Photosynthesis and respiration are discussed in detail in Chapters 9 and 10, respectively.

## Electron Transport Systems

By now it is probably clear to you that energy is transported in electrons and that this energy can be cashed in for ATPs. You may still wonder, however, how the electron transport system strips the electrons of their energy so it can be transferred to ATP. Biologists wonder the same thing. We have solved much of the puzzle, but a few critical questions remain to be answered. We do know that electron transport systems are composed of iron-containing proteins called **cytochromes**. The iron provides an affinity for electrons, giving cytochromes an electron-carrying ability. Unlike the electron-carrying coenzymes NAD⁺ and FAD, cytochromes cannot move from one place to another. A series of different cytochromes are embedded next to one another in the cell's membranes, forming a chain of electron carriers. Each cytochrome accepts a pair of electrons from the previous member of the sequence and donates them to the next cytochrome down the line. Electrons therefore move through the electron transport system in much the same way that water moves along a bucket brigade.

Almost as soon as a cytochrome is reduced, it transfers the electron pair to the next cytochrome. Energy is released during each of these redox reactions. Thus, the electrons move toward a lower energy level each time they are transferred until they emerge from the final cytochrome exhausted of their initial energy holdings. Evidence suggests that the released energy is used to move protons (H⁺) in a single direction across cell membranes, so that they accumulate on one side. In other words, the energy released during the sequential oxidation and reduction of the cytochromes is stored in the form of a concentration gradient.

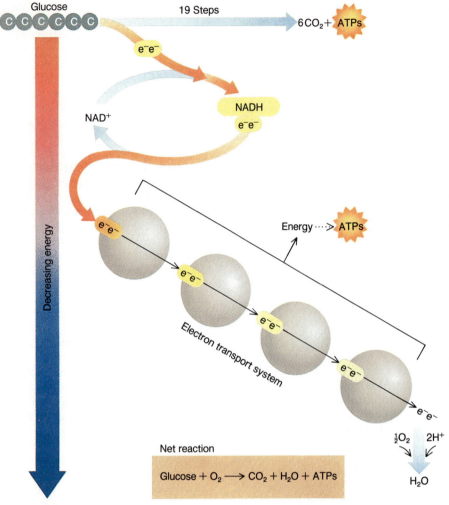

**Figure 8-3**
Most of the energy released by the oxidation of glucose is carried in energy-rich electrons of NADH, which takes them to the electron transport system, where their energy is used to produce ATP. One advantage of this complexity is to release energy in small, manageable bits. Each molecule of glucose releases 12 energized electron pairs. These move through the electron transport system one pair at a time.

the cell's chemiosmotic machinery.

Anchoring the electron transport system to membrane surfaces keeps the cytochromes in their proper sequence. Maintaining the proper cytochrome sequence is critical to the system's performance. To work properly, the cytochromes must be arranged in order of increasing electron affinity, each cytochrome having a greater attraction for electrons than the previous member in the chain. This arrangement allows electrons to release their energy in a series of small steps rather than all at once. If the final (most electron-attractive) cytochrome were allowed to compete with the first member of the chain, it would always win the electrons, and the entire sequence would be skipped over, literally short-circuited. The electrons would be transferred directly from the first member to the last, bypassing the ATP-generating mechanism and releasing their energy in one large unmanageable burst. The energy harvest therefore depends on the cytochromes being aligned so that each is physically separated from all but the two adjacent members of the chain.

## Energy Utilization

Energy is required for a multitude of biological tasks. It is needed to power all types of movement, such as cyclosis, phagocytosis, intracellular packaging of materials, muscular contraction, and locomotion. In addition, enormous amounts of ATP (and energy-rich electrons) are consumed for biosynthesis of new or replacement compounds. Energized electrons are carried to reductive biosynthetic processes by NADPH, the cell's reducing power. Photosynthesis directly generates NADPH and therefore a reservoir of reducing power. Electrons released during oxidation of food molecules are picked up by NAD+ rather than NADP+. Therefore, for humans and other nonphotosynthetic organisms to build a pool of reducing power, electrons must be transferred from NADH to NADP+.

(Recall that gradients represent potential energy.) To harvest this potential, these membranes are equipped with "proton pores," channels that allow the amassed protons to flow back across the membrane, toward the region of lower concentration. The energetics of proton flow through these channels is similar to water flow through bathtub drains that allow the liquid trapped in the tub to escape, to seek a lower, more stable energy level.

Phosphorylating enzymes attached to the region adjacent to each pore use the energy of the flowing protons to manufacture ATP. Although it is not clear exactly how this conversion occurs, enough of this explanation is substantiated to have earned it an official name. It is called **chemiosmosis**, the use of electrical and osmotic gradients drive the formation of ATP. Figure 8-6 shows the relationship between electron transport systems and

$$\text{NADH} + \text{NADP}^+$$
$$\downarrow$$
$$\text{NAD}^+ + \text{NADPH}$$

(reducing power)

Whether a cell leaves electrons in NADH or transfers them to form NADPH depends on the organism's energy resources. When energy is abundant, the production of NADPH is favored, providing an ample supply of electrons needed for biosynthesis of new macromolecules, such as proteins and lipids. These compounds are essential to growth; in addition, their formation is an excellent way to store surplus energy. In short, energy abundance favors a buildup of reducing power to supply electrons for increased biosynthesis. When energy resources are scarce, however, most of the NADH molecules cash in their electrons for ATP, NADP$^+$ gets just enough electrons for the minimal biosynthesis needed to maintain the status quo.

## Directing Metabolic Traffic

Regulating the relative amounts of NADH and NADPH is just one example of the ability of cells to adapt to constantly changing conditions. Cells also direct metabolic intermediates into the pathways that best satisfy the cell's needs at that particular moment. For example, glucose can be directed into several different metabolic routes. It can be dismantled into $CO_2$ and $H_2O$ to generate energy; it can be linked to other glucose molecules to form starch or cellulose in plants, or glycogen in many animals; it can be partially disassembled and the fragments used to build lipids; or it can be modified to form amino acids or nucleotides. Even the simplest cell can accurately "evaluate" these needs and direct the fate of glucose (and the thousands of other metabolic intermediates for which multiple options exist). How can cells "decide" which of these options best satisfies its needs?

The answer is that cells need not (indeed, *can*not) make decisions. The nature of the enzymes themselves regu-

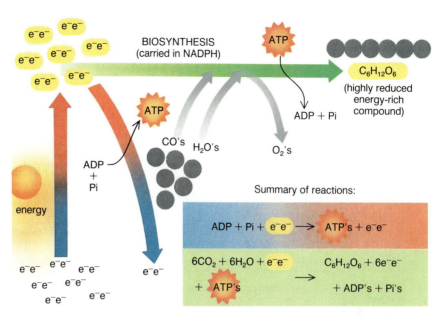

**Figure 8-4**
**Two optional fates of energy-rich electrons.** Energized by sunlight, some electrons provide energy for ATP formation; others are used by the cell as reducing power (NADPH) for biosynthesis (which also requires ATP).

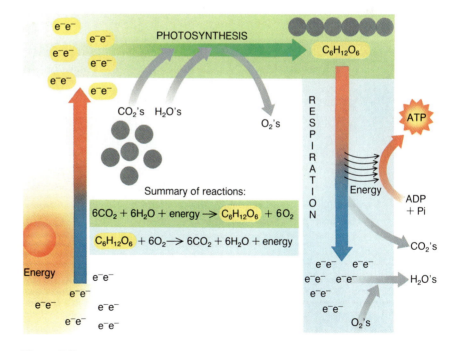

**Figure 8-5**
**Energized electron cycle.** Energy flows from compound to compound, and from organism to organism, in energy-rich electrons. Electrons are originally excited by the sun. This energy is stored during photosynthesis in reduced molecules. Respiration releases the chemical energy stored in the reduced molecule, then discards the energy-poor electrons by giving them (and the accompanying hydrogen ions) to oxygen, forming water.

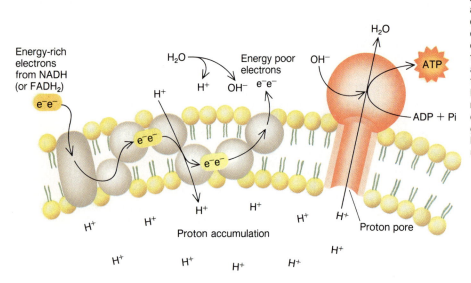

**Figure 8-6**
**Abbreviated scheme of electron transport and chemiosmosis.** Energy is released from electrons each time they are transferred from one cytochrome to the next member in the chain. It is believed that at two or three points in the chain, enough energy is released to propel across the membrane a hydrogen ion (a proton from the spontaneous disassociation of water: $H_2O \rightarrow H^+ + OH^-$). This leaves $OH^-$ ions (the remainder of the disassociated water) on the other side, producing an electrical and chemical gradient. Protons rush back across the membrane through channels, creating the force used by enzymes to generate ATP.

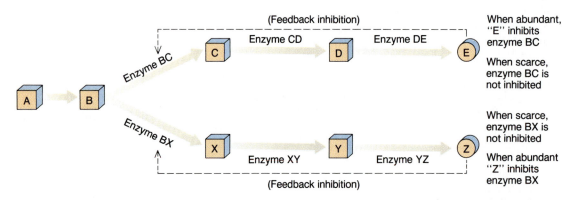

| When end product "E" is . . . | And end product "Z" is . . . | Then enzyme BC is . . . | And enzyme BX is . . . | Which allows production of . . . |
|---|---|---|---|---|
| Scarce | Scarce | Active | Active | E and Z |
| Scarce | Abundant | Active | Inhibited | E only |
| Abundant | Scarce | Inhibited | Active | Z only |
| Abundant | Abundant | Inhibited | Inhibited | Neither E or Z |

**Figure 8-7**
**Regulation of metabolism by feedback inhibition.** By controlling the activities of the two key enzymes ("BC" and "BX"), concentrations of end products determine which of these two optional pathways will be operational. Each condition shown here (for example, high Z concentration; low E concentration) is temporary. As the amount of each end product changes, so does the activity of the corresponding key enzyme. The table shows the activity states that exist for differing concentrations of E and Z.

late a particular pathway according to the cell's needs. Enzymes provide the switch that turns on or off a particular set of reactions. A pathway can be in-hibited either by producing fewer of its key enzymes or by temporarily inacti-vating these enzymes. The opposite situation, enhancement of a pathway, is achieved by increasing the concen-tration or activity of the key enzymes. Such enzyme regulation occurs in re-sponse to the concentration of the pathway's substrates or end products. (These concentrations represent the cell's need for the corresponding set of reactions.) For example, when an end product of a particular pathway is in ample supply, it inhibits one of the first enzymes in the pathway. As a result, the entire sequence of reactions is halted, allowing the cell's resources to be channeled into other pathways.

## Feedback Inhibition

**Feedback Inhibition** (also called *negative feedback*) is a mechanism of regulating enzyme activity by temporarily inactivating a key enzyme in a biosynthetic pathway when the concentration of the end product becomes elevated. In other words, the more end product there is, the more it "feeds back" and inhibits the key enzyme, as illustrated by the hypothetical branching pathway shown in Figure 8-7. In high concentrations, the end product Z binds to and inhibits the activity of the first enzyme unique to that metabolic branch. The substrate is then diverted to other pathways—in this instance, the reaction sequence that produces E. When the situation is reversed, enzyme BC is inhibited and the other branch is temporarily released from its regulatory restraints, so more

Z is produced. In this way, the cell channels its resources where they are needed most: for manufacturing products that need replenishing rather than wasting substrates to make compounds that are in ample supply at that moment.

Not all enzymes respond to feedback inhibition, but those that do are called **allosteric enzymes** (allosteric = shape-changing). These enzymes possess an additional site that makes them responsive to the end product's concentration. This area on the enzyme, called the *allosteric site*, can bind with the end product of the pathway. Such an event distorts the enzyme's active site so that it can no longer recognize the substrate(s) (Figure 8-8). In this way, the enzyme is *temporarily* inactivated while the concentration of the end product is high enough to favor its binding to the allosteric site. When the end product becomes scarce, it separates from the allosteric site and the enzyme regains activity. Since the rate of reaction depends on the number of active enzymes available, the amount of allosteric inhibition is proportional to the amount of end product present. When the concentration of end product is low, most of the enzymes remain active, and the pathway is turned on.

As the concentration of the end product increases, more of the enzymes will be inhibited by allosteric binding, thereby decreasing activity of the pathway. Allosteric enzymes provide one assurance that a biosynthetic pathway operates only when the product of that pathway drops below concentrations needed by the organism.

## Control of Enzyme Synthesis

Although feedback inhibition of enzyme activity provides immediate inactivation of a metabolic pathway it still allows energy to be wasted on synthesizing unneeded enzymes. Organisms possess additional control mechanisms that prevent such metabolic extravagance. The end products of most biosynthetic pathways are capable of blocking the *production* of the pathway's key enzymes as well as the activity of these enzymes. In many catabolic pathways, the opposite situation exists—key enzymes are synthesized only when the corresponding substrate is available. For example, the enzymes for digesting the sugar lactose are not manufactured when there is no lactose to digest. The mechanism by which cells avoid squandering their resources making unneeded enzymes is described in Chapter 28.

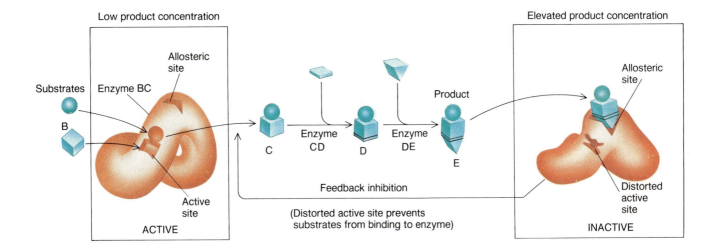

**Figure 8-8**
**Metabolic control by allosteric enzyme inhibition.** When concentrations of the product (E) are low, the first enzyme (BC) is active and the pathway proceeds to completion. As the end product (E) accumulates, the likelihood of it binding to the enzyme's allosteric site increases. When bound the product E shuts down the pathway by changing the first enzyme's shape to an inactive configuration.

Synopsis

## MAIN CONCEPTS

- **Metabolic pathways** change molecules in manageable steps, generating metabolic intermediates in the process. Catabolic pathways are generally endergonic and oxidative (release electrons). Anabolic pathways are exergonic and reductive (absorb electrons).

- **Energy distribution often follows the fate of electrons.** Energized electrons contain energy initially introduced by photoexcitement of electrons; electrons released by oxidizing highly reduced organic compounds temporarily retain this energy. Electron carriers transport energized electrons. $NAD^+$ or FAD carries energized electrons (as NADH or $FADH_2$) to the electron transport system where their energy is used to form ATP by chemiosmotic phosphorylation. In this process energy released from electrons is used to concentrate protons on one side of a membrane. The flow of these protons back across the membrane drives phosphorylation. $NADP^+$ carries energized electrons (as NADPH—reducing power) to electron absorbing biosynthetic pathways.

- **Routing metabolic intermediates** changes a compound's metabolic direction according to the needs of the cell. Feedback inhibition temporarily inactivates a pathway's key enzyme(s) as the product accumulates to surplus. Allosteric enzymes are temporarily inactivated by feedback inhibition when the pathway's product binds to the enzyme and thereby distorts it into a nonfunctional configuration. Regulation of enzyme synthesis avoids the waste of synthesizing enzymes during periods when they are not needed.

## KEY TERM INTEGRATOR

### Metabolic Strategies

| | |
|---|---|
| **metabolic pathways** | A series of chemical changes that alter a compound in small sequential steps. Whether linear or **cyclic pathways**, the **metabolic intermediates** generated represent possible branch points for diversion of surpluses to other pathways. This provides a mechanism for integrating energy-releasing reactions (**catabolism**) and energy-consuming processes (**anabolism**), for example, linking glucose oxidation with the **biosynthesis** of energy storage compounds. |

### Oxidation–Reduction Reactions

| | |
|---|---|
| **redox reactions** | Electron-releasing **oxidation** reaction coupled with an electron-accepting **reduction** reaction. |
| **electron carriers** | Molecules that may briefly contain **photoexcited** electrons or energized electrons released from oxidation of highly reduced compounds. $NAD^+$ and **FAD** may carry their electron cargo to the **cytochromes** of the **electron transport system**, where their energy is used for ATP production via the process of **chemiosmosis**. The electrons carried in NADPH represent a cell's **reducing power**, electrons used for biosynthesis. |

### Directing Metabolic Traffic

| | |
|---|---|
| **feedback inhibition** | Inhibiting the activity of a key enzyme in a metabolic pathway when the pathway's end product is in ample supply. Enzymes regulated in this way are **allosteric enzymes**. |

# Review and Synthesis

1. Describe two advantages of using metabolic pathways versus single-step reactions.

2. In the equation on page 112, glucose is oxidized, and oxygen is ___. Which of the two end products is in a more reduced state? How could you arrive at that answer by simply looking at the chemical formulas $CO_2$ and $H_2O$?

3. Discuss three differences in the way glucose is burned by simple combustion versus its complete oxidation by an organism during respiration.

4. Arrange the following in the proper sequence: oxidation of protein, photoexcitement, reduction of $NAD^+$ to NADH, low-energy electrons leaving cell in water, transfer of electrons from NADH to NADP, transfer of NADH to the electron transport system, ATP production, reduction of $CO_2$ to sugar, biosynthesis of protein.

5. Select one or more items from the following list which connect each pair of terms.

   NADPH ATP electrons NADH energy

   degradation   (_____)   biosynthesis
   anabolism     (_____)   catabolism
   oxidation     (_____)   reduction
   catabolism    (_____)   electron transport system

6. What is the advantage of regulating enzyme synthesis rather than enzyme activity? What is the advantage of regulating enzyme activity? (Hint—it takes time to produce enzymes and for enzymes already synthesized to deteriorate.)

7. What would happen to you (and all other forms of life) if photoexcitement of electrons were somehow prevented?

# Additional Readings

Baker, J., and G. Allen. 1981. *Matter, Energy, and Life.* 4th ed. Addison-Wesley, Reading, Mass. (Introductory.)

Cloud, P. 1983. "The biosphere." *Scientific American* 249:176–89. (Intermediate.)

Hinkle, P., and R. McCarty. 1978. "How cells make ATP." *Scientific American* 238:104–23. (Intermediate.)

Lehninger, A. 1971. *Bioenergetics: The Molecular Basis for Biological Energy Transformations.* 2nd ed. Benjamin–Cummings. (Advanced.)

Readings from Scientific American. 1980. *Molecules to Living Cells.* W. H. Freeman, San Francisco. (Intermediate.)

Chapter 9

# Processing Energy: Photosynthesis

life is powered by the sun. It may not be obvious at first, but the energy that gives a cheetah its incredible speed or a bird its power to fly originally came from the sun. The energy of sunlight, however, becomes energy for life only after it has been transformed into chemical energy through photosynthesis.

Virtually all organisms depend on photosynthesis for supplies of energy-rich food. The cheetah, for example, gets its food by eating an antelope, which in turn ate grasses and other plants that photosynthesized. Although the majority of organisms depend on photosynthesis, only plants, algae, and some protists and bacteria are themselves photosynthetic. The remainder of organisms—more than 80 percent of living species—cannot photosynthesize.

All photosynthesizers are **autotrophs.** That is, they provide themselves with their own nutritional needs. Photosynthetic autotrophs harness the sun's energy and use it to make energy-rich molecules. These molecules are then used either to manufacture ATP through cellular respiration (as discussed in Chapter 10), or as molecular building blocks for constructing other organic compounds, many of which are combined to make new cells and tissues. Although most autotrophs are photosynthetic, there are some autotrophic bacteria that acquire their energy from inorganic substances through a process called chemosynthesis, discussed at the end of the chapter.

Organisms that are unable to carry out either photosynthesis or chemosynthesis are **heterotrophs.** That is, they depend on other organisms (ultimately, the autotrophs) to supply them with their nutritional needs. Most heterotrophs rely on photosynthetic organisms as the critical link in converting energy from the environment into food for life. Thus, photosynthetic autotrophs provide not only themselves with food, but a vast array of heterotrophs as well—you included. For this reason, photosynthesis is considered by most biologists to be the most important set of chemical reactions in all biology (Figure 9-1).

# An Overview of Photosynthesis

At one time scientists believed that plants, the principal photosynthesizers on earth, absorbed all of their food and nutrients from the soil alone. But in the early 1600s, a botanist named Jan Batista van Helmont conducted a simple experiment that disproved this hypothesis of "soil-eating" plants. After planting a 5-pound willow tree in 200 pounds of dry soil, van Helmont added only rainwater to the soil for a period of 5 years. When he reweighed the soil and willow tree, the tree had gained 164 pounds, yet the soil had lost only 2 ounces, not 164 pounds as would be expected of a soil-eating plant. He concluded that the willow's weight gain must have come from the added water, a conclusion we now know is only partly correct.

It is true that a large amount of the willow's weight was from the water in its tissues. But willows, like most other plants, gather the carbon they need to build tissues from the air. Through photosynthesis, the carbon in atmo-

a

b

**Figure 9-1**
**Life is centered around photosynthesis.***(a)* Plants supply energy for virtually all organisms in the biosphere by transforming light energy into chemical energy. Every year plants manufacture about 150 billion metric tons of sugars, a weight equivalent to 54 times the combined weight of all people living on earth today.*(b)* Photosynthetic plants are essential to human existence, as well as virtually all other animals. Although about 3000 species of plants have been cultivated throughout human history, today only 15 plant species supply most of the world's human population with food.

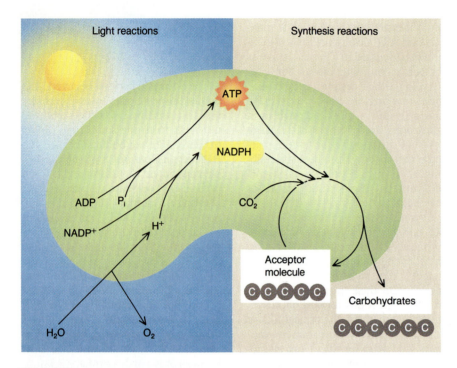

**Figure 9-2**
**Photosynthesis: a two-stage process.** Photosynthesis is divided into light reactions and synthesis reactions. During the light reactions, the energy of sunlight provides the power to generate ATP and NADPH from ADP, $P_i$, $NADP^+$, and $H_2O$. Oxygen gas is given off as a byproduct. In the synthesis reactions, ATP and NADPH from the light reactions provide the energy and electrons to assemble $CO_2$ molecules into carbohydrates. Because light energy is not directly involved in the synthesis reactions, they are sometimes called *the dark reactions.*

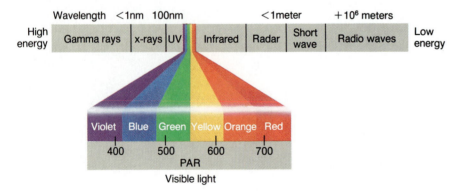

**Figure 9-3**
**Solar energy for life.** The biosphere is constantly bombarded with solar radiation, pulsating waves of high-intensity energy that *radiate* out from the sun. The wavelengths of radiation reaching the earth range from short wavelength (high-energy) gamma rays, to extremely long wavelength (low-energy) radio waves. Our eyes can detect only a small segment of the radiation spectrum, the portion we call visible light (from 380 nm to 750 nm). Photosynthetic pigments are also sensitive to visible light. They selectively absorb wavelengths between 400 nm and 700 nm, known as PAR (*Photosynthetically Active Radiation*).

spheric $CO_2$ is used to form the backbone of new organic compounds which are then fashioned into new cells for growth. Only a few of the minerals needed for growth actually come from the soil. (This accounts for the 2-ounce weight loss measured by van Helmont.)

Photosynthesis takes place inside the chloroplasts of eukaryotic cells (Chapter 6). Within these organelles, clusters of light-absorbing pigments capture the energy of sunlight and use that energy to manufacture energy-rich glucose from carbon dioxide and water. Glucose not only provides a store of chemical bond energy, but also serves as a molecular foundation to produce thousands of other molecules needed by the plant. Oxygen is released as a byproduct.

Expressing the overall process in a chemical equation makes photosynthesis appear rather simple:

$$6CO_2 + 12H_2O \xrightarrow{\text{light energy}} \underset{\text{glucose}}{C_6H_{12}O_6} + 6O_2 + 6H_2O$$

But photosynthesis is not a single pathway as this equation suggests. The entire process has some 20 chemical reactions, including both linear and cyclic pathways.

Botanists have grouped the 20-odd reactions in photosynthesis into two general stages: **light reactions** and **synthesis reactions** (Figure 9-2). During the light reactions, energy from sunlight is absorbed and used to make chemical energy in the form of energy-rich ATP and NADPH. Water is split into $H^+$ (protons), electrons, and oxygen during these reactions. The balanced chemical equation for the light reactions is written as

$$18ADP + 18P_i + 12NADP^+$$
$$+ 12H_2O \xrightarrow{\text{light energy}} 18ATP$$
$$+ 12NADPH + 6O_2 + 6H_2O$$

Since light energy is not directly involved in the synthesis reactions, the second stage of photosynthesis is sometimes called *the dark reactions.* During the synthesis reactions, $CO_2$ is added to another molecule, called a carbon-acceptor, and carbohydrates (ultimately glucose) are synthesized using the energy of ATP, and energized electrons and hydrogens carried by NADPHs from the light reactions. A balanced equation for the synthesis reactions is

$$6CO_2 + 6 \text{ acceptors} + 18ATP \\ + 12NADPH \longrightarrow 1 \text{ } C_6H_{12}O_6 \\ + 18ADP + 18P_i + 12NADP^+ \\ + 6 \text{ acceptors}$$

Note that six carbon-acceptors are regenerated at the conclusion of the synthesis reactions, indicating the presence of a cyclic pathway.

# The Light Reactions

Light reactions take place within the thylakoid membranes of chloroplasts. They begin when the energy in sunlight is captured by light-absorbing photosynthetic pigments.

## Photosynthetic Pigments: Capturing Light Energy

Although solar energy ranges from short (less than $10^{-14}$ meters) to relatively long wavelengths (greater than $10^6$ meters) (Figure 9-3), not all wavelengths are absorbed by photosynthetic pigments. Each pigment absorbs only certain wavelengths of light. But because there are many kinds of pigments, together they absorb light energy with wavelengths between 400 nm (violet light) and 700 nm (red light). This range is referred to as **PAR**, an abbreviation for **P**hotosynthetically **A**ctive **R**adiation (Figure 9-4).

The various kinds of light-capturing pigments are classified into three groups:

- **Chlorophyll pigments.** Several forms of chlorophyll collectively absorb wavelengths of light from 400 to 500 nm (violet and blue light), and from 600 to 700 nm (orange and red light). Because chlorophyll reflects green light (500 to 600 nm), they give plants their green color.

- **Carotenoid pigments.** Carotenoids absorb light mainly between 400 and 500 nm wavelengths. Yellow, orange, and red wavelengths are reflected by carotenoids, giving carrots, oranges, tomatoes, and the leaves of some plants during the fall their characteristic color. Chlorophylls and carotenoids are the light-capturing pigments used by all photosynthesizers. However, two small groups of organisms also use phycobilins.

- **Phycobilins.** Phycobilins are found only in photosynthetic cyanobacteria and red algae. They absorb wavelengths of light between 450 and 650 nm.

## Organization of Photosynthetic Pigments

Capturing light energy and converting it into chemical energy requires very intricate organization. Photosynthetic pigments are embedded in the thylakoid membranes of chloroplasts in precise clusters called **photosystems** (Figure 9-5). Together, these pigments form a *light-harvesting complex* that absorbs light energy throughout the PAR wavelengths. A single chloroplast contains several thousand photosystems.

Each photosystem has a special chlorophyll molecule called the **reaction center** surrounded by 250 to 350 **antenna pigments**. Antenna pigments are so named because they gather light energy and then channel this energy to the reaction center, much as an antenna receives and relays radio and TV waves. In plants, most antenna pigments (between 200 and 300) are chlorophyll molecules, and about 50 are carotenoid molecules.

To date, scientists have identified two kinds of reaction center chlorophylls. One reaction center chloro-

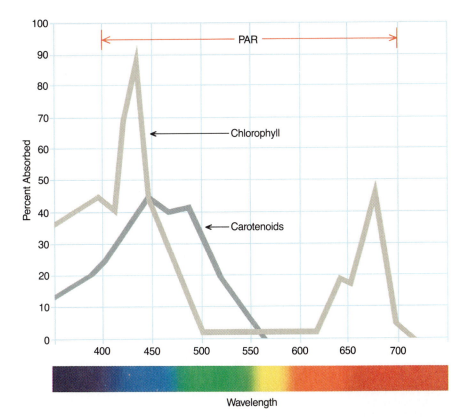

**Figure 9-4**
**Absorption spectrum of photosynthetic pigments.** Most absorption of light energy by photosynthetic pigments takes place at the peaks of each graph; the valleys indicate wavelengths that are reflected. Chlorophyll pigments absorb wavelengths mainly in the violet-blue and orange-red ranges. Carotenoid pigments absorb violet to blue wavelengths. All plants use chlorophyll and carotenoid pigments for photosynthesis.

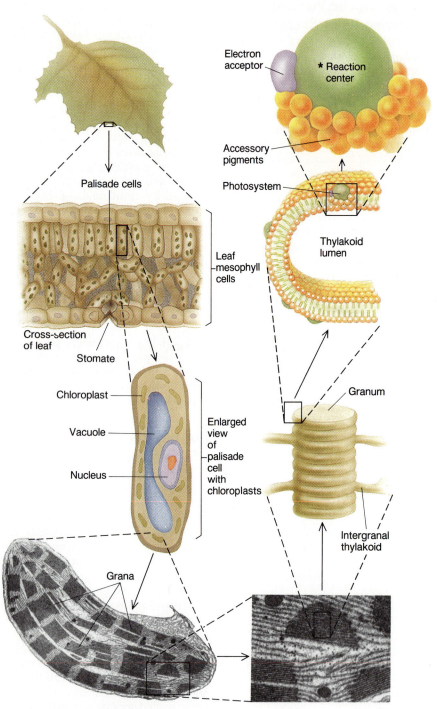

Electron
acceptor
* Reaction
center
Accessory
pigments
Photosystem
Thylakoid
lumen
Palisade cells
Leaf
mesophyll
cells
Cross-section
of leaf
Stomate
Granum
Chloroplast
Vacuole
Enlarged
view
of
palisade
cell
with
chloroplasts
Nucleus
Intergranal
thylakoid
Grana

\* Greatly enlarged for emphasis

**Figure 9-5**
**Organization for photosynthesis.** Photosynthesis takes place in the chloroplasts of plant cells. A leaf cell may contain as many as 60 chloroplasts. The thylakoid system of the chloroplast is the site of the light reactions, whereas the stroma is the site of the synthesis reactions. Photosynthetic pigments are precisely arranged in the thylakoid membranes to form light-harvesting photosystems. The reaction center of each photosystem receives energy from surrounding antenna pigments. When sufficent energy is absorbed by the reaction center, its electrons are boosted to a higher energy level and are then passed to an electron-acceptor molecule. This starts a chain of chemical reactions that leads to the formation of ATP and NADPH, and ultimately energy-rich carbohydrates.

phyll is activated by wavelengths of light energy up to 700 nm and has been aptly named the **P700 reaction center.** (The "P" stands for pigment.) The second reaction center chlorophyll is activated by wavelengths up to 680 nm, so it was named the **P680 reaction center.** Photosystems with a P700 reaction center are named **Photosystem I,** whereas those with a P680 reaction center are called **Photosystem II.**

When the P700 and P680 reaction center chlorophyll receive sufficient energy from surrounding antenna pigments, two of their electrons are boosted to a higher energy level; the electrons become energized. These energized electrons are then passed to an electron-acceptor molecule, one acceptor for the P700 reaction center and another for the P680 reaction center. Each electron acceptor then transfers the energized electrons to other molecules, thereby initiating a flow of electrons that ultimately produces chemical energy—ATP and NADPH, or ATP alone.

## Converting Light Energy into Chemical Energy

The electrons of P700 and P680 remain energized for less than one-billionth of a second. If, during this fraction of an instant, these electrons are not passed to an electron-acceptor molecule, they release their excess energy as heat or light as they return to their original energy levels. But if the photoexcited electrons are transferred to an electron acceptor, the energized electrons are then shuttled along one of two molecular pathways that eventually leads to the formation of chemical energy. One pathway forms a cycle; that is, the electrons return to their original reaction center. The second pathway does not form a cycle. In this noncyclic pathway, energized electrons are ultimately passed to NADP$^+$ and do not return to their reaction center. Since both pathways of electron flow result in phosphorylation (ADP + P$_i$ → ATP) (see Connections: ATP, in Chapter 3), they have been

named **cyclic photophosphorylation** and **noncyclic photophosphorylation** —*cyclic* or *noncyclic* because of the pathway of energized electrons; *photo-* because light energy is used to energize electrons; and *phosphorylation* because ATP is formed. First, we will describe how the simpler, cyclic pathway works.

## CYCLIC PHOTOPHOSPHORYLATION

Only Photosystem I operates during cyclic photophosphorylation (illustrated in Figure 9-6). Light energy is absorbed by the antenna pigments and funneled into the P700 reaction center. Photoexcited electrons from P700 are passed to an electron acceptor, and then are relayed down an electron transport system that returns the electrons to the same P700 reaction center. ATP is synthesized as a result of this cyclic electron journey.

Most likely, cyclic photophosphorylation is the way the earth's first autotrophic bacteria converted light energy into chemical energy. Even today, the only means of making ATP for some prokaryotic bacteria is through cyclic photophosphorylation. But if cyclic phosphorylation were the only available means of generating usable chemical energy, there could be no multicellular autotrophs. ATP is not produced fast enough or in large enough quantities during cyclic photophosphorylation to meet the energy and molecular demands for building and maintaining a multicellular organism. The earth now abounds with more than 350,000 species of multicellular plants, each possessing another electron pathway that meets these requirements. By teaming Photosystem I and Photosystem II during noncyclic photophosphorylation, not only is ATP produced, but NADPH as well.

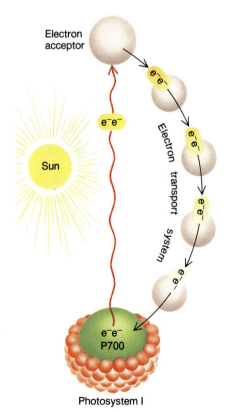

**Figure 9-6**
**Cyclic photophosphorylation.** Photosystem I, its electron-acceptor molecule, and an electron transport system participate in cyclic photophosphorylation. Photoexcited electrons from the P700 reaction center are first passed to an electron acceptor, and then are shuttled down the electron transport system and relayed back to P700. ATP is produced by chemiosmosis.

## NONCYCLIC PHOTOPHOSPHORYLATION

During noncyclic photophosphorylation, photoexcited electrons flow along a linear path. They originate from water, travel through a number of molecules, and ultimately flow to NADP⁺ to form NADPH. As a result of their transfer between molecules, ATP is also synthesized. This more complicated electron course is diagrammed in Figure 9-7.

As in cyclic photophosphorylation, the noncyclic flow of electrons begins when light energy is absorbed by photosynthetic pigments. Both Photosystem I and Photosystem II participate in noncyclic flow, and both photosystems are activated simultaneously. In Photosystem II, absorbed energy is funneled to the P680 reaction center, boosting two electrons to a higher energy level. These energized electrons then pass to Photosystem II's electron-acceptor molecule. The oxidized P680 reaction center now has a very strong attraction for electrons. This attraction pulls electrons from the hydrogen atoms of a water molecule (one electron from each hydrogen), splitting the water. This process is called **photolysis** (photo = light, lysis = splitting). The two electrons that were stripped from the water's hydrogen atoms fill the electron vacancy in P680. The two protons ($H^+$) remain within the thylakoid sac, and liberated oxygen atoms combine to form $O_2$.

Meanwhile, the energized electrons passed from P680 to its electron ac-ceptor are shuttled down an electron transport system that delivers them to the P700 reaction center of Photosystem I. But why would P700 accept electrons? Recall that light strikes both Photosystem I and Photosystem II at the same time. Therefore, as the events just described are taking place in Photosystem II, light energy is also being absorbed by Photosystem I. The energized P700 electrons are transferred to an electron acceptor. This leaves a two electron vacancy in the oxidized P700 reaction center. The electrons routed from Photosystem II fill this vacancy. These are the electrons that originated in P680.

Once the electron-acceptor molecule of Photosystem I accepts energized electrons from P700, it passes

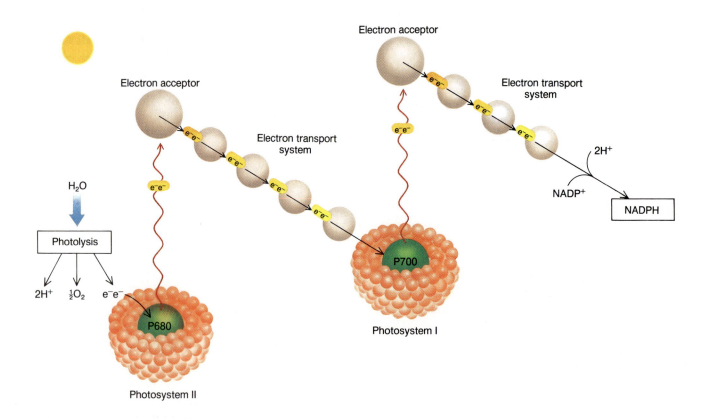

**Figure 9-7**
**Noncyclic photophosphorylation.** Photosystem I, Photosystem II, each of their electron-acceptor molecules, and two electron transport systems take part in noncyclic photophosphorylation. Photoexcited electrons from the P700 and P680 reaction centers are passed to electron acceptors, which in turn transfer energized electrons to an electron transport system. The electron vacancy created by photoexcitement in P680 is filled with electrons from the photolysis of water. The electron vacancy in P700 is filled with electrons that originated in P680. The photoexcited electrons from P700 eventually combine with NADP⁺ and H⁺ to form NADPH. ATP is also produced by chemiosmosis as a result of the noncyclic flow of energized electrons.

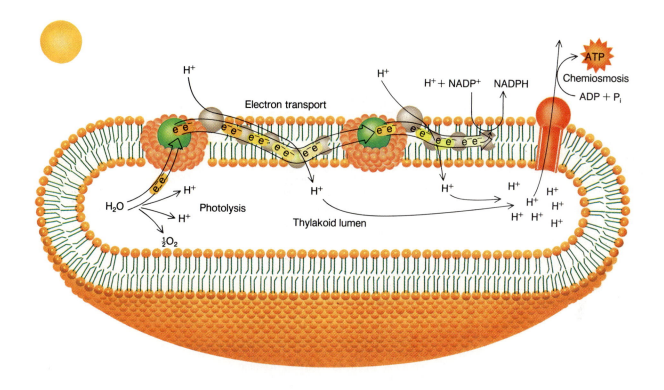

**Figure 9-8**
**Building a proton gradient for chemiosmosis.** The transport of energized electrons along an electron transport system, and the splitting of water during photolysis, concentrate protons inside the thylakoid lumen. This produces a concentration gradient of protons across the thylakoid membrane sufficient enough to drive the phosphorylation of ADP to ATP as protons diffuse through special membrane channels.

them to yet another transport chain. The final molecule in this chain relays the electrons to $NADP^+$ to form NADPH.

As in cyclic photophosphorylation, the flow of energized electrons from one cytochrome to another in an electron transport system sets up a proton gradient across the thylakoid membrane that powers ATP synthesis by chemiosmosis. Cyclic and noncyclic photophosphorylation are summarized in Table 9-1.

## Chemiosmosis: Making ATP

For chemiosmosis to occur, an electrochemical gradient must be established across a membrane (Chapter 8). During the light reactions of photosynthesis, two sources of protons ($H^+$) establish this electrochemical gradient across the thylakoid membranes. One source of $H^+$ is from the transport of electrons along electron transport systems. As illustrated in Figure 9-8, cytochromes are embedded in thylakoid membranes so that the first to receive energized electrons is on the stroma side and the last is nearest the thylakoid lumen, the space inside each flattened thylakoid sac. As negatively charged electrons pass from one cytochrome to the next in the transport chain, protons move from the stroma into the thylakoid lumen. Protons accumulate in the thylakoid lumen during both cyclic and non-cyclic flow of electrons during the light reactions.

A second source of $H^+$ is photolysis. Water is split inside the thylakoid lumen, releasing protons. During noncyclic flow, protons accumulate in the thylakoid lumen from both photolysis and electron transport.

As a reservoir of $H^+$ builds within the thylakoid lumen, a proton gradient is established across the membrane. The protons diffuse through special membrane channels, and, as they exit, energy is provided for phosphorylation (Figure 9-8).

## The Synthesis Reactions

Since both ATP and NADPH are forms of chemical energy, why do photosynthesizers manufacture en-

TABLE 9-1

## Summary of Cyclic and Noncyclic Photophosphorylation

| Cyclic | Noncyclic |
|---|---|
| 1. Light is absorbed by pigments in Photosystem I. | 1. Light is absorbed by pigments in Photosystems I and II. |
| 2. Energized electrons from P700 pass to an electron acceptor. | 2. Energized electrons from P680 of Photosystem II and energized electrons from P700 of Photosystem I pass to separate electron acceptors. |
| 3. Energized electrons are relayed down an electron transport system and returned to the original P700 reaction center. | 3. Photolysis of water fills the electron vacancy in P680, releasing protons and oxygen. |
| 4. ATPs are generated by chemiosmosis as a result of the cyclic flow of electrons. | 4. P680's electron acceptor passes energized electrons to an electron transport system that relays them to the P700 reaction center of Photosystem I. |
| | 5. The electron acceptor for P700 passes energized electrons to another electron transport chain that shuttles them to $NADP^+$, forming NADPH. |
| | 6. ATPs are generated by chemiosmosis as a result of a noncyclic flow of electrons and the release of protons from photolysis. |

ergy-rich glucose molecules during the synthesis reactions? The answer has to do with the stability, transportability, and versatility of these molecules. Neither ATP nor NADPH can be stored or translocated from one part of a plant to another, but glucose can. Transport and storage of energy-rich compounds are critical to the survival of a multicellular plant. For example, glucose produced in leaves must be moved to growing regions of roots and stems where photosynthesis does not take place. In addition, glucose can be used to construct other organic compounds, whereas ATP and NADPH cannot. Thus, the function of the synthesis reactions is to use the energy and electrons of ATP and NADPH to manufacture glucose. $CO_2$ is gathered from air to supply carbon for glucose construction.

Like the light reactions, the synthesis reactions occur in the chloroplasts of eukaryotic cells. But unlike the light reactions, the synthesis reactions take place on the thylakoid surfaces facing the stroma and in the stroma itself.

Three variations of the synthesis reaction have been identified in plants — $C_3$, $C_4$, and **CAM Synthesis.** Table 9-2 lists common plants with $C_3$, $C_4$, and CAM synthesis reactions. All three synthesis reactions share the same cyclic pathway for producing glucose and regenerating the starting molecule (this pathway is called the Calvin Cycle — see "$C_3$ Synthesis" later in this chapter), but differ in the way $CO_2$ is first combined with a carbon-acceptor molecule. As we will see, this seemingly small difference has given some plants a metabolic edge in certain environments.

## $C_3$ Synthesis

About 80 percent of all plants use the $C_3$ synthesis pathway exclusively for

constructing glucose, making it the predominant synthesis pathway among the earth's photosynthesizers.

$C_3$ synthesis, and indeed all three types of synthesis reactions, begin with **$CO_2$ fixation,** the combining of $CO_2$ with a carbon-acceptor molecule. In $C_3$ synthesis, however, $CO_2$ is fixed to a 5-carbon sugar, **ribulose-bisphosphate (RBP),** by the enzyme **RBP carboxylase.** However, the 6-carbon molecule formed by this union is so unstable that it immediately splits into two 3-carbon molecules, **phosphoglycerate (PGA).** Because PGA is the first stable carbohydrate to be produced, the pathway was named **$C_3$ synthesis.**

PGA, however, is not a high-energy molecule. In fact, two molecules of PGA contain less energy than a single molecule of RBP, the starting carbon-acceptor molecule. This explains why $CO_2$ fixation is spontaneous and does not require energy from the light reactions. To synthesize high-energy mole-

TABLE 9-2

| Plants with C₃, C₄, and CAM Synthesis Reactions | | |
|---|---|---|
| **C₃** | **C₄** | **CAM** |
| *Legumes (bean, pea) | *Sugarcane | *Pineapple |
| *Wheat | *Corn | Cactus |
| *Oats | *Sorghum | Jade plant |
| *Barley | Crabgrass | Lilies |
| *Rice | Russian thistle | *Agave (tequila) |
| Bluegrass | Bermuda grass | Some orchids |

*Agricultural crops.

cules, the energy and electrons from ATP and NADPH from the light reactions are used to reduce each PGA to a more energy-rich molecule, **phosphoglyceraldehyde (PGAL).** Two PGAL molecules can then be complexed to form one glucose.

Although this explanation tells us how glucose is formed, the Calvin Cycle is not complete inasmuch as RBP has not been regenerated. Fortunately, the regeneration of RBP and the production of glucose are rather

simple to explain. Instead of one $CO_2$ molecule becoming fixed, consider six. When six $CO_2$ molecules combine with six RBP molecules, not only can one 6-carbon glucose be synthesized, but also the original six RBPs can be regenerated and the cycle can start over again. This is the **Calvin Cycle,** named after Melvin Calvin, the investigator who received a Nobel prize in 1961 for identifying the sequence of reactions. The Calvin Cycle is diagrammed in Figure 9-9.

## C₄ Synthesis

At high temperatures (45° C, 103° F, or higher), plants with $C_4$ synthesis produce as much as six times more glucose than $C_3$ plants. $C_4$ plants also do better than $C_3$ plants in environments in which water is limited and soil nutrients are scarce. What makes $C_4$ plants so different from $C_3$ plants?

The difference is the way they fix $CO_2$. In $C_3$ plants, $CO_2$ is only fixed to RBP by RBP carboxylase. For RBP carboxylase to work, $CO_2$ must be

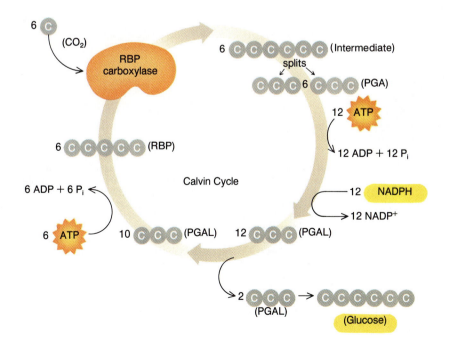

**Figure 9-9**
**C₃ synthesis and the Calvin Cycle.** Glucose is produced by the Calvin Cycle. Six $CO_2$ molecules are combined with six RBPs to form six unstable, 6-carbon carbohydrates. Each carbohydrate immediately splits in two, producing a total of 12 molecules of PGA. The energy and electrons from 12 ATPs and 12 NADPHs generated during the light reactions transform the PGAs into 12 molecules of PGAL. Two PGALs combine to form glucose, whereas the remaining 10 PGALs recombine to regenerate the six molecules of RBP that started the Calvin Cycle.

plentiful. But in $C_4$ synthesis, the enzyme **PEP carboxylase** fixes $CO_2$ to another carbon acceptor, **PEP** (Phosphoenol pyruvic acid). PEP carboxylase has a much higher affinity (attraction) for $CO_2$ than RBP carboxylase. Therefore, $CO_2$ levels can be very low in a leaf of a $C_4$ plant, much lower than normal air concentrations, and $CO_2$ will still be fixed to PEP by PEP carboxylase. RBP carboxylase cannot fix $CO_2$ at levels lower than normal air concentrations, so in $C_3$ plants, glucose synthesis is stalled.

For plants growing in hot, dry environments in which water loss is critical, $C_4$ synthesis offers a major survival advantage. As $CO_2$ enters the leaves of plants through small pores, or *stomates,* precious water vapor exists at the same time. Since $C_4$ plants can extract $CO_2$ even at extremely low concentrations, stomates can be kept very narrow or closed altogether, reducing water loss, yet enabling them to still fix $CO_2$. On the other hand, for $C_3$ plants

to maintain $CO_2$ concentrations high enough to allow fixation by RBP carboxylase, stomates must be kept wide open, allowing great amounts of water vapor to escape from internal tissues. Open stomates could quickly prove fatal to $C_3$ plants in dry environments; therefore, stomates must be kept closed during the heat of the day to conserve precious water. However, closing stomates stops the synthesis reactions for $C_3$ plants by blocking the entrance of $CO_2$.

All $C_4$ plants have leaves with a distinctive characteristic termed **Kranz anatomy** (Figure 9-10). Leaves of $C_4$ plants contain mesophyll cells (cells between the upper and lower epidermis of a leaf), and bundle sheath cells (cells that encircle a leaf's vein). Both of these cell types contain chloroplasts. The bundle sheath cells of $C_3$ and CAM plants do not have chloroplasts; only their mesophyll cells are photosynthetic. Kranz anatomy allows the synthesis reactions to be di-

vided between different parts of the leaf. This division increases the rate of glucose production.

$C_4$ synthesis begins when $CO_2$ is fixed by PEP carboxylase to PEP in the chloroplasts of mesophyll leaf cells. The product of $CO_2$ fixation is oxaloacetate, a 4-carbon acid (hence the name $C_4$). Depending on the plant species, oxaloacetate is converted into malic acid or aspartate. The malic acid or aspartate is then transferred from the chloroplasts of the mesophyll cells to the chloroplasts of the bundle sheath cells, where it splits, releasing $CO_2$ and a 3-carbon pyruvate molecule. The pyruvate cycles back to the mesophyll cells and combines with a phosphate from ATP to regenerate PEP. The released $CO_2$ is fixed to RBP, and glucose is produced through the Calvin Cycle in the same way as in $C_3$ synthesis.

Dividing the synthesis reactions between mesophyll cells and bundle sheath cells allows $CO_2$ concentrations

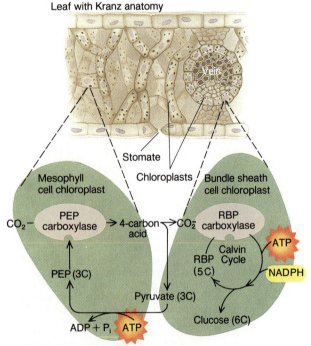

**Figure 9-10**

**Kranz anatomy and $C_4$ synthesis.** Plants with $C_4$ synthesis have chloroplasts in their mesophyll cells and in large bundle sheath cells surrounding the leaf veins. This unique chloroplast distribution enables $C_4$ plants to photosynthesize up to six times faster than $C_3$ plants. The $C_4$ synthesis reactions take place in different leaf cells. $CO_2$ fixation by PEP carboxylase takes place in the chloroplasts of mesophyll cells. The resulting 4-carbon oxaloacetates are converted and moved to the chloroplasts in bundle sheath cells. Here, the 4-carbon acids are broken down to release $CO_2$ and pyruvate. The $CO_2$ is fixed to RBP by RBP carboxylase, starting the Calvin Cycle. The ATPs and NADPHs produced during the light reactions in the bundle sheath cell chloroplasts power the production of glucose and the regeneration of RBP. Pyruvate is transported back to the mesophyll cells and reconverted into PEP with energy from ATP.

to build in the bundle sheath chloroplasts. Since $CO_2$ can be fixed to PEP at low concentrations in the mesophyll cells, $CO_2$ is "imported" from many mesophyll cells throughout the leaf and concentrated in bundle sheath cells to levels high enough to be continually fixed by RBP carboxylase and maintain continuous glucose production. The results for $C_4$ plants are faster growth rates and less water loss per glucose produced than for plants with $C_3$ synthesis.

## CAM Synthesis

CAM, an abbreviation for **Crassulacean Acid Metabolism**, is the least common of the three synthesis reactions. It is found in fewer than 5 percent of plants. Biochemically, CAM synthesis is identical to $C_4$ synthesis except there is no separation of synthesis reactions between mesophyll and bundle sheath cells. Instead, all CAM synthesis reactions take place in the same cell. But the reactions are separated by time, a factor that is very important to the survival of CAM plants in dry environments.

Like $C_4$ synthesis, CAM is a metabolic adaptation to particular environmental conditions. Many CAM plants are found in hot, dry deserts, environments that pose the greatest threats to survival because of the scarcity of water and high temperatures. Huge quantities of water can be lost when desert plants open their stomates to take in $CO_2$ for the synthesis reactions. But CAM synthesis helps conserve precious water by separating the time of $CO_2$ fixation, the light reactions, and the Calvin Cycle.

CAM plants open their stomates at night to take in $CO_2$; $C_3$ and $C_4$ plants open their stomates during the daytime. The diffusion gradient of water vapor between a plant and the air is steeper during the day. Consequently, the rate of water vapor loss is less at night.

Although it is clearly advantageous for a desert plant to have its stomates open at night, how are CAM plants able to use the $CO_2$ to make glucose when energy-rich ATP and NADPH

are produced only during daylight? Unlike in $C_3$ synthesis, $CO_2$ fixation does not occur immediately before the Calvin Cycle during CAM synthesis. At night, when a CAM plant is actively taking in $CO_2$, the $CO_2$ is swiftly fixed to PEP, forming oxaloacetate (Figure 9-11). The oxaloacetate is converted to malic acid, and the malic acid is moved from the chloroplast to the vacuole for storage. Throughout the night, more and more malic acid is produced and stored in the cell's vacuole, building a reservoir of fixed $CO_2$. (Segregating the malic acids in the vacuole is critical, for the cell could not metabolize normally if the acid accumulated in the cytoplasm.) At sunrise, stomates close (preserving water), and sunlight powers ATP and NADPH production in the light reactions. At the same time, malic acids are gradually removed from the vacuole and moved back to the chloroplast. There, malic acid is broken down into $CO_2$ and pyruvate. It is then that $CO_2$ becomes fixed to RBP and the Calvin Cycle begins.

## Photorespiration

Strangely enough, bright sunlight can have a negative impact on photosynthesis. Although photosynthesis proceeds faster in bright sunlight, faster rates quickly deplete $CO_2$ and produce large quantities of oxygen gas. When oxygen concentration is high and $CO_2$ concentration is low, biochemical shifts occur in $C_3$ plants, drastically reducing glucose production.

These shifts occur because both oxygen and $CO_2$ compete for the same active site on the RBP carboxylase enzyme. Hence, as oxygen increases and $CO_2$ decreases, RBP carboxylase fixes $O_2$ to RBP instead of $CO_2$, preventing the Calvin Cycle from operating. In addition, ATP and NADPH from the light reactions are diverted to other reactions, not those that drive the Calvin Cycle. Without the Calvin Cycle, no glucose is produced, and metabolism and growth decline. This detrimental process is called **photorespiration**—

*photo-* because light stimulates this reaction by forming $O_2$ during photolysis, and *respiration* because $O_2$ is consumed and $CO_2$ is released, the same changes that occur during cellular respiration (Chapter 10). Photorespiration can cut glucose production in $C_3$ plants by as much as 50 percent during peak sunlight hours.

Photorespiration is virtually nonexistent in $C_4$ plants. Because PEP carboxylase has such a high affinity for $CO_2$, $CO_2$ fixation continues even at extremely low concentrations. Since $C_4$ plants concentrate $CO_2$ in the chloroplasts of bundle sheath cells (and not $O_2$), $CO_2$ levels there are nearly always high enough to favor $CO_2$ fixation to RBP, starting the Calvin Cycle. Thus, $C_4$ plants are well adapted to the bright sunlight of tropical areas and deserts. Moreover, because they carry out photosynthesis at higher temperatures, in hot environments $C_4$ plants grow faster, more vigorously, and reproduce in greater numbers than their $C_3$ co-inhabitants that are sapped by energy-draining photorespiration. Corn, sugarcane, and sorghum are important $C_4$ food crops that grow rapidly in tropical areas.

So why haven't vigorously growing $C_4$ plants taken over all the earth's habitats? When $C_4$ plants are introduced by people into some environments, the consequences have indeed been swift. But $C_4$ plants do better than $C_3$s only where it is warm or dry. In cold environments, $C_3$ plants do better than $C_4$s because they are less sensitive to cold.

## Chemosynthesis

Although the predominant means of generating energy-rich organic molecules is through photosynthesis, some organisms depend on chemosynthesis for a supply of chemical energy. For example, an unusual community of organisms was recently discovered that relies totally on chemosynthesis. This fascinating community was found in a very unexpected place—around scorching hot vents deep in the

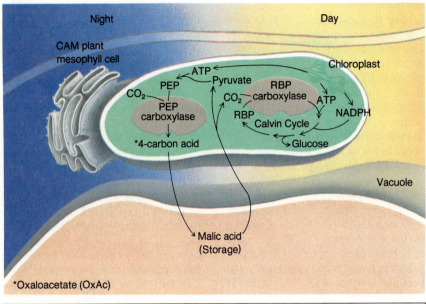

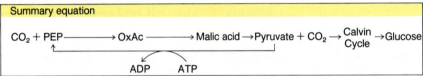

**Figure 9-11**
**CAM synthesis.** CAM synthesis reactions are identical to $C_4$ synthesis reactions except they take place at different times within the same mesophyll cell. At night, stomate pores open to take up $CO_2$. The $CO_2$ is fixed to PEP by PEP carboxylase producing oxaloacetate. The oxaloacetate is then converted to malic acid, which is then stored in the cell's vacuole. $CO_2$ fixation continues throughout the night building a reservoir of malic acid in the vacuole. During the daytime, CAM plants close their stomates, conserving water. The stores of malic acid are gradually moved back to the chloroplasts and split to release $CO_2$ and pyruvate. The released $CO_2$ is then fixed to RBP by RBP carboxylase and is introduced into the Calvin Cycle, while the pyruvate is converted into PEP with energy from ATP.

Pacific Ocean (see Bioline: Living on the Fringe of the Biosphere).

Chemosynthesis and photosynthesis are very similar. Both of these energy-converting reactions require an electron and hydrogen donor, a source of carbon ($CO_2$ is the carbon source), and energy. In photosynthesis, electrons and hydrogen come from the photolysis of water. In chemosynthesis, the electron donor is an inorganic substance, such as hydrogen,

nitrogen, iron, or sulfur compounds. During chemosynthesis, electrons are stripped from these inorganic substances, and the reduced substance supplies the energy for producing ATP and NADPH, which, in turn, provide fuel and reducing power for the synthesis of carbohydrates.

The bacteria that surround the hot vents described in the Bioline are all chemosynthetic autotrophs. One type of chemosynthetic bacterium oxidizes

the hydrogen in hydrogen sulfide to obtain energy for producing carbohydrates. The chemical equation for chemosynthesis of this hot vent bacterium is

$$CO_2 + 2H_2S \longrightarrow CH_2O + H_2O + 2S$$

Only a few species, all of them bacteria, carry out chemosynthesis. Together, photosynthetic and chemosynthetic autotrophs provide the energy-rich molecules and nutrients that support all life in the biosphere.

# B I O L I N E

## LIVING ON THE FRINGE OF THE BIOSPHERE

One of the most extraordinary and productive communities is found at the very fringe of the biosphere—in the pitch-black depths of the Pacific Ocean near the Galapagos Islands. At a depth of more than 2500 meters (8000 feet), strange marine creatures were discovered in 1977 crowded near mineralized chimneys that spewed scalding, black water. These hot, deep-sea vents form at the junction of crustal plates off the coast of Central America. Where the plates meet, frigid seawater seeps into red-hot fissures and becomes super heated to more than 350° C (662° F). The scalding water mixes with hydrogen sulfide released from the fissures. Such a hostile environment is an unlikely place to find a thriving community of organisms, but that's exactly what scientists found.

The torrid jets produce a warm, nutrient-laden habitat that is ideal for the growth of chemosynthetic bacteria. Dense clouds containing trillions of bacteria are eaten by a variety of crabs, barnacles, clams, mussels, and small worms, which, in turn, are eaten by larger crabs, fishes, limpets, leeches, whelks, and other worms. A number of organisms also house enormous communities of chemosynthetic bacteria inside their tissues, thereby getting a direct supply of food while, in turn, providing a habitat for the bacteria.

Many of these vent animals had never been seen by scientists before, or were much larger than their closest known relatives. For example, bright red tube worms (up to 3 meters long) acquire energy-rich carbohydrates directly from chemosynthetic bacteria that live inside their bodies.

To date, 17 high-temperature hydrothermal vent communities have been discovered throughout the world, revealing not only new species, but also new genera, families, and possibly even a new phylum of animals. Because only a very small portion of the world's deep seas have been explored, it is likely that even more new species will be discovered as additional vent communities are explored.

# Synopsis

## MAIN CONCEPTS

- **Radiant energy** from the sun is the main source of energy for life on earth.

- **Through photosynthesis,** the energy in sunlight is captured and converted into chemical bond energy.

- **Except for a few organisms that rely on chemosynthesis,** all organisms either directly or indirectly depend on photosynthesis for a supply of energy-rich organic molecules.

- **Photosynthesis is divided into two phases.**
  1. During the light reactions, the energy of sunlight is absorbed by pigments and converted into energy-rich ATP and NADPH. ATP is produced by chemiosmosis, whereas $NADP^+$ gains protons and energized electrons during noncyclic photophosphorylation to produce energy-rich NADPH.
  2. During the synthesis reactions, $CO_2$ is fixed to a carbon-acceptor molecule, and energy-rich carbohydrates are produced using the energy and electrons from ATP and NADPH.

- **Of the three types of synthesis reactions,** $C_4$ and CAM synthesis aid survival in certain environments over plants with $C_3$ synthesis.

- **In bright sunlight, photorespiration in plants** with $C_3$ synthesis reduces glucose production by as much as 50 percent. This, in turn, retards growth and reproduction.

- **Virtually all autotrophs are photosynthetic.** However, a few types of bacteria oxidize inorganic molecules, using the electrons and hydrogen to ultimately produce energy-rich organic molecules through a process called chemosynthesis. Chemosynthetic bacteria not only supply themselves with their own nutrition, but also form the food base for unique deep-sea vent communities.

## KEY TERM INTEGRATOR

Organization for Photosynthesis

| | |
|---|---|
| **photosynthetic pigments** | **Chlorophylls, carotenoids,** and **phycobilins** |
| **photosystem** | **Reaction center + antenna pigments** |
| **Photosystem I** | P700 reaction center + antenna pigments |
| **Photosystem II** | P680 reaction center + antenna pigments |

Photosynthetic Processes

| | |
|---|---|
| **Light reactions** | |
| cyclic photophosphorylation | Uses Photosystem I, produces ATP only. |
| noncyclic photophosphorylation | Photosystem I and II, produces ATP and NADPH, water is split during **photolysis.** |
| **Synthesis reactions** | |
| $CO_2$ fixation | Combines $CO_2$ with a carbon-acceptor molecule |
| the Calvin Cycle ($C_3$ Synthesis) | **RBP carboxylase** fixes $CO_2$ to **RBP (Ribulose-bisphosphate);** glucose is produced and RBP is regenerated. |
| $C_4$ synthesis | **PEP carboxylase** fixes $CO_2$ to **PEP (phosphoenol pyruvic acid)** in mesocells; the resulting 4-carbon acid is |

| | |
|---|---|
| | moved to bundle sheath cells and releases $CO_2$ for the Calvin Cycle. |
| **CAM synthesis** | PEP carboxylase fixes $CO_2$ to PEP at night; during the day, the 4-carbon acid releases $CO_2$ for the Calvin Cycle. |
| **photorespiration** | In $C_3$ plants, oxygen is fixed to RBP instead of $CO_2$; reduces glucose production. |
| **chemosynthesis** | Oxidation of inorganic molecules provides energized electrons and hydrogens for forming carbohydrates. |

## Review and Synthesis

1. Rearrange the order to match the correct sequence of reactions during photosynthesis.

   | | |
   |---|---|
   | glucose formation | PGA production |
   | photolysis | chemiosmosis |
   | $CO_2$ fixation | absorption of |
   | noncyclic flow |    light |
   | cyclic flow | PGAL production |

   Are there any missing reactions? If so, place them in the appropriate location.

2. Arrange the following so that they are in proper sequence for cyclic photophosphorylation, and then arrange them for noncyclic photophosphorylation.

   electron transport system
   P680
   $H_2O$
   P700
   $NADP^+$
   electron-acceptor molecule

3. Why is photosynthesis described as having only two phases, the light reactions and the synthesis reactions, when there are more than 20 reactions in all?

4. How many RBPs and $CO_2$ are needed for the Calvin Cycle to produce six glucose molecules and still regenerate the original number of RBPs?

5. What would happen to other forms of life on earth if plants photosynthesized only as much glucose as they required for their own growth, development, and reproduction?

6. Consider an environment that is warm and dry for half the year, and then hot and wet the other half. Would $C_3$, $C_4$, or CAM synthesis be more adaptive? Why?

## Additional Readings

Baker, J., and G. Allen. 1970. *Matter, Energy and Life: An Introduction for Biology Students.* Addison-Wesley, Reading, Mass. (Introductory.)

Bjorkman, O., and J. Berry. 1973. "High-efficiency photosynthesis." *Scientific American* 229:80–93. (Intermediate.)

Childress, J., H. Felbeck, and G. Somero. 1987. "Symbiosis in the deep sea." *Scientific American* 256:114–21. (Introductory.)

Galston, A. W. 1964. *The Life of the Green Plant.* Prentice-Hall, Englewood Cliffs, N.J. (Introductory.)

Hinkle, P., and R. McCarty. 1978. "How cells make ATP." *Scientific American* 238:104–23. (Intermediate.)

Levine, R. 1969. "The mechanism of photosynthesis." *Scientific American* 221:58–70. (Intermediate.)

# Chapter 10

# Processing Energy: Respiration and Fermentation

Every breath you take sustains your life a little longer. It draws in life-supporting oxygen and eliminates the waste, carbon dioxide. Yet, few of us realize why we need oxygen. Oxygen releases the energy stored in the chemical fuel we consume as food. Without it the metabolic furnace that powers our life processes is extinguished. These oxygen-dependent processes bear a name that is often used synonymously with "breathing." This term, *respiration,* means much more than merely inhaling and exhaling atmospheric gases. Nearly all organisms, even those that have never taken a breath, rely on respiration as their primary energy-harvesting strategy. These organisms include all animals, plants, algae, protozoa, fungi, and most bacteria—every organism on the planet with the exception of a few prokaryotes. Clearly, respiration implies more than simply breathing.

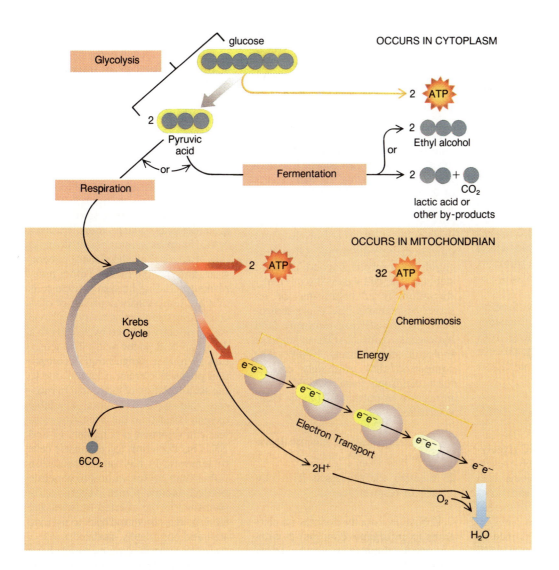

**Figure 10-1**

**Fermentation and respiration,** two optional pathways for harvesting the energy of glucose. Both strategies begin with glycolysis, the splitting of glucose into two molecules of pyruvic acid. (Each ball represents a carbon atom. *Fermentation* converts pyruvic acid to organic byproducts that still retain the carbon, electrons, and hydrogens (and therefore most of the energy) from glucose. *Respiration* completely oxidizes the carbons of glucose to $CO_2$ and extracts energy from the released electrons by sending them through the electron transport system. Respiration is a three-part operation: (1) glycolysis, (2) oxidation of pyruvic acid by the Krebs Cycle, and (3) phosphorylation by chemiosmosis at the electron transport system. The strategy selected depends on whether the organism possesses respiratory equipment and on the availability of oxygen (or another terminal electron acceptor for anaerobic respiration, which is not shown here).

## Preview of the Strategies

The basic strategy in energy-harvesting catabolism depends on converting food molecules to glucose or to a by-product of glucose degradation. The extent to which glucose is dismantled to release its energy is the basis for distinguishing between two fundamental processes, *fermentation* and *respiration*. The precise definition of these terms is best deferred until after you are briefly introduced to their differences. For now, we will refer to fermentation as the incomplete degradation of glucose without the participation of either oxygen or an electron transport system. Unlike fermentation, respiration completely oxidizes glucose to $CO_2$ and $H_2O$ using an electron transport system to extract energy from electrons and store it in the energy-rich bonds of ATP. This process usually requires the participation of oxygen and is therefore called **aerobic respiration** (aerobic = oxygen dependent). Some bacteria, however, are capable of **anaerobic respiration**, which also depends on an electron transport system but proceeds in the absence of oxygen.

Regardless of the strategy (fermentation, aerobic respiration, or anaerobic respiration), stripping energy from glucose begins with the same process: the splitting of glucose by glycolysis (Figure 10-1).

## Glycolysis

The multistep pathway called **glycolysis** (glyco = sugar, lysis = split) severs the 6-carbon glucose molecule into two molecular fragments, each containing three carbons. Each of the enzyme-mediated steps in this pathway redistributes the energy contained in the original glucose molecule until much of it is concentrated in four high-energy phosphate bonds attached to the metabolic intermediates. These phosphates, along with their chemical bond energy, are then transferred to ADP to generate four molecules of ATP. An abbreviated version of glycolysis appears in Figure 10-2.

The *net* ATP yield for glycolysis is two rather than four, because two ATPs must be spent just to get glycolysis "rolling," a process analogous to priming a pump. The input of this energy destabilizes the chemical bonds of glucose, decreasing their resistance to the changes dictated by the glycolytic enzymes. In addition to a net gain of two ATPs, glycolysis also generates two pairs of energized electrons (carried by two NADH s). The molecular remains of glucose following glycolysis are two molecules of pyruvic acid, each containing three carbons. This is the point where the three strategies diverge. They differ in the way they solve a common problem — how to recycle electron carriers ($NAD^+$).

In the absence of $NAD^+$, neither glycolysis nor the subsequent energy-generating processes can proceed. No other compound can accept the electrons released during the early stages of glycolysis. Yet, cells are equipped with only a limited amount of these coenzymes. The amount is so little that a normal rate of glycolysis would quickly deplete the supply. Glycolysis could not proceed beyond its third step, and the cell would die of energy starvation. In all organisms, the reserves of available $NAD^+$ are replenished by transferring electrons (and hydrogens) from NADH to another molecule, regenerating $NAD^+$. The compound that ultimately accepts these electrons from NADH determines two important metabolic characteristics:

1. The end products of glucose degradation.

2. Whether the process is fermentation or respiration.

If NADH directly donates its electrons to an organic compound, without passing them to an electron transport system, the process is called **fermentation**. If the final electron acceptor is a small molecule, like oxygen, that accepts electrons as they have traveled through a chain of cytochromes (an electron transport system), the process is termed **respiration** (sometimes referred to as *cellular respiration* to distinguish it from oxygen-

acquiring processes such as breathing). This distinction between fermentation and respiration may seem a bit bewildering to you now, so let's examine these molecular processes in enough detail to clarify the differences between them.

## Fermentation

In some fermentative organisms, the end product of glycolysis (pyruvic acid) accepts the electrons and hydrogen from NADH. In others, pyruvic acid is converted to an intermediate product before oxidizing NADH. In either case, $NAD^+$ is regenerated for reuse in glycolysis, and pyruvic acid is converted to another compound. These compounds, the end products of fermentation, vary according to the organism (Figure 10-3).

In the most familiar type of fermentation, yeast converts pyruvic acid to an intermediate compound, called acetaldehyde, which then oxidizes NADH and, as a result, is transformed into ethyl alcohol (the alcohol in beer, wine, and spirits). In some other organisms (including human muscle cells that are temporarily deprived of oxygen during strenuous activity), pyruvic acid is directly reduced by NADH to a 3-carbon acid (lactic acid). The sensation of muscle fatigue that accompanies physical exercise is largely due to the accumulation of lactic acid in muscles that have consumed available oxygen faster than the circulatory system can replenish it (Chapters 17 and 20). This depletion forces muscle cells to temporarily rely on fermentation rather than on respiration to generate needed energy. (Once oxygen is replenished in the affected muscles, the lactic acid is converted back to pyruvic acid and completely oxidized.)

For thousands of years, people have been using the end products of fermentation. The organisms that we rely on to supply these valuable products are our frequent allies, the microorganisms (see Bioline: The Fruits of Fermentation).

Regardless of the end product, all fermentations are *anaerobic* (oxygen-independent) processes, requiring nei-

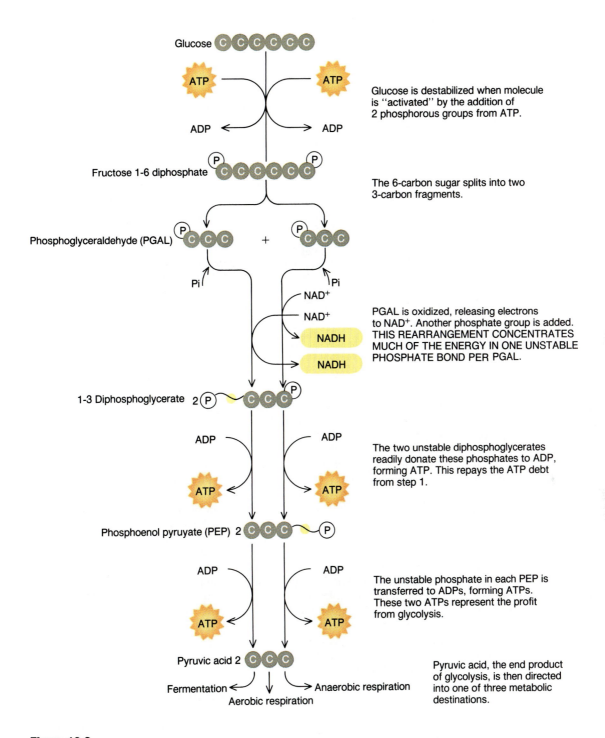

Glucose

Glucose is destabilized when molecule is "activated" by the addition of 2 phosphorous groups from ATP.

Fructose 1-6 diphosphate

The 6-carbon sugar splits into two 3-carbon fragments.

Phosphoglyceraldehyde (PGAL)

PGAL is oxidized, releasing electrons to $NAD^+$. Another phosphate group is added. THIS REARRANGEMENT CONCENTRATES MUCH OF THE ENERGY IN ONE UNSTABLE PHOSPHATE BOND PER PGAL.

1-3 Diphosphoglycerate

The two unstable diphosphoglycerates readily donate these phosphates to ADP, forming ATP. This repays the ATP debt from step 1.

Phosphoenol pyruyate (PEP)

The unstable phosphate in each PEP is transferred to ADPs, forming ATPs. These two ATPs represent the profit from glycolysis.

Pyruvic acid

Fermentation ← ↓ → Anaerobic respiration
Aerobic respiration

Pyruvic acid, the end product of glycolysis, is then directed into one of three metabolic destinations.

**Figure 10-2**
**A condensed version of glycolysis.** Although some of the 10 reactions have been omitted for clarity, the overall activities of the pathway are evident—the oxidation of glucose to two molecules of pyruvic acid with the release of two NADH and a profit of two ATPs. The number of carbon atoms at each step in the reaction sequence is six, the six carbons provided by glucose. Net reaction: Glucose ⟶ 2 pyruvic acid + 2 ATPs + 2 NADH

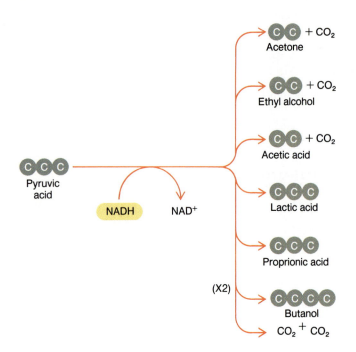

**Figure 10-3**
**Some metabolic products of fermentation.** Pyruvic acid may be converted to several end products, depending on the organism and, for some organisms, the absence of oxygen. In all cases, the final compound contains electrons (and hydrogen) donated by NADH. This recycles the coenzyme, allowing it to continue its essential role as electron acceptor in glycolysis. The usefulness of these fermentation products is discussed in the Bioline.

ther molecular oxygen nor an electron transport system. For many organisms that live in environments lacking oxygen, this is the only method of obtaining energy. Their total ATP yield is a mere two molecules per glucose oxidized. Most of the potential energy of glucose is left in the waste products, where it remains untapped at the end of fermentation. The energized electrons in NADH, which represent a potential for ATP production, are absorbed into this final fermentation waste product. An even greater reserve of energy remains in chemical bonds of pyruvic acid. This energy also ends up in the end product. In fact, more than 90 percent of glucose's chemical energy is untapped by glycolysis. (The flammability of ethyl alcohol, which is used as a fuel in automobiles, testifies to the high energy content of the molecule.)

An organism that could completely oxidize the two pyruvic acids generated by glycolysis could harvest this energy and obtain *an additional* 30 ATPs, plus 6 more from the two molecules of NADH. Compare this value to the 2 ATPs generated by glycolysis alone. Acquiring this enormous energy profit requires that electrons be processed by an electron transport system. Hence, only organisms equipped

with cytochromes can collect this additional profit. Yet this additional equipment is well worth the metabolic expense to make cytochromes. Respiration offers a 19-fold increase in energy efficiency over fermentation. Let's examine where all those additional ATPs came from.

## Respiration

Like fermentation, respiratory oxidation of glucose begins with glycolysis and generates two molecules of pyruvic acid. Unlike fermentative organisms, however, cells capable of respiration can harvest the energy in the two NADH molecules generated during glycolysis. They do this by channeling the reduced coenzyme's energy-rich electrons through the electron transport system, which coupled with chemiosmosis, generates ATP. In prokaryotes, respiration yields an additional six ATPs from the two NADH molecules released during glycolysis. In eukaryotic cells, however, the yield is less. Because glycolysis occurs in the cytoplasm, the energized electrons must be transported through the mitochondrion membranes to reach the electron transport system. This shuttling of electrons requires energy and, in most eukaryotic

the three ATPs that can be produced by the oxidation of each NADH. The two NADH molecules from glycolysis therefore provide an additional four molecules of ATP in these cells.

But this is just the tip of the efficiency iceburg—respiring organisms can then harvest the huge energy reserves in the two pyruvic acids that remain at the end of glycolysis. They do so by completely oxidizing the two pyruvic acids to six molecules of $CO_2$, using a cyclic metabolic pathway that, in eukaryotes, occurs in the mitochondria. This pathway is called the **Krebs Cycle**.

### Krebs Cycle

Before pyruvic acid enters the Krebs Cycle, one of its carbons is oxidized to $CO_2$, leaving a 2-carbon molecule, acetic acid (Figure 10-4, step 1). This process releases hydrogen and high-energy electrons that are picked up by $NAD^+$, forming NADH. In the process, acetic acid is temporarily coupled to a carrier molecule, coenzyme A (CoA), forming the complex called *acetyl coenzyme A* (acetyl CoA). The coenzyme forms a metabolic bridge between glycolysis and the Krebs

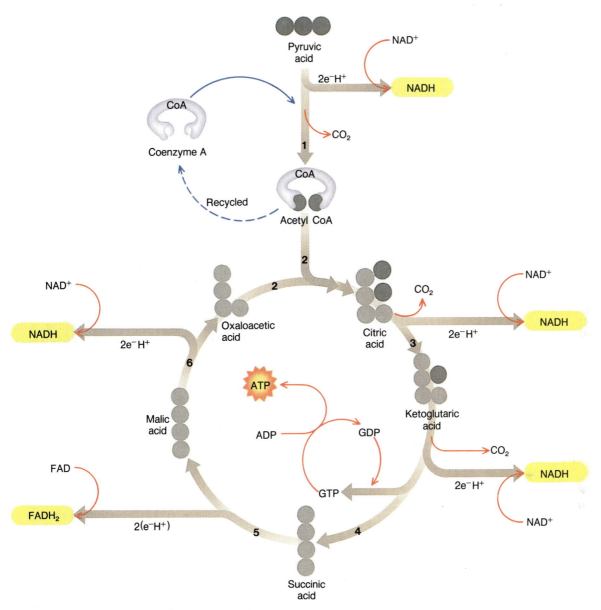

**Figure 10-4**
**An abbreviated version of the Krebs Cycle.** (See text for description of each step.) The overall function of the cycle is to completely oxidize pyruvic acid to $CO_2$ with the production of NADH and $FADH_2$, and one ATP directly. The energy in the coenzymes can be cashed in for ATPs at the electron transport system (via chemiosmosis). Because glycolysis generates *two* pyruvic acids, oxidation of a single glucose produces *twice* the amounts shown here. (Carbons from pyruvic acid are distinguished from those of oxaloacetate by shading.)

Cycle. (To minimize confusion, all future references to the Krebs Cycle will include this preparatory step.)

The Krebs Cycle itself begins when the two-carbons of acetyl CoA are joined with a 4-carbon molecule, oxaloacetic acid (step 2 in Figure 10-4). The resulting 6-carbon compound is *citric acid,* the substance that gives the Krebs Cycle its alternative name, the

"citric acid cycle." One of citric acid's two new carbons is then oxidized to a molecule of $CO_2$ in another NADH-producing reaction (step 3). The product of this reaction (keto-glutaric acid) contains the last carbon that remains of the original pyruvic acid. This carbon is removed in step 4, producing a 4-carbon molecule (succinic acid), another NADH, and the only ATP gen-

erated directly from the Krebs Cycle.*
Even though succinic acid contains no more carbons from pyruvic acid, it still retains two pairs of electrons and hydrogens from pyruvic acid that were

---

* Actually, a related compound, GTP, carries the energy to the ATP. GTP is guanine triphosphate, a high-energy compound that uses a different nucleotide base, guanine, instead of adenine.

not released when the $CO_2$s were removed. These electrons contain considerable energy. Oxidizing succinic acid to oxaloacetic acid releases these electrons, forming $FADH_2$ and NADH (steps 5 and 6). Oxaloacetic acid is then free to pick up another acetic acid from acetyl CoA to again form citric acid, starting the cycle again.

Each "turn" of the Krebs Cycle oxidizes pyruvic acid to its simplest, most oxidized form, $CO_2$, gradually releasing the usable energy that was in the original glucose molecule. The majority of this energy is transferred to several molecules of NADH and $FADH_2$. In most respiratory organisms, the key that unlocks this treasury of energy is a small, electron-attracting molecule — oxygen.

## The Value of Oxygen
The remarkable efficiency of aerobic respiration compared to fermentation reveals the advantage of using oxygen instead of pyruvic acid to relieve NADH of its electron cargo. Oxygen is an immensely more powerful oxidizing agent than is pyruvic acid or any

other organic compound. This means that much more energy will be released by oxidation with oxygen than with an organic compound. Oxygen returns the energized electrons to their lowest energy state, whereas fermentation leaves the electrons in a much more energized form, just as water falling into a pool close to a mountain's peak still retains most of its potential energy. Once a pair of electrons (and hydrogens) has been captured by oxygen, they have achieved their lowest energy state. (In the same way falling water has reached its lowest potential energy state once it reaches sea level.)

Cytochromes help harvest the energy released when electrons fall from their high-energy states in NADH or $FADH_2$ to their "sea level" state when accepted by oxygen. The released energy is coupled with chemiosmotic phosphorylation, producing several molecules of ATP (Chapters 8 and 9). But without an electron transport system, the electrons would have no biological means of energetically descending to oxygen. They would remain at a high-energy level, trapped in the byproducts of fermentation.

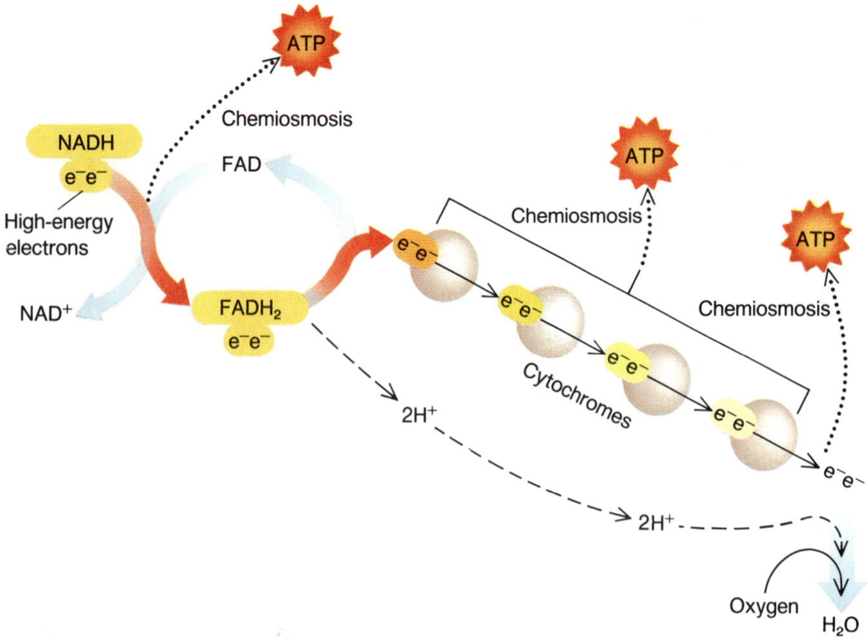

Whereas fermentation taps less than 3 percent of the energy stored in glucose, respiration captures about 40 percent of it. This 40 percent is enough energy to form 36 or 38 molecules of ATP for each glucose molecule oxidized. (The rest is lost as unusable energy, increasing entropy in the universe.) All but four of these ATPs come from energized electrons released during glucose oxidation and processed by the electron transport system. As illustrated in Figure 10-5, each NADH that donates electrons to the electron transport system generates three ATPs. The first of these phosphorylation events occurs with the energy released during the initial transfer, when NADH reduces FAD to $FADH_2$. Therefore, a pair of electrons introduced into the cytochrome chain by $FADH_2$ skips this first phosphorylation step and yields only two ATPs. It is important to remember this fact when we audit the cell's energy expenses and profits to pinpoint the transactions that generate such an impressive profit margin.

Oxygen acts as the **terminal electron acceptor** of the electron transport system. Without such a molecule to remove the electron pair from the final cytochrome, the entire respiratory chain would become choked with a backup of electrons, and the clogged system would grind to a halt. This is why you breathe — to provide a powerful oxidizing agent that will remove the electrons from the last cytochrome and keep the system functioning. Oxygen is essential to the respiration of all aerobic organisms, from bacteria to humans. Without it, ATP production drops to practically nothing. This quickly leads to energy starvation and fatal deterioration of biological order, unless the organism can switch to fermentation to provide the ATP needed to sustain a much slower pace of biological activity. Although such metabolic dexterity is common among microorganisms, fermentation does not provide enough energy to sustain the vast energy demands of larger multicellular organisms. In these organisms, the efficiency of respiration has created a biological dependency on it. We either respire or we die.

To better appreciate the efficiency

**Figure 10-5**
**Extraction of energy from electrons by the electron transport system and chemiosmosis.** Energy-rich electrons, introduced by NADH or $FADH_2$, are channeled through a sequence of cytochromes. Energy released from electrons during their journey to a lower energy state is used for chemiosmotic production of ATP (Chapter 8). The process yields three ATPs for each NADH oxidized or two ATPs for each $FADH_2$ oxidized.

## B I O L I N E

### THE FRUITS OF FERMENTATION

Biological events that take place in the absence of oxygen have been creating fortunes and livelihoods for people since the dawn of recorded history. The metabolic byproducts of fermentation make our diets more interesting, protect us from disease, retard food spoilage, and provide the mixed blessing of alcoholic beverages. Alcoholic drinks are the product of the conversion of pyruvic acid to ethyl alcohol by brewer's yeast. Alcoholic fermentation also releases $CO_2$, which accounts for the bubbling that accompanies the process. If this gas is not allowed to escape, it becomes trapped in the liquid, producing the natural carbonation of beer and champagne. The same process, alcoholic fermentation by yeasts, elevates dough during the leavening of bread and pastries. In fact, until the mid-1800s, commercial bakeries used yeast left over from the brewing of beer to leaven their pastries. The dried baker's yeast familiar to modern cooks contains living cells that quickly convert the sugar in dough to ethyl alcohol and $CO_2$. The $CO_2$ forms the expanding bubbles that cause the dough to rise; the dough literally inflates with the gas. The alcohol is driven off by baking, explaining why you can eat bread without fear (or hope) of inebriation.

Bacteria generate a wide variety of valuable fermentation products. Lactic acid is perhaps the most common of these. This slightly sour acid imparts flavor to yogurt, rye bread, and a few cheeses. It also lowers the pH of food below that which is tolerable by many spoilage microbes. Lactic acid-containing dairy products are therefore more resistant to spoilage than the milk from which they were made.

Swiss cheese is produced by other types of bacteria that convert pyruvic acid to propionic acid and $CO_2$. The acid imparts a nutty flavor to the cheese, while the gas creates the large bubbles that form its characteristic holes. Vinegar is a product of acetic acid-producing bacteria that grow in apple cider or grape mash. The tangy taste of pickled cabbage (sauerkraut), cucumbers (pickles), and olives reflects the presence of lactic and acetic acids, microbial byproducts that provide a flavorful twist to these foods. Soy sauce requires an eight- to twelve-month fermentation period before the mash of soy beans acquires its characteristic flavor.

Acetone and other industrially produced solvents are also byproducts of fermentation, as is isopropyl (rubbing) alcohol. Fermentation may be an inefficient means for an organism to harvest biological energy, but it has proven a very efficient source of valuable products.

---

of respiration, consider the end products of the process, $CO_2$ and $H_2O$. Carbon dioxide is the most oxidized form of carbon, which means that most of the chemical energy in the highly reduced glucose molecule has been released. More than 90 percent of the usable energy is harvested by oxygen-dependent respiration.

The water formed as a respiratory end product is generated as oxygen accepts two electrons from the last cytochrome:

$$\text{two electrons} + 1/2\ O_2 \longrightarrow O^{-2}.$$

The negative charges of this oxygen ion are immediately neutralized by two $H^+$:

$$O^{-2} + 2\ H^+ \longrightarrow H_2O.$$

The electron pair has been stripped of its energy, forming water, an extremely stable, unreactive compound.

### The Role of Mitochondria

Processing NADH and $FADH_2$ by the electron transport system is the final act of the three-act respiratory drama. (Glycolysis represents the opening event, and the Krebs Cycle the second act.) The relationship of these three steps and the location of their occurrence are depicted in Figure 10-6. The phosphorylation events in the third step require the peculiar double-membraned structure of mitochondria. The cytochromes are embedded in the inner membrane of this energy-generating organelle. The outer compartment serves as the holding tank for protons ($H^+$) pumped across the inner membrane from the matrix (the inner most compartment). Proton pores in the inner membrane provide channels

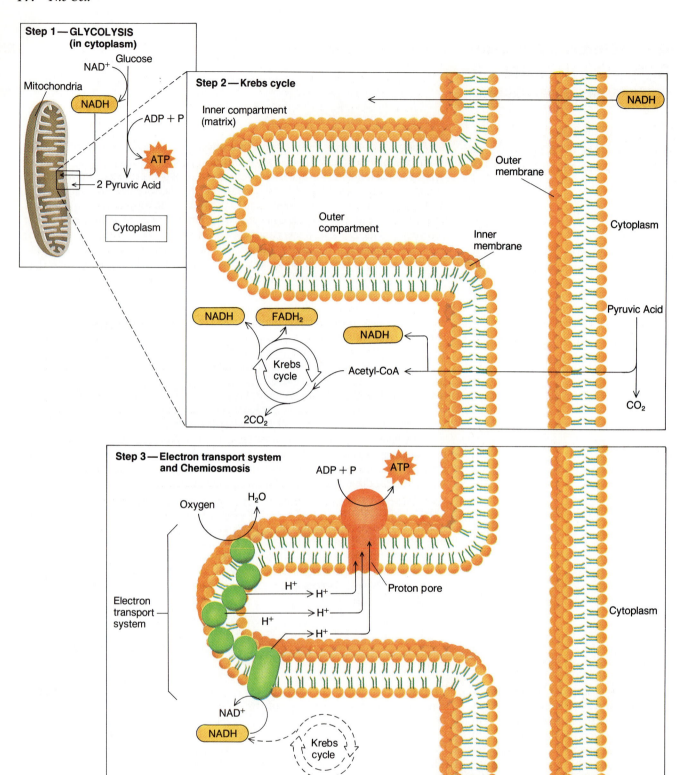

**Figure 10-6**

**The three parts of respiration.** Step 1. Glycolysis converts glucose to pyruvic acid in the cytoplasm, generating NADH and a small amount of ATP. Step 2. Pyruvic acid is moved into the mitochondrion as it is converted to acetyl CoA, which is then completely oxidized by the Krebs Cycle to $CO_2$, generating several molecules of NADH and one $FADH_2$. The NADH molecules from glycolysis also move into the mitochondrion. Step 3. The energized electrons carried by NADH and $FADH_2$ are processed by the electron transport system and coupled with chemiosmotic production of large amounts of ATP.

through which these protons stream back into the matrix to balance the concentration inequity. The power generated by this proton flow is used to generate ATP in the mitochondria. In prokaryotic cells, which lack mitochondria, the cytochromes and the chemiosmotic machinery are embedded in the plasma membrane.

## Balancing the Metabolic Books

We have already discussed the number of ATPs generated by glycolysis (page 139). Let us now examine the energy produced by the Krebs Cycle, the pathway that represents the most productive metabolic marketplace for generating energy profits. The Krebs Cycle directly manufactures only two ATPs, one for each pyruvic acid oxi-

dized. (Remember, each glucose produces *two* pyruvic acids.) Most of the sugar's energy is released in the form of reduced electron carriers, NADH and $FADH_2$. These reduced coenzymes are generated in the mitochondria where they can be cashed in for their full ATP values of three for each NADH and two for each $FADH_2$. Each pyruvic acid yields a *potential* of 15 ATPs. Twelve of these come from the four NADH molecules released by the Krebs Cycle (including the NADH released during the conversion of pyruvic acid to acetyl CoA), two more ATPs from $FADH_2$, and one produced directly from the Krebs Cycle. The total profit from oxidizing both pyruvic acids is therefore 30 ATPs. Add to this the six or eight ATPs from glycolysis (page 140), and we find that the aerobic oxidation of one glucose molecule liberates enough energy to

produce 36 or 38 molecules of ATP. More than 90 percent of these ATPs depend on the availability of oxygen as the terminal electron acceptor. A metabolic ledger pinpoints the source of each ATP in Figure 10-7.

This accounting represents the *potential* number of ATPs that each molecule of glucose provides an organism. But electrons are also needed for other purposes, such as biosynthesis or, in a more unusual example, generating light (Figure 10-8). Although these activities reduce the number of ATPs produced, their energy is not wasted, since it is absorbed into the reduced molecule or used to do biological work.

## Anaerobic Respiration

Some microorganisms that live in oxygen-devoid environments still make use of their respiratory machinery, converting the energy carried by the electrons of NADH into ATP. These organisms utilize a terminal acceptor other than oxygen (for example, nitrate or sulfate) as the oxidizing agent that removes spent electrons from the final cytochrome. **Anaerobic respiration**, as this process is called, is less efficient than aerobic respiration because none of these alternative terminal electron acceptors is as powerful an oxidizing agent as oxygen. Consequently, the electrons never achieve their lowest energy level, and they leave the cell with some of their energy still untapped. Nonetheless, the process is efficient enough to generate one or two ATPs for each NADH oxidized by the cytochromes, providing a definite improvement over fermentation. Many of these organisms have the additional capability to switch to aerobic (oxygen-dependent) respiration the instant oxygen becomes available. For some of these organisms, however, exposure to oxygen is fatal (see Bioline: Oxygen).

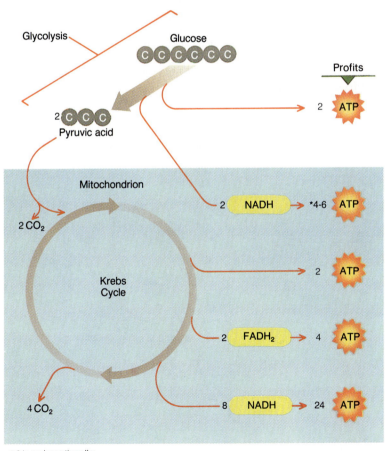

Glycolysis

Glucose

C C C C C C

Profits

2 ATP

2 C C C
Pyruvic acid

Mitochondrion

2 CO₂

Krebs Cycle

4 CO₂

2 NADH → *4-6 ATP

2 ATP

2 FADH₂ → 4 ATP

8 NADH → 24 ATP

Total = 36-38 ATP

* 6 in prokaryotic cells;
4 in eukaroyotic cells
(except heart and liver
which yield 6)

**Figure 10-7**
**A metabolic ledger** reveals the energy profits from oxygen-dependent respiratory oxidation of one glucose molecule. More than 90 percent of the ATPs are collected at the electron transport system. Strictly fermentative organisms lack not only an electron transport system, but also the Krebs Cycle. They therefore obtain only two ATPs per glucose fermented.

**Figure 10-8**
**Bioluminescence—a glowing tribute to respiration.** In some organisms, as much as 20 percent of the NADH donate their energized electrons to an electron transport system that releases the energy as light. Although the most familiar of these self-illuminating organisms is the firefly, other examples are illustrated here. *(a)* Bioluminescent microorganisms live in these corals, providing the animal its glow. *(b)* Other bioluminescent microorganisms are sometimes so common in sea water they cause the surface of the ocean to glow with an eerie light.

# Coupling Respiration with Digestion

So far, we have limited our discussion of energy to that obtained by glucose oxidation. That must seem enormously oversimplified to you, since you have probably never sat down to a meal of pure glucose. What about the hamburger and french fries that may be fueling your biological processes right now?

The central respiratory pathway we have described (glycolysis, Krebs Cycle, electron transport system) not only extracts energy from glucose, but also accepts molecules from other pathways. All digestible food molecules can be broken down to either glucose or to intermediates that can be channeled into either glycolysis or the Krebs Cycle for complete oxidation to $CO_2$ and water (Figure 10-9). For example, the protein in your hamburger is hydrolyzed to amino acids, which are absorbed into your cells and, if not needed for building new proteins in these cells, may be converted to various intermediates of glucose degradation. The 3-carbon amino acid alanine, for example, can be converted to pyruvic acid simply by removing its amino group. The pyruvic acid can then be introduced into the Krebs Cycle for complete oxidation. Removal of the amino group from glutamic acid (a 5-carbon amino acid) produces keto-glutaric acid, an intermediate of the Krebs Cycle. Each intermediate of glycolysis and the Krebs Cycle is treated by the corresponding enzyme exactly the same whether they came from glucose degradation or from the fat, protein, and starch of your hamburger. Thus, any food molecule that can be degraded or converted to an intermediate of these central respiratory pathways can be completely oxidized for their energy.

This principle also works in reverse. That is, virtually any compound which an organism can synthesize can be manufactured by diverting intermediates of glycolysis or the Krebs Cycle into the appropriate biosynthetic pathway. For example, much of the surplus acetyl CoA produced when glucose is abundant will be diverted to lipid synthesis for energy storage. In this way, an organism integrates its metabolic processes into a managed traffic pattern that can be directed according to the demands of the cell.

**Figure 10-9**
**Some common denominators in metabolism.** Virtually all the digestive pathways are linked to oxidative glucose metabolism by shared intermediates (shown here in squares). These intermediate compounds provide intersections that allow the digestive products of proteins, lipids, carbohydrates, and nucleic acids to be channeled into glycolysis or the Krebs Cycle for complete oxidation. Reversing the strategy provides the raw materials for biosynthesis by diverting these intermediates into pathways for constructing macromolecules when needed.

# B I O L I N E

## OXYGEN: THE EARTH'S DEADLY ENVELOPE

Most of us think of oxygen as a precious, life-sustaining substance. You probably remember brief experiences without it, moments of panic when you were submerged under water too long or when overexertion depleted your oxygen reserves to the extent that it seemed impossible for you to "catch your breath." Yet oxygen wasn't originally a welcome addition to the earth's atmosphere. In fact, its gradual appearance between 2 and 3 billion years ago may have been the most catastrophic single occurrence in the history of life on earth. It spelled extinction for millions of species. Creatures that had prevailed for millions of years disappeared forever, banished by oxygen, one of the most universally poisonous substances the world had ever known. Where did it come from? It was produced by a new type of microorganism, an evolutionary invader that changed the face of the planet even more dramatically than would the eventual reign of humans billions of years later. These devastating microbes were cyanobacteria (Chapter 31), the prokaryotes that "invented" oxygen-evolving photosynthesis.

Up to that time, all the earth's organisms were **anaerobes**, bacteria that depended on oxygen-free metabolic processes to obtain energy and nutrients. Bacterial photosynthesis, for example, utilizes hydrogen sulfide ($H_2S$) instead of water as a source of hydrogens and produces sulfur rather than oxygen as waste. This was the ancient order on the early oxygen-devoid earth. Following the rise of the cyanobacteria, however, so much oxygen infused the oceans and atmosphere that the old order was overthrown, inexorably altering the history of life on earth. The success of the strategy catapulted cyanobacteria to the top of the evolutionary hierarchy of their day, a domination that lasted more than 2 billion years.

Eventually, the prevalent forms of life not only withstood the poisonous effects of oxygen, but actually become dependent on it, using it to achieve enormous increases in their metabolic efficiency. These evolutionary innovators were the first aerobes — the first users of oxygen. They were in fact probably the cyanobacteria themselves. They had electron transport systems that took advantage of oxygen's powerful attraction for electrons to extract additional energy from their food. This revolutionary development, aerobic respiration, led to an ability to utilize not only the food they manufactured by photosynthesis, but the waste ($O_2$) as well. Respiration worked so well that some of these early aerobes lost their photosynthetic machinery and consumed the food produced by other organisms, using their new respiratory powers to harvest the energy. In the process, they generated wastes ($CO_2$ and $H_2O$) which photosynthetic organisms needed to produce food, food for themselves, and food for virtually all the organisms on earth. This flow of carbon between autotrophs and oxygen-using heterotrophs established the fundamental cycle that allowed such a generous diversity of life to eventually enrich the earth. In addition, these early aerobes started an evolutionary line that ultimately produced all oxygen-using heterotrophs, including humans.

Living with oxygen is still a dangerous business. It is a little like living with fire: you need special protection to prevent it from burning you up. In many biological reactions, oxygen is converted to even more powerful oxidizing agents, called *superoxide radicals* ($O^-$), that combine with essential cellular biochemicals in the cytoplasm. Such uncontrolled oxidation of cytoplasm is always fatal. The survival of all aerobes depends on protective enzymes that disarm superoxide radicals by converting them to water the instant they are formed.

Early bacteria didn't need protection from oxygen, since virtually none of the gas was around at the time. Some of their descendants still survive in a few oxygen-devoid environments, such as aquatic sediments, hot sulfur pools, and the interior recesses of your intestinal tract. But most of them died with the appearance of oxygen, which favored the survival of organisms with the enzymatic endowment for quickly defusing oxygen's powerful destructive tendencies. Without this protection, which you inherited from these early prokaryotes, the oxygen in your next breath would bring certain death instead of sustaining your life.

## Synopsis

### MAIN CONCEPTS

- **Glycolysis and the Krebs Cycle** provide the metabolic means of using the food produced by photosynthesis and chemosynthesis. They harvest energy by breaking down food to glucose or to an intermediate of glycolysis or the Krebs Cycle.

  *Glycolysis:* the first (and in some organisms the only) pathway for glucose metabolism. Located in cytoplasm, it converts glucose to two molecules of pyruvic acid (3 carbons), 2 ATP's and 2 NADH (a potential yield of 8 ATP's).

  *Krebs Cycle:* the pathway in respiratory organisms that completely oxidizes the two pyruvic acids from glycolysis to 6 $CO_2$ molecules. Located in mitochondria in eukaryotes (cytoplasm in prokaryotes), its major function is to generate reduced coenzymes for biosynthesis and energy production. It yields 2 ATP, 8 NADH and 2 $FADH_2$ per glucose oxidized, or a potential yield of 30 ATP's.

- **Fermentation** is the partial oxidation of sugar (usually glycolysis) and no electron transport system. The two pairs of high-energy electrons generated by glycolysis are discarded without extracting the energy. These electrons are donated directly to an organic molecule, which is usually the terminal electron acceptor. At the end of fermentation, more than 90 percent of glucose's chemical energy remains in discarded end products.

- **Respiration** is the complete oxidation of sugar to $CO_2$ and $H_2O$ using an electron transport system and a small (usually) inorganic molecule such as oxygen, as the terminal electron acceptor. It is a three-part operation—glycolysis, the Krebs Cycle, and the electron transport system. Glycolysis and the Krebs Cycle oxidize glucose to $CO_2$. In addition, a few ATPs are produced by these two pathways, but most of glucose's chemical energy is released as high-energy electrons in coenzymes, which are processed at the electron transport system. This system extracts energy from the electrons brought there by NADH or $FADH_2$ and, coupled with chemiosmosis, uses it to produce ATP (3 per NADH; 2 per $FADH_2$). The resulting low-energy electrons are removed from the final cytochrome by oxygen in aerobic respiration or by an acceptor molecule other than oxygen in anaerobic respiration.

### KEY TERM INTEGRATOR

| | |
|---|---|
| **glycolysis** | The splitting of glucose into two molecules of *pyruvic acid*. It is the first step in energy-harvesting catabolism. **Fermentation** uses the partially disassembled sugars to accept the electrons released during oxidation. |
| **respiration** | Completely oxidizes glucose, using a combination of glycolysis and the **Krebs Cycle**, processing the released electrons by running them through an electron transport system for phosphorylation. In **aerobic respiration**, these electrons are finally accepted by molecular oxygen. **Anaerobic respiration** also uses an electron transport system for phosphorylation, but a small molecule other than oxygen is the **terminal electron acceptor**. |

# Review and Synthesis

1. Rank the following substances as oxidizing agents, going from the least electron-attracting to the most: $NAD^+$, oxygen, FAD, glucose, cytochromes.

2. Why would an organism with respiratory equipment ever resort to fermentation? Why can't you switch to fermentation to sustain yourself indefinitely in similar circumstances?

3. Each ATP captures about 1 percent of the energy released by the complete oxidation of a glucose molecule. Calculate the energy efficiency of respiration, assuming a maximum of 36 ATPs are produced. What happens to the rest of the energy? How many times more energy efficient is cellular respiration than an automobile (cars convert only 25 percent of gasoline's released energy into the mechanical energy of motion)?

4. Remembering the details of electron transport and chemiosmotic production of ATP, why is the mitochondrion a double-membraned structure?

5. If half a cell's NADH is oxidized by $NADP^+$ to be used for biosynthesis, how many ATPs could be produced by complete oxidation of glucose to $CO_2$ and $H_2O$?

6. Fermentation is one solution to the problem of relieving NADH of its electron cargo, so that the coenzyme can be recycled as an electron acceptor in glycolysis. How has respiration turned this problem into an enormous asset?

7. Although neither anaerobic respiration nor fermentation requires oxygen, why is anaerobic respiration considered true respiration, whereas fermentation is not?

8. Cyanide, the respiratory poison used in the gas chamber, binds to one of the cytochromes and blocks the flow of electrons. The lethal effects of cyanide are therefore similar to all but one of the following: drowning, oxygen deprivation, suffocation, oxygen toxicity. Circle the exception.

# Additional Readings

Dickerson, R. 1980. "Cytochrome *c* and the evolution of energy metabolism." *Scientific American* 242:136–53. (Intermediate.)

Hinckle, P., and R. McCarty. 1978. "How cells make ATP." *Scientific American* 238:104–23. (Intermediate.)

Lechtman, M., B. Roohk and R. Egan. 1978. *The Games Cells Play: Basic Concepts of Cellular Metabolism.* Benjamin–Cummings Co. (Intermediate.)

Lehninger, A. 1982. *Principles of Biochemistry.* 2nd ed. Worth Publishing Co. (Advanced.)

Tribe, M., and P. Whittaker. 1972. *Chloroplasts and Mitochondria.* Crane, Russak and Co. (Advanced.)

Chapter 11

# Cell Division: Mitosis

Every organism starts life as one cell. Once you were just a single fertilized egg, the product of the union of gametes, a sperm from your father and an egg from your mother. Of course, unicellular organisms live their entire lives as one cell. But in a multicellular organism like yourself, countless divisions of the original fertilized egg build the organism to an astonishing level of cellular complexity and organization. During this period of growth, many cells embark on a course of specialization that commits them to perform specific functions. Some cells function in cell division—either they divide to produce gametes for sexual reproduction, or they divide to make new cells for growth or to replace old and damaged cells. Thus, cell division is at the very core of life itself. It enables organisms to grow, reproduce, and repair damaged and worn tissues—three fundamental activities of life.

New cells originate only from other living cells. The dividing cell is called a **mother cell**, and its descendants are called **daughter cells**. There is a reason for using these "familial" terms. The mother cell transmits copies of its hereditary information (DNA) to its daughter cells, the next cell generation. In turn, the daughter cells can become mother cells to their own daughter cells, passing along the same genes as the original mother cell to yet another generation of cells.

For hereditary information to be perpetuated from generation to generation, DNA, the carrier of the genetic information, must be *replicated* (duplicated) before the cell divides so that each new daughter cell receives a complete copy of the hereditary instructions. (How DNA replicates is described in Chapter 27.) Since DNA is part of a cell's chromosomes (Chapter 6), the chromosomes duplicate as well. After chromosomal duplication, the remainder of the division activities proceed in a way that ensures each daughter cell receives the same share of genetic instructions, as well as a portion of the cell's cytoplasm and organelles.

## Types of Cell Divisions

Simple prokaryotic cells divide by a process called **binary fission** in which the replicated DNA and the mother cell's cytoplasm become equally split between two new daughter cells (Figure 11-1). Since prokaryotes lack a nuclear membrane, their DNA is attached directly to the cell's plasma membrane. At the beginning of binary fission, DNA replicates, with the two DNA molecules attached to slightly different points on the plasma membrane. A new section of plasma membrane grows between these attachment points, separating the DNA molecules. A partitioning plasma membrane and cell wall then develop between the DNA molecules to form the two prokaryotic daughter cells.

There are four reasons why cell division is more complicated in eukaryotic cells than binary fission in prokaryotic cells.

1. Eukaryotic cells house greater amounts of DNA than do prokaryotic cells. The quantity of DNA in a human cell, for example, is 700 times greater than that in any prokaryotic cell. Eukaryotic cells have greater stores of DNA because each eukaryotic cell in a multicellular organism is descended from the same mother cell (the fertilized egg) and therefore contains the full repertoire of genetic instructions for growing an entire organism. A cell in your big toe, for instance, has all the genes for coding for a completely new you.

2. Eukaryotic cells contain anywhere from 2 to more than 1000 chromosomes (compared to a single strand of DNA in prokaryotic cells). In order to ensure that each daughter cell receives an exact share of chromosomes during cell division, eukaryotic cells construct a complicated rigging structure called the **spindle apparatus** for organizing and separating duplicated chromosomes.

3. Chromosomes are housed in a membrane-bound nucleus in eukaryotic cells, and a new nucleus must be manufactured for each daughter cell during cell division.

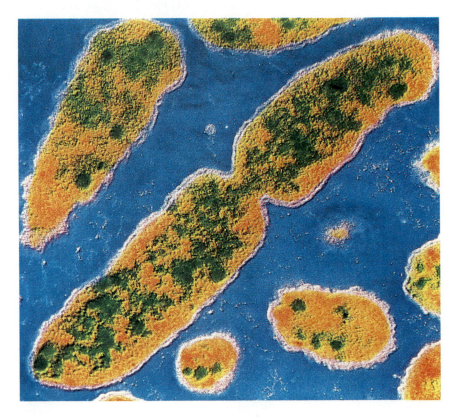

**Figure 11-1**
**Binary fission of a prokaryotic cyanobacterium** Duplicated DNA has already been separated, and the crosswall that will eventually finalize cell division is partially manufactured.

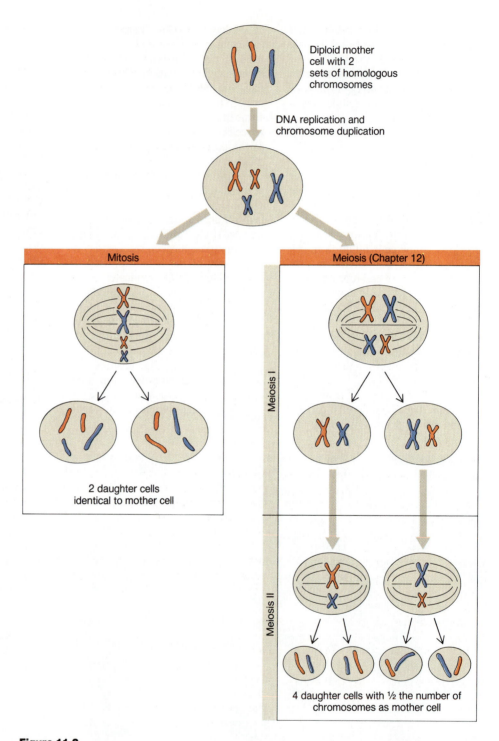

Diploid mother
cell with 2
sets of homologous
chromosomes

DNA replication and
chromosome duplication

**Mitosis**

**Meiosis (Chapter 12)**

Meiosis I

Meiosis II

2 daughter cells
identical to mother cell

4 daughter cells with ½ the number of
chromosomes as mother cell

**Figure 11-2**
**Mitosis versus meiosis.** Mitosis results in two daughter cells that have exactly the same number
of chromosomes as the mother cell. However, in meiosis, the mother cell produces four daughter
cells with half the number of chromosomes. (Meiosis is discussed in Chapter 12.)

4. The organelles in the cytoplasm of eukaryotic cells must be duplicated and then sorted more or less equally between the new daughter cells.

Because of these complications, biologists separate the events in eukaryotic cell division into two stages —karyokinesis and cytokinesis. During **karyokinesis** (karyo = nucleus, kinesis = division), the nucleus and its contents (including duplicated chromosomes) divide to form new nuclei. The cell's cytoplasm and the new nuclei are then partitioned into separate daughter cells during **cytokinesis** (cyto = cytoplasm).

Dividing eukaryotic cells undergo one of two types of karyokinesis, mitosis and meiosis (Figure 11-2). (Although mitosis and meiosis specifically refer to division of the nucleus, like most biologists we use these terms for the entire process of cell division, including cytokinesis.) **Mitosis** produces *two* daughter cells with exactly the same number of chromosomes as the mother cell. The daughter cell chromosomes are precise copies of those of the mother cell, giving them exactly the same genetic potential as the mother cell. Thus, mitosis is the type of cell division that maintains the chromosome number and generates new cells for growth, maintenance, and repair of an organism.

How often does mitosis take place? Some estimate that in each person's body more than 25 million mitotic cell divisions occur *each second*. In some organisms, including many plants, mitosis is also a means of producing an entirely new organism through **asexual reproduction**, that is, reproduction without the union of male and female gametes. The offspring produced by asexual reproduction have exactly the same genes as their parent.

**Meiosis** produces *four* daughter cells, each with one-half the number of chromosomes as the mother cell (Figure 11-2). Cells with one-half the number of chromosomes are called **haploid** or **1n** cells. The haploid daughter cells eventually form egg or sperm gametes for sexual reproduction. When these sex gametes fuse during fertilization, a **diploid (2n)** cell, called a zygote, is formed with a different genetic potential than that of either parent. The importance of sexual reproduction and the details of meiosis are discussed in the next chapter.

# The Mechanics of Eukaryotic Cell Division

The earth has cradled life for more than 3.5 billion years. For the first 2.5 billion years (almost three-quarters of the total time there has been life on earth), prokaryotes were the only form of life and binary fission was the only form of cell division. But two critical breakthroughs, both of which involved cell division, sparked the evolution of eukaryotic cells and the eventual proliferation of multicellular organisms. First, DNA became complexed with proteins to form eukaryotic chromosomes. The evolution of chromosomes provided a means to package huge stores of hereditary information needed for constructing a complex, multicellular organism. It also allowed these large stores to be condensed so that they could be moved during cell division. The second breakthrough was the evolution of a spindle apparatus, the cellular rigging for attaching and moving chromosomes during cell division. Without a spindle apparatus, precise separation of duplicated chromosomes would be virtually impossible.

## Chromosomes

Recall from Chapter 6 that chromosomes are visible only during cell division; the remainder of the time an organism's hereditary information appears as a tangle of threadlike material called chromatin. In preparation for division, the chromatin coils, compacting enormous lengths of DNA into chromosomes that can be separated and moved during cell division. The importance of coiling becomes clear when you consider that the DNA of a single human chromosome is 30,000 times longer than it is when it is coiled. Imagine the molecular chaos inside a human cell during the separation of 92 uncoiled chromosomes (46 chromosomes and their duplicates). Inevitable tangling would pull the uncoiled DNA to pieces, forming a jumble of DNA fragments that would be unevenly sorted into the daughter cells. The result would be daughter cells with an incomplete set of genetic blueprints. No species could survive disorder of this magnitude.

### THE STRUCTURE OF CHROMOSOMES

Chromosomes are complexes of DNA, which carries the genetic blueprints for an organism, and proteins (Figure 11-3). The proteins play two important roles. Some help to package and condense large amounts of DNA, whereas others control whether or not hereditary information is expressed. (How genes are expressed is covered in Chapter 28.)

At the beginning of cell division, each chromosome appears as two thick "rods," pinched together at one point (Figure 11-3b). Each "rod" is called a **chromatid**, and the point of attachment is the **centromere** (Figure 11-3c). Chromatids are genetic duplicates of each other, exact copies that were constructed during DNA replication. When a cell divides, the centromere splits, allowing chromatids to disconnect into two "sister" chromosomes. Each "sister" chromosome moves to an opposite end of the cell and becomes partitioned into a new nucleus and finally into a separate daughter cell.

### CHROMOSOME PAIRS

In eukaryotic cells, chromosomes occur mostly in pairs, each chromosome having a partner that is almost identical in appearance (Figure 11-3b). The two similar-shaped chromosomes are called **homologues** (homo = same, log = writing) because each has exactly the same sequence of genes along its length as the other. Homologues are not the same as chromatids. Chromatids are exact duplicates of the *same* chromosome, whereas homologues are different chromosomes.

Cells that contain homologous pairs of chromosomes are **diploid** cells (di = two, ploid = multiple). Diploid

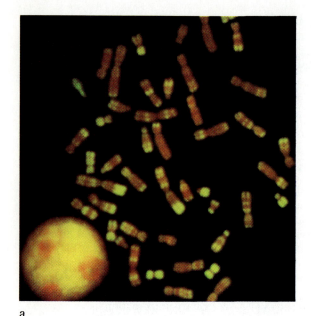

a

b

Duplicated chromosome

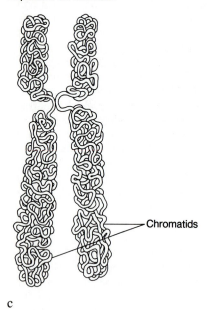

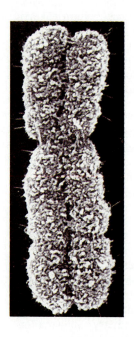

Chromatids

c

**Figure 11-3**
**Human chromosomes.**
(a) Each body cell contains 46 chromosomes in its nucleus.
(b) Each chromosome has a lookalike called a *homologue.* The 46 human chromosomes can be paired, homologue to homologue, producing 23 pairs of *homologous chromosomes.*
(c) Before a cell divides, each chromosome duplicates itself. A duplicated chromosome is made up of two identical chromatids, held together by a centromere.

**6.** What two innovations related to cell division led to the evolution of eukaryotic cells and the rapid increase in multicellular organisms? Why were these events so important?

**7.** Using pieces of colored yarn or crayons, set up the mitosis template (below) on a large sheet of paper and then physically run through the phases of mitosis. This will not only improve your understanding of the activities that take place in each step of mitosis, but will also help you learn and remember the details.

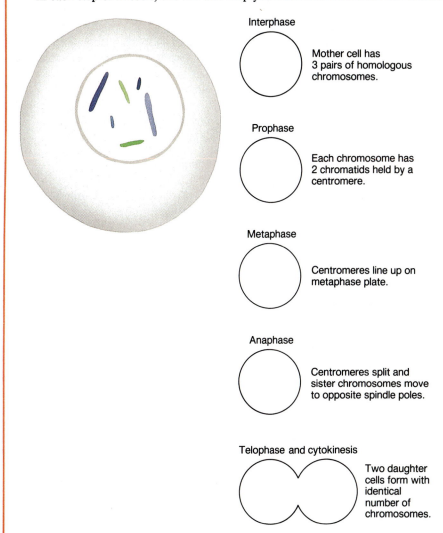

Interphase

Mother cell has 3 pairs of homologous chromosomes.

Prophase

Each chromosome has 2 chromatids held by a centromere.

Metaphase

Centromeres line up on metaphase plate.

Anaphase

Centromeres split and sister chromosomes move to opposite spindle poles.

Telophase and cytokinesis

Two daughter cells form with identical number of chromosomes.

## Additional Readings

John, P. (ed.). 1981. *The Cell Cycle.* Cambridge University Press, Cambridge, England. (Advanced.)

Maiza, D. 1974. "The cell cycle." *Scientific American* 230:54–64. (Introductory.)

Readings from *Scientific American.* 1986. "Cancer biology." W. H. Freeman, San Francisco. (Introductory.)

Sloboda, R. 1980. "The role of microtubules in cell structure and cell division." *American Scientist* 68:290–98. (Intermediate.)

Chapter 12

# Cell Division: Meiosis

*E*ach organism is a living showcase displaying thousands of inherited features. When you glance in a mirror, you see hundreds of inherited characteristics —the length and thickness of your eyelashes, the width of your smile, the depth of a dimple. All the genetic instructions that created you are filed in only 46 chromosomes. Actually, two complete sets of genetic instructions are stored in every cell of your body. One set (in 23 chromosomes) is derived from your father's sperm, and the other set (the remaining 23 chromosomes) originated from your mother's egg. At conception, the sperm nucleus fused with the egg nucleus to produce a cell (zygote) with 46 chromosomes. This cell and all of its progeny have dutifully duplicated those 46 chromosomes each time they divided, eventually growing into a remarkable cell collection—you.

But since your parents are also collections of cells containing 46 chromosomes, how did they manufacture sperm and egg gametes with only 23 chromosomes—exactly half the number of chromosomes? Not by mitosis, for mitosis produces daughter cells with exactly the same number of chromosomes as the mother cell (see Figure 11-2). The type of cell division that divides the number of chromosomes in half and leads to the formation of gametes for sexual reproduction is meiosis.

## Meiosis and Sexual Reproduction

Reproduction is one of the fundamental characteristics of life (see Connections: Asexual Versus Sexual Reproduction). The sequence of events during the lifetime of an organism, from the moment it is formed to when it reproduces, forms what biologists call the **life cycle** of an organism. For unicellular bacteria that reproduce asexually, their life cycle is very simple. When they grow to a certain size, they divide by binary fission to form two new individuals, each of which then begins a new life cycle.

On the other hand, most multicellular organisms reproduce sexually.

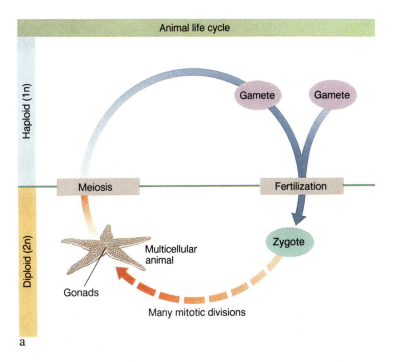

a

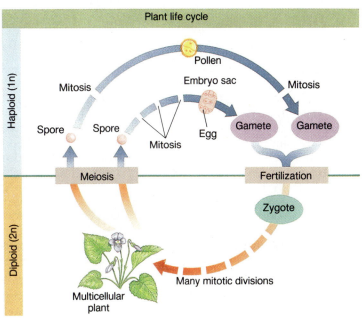

b

**Figure 12-1**
**Sexual life cycles.** Sexual reproduction requires both meiosis and fertilization. Meiosis reduces the number of chromosomes in half to form haploid cells that develop into gametes. Gametes fuse during fertilization to form a zygote (sometimes called a "fertilized egg").
*(a) Animal life cycle.* In animals, the haploid daughter cells produced by meiosis directly form gametes. After fertilization, numerous mitotic divisions of the zygote produce a multicellular adult with gonads containing germinal tissues that divide by meiosis. The life cycle of most animals is predominantly diploid.
*(b) Plant life cycle.* In plants, the haploid daughter cells produced by meiosis form spores. The spores divide by mitosis to form a multicellular, haploid structure (in flowering plants, a pollen grain and embryo sac) that houses the sperm and egg gametes. Following fertilization, many mitotic divisions of the zygote produce a multicellular diploid plant with germinal tissues that, like in animals, divide by meiosis, completing the cycle. Unlike animals, however, the life cycle of plants alternates between two distinct multicellular generations. (We explore alternation of generations in plants more in Chapters 21 and 33.)

They form haploid gametes that combine during fertilization to form a single, diploid zygote. Thus, meiosis and fertilization are critical events in the sexual life cycles of organisms because both processes change the number of chromosome sets in cells. Meiosis generates haploid cells that eventually form gametes, and fertilization produces a diploid zygote. Alternating between a haploid phase and a diploid phase during a sexual life cycle is known as **alternation of generations**.

In organisms that reproduce sexually, meiosis always precedes the formation of gametes. Meiosis takes place only in diploid **germinal tissues**, which are so called because they produce **germ cells** (or **gametes**). (In this case, the word "germ" does not refer to a disease-causing organism.) Germinal tissues are found in the **gonads** of animals—the female ovaries and the male testes. In most plants, germinal tissues are located in the ovary and anthers of flowers. All remaining cells that make up the body of an animal or plant (all cells except the germinal cells) are **somatic cells**. Somatic cells often divide by mitosis to produce identical daughter cells for growth or repair, but they never divide by meiosis.

The way gametes are produced is different in animals than in plants. In animals, meiosis produces daughter cells that directly specialize to form gametes (Figure 12-1a). In plants, meiosis produces haploid daughter cells that form *spores* (Figure 12-1b). The plant spores then divide by mitosis, growing into a multicellular haploid plant that produces the sex gametes. Sexual reproduction in plants and animals is discussed further in Chapters 21 and 23, respectively.

## Importance of Meiosis

If gametes contained the same number of chromosomes as other body cells, chromosome number would double with each generation. For humans, this would mean that the first generation of offspring would have 92 chromosomes (46 from the sperm plus 46

CONNECTIONS — Formation of gametes (Chapter 21)
Alternation of generations (Chapter 33)
Animal Reproductive strategies (Chapter 23)

## ASEXUAL VERSUS SEXUAL REPRODUCTION

All organisms reproduce, but why are there two forms of reproduction—asexual and sexual reproduction? Why don't all organisms simply produce an exact, yet younger, version of themselves? This is, in fact, what happens during asexual reproduction. Offspring produced by asexual reproduction result from repeated mitotic divisions of a single parent. Because each daughter cell receives an exact copy of the mother cell's chromosomes during mitosis, asexually produced offspring are genetically identical to their parent and, as a result, develop identical inherited traits. Logically, if a parent is successful in surviving and reproducing in a particular habitat, identical offspring should also be successful. For example, when a single aphid lands on an unoccupied plant, it can quickly reproduce hundreds of identical offspring, each of which is equally capable of exploiting the new food resource. In addition to being an effective means of quickly populating an area, asexual reproduction is a successful means of reproducing in uniform environments where conditions do not change much from one place to another, or remain stable over long periods of time.

Most of the earth's habitats are not uniform, however. Conditions not only change over time (the repeated ice ages, for instance), but they also vary from one location to another. A stroll through a forest, for example, reveals very different environmental conditions between cool, shady areas beneath dense trees, and the very warm, bright conditions of open meadows with no trees. In variable environments, reproducing identical offspring by asexual reproduction is risky, for if the parents are ill equipped for a change, then all the identical offspring produced by asexual reproduction will be similarly ill equipped. A change in the environment could possibly kill all individuals, causing the species to disappear from earth forever.

But if each member of a species has slightly different traits, chances are improved that some individuals may possess characteristics that enable them to survive and reproduce in a new habitat or following a change in the environment. But what causes this variation among members of the same species?

Since genes code for inherited characteristics, the greater the variation in genes among members of a species, the greater the variation in their characteristics. Sexual reproduction increases genetic variation for three reasons.

1. In meiosis, new combinations of genes are produced when corresponding chromosome segments exchange places (crossing over).

2. Homologous chromosomes separate independently during Metaphase I of meiosis, mixing maternal and paternal chromosomes in new combinations (independent assortment).

3. During *fertilization*, a sperm fuses with an egg to form a genetically unique zygote.

The combination of crossing over, independent assortment, and fertilization quickly builds genetic variability to enormous levels. Genetic variation is the foundation of evolutionary change, enabling species to adapt to new environmental conditions, which is a fundamental prerequisite for the continuation of life on earth.

from the egg), the second generation 184, the third 368, and so on. As you can imagine, cells would become so overloaded with surplus chromosomes after just a few generations that they would soon stagnate and die. Thus, for organisms that reproduce sexually, meiosis is crucial for the perpetuation of life. By reducing the number of chromosomes in half (to one complete set, instead of two sets as in diploid cells), meiosis ensures that the chromosome number of a species remains the same from generation to generation.

But the importance of meiosis goes beyond maintaining chromosome number from one generation to the next. Meiosis also builds *genetic variability* in a species. That is, meiosis increases the variation of genes, which in turn increases the range of characteristics in individuals that make up the species. A glance at a crowd of people instantly reveals the enormous variability that can exist within the human species. A greater range of characteristics improves the chances that some individuals will survive environmental changes. Throughout the history of life on earth, there have been a great many environmental changes, from worldwide shallow seas, to mountain building, to repeated ice ages. (The effects of environmental changes on the perpetuation of species are discussed in Chapters 29 and 30.)

Genetic variability increases during meiosis because chromosomes and genes are reshuffled and then distributed to haploid daughter cells in new combinations. As a result, each of the resulting gametes has a different genetic mixture. When gametes combine during fertilization, the organism that is produced is different from all others of the species. This explains why you *resemble* members of your family, yet do not look identical to them. In fact, because of meiosis, the odds against two people being genetically identical are astronomically slim, approximately 1 chance in $1 \times 10^{9031}$ (1 followed by 9,031 zeros; there isn't even a word for a number that large!) The one exception to this is identical twins. Since identical twins develop

from the same zygote, they share identical genes and therefore look virtually alike.

# Phases of Meiosis

There are many similarities between meiosis and mitosis (Chapter 11). Even the names of the stages are the same (prophase, metaphase, anaphase, and telophase). Although the basic division process is similar, there are clear differences, however. We would expect these differences, knowing that mitosis maintains chromosome number while meiosis divides the number in half.

Like mitosis, meiosis is a continuous process. Biologists divide this process into phases, each of which is marked by changes in chromosome position or in the characteristics of the dividing cell.

One fundamental difference between mitosis and meiosis is in the number of cell divisions. Instead of one division as in mitosis, meiosis has two: **meiosis I** and **meiosis II** (see Figure 11-2).

Both meiosis I and II are subdivided into prophase, metaphase, anaphase, and telophase (identified as prophase I, metaphase I, etc., and prophase II, metaphase II, etc.). The interphase period separating meiosis I and II is usually short, for unlike the interphase that precedes prophase I or that of mitosis, new chromosomes are not synthesized during this interphase. The entire sequence of meiotic phases is shown in Figure 12-2.

## Meiosis I

Meiosis I is called the **reductional division** of meiosis because the daughter cells produced at the end of telophase I contain half the number of chromosomes as the mother cell. This reduction in number of chromosomes is achieved by separating the homologues of each pair of homologous chromosomes. As a result, the daughter cells contain only one complete set of chromosomes instead of two sets as in the diploid mother cell. But before homologues become separated, some

very important chromosome choreography takes place during prophase I. This choreography increases genetic variability.

### PROPHASE I

Unlike the prophase of mitosis which usually lasts only a few minutes, prophase I can take hours, days, and sometimes even years. During this time, homologues join partners, and, while connected, often exchange segments of their chromosomes with each other. How do these chromosome interactions come about?

Before prophase I begins, chromosomes have already duplicated during interphase to form two chromatids united by a centromere. When chromosomes first appear during the early stages of prophase I, the chromatids are bunched together so closely that they appear to be a single strand rather than two joined chromatids (Figure 12-2a). The cell churns the chromosomes, tumbling homologues over one another. When homologous partners bump into each other in just the right way, they pair off (Figure 12-2b). This pairing of homologous chromosomes is called **synapsis**. Since each homologue is made up of two chromatids, a synapsed pair of homologous chromosomes forms a **tetrad**, which is a unit of four chromatids (Figure 12-3).

It is during synapsis that homologues exchange segments with each other. This process is called **crossing over**. During crossing over, one chromatid of one homologue overlaps with a chromatid of the other homologue. Since chromatids generally overlap in more than one spot, they form cross-shaped junctions called **chiasmata** (Figure 12-4). At each chiasma, regions of homologous chromatids containing matching genes may fragment into sections and then switch places with one another.

Exchanging chromatid pieces during crossing over can produce new gene combinations on a chromosome. Recall from Chapter 11 that homologous chromosomes have identical sequences of genes along their length. But each homologue may not necessarily code for the same characteristic

**Figure 12-2**
**The stages of meiosis.**

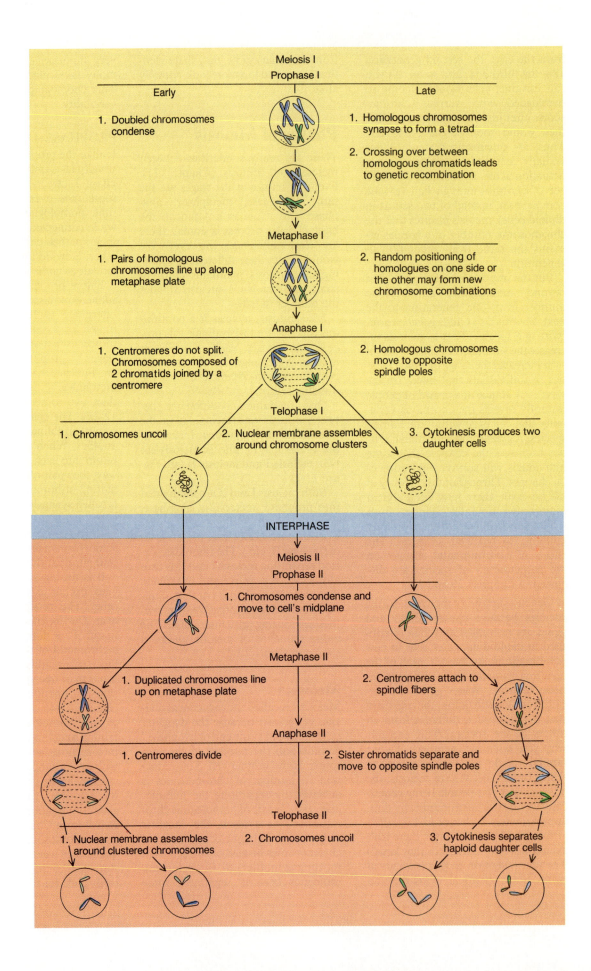

for a particular gene. For example, consider the gene for height in plants. One homologue may code for a tall individual, whereas the other codes for a short plant. Both chromosomes have a gene for height at the same location, but each produces a different trait.

Chromosomes carry thousands of genes, not just one. For this reason, crossing over between homologous chromatids can produce different gene combinations than were originally found in either chromatid. Let's see how crossing over can produce new gene combinations in homologous chromosomes with genes for height at one location and genes for flower color at another. We will represent these chromosomes as follows:

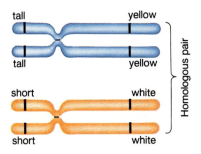

Here, one homologue (on top) directs the development of a tall individual with yellow flowers, whereas its homologue produces a short plant with white flowers. Since the gene for height is physically linked with the gene for flower color by being on the same chromosome, all tall offspring will have yellow flowers and all short ones will have white flowers. Where one gene goes, the other one follows when both occur on the same chromosome.

Crossing over can reshuffle gene combinations, a process known as **genetic recombination**. In our example of plant chromosomes, crossing over and recombination team the gene for a tall plant with the gene for white flowers on one chromatid, and the gene for a short plant with the gene for yellow flowers on the homologous chromatid shown below.

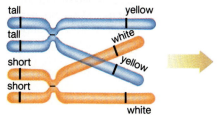

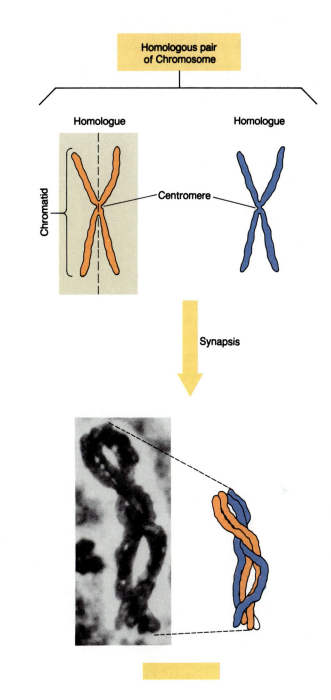

**Figure 12-3**
**Homologous chromosomes forming a tetrad during synapsis.**

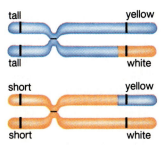

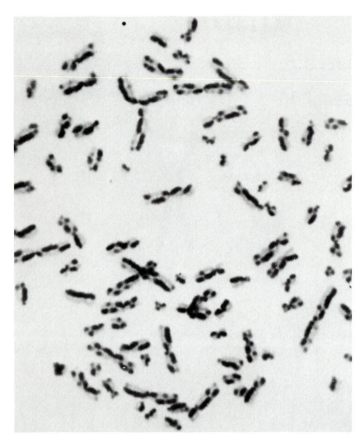

**Figure 12-4**
**Synapsed homologous chromosomes.** Chiasmata, points where chromatids cross each other, are indicated by arrows. Exchanging chromosome segments at chiasmata results in genetic recombination, which in turn may produce new gene combinations on a chromatid, thereby increasing the genetic variability of the species.

Notice that the recombined genes are found on only two chromatids. The original gene combinations remain on the two chromatids that did not participate in crossing over.

Typically, multiple exchanges take place between coupled homologues. During prophase I in human meiosis, for example, an average of 10 crossing-over exchanges occur for each synapsed pair of homologous chromosomes.

After crossing over and recombination are complete, synapsed chromosomes begin repelling each other. Additional coiling also forces the homologues further apart, sliding chiasmata down to the ends of the chromosomes. The final prophase I activities include completing con-

struction of the spindle apparatus and disassembling the nuclear membrane and nucleolus. These activities set the stage for metaphase I.

## METAPHASE I
At the beginning of metaphase I, paired homologous chromosomes line up on the metaphase plate. Each chromosome is flanked by its homologue on the opposite side of the plate (Figure 12-2c). The centromeres of each duplicate homologue attach to a spindle fiber that leads to an opposite spindle pole.

When homologous chromosomes line up on the metaphase plate, each homologue can line up on one side or the other of the metaphase plate, de-

pending entirely on random chance. Hence, all the chromosomes from one parent do not necessarily line up on the same side of the metaphase plate, so they don't always end up in the same daughter cell. Put another way, the chromosomes of each parent are sorted independently during metaphase I. This **independent assortment** of homologues usually shuffles homologous chromosomes to form new chromosome combinations in the daughter cells which later produce offspring with random mixtures of traits from both parents. Depending on the number of chromosomes, independent assortment can produce an enormous potpourri of possible mixtures (Figure 12-5). For a human, this chromosome potluck works out to

(a) Distribution of 2 pairs of homologous
chromosomes by independent assortment

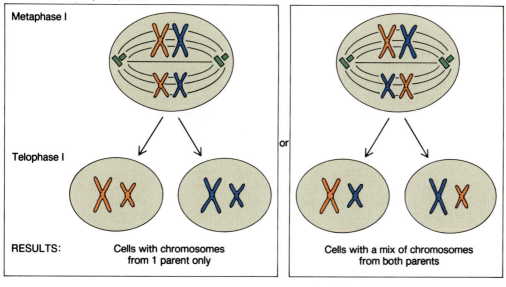

(b) Distribution of 3 pairs of homologous
chromosomes by independent assortment

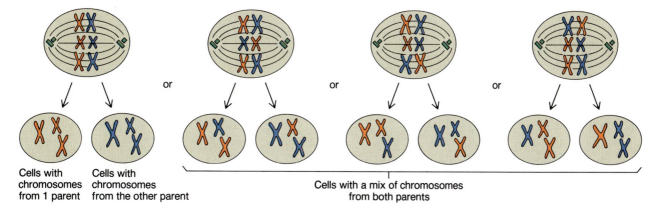

**Figure 12-5**
**Independent assortment of homologous chromosomes during metaphase I.** Paired homologous chromosomes line up at random along the metaphase plate. Because different pairs of homologous chromosomes line up independently, new chromosome combinations are often produced in the daughter cells, leading to increased genetic variability in a species. (To follow independent assortment, the homologues of one parent are colored blue, and those from the other parent are red.)
(a) In a cell with two pairs of homologous chromosomes, there are two possibilities for chromosome alignments during meiosis I. Either both chromosomes from the same parent go into the same daughter cell, or chromosomes are mixed and each daughter cell receives one chromosome from one parent and the other chromosome from the other parent.
(b) In a cell with three pairs of chromosomes, there are eight possible combinations. Because of independent assortment during meiosis I, the number of possibilities of chromosome alignments increases with each additional chromosome. In human cells with 23 pairs of chromosomes, there are about 8.5 million possible chromosome combinations.

more than 8 million ways the 23 chromosomes from your father can combine with the 23 chromosomes from your mother during metaphase I. Add to this the gene exchanges that occur during crossing over, and it is easy to see how meiosis vastly increases the genetic variability of a species.

### ANAPHASE I

The events of anaphase I are much simpler than those of prophase I and metaphase I. The function of anaphase I is to separate the homologues of each pair of homologous chromosomes, thereby reducing the number of chromosome sets in half in the daughter cells that form during telophase I.

Anaphase I begins when the spindle fibers attached to the centromeres of each chromosome shorten and pull duplicated homologues to opposite spindle poles (Figure 12-2d). Unlike anaphase of mitosis, the centromere does not split during anaphase I of meiosis. Each homologous chromosome is therefore still made of two chromatids joined by a centromere. When separated homologues cluster near their spindle pole, anaphase I is completed and telophase I begins.

### TELOPHASE I AND INTERPHASE

Telophase I is almost identical to the telophase of mitosis. In most organisms, the chromosomes uncoil and a nuclear membrane assembles around each chromosome cluster. Cytokinesis divides the cytoplasm into two independent daughter cells, each of which has one set of homologues (Figure 12-2e). Both daughter cells then begin interphase.

Because DNA does not replicate during interphase between telophase I and prophase II, the period separating meiosis I and II is usually very brief. Each chromosome enters prophase II as two chromatids united by a centromere. If centrioles are present, they migrate to opposite poles during this interphase in preparation for the more rapid, straightforward second division, meiosis II.

## Meiosis II

Since sister chromatids become separated during meiosis II, the events in meiosis II are very similar to those in mitosis. During **prophase II**, chromosomes shorten and thicken as they journey to the cell's equator to begin metaphase II (Figure 12-2f). At **metaphase II**, the centromeres line up on the metaphase plate (Figure 12-2g). **Anaphase II** begins when the centromeres split and chromatids separate into sister chromosomes. The sister chromosomes are then pulled toward opposite spindle poles (Figure 12-2h). When the chromosomes reach opposite poles, anaphase II is completed and telophase II begins. During **telophase II**, a nuclear membrane is assembled around each of the four clusters of chromosomes. The cells then divide by cytokinesis to produce four haploid daughter cells (Figure 12-2i). Each daughter cell contains one chromosome from each homologous pair of the original diploid cell and, therefore, has *one* complete set of hereditary instructions.

# Synopsis

## MAIN CONCEPTS

- **Cells divide either by mitosis or meiosis.** Unlike mitosis which maintains chromosome number, meiosis reduces the number of chromosomes in half by separating the homologues of a diploid mother cell into four haploid daughter cells.

- **Meiosis produces cells that ultimately form gametes for sexual reproduction.**

- **Because meiosis produces haploid cells,** the number of chromosomes of species is maintained from generation to generation.

- **Organisms that reproduce sexually** alternate between haploid and diploid generations during their life cycles. Meiosis and fertilization change the number of chromosome sets in cells and thus mark the transitions between these generations.

- **Genetic recombination and independent assortment** during meiosis increase the genetic variability of a species, enhancing the chances that some members of the species will survive inevitable environmental changes.

## KEY TERM INTEGRATOR

### Sexual Life Cycles

| | |
|---|---|
| **diploid zygote** | Formed when haploid gametes fuse during fertilization. |
| **alternation of generations** | Alternation between haploid and diploid phases during a sexual **life cycle** resulting from meiosis and fertilization. |
| **germinal tissues** | Cells that divide by meiosis to form **germ cells**, or gametes. Germinal tissues are located in the **gonads** of animals. |
| **somatic cells** | All body cells that do not divide by meiosis. |

### Steps of Meiosis I and II

| | |
|---|---|
| **prophase I** | Homologous chromosomes form a **tetrad** during **synapsis**. Chromatids of homologous chromosomes overlap, forming **chiasmata**. **Crossing over** of chromatid segments may occur at chiasmata, leading to **genetic recombination** and increased **genetic variability**. |
| **metaphase I** | **Independent assortment** of paired homologues as they line up on the metaphase plate often increases genetic variability of a species. |
| **anaphase I** | Paired homologues separate and move to opposite poles. Centromeres of duplicated chromosomes do not split. |
| **telophase I and cytokinesis** | Separated homologues are partitioned into two daughter cells. |
| **Interphase** | Brief period with no replication of genetic material. |
| **prophase II** | Chromosomes condense and move to center of cell. |
| **metaphase II** | Centromeres line up on the metaphase plate. |
| **anaphase II** | Centromeres split, and sister chromatids move to opposite poles. |
| **telophase II and cytokinesis** | Four haploid daughter cells form, each with exactly half the number of chromosomes as the mother cell. |

# Review and Synthesis

1. Complete the sexual life of a squirrel by filling in the following terms.

   zygote           meiosis
   sperm            germinal tissues
   egg              many mitotic divisions
   fertilization

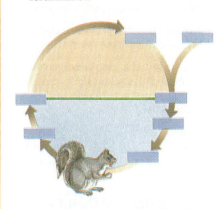

2. Match the activity with the phase of meiosis in which it occurs.

   ___ a. synapsis                           1. prophase I
   ___ b. crossing over                      2. metaphase I
   ___ c. centromeres split                  3. anaphase I
   ___ d. independent assortment             4. prophase II
   ___ e. homologous chromosomes separate    5. anaphase II
   ___ f. cytokinesis                        6. telophase II

3. How do crossing over and independent assortment increase the genetic variability of a species?

4. Why is only meiosis I referred to as the reductional division?

5. Circle the activities that occur during prophase I which explain why this phase takes hours or even days to be completed.

   formation of spindle apparatus    nucleolus forms
   synapsis                          cleavage
   independent assortment            crossing over
   chiasmata form                    DNA replication
   nuclear membrane disassembled     cell plate from

6. Why do we need both meiosis and mitosis for survival of our species?

7. Using colored pencils, pieces of yarn, crayons, or any other rod-shaped object to represent chromosomes, set up the following template on a large sheet of paper and then physically run through the phases of meiosis. The exercise involves a mother cell with three pairs of homologous chromosomes (six chromosomes). Make homologous chromosomes equal in length, but use one color for the three homologues that originated from the father and another color for homologues from the mother. This exercise not only improves your understanding of the activities that take place in each of the steps of meiosis, but also helps you learn and remember the details.

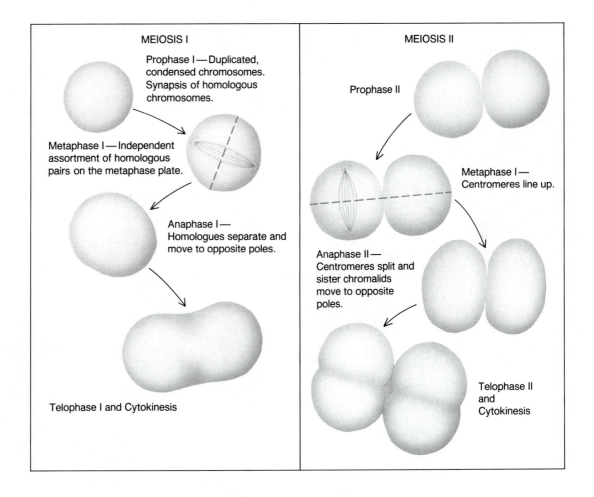

**MEIOSIS I**

Prophase I—Duplicated, condensed chromosomes. Synapsis of homologous chromosomes.

Metaphase I—Independent assortment of homologous pairs on the metaphase plate.

Anaphase I— Homologues separate and move to opposite poles.

Telophase I and Cytokinesis

**MEIOSIS II**

Prophase II

Metaphase I— Centromeres line up.

Anaphase II— Centromeres split and sister chromalids move to opposite poles.

Telophase II and Cytokinesis

**Additional Readings**

John, B. 1976. "Myths and mechanisms of meiosis." *Chromosoma* 54:295–325. (Intermediate.)

Leff, D. 1983. "The way of a sperm with an egg." In *The Cell. Inter- and Intra-relationships. NSF Mosaic Reader.* Avery Publishing, Wayne, N.J. (Introductory.)

Maiza, D. 1961. "How cells divide." *Scientific American* 205:100. (Introductory.)

# 4

# *Multicellular Organization of Plants*

Flowers capture the attention of passing animals (mostly insects). As animals gather nectar, they transfer sperm-containing pollen from flower to flower for sexual reproduction.

Chapter 13

# Plant Tissues and Organs

A natomically, plants are less complex than most animals. However, less complexity does not mean inferiority. The simpler anatomy of plants is a perfect example of how form is often exquisitely adapted to an organism's way of life. Because a plant's energy source (sunlight) flows to them, plants do not need all of the complex anatomical equipment to lure, hunt, chase, seize, dismember, or digest their energy source that animals do. The abundance of worldwide plant life confirms that plants are perfectly suited for their autotrophic lifestyle. The dependence of animals on plants for food reaffirms that in the biological world greater complexity does not mean increased superiority; survival is what counts.

## The Basic Plant Design

The earth's 400,000 plant species display a wide range of **morphology** (external body form) and **anatomy** (internal cell structure). A plant can be as simple as a mat of nearly identical cells, or as complicated as a giant redwood composed of billions of specialized cells. With such a broad range of structural variation, there is no "typical" plant design. In this chapter, we focus our discussion on the body plan of flowering plants, which are the most familiar plants and the plant group with the largest number of species. (We describe the other plant groups in Chapter 33.) Flowering plants are also the most complex plants. All are **vascular plants**, that is, they have specialized conducting cells that form a circulatory system of vessels and tubes for transporting water, minerals, and food and other organic molecules from one region to another.

Botanists subdivide flowering plants into two main groups: the **dicotyledons** or dicots (di = two, cotyledon = embryonic seed leaf) and the **monocotyledons** or monocots (mono = one). Table 13-1 illustrates several morphological and anatomical differences that distinguish monocots from dicots. (New terms in the table are explained later in the chapter.)

**TABLE 13-1**

| **Dicot and Monocot Comparison** | | |
|---|---|---|
| | **DICOT** | **MONOCOT** |
| Embryo | 2 Cotyledons | 1 Cotyledon |
| Flowers | Parts in 4s or 5s | Parts in 3s |
| Leaves | Net veined | Veins parallel |
| Roots | One main root (tap root system) | Many main roots (fibrous root system) |
| Stem anatomy | Vascular bundles in rings (Vascular bundle) | Scattered vascular bundles |
| Root anatomy | Xylem in center | Pith |
| Secondary growth | Yes | No |

Monocots and dicots share the same general body plan composed of two main parts: a **shoot** and a **root** (Figure 13-1). The shoot system consists of stems, leaves, flowers, and fruits. Interconnecting vascular tissues transport materials throughout the shoot and match up with the root's vascular tissues, so that absorbed water and minerals can be conducted to shoot tissues, and food produced by the shoot can be transported to nourish root tissues.

Leaves are attached to stems at **nodes**; the leaf-free area between two nodes is an **internode** (Figure 13-1).

Directly above each leaf on the stem is always an **axillary bud**, an undeveloped shoot. Axillary buds either develop into flowers or grow into new stems bearing additional leaves and more axillary buds. Generally, leaves have a large, flattened **blade** for collecting sunlight for photosynthesis, and a **petiole**, a stock that connects the blade to the stem.

The root system is made up of hundreds, and sometimes thousands, of branching roots. Water and minerals are absorbed through **root hairs**, elongated surface cells near the tip of each root. Once absorbed, water and minerals are conducted through vascular tissues to the shoot. In addition to absorption and conduction, the root system often helps physically support the shoot and may act as a depository for stored food.

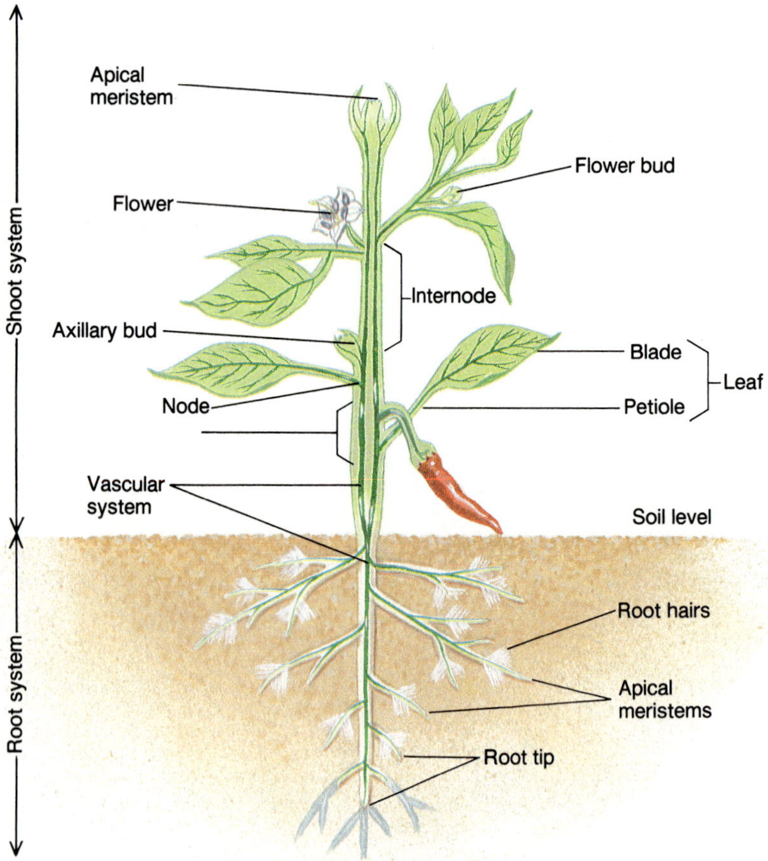

## Annuals, Biennials, and Perennials

The longer a plant lives, the more complicated its anatomy. **Annuals** are plants, such as grasses and herbs, that live for one year or less. They are anatomically simpler than longer lived **biennials**, plants that live for two years, or **perennials**, plants that live longer than two years. Biennials and perennials can live longer because they produce new cells to replace those that cease functioning or die. Because each plant cell has a surrounding wall (Chapter 6), old plant cells do not just disappear when they die. They leave cell "skeletons" where they once lived. Although these persistent cellular remains may complicate anatomical architecture, many perform special functions even after they die. Cells that conduct water and minerals, for example, or those that insulate and protect delicate living tissues, can do so only after they die.

## Plant Tissues

As in other many-celled organisms, the cells of a multicellular plant work together in cooperative groups to form **tissues**. **Simple tissues** are made up of the same type of cells, whereas **complex tissues**, such as the vascular tissues described later in this chapter, are composed of different kinds of cells working together to carry out a particular function.

Botanists recognize four major categories of simple tissues in plants— *meristems, parenchyma, collenchyma,* and *sclerenchyma.*

## Meristems

**Meristems** are clusters of cells that divide by mitosis to produce new cells for growth. Each time a meristematic cell divides, at least one daughter cell remains meristematic, preserving the meristem. The other daughter cell

**Figure 13-1**
**Basic plant design.** A plant body has an aerial shoot system and subterranean root system. The highly branched root system absorbs water and minerals from the soil, while the shoot produces stems and leaves for gathering sunlight and extracting $CO_2$ from the air for photosynthesis. The shoot also produces flowers for sexual reproduction and fruits to help disperse seeds. All flowering plants have a network of interconnecting vascular tissues for transporting water, dissolved minerals, and organic molecules.

may continue to divide for a time but eventually differentiates into a specialized cell.

All shoots, axillary buds, and roots have **apical meristems**, groups of meristematic cells located at their apexes (tips) (review Figure 13-1). Divisions of apical meristem cells increase the length of these structures. Growth from apical meristems is defined as **primary growth**, and the cells and tissues produced by apical meristems form the plant's **primary tissues**. Annual plants have only primary growth and thus are composed entirely of primary tissues. These plants are called **herbaceous**.

Like annuals, biennials and perennials have apical meristems for primary growth. But in addition to apical meristems, biennials and perennials develop **secondary meristems** or **cambia** (singular = cambium), rings and clusters of meristematic cells that increase the width of stems and roots when they divide. Growth from cambia is called **secondary growth**, and the cells produced by these secondary meristems form the plant's **secondary tissues**. Thus, a pine or oak tree (or any other perennial) has both primary growth from apical meristems at the tips of stems and roots, and secondary growth from cambia in between (Figure 13-2).

Long-lived plants form two types of cambia: a vascular cambium and cork cambium. The **vascular cambium** produces additional vascular tissues that (1) increase the transport capacity of the vascular system in order to accommodate the plant's increase in size, and (2) replace aging or damaged conducting cells. The **cork cambium** produces cork cells that replace outer protective tissues damaged as stems and roots expand during secondary growth. Most secondary tissues are hard and woody because the cells are compact and have walls impregnated with lignin, a hardening agent.

Together, apical and secondary meristems impart potential immortality to perennial plants. Theoretically, a perennial can grow forever by continually substituting new cells for aging and nonfunctioning cells, an ability animals do not have. This potential for continuous growth is called *open*

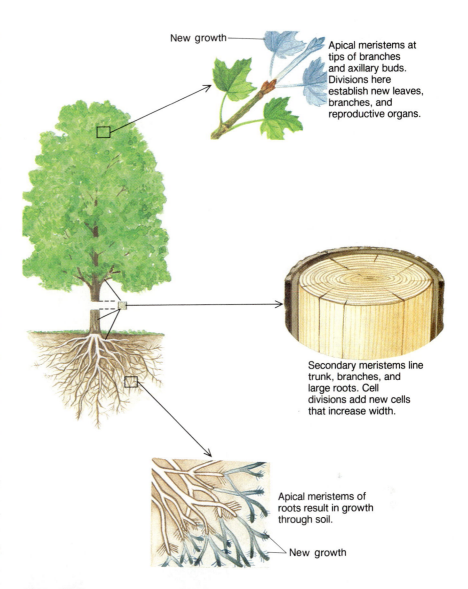

New growth

Apical meristems at tips of branches and axillary buds. Divisions here establish new leaves, branches, and reproductive organs.

Secondary meristems line trunk, branches, and large roots. Cell divisions add new cells that increase width.

Apical meristems of roots result in growth through soil.

New growth

**Figure 13-2**
**Growing taller and wider at the same time,** perennial trees and shrubs have both apical meristems and secondary meristems. Growth from apical meristems at the ends of stems and roots produces new primary tissues, while secondary meristems replace worn and damaged tissues, as well as increase the transport capacity of the vascular system.

*growth.* It enables some plants to live for hundreds and even thousands of years (see Bioline: Nature's Oldest Living Organisms).

## Parenchyma

The most prevalent type of cell in an herbaceous plant is **parenchyma** (Figure 13-3). Parenchyma cells form expanses of simple tissues and team up with other kinds of cells to form complex tissues.

Parenchyma cells carry out many plant functions, including

- Photosynthesis.
- Storage of food, water, and pigments.
- Transport of nutrients and other materials between certain areas.
- Formation of replacement tissues for healing of wounds.
- Generation of new roots, stems, and other plant parts.

With such a variety of functions, it's not surprising that parenchyma cells vary a great deal in size, shape, and even cell wall characteristics, although the majority are generally roundish and have thin walls. Despite such variation, parenchyma cells are distinguished from all other plant cells in their ability to differentiate into any other kind of plant cell. Thus, parenchyma cells are one of the most versatile cells in the plant.

## Collenchyma

**Collenchyma** cells are elongated, cigar-shaped cells with unevenly thickened cell walls (Figure 13-3). Collenchyma cells are found almost exclusively in primary tissues where they help support growing plant parts. In stems, collenchyma tissues lie just beneath the epidermis, helping to keep the stem upright.

## Sclerenchyma

The thick secondary walls of **sclerenchyma** cells make them an even stronger strengthening and supportive cell than collenchyma. Sclerenchyma cells come in a variety of sizes and shapes. Some are long and thin, whereas others are shaped like dumbbells, hourglasses, and even stars. Since they provide support or physical protection, the critical component of a sclerenchyma cell is its cell wall, not the living protoplasm. Sclerenchyma cells therefore continue to function even after the cell dies.

Sclerenchyma cells are divided into two categories: fibers and sclereids. All **sclerenchyma fibers** are extremely long, narrow cells with uniformly thickened walls (Figure 13-3). The cell wall allows fibers to bend and then spring back to their original position. When flexible sclerenchyma fibers are strategically positioned in stems, petioles, leaves, and flowers, these structures can bend or flutter, yet immediately reorient. Such elasticity is particularly important in leaves. When winds die down, for example, leaves quickly reorient, exposing the

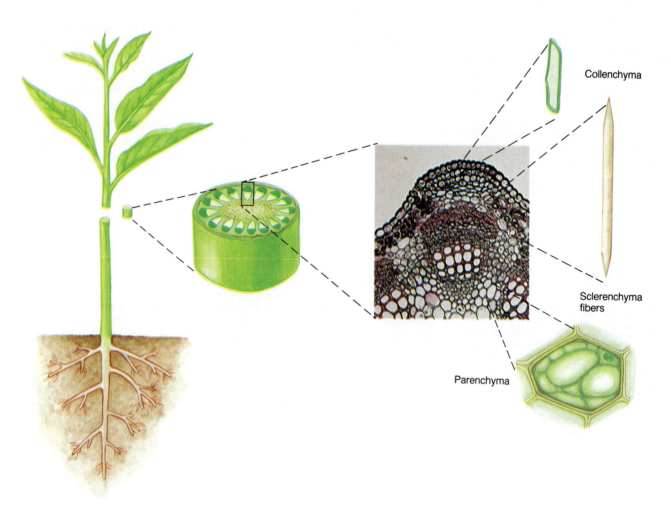

**Figure 13-3**
**Simple primary tissues.** Roundish parenchyma cells fill most of the stem. Elongate collenchyma cells with thickened walls are found immediately under the epidermis, and dense collections of sclerenchyma fibers form a cap of supportive cells over each vascular bundle.

# B I O L I N E

## NATURE'S OLDEST LIVING ORGANISMS

a

b

Imagine living for 5000 years. If this were possible, you would have been born around 3000 BC at the beginning of human civilization itself. Then, our ancestors were just starting to practice agriculture, build cities, develop art, and scratch out the first forms of writing. During your 5000-year lifetime, you would have lived through the grandeur of the Roman Empire and seen it fall into ruin; you would have survived the constraints of the Middle Ages to experience the glory of the Renaissance; you would have witnessed the discovery of North America and survived tragic world wars; and you would have "heard" the footsteps of people walking on the moon and experienced the beginning of an age of computer technology.

But humans cannot live that long; vital cells simply wear out and cannot be replaced. (Human nerve cells, for example, live for about 100 years; they cannot be replaced.) For the same reason, other animals live no longer than a few hundred years. But many plants have an ability no animal has — theoretical immortality. Because plants are able to continually replace worn-out cells with new ones, they maintain a constant supply of young, efficient cells, providing themselves potential agelessness.

For many years, the reigning "Oldest Living Organism" has been the Bristlecone pine *(Pinus longaeva)* with a documented age of 4900 years (see photo inset *a*). In fact, it was a Bristlecone pine that lived through all of those milestones in the development of human civilization. Strangely, these ancient plants do not live in a rich environmental setting. Instead, they grow on windbeaten, desertlike slopes of the White Mountains in eastern California.

But the reign of the Bristlecone pine may soon be coming to an end: it is being threatened by an unimposing, common neighbor, the creosote bush *(Larrea divaricata)* (see photo inset *b*). Less than 250 miles away from the kingdom of Bristlecone pines, some desert creosote bushes have been growing clones of themselves for an estimated 11,000 years, making Bristlecone pines middle-aged by comparison. But why should clones be considered part of the same living plant?

After about 100 years, a creosote bush begins to form rings of satellite clones, each developing form rooting branches that contact the soil. Eventually, the parent creosote bush dies, but its clones live on to produce even more clones.

Although the botanical details are slightly different, natural cloning in the creosote bush is analogous to the way the Bristlecone pine produces rings of secondary tissues year after year. The only living Bristlecone pine tissues are young cells generated by apical meristems and cambia. The Bristlecone pine, like the creosote bush, does not have cells that are thousands of years old. Both forms of growth, cloning and secondary growth, accomplish the same thing — perpetuation, over thousands of years, of cells that originated from one individual plant. For this reason, as soon as precise aging studies are completed on creosote clones, the common creosote bush will likely assume the throne as the "Oldest Living Organism" on earth.

---

maximum surface to sunlight for photosynthesis.

**Sclereids**, a group of sclerenchyma cells with varied shapes and functions, all have hard, thick cell walls. The shape of the sclereid determines its function. For example, sharp, star-shaped sclereids (Figure 13-4) not only help support loosely arranged plant tissues, but also quickly discourage even the hungriest browsing animal. The sclereids puncture the animal's gums and tongue. Densely packed rectangular-shaped sclereids form the "shell" of a walnut, protecting its em-

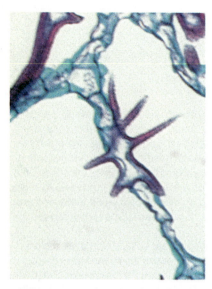

**Figure 13-4**
**Sclereids** come in various shapes. Star-shaped sclereids in the giant leaves of water lilies help support loosely packed cells, maintaining large airspaces to keep the leaf afloat. Many herbivores learn to avoid plants with flesh-piercing star-shaped sclereids.

bryo from injury as it falls from a tree or tumbles along the ground. The gritty texture of a pear is caused by clumps of stone-shaped sclereids that help support huge parenchyma cells engorged with sweet pear juices.

## Plant Tissue Systems

Tissues that run throughout a plant's root, stem, leaf, and flower form a **tissue system**. Plants have three such systems, and each is formed by a different **primary meristem**, groups of meristematic cells produced by the apical meristems that develop all primary tissues. The three primary meristems and the tissue systems they produce are as follows:

• The **protoderm**, which forms the **dermal tissue system** or **epidermis**, the outer cell covering of the plant.

• The **procambium**, which produces the **vascular tissue system** of conducting cells to transport food, water, and minerals throughout the plant.

• The **ground meristem**, which produces all remaining cells and tissues, forming the **ground tissue system**.

The three tissue systems of an herbaceous plant are illustrated in Figure 13-5.

## The Dermal Tissue System

The dermal tissue system is usually just one cell layer thick. This single sheet of cells is the only thing that stands between the fragile interior cells and the often hostile, disruptive elements of the outside world. For protection, the outer cell walls of the shoot epidermis are covered by a protective **cuticle**, a waxy layer that retards water vapor loss and helps prevent dehydration. The cuticle is such an effective water barrier that we use the water-repelling cuticle of some plants to protect some of our own prized possessions. For example, the cuticle from the Brazilian wax palm is harvested to make carnuba wax for safeguarding furniture and cars.

The dermal tissue system is composed of many kinds of cells, including epidermal cells, guard cells, trichomes, epidermal hairs, root hairs, and the periderm in secondary tissues.

### EPIDERMAL CELLS
**Epidermal cells** are all cells in the dermal tissue system other than guard cells, root hairs, and other dermal cells with highly specialized shapes and functions. Unlike the cells in many other plant tissues, all epidermal cells function while alive. Some, such as those in begonias, become meristems, asexually producing buds that grow into complete new plants. In the leaves of red cabbage, the petals of flowers, and many other colorful plant parts, the vacuoles of epidermal cells are engorged with water-soluble pigments (Figure 13-6). Brightly colored flower petals and fruits attract animals that play a crucial role in the plant's sexual reproduction (Chapter 21).

### GUARD CELLS AND STOMATES
The epidermis of the shoot system is riddled with microscopic pores called **stomates**, or **stomata** (Figure 13-7). Stomates allow gases to be exchanged between the plant and the external environment; $H_2O$ vapor, $CO_2$, and $O_2$ enter and exit by diffusion. A stomatal pore is formed between two specialized **guard cells**. When internal turgor pressure changes in guard cells, they change shape, which in turn changes the size of the stomatal pore and alters the rate of gas diffusion. The ability of guard cells to regulate internal turgor

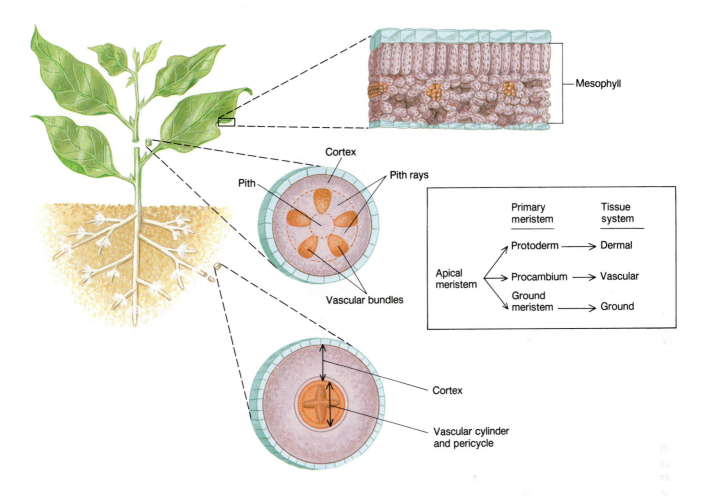

| | Primary meristem | Tissue system |
|---|---|---|
| Apical meristem | Protoderm → | Dermal |
| | Procambium → | Vascular |
| | Ground meristem → | Ground |

**Figure 13-5**
**Plant tissue systems.** Apical meristems produce three primary meristems. Primary meristems differentiate into three continuous tissue systems that make up the body of a plant. The protoderm produces the dermal tissue system, the procambium the vascular tissue system, and the ground meristem the ground tissue system. The ground tissue system includes the cortex, pith, and pith rays in stems, the mesophyll in leaves, and the cortex in roots.

**Figure 13-6**
**A natural cobblestone** of epidermal cells on an alyssum flower petal. The sculptured epidermis gives alyssum petals a velvety texture. Bright pigments stored in the vacuoles of these epidermal cells give the petals their brilliant color.

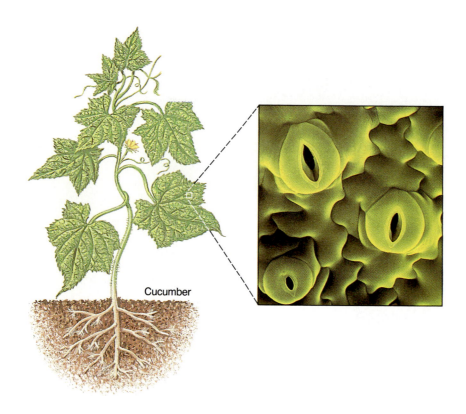

Cucumber

**Figure 13-7**
**Stomates** are formed between adjacent guard cells. Stomatal pores allow gases to diffuse into and out of airspaces between internal cells. $CO_2$ mainly diffuses inward for photosynthesis, and water vapor usually diffuses out. As internal turgor pressure increases or decreases, guard cells change shape, thereby changing the size of the pore, permitting more or less gas to enter or leave the leaf. This scanning electron micrograph shows a stomate of a cucumber leaf.

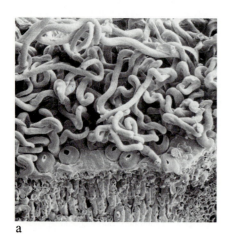

a

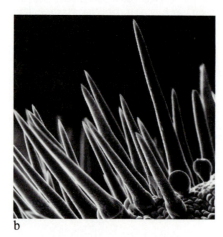

b

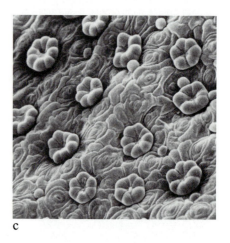

c

**Figure 13-8**
**Epidermal hairs and trichomes.**
*(a)* This scanning electron micrograph shows the thick layer of intertwining epidermal hairs on the lower surface of a New Zealand Christmas tree leaf. These hairs not only snare plant-eating insects, but also help reduce water vapor loss to the atmosphere.
*(b)* Saberlike trichomes on a cucumber leaf impale insects that might otherwise eat the plant's tissues.
*(c)* Bright orange glandular hairs dot the surface of a Swedish ivy leaf. Their secretions create a sticky mire that strands crawling insects.

pressure is critical to the survival of many plants, particularly those growing in dry environments where excessive loss of water vapor can quickly lead to dehydration and death.

The number of stomates varies between different species and between different parts of the shoot. The density of stomates on an apple leaf, for instance, is about 20,000 stomates per each square centimeter (roughly 120,000 stomates per square inch). This means that a single apple leaf has more than 3.5 million stomates. Some plants have three and four times that number.

## TRICHOMES AND EPIDERMAL HAIRS

Like guard cells, **trichomes** (sharply pointed dermal cells) and **epidermal hairs** (tubular or gland-tipped dermal cells) are found only in the shoot. Your body may be enveloped in epidermal hairs at this moment, since the cotton "fibers" in clothes are actually long epidermal hairs formed on the surfaces of cotton seeds.

There is a great variety of epidermal hairs and trichomes (Figure 13-8), most of which function to help retard water vapor loss or to protect plants from being eaten by herbivores. Some epidermal hairs secrete a sticky substance that stops plant-eating insects in their tracks; eventually, these insects die of exhaustion or starvation. The epidermal hairs of a sundew plant not only hold an insect in place, but also secrete enzymes that digest the captured insect, providing needed nutrients for the plant (Figure 13-9).

**Figure 13-9**
**Lunch on the wing.** The bright red epidermal hairs on the leaves of this sundew plant not only successfully attract an unwary damselfly, but also hold the fly in place as they curve inward and press it flat against the leaf blade. The blade secretes enzymes that digest the victim for nutrients.

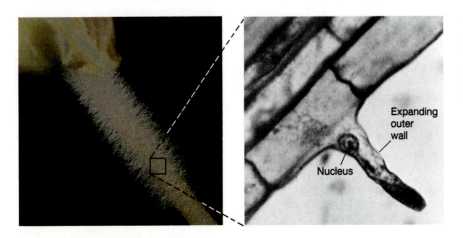

Expanding outer wall

Nucleus

**Figure 13-10**
**A two-day-old radish seedling** sends a root downward. Many root hairs give the root its fussy appearance. The enlargement shows the cell wall of a single cell beginning to expand to form a root hair for absorbing water and minerals from the soil.

Like some epidermal hairs, trichomes can also be death-wielding structures, stabbing insects with their pointed tips. Others entangle or trap victims in a massive maze; unable to escape, the snared insect eventually dies—the end of another plant eater.

## ROOT HAIRS

Ironically, the basic function of the shoot epidermis is exactly opposite that of the root, even though both are part of the same tissue system. Whereas the outer walls of all shoot dermal cells are covered with a waxy cuticle to reduce water vapor loss to the atmosphere, there is no cuticle on the epidermis of the root system. Such a water-repellent layer would fatally block roots from absorbing water and minerals from the soil.

Plants absorb water and dissolved minerals from the surrounding soil through **root hairs**. These hairs are specialized cells with elongated external walls that create a large surface area for absorption (Figure 13-10). (The roots of most plants also have cooperative fungi that grow around and into root cells, increasing their ability to absorb water and minerals. This partnership is discussed more fully in the next chapter.)

## THE PERIDERM

The **periderm** is a secondary tissue that takes over the protective and regulating functions of the epidermis when the epidermis has become damaged. During secondary growth, for example, the enlarging diameters of stems and roots split the epidermis. This disruption stimulates groups of parenchyma cells beneath the torn epidermis to become meristematic, forming a cork cambium. The newly formed cork cambium produces layers of compact cork cells toward the outside. Eventually, the entire epidermis is replaced by layers of cork cells as secondary growth progresses. Cork is only one part of the periderm. Technically, the periderm is composed of the cork cells, the cork cambium, and a layer of parenchyma cells inside the cork cambium (Figure 13-11a).

Like sclerenchyma, cork cells function when they are dead. Each cell has

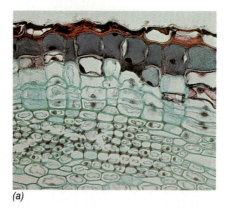

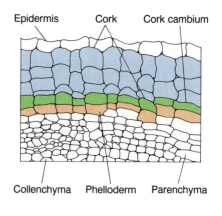

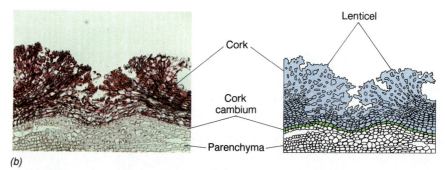

(a)

(b)

**Figure 13-11**
**The periderm:** Protection and gas exchange in secondary dermal tissues.
(a) A periderm begins forming beneath a split in the epidermis.
(b) Gases are exchanged through lenticels in mature periderm.

many secondary wall layers impregnated with a waterproof wax, suberin. As wall layers accumulate during the maturation of a cork cell, supplies of water and nutrients are gradually reduced, until movement to the cell's cytoplasm is blocked altogether and the cell dies. But it is precisely because of these waxy cell layers that cork is able to seal off internal tissues, protecting them against excessive water vapor loss, disease, extreme weather, and foraging animals. In other words, young, living cork cells don't function as effectively as fully differentiated, dead cork cells.

But how do living cells (the cork cambium, for instance), sealed within the compact corky layers, get $O_2$ from the atmosphere for life-sustaining respiration? Gases are exchanged through **lenticels**, random eruptions in the periderm that create air channels for transferring $CO_2$ and $O_2$ (Figure 13-11b).

Many of us commonly refer to the periderm as **bark**. Actually, the periderm only forms part of bark, which is a collective term for all plant tissues outside the secondary xylem (bark includes the vascular cambium, secondary phloem, parenchyma cortex cells, the cork cambium, and cork). Since bark contains critical living tissues, stripping it off will quickly kill a shrub or tree.

## The Ground Tissue System

The ground tissue system comprises all plant tissues except those in the dermal and vascular tissue systems (see Figure 13-5). It includes cells in the **cortex** (the region between the epidermis and the vascular tissues), **pith** (the area from inside the vascular tissues to the center of the stem or root), and **pith rays** (areas between bundles of vascular tissue).

In herbaceous plants, ground tissues form the bulk of the plant. (Plants with secondary growth are mostly made up of secondary vascular tissues.) The photosynthetic cells between the upper and lower epidermis of a leaf are also part of the ground tissue system. Thus, ground tissues are the principal sites of photosynthesis, and food and water storage. Ground tissues also contain collenchyma and sclerenchyma that provide support for shoot structures.

## The Vascular Tissue System

The vascular tissue system forms an internal circulatory network, the "veins and arteries" of a vascular plant. In the shoot, vascular tissues are grouped into **vascular bundles**, whereas in the root they form a central **vascular cylinder** (Figure 13-5).

The vascular tissue system contains two complex tissues: **xylem and phloem**. Water and dissolved minerals are carried mostly through the xylem, whereas food (carbohydrates produced during photosynthesis) and other organic substances (such as hormones, amino acids, and proteins) move through the phloem. Although xylem and phloem are made up of different cell types, only some actually transport materials.

### XYLEM

Water and minerals are transported through **tracheids** and **vessel members**, both of which are the conducting cells of the xylem (Figure 13-12). Primary xylem (xylem produced by the procambium) contains vessel members with distinctive wall patterns caused by stretching as the stem elongates during growth. In addition to tracheids and vessel members, xylem contains sclerenchyma fibers, sclereids, and parenchyma cells. With the exception of parenchyma, xylem cells have strong cell walls, so that the xylem also functions to help support aerial structures. The presence of parenchyma permits food storage and lateral conduction.

Secondary xylem is commonly called **wood**. New secondary xylem is produced by divisions of the vascular cambium, mostly in yearly intervals. Toward the end of one growth period, the inside of xylem tracheids and vessel members gets progressively smaller owing to decreasing supplies of water. This phenomenon generally occurs during drier summer months, although it may occur at any time when water is scarce. When favorable temperatures return and water again becomes plentiful (usually in spring), the vascular cambium produces new tracheids and vessel members with large diameters. The abrupt transition from xylem cells with narrow diameters to cells with large diameters forms a distinct line, which constitutes the border of a **growth ring** (Figure 13-13). Because one growth ring is *usually* produced each year, the number of growth rings is often used to estimate the age of a tree. Nevertheless, many botanists have abandoned the term "annual ring" because there can be more than one growth spurt, and therefore more than one growth ring, in a single year.

Tracheids and vessel members function when dead. Water and dissolved minerals move passively through their hollow interiors. The walls of each conducting cell are riddled with pits (Chapter 6), which are small holes in the secondary wall adjacent to thin, water-permeable depressions in the primary wall (Figure 13-14). Xylem-conducting cells line up, so that the pits of one match up exactly with those of a neighbor, forming a pit pair. Water flows from one cell to the next through pit pairs.

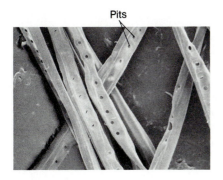

a Tracheids

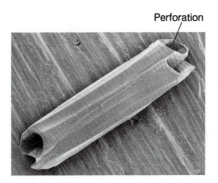

b Vessel members (secondary)

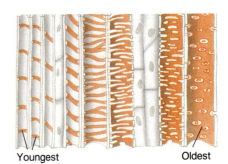

Youngest     Oldest

c Vessel members (primary)

**Figure 13-12**
**Xylem-conducting cells:** Transporting water and dissolved minerals.
*(a)* Tracheids are long and thin. In this preparation, the tracheids have been separated. In functioning xylem, tracheids are precisely aligned so that the openings (pits) of one match up with those of other tracheids.
*(b)* Vessel members are shorter and wider than tracheids, and have perforation plates or completely open ends. Columns of vessel members form open vessels for translocating water and minerals.
*(c)* Primary xylem is composed of vessel members with characteristic cell wall patterns. These patterns form as vessel members become stretched during growth of a stem or root.

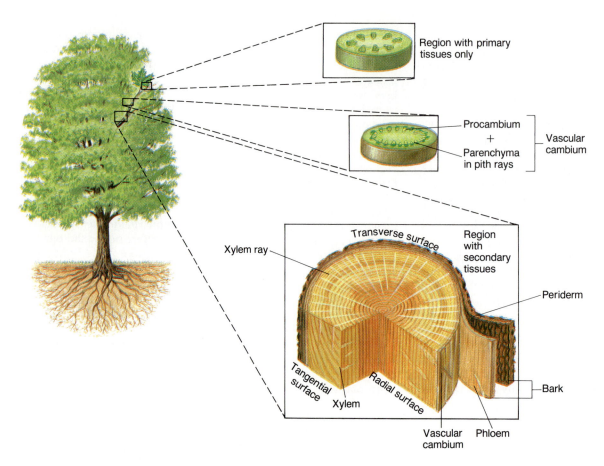

**Figure 13-13**
**Section of the trunk of an oak tree.** Although primary growth continues at the tips of stems, most of the tissue in a tree is secondary xylem or wood. In regions with secondary growth, the vascular cambium divides to produce secondary xylem toward the inside and secondary phloem toward the outside.

In addition to having pits, opposite ends of vessel members are either completely open or have a **perforation plate**, an area where large portions of the end wall are absent (Figure 13-15). Vessel members are stacked one on top of another to form **vessels**, that is, open conduits through which water readily travels (Figure 13-15). Because longer, narrower tracheids do not have perforated or open ends, water is conducted only through pit pairs.

In conifers, ferns, and other lower vascular plant groups, tracheids are the only water- and mineral-carrying cells. However, the xylem of flowering plants contains both tracheids and more efficient vessel members.

PHLOEM
Running parallel with the xylem is phloem, the complex tissue through which dissolved food and other

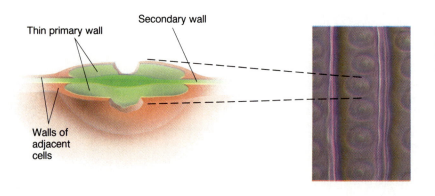

**Figure 13-14**
**Pits.** Water and dissolved minerals are moved through connecting pits of adjacent xylem-conducting cells.

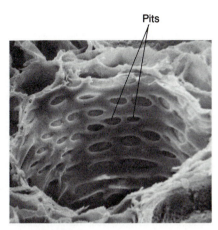

**Figure 13-15**
**A vessel pipeline** is formed by stacks of vessel members. This inside view of a cucumber vessel member shows the pits in the sides of the secondary wall.

organic materials are transported throughout a plant. But unlike tracheids and vessel members, phloem-conducting cells function when living.

All phloem-conducting cells are called **sieve elements** because they have clusters of pores in their cell walls that resemble a strainer or sieve. There are two types of sieve elements — **sieve cells** in conifers and lower vascular plants, and **sieve-tube members** in flowering plants (Figure 13-16). Sieve-tube members have **sieve plates**, end walls that contain large pores. A row of sieve-tube members, stacked sieve plate to sieve plate, forms a **sieve tube**. Although sieve cells lack sieve plates, they have clusters of small pores through which food is moved from sieve cell to sieve cell.

Although sieve-tube members and sieve cells are *technically* alive, neither has a nucleus or most of the other cellular organelles of a typical living cell. As these conducting cells mature, their organelles degenerate, leaving only a veneer of cytoplasm adhering to the inside wall and a delicate mesh of **p-protein** (an abbreviation for "phloem protein") extending between sieve elements. To remain alive, sieve cells and sieve-tube members are continually nourished by special parenchyma cells that lie adjacent to them. The specialized parenchyma cell that supports a sieve cell is called an **albuminous cell**,

and the cell associated with a sieve-tube member is a **companion cell**. Sieve-tube members and companion cells are interdependent; if one dies, its partner dies soon after.

# Plant Organs

Plant cells, tissues, and tissue systems are organized into four plant organs — the **stem**, **root**, **leaf**, and **flower**.

## The Stem

The stem is an exquisite example of nature's bioengineering. Not only do stems physically support all the leaves, flowers, fruits, and even other stems, but they actually produce the very structures they support.

### PRIMARY STEM TISSUES

At the tip of every stem is a dome-shaped apical meristem (Figure 13-17). The ends of stems are divided into three regions:

- The **meristematic region** contains meristematic cells of the apical and primary meristems. Young leaves and juvenile axillary buds begin developing in this region.

- The **region of elongation** is just below the meristematic region. In this region, cells enlarge and lengthen as they differentiate into the plant's primary tissues.

- The **region of maturation** lies immediately below the region of elongation. Here, differentiation of nearly all cells is complete.

Monocot and dicot flowering plants differ from each other in the way their primary tissues are arranged (see Table 13-1). In dicots, vascular bundles are organized in discrete rings (Figure 13-18), whereas more numerous mon-

a

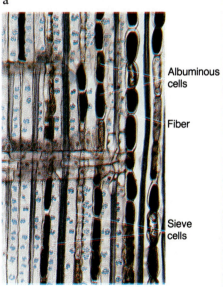

Albuminous cells

Fiber

Sieve cells

b

**Figure 13-16**
**Phloem sieve elements:** Conducting organic molecules.
*(a)* Sieve cells and albuminous cells are found in conifers and lower vascular plants.
*(b)* Sieve-tube members and companion cells occur in flowering plants. Sieve-tube members have distinct sieve plates at opposite ends.

Companion cell

Sieve plate

50 μm

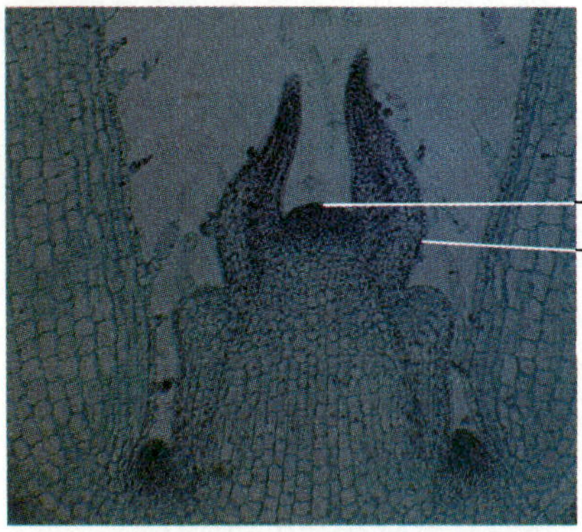

---- Apical meristem

---- Juvenile leaf

**Figure 13-17**
**Stem apex.** A longitudinal cut down the center of a stem tip. The apical meristem is the dome-shaped mass of meristematic cells. The upward arched "arms" are juvenile leaves, and the mass of meristematic cells above each young leaf is a developing axillary bud.

ocot vascular bundles are scattered throughout the ground tissues.

Dicots and monocots differ in another important way. Monocots do not form cambia for secondary growth, whereas perennial and biennial dicots do. (Perennial monocots, such as a palm tree, have a different form of growth in which the apical meristem continually forms new vascular bundles, not rings of secondary vascular tissues from a vascular cambium.) When secondary growth is initiated in dicot stems, procambium between the primary xylem and phloem, and parenchyma cells in the pith rays differentiate to form a continuous ring of vascular cambium cells (Figure 13-13).

## SECONDARY STEM TISSUES

The tallest organisms on earth (see Bioline: Nature's Largest Organisms) are plants that have secondary growth. These plants can grow so large because their diameters (and thus the plant's support strength) increase as they get taller. (Plants that elongate without getting thicker fall over if they grow too tall.) In other words, as primary growth from apical meristems elongates stems, vascular and cork cambia increase support and transport capacity by adding new cells and replenishing damaged older cells. In addition, the new secondary vascular tissues match up with the primary xylem and phloem to maintain a continuous connection of vascular tissues throughout the plant, so even the cells at the plant's tip receive water and can transport the organic molecules they produce.

The vascular cambium always lies between secondary xylem and secondary phloem, producing new conducting and supporting tissues. When a vascular cambium cell divides, one cell remains meristematic, whereas the other enlarges and differentiates into either a secondary xylem or secondary phloem cell. The combination of cell division and cell enlargement causes tissues outside the vascular cambium to be pushed out, crushing or rupturing some of the more delicate cells in the process, such as phloem cells. Vascular cambium cells also divide along a radius to produce additional vascular cambium cells that accommodate increases in circumference as the stem enlarges during secondary growth.

The vascular cambium produces cells that move substances up and down the stem, and other cells that conduct materials laterally, through secondary xylem and phloem rays (Figure 13-13).

## The Root

For many plants, most of its tissues lie hidden in the ground. The extensiveness of a plant's subterranean root system is illustrated by a tiny rye plant. A

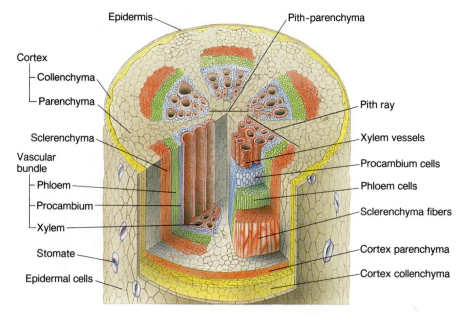

Epidermis — Pith-parenchyma
Cortex
— Collenchyma
— Parenchyma
Pith ray
Sclerenchyma — Xylem vessels
Vascular bundle — Procambium cells
— Phloem — Phloem cells
— Procambium — Sclerenchyma fibers
— Xylem
Stomate — Cortex parenchyma
Epidermal cells — Cortex collenchyma

**Figure 13-18**
**Arrangement of primary tissues in a dicot stem.** Cells are tiered for a three-dimensional view.

# BIOLINE

## NATURE'S LARGEST ORGANISMS

A tour of biology takes us from exceptionally small, microscopic organisms to colossal giants that make *us* feel miniscule. The ancient animal giants were the dinosaurs. The world's largest meat-eater was *Tyrannosaurus rex* with a length of 15 m (50 ft); just the head of this prehistoric giant was more than 1 m (3.3 ft) long. Today, the Blue whale *(Balaenoptera musculus)* is by far the largest animal, measuring 33 m (109 ft) and weighing more than 150 tons. Impressive? Certainly. But these animals are themselves dwarfed by the true mammoths of nature, plants.

The Amazon water lily, *Victoria amazonica,* produces leaves that are twice the size of the head of *Tyrannosaurus rex.* Even a flower of the corpse lily, *Rafflesia arnoldii,* is larger than the head of the once mighty *Tyrannosaurus.*

Some plants are more than three times the size of the largest Blue whale. These are the giant redwoods *(Sequoia sempervirens),* plants that tower more than 112 m (370 ft) into the air (see photo). Some other plants, while less impressive than redwoods, are huge nonetheless. For example, "giant" bamboo grows about 50 m (164 ft) tall and "giant" kelp, *Macrocystis,* reaches lengths of only 65 m (214 ft).

Does any other plant match the height of the giant redwoods? Possibly. In fact, one plant is reported to surpass the redwood's enormous height. An immense eucalyptus tree taller than 140 m (460 ft, almost half the height of the Empire State building) was reported in Australia. Unfortunately it was cut into sections before being measured, so the combined length may have been more than one individual. Until another of these giants is found and measured in tact, the giant redwoods remain nature's "Largest Living Organism."

---

rye plant only 25 cm tall (10 in) may have more than 14 million roots, which if placed end to end would stretch over 609 km (380 miles). It's logical to assume that a massive plant like a giant sequoia would have even more roots than the rye, but surprisingly it has fewer. Although the sequoia has fewer roots, its root system is enormously larger than that of the rye plant. The unexpected disparity in numbers of roots exists because the rye and giant sequoia have different types of root systems. The giant sequoia has a **tap root system** (Table 13-1), which is similar to a carrot with its one main root and many smaller **lateral roots** (Figure 13-19*a*). Conifers (cone producing plants) and most dicots have a tap root system. The rye, like most other monocots, has a **fibrous root system** (Figure 13-19*b*) with many main roots (more than 140 for rye). In both root systems, main roots produce lateral roots, lateral roots form more

lateral roots, and so on until you can't follow their origin any further. The repeated branching of roots generates many root tips, which are locations for water and mineral absorption.

Some roots arise from unexpected places—stems or even leaves. Roots that develop from shoot tissues form a third root system category, the **adventitious root system**. In corn, adventitious roots form at the base of the stem (Figure 13-19*c*), helping to support the shoot when burdened with heavy clusters of fruits (or what we call a "corn cob").

## PRIMARY ROOT TISSUES

The root apical meristem, unlike the stem apical meristem, is not located at the very tip (Figure 13-20). Instead, at the tip of a root is a **root cap**, a protective cellular helmet that surrounds delicate meristematic cells and shields them from abrasion as the root grows

through the soil. The root's apical meristem continuously regenerates its root cap as root cap cells are worn away or pierced by sharp soil particles. Root cap cells manufacture and secrete a gelatinlike substance that helps to lubricate the root, easing penetration through the soil. This secretion also creates a favorable habitat for certain soil bacteria; these microorganisms supply the plant with essential elements, especially nitrogen. In addition, root cap cells produce growth hormones, chemicals that control the pattern and rate of cell divisions in the apical meristem so that growth continues downward. (Plant hormones are discussed in Chapter 22.)

In addition to forming its own cap, the apical meristem of the root also produces cells behind itself that differentiate into the root's primary tissue systems. As in stems, roots have a *meristematic region,* a *region of elongation,* and a *region of maturation* (Fig-

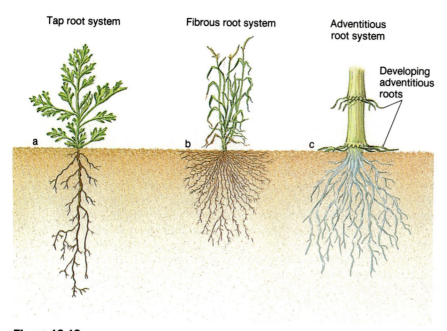

**Figure 13-19**
**Three types of root systems.**
*(a)* A *tap root system* consists of one, generally deep, main root with many branching lateral roots.
*(b)* A *fibrous root system* is generally shallow and spreads out from the stem base. Fibrous root systems have more than one main root, each with many branching lateral roots.
*(c)* In an *adventitious root system,* roots arise from plant organs other than the root itself, mainly stems or leaves. Adventitious roots form when plant cuttings are made.

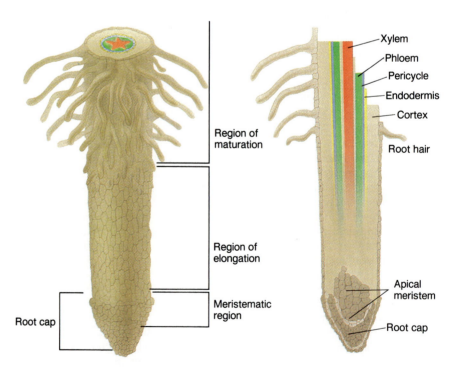

**Figure 13-20**
**The root apex:** External morphology and internal anatomy. As in stems, the root apex includes three regions of growth—the meristematic region of active cell division, the region of elongation where cells enlarge (in length as well as width) and begin differentiating, and the region of maturation defining the area where most cells, including the root hairs, have completed differentiation. Water and minerals are absorbed in the region of maturation.

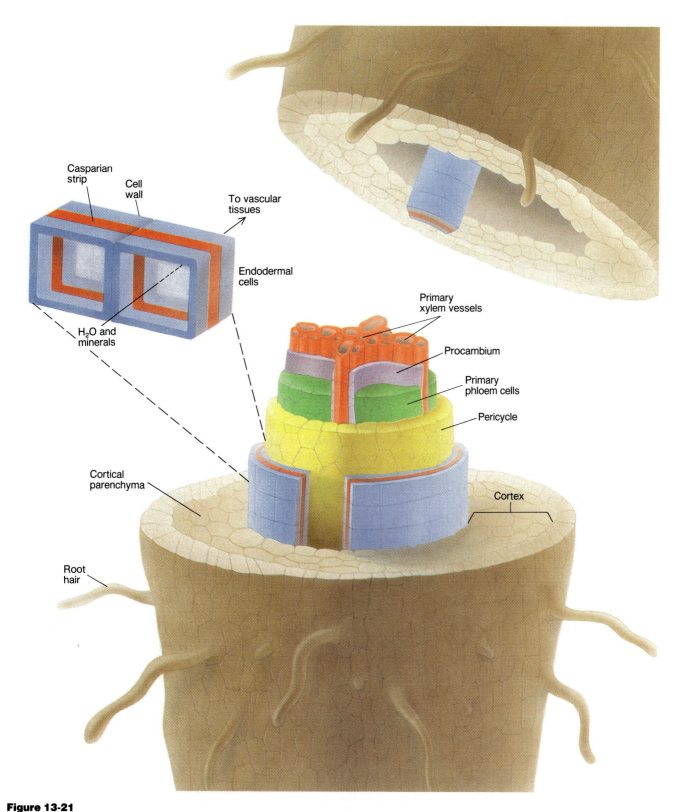

**Figure 13-21**
**A dicot root.** (Cells are tiered for a three-dimensional view.) The innermost layer of cortex cells is the endodermis. A casparian strip encircles each endodermal cell, preventing water from entering the vascular cylinder by moving along cell walls or between cells. In this way, the movement of water and dissolved minerals is controlled by the endodermis. The pericycle forms a cylinder of cells immediately inside the endodermis. Pericycle cells divide to form lateral roots or form part of the vascular cambium for secondary growth.

ure 13-20). Root hairs form in the region of maturation.

Primary xylem takes on different arrangements in different species. In the dicot root illustrated in Figure 13-21, the primary xylem forms a four-pointed star in the center. Primary phloem is clustered between the arms of the xylem star. The arrangement is different in monocot roots. In monocots, xylem and phloem alternate with each other around a central pith (Table 13-1).

Roots contain two important tissues that are not found in stems: an endodermis and a pericycle. Both are cylinders of cells that encircle the vascular tissues (Figures 13-20 and 13-21). The **endodermis** is the innermost layer of cortex cells. Each endodermal cell is itself encircled by a band of waxy suberin, forming what is called the **casparian strip**. Water-repelling casparian strips of adjacent endodermal cells are aligned, creating a hydrophobic barrier that prevents water and dissolved minerals from entering or exiting the root vascular tissues by moving between cells or along cell walls. The only way these nutrients can enter or exit the xylem, then, is by passing *through* living endodermal cells. By using active and passive transport to change solute concentration, endodermal cells can control the uptake of water and minerals, and prevent water from leaking out of the root and back into the sometimes dry or salty soil. Without such regulation, plants growing in somewhat dry or salty soil would never be able to create a gradient steep enough for water to enter their roots and would quickly lose water to the soil.

Like the endodermis, the **pericycle** is one cell layer thick. It lies immediately inside of the endodermis and encircles the vascular tissues (Figure 13-21). Mitotic divisions of pericycle cells produce lateral roots. If the plant is a biennial or perennial dicot, the pericycle forms part of the root's vascular cambium for secondary growth.

## SECONDARY ROOT TISSUES

If there is secondary growth in the stem, there is secondary growth in the root. Secondary tissues are so similar in roots and stems that, except for the absence of a pith in dicot roots, after a few years of secondary growth it is difficult to distinguish a cross section of a stem from that of a root (Figure 13-22).

## The Leaf

Leaves are the plant's "solar-collecting units," "energy generators," and "energy transmission lines," all rolled up into one efficient structure. Leaves capture solar energy, house chloroplasts for converting the energy in sunlight into chemical bond energy during photosynthesis, and have an extensive network of interconnecting vascular tissues for transporting materials.

In general, leaves only have primary tissues (Figure 13-23). The protoderm produces an upper and lower epidermis. The ground meristem produces the leaf's **mesophyll**—the layers of cells between the upper and lower epidermis. In dicot leaves, densely packed, column-shaped **palisade parenchyma** cells lie next to the upper epidermis. Below the palisade parenchyma cells are loosely packed **spongy parenchyma** cells. Monocot leaves generally have only spongy paren-chyma. The mesophyll contains airspaces between the cells, which form corridors for diffusion of $CO_2$ needed for photosynthesis.

The extensive vascular network in leaves is produced by the procambium. Leaf vascular tissues form **veins,** which create a characteristic **venation** pattern for different plants. Monocots have leaves with parallel veins, whereas dicots have highly branched, or netted, venation (Table 13-1). The vascular tissues of a leaf are so extensive that each mesophyll cell is no more than a few cells away from a xylem or phloem cell (Figure 13-24). Each vein is surrounded by a **bundle sheath,** parenchyma cells that regulate the uptake and release of materials between the vascular tissue and the mesophyll cells.

## The Flower

The balance between form and function is displayed most elegantly in flowers. Flowers are the site of sexual reproduction for a plant. Each flower part participates in the process, either producing gametes or influencing how gametes are transferred so that fertilization can take place.

A flower is a very short stem with highly modified leaves. Some modified leaves contain germinal tissues that divide by meiosis to produce gametes. Following fertilization, the flower develops (ripens) into a fruit that contains seeds with newly developed plant embryos. The fruit protects these embryos and helps disperse them to new habitats. Because of their complexity and central role in reproduction, an entire chapter (Chapter 21) is devoted to the structure and functions of flowers and fruits.

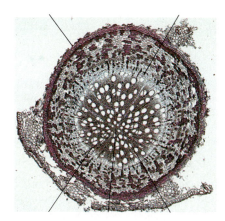

**Figure 13-22**
**Is this a stem or a root?** Since the center is filled with primary xylem here, this must be a root with secondary growth.

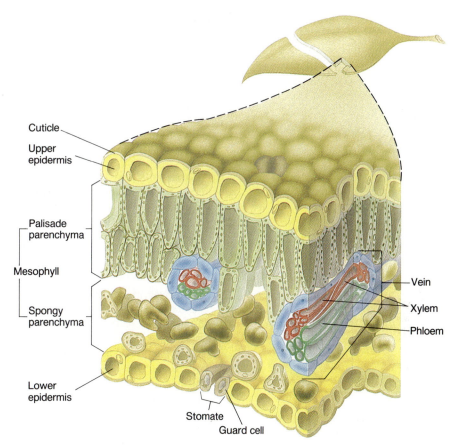

Cuticle

Upper epidermis

Palisade parenchyma

Mesophyll

Spongy parenchyma

Lower epidermis

Stomate

Guard cell

Vein

Xylem

Phloem

**Figure 13-23**
**Dicot leaf.** Although stems generally contain photosynthetic cells, most photosynthesis takes place in a plant's leaves. Chloroplasts for photosynthesis are found in both palisade and spongy parenchyma cells. Gases diffuse in and out through stomates. Vascular bundles form a network of leaf veins, with xylem toward the upper surface and phloem below.

**Figure 13-24**
**A leaf cleared** of all tissues except xylem shows the extensive vascular network in a magnolia tree leaf.

## Synopsis

### MAIN CONCEPTS

- **Although less complex than animals, a plant's simpler anatomy is well adapted for its autotrophic life.** The plant's highly branched root system absorbs and conducts water and dissolved minerals, as well as helps support the aerial shoot system. The plant's shoot produces both stems with leaves for photosynthesis and axillary buds that develop into more stems or produce flowers for sexual reproduction.

- **Most plants have specialized vascular tissues** for conducting water and minerals (xylem), and for transporting food and other organic compounds (phloem).

- **Long-lived plants have two forms of growth** that occur simultaneously: primary growth from apical meristems and secondary growth from cambia. Plants that live for one year or less generally only have primary growth.

- **Long-lived plants are able to continually produce new cells to replace old or damaged cells from meristems and cambia.** As a result, many plants can live for hundreds of years and some for thousands of years.

- **Many plant cells perform their specialized functions when dead.** These include xylem tracheids and vessel members, cork cells, sclerenchyma fibers, and sclereids.

- **Living parenchyma cells are not only the most prevalent type of cell in many plants,** but also the most versatile cell in the plant body. They can change into any plant cell, including meristematic cells.

- **Plants have evolved many ways of defending themselves against plant-eating animals,** including swordlike trichomes that stab or catch insects, epidermal hairs that ensnare insects, and sticky glandular hairs that hold insects until they die of exhaustion or starvation.

### KEY TERM INTEGRATOR

**The Plant Body**

| | |
|---|---|
| **morphology and anatomy** | External form and internal architecture |
| **vascular plants** | Contain **xylem** and **phloem** tissues for conducting water, dissolved minerals, and organic molecules. |
| **monocotyledons dicotyledons** | Two classes of flowering plants, the most diverse plant group. Both monocots and dicots have a **shoot** and a **root**. Leaves are made up of a **blade** and **petiole**, and **axillary buds** are attached to the shoot at **nodes**. Leafless areas between nodes on the stem are **internodes**. The root produces **root hairs** for absorbing water and minerals from the soil. |
| **annuals** | Plants that live for one year or less. |
| **biennials** | Plants that live for two years. |
| **perennials** | Plants that live longer than two years. |

## Plant Tissues

| | |
|---|---|
| **tissues** | Similar cells are grouped into **simple tissues**, whereas **complex tissues** are made up of more than one kind of cell. Simple plant tissues include **parenchyma, collenchyma, sclerenchyma** (including **sclerenchyma fibers** and **sclereids**), and **meristems** (apical meristems and cambia). |
| **apical meristems** | **Produce primary growth** and **primary tissues. Herbaceous** plants have only primary tissues. |
| **secondary meristems (cambia)** | Cell divisions of cambia (**vascular cambium** and **cork cambium**) result in **secondary growth** and produce **secondary tissues.** Secondary tissues include **wood** (secondary xylem), secondary phloem, and the **periderm** containing **cork** and **lenticels.** |

## Tissue Systems

| | |
|---|---|
| **primary meristem** | Forms the plant's **tissue systems.** The three primary meristems are the **protoderm**, the **procambium**, and the **ground meristem**. Three continuous tissue systems extend throughout a plant's body. |
| **dermal tissue system (epidermis)** | Contains **epidermal cells, guard cells** (which form **stomates**), **trichomes, epidermal hairs,** and the periderm in secondary tissues. The **cuticle** helps prevent dehydration. |
| **ground tissue system** | Includes cells and tissues that make up the **pith, cortex,** and **pith rays.** |
| **vascular tissue system** | The network of complex conducting tissues arranged in **vascular bundles** in stems and a **vascular cylinder** in roots. *Xylem*-conducting cells include **tracheids** and **vessel members** with **perforation plates** or open-end walls. Stacks of open-ended vessel members form **vessels.** Secondary xylem forms **growth rings.** Phloem-conducting cells, or **sieve elements**, include **sieve-tube members** with **companion cells** and **sieve cells** with **albuminous cells.** Sieve-tube members are connected by **sieve plates** and contain **p-protein.** |

## Plant Organs:  stem, root, leaf, flower

| | |
|---|---|
| **stem** | Divided into the **meristemic region**, the **region of elongation**, and the **region of maturation.** |
| **root** | A **tap root system** has one main root, a **fibrous root system** has many main roots, and an **adventitious root system** develops from a stem or leaf. The apical meristem of a root is protected by a **root cap.** Unlike stems, roots have an **endodermis** with each cell surrounded by a **casparian strip**, and a **pericycle** that produces **lateral roots** or forms part of the root's vascular cambium. |
| **leaf** | The **mesophyll** between the upper and lower epidermis of a leaf contains **palisade parenchyma** and/or **spongy parenchyma.** Vascular tissues form **veins** that create characteristic **venation** patterns. Each vein is surrounded by a **bundle sheath.** |

# Review and Synthesis

1. Using this list of the key terms, complete the following.

| | | |
|---|---|---|
| apical meristem | node | collenchyma |
| axillary bud | stem | sclerenchyma |
| bundle sheath | leaf | palisade |
| casparian strip | root | adventitious |
| endodermal | flower | root system |
| root cap | spongy | |

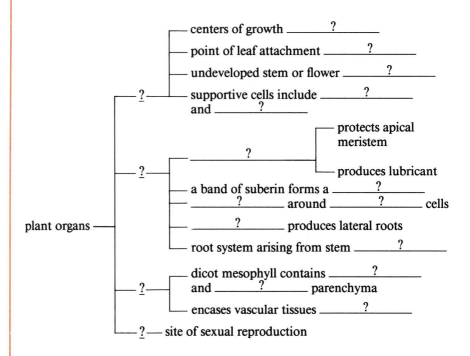

```
                          ┌── centers of growth ____?____
                          ├── point of leaf attachment ____?____
                          ├── undeveloped stem or flower ____?____
                   ?──────┤── supportive cells include ____?____
                          │   and ____?____
                          │
                          │                    ┌── protects apical
                          │        ____?____ ──┤   meristem
                   ?──────┤                    └── produces lubricant
                          │
                          ├── a band of suberin forms a ____?____
plant organs ──────┤      ├── ____?____ around ____?____ cells
                          ├── ____?____ produces lateral roots
                          └── root system arising from stem ____?____
                          │
                          │   ┌── dicot mesophyll contains ____?____
                   ?──────┤   │   and ____?____ parenchyma
                          │   └── encases vascular tissues ____?____
                          │
                   ?── site of sexual reproduction
```

2. Identify (with an "X" for xylem and a "P" for phloem) which of the following cells is included in the xylem, in the phloem or both. Underline those that actually do the conducting.

| | |
|---|---|
| parenchyma | tracheid |
| sieve-tube member | sclereid |
| companion cell | vessel member |
| sieve cells | albuminous cells |

3. How are the terms "periderm" and "bark" related? Why do most botanists avoid using the term "bark"?

4. Why are parenchyma cells sometimes called "the most important type of cell in a plant"?

5. Sketch a transverse (cross) section of a dicot stem with three seasons of secondary growth. Label the following:

| | |
|---|---|
| vascular cambium | secondary xylem |
| secondary phloem | crushed primary phloem |
| primary xylem | pith |
| xylem and phloem rays | cork cambium |
| cork | lenticels |

6. You find an herbaceous plant that has leaves with parallel veins and flowers with 12 colorful petals. There are no fruits or seeds, so it's impossible to determine whether the embryo has one or two cotyledons. Nevertheless, you are confident that it is a monocot. Why? What other characteristics could you look for to verify your hypothesis?

7. Plants are either annuals, biennials, or perennials. What is the adaptive value for each of these life span strategies? In which environment (desert, seashore, mountain, lake, and rainforest) would one strategy aid in survival and reproduction over the others? Why are all three usually found together, regardless of environment?

## Additional Readings

Esau, K. 1977. *Anatomy of Seed Plants.* John Wiley and Sons, New York. (Advanced.)

Evans, M., R. Moore, and K. Hasenstein. 1986. "How roots respond to gravity." *Scientific American* 255:112–19. (Intermediate.)

Galston, A., P. Davis, and R. Satter. 1980. *The Life of the Green Plant.* Prentice-Hall, Englewood Cliffs, N.J. (Introductory.)

Hohn, R. 1980. *Curiosities of the Plant Kingdom.* Universe Books, New York. (Introductory.)

Simpson, B., and M. Conner-Ogorzaly. 1986. *Economic Botany: Plants in Our World.* McGraw-Hill, New York. (Introductory.)

Chapter 14

# Circulation in Plants: Transporting Water, Minerals, and Food

Sometimes even our language gets in the way of gaining an understanding of biology. For example, being "animated" is to be energetic and lively, like an animal. On the other hand, being sluggish, dull, and passive is to "vegetate" —in other words, to be plantlike. But plants are not inactive. Hidden behind their immobility and stillness is not only the biochemical bustle of cellular metabolism, but also a number of surprisingly "animated" behaviors. For example:

- Plants join in battle with other plants for space and environmental resources, similar to the competition found between many animals. Some plants employ biochemical warfare tactics to outcompete their rivals.

- Although most plants respond more slowly than animals, some, like the Venus fly-trap, move fast enough to catch flies.

- Like warm-blooded animals, some plants store the heat produced from their own metabolism and use it to melt snow around them for water, or to entice insects to their flowers.

- Some plants, like hibernating animals, become dormant during cold periods, and then resume growth and reproduction when favorable conditions return.

- Like animals, most plants have a complex circulatory system. In vascular plants, labyrinths of xylem and phloem tissues circulate needed water and nutrients throughout the plant body.

The force that powers the circulatory system of a plant, however, is very different from that found in animals. Plants have no muscles or hearts to create the pressure that pumps (pushes) water and nutrients through their circulatory system. Even so, food and other organic molecules are indeed *pushed* from regions of high pressure to regions of low pressure through connecting sieve elements in the phloem.

The situation is reversed in the xylem. Instead of raising pressure at one end of a vascular tube as in the phloem, very low or negative pressure is generated at one end of the xylem. This pressure creates a strong tension that *pulls* water and minerals through

columns of lifeless conducting cells.

A plant's vascular system is able to withstand steep pressure gradients because plant cells have strong cell walls. Plant cell walls not only resist stretching under high pressure as materials are pushed through phloem-conducting cells, but also resist collapsing under low or negative pressures as water and minerals are pulled up through the xylem. Use of negative pressure for circulation is impossible for animals because animal cells lack pressure-resistant cell walls.

## Xylem: Water and Mineral Transport

Huge amounts of water are transported through the xylem of a vascular plant. A single corn plant, for example, conducts more than 206 liters (43 gallons) of water during a four-month growing period. At that rate, an acre of corn transports enough water in four months to fill more than 16 backyard swimming pools (about 68,000 liters or 324,000 gallons).

Plants take in and conduct far greater amounts of water than ani-

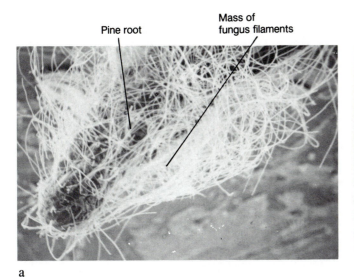

a                    b

**Figure 14-1**
**Close encounters with mutual benefits.** *(a)* **Mycorrhizae.** The filaments of a soil fungus join with a plant root to form mycorrhizae. The fungus absorbs some food (carbohydrates) stored in the root, while the plant receives water and minerals absorbed by widespread fungal filaments. *(b)* **Root nodules.** Some plants house nitrogen-fixing bacteria in lumplike root nodules. As with mycorrhizae, the alliance benefits both. Bacteria harvest some of the plant's carbohydrate stores, and the plant receives nitrogen in a form it can use.

mals. A sunflower plant, for instance, will imbibe more than 17 times as much water as you will in one day. Plants demand such large amounts of water because they lose up to 98 percent of the water they absorb to the air as water vapor. Water vapor is lost when plants expose moist leaf cells to the air to extract gaseous carbon dioxide for photosynthesis. When water is exposed to air, it evaporates; the drier the air, the faster the evaporation rate.

Water vapor loss from plant surfaces is called **transpiration.** Most water vapor is transpired from leaves through open stomates. Unless the plant's roots absorb enough water to replace the amount lost during transpiration, the plant will wilt and eventually die from dehydration.

## Absorption

In the previous chapter we learned that plants absorb water and minerals through root hairs found near root tips (refer to Figure 13-10). Since water and minerals are widely dispersed in the soil and are usually available only in small amounts, plants develop extensive root systems with many root tips to mine the soil for these scant resources. In addition, more than 90 percent of vascular plants form root partnerships called **mycorrhizae** with soil fungi (Figure 14-1*a*). A mycorrhizae association increases a plant's ability to extract water and minerals from soil because the filaments that make up the fungus extend further into the surrounding soil than the plant's root system. This creates an enormous surface area for water and mineral absorption. In fact, mycorrhizae associations are so critical that the young seedlings of some plants will quickly perish from dehydration or nutrient deficiency if they do not develop mycorrhizae (see Bioline: Mycorrhizae and Our Fragile Deserts). Mycorrhizae are mutually beneficial. The plant benefits by receiving greater supplies of water and minerals than it could absorb on its own, and the heterotrophic fungus benefits by absorbing food stored in the plant's root and using it for its own metabolism.

In addition to forming cooperative alliances with fungi, some plants develop **root nodules** that house nitro-gen-fixing bacteria. In exchange for food, water, and living space, resident bacteria supply the plant with nitrogen, a mineral essential for normal growth and development (Figure 14-1*b*).

## AVENUES OF WATER UPTAKE

Water enters roots by osmosis. Once inside, water diffuses through the root cortex either along cell walls, through the airspaces between cortical parenchyma cells, or from cell to cell via plasmodesmata (Figure 14-2). But once water reaches the endodermis, aligned casparian strips encircling each endodermal cell form a barrier that prevents water from moving between cells or along walls (see Figure 13-21). The only way water gets into

the vascular cylinder of the root is by moving through the protoplasm of an endodermal cell as minerals are actively transported into and out of the cells. Water continues to follow the inbound path of actively transported minerals to the pericycle, and finally into a xylem-conducting cell. Inside the xylem, water and minerals join a *transpiration stream* that swiftly conducts them through the plant.

## MINERAL ABSORPTION AND REQUIREMENTS

For most plants, mineral ions must be dissolved in water before they can be absorbed by root hairs or mycorrhizae filaments. When the mineral concentration in the root is less than that in

PATHWAYS OF WATER AND MINERAL MOVEMENT

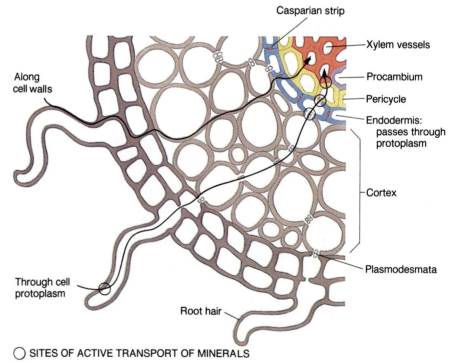

○ SITES OF ACTIVE TRANSPORT OF MINERALS

**Figure 14-2**
**Roadmap to the xylem: Pathways of water and mineral transport through root tissues.** Water enters roots by osmosis, whereas most dissolved mineral ions are actively transported into root hairs using the energy of ATP. Water moves through the cortex of living cells, between the cells, or along cell walls, but is prevented from moving between endodermal cells by the casparian strips. Minerals are actively transported through the endodermis, the pericycle, procambium, and into xylem-conducting cells, creating an osmotic gradient. Water flows along this gradient.

# D I S C O V E R Y

## MYCORRHIZAE AND OUR FRAGILE DESERTS

The desert—hot, dry, and desolate. That is the image most of us have of a desert. In fact, Webster's New World Dictionary defines a desert as a barren wasteland, incapable of supporting almost any plant and animal life. Certainly, the "barren" discriptor fits some areas of the Sahara and Namib deserts in Africa, but even a brief scouting trip through the deserts of North America will reveal a multitude of desert organisms. The California deserts, for example, support about 1200 plant species, 200 species of vertebrate animals, and numerous insects and other invertebrates. The conception of a desert as an uninhabited wasteland is not correct.

Another common misconception which many of us have is that, because desert plants and animals can withstand unusually harsh desert conditions, they are "tough" enough to withstand virtually any type of disturbance, even hordes of racing off-road vehicles and motorcycles, or the effects of constructing thousands of miles of transmission lines to link power plants to major population centers. In fact, the opposite is true; the desert is easily damaged and is very, very slow to recover. Thousand-year-old Indian trails and campsites are still clearly visible in aerial photographs, proof that virtually no natural recovery has occurred in thousands of years. One reason why North American deserts are so very slow to recover from human disturbances is that the most common plant, the creosote bush *(Larrea divaricata),* lives for thousands of years. It would take that long to replace a damaged creosote bush with a new one (see Bioline: Nature's Oldest Living Organisms, Chapter 13). Unfortunately, the misconception that the desert is a tough ecosystem has led to a great deal of damage to the deserts in southern California.

The California deserts are a valuable recreational, industrial, and biological resource. The Department of Interior estimates that 11 million people use the desert for recreation every year. One-half million people live in 100 desert cities, some of whom work in mines, military bases, or electrical power plants. As a biological resource, more than 100 of the animal and plant species found in the California deserts are listed as rare, endangered, or threatened with extinction. With continued fast-paced development in the deserts, can these species, and the new species that become endangered because of further destruction, be saved?

When a local power company built a power transmission line near Barstow, California, in 1970, the new line divided one of the few remaining populations of the Mojave ground squirrel, an endangered species. The powerline and access road prevented individuals on one side of the powerline from breeding with individuals on the other side. This separation lowered the reproduction rate, causing the number of Mojave ground squirrels to plummet. Could anything be done to help regrow new native plants for food and cover so that separated Mojave squirrels would once again reach each other and interbreed?

During a three-year study, all attempts to seed the damaged areas failed, even when seeds were repeatedly irrigated. Attempts to transplant growing seedlings also failed. When the investigators analyzed the soil, they discovered there were no mycorrhizal fungi in the disturbed soil. As a result, the roots of seedlings could not develop mycorrhizae. Without mycorrhizae, these young plants quickly died of dehydration, incapable of forming root systems large enough to gather sufficient supplies of water even when irrigated. However, when the researchers artificially added mycorrhizae fungi to the disturbed soil, many seedlings lived. The means of rehabilitating this sensitive desert habitat had therefore been discovered, but to no avail: the costs of injecting the disturbed soil are astronomical. Thus, this population of Mojave ground squirrels will have to survive without the help of scientists.

Of course, some important lessons have been learned from this study. The power company will no longer scrape desert soil off to the side when it constructs power transmission lines through the desert. It will also be more careful in the future not to damage the habitat of any endangered desert species, for the cost of reconstructing the habitat is prohibitive.

the soil, minerals enter the root passively. But more often, minerals are actively transported into the root, building concentrations higher in the root than in the surrounding soil. Some root cells contain as much as 10,000 times more potassium than the soil. Once inside the root, dissolved minerals are actively transported through cortical parenchyma cells, through the endodermis, and finally into the xylem (Figure 14-2). Because of this one-way active transport path to the xylem, minerals are concentrated in the xylem and are prevented from leaking back into the cortex or soil by the endodermis.

While moving with the transpiration stream, some minerals diffuse into surrounding root and stem tissues. Others continue to flow with the transpiration stream to the leaves. Minerals not used in the leaf are discharged into the phloem and exported along with food. In the phloem's descending stream, minerals may travel back to the root again where they are either used by root cells or reenter the ascending transpiration stream again and recycle back to the shoot system.

Without adequate supplies of minerals, plants cannot grow normally. Plants use minerals to build complex organic molecules (amino acids, nucleic acids, ATP, chlorophyll and other pigments, enzymes, and coenzymes) and to help construct the middle lamella that cements adjoining plant cells.

Sixteen minerals are essential for normal plant growth (Table 14-1). These **essential nutrients** are divided into two groups, depending on the quantity required: **macronutrients,** which are needed in very large amounts for normal growth, and **micronutrients,** which are needed in small amounts.

TABLE 14-1

## Primary Uses of the 16 Essential Plant Nutrients

**Macronutrients**

| | |
|---|---|
| Carbon (C) | Major component of organic molecules |
| Oxygen (O) | Major component of organic molecules; cell respiration |
| Hydrogen (H) | Major component of organic molecules |
| Nitrogen (N) | Component of amino acids, nucleic acids, chlorophyll, and coenzymes |
| Potassium (K) | Components of proteins and enzymes; helps regulate opening and closing of stomata |
| Calcium (Ca) | Component of middle lamella; enzyme cofactor; influences cell permeability |
| Phosphorus (P) | Component of ATP, ADP, proteins, nucleic acids, and membrane phospholipids |
| Magnesium (Mg) | Component of chlorophyll; enzyme activator; cofactor in amino acid and vitamin formation |
| Sulfur (S) | Component of proteins and coenzyme A |

**Micronutrients**

| | |
|---|---|
| Iron (Fe) | Catalyst for chlorophyll synthesis; component of cytochromes |
| Chlorine (Cl) | Osmosis and photosynthesis |
| Copper (Cu) | Enzyme activator and component; chlorophyll synthesis |
| Manganese (Mn) | Enzyme activator; component of chlorophyll |
| Zinc (Zn) | Enzyme activator; influences synthesis of hormones, chloroplasts, and starch |
| Molybdenum (Mo) | Essential for nitrogen fixation |
| Boron (B) | Important in flowering, fruiting, cell division, water relations, hormone movement, and nitrogen metabolism |

\*\*Memory Joggers\*\*
*To remember the macronutrients, use the phrase "C. HOPK(i)NS Car is an MG"*
*To remember the micronutrients, use the phrase "A Festive MoB comes in (=CuMnZn) Clapping"*

## ROOT PRESSURE AND GUTTATION

At night, water pressure can build in the roots of some plants as roots continue to actively pump minerals into the xylem, drawing more water in. The continuous influx of water produces a positive **root pressure** that pushes water and dissolved minerals up the xylem. In some cases, root pressure can build high enough to force water and minerals completely out the tips of leaves. This process is called **guttation** (Figure 14-3). Root pressure falls and guttation stops as soon as transpiration begins in the morning.

Root pressure cannot build during the daytime because of transpiration. As a result, root pressure accounts for

**Figure 14-3**
**Forcing the issue.** During guttation, glistening droplets of water and dissolved minerals are forced out leaf tips by positive root pressure.

the movement of water and minerals only for certain plants, and only when transpiration is stopped and soil water is plentiful. For most plants, the principal means of water and mineral transport is by **transpiration pull,** a process initiated by transpiration.

## Transpiration

As water vapor is lost during transpiration, the water deficit generates a negative pressure (tension) in the xylem strong enough to pull more water to the top of even the tallest tree (Figure 14-4). The faster the transpiration rate, the stronger the tension, and the more accelerated the flow of the transpiration stream.

### TRANSPIRATION PULL

Transpiration pull is the result of a chain of events that starts when leaves absorb solar radiation. Sunlight heats leaves, causing water to evaporate from moist cell walls. The evaporated water is immediately replaced, first with water from inside the cell, then with water from neighboring cells

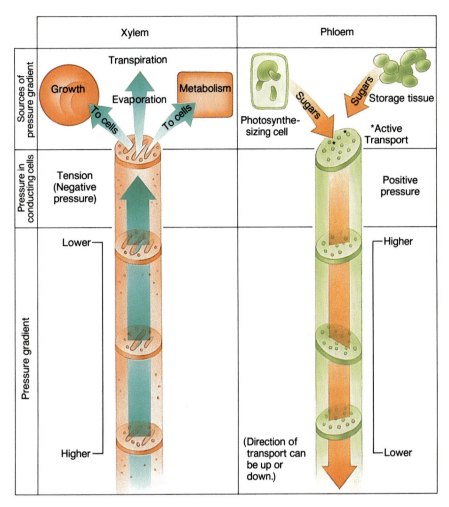

**Figure 14-4**
**Xylem versus phloem transport.** Pressure gradients move materials through both the xylem and phloem. In the xylem, tension, or negative pressure, builds as water evaporates during transpiration and as water is used by cells for metabolism and growth. As water exits the xylem, cohesively bonded water molecules are pulled up. In the phloem, sugars and organic molecules are actively loaded into sieve elements, creating an osmotic gradient that draws water into the phloem, thereby increasing its internal pressure. Pressure falls at sinks as assimilates are actively unloaded and water flows out of the phloem. This pressure gradient pushes materials through the phloem from sources (high pressure) to sinks (low pressure).

deeper in the leaf, and, finally, with water in the xylem.

In the xylem, water molecules cling to each other by strong cohesive bonds (Chapter 3), forming an unbroken water column that extends all the way from the leaf vein, through the stem, and into the root. Because of these bonds, when one water molecule exits the xylem to replace transpired water, it exerts a pull on the adjacent water molecules. This tension is transmitted down the entire water column. As water continually exits the xylem to

replenish lost water, cohesive bonds pull water up. This is transpiration pull.

Why doesn't increasing tension in the xylem simply snap cohesive bonds, pulling water molecules apart and breaking the continuous water column necessary for transpiration pull to work? The reason has to do with the strength of bonds and the width of xylem tracheids and vessel members. In narrow xylem cells, the strength of attraction between water molecules is comparable to steel wire of the same diameter. As a result, in a narrow tube a column of water can withstand tension 10 times that needed to pull water to the top of the tallest redwood tree, the tallest organism on earth.

The tremendous tension produced in the xylem can even shrink the diameter of a large tree trunk! Because some water molecules are adhesively bonded to xylem cell walls, as well as cohesively bonded to other water molecules, the tension in the xylem can actually pull tracheid and vessel walls inward, enough so that it can measurably reduce the diameter of a tree trunk.

## REGULATING TRANSPIRATION RATES

The difference in concentration of water vapor between the airspaces *inside* a leaf and the concentration of water vapor of the outside air determines the rate of transpiration. The greater the difference, the faster the transpiration rate, and the smaller the difference, the slower the rate of transpiration (Figure 14-5).

Because their outside surfaces are covered with a waxy cuticle, less than 10 percent of transpired water is lost directly through the cells that make up the epidermis. The main avenue of water vapor loss is through open stomates, the pores formed between adjacent guard cells.

Guard cells are able to change the size of a stomatal pore, thereby changing the rate of transpiration. Because of their shape and cell wall construction, influxes of water increase internal turgor pressure and cause guard cells to bend. As facing guard cells bow out in opposite directions, the stomate pore enlarges. When water exits guard cells, turgor pressure falls, guard cells straighten, and the pore narrows.

Guard cells control the influx and efflux of water by actively transporting potassium ions ($K^+$). When $K^+$ is actively transported into a guard cell, an osmotic gradient is established that draws water in, increasing turgor pressure and enlarging stomatal pores. When $K^+$ is transported out, turgor pressure falls and stomatal pores narrow. This is one way plants are able to regulate rates of water vapor loss, re-

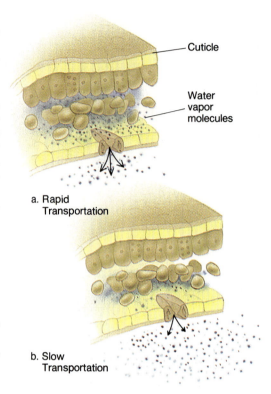

a. Rapid Transportation

b. Slow Transportation

**Figure 14-5**
**Transpiration rate** is directly proportional to the difference in concentration of water vapor molecules inside a leaf and the outside air. Since the water vapor is always high inside the leaf, the humidity of the air greatly affects the transpiration rate.

ducing the possibility of wilting or becoming dehydrated.

Why don't plants, especially those growing in hot, dry environments, just keep their stomates closed all of the time so that they won't lose large amounts of water? Remember, plants must open their stomates to allow $CO_2$ to diffuse into the leaf for photosynthesis, and when stomates open, water vapor races out. In addition to changing internal turgor pressure in guard cells, many plants have evolved other adaptations that juggle the cost of water loss with the amount of $CO_2$ gained. They have reached different physiological and anatomical compromises that help balance the need for $CO_2$ with the danger of dehydration. Some of these adaptations are illustrated in Figure 14-6.

## ENVIRONMENTAL FACTORS THAT AFFECT TRANSPIRATION

Although the opening and closing of stomates often has the greatest impact

on transpiration rates, the following environmental factors also affect the rate of transpiration. They either influence guard cell action or change the concentration gradient of water vapor between the air inside a leaf and the outside atmosphere.

- *Radiation.* Absorbed solar radiation raises the temperature of a leaf, causing more water to evaporate. For every 10° C rise in leaf temperature, the evaporation rate doubles. The higher the leaf temperature, the greater the difference in the concentration of water vapor between the airspaces inside a leaf and the outside air. This leads to faster transpiration rates. In most plants, stomates close when temperatures exceed 30° C.

- *Air humidity.* The air inside a leaf is always saturated with water vapor. When atmospheric humidity is low, a steep concentration gradient exists that leads to rapid rates of transpiration. On humid days, the vapor gra-

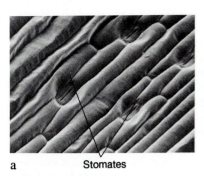
a    Stomates

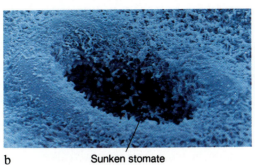
b    Sunken stomate

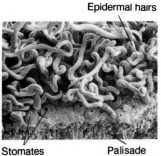
Epidermal hairs

Stomates    Palisade mesophyll of leaf
c

**Figure 14-6**
**Adaptations that affect transpiration rates.** *(a)* **Number of stomates.** Plants can have many thousands of stomates per each square centimeter of leaf surface. Plants adapted to moist habitats have stomates on both the upper and lower epidermis. On the other hand, plants adapted to arid areas have fewer stomates that are restricted to the cooler lower surfaces where the vapor gradient between the inside and outside of the leaf is less. *(b)* **Position of stomates.** Some plants conserve water by recessing stomates in sunken crypts. This placement doesn't affect $CO_2$ diffusion, but retards water vapor loss by forming small pockets of high humidity. Sunken stomates also protect the plant from the drying affects of wind. *(c)* **Epidermal hairs.** Dense epidermal hairs reflect light, reducing the leaf's heat load and slowing transpiration. Epidermal hairs also divert drying winds away from stomate openings.

dient is less, reducing transpiration rates and overall water loss.

• *Wind.* Wind increases transpiration by blowing away water vapor molecules at the leaf's surface. This creates a steep vapor gradient, which in turn increases transpiration.

• *Carbon dioxide concentration.* Generally, stomates open when the carbon dioxide concentration in leaf airspaces falls below a critical concentration. However, when a leaf begins to dry out, stomates close no matter what the internal level of $CO_2$ is, thereby curbing further water

loss. (Plants can survive $CO_2$ shortages much longer than they can dehydration.)

## Phloem: Food Transport

Substances continually move through the vascular tissues of a plant. The rate

RADIATION

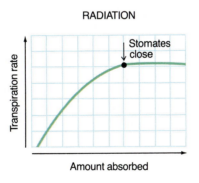

TEMPERATURE

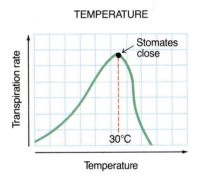

HUMIDITY

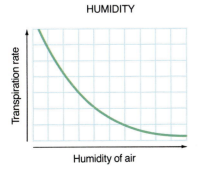

WIND

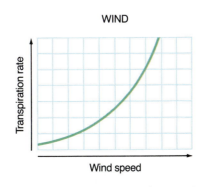

$CO_2$ CONCENTRATION

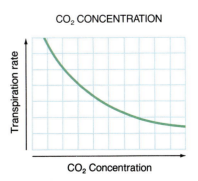

that food moves through the phloem can reach speeds of up to 100 cm/hour, or roughly 0.0013 miles/hour, the speed of a tired snail (see DISCOVERY: Aphids and phloem research). Although this is only about ⅕ the rate that water and minerals move through the xylem, it is much faster than moving substances using cytoplasmic streaming or diffusion from cell to cell. Such rates are too slow to distribute adequate supplies of nutrients throughout a plant.

## Translocation Through the Phloem

Since many of the solutes in the phloem are products of assimilation (biochemical construction of molecules by the plant), moving phloem sap is called an **assimilate stream.** Unlike the xylem's transpiration stream, the assimilate stream is pushed through columns of living sieve elements (Figure 14-4).

Assimilates always flow from *sources* to *sinks.* Sources are the parts of a plant where assimilates are synthesized (such as a photosynthesizing leaf cell) or stored (such as in parenchyma tissues in stems and roots). Sinks are plant tissues that use assimilates, such as actively growing meristematic tissues of roots and shoots, developing fruits and seeds, and storage tissues. (Notice that storage tissues can be both sources and sinks; they are sources when they export assimilates and sinks when they import assimilates to build reserves.)

Because phloem cannot be studied without causing some damage to delicate conducting cells, scientists are still debating how materials are translocated through the phloem. Various hypotheses (models) have been proposed for phloem transport. These models can be classified into two groups: those requiring active transport, and those based on passive transport only.

In active transport models, assimilates are actively transported along fibers of p-protein that run through adjoining sieve elements. The energy in ATP is used to contract p-protein fibers or cause them to undulate. Either way, the movement of p-protein fibers pumps assimilates through sieve elements. Although there is some evi-

dence to support such active transport models, there is more evidence for a passive transport mechanism, in particular the pressure flow hypothesis.

## Pressure Flow Hypothesis

In the **pressure flow hypothesis** of phloem transport, higher pressures are created at sources when assimilates are actively transported into the phloem-conducting cells. This process is called **phloem loading.** Phloem loading increases the assimilate concentration inside the sieve elements and creates an osmotic gradient that causes water to flow in from nearby xylem cells. At the sink end of a sieve tube, assimilates

are actively transported out of the phloem. This process is called **phloem unloading.** Phloem unloading reduces assimilate concentration in the phloem and increases concentration in surrounding cells, causing water to flow out of the phloem into the sink or back to the xylem. Since sources and sinks are connected by columns of sieve elements, water pressure builds at a source and drops at a sink. This pressure difference pushes the assimilate stream from the source to the sink (Figure 14-7). Companion and albuminous cells are required for pressure flow because they supply ATP to sieve elements for phloem loading and unloading.

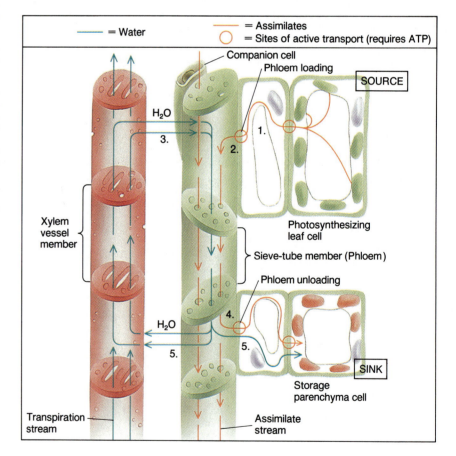

**Figure 14-7**
**Phloem transport by pressure flow.** Assimilates always flow from sources to sinks. (1) A photosynthesizing leaf cell (the source) produces sugars (assimilates). (2) Sugars are actively transported into a sieve tube during phloem loading. (3) As sugar concentration increases in the sieve tube, water is drawn in from nearby xylem vessels. This influx of water raises the pressure in the sieve tube. (4) At sinks (storage parenchyma cells in this example) sugars are actively unloaded, lowering sugar concentration in the sieve tube and (5) causing water to be recycled back to the xylem or be drawn into surrounding cells. These pressure differences between sources and sinks initiate the flow of an assimilate stream.

# D I S C O V E R Y

## APHIDS AND PHLOEM RESEARCH

Until recently, investigators could only speculate on what was being transported through the phloem. Regardless of how carefully they tried to extract sap from phloem, the sieve elements collapsed and plugged up, preventing them from getting a sample. Ironically, a common plant pest provided scientists with a "tool" to collect pure phloem sap, enabling them to not only verify once and for all that food and other organic substances were actually transported through the phloem, but how fast these substances move. These unusual "research assistants" were aphids.

Aphids crowd the growing stem tips and leaves of many plants in early spring. Their hollow, needlelike mouthparts (stylets) enable them to bore directly into the phloem and feed on the plant juices flowing through the phloem conducting cells; aphids could do what scientists were unable to do — tap delicate phloem cells without damaging them. The moment the aphids stylets penetrate a sieve element, their bodies balloon out as sap gushes from transport cells; sometimes the pressure in the sieve elements is high enough to force sap completely through the insect's digestive tract and out its anus, forming small drops of "honeydew" (photo a). To determine the content and transport rate of phloem sap, investigators simply allow aphids to tap the phloem, then anesthetize them and cut off their bodies. The pure sap that exudes from the remaining stylets is then collected (photo b). Analysis of the sap reveals it to be a very concentrated syrup of 10–25% sugar (mainly sucrose), with small amounts of amino acids, other nitrogen-containing substances, hormones, and a few dissolved mineral ions.

Stylet          Honeydew drop

a

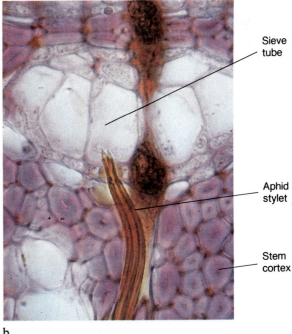

Sieve tube

Aphid stylet

Stem cortex

b

*(a)* The stylet mouthparts of aphids enable them to penetrate phloem cells without damaging delicate conducting cells. Once the phloem is tapped, the aphid feeds on the plant's food supply. Positive pressure in the phloem forces phloem sap through the digestive tract of an overfed aphid, forming a honeydew drop.
*(b)* Aphids provided investigators with the perfect tool for studying phloem content and transport. When the aphid's body is removed, assimilates flow out the insect's stylet for analysis.

## Synopsis

### MAIN CONCEPTS

- **A vascular plant's circulatory system includes two sets of conduits:** (1) water and dissolved minerals, which are conducted through the xylem, and (2) food and other organic molecules, which are translocated through the phloem.

- **Extremely large quantities of water are pulled up through nonliving xylem tracheids and vessel members by transpiration pull.**

- **Water and dissolved minerals** are absorbed through root hairs or through the filaments of fungi that form mycorrhizae with plant roots.

- **Lump-shaped root nodules on some plants house nitrogen-fixing bacteria** that directly supply roots with usable nitrogen.

- **The casparian strips surrounding endodermal cells in the root funnel** dissolve minerals and water into the xylem vascular tissues and prevent water from leaking out as the soil dries.

- **Most of the water absorbed by a plant** evaporates from cell surfaces and exits through stomates during transpiration.

- **The rate of transpiration** is directly proportional to the difference in the concentration of water vapor in the airspaces inside a leaf and the outside air.

- **Products of biochemical reactions are actively loaded into the phloem for translocation,** generating pressure that pushes assimilates to sinks by pressure flow. Assimilates are actively unloaded into tissues for immediate use or storage.

### KEY TERM INTEGRATOR

Absorption of Water and Minerals

| | |
|---|---|
| **mycorrhizae** | Partnerships in which the roots of most plants develop with soil fungi, improving water and mineral absorption over that which is possible only through their own root hairs. |
| **root nodles** | House nitrogen-fixing bacteria in some plants. |
| **essential nutrients** | Needed for normal plant growth. **Macronutrients** are essential nutrients needed in large amounts, and **micronutrients** are needed in small amounts. |

Water and Mineral Transport Through the Xylem

| | |
|---|---|
| **transpiration** | Water vapor loss from plant surfaces. During this process, cohesively bonded liquid water molecules are pulled up the xylem by **transpiration pull**. Dissolved minerals move with the transpiration stream. |
| **root pressure** | Positive pressure that in some plants is high enough to cause **guttation** of liquid water and minerals out leaf tips. |

Transport of Organic Molecules Through the Phloem

| | |
|---|---|
| **assimilate stream** | Moving stream of sugars and other organic molecules (assimilates). |
| **pressure flow hypothesis** | A theory explaining the transport of organic molecules in the phloem. Water pressure builds at sources of assimilates as they are actively transported into phloem-conducting cells during **phloem loading**. At sinks, water pressure falls as assimilates are actively transported out of the phloem by **phloem unloading**. The pressure difference pushes the assimilate stream through the phloem. |

# Review and Synthesis

1. Complete the following graphic organizer using the terms below:

| | | |
|---|---|---|
| phloem loading | sugars | organic molecules |
| phloem unloading | xylem | micronutrients |
| minerals | water | macronutrients |
| pressure flow | phloem | transpiration pull |
| root pressure | guttation | |

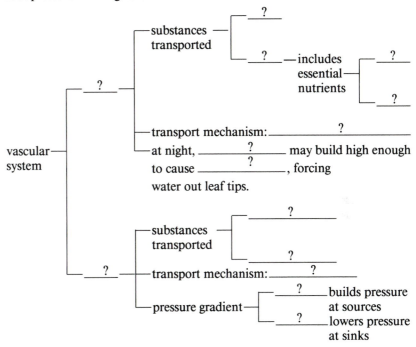

2. Where appropriate, put an "R" for root nodule or an "M" for mycorrhizae.
___ a. root and fungus partnership
___ b. directly supplies nitrogen to plants
___ c. root and bacteria partnership
___ d. creates a large surface area for water and mineral absorption

3. How can you determine whether a drop of water on a leaf tip is dew (condensation from the air) or water forced out by guttation?

4. List the essential nutrients that are
   a. components of organic molecules
   b. enzyme cofactors
   c. needed for chlorophyll formation
   d. needed for constructing the middle lamella
   e. enzyme activators

5. Which moves faster, the assimilate stream or the transpiration stream? Why?

6. How are root pressure and pressure flow similar?

# Additional Readings

Biddulph, O., and S. Biddulph. 1959. "The circulatory system of plants." *Scientific American* 200:26. (Introductory.)

Brum, G., R. Boyd, and S. Carter. 1983. "Recovery rates and rehabilitation of powerline corridors." In *Environmental Effects of Off-Road Vehicles,* R. Webb and H. Wilshire (eds.). Springer-Verlag, New York. (Intermediate.)

Epstein, E. 1973. "Roots." *Scientific American* 228:48–58. (Introductory.)

Galston, A., P. Davis, and R. Satter. 1980. *The Life of the Green Plant.* Prentice-Hall, Englewood Cliffs, N.J. (Introductory.)

Zimmerman, M. 1963. "How sap moves in trees." *Scientific American* 208:132. (Introductory.)

# 5

## *Multicellular Organization of Animals*

Chapter 15

# Orchestrating the Organism — The Nervous System

Most animals are more complex than plants. Almost all animals possess muscles for movement. Unlike plants, all must obtain and digest organic nutrients and, with few exceptions, eliminate solid wastes. They must also possess more complex systems for supplying oxygen to all their tissues. Moreover, many are active mate-seekers. Humans and other mammals have 11 organ systems for accomplishing these operations (see Connections: Multicellular Organization in Animals). Such complexity requires an internal communications network that coordinates the activities of each system with those of the others and with environmental conditions. In this way, all parts of the body can be co-regulated according to the animal's needs.

In vertebrates, there are actually two great communications networks. The first is the complex of chemical messengers, called *hormones,* secreted by endocrine glands (Chapter 16). The second network, the nervous system, speeds electrical messages over biological "wires and cables" (nerve cells and nerves), triggering immediate responses in the cells that receive the command. These messages enable an animal to quickly detect and respond to stimuli, that is, changes in the organism's external or internal environment that elicit a response. Neurological messages regulate heart rate, breathing, digestive activity and coordinate movement and hundreds of other processes. The nervous system also enables some animals to store and recall information, to think, and to reason. This chapter seeks to stimulate these latter abilities by challenging your nervous system to understand itself.

## Neurons and Targets

Each nerve cell, called a **neuron**, delivers a message to one or more "target cells" that respond to the neurological impulse. The effect of the impulse depends on the nature of the neuron and the type of target cell.

- *Nature of the neuron.* Some neurons, called **excitatory neurons**, stimulate their target cells into activity; others, the **inhibitory neurons**, oppose a response, encouraging the target to remain at rest.

- *Nature of the target cell.* Neurological activation generates different results according to whether the target is
  —a muscle fiber (contracts and exerts force),
  —part of a gland (secretes, that is, produces and discharges a particular substance), or
  —another nerve cell (propagates an impulse of its own).

Muscle fibers and glands are called **effectors** because their activation produces a specific effect. Most impulses travel through a circuit of many neurons before arriving at an effector. Each neuron along the way is the target of the previous nerve cell and relays the neurological impulse to the next cellular target.

Your existence, second by second, depends on continuous neurological activity. Without orders from the nervous system, you would not be able to contract the muscles that draw air into your lungs. You would not be able to generate neurological messages to activate movements needed to obtain food, chew it, and swallow it. Even if food was to somehow enter your stomach, your digestive glands would not release many of the enzymes needed to digest it. Nervous control prevents internal chaos by providing a means of functionally plugging the body's 60 trillion cells into a command center that integrates their activities, allowing an animal to maintain homeostasis in a changing environment (see Connections: Homeostasis, Chapter 20). These are just a few of the activities that require continuous neurological control.

## The Roots of the Nervous System

People tend to believe that a large and sophisticated brain is the ultimate achievement of evolution, with humans the benefactors of this "evolutionary pinnacle." Yet the nervous systems of simpler animals, some of which lack even the earliest traces of a brain, are just as effective in enabling these organisms to survive and reproduce as is the large brain of humans. Our "advanced" brains are merely one adaptive strategy. Some animals, notably sponges, succeed without a single neuron, as do all plants, fungi, protists, and monerans.

The simplest nervous system is a network of overlapping nerves that

**Figure 15-1**
**Simple but effective.** The anemone's predatory skills rely on a simple system of sensory neurons that launch a feeding response when they detect a fish in its tentacles. Stinging darts quickly paralyze the victim, which is then moved by the tentacles into the mouth.

## MULTICELLULAR ORGANIZATION IN ANIMALS: AN OVERVIEW

Animal cells are bound by many of the same limitations that restrict the size and complexity of plant cells (Chapter 6). Like plants, animals overcome these limitations by forming multicellular complexes that themselves constitute an individual organism. One trillion cells, differentiated into 200 different cell types, create a human being. Humans share the same multicellular heirarchy as other animals. In virtually all animals, specialized cells collaborate to form tissues. Some of these tissues are *germinal,* producing gametes (sperm and eggs) for sexual reproduction. The rest are *somatic,* the tissues that make up the animal's body and perform all the functions needed for survival of the individual. In vertebrates, somatic tissues fall into one of four tissue types.

- *Connective tissue* — protects, supports, and holds together the structures of the body; also includes blood cells produced by these tissues:

  1. *Tendons and ligaments:* secure muscle to bones, bone to bone, and hold internal organs in place.

  2. *Bone, bone marrow, and cartilage.*

  3. *Blood* — red blood cells produced by bone marrow.

  4. *Adipose (fat-storing) tissue.*

- *Muscle tissue* — bundles and sheets of contractible cells that shorten when stimulated, providing force for controlled movement.

- *Nervous tissue* — excitable cells that receive stimuli and, in response, transmit an impulse to another part of the animal.

- *Epithelial tissue* — continuous sheets of tightly packed cells that cover the body, line the digestive and respiratory tracts, and package internal organs. Glands (structures that secrete a specific substance) are also forged from specialized epithelial tissue.

Tissues combine with one another to form sheets, layers, tubes, bundles, and other arrangements that constitute *organs.* Organs are specialized structures found in all but the simplest animals. From these are built an *organ system,* a cooperative of different organs committed to the same task, such as disposal of wastes (excretory system), movement and support (muscular and skeletal systems), or distribution of nutrients and oxygen to cells in the body (circulatory system). The 11 systems that make up the vertebrate body (nervous, endocrine, digestive, immune, integumentary, reproductive, respiratory, plus the excretory, circulatory systems, muscular and skeletal are the focus of this and the next five chapters on animal anatomy, physiology, and reproduction. We do not attempt to discuss every system in every type of animal. Rather, we supply selected examples that best illustrate the structure and function of each system, including a few simpler strategies that reveal trends in the evolutionary development of each system.

lace the body of jellyfish, sea anemones, and hydras. Their *nerve nets* give these simple animals the ability to slowly move around, and to seize and paralyze prey with their stinging tentacles, and then shove the incapacitated victim into the mouth (Figure 15-1).

Compare the anemone's nerve net with the primitive nervous system of the flatworm in Figure 15-2. Although the flatworm's system is only slightly more complicated, it represents an evolutionary breakthrough — **cephalization**, the trend toward neuron clustering at the front of the animal (cephalo = head). These neuron clusters, called *ganglia,* become more pronounced as animals increase in complexity. This tendency toward centralizing nervous control integrates the responses of various parts of a complex animal.

As control centers, however the ganglia of flatworms and roundworms provide so little control that the worm can lose its head and continue to live while growing a replacement. The major ganglia of insects, lobsters, and shrimp are considered to be brains and may be capable of learning. The octopus (which belongs to the same phylum as snails, clams, and slugs) possesses the most complex brain of any invertebrate. Discrete regions of the octopus brain control specific functions (such as eye movements and respiratory activities) and an impressive amount of "intelligence." Neurological centralization culminates with the vertebrates, animals that protect their central nervous systems (the brain and spinal cord) by encasing them in special bony or cartilagenous compartments.

The nervous systems of all these animals, from jellyfish to humans, are composed of individual neurons that employ the same mechanism for conducting impulses. Let's examine how a single nerve cell can relay the messages that are so essential to the continued existence of practically every animal on earth.

## Anatomy of a Neuron

Describing a typical nerve cell is impossible; there are more than 100 billion of them in your body, and no two are exactly alike. Nonetheless, they are all composed of the same four basic parts (Figure 15-3). (1) **Dendrites** are extensions of the cell that carry impulses from the area of stimulation (the cell's "input regions") to the cell body. The (2) **cell body** relays impulses from dendrites to the (3) **axon**, a usually long, sometimes branching extension that terminates in a group of (4) **synaptic knobs** (the cell's output regions). These knobs deliver the neurological message to the target cell. Neurological impulses always travel in one direction—from dendrites to synaptic knobs.

## Dendrites

Stimuli impinge on nerve cells at the dendrites. Most dendrites are multi-branched "sensors" that extend in many directions. When a dendrite detects a change in conditions, it alerts the rest of the neuron of the disturbance. Some dendrites receive their activating messages from the external environment, as do pain and pressure receptors in the skin. Stimuli may also originate from other cells (such as other neurons) adjacent to dendrites or from chemical or physical changes in tissue fluids. In general, the more dendrites that are stimulated, the greater the sum of impulses that enter the membrane of the neuron's cell body.

## Cell Body

The nucleus and most of the neuron's other life-sustaining organelles reside in the cell body. The Golgi complex in this portion of the cell is particularly active, manufacturing vesicles that contain the chemicals that ultimately transfer the neuron's impulse to its target cell. The cell body is also the neurological "decision maker." It receives input not only from dendrites, but from other neurons as well. These neurons deliver their impulses directly

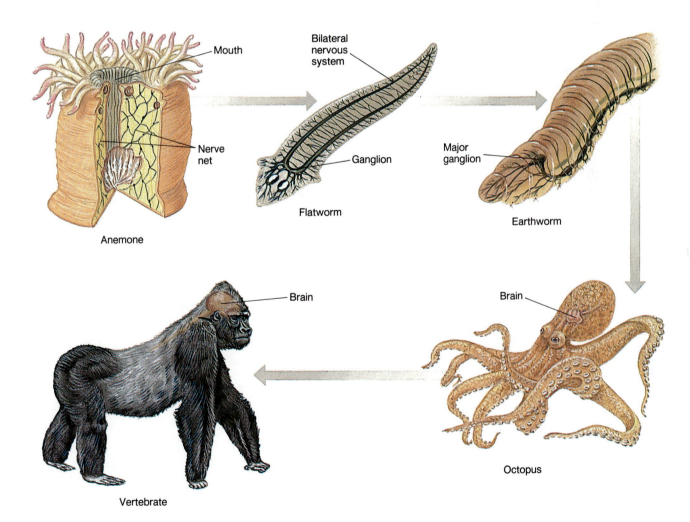

**Figure 15-2**
**Evolution of the nervous system** is revealed by these organisms, presented in their order of evolutionary complexity. The trend toward cephalization (the development of control centers in the head) first appears in the flatworms.

to the cell body's membrane. The sum of these impulses from dendrites and from the neurons determines whether the total stimulation will elicit a response. Of course, the "decision" is not a conscious one, but simply a matter of the neuron membrane's resistance to change. If the total stimulus is strong enough to exceed the neuron's *threshold* resistance, the cell body relays the impulse to the axon for transmission to the target cell. Weaker stimulation simply decays at the cell body, and the target cell receives no impulse.

A neuron acts in an *all-or-none* fashion—it either relays an impulse or it doesn't. Once a particular neuron's threshold is reached, the intensity of the resulting impulse is the same regardless of the strength of the stimulus.

## Axon

The length of the axon provides one of the adaptive advantages of having neurons. Axons can transmit neurological impulses over distance, sometimes to targets several meters away from the cell body (although the axons of some neurons are very short). The neurons that carry impulses from a giraffe's brain to its legs exceed 3 meters in length; these are the longest cells in the animal world. The problem with transmitting impulses over such distances lies in the speed needed to get the signal to the target fast enough to trigger a quick response in the effector. When you accidentally stumble, the only thing that keeps you from falling is your ability to thrust your foot and leg forward fast enough to arrest your forward plunge. This action requires extremely rapid transmission of neurological impulses over neurons that span the distance between the central nervous system and the muscles of the leg.

One way to increase the velocity of the signal is to increase the diameter of the neuron, but this would occupy large amounts of valuable body space. For vertebrates, this problem is solved by covering the axons of high-velocity neurons with a **myelin sheath**. This jacket increases the speed of a neurological impulse to 120 meters per second (250 miles per hour), about 20 times faster than the impulses of an unmyelinated neuron of the same diameter.

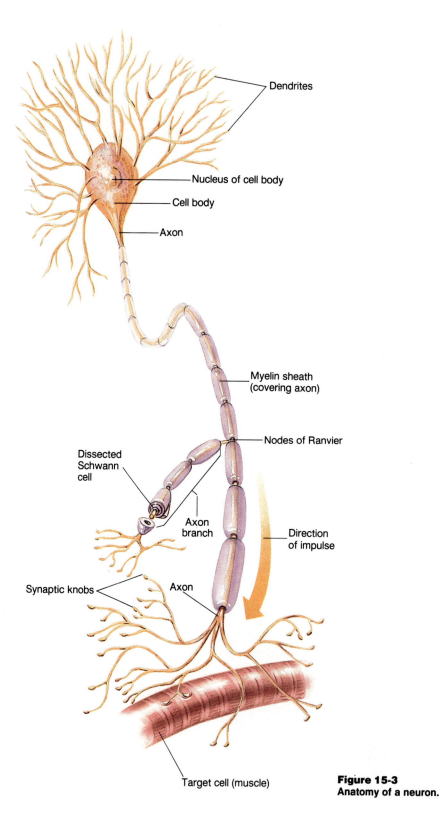

**Figure 15-3**
**Anatomy of a neuron.**

Dendrites

Nucleus of cell body

Cell body

Axon

Myelin sheath (covering axon)

Nodes of Ranvier

Dissected Schwann cell

Axon branch

Direction of impulse

Synaptic knobs

Axon

Target cell (muscle)

The myelin sheath is composed of cells, called **Schwann cells**, that resemble linked sausages (review Figure 15-3). Each Schwann cell is wrapped around the axon, forming a spiral of flattened membrane layers. Cell membranes consist predominantly of lipids (Chapter 6), and lipids are poor conductors of electricity. Thus, the myelin sheath also functions as living "electri-

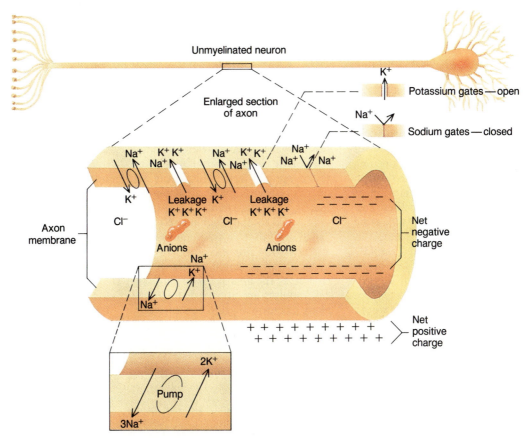

**Figure 15-4**
**Polarization in the resting neuron.** An ATP-driven sodium-potassium pump maintains membrane polarity by pumping sodium ions ($Na^+$) out of the cell, leaving behind chloride ions ($Cl^-$). Each sodium exported is accompanied by the import of a potassium ion ($K^+$). Unlike large sodium ions however, the potassium ions are small enough to leak back through the membrane, diffusing out of the cell where the $K^+$ concentration is lower. The result is an overall positive charge outside the cell ($Na^+$ and $K^+$), and a negative charge inside the cell. Large organic anions (negatively charged ions) cannot escape the cell, so they remain in the cytoplasm, intensifying the ionic gradient. The strength of the electrical potential produced by this gradient is measured as voltage.

cal tape" that insulates the axon against leakage of the impulse, thereby preventing neurological short-circuiting between adjacent nerve cells. The insulation is not complete, however, as tiny naked gaps remain exposed between adjacent Schwann cells. These uninsulated gaps, called the **nodes of Ranvier,** help accelerate neurological impulses by a mechanism presented in the upcoming section, "Neurological Impulses."

### Synaptic Knobs

Neurological impulses are transferred from neurons to target cells by the synaptic knobs. Neurons that have had their synaptic knobs removed by research scientists can no longer elicit a

response. These experimentally crippled cells can carry the message to their targets but cannot deliver the command without the knobs.

## Neurological Impulses

Transmission of an impulse along a neuron is often compared to an electrical signal conducted along a wire. But this analogy fails to reveal the source of these electrical impulses. How can planting a sewing needle into your thumb generate the electric charge that tells the brain to experience the pain? The source of this electricity is not the needle; it is the nerve cell itself.

### Electrical Signals

Even in their "resting" states (when they are not carrying an impulse), nerve cells are generating an electrical charge. In actuality, a resting neuron is not resting at all but is steadily maintaining its *excitability,* that is, the capacity to conduct an impulse in response to a stimulus. The neuron primes itself by building an ionic gradient along its plasma membrane, as described in Figure 15-4. With its net positive charge on the outside ($Na^+$ and $K^+$) and a buildup of negative charges on the inside ($Cl^-$ and large anions), the "resting" neuron has generated electric potential, a reservoir of energy stored as voltage. This voltage has been measured by inserting microscopic electrodes into a nonconducting nerve cell, revealing the axon's **resting potential** to be about − 70 millivolts. As long as the membrane provides a barrier that maintains this *polarity* (separation of charges), the electric potential is maintained.

Any stimulus that exceeds the neuron's threshold triggers the membrane to reverse its polarity at the point of stimulation. It does so by opening pores in the membrane, pores through which sodium can flow. As soon as these *sodium gates* are open, these positive ions stream through. (At the same time, potassium gates are closed.) In theory, the charges would rapidly equalize on both sides of the membrane, briefly depolarizing it. In reality, however, the sodium rushes through so fast that the charge would

be momentarily reversed. The momentum of the ionic flow would cause the rushing ions to "overshoot" the mark of exact depolarization, much like a swinging pendulum overshoots its lowest energy point. The resulting polarity reversal is the upward spike on the graph in Figure 15-5. The sodium gates then slam shut and the potassium gates open, allowing the original polarity of the resting membrane to be reestablished. This entire sequence of events, called an **action potential**, takes place in about one-thousandth of a second. Such neurological "sparking" (the action potential) starts a *neurological impulse.*

An action potential occurs in one spot along the neuron. Yet neurological impulses are not fixed; they race along a neuron. So how can an action

potential at one point excite a neuron at some distant point? The answer is that an action potential is also a stimulus. It disturbs the adjacent resting membrane, causing its sodium gates to spring open and initiate an action potential at this new point. (By now the preceding action potential has subsided.) This new action potential then initiates another next to it, which produces a wave of action potentials that travel along to the end of the excited neuron. Because the energy for relaying action potentials is generated equally along the neuron's plasma membrane, the impulse arrives at its target cell as strong as it was at its point of origin.

Axons that are insulated with myelin sheaths can reverse their membrane polarity only at the nodes of

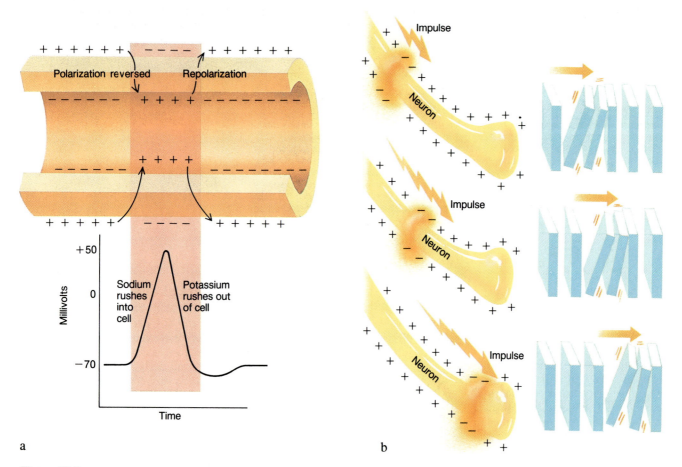

a                                                                          b

**Figure 15-5**

**An action potential** occurs when the membrane's sodium gates open and allow the charge imbalance to briefly equalize. (The peak in the action potential graph is due to overshoot.) A moment later, simultaneous closure of the sodium gates and opening of the potassium gates allow K⁺ to diffuse across the membrane and reestablish the resting potential at that location. The disturbance, however, has triggered the adjacent area to briefly open its sodium gates, which repeats the performance. In this way, the impulse races along to the end of the neuron, much like falling dominoes, but with an important difference—each "domino" along the neuron rights itself immediately after falling.

Ranvier (the uninsulated spaces between Schwann cells). An action potential in one node triggers polarity reversal in the next node, so that the impulse skips from node to node rather than taking the slower continuous route through the membrane, in what is referred to as *saltatory propagation* (saltare = to leap) of impulses. Such propagation accelerates the velocity of transmission along high-speed neurons, such as those that lead from the spine to the foot.

## Chemical Signals — Jumping the Synaptic Gorge

Neurons articulate with their target cells at neurojunctures called **synapses**. The neuron is separated from its target by a gap called the **synaptic cleft** (Figure 15-6). The cleft in some invertebrates is narrow enough for direct electrical transmission of an impulse to the target cell. In most of the neurojunctures of vertebrates,

however, the cleft is too great for the incoming action potential to jump to the target cell. Instead, the message is transmitted across the synaptic cleft by chemicals called **neurotransmitters**. These substances are stored in vesicles packed in the synaptic knob, where they have little effect as long as they remain locked away inside these vesicular containers (Figure 15-7). Receptor molecules on the postsynaptic membrane of the target cell can join and react with the neurotransmitter chemical when it is released from the neuron's vesicles. But as long as the cleft contains no neurotransmitter, the receptor molecules remain unbound, and the target cell continues to rest.

When an impulse reaches the end of a neuron, the membranes of vesicles fuse with the plasma membrane of the synaptic knob. In other words, the neurological impulse triggers exocytosis (Chapter 7), and the synaptic knob discharges its neurotransmitter into the cleft. The released chemicals

bind to receptor sites on the target cell's membrane. Binding opens nearby ion pores, leading to a change in membrane polarity. (Resting effector cells generate membrane potential just as neurons do.) Neurons that excite their targets affect receptors that open the sodium gates. The flow of sodium ions reverses the polarity in the target cell's membrane, which causes it to secrete or contract, or conduct another neurological impulse to its target cell. In contrast, neurons that inhibit their targets affect different receptors, those that open potassium gates. Opening these gates *hyper*polarizes the postsynaptic membrane, increasing its activation threshold so that more excitatory impulses are required to reverse the membrane's polarity and excite the target cell.

If the synaptic story ended here, the target cell would be unable to recover from the chemical message. Neurochemical stimulation at the synapse would lock an activated target into a

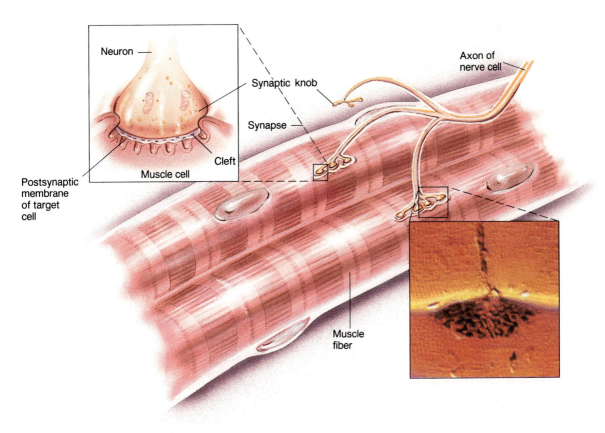

**Figure 15-6**
**Neurojuncture between neuron and effector cell,** in this case a muscle. The target cell could also be the cells of a gland or another neuron.

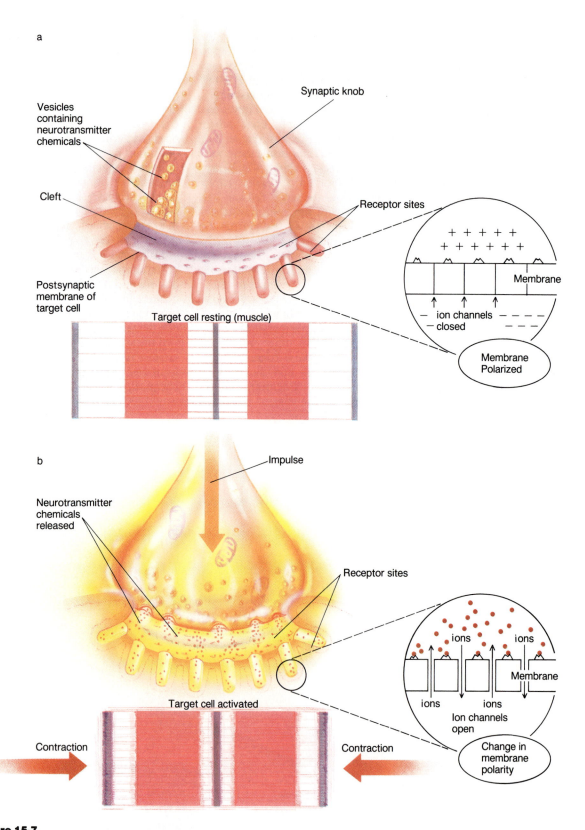

a

Synaptic knob

Vesicles
containing
neurotransmitter
chemicals

Cleft

Receptor sites

Postsynaptic
membrane of
target cell

+ + + + +
+ + + + + +

Membrane

− ion channels −
− closed −

Membrane
Polarized

Target cell resting (muscle)

b

Impulse

Neurotransmitter
chemicals
released

Receptor sites

ions        ions

Membrane

ions      ions

Ion channels
open

Target cell activated

Change in
membrane
polarity

Contraction

Contraction

**Figure 15-7**
**Transmission of nerve impulse at the synapse.** In a resting neuron *(a)* the cleft contains no
neurotransmitter chemicals, the receptor sites remain empty, and the target cell stays relaxed. *(b)*
Arrival of an action potential at the synaptic knob promotes fusion between membranes of
vesicles and cell, leading to exocytosis. Released neurotransmitter saturates the synaptic cleft
and binds to receptors on the target cell membrane, which open ion gates in response. If enough
gates are open, the polarity of the target's membrane changes, generating a response in the
target cell.

perpetually excited state. Yet a muscle fiber can relax the instant after an impulse is delivered. Within one-five-hundredth of a second, some types of neurotransmitters are destroyed by enzymes on the target cell's membrane. These enzymes disassemble the neurotransmitter almost as soon as it reacts with the receptors. The duration of the activation is therefore very brief. Other types of neurotransmitters rapidly diffuse out of the synaptic cleft or are immediately reabsorbed by the synaptic knob. The effect of each impulse therefore lasts no more than a few milliseconds.

A target cell's threshold is generally greater than can be exceeded by a single impulse. In fact, most target cells remain resting until they receive at least 50 simultaneous excitatory impulses. This threshold can be further elevated by the presence of **inhibitory neurotransmitters**, substances released from inhibitory neurons where they synapse with the target cell. Impulses from inhibitory neurons open the potassium gates and render the effector less responsive to the commands of the excitatory neurons. A target cell may be influenced by impulses from hundreds of different excitatory and inhibitory neurons (Figure 15-8). The effector therefore responds to a complex variety of environmental changes. If the combined input from all the incoming synapses is enough to trigger a response, the cell performs its activities. Until then, it continues to rest.

About 30 chemicals are believed to be neurotransmitters, a few of which are described in Table 15-1. Among the most recently discovered are the *endorphins* and *enkephalins,* natural opiates (morphinelike substances) that affect neurons in the brain and spinal cord. These nerve cells have receptors to which the internal opiates attach. If the neuron is part of a pain-transmitting pathway, attachment of endorphins and enkephalins reduces the relay of information to the brain, thereby blocking the sensation of pain. The pain-killing effectiveness of endorphins is equal to that of morphine, the most potent pain medicine known. Other cells, notably neurons associated with mood and emotion, also have receptors that, when occupied by endorphins (or the synthetic opiates morphine, heroine, and codeine), produce sensations of euphoria (well-being). As a result, these pain killers also generate a "high" that can lead to abuse and addiction. Even physical exercise may be addictive due to the release of endorphins and enkephalins during strenuous or painful workouts. This phenomenon is commonly known as "runner's high."

Unfortunately, neurons do not always function properly. Things can go dreadfully wrong at the synapse—for example, when the cleft is occupied

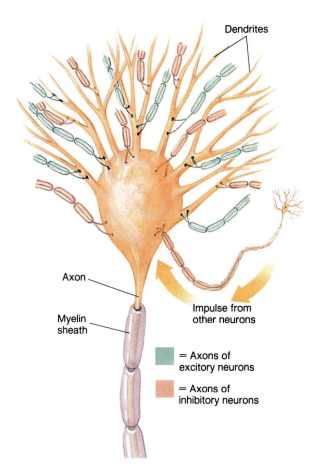

Dendrites

Axon

Myelin sheath

Impulse from other neurons

= Axons of excitory neurons

= Axons of inhibitory neurons

**Figure 15-8**
**Multiple synapses on a single neuron** can run into the thousands. Whether the neuron relays the message depends on the combined stimulation of the excitatory and inhibitory neurons exceeding the membrane's threshold.

TABLE 15-1

## Some Neurotransmitters and Their Effects

| Neuro-Transmitter | + or − [a] | Most Common Target Cells | Predominant Effect |
|---|---|---|---|
| Acetylcholine | + | Voluntary muscles | Stimulates muscle contraction. |
| | − | Heart muscle | Increases threshold of contraction. |
| Glycine | − | Motor neurons to voluntary muscles | Raises threshold of excitation, checking uncontrolled muscle contraction. |
| Dopamine | − | Neurons that produce acetylcholine | Prevents overactivity of neurons that activate muscles. (Deficiencies result in uncontrolled muscle contractions of Parkinson's disease.) |
| Norepinephrine (noradrenaline) | + | Neurons of central nervous system responsible for arousal, attention, and mood; involuntary muscles (e.g., heart); glands | Increases alertness and attention; heightens readiness for muscular activity. |
| GABA | − | Motor neurons to voluntary muscles | Prevents uncontrolled muscle contraction. |
| Serotonin | − | Neurons in the brain that maintain wakefulness | Induces sleep; may modulate mood. |

[a] "+" = excitory; "−" = inhibitory.

by chemicals that interfere with neurotransmitters or their immediate removal (see Bioline: Deadly Meddling at the Synapse).

## Interpreting the Impulses

Different types of neurons respond to different types of stimuli, some of which are from external sources and others from internal changes. **Sensory neurons** detect changes in the external and internal environments, generating impulses that travel *toward* the central nervous system. **Interneurons** receive neurological impulses and relay to other neurons. Interneurons are therefore the neurological switchtracks that route incoming or outgoing impulses, integrating the millions of signals racing through the system. Most of the neurons in the brain and spinal cord are interneurons. Outgoing impulses are carried by **motor neurons**, the nerve cells that affect muscles or glands.

All these neurons transmit impulses in exactly the same way and with the same degree of strength. A neuron is "all-or-none"—it either fires an impulse or it doesn't. It does not conduct a stronger signal in response to increased intensity of stimulation. Yet we are capable of detecting quantitative differences in stimuli. Our sensory neurons do not report temperature as simply "hot" or "not hot," but readily perceive a range of thermal differences, from unpleasantly cold to scalding hot. This ability may be due in part to the presence of many neurons that have different thresholds. Hence, they

# B I O L I N E

## DEADLY MEDDLING AT THE SYNAPSE

Nerve poisons! Few substances have acquired such notorious reputations as this group of chemicals, and none deserves the reputation more. In general, they are the most toxic substances on the planet. In minute quantities they can arrest muscle activity by interfering with *acetylcholine,* the neurotransmitter released by motor nerves to activate their target muscle cells. The following examples illustrate opposite mechanisms by which nerve poisons sabotage the normal function of the synapse.

### Curare

For centuries hunters in South America enhanced their predatory skills by dipping their weapons in extracts from various species of tropical plants. A small dart from a blowgun becomes a missile of certain death when coated with curare, the powerful nerve poison contained in these botanical extracts. Curare blocks the neurotransmitter receptor sites on the postsynaptic membrane of muscle cells. As a result, muscles fail to respond to neurological commands to contract, even though the synapse is saturated with acetylcholine. Muscles are paralyzed in a state of relaxation, a phenomenon called *flaccid paralysis* (flaccid = floppy). Death occurs quickly once the muscles used for breathing stop functioning. The victim suffocates.

### Organophosphates and Nerve Gas

Some neurotoxins don't interfere with acetylcholine, but rather inhibit its enzymatic destruction. In the presence of these toxins, the neurotransmitter remains in the synapse, causing perpetual muscle contraction. These victims suffer *spastic paralysis,* an inability to relax the muscles. When inhaled, nerve gas causes this type of paralysis. So do organophosphate pesticides, which are deadly for humans as well as for the insect pests for which they are intended.

### Tetanus

Another type of spastic paralysis is associated with a simple bacterial infection of wounds. This disease is *tetanus.* When growing in wounds, tetanus bacteria release a powerful neurotoxin that seeps into the bloodstream and invades neuromuscular synapses throughout the body. Only inhibitory synapses are affected, however. The toxin blocks the attachment of *glycine,* an inhibitory neurotransmitter, to its receptors on the postsynaptic membrane of muscles. The result is a lowered threshold of excitation of the muscle, so that even low levels of activating impulse set off unchecked muscle spasms throughout the body. The muscles become locked in a state of complete "tetanus" (a word meaning muscle contraction). Although the jaw muscles are the first to be affected (tetanus is commonly known as "lockjaw"), the toxin eventually hits all the body's voluntary muscles. The back is bent into an exaggerated arch by muscle spasms strong enough to snap the backbone before the victim eventually loses the ability to breathe and dies of asphyxiation.

respond to different levels of the same type of stimulus. Low-level stimulation activates the few neurons with the lowest thresholds. A stronger stimulus triggers higher threshold nerve cells leading to slightly different parts of the brain that interpret the stimulus as "hotter." A stronger stimulus also activates more nerve cells, increasing the sum of the impulses that reach the brain. Similarly, stimulus strength may also influence how frequently action potentials are launched down a neuron. The greater the frequency, the stronger the response in the target cells.

Let's return to the unpleasant experience of the needle in your finger. Before you even look at the damage, you know exactly what has happened. Your brain has been notified of the disturbance and translates the incoming neurological signals as "sharp pain in the tip of the left index finger." The action potentials triggered by this stimulus, however, carry no such message. The group of neurons that carries these impulses from the finger terminate at a portion of the brain that alerts the consciousness of the exact location and nature of the disturbance. Any time the cells in this portion of the brain are activated, the same sensation is created in the same location in the body. Therefore, it is the brain that perceives the sensation, not the finger. This mechanism is dramatically demonstrated during open skull surgery. Because the brain has no sensory receptors of its own, electrodes can be painlessly inserted into the exposed brain of a fully conscious patient. If the electrode stimulates the previously described portion of the brain, the patient will experience a very real sensation of a needle piercing his or her unassaulted index finger.

# Neurological Teamwork —The Nervous System

The simple all-or-none activities of a single neuron could hardly provide the adaptability needed for complex animals to adjust to constant changes in internal and external environments.

In the case of humans, such adaptability requires the participation of billions of neurons, each making its "fire-or-don't-fire" decision. Twelve billion such decisions are occurring in your body at this instant, the sum of which allows you to conduct all the fundamental biological processes that sustain your life. As mentioned earlier, organizing and coordinating such a biologically ambitious project as a human being, a lobster, or a mountain goat require centralization of control. Let's examine the simplest example of centralization, the reflex arc.

## The Reflex Arc

The quickest motor reactions to stimuli are reflex responses. Their speed is due to the direct delivery of the impulse to the nearest part of the central nervous system, and the immediate return of an impulse for triggering a response. This neurological arc can be illustrated by one such pathway, the one responsible for your foot jumping forward when the doctor strikes below your knee with a rubber hammer. The hammer's impact on the tendon stretches the attached thigh muscle, which reacts by automatically shortening. This type of arc is called the *stretch reflex*. Special sensory neurons, called *stretch receptors*, are embedded in muscles. Whenever the muscle stretches beyond a certain length, stretch receptors fire an impulse to the interneurons of the spinal cord. The message is immediately relayed to the motor neurons that lead back to the muscle, triggering it to contract. In this way, the muscle maintains approximately the same length and so enables posture to be maintained (Figure 15-9). The circuit is called a "reflex" because it is automatic and is not under deliberate control.

Many reflexes occur without even conscious awareness of the stimulus or the reaction. The stretch reflex automatically helps you maintain your posture and balance but can be overridden by stronger input from other motor neurons. Examples of such neurons are those that initiate the muscle changes needed for walking or

for compensating for a brief loss of footing.

The reflex arc is not limited to maintaining posture. It also provides the ability to speedily withdraw from dangerous stimuli. Your day is full of reflex responses. An overambitious sip of steaming coffee, for example, triggers a reflex activation of the muscles in the tongue, jaw, and mouth. The tongue automatically jumps to the back of the mouth (which may open and discharge the liquid before your brain can consider the opinion of the other people in the restaurant). Other reflexes include regulation of blood pressure, control of pupil size in response to changes in light intensity, and withdrawal of the hand when it encounters a sharp or hot object. These are involuntary responses that occur very rapidly in a *stereotyped* manner. That is, the reflex is the same every time the same simple stimulus is encountered. Awareness of the stimulus comes an instant later, when the information reaches the brain via other neurons that have been fired in response to the original stimulus or to the changes induced by the reflex reaction. Reflexes are the simplest form of animal behavior (see Animal Behavior I: The Neurological Basis of Behavior).

## Nerves and Neural Clusters

*Nerves* are "living cables" composed of many individual neurons bundled together in parallel alignment, plus their supporting cells (Figure 15-10). Neurons also cluster together to form control centers. In vertebrates, these centers are the brain, spinal cord, and ganglia. The neurons of nerves and control clusters are supported by nonconducting *glial cells*. Some glial cells (the Schwann cells) accelerate neurological impulses. Others are believed to nourish the nerve, protect it, and supply additional services. The glial cells outnumber the neurons by about 10 to one.

In most animals, the nerves of vertebrates are organized into a nervous system that consists of two fundamental divisions, the central and peripheral systems. In vertebrates, the **central**

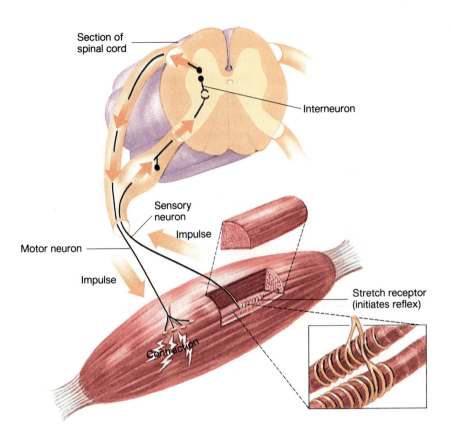

**Figure 15-9**
**The reflex arc** is illustrated here by the stretch reflex.

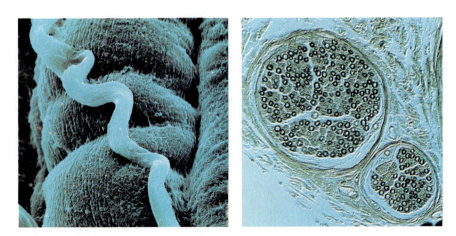

**Figure 15-10**
**A motor nerve** winding across the surface of muscle fibers. In cross section, the individual neurons in the nerve are seen, bound together in a sheath of connective tissue.

nervous system includes the brain and spinal cord, which are neuron clusters encased in the skull and backbone. The neurons that permeate the rest of the body comprise the **peripheral nervous system**. These two divisions of the nervous system constantly interact with one another. The peripheral nervous system supplies information to the central nervous system, which processes the input and directs impulses to the appropriate muscles and glands. These instructions are delivered to target cells by motor neurons of the peripheral system.

## The Central Nervous System

### THE BRAIN

Although the **brain** constitutes only about 2.5 percent of your body weight,

it consumes 25 percent of your body's oxygen supplies to generate the ATP needed to fuel its activities. If not replenished, the brain's oxygen supplies would be exhausted in 10 seconds.

The brain is not the firm object that many people imagine it to be, but a rather delicate 3-pound (about 1.5 kilogram) mass of gelatin-like tissue. It consists of two distinct regions that differ in color. The darker region, called **gray matter**, contains the cell bodies and dendrites of the brain's neurons. In other words, gray matter is rich with neuron-to-neuron synapses, places where neurological associations are made. The other region is called **white matter** to denote its lighter color. Myelinated axons dominate the white matter, its whiteness provided by the myelin sheaths. Enveloped in the

white matter lie hollow chambers filled with a liquid, the **cerebrospinal fluid**. This fluid also surrounds the delicate brain (and spinal cord), cushioning it against injury. To keep the brain afloat, the entire surface of the central nervous system is encased in a watertight membrane, the *meninges,* that prevents the cerebrospinal fluid from seeping out into the surrounding tissues.

The vertebrate brain is divided into three principal regions, the **forebrain**, **midbrain**, and **hindbrain** (Figure 15-11). In humans, the forebrain dominates the brain and is the region responsible for such cognitive processes as conscious thought, reasoning, memory, and language. The hindbrain and the tiny midbrain are small in humans. Together, these two smaller

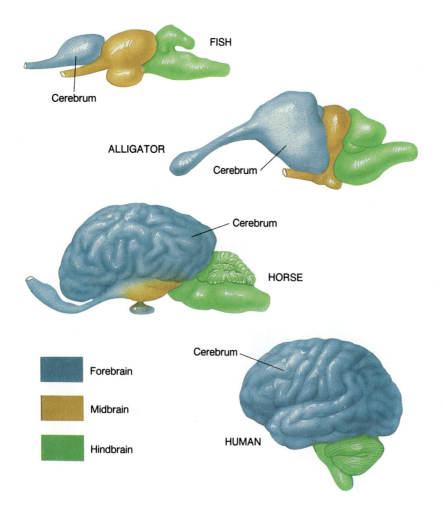

FISH

Cerebrum

ALLIGATOR

Cerebrum

Cerebrum

HORSE

Cerebrum

Forebrain

Midbrain

Hindbrain

HUMAN

**Figure 15-11**
**Evolutionary changes in the brains of vertebrates.** These animals, shown here in their order of evolutionary complexity, reveal the tendency toward increased cerebrum size and reduction in midbrain importance. In lower vertebrates, the midbrain processes auditory and optical input (sound and vision). The midbrain is diminished in mammals, which relay impulses from eyes and ears directly to the forebrain.

# ANIMAL BEHAVIOR I

## THE NEUROLOGICAL BASIS OF BEHAVIOR

We often find ourselves fascinated, horrified, and even inspired by the behavior of other animals. We admire the courage of the mother bird who risks her life feigning a broken wing to lure a cat away from her babies. We wince at the greed of sharks. A couple of snow monkeys "lovingly" grooming one another touches us with their apparent selfless devotion to each other. We tend to see anthropomorphic (human-like) motives in animal behavior. While that may make animals seem more understandable, it does little to help us understand the real reason animals act as they do.

Many behaviors, or the potential to perform them, are as much a part of an animal's inheritance as wings and fur color — they are gene-determined traits that an animal inherits from its parents. Some behaviors are programmed into an animal's neurological circuits. When that circuit is activated by a particular stimulus, the program "plays" that behavior — always the same behavior. Most of these are adaptations as essential to an animal's survival as are its structural features. A frog's tongue may be anatomically well-suited for catching insects, but without "knowing" when and how to perform the tongue-flick, the frog would starve. The tongue-flick is not a learned behavior. The action is ingrained in the animal's nervous system.

Each behavior is an animal's coordinated neuromuscular response to an internal or external stimulus. As the word "neuromuscular" implies, behavior is determined by the nervous system. Muscle activity drives the movement that characterizes the behavior, but neural activity determines which muscles will be activated and the sequence of activity. Swallowing behavior forces a bit of food in the proper direction only if the muscles in the tongue, throat, and airways act in the proper sequence. Should the sequence be wrong, food will be moved backward (or down the wrong tube into the respiratory tract). The swallowing action normally works well because each muscle automatically contracts immediately after the preceding muscles contract, squeezing the food in the correct direction. This automatic neuron-coordinated behavior is an example of a relatively complicated reflex. Even more complex behavior, such as a cheetah chasing and bringing down an impala, is also coordinated by neurological activity.

Behaviors, and their corresponding neural circuits, are naturally selected like any other genetically acquired adaptation. Thus, the study of animal behavior, a science called **ethology**, is another area of biology in which natural selection is the underlying principle. An animal either behaves appropriately according to the circumstances (to escape a predator, for example) or it dies, leaving fewer offspring than an animal whose behavior allows it to survive (and reproduce) longer.

Behavioral biologists avoid the temptation to ascribe human-like motives to a particular behavior. That "fast-thinking courageous" mother bird we admired earlier is not being either smart or brave. She is simply following automatic neurological instructions, a genetically acquired program that increases the likelihood that her chicks will survive. The female chicks that survive thanks to their mother's automatic diversion behavior will pass on this behavioral trait to their progeny, and the likelihood of species success is improved.

Nonetheless, courage, love, and devotion do influence some behaviors in at least one animal species — humans. Very few human behaviors are believed to be programmed for automatic play following a simple stimulus. Perhaps our greatest behavioral adaptation is an ability to learn new behaviors that suit the conditions. Other animals are compelled more by inherited instructions that direct **innate** (inborn) behaviors, also called *instincts*. Whether innate or learned, however, all behavior has its roots in the nervous system.

## Neurology of Innate Behavior

The neural circuits that direct the enactment of a particular behavior are generally believed to be analogous to the circuits that instruct a computer to perform a sequence of automatic activities. With a particular program, the same command always stimulates the same response from the computer. With a different program, that stimulus would trigger another type of response. The predetermined nature of certain behaviors is reflected in the terms behavioral biologists use for neurological circuits. They call them *genetic programs*.

## Closed Genetic Programs

Innate behaviors that are not modifiable by experience are called **closed genetic programs**. They compel an animal to respond with **stereotyped behavior**, that is, behavior that is automatically launched in response to the **releaser** (the stimulus that provokes a particular behavior). Innate stereotyped behavior is the same regardless of the animal's experiences. A garden spider always weaves the same web pattern. A cowbird sings its mating song flawlessly without having to either practice or copy the song of another cowbird. Air-breathing animals cannot be taught to inhale and exhale — breathing behavior must be automatic and effective from the moment its first breath is needed.

Animals come equipped with a stock of innate stereotyped behaviors, each involving the precisely timed mobilization of many muscles. They include simple reflexes (withdrawal from pain, for example) or more complicated patterns, such as some specific predatory behaviors (the frog's tongue-flick) and maneuvers to escape a predator. Because each innate stereotyped behavior is virtually identical every time it is performed, it is called a **fixed action pattern**.

Fixed action patterns are essential for behaviors that must work perfectly the first time they are tried. A bird taking its initial flight from its cliffside nest cannot take a practice run. Getting it wrong even once is fatal. Similarly, an escape behavior triggered by a predator's attack must be performed correctly the first time or the animal ends up as dinner. These and other innate stereotyped behaviors are programmed into the nervous system, so the animal doesn't have to learn them. The presence of a particular stimulus, the releaser, excites a chain of neurons that bring about the behavior. It is automatic; once started, the behavior always proceeds to completion unless another stimulus (or injury) interrupts it. Many of these behaviors are programmed into the animal during embryonic development.

## Open Genetic Programs

A bird's flight may be a fixed action pattern, but landing is another story altogether. Safely stopping the flight is a more complex behavior than is flying (ask any airplane pilot), one that requires considerable adjustment to existing conditions. A bird landing on a branch is a maneuver that changes each time it happens because the target and the conditions (wind, size of branch, angle of approach) are different. In other words, it requires the bird to respond to feedback from the sense organs. If it fails to adjust its landing behavior according to the existing conditions, the bird will crash. Flight and other behaviors that must change moment by moment are called **open genetic programs**. The basic program is inherited, but the overall response to a particular stimulus is modified by the conditions of the moment and by *learning.* With open genetic programs, the behavior pattern is *nonfixed;* rather it is the coproduct of inherited potential and modification of that potential by experience. The development of innate and learned behavior are discussed in more detail in Chapter 24 (*Animal Behavior Box II*).

## Behavioral Thresholds

An animal doesn't always respond to a releaser. Several neural mechanisms establish a hierarchy of responsiveness, so an animal can respond when two releasers stimulate conflicting behaviors simultaneously. In general, releasers that trigger escape behaviors take precedent, so an animal confronted with a predator and a potential mate at the same time will escape the threat and ignore the mate. Even this hierarchy can be modified, however. An animal's threshold of sensitivity to a releaser varies according to the conditions that exist at that moment. Two such variables are *drive* and *motivation.* Hunger, thirst, and reproductive urges change an animal's responsiveness to a particular stimulus. An animal may ignore a food stimulus when satiated, whereas the same stimulus will trigger an automatic feeding response when it is hungry, even to the point of ignoring a predatory threat. In vertebrates, such variations in motivation and drive are controlled largely by the hypothalamus. Sexual motivation, hunger, thirst, etc, all influence responsiveness stimuli that lead to behaviors that satisfy these short-term needs. The hypothalamus is part of the *reticular formation,* which regulates the general state of neurological arousal and thus receptiveness to stimuli (this part of the brain is discussed in the next section of this chapter). Although the reticular formation has probably filtered out the sensation of your chairseat, a sharp skin-piercing object there would arouse the system, and you would automatically launch into an embarrassing "leap up and scream" response.

This first special feature in a series on animal behavior reveals that behavior, like other traits, is a biologically determined product of natural selection. Later reports in this series focus on embryonic and post embryonic development of behavioral traits and the ways these behaviors influence reproductive success and community cooperation.

regions make up the **brainstem**. The brainstem coordinates most of the automatic, involuntary body processes, such as breathing, heart beat, balance, and muscle coordination.

The major evolutionary change in the vertebrate brain is clearly revealed by the steady increase of forebrain size, particularly the **cerebrum**. In humans, it has become the most predominant part of the brain (Figure 15-12). Its two halves, called cerebral hemispheres, are generally associated with "higher" brain functions, such as speech and rationale. Actually, these higher functions are attributes of the outer, highly wrinkled layer of the cerebrum, the ce-

rebral cortex (cortex = rind). Every cubic inch of this one-tenth of an inch thick layer of gray matter contains 10,000 miles of connecting neurons. The convolutions (wrinkles) in the cerebrum increase the surface area of the cortex without enlarging the space required to house it. These convolutions are generally signs of higher cerebral capabilities, such as intellect, artistic and creative abilities, and a greater capacity for learning and memory compared to the smooth-brained lower vertebrates. Each cerebral hemisphere is composed of four lobes: the frontal, parietal, temporal, and occipital (see Figure 15-13).

Four narrow strips of cortex on either side of the *fissure of Rolando* (Figure 15-14) house the brain's motor and sensory regions. Notice that the amount of brain space devoted to each body part is proportional to the degree of neurological control over that part, and not to the size of the body part. For example, the intricate manipulations of the human hand require much more brain control than does regulating the relatively coarse movements of the legs, even though legs are much larger than hands.

Most of the sensory information streaming into the cerebrum is routed through the thalamus, which also coordinates outgoing motor signals. The thalamus contains part of the *reticular formation,* the rest of which is located in the brainstem. The reticular formation selectively arouses brain activity, producing a state of wakeful alertness. These neurons intercept incoming impulses from sensory nerves and, in response to some of these send a burst of neurological stimulation to the cerebral cortex. The resulting cerebral arousal translates into wakefulness and an ability to concentrate thoughts and attention. The reticular formation can be activated by any number of internal factors or external stimuli — the sound of a horn blasting, a flashlight shining in your eyes, or the middle-of-the-night memory of a forgotten biology exam tomorrow morning. When the reticular system fails to maintain arousal, however, the brain falls asleep and stays sleeping until the reticular formation is activated, for instance, by sensory input from an alarm clock.

The reticular formation is small; it is no larger than your little finger. None-

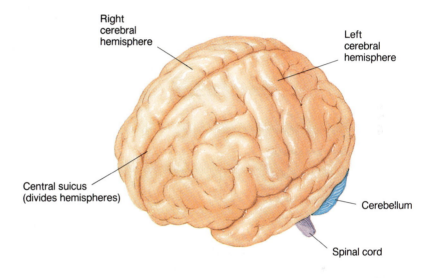

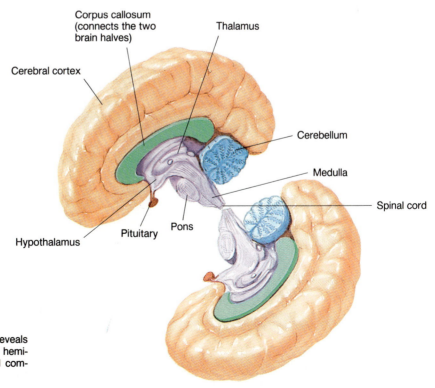

**Figure 15-12**
**Two views of the brain.** The intact human brain reveals mostly cerebrum. Separating the two cerebral hemispheres reveals some of the brain's structural complexity.

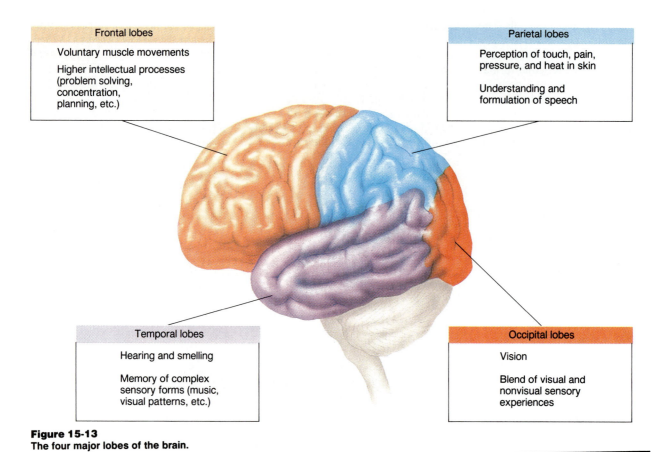

**Frontal lobes**

Voluntary muscle movements

Higher intellectual processes (problem solving, concentration, planning, etc.)

**Parietal lobes**

Perception of touch, pain, pressure, and heat in skin

Understanding and formulation of speech

**Temporal lobes**

Hearing and smelling

Memory of complex sensory forms (music, visual patterns, etc.)

**Occipital lobes**

Vision

Blend of visual and nonvisual sensory experiences

**Figure 15-13**
**The four major lobes of the brain.**

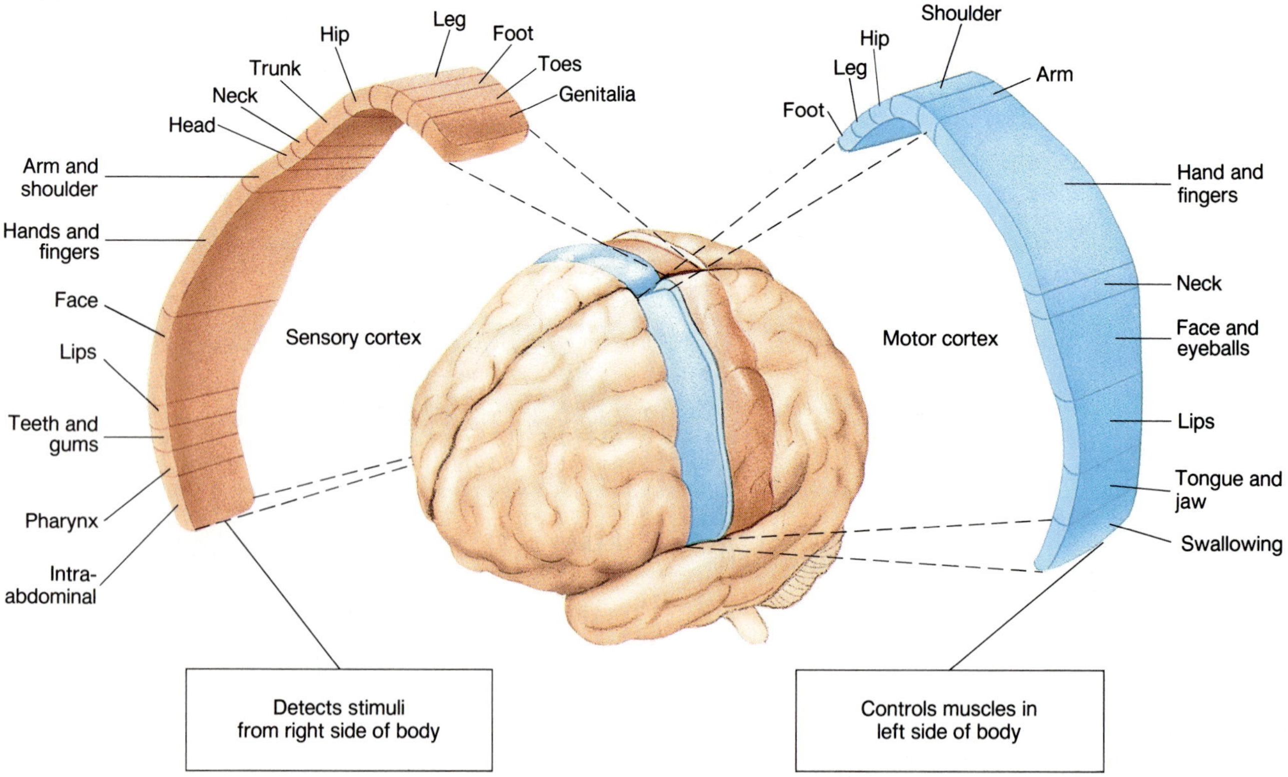

Hip
Leg
Foot
Trunk
Toes
Neck
Genitalia
Head

Shoulder
Hip
Arm
Leg
Foot

Arm and shoulder

Hand and fingers

Hands and fingers

Neck

Face

Face and eyeballs

Lips

Lips

Teeth and gums

Tongue and jaw

Pharynx

Swallowing

Intra-abdominal

Sensory cortex

Motor cortex

Detects stimuli from right side of body

Controls muscles in left side of body

**Figure 15-14**
**The motor and sensory cortex** is a narrow strip that spans the brain like headphones. The motor cortex activates various muscle groups for motion. The sensory cortex registers the sensations of touch, pain, pressure, heat, cold, and body position.

theless, it helps you cope with the 100 million impulses that assault the brain every waking second. It screens out the trivial while allowing vital or unusual signals through to alert the mind. Stop reading for a minute and notice the efficiency of the system. Listen to the sounds that a moment ago weren't in your awareness. Feel the contact of your clothing against your skin. Taste the flavor of your own mouth. The reticular system really works!

Extending from the bottom of the thalamus is a portion of the brain that is only about the size of your thumb tip, but without it homeostasis would be impossible. The **hypothalamus** (hypo = below) regulates body temperature, keeping it around 37° C (98.6° F). It also helps control blood pressure, heart rate, and the body's urges, such as hunger, thirst, and sex drive. The hypothalamus and the thalamus comprise part of an interconnected group of brain structures buried in the core of the forebrain. This complex is called the **limbic system** (Figure 15-15). Its different components are associated with pleasure and joy, pain and fury, memory, and an ability to balance emotions. Activating the limbic system can generate anger or euphoria, sexual arousal, or deep relaxation. One part of the system, the *hippocampus,* seems to be essential for short-term memory; without it, you would have already forgotten how this sentence began.

The major components of the brain and their functions are summarized in Table 15-2.

The human brain is connected to the peripheral nervous system in two ways. Twelve pairs of **cranial nerves** enter and leave the brain through holes in the skull. These nerves link the brain to sensory organs located in the head, such as those for seeing, hearing, tasting, and smelling. Most neurological information, however, enters and leaves the brain through the spinal cord.

### THE SPINAL CORD

The brainstem merges with the other major component of the central nervous system, the **spinal cord**. The spinal cord not only provides a centralized mass of neurons for processing neurological messages (as in the reflex arc), but also links the brain with that part of the peripheral nervous system not reached by the cranial nerves. Like the brain, the spinal cord is composed of white matter (myelinated axons) and gray matter (dendrites and cell bodies). However, in the spinal cord the arrangement is reversed—the white matter surrounds the gray matter (Figure 15-16). Bones along the spinal column, the *vertebrae,* articulate with one another to form flexible armor around the spinal cord. Through the spaces between the vertebrae, 31 pairs of **spinal nerves** extend from the spinal cord and branch into successively smaller nerve bundles that supply the limbs and torso with sensory and motor neurons.

In general, sensory impulses enter the spinal cord through *dorsal roots,* which are the branches of the spinal nerves that protrude from the back (dorsal) side of the spinal cord. Motor impulses leave the spinal column through the *ventral roots* that extend from the front (ventral side) of the spinal cord. The two roots join to form a single spinal nerve.

## The Peripheral Nervous System

The peripheral nervous system provides the neurological bridge between the central nervous system and the environment, and between the central nervous system and effectors. Peripheral neurons connect the central nervous system with virtually every part of the body. Signals enter and leave the brain and spinal cord through *afferent* (incoming) and *efferent* (outgoing) peripheral nerves. The **afferent nerves** supply our central nervous systems with the input needed for conscious awareness of and responsiveness to conditions in the body and in the world around us. The **efferent nerves** carry messages to the effectors, the muscles and glands. These outbound nerves are divided into two systems: the somatic and the autonomic nervous systems. The **somatic nervous system** carries messages to the muscles that move the skeleton. These movements generally follow voluntary orders from the brain, although stereotyped reflex movements are also under the control of the somatic nervous system. The **autonomic nervous system**

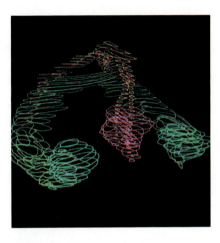

**Figure 15-15**
**The limbic system as envisioned by a computer.** The representation reveals how the system flares like a wishbone in the center of the brain.

controls the involuntary activities of the internal organs. The level of activity in each internal organ is influenced by the individual's energy supplies and state of stress at any given time. For example, when you are under extreme stress, exertion, or fear, heart and circulatory functions increase and digestive activities subside until calmer conditions are restored. To facilitate these adjustments the autonomic nervous system is really two distinct systems: (1) the **parasympathetic system**, which is active during relaxed normal activity, and (2) the **sympathetic system**, which takes over in times of emergency or prolonged exertion. Figure 15-17 depicts how these systems interplay with the central nervous system and with each other.

The activities of the sympathetic and parasympathetic autonomic systems are revealed in Figure 15-18. Notice that many homeostatic functions are stimulated by the parasympathetic neurons. In emergencies, these parasympathetic functions are temporarily inhibited by the sympathetic system and replaced by those functions that favor heightened awareness, alert responses, and strength. During "fight or flight" reactions, induced by fear or alarm, the adrenal gland releases hormones that further prepare the body to meet the emergency. This teamwork illustrates the interrelatedness of the nervous and endocrine systems.

TABLE 15-2

| Components of the Human Brain | | | |
|---|---|---|---|
| **Structure** | **Location** | **Function** | **Comments** |
| Brainstem | Hindbrain and midbrain | Controls breathing, heart beat, gastrointestinal activity | Automatic, involuntary functions; contains the major part of the reticular formation |
| Reticular formation | Brainstem and thalmus | Arouses brain activity often in response to incoming stimuli; screens out "trivial" sensory signals; coordinates voluntary muscle activity; houses respiratory and cardiovascular centers | Loss of reticular function leads to coma. |
| Cerebellum | Hindbrain | Regulates balance and muscle coordination | |
| Cerebrum: —cortex (gray matter | Forebrain | Controls conscious thought, speech, "intelligence," rationale, personality, sensory-motor coordination | This thin (3mm) outer layer is the origin of all cerebral functions. |
| —White matter | Forebrain | Relays impulses between cortex and rest of the brain and spinal cord | Controls no specific brain functions |
| Thalamus | Forebrain | Routes all incoming sensory information (except smell) to the cortex; also handles outgoing motor impulses | Contains part of the reticular formation |
| Hypothalamus | Forebrain | Maintains homeostasis; regulates body temperature, hunger, thirst, sex drive, blood pressure, and heart rate | |
| Limbic system | Dispersed throughout forebrain | Influences emotion, memory | Comprised of the thalamus, hypothalamus, amygdala hippocampus, and parts of the cortex |
| Corpus callosum | Between the cerebral hemispheres | Connects the two halves of the cerebrum | Informs each hemisphere of the other's activities |

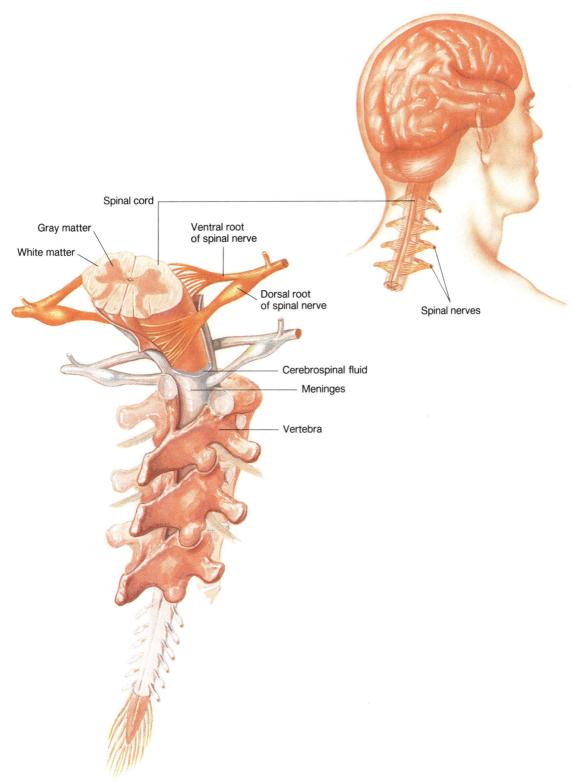

Spinal cord

Gray matter

White matter

Ventral root
of spinal nerve

Dorsal root
of spinal nerve

Cerebrospinal fluid

Meninges

Vertebra

Spinal nerves

**Figure 15-16**
**The spinal cord** extends through hollows in the vertebrae (the individual bones that make up the backbone) and joins with the brain through a hole in the base of the skull.

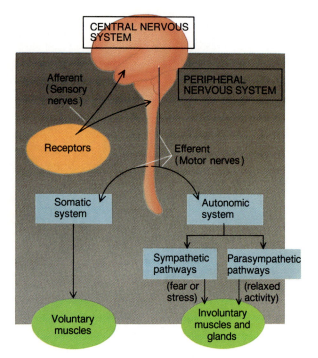

**Figure 15-17**
**Interplay between central and peripheral nervous systems.** Receptors relay information about the internal and external environment to the central nervous system by way of afferent (incoming) nerves. The central nervous system responds to this input by sending impulses to target cells through efferent (outgoing) nerves. Efferent nerves are divided into two systems. (1) the somatic system that acts on voluntary muscles, and (2) the autonomic system that either excites or inhibits involuntary muscles and glands. The exact effect of autonomic activity depends on whether the sympathetic or parasympathetic pathways are activated. (These effects are described in Figure 15-18.)

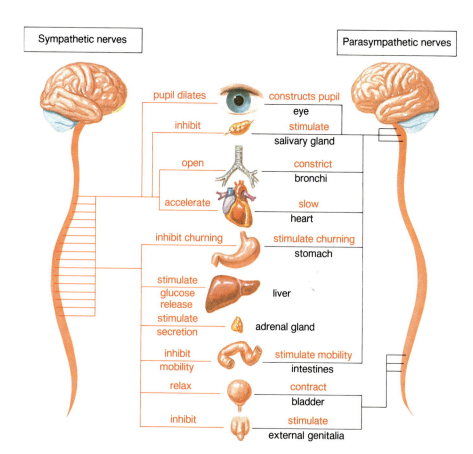

**Figure 15-18**
**The two divisions of the autonomic nervous system** control most of the same internal organs, but under contrasting circumstances. The effects of the parasympathetic system predominate during normal relaxed activity. During emergencies (fight or flight situations), the sympathetic system is dominant, eliciting physiological changes that increase the level of performance to better adapt to the emergency.

# The Senses — Monitoring the Environment

Awareness of your surroundings is provided by five specialized types of sensory structures. These structures sample the environment and rush their findings to the brain, which translates the impulses into pictures, smells, sounds, tastes, and sensations of pain, pressure, heat, and motion.

Examining an animal's sensory structures reveals a great deal about the environment to which it is adapted. Animals that live in total darkness, for example, rely less on vision and more on smell, touch, or sounds. Some deep-sea fishes and cave-dwelling reptiles lack eyes altogether. Bats are well adapted to their nocturnal lifestyle. Although their vision is poor, they adroitly fly through the night, finding food and avoiding obstacles. Bats rely predominantly on *echo location* to "see" where they are going. In other words, they use biological sonar, emitting sounds that return after bouncing from objects. This sensory adaptation allows the bat in Figure 15-19 to know the precise location of its prey and to avoid crashing into objects.

Although humans have poor echo location abilities, our specialized sensory organs collect other types of stimuli, sometimes amplifying these stimuli to increase the sensitivity of the senses. Each sensory structure is characterized by a specialized sensory receptor. These are listed below, followed by the type of stimulus they detect.

- *Mechanoreceptors* — mechanical energy, such as motion, touch, pressure, and sound.

- *Thermoreceptors* — heat and cold.

- *Chemoreceptors* — taste, smell, osmotic strength of blood, concentration of critical nutrients (glucose, amino acids), and gases ($O_2$ and $CO_2$) in blood.

- *Nociceptors* — pain induced by excess heat or pressure and by irritating chemicals (or chemicals released from damaged or inflamed tissue).

- *Photoreceptors* — visible light.

The following brief discussion of how each type of receptor contributes to sensory structures in humans provides a model for understanding analogous organs in other animals.

## Heat, Cold, Pressure, and Pain Receptors

Although the skin is not considered a specialized sensory organ, it is generously endowed with several types of sensory receptors. Many types of mechanoreceptors detect touch and pressure, nociceptors sense pain, and thermoreceptors distinguish temperatures. Neurons for detecting each type of sensation in skin and internal organs are organized into specialized receptor structures (Figure 15-20). Keep in mind that the sensations are actually perceived in the brain and not in the sensory structure itself.

## Chemoreceptors

Several types of chemoreceptors detect various substances dissolved in secretions of the nose and mouth or in tissue fluids. For example, the *olfactory* receptors on the nasal epithelium (olfaction = sense of smell) and the *taste buds* on the upper surface and edges of the tongue are chemoreceptors. Olfactory chemoreceptors form hairlike extensions that protrude from the epithelial cells lining the upper surface of the nasal cavity. (Some animals, however, use their skin, antennae, or other body parts to detect odors; see Figure 15-21.) The odor of dissolved substances is detected by

**Figure 15-19**
**Sonar-directed attack.** Bats depend on echo location to guide them to their prey in the dark.

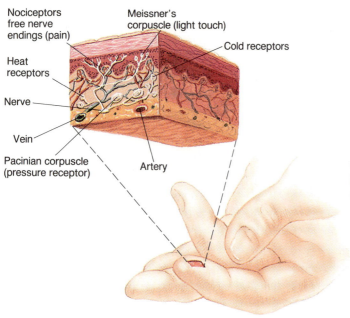

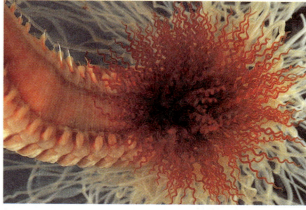

a

b

**Figure 15-20**
**Sensory receptors.** *(a)* In human skin. *(b)* The luxuriant mane of the spaghetti worm contains millions of sensory receptors that supply information about its environment.

**Figure 15-21**
**Chemoreceptors** in olfactory organs (antennae) of a male prometheus moth. These huge antennae can smell a female prometheus moth from a distance of 7 miles (11 km).

chemostimulation of these neurons, which fire their impulses to the olfactory bulb of the brain. Here the neural excitement is translated into a particular smell. Much of what we refer to as "taste" is really a combination of taste and smell. The taste buds are capable of registering only four fundamental tastes—sweet, sour, salty, and bitter. The receptors for each of these tastes reside in different regions of the tongue.

## Photoreceptors

To most people, vision is the richest form of sensory input. The complexity of our visual system lends support to the evolutionary importance of this sense. A carpet of photoreceptors provides the screen on which visual images are projected. But the system also depends on the participation of several accessory structures to regulate the amount of incoming light, direct and focus it, and maintain the proper distance between the pupil and the "projection screen." The eyeballs of vertebrates (and some invertebrates) embody all the structures needed to execute these complex visual tasks (Figure 15-22). Some variations in vi-

sual equipment are shown in Figure 15-23.

## Mechanoreceptors for Hearing and Balance

The neurons in the ear not only detect movement (vibrations) in air molecules for translation into sound, but they also help maintain equilibrium and balance. These structures, depicted in Figure 15-24, are distributed in three regions of the ear—the outer, middle, and inner ear.

### SOUND RECEPTION

The auditory components of the ear enable us to hear sounds. Sound vibrations travel outward through air from the source, much like ripples in a pond. These vibrations impinge on the *tympanic membrane* (the eardrum), causing it to vibrate gently. The quivering eardrum moves the tiny hammer-shaped bone that presses against it. The vibrations then travel to the next bone in the sequence, the "anvil." This exchange amplifies the magnitude by a factor of 20 before passing the vibrations to the last bone in the series, shaped like a stirrup. From here the vibrations enter the snailshell-

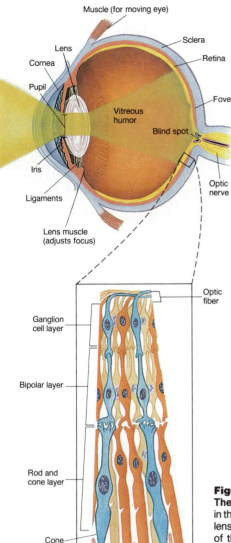

Muscle (for moving eye)

Sclera

Lens

Retina

Cornea

Pupil

Fovea

Vitreous humor

Blind spot

Iris

Optic nerve

Ligaments

Lens muscle (adjusts focus)

Optic fiber

Ganglion cell layer

Bipolar layer

Rod and cone layer

Cone

Rod

**Figure 15-22**

**The human eye.** Light enters through the cornea and then travels through the pupil, the dark hole in the middle of the colored iris. Pupil size changes to regulate the amount of light that enters. The lens projects the inverted image on the retina, the photosensitive blanket of cells that line the back of the eyeball. The retina has two kinds of photoreceptors: rods and cones. The acute light sensitivity of rods allows us to see in very dim light. Cones allow us to see color. Rods and cones relay their neurological signals to the brain by way of bipolar cells and the optic nerve. The hole through which the optic nerve exits the inner surface of the eye lacks photoreceptors and creates a small ''blind spot'' in the visual field.

shaped cochlea and generate movement in "hair cells." When these sensitive cells (mechanoreceptors) brush against the overlying membrane (the tectorial membrane), the movement stimulates an impulse that is sent to the brain and interpreted into sound.

BALANCE

The fluid-filled *semicircular canals* and associated structures (saccule and utricle) in the inner ear provide us with

the ability to maintain balance, to detect up from down (even with our eyes closed), and to feel changes in our body's position. Sensory hairs in these structures relay information about the direction of gravitational pull and any body movements that disturb the fluid in these structures. When you turn your head, the semicircular canals move faster than the fluid they contain. This bends the sensory hairs and informs the brain of the motion. Be-

cause the three canals are perpendicular to one another, motion in any direction can be detected. The position of the head is ascertained by the saccules and utricles, which contain tiny grains of calcium carbonate that lie on the sensory hairs. The direction of pressure on these hairs tells us whether we are upright, reclined, or upside down. Movement of the crystals provides us with the feeling of acceleration or deceleration.

**Figure 15-23**
Some striking photoreception adaptations. *(a)* Stalked eyes of a mantis shrimp. *(b)* Compound eyes of robber fly are made of hundreds of individual lenses that give them an extremely broad visual field. *(c)* Only their eyes batray the silent nocturnal approach of this flotilla of alligators closing in on an observer's spotlight. A special layer of tapetum cells in the retina reflects, light, creating the telltale orange-red glow. (Inset—an alligator eye in daylight.)

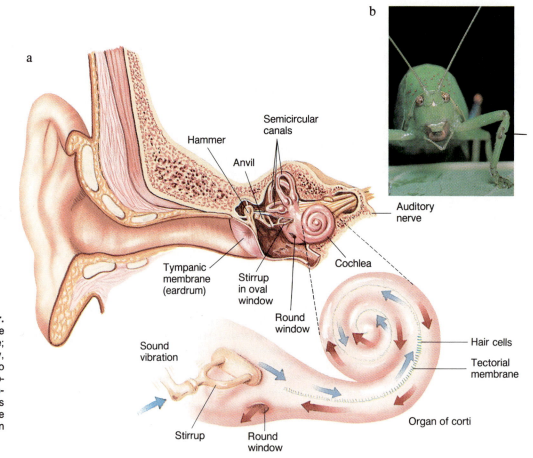

**Figure 15-24**
*(a)* **Anatomy of the human ear.** The internal structures in blue are concerned with balance; those in red receive, amplify, and relay sound vibrations to the brain (see text for description of function). *(b)* **Less sophisticated but effective.** This katydid displays its simple hearing apparatus located on the legs.

# Synopsis

## MAIN CONCEPTS

- **The outcome of neurological activity** depends on whether the neuron is excitatory or inhibitory, and on the nature of the target cell. Excitatory neurons elicit the following from their effectors:

  Glands secrete.

  Muscles contract.

  Neurons relay the action potential to another target.

  Inhibitory neurons elevate the threshold of the effector, increasing the excitatory stimulation required to activate the target.

- **Neurological impulses**

  *Resting neurons* pump ions across the plasma membrane, building electrical potential along the neuron's surface.

  *Action potential* is the reversal of this polarity.

  *Impulses* travel from dendrites to the axon's synaptic knobs, where neurotransmitters are released into the synapse.

  *Synaptic transmission* occurs when these chemicals bind to the target cell's membrane. Depending on the neurotransmitter, the target is either excited or inhibited.

- **Neurological coordination and control centers**

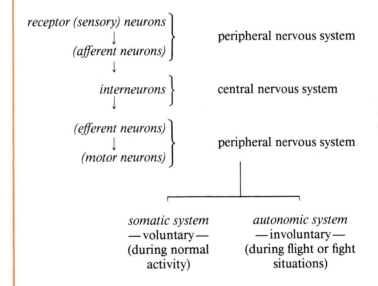

- **Central Nervous System**

  *The brain*—in humans, the major center for integrating and responding to neural input.

  *The cerebral cortex* controls voluntary movement, receives and perceives sensory input, and is rich with interneurons for complex associations needed for learning, language, and thinking.

  *The brainstem* contains structures that regulate waking and sleeping, codination, and involuntary or semivoluntary activities, such as body urges, emotion, breathing, and heart beat.

  *The spinal cord*—the relayer of many reflex impulses; conduit for neurological impulses into and out of the brain.

## KEY TERM INTEGRATOR

### Neurons and Targets

| | |
|---|---|
| **neurons** | Respond to *stimuli* and deliver an impulse to their target cells. **Excitatory neurons** stimulate the target, and **inhibitory neurons** oppose target stimulation. The target may be an **effector** (muscle or nerve cell) or another neuron. |

### Neuron Structure and Function

| | |
|---|---|
| anatomy | Impulses arise at the **dendrites** and sweep into the **cell body**. From there they are relayed down the **axon** to the **synaptic knobs** for delivery to their targets. **Schwann cells** surround some axons or dendrites with a **myelin sheath** that speeds the impulse, which jumps from one **node of Ranvier** to the next. |
| neurological impulse | A neuron maintains a polarity of charges along its membrane, a charge buildup called **resting potential**. Stimulation triggers an **action potential** that is conducted to the **synapse**. **Neurotransmitters** released into the **synaptic cleft** either activate or inhibit the target, depending on whether they are excitatory or **inhibitory neurotransmitters**. |

### The Nervous System—Integration, Association, and Regulation

| | |
|---|---|
| **reflex arcs** | Enable an animal to react rapidly to a sudden stimulus by routing the impulse through the shortest loop—**sensory neurons** (detectors of stimulus) to **interneurons** (associators) to **motor neurons** (effectors of a response). |
| **central nervous system** | Includes the **brain** and **spinal cord**, both of which are composed of association-rich synapses between neurons (the **gray matter**) and myelin-rich axons (the **white matter**). Brain and spinal cord are bathed in **cerebrospinal fluid**. In humans and other mammals, the **brainstem**, the complex of the **midbrain**, and **hindbrain**, is small compared to the **forebrain**, which contains the **cerebrum** and the deeply wrinkled **cerebral cortex**. The forebrain also includes the **hypothalamus** and thalamus, which constitute much of the brain's **limbic system**, the seat of human emotions. |
| **peripheral nervous system** | Incoming impulses travel toward the central nervous system in **afferent nerves**. They enter the spinal cord in **spinal nerves** and enter the brain either in **cranial nerves** or up from the spinal cord. **Efferent nerves** carry the impulse from the central nervous system to effectors. Some of these *outbound* nerves, those of the **somatic nervous system**, control voluntary automatic activities that continue with no conscious intervention. The autonomic system has two branches—the **parasympathetic system**, which maintains normal calm activity and the **sympathetic system**, which temporarily takes over in times of emergency. |

# Review and Synthesis

1. Describe a safeguard that prevents a nerve from carrying a neurological impulse in the wrong direction. What would be the effect of "two-way traffic" along a neurological tract?

2. What is the source of the electrical energy needed to fire an impulse along a neuron? Why doesn't the strength of the impulse diminish as it travels along the neuron?

3. How would each of the following influence the indicated target cell's threshold of activation?
   glycine — voluntary muscles
   tetanus toxin — voluntary muscles
   endorphins — pain receptor neurons
   serotonin — brain centers that maintain wakefulness
   dopamine — acetylcholine-producing neurons

4. Some persons who have lost a limb experience severe and very real pain in the missing extremity. Apply what you've learned of sensory perception to explain such "phantom limb" pain.

5. The sight or smell of food triggers automatic salivation in hungry dogs and people, but not in satiated individuals. What type of neurological pathway is automatic salivation? How might a full stomach inhibit this pathway?

6. What are two links between the brain and the peripheral nervous system?

7. Speculate on the effect of severe injury to each of the following parts of the central nervous system: medulla of the brainstem; reticular formation; the spinal cord when severed approximately one-third of the way down the back.

8. Name the sensory structure(s) in humans that contain(s): photoreceptors; mechanoreceptors; thermoreceptors; chemoreceptors; nociceptors; cold receptors; olfactory receptors; calcium carbonate crystals; tympanic membrane; heat receptors; and tectorial membrane and neurologically sensitive hair cells.

# Additional Readings

Bullock, T., R. Orkand, and A. Grinnel. 1977. *Introduction to Nervous Systems.* W. H. Freeman, San Francisco. (Intermediate.)

Fincher, J. 1981. *The Human Body — The Brain.* U.S. News Books. (Introductory.)

Iversen, L. 1979. "The chemistry of the brain." *Scientific American* 241:134–49. (Intermediate.)

Keynes, R. 1979. "Ion channels in the nerve-cell membrane." *Scientific American* 240:126. (Intermediate.)

Vander, A., J. Sherman, and D. Luciano. 1986. *Human Physiology.* 4th ed. Chapters 8, 18, 19, and 20. McGraw-Hill, New York. (Intermediate to advanced.)

Wurtman, R. 1982. "Nutrients that modify brain function." *Scientific American* 246:50–59. (Intermediate.)

Chapter 16

# Orchestrating the Organism— The Endocrine System

T hings could have gone terribly wrong. For days the child received no calcium in her diet. The absence of this essential mineral in the blood could trigger a number of life-threatening breakdowns. Neurons would fail to release their neurotransmitters. Muscles would lock up in painful spasms. Blood would clot reluctantly, if at all, resulting in fatal hemorrhages. But these disasters never materialized, prevented by four small glands buried in the child's neck. These pea-sized glands sensed the drop in blood calcium and sounded a molecular alarm, releasing tiny amounts of a chemical, a hormone, that stirred the body into corrective action. Some calcium was withdrawn from the bones (the body's calcium bank) and poured into the bloodstream. At the same time the hormone slowed excretion of the mineral while enhancing its absorption from the intestine. The deficit was managed without misfortune.

Yet the danger was not over yet, for a backlash from these calcium-boosting measures threatened to shoot the concentration of calcium to dangerously high levels in the blood. Fortunately, before that could happen, another gland began releasing its hormone, one that prevents calcium excesses by reversing the effects of the first hormone. The constant turning on and off of these two glands maintains a stable concentration of calcium in the blood, even during temporary deficiencies in its supply. Without such a system, wild vacillations in calcium concentrations would quickly erase any chance of life continuing.

Animal bodies face a constant threat of things "going wrong." Fluctuations in hundreds of critical chemicals occur minute by minute and must be corrected quickly before they disrupt homeostasis and interfere with the life processes. Life simply cannot tolerate extreme fluctuations in the internal environment. One of the jobs of the endocrine system is to detect such variations and release **hormones**— chemical messengers that direct target tissues to change, to alter their activities (or sometimes their form). The endocrine system's glands help maintain a suitable internal environment within the narrow ranges that allow life to continue. And this is only one of its jobs.

Hormones also regulate long-term growth. They speed up or slow down the body's overall metabolic rate. They tell animals when to become reproductively mature, when to produce sperm, when to release eggs, when to expel a fully developed fetus, and when to produce milk (in mammals).

Some hormones enhance the aggressiveness of the body's defenses against disease; others temper these defenses to minimize the chances of self-inflicted injury from an overactive immune system. Hormones supervise the pressure and volume of blood and enhance the formation of blood cells. Hormones can transform an animal until it bears little resemblance to its former appearance (Figure 16-1).

a

b

Juvenile Emperator

Adult Emperator

**Figure 16-1**
**Some effects of hormones.** *(a)* Molting of this insect is chemically initiated by a hormone *(ecdysone). (b)* This emperor fish broadcasts its sexual maturity by hormone-induced changes in coloration.

In humans, the body is harmonized by the coordinating activities of hormones, many of which are produced by nine **endocrine glands** ("ductless" glands). These glands secrete their products directly into surrounding tissue fluids and blood vessels for distribution by the circulatory system to the rest of the body. Hormones regulate activities in target tissues throughout the body, not just in those that are adjacent to the endocrine gland. (Another group of glands, the **exocrine glands**, discharge their secretions through ducts directly to their sites of action. They include the glands that secrete sweat, oil, tears, stomach acid, and digestive enzymes. These glands are not part of the endocrine system and are discussed with their associated organ system in later chapters.)

In some cases the endocrine "gland" is no more than individual cells residing in various nonendocrine organs. The brain, for example, acts as a gland, secreting some 45 separate hormones, such as *enkephalins* that act as natural opiates for pain relief (Chapter 15). Most "brain hormones" are neurotransmitters that also have hormonal activities. For example, the intestinal hormone cholecystokinin not only promotes release of digestive enzymes in response to stomach contents entering the small intestine, but is also a neurotransmitter produced in the brain, where it acts as an appetite suppressor. The discovery of new human hormones is growing at an astounding rate. In 1970 fewer than 30 of these chemicals were identified. Today some 200 hormones produced by various cells scattered throughout the body are being investigated.

## The Neuroendocrine Connection

The ultimate effect of a hormone depends on its target cell's specialty. This specialized activity is performed (or in some cases inhibited) when the hormone attaches to specific receptors on the target cell's plasma membrane. This may sound familiar, much like the influence exerted by neurotransmitter chemicals on the target cells of neurons discussed in the previous chapter. Unlike neurons, however, endocrine cells often take longer to affect their targets, which may be a meter or more away rather than just across a microscopic synapse. Hours, or even days, may elapse before a secreted hormone begins to exert its effect. Hormones also persist longer than do neurotransmitters, which are instantly degraded or removed from the synapse. Yet in spite of these differences, hormones and neurotransmitters share much in common, as do the systems that produce them.

Both systems work hand in hand to achieve a common goal: supervising the enormous complexity of an animal's body so that each part works as part of an integrated team. In tackling this massive task, the two systems collaborate on some functions and work independently on others. Their collaborative efforts are coordinated in two ways. First, some neurotransmitters activate glands to secrete hormones that, in turn, stimulate the autonomic nervous system. Mobilization of the sympathetic neurons during a fight or flight response, for example, triggers the release of hormones from the core of the adrenal glands. These hormones enhance the temporary dominance of sympathetic neurons over the parasympathetic system during emergencies (Chapter 15).

But the most profound neuroendocrine link is between the hypothalamus and the pituitary (Figure 16-2). Part of the brain, the hypothalamus, is physically connected to the pituitary and functionally linked to it by blood vessels and by neurons. The pituitary's job is to release hormones, most of which activate other glands. Centralizing much of the endocrine system under the command of a single control center coordinates glandular activity. The pituitary determines which hormones will be secreted to best suit the needs of the whole organism at any particular moment. Coordinating this network of hormones, requires information about the state of affairs throughout the organism, information routed through the hypothalamus. Sensory neurons scattered throughout the body detect local conditions and relay this information to the hypothalamus, which responds by activating the pituitary to release the appropriate hormones.

## "Reins and Spurs"— Regulating Hormone Production

Hormones exert their effects in extraordinarily tiny amounts, and their targets are acutely sensitive to even slight endocrine variations. The levels of hormones and tissue fluids must be tightly controlled to avoid the disruptive effects of either deficits or excesses. A number of factors regulate the precise concentrations of a hormone.

- *Negative feedback.* In many cases, the concentration of a hormone directly influences its own production. Some endocrine cells produce hormones only when the levels of those hormones in the blood drop below "normal" (Figure 16-3). Deficits in the hormone spur the gland to secrete more, and the concentration in the blood begins to rise. Elevated concentrations have the opposite effect; they pull the reins on hormone secretion before an excess can accumulate. This regulatory mechanism, called **negative feedback**, is similar to the way you maintain a constant driving speed in a car. Your speedometer alerts you to excessive speed, and you reduce your velocity. When you notice that you are not producing enough speed, you apply more pressure on the gas pedal. Many glands are equipped with sensors that, like the speedometer, inform them of their endocrine "speed." These glands constantly self-correct to maintain stable hormone concentrations.

- *Positive feedback.* Some hormones enhance their own production by activating the glands that produce them. The result is just the opposite of negative feedback. With positive feedback, elevated hormone concentrations *increase* hormone release. Although, in most situations, such self-perpetuating escalation would disrupt homeostasis, some situations demand rapid increases in

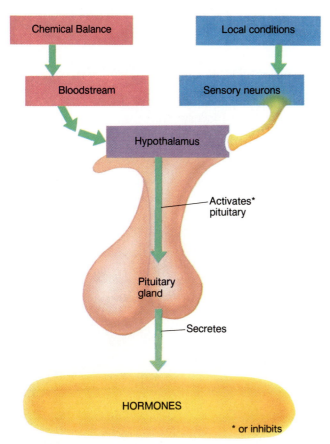

**Figure 16-2**
**The major neuroendocrine connection** (see text for explanation).

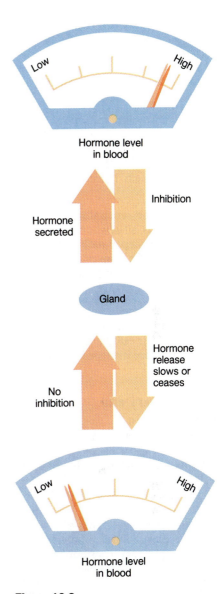

**Figure 16-3**
**Negative feedback** provides one way to regulate hormone concentration in the blood. Higher concentrations of hormone inhibit the gland that secretes it, which resumes activity again as levels decrease.

hormone concentration, for example, to promote expulsion of the fetus during childbirth.

- *The condition being regulated.* Some glands are sensitive to the changes their hormones elicit. This chapter opened with an example of this type of regulation, the calcium-sensitive glands that stabilize levels of this mineral in the blood. In this form of negative feedback, the gland responds to the eventual effect (concentrations of calcium in this example) rather than to concentrations of the hormone itself.

- *Autonomic neurons.* Sensory input may activate the autonomic nervous system either to promote or to inhibit hormone secretion. Blood pressure is partially regulated by such a mechanism. When neurons sense a drop in blood pressure, the nervous system orders the secretion of hormones that increase blood volume and constrict blood vessels enough to maintain circulatory pressure.

- *Inactivation or removal.* Hormones do not persist indefinitely. Most are removed by the kidneys and excreted. Others are enzymatically inactivated or transformed into other compounds. If not replaced by continued secretion, the hormone slowly diminishes in concentration.

## A Closer Look at the System

Hormones are produced either by distinct endocrine glands or by specialized cells embedded in nonendocrine organs (Figure 16-4). Table 16-1 describes a few of these chemical messengers in humans.

All hormones are organic compounds that, with few exceptions, fall into two fundamental categories:

1. Most hormones are amino acids, derivatives of amino acids, or *peptides* composed of more than one amino acids. (This category includes protein hormones such as insulin.)

2. Most of the remaining hormones are *steroids,* all of which possess the same ring-shaped molecular skeleton, but differ from one another in the chemical groups attached to the rings (refer to Figure 4-7).

Peptide hormones and steroid hormones employ different approaches to activating their cellular targets. These mechanisms are discussed later in this chapter.

## The Pituitary and Hypothalamus

The pituitary has been traditionally called the "master gland" because it controls the activities of so many other endocrine elements. The master gland, however, has a master of its own, the hypothalamus. The hypothalamus exerts two types of control over the double-lobed pituitary. The **posterior pituitary** (posterior = rear) manufactures no hormones of its own, but receives those produced by the cell bodies of neurons in the hypothalamus. The chemicals migrate down the axons of these neurons, which terminate in the posterior pituitary. This pituitary lobe, therefore, serves as a storage depot for hypothalamic hormones (Figure 16-5a), which are released on command from the hypothalamus. The **anterior pituitary** (anterior = front) is a true endocrine gland, manufacturing six hormones that are released when the hypothalamus senses a need for them and discharges chemicals called *releasing factors* (Figure 16-5b). The releasing factors diffuse into nearby blood vessels and are transported directly to the anterior pituitary, where they trigger the release of hormones. The hypothalamus also produces inhibiting factors that pre-

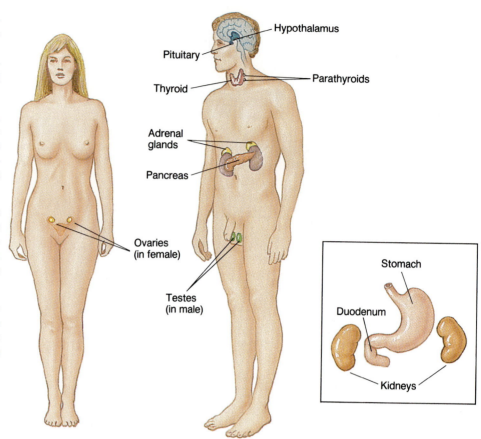

**Figure 16-4**
**The major endocrine glands**, plus some endocrine cells that are dispersed in nonglandular tissue (inset).

vent excessive secretion of hormones by the anterior pituitary.

## HORMONES OF THE POSTERIOR PITUITARY

Two hormones are released by the posterior pituitary: **antidiuretic hormone** (ADH), which reduces water lost during urine formation and, in females, **oxytocin**, which triggers uterine contractions during childbirth. Oxytocin also stimulates the release of milk when the breast is stimulated by nursing. Sensory impulses relayed by nerves from the stimulated nipple provoke action potentials in the oxytocin-producing neurons of the hypothalamus. When the action potentials reach the posterior pituitary, the neurons release oxytocin by exocytosis. Milk is then released the instant the oxytocin-rich blood reaches the breasts.

As its name implies, ADH prevents *diuresis* (an increase in the water content of urine) by promoting water recovery from the kidneys. Water is drawn back into the blood, which becomes more dilute as a result. Neurons, called *osmoreceptors,* in the hypothalamus sense when blood is becoming too dilute and temporarily inhibit the release of ADH. As ADH concentration decreases, less water is recovered from urine, and blood becomes more concentrated. When osmoreceptors sense that the blood has become too concentrated, they again trigger the release of ADH. This hormone also has the effect of elevating blood pressure; since water is not eliminated, the volume of blood increases, raising its pressure against the vessels. Because of its dual effects, ADH is also known by the name *vasopressin.*

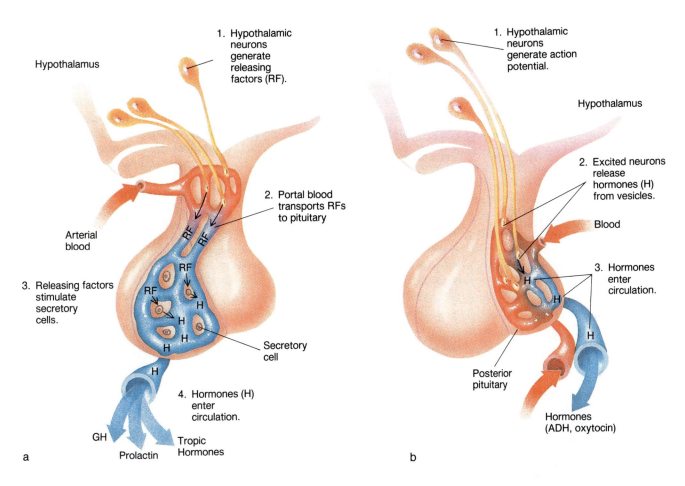

**Figure 16-5**
**The pituitary–hypothalamic linkage.** The hypothalamus is functionally linked by neurons to the posterior pituitary *(a)* and by the bloodstream to the anterior pituitary *(b)*.

## HORMONES OF THE ANTERIOR PITUITARY

Of the six known hormones produced by the anterior pituitary (Figure 16-6), four regulate the activity of other endocrine glands. The remaining two are direct-acting hormones. One of these, **prolactin**, promotes the production of milk by mammary glands in the breast. The other encourages body growth by stimulating bone elongation and accelerating protein synthesis. This compound, called **growth hormone (GH)**, is produced predominantly during childhood and adolescence, and determines the eventual size of the individual. When the pituitary ceases its GH output (as ordered by the hypothalamus), overall growth stops. This occurs sooner in girls than in boys, accounting for the shorter average height of women. Interestingly, most growth takes place while sleeping, when bursts of GH secretions usually occur. Extremes in height (giantism or dwarfism) are due to overproduction of or deficiencies in GH during childhood.

The remaining four anterior pituitary hormones stimulate the activities of other endocrine glands. **Adrenocorticotropic hormone (ACTH)**, for example, stimulates the cortex (outer layer) of the adrenal glands to secrete its hormones. Another activates secretion by the thyroid gland and is called **thyroid-stimulating hormone (TSH)**. The other two, which act on the gonads (testes in males, ovaries in females), are appropriately called *gonadotropins*. These two gonadotropins, **follicle-stimulating hormone (FSH)** and **leutinizing hormone (LH)**, promote gamete development and stimulate gonads to produce sex hormones. Some details of how these four **tropic hormones** influence their target glands are provided in the following sections.

## The Adrenal Glands

Perched atop each kidney, the adrenal glands (ad = upon, renal = kidney) respond to chemical orders from two directors—hormones from the anterior pituitary and neurological impulses from the sympathetic nervous system. Each adrenal gland is essentially two independent glands that pro-

TABLE 16-1

## The Major Endocrine Glands and Their Hormones

| Gland | Hormone | Regulates |
|---|---|---|
| Anterior pituitary | Growth hormone (GH) | Growth; organic metabolism |
| | Thyroid-stimulating hormone (TSH) | Thyroid gland secretions |
| | Adrenocorticotropic hormone (ACTH) | Adrenal cortex secretions |
| | Prolactin | Milk production |
| | Gonadotropic hormones: Follicle-stimulating hormone (FSH); Leutinizing hormone (LH) | Production of gametes and sex hormones by gonads |
| Posterior pituitary | Oxytocin | Milk secretion; uterine motility |
| | Antidiuretic hormone (ADH) or vasopressin | Water excretion; blood pressure |
| Adrenal cortex | Cortisol | Organic metabolism |
| | Androgens | Growth and, in women, sexual excitability |
| | Aldosterone | Sodium and potassium excretion |
| Adrenal medulla | Epinephrine and norepinephrine | Organic metabolism; cardiovascular function; response to stresses |
| Thyroid gland | Thyroxine (T-4) Triiodothyronine (T-3) | Energy metabolism; growth |
| | Calcitonin | Calcium in blood |
| Parathyroids | Parathyroid hormone | Calcium and phosphate in blood |
| Gonads | | |
| Ovaries | Estrogens and progesterone | Reproductive system; growth and development; secondary sex characteristics of females |
| Testes | Testosterone | Reproductive system; growth and development; male secondary sex characteristics; sex drive |
| Pancreas | Insulin and glucagon | Organic metabolism; blood glucose concentration |
| Kidneys | Renin | Adrenal cortex; blood pressure |
| | Erythropoietin | Erythrocyte production |
| | 1,25-Dihydroxy vitamin $D_3$ | Calcium balance |
| Gastrointestinal tract | Gastrin | GI tract; liver; pancreas; gall bladder |
| | Secretin | |
| | Cholecystokinin | Release of digestive enzymes |
| | Gastric inhibitory peptide | Mobility of stomach contents |
| Thymus | Thymosin | Lymphocyte development |
| Pineal gland | Melatonin | Biological cycles; sexual maturation |

*Modified from Vander et al., 1986, Human Physiology, 4th ed., McGraw-Hill, New York.*

duce very different hormones (Figure 16-7). The gland's outer layer, the **adrenal cortex**, secretes hormones in response to ACTH from the anterior pituitary. The core of each adrenal gland, the **medulla**, is linked by neurons to the hypothalamus, which stimulates the medulla to secrete its hormones when the sympathetic nervous system is aroused.

## THE ADRENAL CORTEX

The adrenal cortex secretes a family of steroid hormones, called **corticosteroids,** that fall into three functional categories.

- **Glucocorticoids**, which regulate sugar and protein metabolism and elevate blood sugar levels by encouraging the conversion of amino acids to glucose. These hormones also moderate the protective inflammatory response, tempering this defensive mechanism so that it is less likely to injure the individual it is there to protect.

- **Mineralocorticoids**, which regulate the level of sodium and potassium in the blood. The most important of these is *aldosterone,* which stimulates the kidney to resorb sodium back into the blood. Sufficient sodium in the blood is essential for water recovery. Too little of the mineral results in water loss and a drop in blood pressure. Too much sodium in the blood has the opposite effect, contributing to hypertension (high blood pressure).

- **Sex hormones**, which influence the production of gametes and the development of male or female sex characteristics. (The primary source of these hormones, however, are the testes and ovaries.)

The adrenal cortex is essential to life. Without aldosterone, for example, the body quickly loses sodium and potassium. These minerals are needed for neuron activity, muscular contraction, blood pressure stability, and transport of substances across the plasma membranes. Without glucocorticoids, glucose concentrations in the blood plunge, crippling cellular respiration.

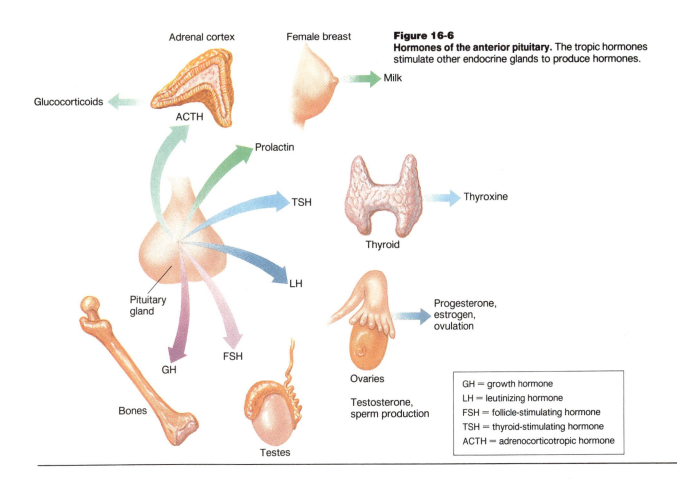

Adrenal cortex

Female breast

Glucocorticoids

ACTH

Milk

Prolactin

TSH

Thyroxine

Thyroid

LH

Pituitary gland

Progesterone, estrogen, ovulation

GH

FSH

Ovaries

Bones

Testosterone, sperm production

Testes

**Figure 16-6**
**Hormones of the anterior pituitary.** The tropic hormones stimulate other endocrine glands to produce hormones.

GH = growth hormone
LH = leutinizing hormone
FSH = follicle-stimulating hormone
TSH = thyroid-stimulating hormone
ACTH = adrenocorticotropic hormone

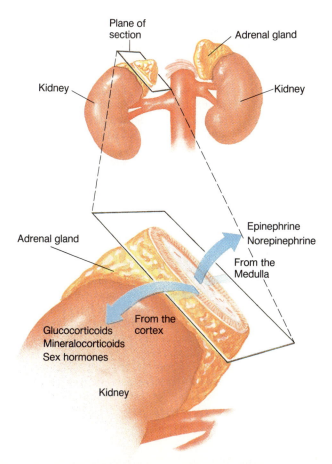

Plane of section

Adrenal gland

Kidney

Kidney

Epinephrine
Norepinephrine

From the Medulla

Adrenal gland

Glucocorticoids
Mineralocorticoids
Sex hormones

From the cortex

Kidney

A partially impaired adrenal cortex results in varying degrees of these symptoms, a condition known as *Addison's disease.* The disease may sometimes be treated by administering corticosteroids to supplement the deficiency. Corticosteroids are also used to combat persistent inflammation and severe allergies, and to suppress the immune response in patients who have received organ transplants so they will not reject the foreign tissue. These hormones (or their synthetic derivatives) are also used to accelerate healing after trauma, such as surgery. Like most drugs, however, corticosteroids have some severe side-effects that argue against their indiscriminate use (see Bioline: Chemically Enhanced Athletes).

## THE ADRENAL MEDULLA

One of the "fight or flight" responses discussed in the previous chapter was the activation of the **adrenal medulla** to secrete its two hormones, **epinephrine** (also called *adrenalin*) and **norepinephrine** (or *noradrenalin*). These hormones jolt the body into readiness to escape or confront emergencies. Oxygen consumption increases, the heart beat accelerates, more glucose and oxygen are shunted to voluntary muscles (increasing muscular performance), blood vessels to the skin constrict (helping to reduce blood lost through injured tissues), and red blood cells pour into the bloodstream from their reserves in the spleen (enhancing oxygen transport and replacing blood cells lost during bleeding). These hormones also inhibit activities (such as contraction of digestive muscles) that are of little help in an emergency. Enzymes that destroy these hormones assist in recovery from the effects of adrenal medulla activity once the threat is over and the sympathetic nerves relinquish their control to the parasympathetic system.

## The Thyroid Gland

Early in this century, many people were afflicted with enormous swellings in the front of their necks. This condition, known as *simple goiter,* is an enlargement of the **thyroid gland** (Figure 16-8), a butterfly-shaped gland that

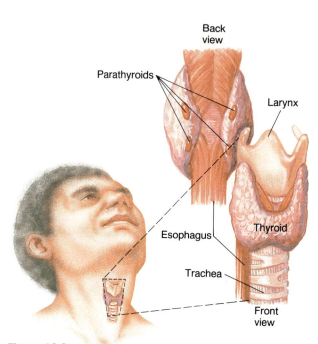

**Figure 16-8**
**The thyroid and parathyroid glands.**

lies just in front of the windpipe in people. (Most other vertebrates have two thyroid glands.) The swelling is often accompanied by lethargy, hair loss, slowed heart beat, and mental sluggishness. Goiters occurred frequently in geographic regions where natural sources of dietary iodine were scarce. The link between iodine deficiencies and the disease was not established until 1916, however. When iodine was proven to prevent the condition, it was added to salt, virtually eliminating the condition from the United States and Europe.

Without iodine, the thyroid gland fails to manufacture two metabolism-regulating hormones. A person's *basal metabolic rate* is enhanced by these two hormones, **thyroxin** and **triiodothyronine**, which increase the energy-generating oxidation of carbohydrates. Normally, their concentrations are regulated by negative feedback. The hypothalamus turns off the release of TRF (thyroid-releasing factor) when the blood concentrations of the two hormones are sufficient. But in the absence of iodine, neither of the hormones is produced, and the person

suffers from energy deprivation. Since the hypothalamus detects no thyroxine in the blood, it continues to pump out the releasing factors that order the pituitary to release *thyroid-stimulating hormone.* The overactive thyroid gland enlarges as a result of this perpetual stimulation. Dietary iodine restores the thyroid hormones, thyroid activity and metabolism normalize, and the gland returns to its unobtrusive size.

Some persons have defective thyroid glands that produce no hormone even when iodine is plentiful. The effects of the resulting *hypothyroidism* (too little thyroxine) are much more severe when it strikes children, who develop *cretinism* as a result. The physical and mental retardation caused by this condition can be prevented by early treatment with thyroxine supplements. The same treatment is effective in correcting the sluggish basal metabolic rate of adult hypothyroid victims.

An excess of thyroxine, that is, *hyperthyroidism,* also enlarges the thyroid gland. In these persons, hyperactivity, weight loss, nervousness, in-

## B I O L I N E

## CHEMICALLY ENHANCED ATHLETES

During the 1988 Olympic Games in Seoul, Korea, the sports world was shocked by the disqualifications of some athletes that had already received their medals. The reason—steroids were found in the athletes' urine. During the Pan American Games of 1983, highly sensitive tests were used to detect illegal drug usage by athletes. Topping the list of banned substances were steroids, commonly referred to as "anabolic" steroids because they tend to increase biosynthesis (anabolism), especially protein synthesis. The most common steroid is the male sex hormone testosterone and its synthetic derivatives. Many athletes believe these drugs help build muscle, restore energy, and enhance aggressiveness. Their use has grown so common among athletes that suppliers of black market drugs distribute printed advertising brochures and order forms at gyms.

The dangers of abuse, however, are considerable. Through negative feedback, steroid surpluses in the blood inhibit the production of leutinizing hormone by the pituitary. In men this lowers sperm counts and may cause atrophy (wasting) of the testicles.

When a man's body is flooded with testosterone, some of it is converted to the female hormone estrogen, often stimulating breast enlargement. In women, testosterone-like steroids stimulate growth of facial hair and the development of other masculine features. Steroid supplements also increase the likelihood of liver cancer and promote growth of existing prostate cancer. Prolonged exposure to steroids suppresses the immune system, compromising the body's ability to defend itself against many infectious diseases.

The victims of steroid abuse are not limited to athletes. *Cortisone,* one of the glucocorticoids, is frequently used to suppress the symptoms of inflammation. In some cases this usage is warranted, for example, to reduce swelling of inflamed tissues that may otherwise close the airways and prevent breathing. Cortisone also reduces pain associated with inflammation and suppresses the symptoms of some allergies. Although cortisone makes the patient feel better temporarily, it can cripple the body's immunity and lead to severe, even fatal, infections.

somnia, and irritability are usually accompanied by the disease's most striking external symptom—exophthalmia (eyeballs bulging from the sockets). Hyperthyroidism is also called *exophthalmic goiter.*

A third thyroid hormone, called **calcitonin**, regulates blood calcium levels in cooperation with another set of glands, the parathyroid glands.

### The Parathyroid Glands

The drama that opened this chapter is being played out in your body at this very moment. Balancing the calcium concentrations in the blood requires the cooperation of the thyroid gland and four tiny glands, called **parathyroid glands**, attached to it. Parathyroid glands secrete **parathyroid hormone (PTH)**, which reverses the effects of calcitonin (secreted by the thyroid gland). PTH, secreted when blood calcium levels are low, acts on the bones, kidneys, and intestines to restore and maintain normal calcium concentrations in the blood (Figure 16-9, *top*). Under PTH influence, bones are decalcified, and the released mineral is absorbed into the bloodstream; kidneys retain calcium rather than excreting it; and calcium absorption from the intestines is enhanced. Negative feedback turns off PTH secretion when concentrations of the mineral normalize. Calcitonin, secreted by the thyroid gland when calcium accumulates in abnormally high concentrations in the blood, exerts the opposite effects of PTH (Figure 16-9, *bottom*).

### The Pancreas

The pancreas is predominantly an exocrine gland that produces digestive enzymes rather than hormones. Embedded in the pancreas, however, are several endocrine centers called **islets**

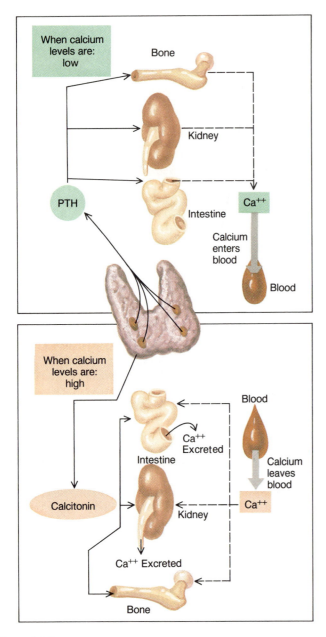

**Figure 16-9**
**Balancing calcium**. A low calcium concentration in the blood stimulates PTH secretion by parathyroid glands (shown in green) and inhibits calcitonin secretion by the thyroid gland (in red). A high calcium concentration have the opposite effects.

Glucose stored as glycogen does little unless these reserves can be tapped and distributed by the bloodstream to needy tissues. When the concentration of blood sugar begins to drop below normal, the islets of Langerhans alter their secretory priorities and begin releasing **glucagon** instead of insulin (Figure 16-10, *right frame*). This hormone promotes glycogen breakdown and elevates glucose concentration in the blood, especially during times of stress when increased cellular and physical activity is likely to require greater energy expenditures.

People with insulin deficiencies suffer from diabetes or, more correctly, *diabetes mellitus*. The loss of glucose in urine provides a convenient way of detecting diabetes by simply testing the urine for excess sugar. Diabetics tend to become dehydrated because their excess urinary sugar interferes with the osmotic recovery of water by the kidneys. Even though the blood is saturated with glucose, it is utilized poorly because, without enough insulin, it is not absorbed into cells. Cells must turn to other sources of energy, such as protein and fat reserves, ultimately causing the diabetic to become emaciated. The immune system also suffers, and as a result normally harmless microbes can cause serious infections in diabetics. The treatment for this type of diabetes is straightforward—periodic injections of insulin to replace the missing hormone.

## The Ovaries and Testes

When stimulated by pituitary gonadotropins, the **ovaries** and **testes** secrete the powerful steroid hormones **testosterone** (the male sex hormone) and **estrogen** (the female sex hormone). These substances promote the development of secondary sex characteristics, those nongenital attributes that distinguish genders (such as deeper voices and facial hair in men and enlarged breasts in women). The activities and regulation of the sex hormones are covered in more detail in Chapter 23).

## Prostaglandins

**Prostaglandins** are secreted by endocrine cells scattered throughout the body, and their activities seem as diverse as their sources. They are pre-

of **Langerhans**, which secrete two hormones. Together, these hormones maintain proper glucose concentrations in the blood. One of these, **insulin**, is secreted when the concentration of glucose in the blood begins to exceed normal, usually as sugar floods the bloodstream following a meal (Figure 16-10, *left frame*). Insulin stimulates the uptake of glucose by the cells in the body, a necessary first step in the utilization of the sugar. Insulin also directs the conversion of excess glucose to glycogen for storage in liver and muscles. Excessive glucose would otherwise remain in the bloodstream and be excreted in urine. Insulin prevents wasting precious chemical energy by reducing the amount of glucose in the blood.

Insulin can do too good a job, however, and deplete the blood of glucose.

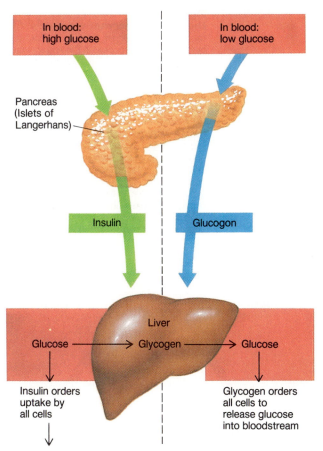

**Figure 16-10**
**Control of blood sugar concentration by the islets of Langerhans in the pancreas**. When blood glucose is high, insulin is secreted, promoting glucose uptake by all cells and storage as glycogen in liver and muscle. Glucagon, produced when blood glucose is low, has the opposite effect.

dominantly found in **semen**, the fluid that is discharged during male orgasm. These prostaglandins cause uterine muscles to contract, which may promote fertilization by facilitating uptake of semen (and sperm cells). Other prostaglandins help protect the body by promoting *inflammation*. The inflammatory response localizes and destroys invading microorganisms and helps heal wounds. These hormones also stimulate nociceptors (pain sensors), thereby contributing to the sensation of pain. The pain-relieving and anti-inflammatory effects of aspirin are due to its ability to inhibit prostaglandins.

The spectrum of prostaglandin activities affects reproduction, digestion, respiration, neurological function, and blood clotting. One type of prostaglandin protects against blood loss by encouraging clotting and blood vessel constriction. Another type of prostaglandin does just the opposite, keeping the blood channels free of clots and obstructions.

## The Pineal Gland

The **pineal gland**, a tiny protuberance in the brain, is perhaps the most mysterious endocrine gland. Although its function in humans remains a mystery, in some other vertebrates it may regulate the darkening of skin in response to changes in light. In these animals, light triggers the pineal gland to release *melatonin,* a hormone that directs certain skin cells to concentrate their pigment. In this way, some vertebrates change color with the seasons, using photoperiod (length of daylight) as the indicator of the season.

The pineal gland is believed to regulate *biorhythms* in mammals as well, perhaps by releasing melatonin in response to changes in photoperiod. (Biorhythms are cyclic changes in physical characteristics or behavior.) Light entering the eyes arouses the central nervous system to inhibit the activity of the pineal gland. Therefore, the longer the day, the less melatonin is released. Melatonin usually inhibits the release of gonadotropins and, therefore, reproductive activity. As winter approaches, days grow shorter, more melatonin is released, and the animal enters a nonreproductive phase. Even though we humans experience no seasonal changes in reproductive activity, we nonetheless are governed by a daily rhythm that may be influenced by the pineal gland. Some researchers are examining the "jet lag" phenomenon, relating it to the time required to reset the pineal "clock."

## Action at the Target

Some target cells can be triggered without the hormone even entering the cell. Most peptide hormones, in fact, exert their influences in an indirect way. By merely binding to a **receptor site** on the target cell's plasma membrane, the hormone activates *adenyl cyclase.* This enzyme converts ATP to one of the most universally important molecules associated with regulation, **cyclic AMP (cAMP)**. This molecule is a ring-shaped version of ATP (minus two phosphates) that can activate resting enzymes or help turn on suppressed genes.

An increase in cAMP in the cytoplasm can dramatically alter a cell's activities (Figure 16-11a). For example, when glucagon, the insulin antagonist, attaches to liver cells, it activates adenyl cyclase, which in turn increases the concentration of cAMP. The cAMP activates the enzyme responsible for exporting glucose from the cells, eventually elevating the concentration of glucose in the blood.

This mechanism requires two chemical "messengers" to activate a target cell. The hormone is the *first messenger,* delivering the instructions to the cell, and cAMP is the *second messenger,* the chemical that stimulates the resulting effect.

But how does such a model explain the specificity of hormones? If glucagon, calcitonin, parathyroid hormone, epinephrine, the pituitary-releasing

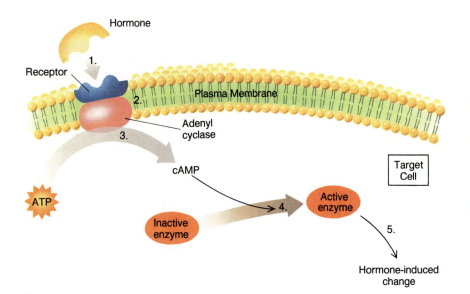

a

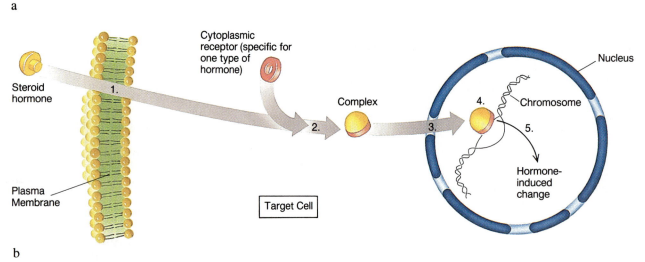

b

**Figure 16-11**
**Mechanisms of hormone action**.
*(a) Second messenger model:* 1. Hormone binds to surface receptor. 2. Binding activates adenyl cyclase. 3. Activated adenyl cyclase converts ATP to cAMP ("second messenger"). 4. cAMP activates (or inhibits) specific enzymes. 5. Activated enzymes catalyze specific changes in the cell.
*(b) Steroid hormones:* 1. Hormone readily passes through plasma membrane and 2. reacts with receptor molecule. 3. The hormone–receptor complex enters the nucleus and 4. attaches to a specific site on the chromosome. 5. Attachment activates gene(s) responsible for the hormone-induced change (for example, production of a hair shaft by a previously dormant hair follicle on an adolescent boy's chin).

factors, and several other hormones do no more than increase cytoplasmic cAMP concentration in target cells, why don't they all cause the body to change in the same way? The key to hormone specificity is the *receptor site* to which the hormone binds. These surface sites on the plasma membrane allow the attachment of only one kind of hormone. A cell that possesses receptors only for epinephrine will bind with no other hormone. Thus, epinephrine is the only hormone that can elevate cAMP levels in this cell. Fur-

thermore, cAMP has different effects in different target cells. Whereas in the liver it stimulates the export of glucose, in cells of the adrenal cortex it activates the glucocorticoid-producing enzymes. Those target cells that respond to multiple hormones have surface receptors for each of them. Liver cells, for example, can bind epinephrine, glucagon, and several other peptide hormones.

Steroid hormones take a much more direct approach (Figure 16-11*b*). These hormones actually enter their

target cells and complex with a free receptor molecule specific for the hormone. The receptor-hormone complex then enters the nucleus, where it attaches to the chromosomes and activates the genes responsible for the hormone-induced changes. Again, the specificity of the receptor molecule determines the hormone(s) to which a target cell responds. The genes that are affected in the cell determine the outcome of hormonal influence.

# Synopsis

## MAIN CONCEPTS

- **Hormones are chemical messengers,** many of which are secreted by ductless glands of the endocrine system. Hormones act in minute amounts on targets anywhere in the body. They reach their targets via the bloodstream or tissue fluids, eliciting changes in their specific target cells.

- **Hormone concentrations are regulated by negative or positive feedback.** The effects of hormones are short-lived, as hormones are excreted or enzymatically converted to other chemicals. Most fall into one of two chemical categories, the peptide hormones and the steroid hormones.

- **The control center — the hypothalamus —** coordinates the secretion of many hormones, especially those released by the double-lobed pituitary and, ultimately, the hormones triggered by the pituitary tropic hormones.

- **Hormones of the posterior pituitary** help prevent excessive water loss during urine formation, and triggers uterine contractions during childbirth and the release of milk from mammary glands when the breast is stimulated by suckling.

- **Hormones of the anterior pituitary** regulate formation of breast milk, stimulate overall body growth, and influence hormone secretion by other endocrine glands.

- **Hormones of the adrenal cortex** regulate sugar and protein metabolism, stabilize sodium and potassium concentrations in blood, and, to a small degree, influence development of gametes and secondary sex characteristics.

- **Hormones of the adrenal medulla** boost metabolic activity, divert energy and oxygen resources to voluntary muscles, and generally prepare the body to cope with perceived danger, emergencies, or stressful situations.

- **Thyroid hormones** regulate the body's overall rate of metabolism and lower the concentration of calcium in the blood. This effect is countered by the parathyroid hormone, which helps balance blood calcium concentrations at a stable level.

- **Pancreatic hormones** promote glucose absorption by cells and conversion to glycogen, decreasing blood glucose when in abundance. When glucose is scarce, it is released from cells and glycogen is broken down, increasing blood glucose.

- **Prostaglandins** are produced by secretory cells scattered throughout the body. They regulate activities associated with reproduction, blood clotting, uterine contraction, inflammation, and the perception of pain.

- **Hormone action at the target.** *Peptide hormones* bind to specific receptor molecules on the target cell's surface. Binding stimulates adenyl cyclase to produce cAMP, which unleashes the cell's potential by activating the enzymes responsible for the hormone-induced change.
  *Steroid hormones* bind to receptor molecules in cytoplasm. The resulting complex attaches to the chromosome and turns on the specific genes responsible for the response to hormone stimulation.

# KEY TERM INTEGRATOR

## Hormones and Their Regulation

| | |
|---|---|
| **hormones** | Chemicals released from **endocrine glands** (as opposed to **exocrine glands**) or from other cells that provoke changes in nearby or distant target cells. **Negative feedback** helps stabilize hormone concentrations. |

## Activities of the Major Endocrine Glands

| | |
|---|---|
| **posterior pituitary** | Releases **antidiuretic hormone** and **oxytocin**, two hormones transferred there from the hypothalamus where they were produced. |
| **anterior pituitary** | Releases two hormones, **prolactin** and **growth hormone**, that directly elicit their ultimate effects. Four others are **tropic hormones (adrenocorticotropic hormone, thyroid-stimulating hormone, follicle-stimulating hormone,** and **leutinizing hormone)** that work indirectly by stimulating other glands to release their hormones. |
| **adrenal cortex** | Produces three groups of **corticosteroids**—**glucocorticoids, mineralocorticoids,** and the **sex hormones** estrogen and testosterone. |
| **adrenal medulla** | Responds to alarming situations by releasing **epinephrine** and **norepinephrine**, hormones that promote the fight or flight response. |
| **thyroid gland** | Regulates metabolic rate by controlling the release of **thyroxine** and **triiodothyronine**; also releases the calcium-eliminating hormone **calcitonin**. |
| **parathyroid glands** | Respond to declines in calcium by releasing **parathyroid hormone** which promotes calcium-conserving measures. |
| **islets of Langerhans** | Located in the pancreas, they release a pair of antagonistic hormones, **insulin** or **glucagon**. |
| **testes and ovaries** | Produce the major sex hormones **testosterone** (male) and **estrogen** (female). |
| **prostaglandins** | Found primarily in **semen**, but have diverse functions, including healing properties and blood clotting. |
| **pineal gland** | Releases melatonin, the hormone that elicits seasonal changes in skin color. |

## Hormones—Mechanism of Action

| | |
|---|---|
| **cAMP** | Second messenger whose increase in concentration evokes a change in the cell when peptide hormone attaches to specific *receptor sites* on the target cell. |

# Review and Synthesis

1. Identify a hormone that exemplifies each of the following regulatory mechanisms: negative feedback, positive feedback, changes in the condition being regulated.

2. Norepinephrine is both a neurotransmitter and a hormone released by the adrenal medulla. Describe two other links between the nervous and endocrine systems.

3. Describe the difference in the way oxytocin and prolactin regulate the production and secretion of milk.

4. For many hormones there are antagonistic hormones that reverse their effects. Name an antagonist for each of the following hormones and describe how each pair works together: parathyroid hormone; insulin; and prostaglandin (as a promoter of inflammation).

5. Why is cAMP called a "second messenger"? What is the first messenger?

6. Insulin deficiencies cause the symptoms of diabetes. What would happen to a diabetic following an insulin overdose? These effects would simulate a deficiency in what hormone?

7. Describe the sequence of hormonal events that would likely occur in your body during a frighteningly strong earthquake. Consider all the glands involved.

# Additional Readings

Guillemin, R. and R. Burges. 1972. "Hormones of the hypothalamus." *Scientific American* 227:76. (Intermediate.)

Mason, S. 1983. *Human Physiology.* (Chapter 10.) Benjamin–Cummings Book Co. (Intermediate.)

Pastan, I. 1972. "Cyclic AMP." *Scientific American* 227:97–105. (Advanced.)

Vander, A., J. Sherman, and D. Luciano. 1986. *Human Physiology. 4th ed.* (Chapters 7 and 9.) McGraw-Hill, New York. (Intermediate to advanced.)

Witzmann, R. 1981. *Steroids: Keys to Life.* Van Nostrand-Reinhold Co., Princeton, N.J., (Intermediate.)

Chapter 17

# Support and Movement — The Skeletal and Muscular Systems

Few people need to be convinced of the importance of their bones and muscles. They constitute more than 65 percent of your body mass and largely determine you physical appearance, from body stature to facial features. Together, they oppose the effects of gravity and inertia, two natural forces that tend to flatten and immobilize living things. Against these forces animals elevate themselves, maintain their functional form, and, for most, move toward food and away from danger.

The power needed to battle gravity and inertia comes from the forces generated by muscles. Muscles also provide the force that maintains the structure of animals that have no firm skeleton. But most animals possess a more rigid form of support, either completely surrounding the body with a protective encasement or forming a system of living girders inside the animal. These living girders are the bones that comprise your skeleton.

## An Animal's Substructure

### Hydrostatic Skeletons

Not all animals are supported by bones or a tough outer skeleton. Some use muscles and the natural supporting properties of water to resist being flattened by gravity. The pressure inside a closed, fluid-filled chamber can be increased by contracting muscles that encircle the chamber. The hydrostatic pressure in the chamber supports the animal against gravity and most other physical forces that would otherwise distort its shape, in much the same way that air in a tire supports the weight of an automobile. Although such **hydrostatic skeletons** are found in some animals that live under water, such as anemones and the individual polyps of a coral (Figure 17-1), earthworms and a few other land organisms successfully use this strategy in terrestrial habitats.

### Exoskeletons

Anyone who has eaten a lobster, found the discarded husk of an insect, or examined a clam shell is familiar with **exoskeletons**. These are hard external coverings used for protection, support, or both. Some animals, notably snails, clams, and scallops, are enclosed in a protective rocklike shell of calcium carbonate. Another type of exoskeleton is a cuticle, a hardened layer that surrounds, supports, and protects some animals (Figure 17-2). These external support systems are found on insects, spiders, crabs, and other *arthropods* (animals with jointed appendages and segmented body parts —Chapter 34). In addition to support and shelter for internal organs, arthropod exoskeletons provide a firm surface against which muscles exert their force, creating a system of rigid levers for movement. Without this rigidity, the animal could not move its limbs at the flexible joints; muscular contraction would merely distort the soft tissue, and the animal would go nowhere.

Arthropod exoskeletons are secreted as a liquid that coats the animal's skin. Chitin, proteins, and salts in this thin layer quickly harden. (Chitin is a polysaccharide discussed in Chapter 4.) The resulting exoskeleton is an armor-hard coat that covers the animal's exterior everywhere except the joints, which remain soft and flexible. The rest of the exoskeleton is so hard, however, that it cannot expand as the arthropod grows larger. The whole apparatus must be periodically discarded and replaced with a larger version as the animal grows. This process is known as **ecdysis**, or molting. The new exoskeleton hardens around a body inflated with air or water. The new exoskeleton is larger than the animal, thereby providing the animal room for growth before the next molting is necessary. Molting is controlled by the hormone *ecdysone*.

### Endoskeletons

Internal support structures constitute an animal's **endoskeleton** (Figure 17-3). Although found in a few invertebrates (sponges and sea stars, for example), the endoskeletons of vertebrates are the most complex and versatile. All vertebrates are supported by internal skeletons, although not all have bones. The skeletons of sharks

**Figure 17-1**
**Stretching tall against the pull of gravity,** these coral polyps use a hydrostatic skeleton to prevent collapsing. By contracting ring-shaped muscles that encircle the body, the coral squeezes into a narrower cylinder, generating hydrostatic pressure that elongates the animal.

a

**Figure 17-2**
**A sampling of exoskeletons.**    The mobility of this spider is provided by dozens of joints in its exoskeleton. Attached muscles and hydrostatic pressure move the appendages, which operate like a series of moving levers.

a                                    b

c                                    d

**Figure 17-3**
**Some organisms with endoskeletons.** *(a)* Sponges are reinforced with a network of hardened *spicules,* branching spikes embedded in the otherwise soft tissues of the animal. *(b)* Hard calcified plates form the endoskeletons of sea stars and sea urchins. *(c)* With not a bone in their bodies, sharks and rays have soft skeletons composed of cartilage. All other vertebrates have bony skeletons, as do amphibians, reptiles, birds, mammals, and these fire gobies *(d).*

and their relatives consist entirely of rubbery **cartilage**, a firm, yet flexible connective tissue that holds its shape while remaining elastic. Before birth, you too were supported by a skeleton of cartilage, but by the time you were born, much of the cartilage had been transformed into 350 **bones**. As you grew, many of these bones fused with one another, so that your adult skeleton consists of 206 individual bones, supplemented by cartilage. (Your ears and nose tip, for example, are sculpted of cartilage.)

In addition to supporting weight, the skeleton shields internal organs and provides rigid surfaces for muscle attachment so that muscle power can be transmitted across movable joints. Bones also provide reserves of calcium and phosphorus that can be tapped when needed. The hollow spaces in bones are filled with soft tissue *(marrow)* that produces blood cells or stores fat.

The skeleton is divided into two functional groups of bones: the axial skeleton and the appendicular skeleton (Figure 17-4). We use the human skeleton as a representative model.

### THE AXIAL SKELETON
Bones aligned along the long axis of the body comprise the **axial skeleton**. They include the skull, vertebral column, and ribcage. The skeleton's protective functions are assumed largely by the axial skeleton. The precious three-pound mass of nervous tissue in your head is enclosed in the **skull**, an unyielding vault of 28 bones. At birth, these individual bones are held together by flexible membranes, allowing the skull to compress a bit so that the head can pass through the birth canal. These membranous connections are called *fontanels,* the baby's vulnerable "soft spots." Midway through the end of a child's second year, the fontanels are replaced by strong interlocking lines of fusion between adjacent plates of the skull (Figure 17-5).

The spinal cord is protected by the **vertebral column**, the backbone. The flexible backbone consists of 33 bones, called **vertebrae**, arranged in a gracefully curved line and cushioned from one another by disks of cartilage.

These disks take on even greater importance in upright animals that walk on two legs because the backbone must bear the entire weight of the upper body. If not for these disks, the vertebrae would grind each other to dust. Nonetheless, years of bearing this weight eventually takes its toll, usually in the form of back pain, one of the most common drawbacks of being human.

The ribcage embraces the chest cavity and protects its vital organs. Ribs extend from the vertebrae and form a "cage" by attaching to the **sternum** (the breastbone) in the front. Only 10 pairs of ribs, however, attach to the sternum. The lower two pairs, called "floating ribs," remain free at their outer ends. (Contrary to popular belief, men and women have the same number of ribs.) Expansion and compression of the ribcage help draw air into and force air out of the lungs during breathing. The marrow of ribs is one of the most prolific producers of red blood cells.

### THE APPENDICULAR SKELETON
The movable appendages attached to the axial skeleton comprise the **appendicular skeleton**. The bones of the pectoral (shoulder) girdle, arms, legs, and pelvic girdle provide a system of levers for mobility and dexterity. The **pectoral girdle** holds the arms to the axial skeleton. It consists of four bones, two **scapulae** (shoulder blades), and a pair of **clavicles** (collarbones). The ability of the scapulae to skate over the surface of the ribcage contributes to the impressive dexterity of the human arm. Having two bones (the *ulna* and *radius*) in the forearm instead of just one allows you to rotate your hand. (The wrist cannot twist in a circle but can only move the hand up and down or side to side.) One of the most extraordinary collection of bones lies at the end of the wrist of humans, monkeys, and other higher primates. The hand combines strength with a dexterity that allows it to manipulate objects precisely.

The pelvic girdle receives the upper body's weight from the vertebral column and transmits it to either the legs or the surface on which you are now

sitting. In women, the central opening of the pelvis is larger to accommodate childbirth. In both sexes, the back of the pelvis is actually part of the vertebral column in which the vertebrae have fused into a single unit. The pelvis is designed for sitting and standing. When sitting, the body's weight rests on the *ischium* (review Figure 17-4). When standing, the load is transmitted to the legs—first the *femur* (the largest bone in the body) and then the two bones of the calf, the *tibia* and *fibula*. Although not as dextrous as the hands, the 26 bones in each foot are arched to withstand tremendous forces. (An arch is the most efficient structure for supporting weight.)

### JOINTS
The skeleton would be stronger if it were a single solid piece of bone, but movement would be impossible. Strength must be compromised somewhat to provide mobility, producing weaker points that are capable of movement. These areas where adjacent bones articulate are the **joints**, and generally the more mobile the joint, the weaker it is. Joints that join the plates of the skull, for example, are very strong but are virtually immobile after the age of two. Compare this to the most mobile joint of the body, the shoulder joint. Any athlete with a shoulder dislocation can painfully testify to its vulnerability.

Mobile joints provide various forms of movement. A *hinge joint,* such as that of the elbow and knee, allows back-and-forth movement, much like the hinge on a door. A *ball-and-socket joint* is the most mobile of all. Two such joints provide the 360° shoulder motion that allows you to swing your arms in vertical circles. In other joints, bones glide along each other's surface or move in two perpendicular arcs, as would two saddles fitted together. The bones in all these joints are held together by strong straps of connective tissue, called **ligaments**. Highly movable joints also need lubrication and cushioning to prevent the bones from crunching against each other. This is the function of the **synovial cavities**, such as are found in the knee (Figure 17-6). The fluid in these cavities lubri-

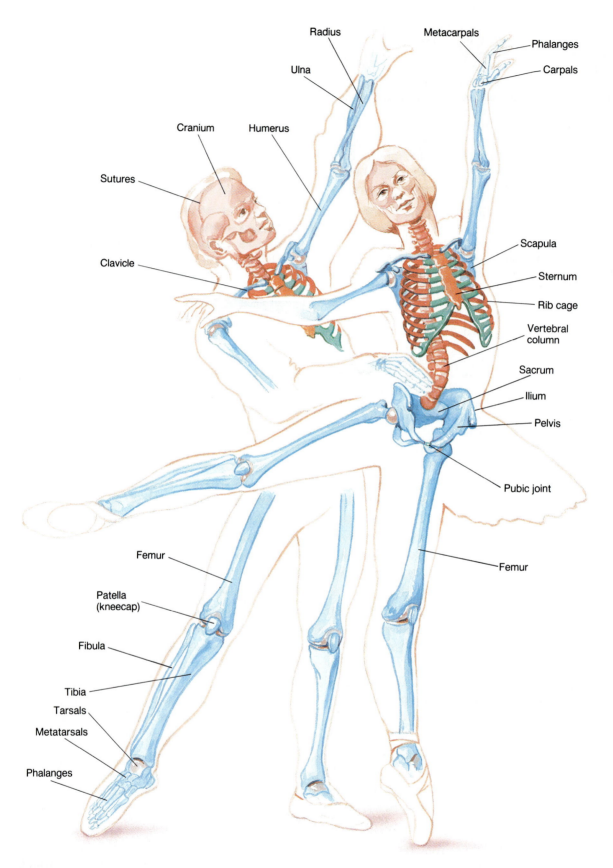

**Figure 17-4**
**The human skeleton.** Bones of the axial portion are shown in (red); those of the appendicular portion are in (blue).

**Figure 17-5**
**Meandering lines of fusion** (sutures) between interlocking bones in the skull increase the strength of the connections.

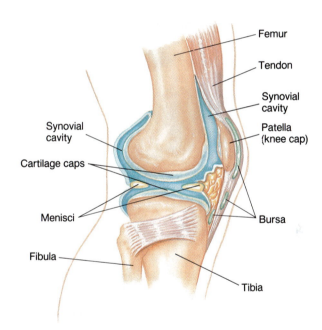

Femur
Tendon
Synovial cavity
Patella (knee cap)
Synovial cavity
Cartilage caps
Menisci
Bursa
Fibula
Tibia

**Figure 17-6**
**Lubricated for life.** The synovial cavity and 13 fluid-filled bursae ease the friction between bones, tendons, and ligaments in the human knee.

cates and separates articulating surfaces. Caps of glistening, smooth cartilage on each surface help cushion and reduce friction. The knee and a few other joints are equipped with additional bags of lubricant called *bursae.* These sacs may become inflamed, producing the pain of *bursitis.* But the major joint problems occur when the joint itself becomes inflamed, eventually leading to degeneration of the cartilage joint caps. Deteriorated caps are replaced with bony outgrowths that rasp painfully against the articulating surface and stiffen the joint, sometimes to the point of crippling im-

mobility. This syndrome is the most common form of *arthritis,* which is an inflammation of the joints (arthr = joint; itis = inflamed).

## The Architecture of Bone
Bone combines two properties that are rarely found in a single material: strength and light weight. Pound for pound, bone is stronger than steel; yet it weighs less than one-fourth as much. One cubic inch of bone can withstand a pressure of 19,000 pounds, the weight of 10 compact automobiles. Bone has been called "living concrete," but it is four times stronger

than concrete and much lighter. A 145 pound person's skeleton weighs only 20 pounds, adding a mere 14 percent to the total weight of the body.

How can something so light be so strong? Woven from flexible protein fibers and hardened by mineral salts, bone is not only hard but also resilient. The protein in bone is **collagen,** the matrix of so many other connective tissues (see CONNECTIONS: Multicellular Organization, Chapter 15). The flexible collagen is hardened by minerals, especially calcium, which comprise 70 percent of the bone's weight. Each of these components strengthens the other. Collagen acts as flexible reinforcing rods that help hold the hard calcium "cement" together. Without collagen, your bones would be so brittle that they would shatter under the body's weight. Without calcium, your bones would bend as though they were made of rubber.

Bones are hollow; this property actually increases their strength. Solid bone would not only be extremely heavy, but would also fracture more easily. Each bone does have a solid, rocklike portion, called **compact bone**, but this usually surrounds a honeycombed mass of **spongy bone** (Figure 17-7). The hollows within spongy bone are filled with **red marrow**, the soft tissue that produces red blood cells. The hollow core of long bones is called the **medullary cavity** in which

**yellow marrow** stores fat.

There is another striking difference between bone and reinforced concrete: bone is alive, even in its most rocklike parts. Within the solid mass of mineral and collagen, living bone cells thrive. These cells, called **osteocytes** (osteo = bone, cyte = cell), are engulfed in the calcified matrix that they manufacture. This matrix is deposited as concentric cylinders that form laminated layers, called **lamella**. The result is a greatly strengthened structure. Osteocytes obtain their nutrients from blood vessels threaded through the **Haversian canals**, the lifeline of compact bone. The tiny chambers that house the osteocytes are connected by **canaliculi**—microscopic canals through which osteocytes touch each other with long

extensions of their surfaces. These interconnections channel nutrients and growth-regulating hormones to every osteocyte in the Haversian system and evacuate their metabolic wastes.

The manufacture of new bone is called **ossification**. In growing bones, the **epiphyseal plates** are the most active centers of ossification (Figure 17-7). Ossification at these plates elongates and thickens bones as the animal grows to adulthood.

Although the bones of adults no longer increase in size, bone is constantly being disassembled and rebuilt. When parathyroid glands release hormones in response to decreases in blood calcium levels, for example, the hormone initiates the decalcification of bone (Chapter 16). It does so by

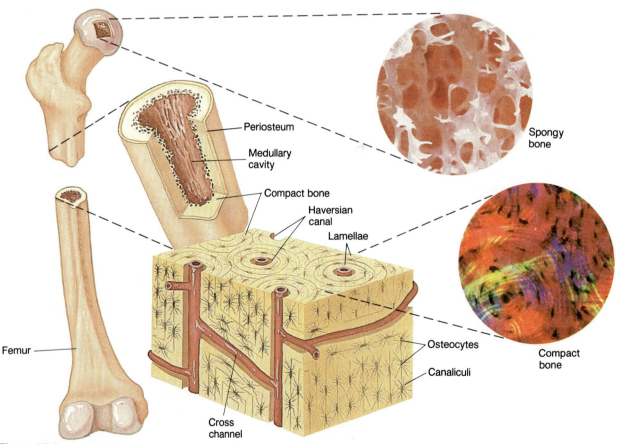

**Figure 17-7**
**The architecture of bone.** Solid-looking compact bone is actually permeated by Haversian canals that contain blood vessels, the osteocytes' lifelines. Oxteocytes manufacture the bone matrix, leaving cross channels and canaliculi so that each cell has access to blood. Spongy bone reinforces the hollow interior, which contains the marrow.

activating another type of bone cell, the **osteoclast**, which attacks bone and removes calcium, dumping it into the bloodstream for use by the body. The bone-destroying action of osteoclasts works in conjunction with bone-building osteocytes to remodel bone, removing older areas and replacing them with new material. When a bone is fractured, osteocytes proliferate and repair the damage with a fresh deposit of ossified material that cements the pieces together. No piece of reinforced concrete can do that.

The constant interplay between bone building and disassembly provides our bodies with an ability to strengthen those bones that receive the most use and to diminish the size of those whose services are in less demand. In other words, physical activity produces larger, stronger bones, just as it enlarges and strengthens muscles. Disuse does just the opposite: it causes *atrophy* (wasting). The speed with which disuse atrophy strikes bones was dramatically illustrated by the 1965 Gemini V space mission, during which American astronauts Cooper and Conrad lost more than 20 percent of the mass in some of their bones. Even enclosing a leg in a cast can diminish the size of a bone by 30 percent in just a few weeks.

# Creating Movement— The Muscles

Without muscles, a skeleton can do nothing more than support an animal. But muscles do much more than move bones and exoskeletons around. Muscles move eyelids and tongues, pump internal fluids through circulatory pipelines, propel nutrients through the digestive thoroughfare, discharge wastes, squeeze secretory products out of glands, suck oxygen into the lungs of air-breathing animals, and generate powerful jets of water that can rocket an animal out of danger.

Although their tasks may differ, all muscles have several things in common. They require tremendous amounts of ATP to fuel their activities (see Connections, Chapter 3). They are superbly efficient in the use of this ATP, converting 35 to 50 percent of

the energy intake into mechanical energy, making them about five times more efficient than an automobile engine. Muscles accomplish their jobs by shortening their lengths when activated. The bulk of their mass consists of long filaments of **contractile proteins** that work like pistons sliding into cylinders, pulling the unit into a shorter configuration. When the protein relaxes, the pistons slide out again, returning to their original length.

Vertebrates are equipped with large amounts of muscle. In fact, about half your body weight consists of three types of muscle—*skeletal, smooth,* and *cardiac muscle.* These muscles differ in physical appearance, the types of jobs they perform, the tissue to which they are attached, their speed of contraction, and the manner in which they are excited into action. Of these three types, the most familiar are the muscles that bulge beneath the skin and give body-builders their characteristic contours. These are skeletal muscles.

## Skeletal Muscle

The majority of muscle tissue in your body is skeletal muscle, averaging about 40 percent of a man's body and 23 percent of a woman's. These muscles are known for their speed, contracting and relaxing very rapidly. (A muscle "twitch" lasts only a fraction of a second.) **Skeletal muscle** derives its name from an important characteristic—most of them are anchored to the bones they move. Skeletal muscles also contract when you consciously command them to do so, that is, they are under voluntary control. For example, you can move your forearm toward your upper arm by ordering your biceps to contract. The voluntary muscle shortens and forces the bone to move upward on its hinge at the elbow (Figure 17-8).

Yet, the biceps cannot move the forearm away from the upper arm. Muscles only *pull;* they cannot push. Another muscle must be used to provide the opposite movement, so that the limb is not stuck in one position. The biceps is paired with the triceps, which contracts to straighten the arm at the elbow. Most skeletal muscles are arranged in such *antagonistic pairs,*

with one reversing the effects of the other. Antagonistic muscle pairs generally don't contract simultaneously. To do so would pit the two in a futile tug of war that would neutralize the effectiveness of both. Neurological orders commanding a muscle to contract are accompanied by simultaneous inhibitory instructions to its antagonist. Thus, when one contracts, the other automatically relaxes. (The exception occurs when you *choose* to contract opposing muscles simultaneously, such as stretching after hours of immobility.)

In spite of its name, not all skeletal muscles move bones. Some nonskeletal activities, such as closing the eyelids and restricting the flow of urine out of the bladder, require the voluntary control afforded by skeletal muscle. Most of the body's 600 skeletal muscles, however, mobilize parts of the bony substructure (Figure 17-9).

### CONTRACTIBLE CELLS
Skeletal muscle consists of bundles of parallel muscle cells. Each bundle is enclosed in a sheath of connective tissue that separates it from adjacent bundles (Figure 17-10). All the bundles in a single muscle are housed together in a resilient wrapper that is also made of connective tissue. At the tips of the muscle, this outer sheath tapers and joins the connective tissue extending from the interior bundles to form the **tendon** that attaches the muscle to bone.

Each multinucleated muscle "cell" in a bundle is the product of several premuscle cells that fused into a single fiber during embryonic development. Because they possess many nuclei, these composite cells are usually referred to as **muscle fibers**. Skeletal muscle fibers bear some distinctive markings. Encircling each one are dark bands called *striations* that give the fiber a striped appearance. For this reason, skeletal muscle is said to be **striated**. The striations are the product of the **myofibrils** that lie parallel to each other and constitute the bulk of the cell's interior. Myofibrils power the muscle. They are the contractile portion that comprises about 80 percent of the fiber's mass. Depending on its thickness, a single muscle fiber may

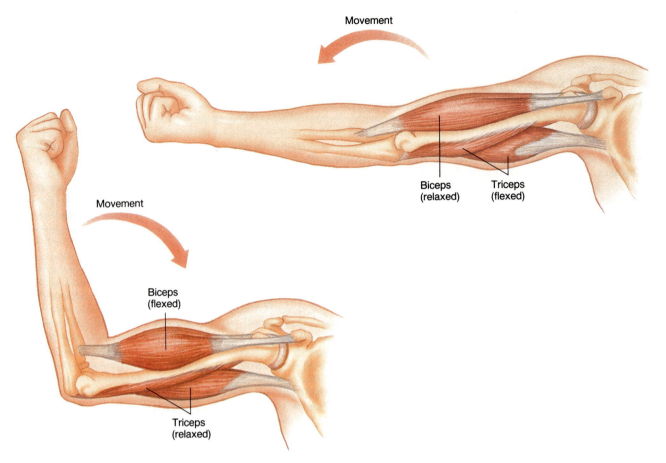

**Figure 17-8**
**Biceps and triceps: Muscles in opposition.** Antagonistic pairing of muscles allows the motion of one to be reversed by the action of the other.

have hundreds or thousands of myofibrils. Improved strength in a muscle is the result of more myofibrils in each muscle fiber.

The striations of a myofibril are arranged in contractible units called **sarcomeres**. Each sarcomere is endowed with a peculiar pattern of bands and lines (Figure 17-10). The darkest line, called Z line, is a membrane disk that separates adjacent sarcomeres. Between the Z lines are several dark bands and light zones. When the myofibril contracts (shortens), the dark bands grow larger, the light zones get smaller, and the Z lines are drawn closer together. These observations have provided major clues in unraveling the molecular mystery of muscle contraction.

## SLIDING FILAMENTS AND MOLECULAR CONTRACTION
The dark bands in a sarcomere represent areas where linear molecules overlap. These overlapping molecular

strands consist of two proteins, called **actin** and **myosin**, that team up to provide living tissue the ability to forcefully contract. The protein complex they form is called **actomyosin**. If all the water were removed from your skeletal muscles, 90 percent of the dried material would be actomyosin.

The mechanism of contraction is a simple one, explained by "sliding filaments." Actin forms thin filaments that are aligned parallel to each other in a circular orientation. This leaves a hollow cylinder. A thicker filament of myosin protrudes slightly into this space (illustrated below). When a

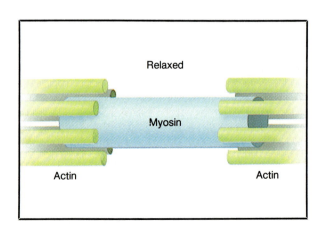

Relaxed

Myosin

Actin                                    Actin

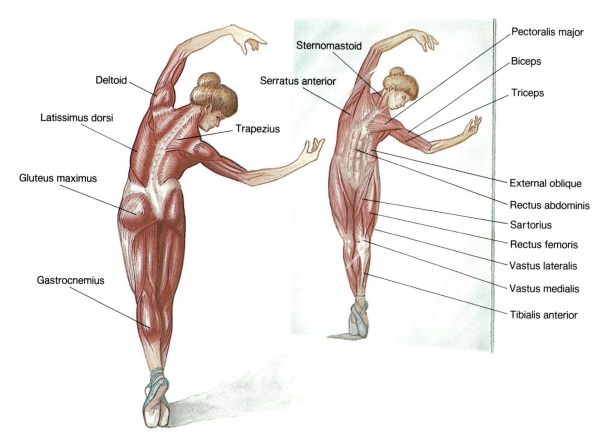

**Figure 17-9**
**600 "engines."** A few of the 600 skeletal muscles that power voluntary movement in the human body.

myofibril is stimulated, the protein rods slide over one another, and the actin "cylinder" is pulled over the myosin "piston." As these filaments slide toward each other, the length of actomyosin is reduced, pulling the ends closer together (illustrated below). In this shortened state, the area of overlap increases (the dark zones get wider), and the Z bands are drawn closer together. This explains the changes in striation that accompany myofibril contraction.

The "molecular forces" that propel the sliding filaments do so in one direction only—that which shortens the

protein complex. The nature of this force remained a mystery until the discovery of rounded bulbs projecting from the ends of the myosin filaments. When a muscle fiber contracts, these projections attach to the nearest actin molecule, forming a "cross-bridge" (Figure 17-11). No cross-bridging occurs in the resting complex. Each myosin bulb however, is loaded with a molecule of ATP, energetically "cocked" like a spring ready to be triggered. When the command arrives, the bulbs attach to the actin molecule and forcibly bend in the direction that slides the actin filaments over the myosin. The bulbs immediately release the actin and snap back to their original positions, recocked by the energy of another ATP. They reattach to the actin at a new site and generate another power stroke in the same direction as the first. A fully contracted muscle shortens by about 20 percent, requiring each cross-bridge to repeat this process thousands of times within

Contracted

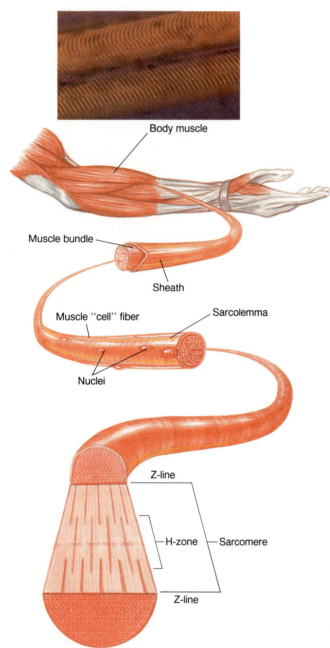

Body muscle

Muscle bundle

Sheath

Muscle "cell" fiber

Sarcolemma

Nuclei

Z-line

H-zone — Sarcomere

Z-line

**Figure 17-10**
**Skeletal muscle** is composed of bundles of parallel multinucleated cells (muscle fibers). Each muscle fiber is covered by the sarcolemma and packed with contractible myofibrils. The contracting unit in myofibrils (the sarcomere) contains overlapping protein filaments that account for the striated appearance of the muscle fiber. (Inset: Striations and nuclei are evident in this magnified view of abdominal muscle fibers.)

milliseconds. This ratchet mechanism for explaining actomyosin contraction works like a team of rowers propelling a boat with a power stroke and lifting their oars out of the water for the return stroke.

It has long been known that calcium is needed for muscle contraction. The mineral apparently promotes the formation of cross-bridges. Resting myofibrils are exposed to virtually no cal-cium. In this state, an inhibitory complex of two proteins, *troponin* and *tropomyosin,* covers the attachment site on the actin, preventing formation of cross-bridges. When calcium floods the myofibril, it briefly binds to the inhibitory complex and distorts its shape, exposing the attachment sites, to which the myosin bulbs immediately connect. That is all it takes. As soon as cross-bridges form, actomyo-sin begins to contract, and remains contracted as long as calcium and ATP are present. The fiber relaxes when calcium is actively evacuated from the actomyosin portion of the muscle fiber.

In the absence of ATP, the bulbs cannot be recocked. They remain attached to the actin filament, locking the protein complex in a state of prolonged rigidity. Shortly after death,

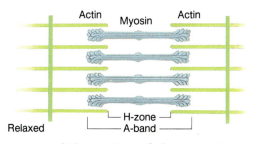

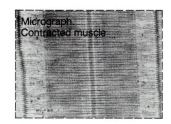

Relaxed

Micrograph Contracted muscle

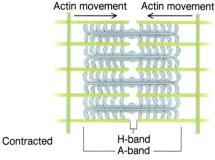

Contracted

**Figure 17-11**
**Sliding filaments of actin and myosin** allow the actomyosin molecule to change length. In its relaxed, longer state, myosin remains unattached to actin. Contraction occurs when "side-arms" projecting from myosin form crossbridges by attaching to actin, and then bend, sliding the two actin subunits toward each other. The action is repeated in a ratchet fashion until the sarcomere is fully contracted to about 20 percent of its original length. The action is fueled by ATP.

when cells can no longer generate ATP, all the muscles in the body stiffen, producing the state of *rigor mortis.* Such prolonged cross-bridging lasts for one or two days after death.

## UNLEASHING THE FIBER'S POTENTIAL

Since calcium triggers contraction, the mineral must somehow quickly flood actomyosin when a stimulatory motor impulse arrives and then be pumped out again, all in a fraction of a second. The anatomy of the **motor unit** (muscle fiber plus motor neuron) reflects the adaptations that make this possible (Figure 17-12). The **sarcolemma** (the muscle fiber's plasma membrane) forms a juncture with synaptic knobs of motor neurons that stimulate (or inhibit) contraction. Impulses from excitatory neurons trigger an action potential in the muscle fiber's sarcolemma. A wave of action potentials sweeps down special branches of the sarcolemma that reach into the core of the cell. These extensions, called **transverse tubules (T-tubules)**, transmit the impulse to the internal membranes of the **sarcoplasmic reticulum (SR)**. The SR is a modified version of the endoplasmic reticulum that stores calcium ions. The arrival of an action potential at the SR triggers it to release calcium ions, which flood the myofibrils and excite them into

contracting. If neurological stimulation is not continued, the SR's membrane immediately reestablishes its resting polarity. Calcium is pumped from the cell's interior back into the chambers of the sarcoplasmic reticulum, allowing the fiber to relax.

Although most of the cell's interior is taken up by myofibrils, there is enough room for the rich supply of mitochondria needed to generate the ATPs to fuel contraction. The nuclei of the fiber are relegated to an odd location—just beneath the sarcolemma near the surface of the cell.

## REGULATING CONTRACTION STRENGTH

Like a neuron, a muscle fiber responds in an all-or-none fashion. It is either completely contracted or fully relaxed. Yet somehow we temper the strength of muscle contraction. The same muscles that generates enough power to help lift a 400-pound barbell can also gently elevate a newborn baby without throwing it against the ceiling. Such control stems from the ability to regulate how many fibers in a muscle contract. Each fiber is equipped with at least one motor neuron. Firing more neurons recruits more fibers to participate, strengthening the magnitude of contraction in the muscle. This additive phenomenon, called **recruitment**, is why you can affectionately slap

friends on the back without knocking them to the floor.

A single nerve impulse triggers a very rapid contraction–relaxation sequence, causing the fiber to *twitch.* When in use, however, muscle fibers usually are flooded with a stream of impulses, and the twitches merge to produce a continuous sustained contraction. The enormous amounts of ATP required to sustain muscle contraction are provided by aerobic respiration. When oxygen is abundant, phosphate is transferred from *creatin phosphate* (a high-energy fuel storage compound in muscles) to ADP to form the ATP. Sustained activity, however, can quickly exhaust the muscle's oxygen supplies and stored energy reserves. At this point, muscle cells shift into anaerobic metabolism, extracting ATP from glucose by glycolysis alone. (See "Fermentation," Chapter 10.) Not only is this a low-yield process, but it also generates lactic acid, which accumulates in the muscle and creates the sensation of fatigue.

Muscles that are frequently contracted to at least 75 percent of their full strength tend to *hypertrophy,* growing in both strength and size. Yet the number of fibers in a muscle is fixed at birth. You are born with all the skeletal muscle "cells" you will ever have, and no amount of exercise will

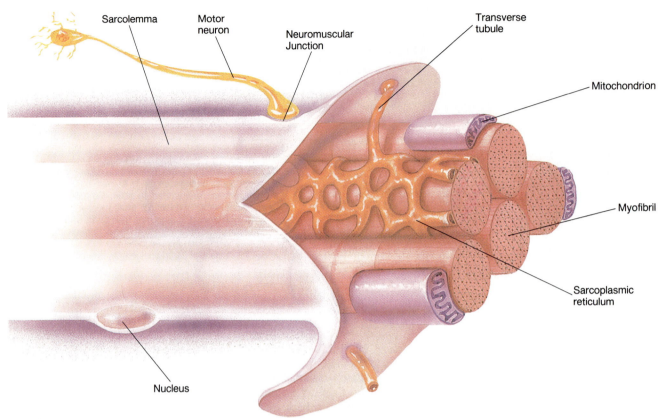

**Figure 17-12**
**The anatomy of a motor unit.** Calcium is housed in the elaborate system of internal membranes (T-tubules and sarcoplasmic reticulum). When an impulse is delivered by a stimulatory motor neuron, the calcium escapes and saturates the myofibrils, causing the fiber to contract.

increase this number. The myofibril content of each fiber, however, reflects the muscle's usage. Well-used muscles develop more myofibrils, increasing the diameter of the muscle and the force it can exert. Exercise also increases metabolic efficiency in muscle, improving endurance as well as strength. Poorly used muscles are subject to the opposite fate; they *atrophy.* If muscle inactivity persists long enough, the diminished size and strength of a muscle can be more or less permanent. Muscle fibers die and are replaced with connective tissue or fat. Even though dead fibers cannot be replaced, exercise can stimulate the surviving fibers to produce more actomyosin and grow in diameter, gaining enough strength to functionally re-

place the lost cells and restore the muscle to full capacity.

## Smooth Muscle

Unlike skeletal muscle, **smooth muscle** is neither striated nor arranged in parallel bundles, nor is it anchored to bone. Smooth muscle fibers are uninucleate cells that interlace, forming sheets that surround the body's hollow organs (Figure 17-13). Actually, this best describes the major type of smooth muscle, called **visceral muscle**. Fibers of the other major category, **multiunit smooth muscle**, are found at nonvisceral locations, such as around blood vessels, in the eyes, and at the base of hairshafts.

Both types of smooth muscle contract more slowly than skeletal muscle

and remain contracted longer. They also respond to *involuntary* commands, contracting without any conscious intervention. Smooth muscle in your urinary bladder, for example, automatically contracts in response to the internal pressure exerted on the walls of a full bladder. Fortunately, we have a backup skeletal muscle under voluntary control that closes off the exit to prevent voiding urine at inopportune moments. These muscles that close off various tubes in the body are **sphinctors**; some of them are composed of smooth muscles, and others of skeletal muscle.

## Cardiac Muscle

Found only in the heart, **cardiac muscle** provides lifelong pumping of blood through the circulatory system. These muscle fibers share characteristics with both skeletal and smooth muscle. Like skeletal muscle, they form cylindrical, striated fibers, but like smooth muscle, they are under involuntary control. Unlike both other types, however, cardiac muscle fibers consist of branched chains of individual cells (Figure 17-14). The cells are tightly fused with their neighbors by **intercalated disks**—folded membranes that readily conduct impulses directly from cell to cell.

Cardiac muscle cells contract automatically, even when no external message commands them to do so. Cardiac muscle cells grown in tissue culture, isolated from contact with any other cells, continue to beat rhythmically. In the heart, intercalated disks help coordinate contraction, so that the cells contract and relax in unison. The impulse that initiates a heart beat originates in the sinoatrial node, the heart's natural pacemaker (Chapter 20).

In this chapter, you have seen how the muscular and skeletal systems are closely parallel in function and importance. They are also closely aligned with another system, the skin, which helps protect and support the animal. One of the major tissues found in skin, the *epithelium,* is discussed in Connections: "The Versatile Epithelium."

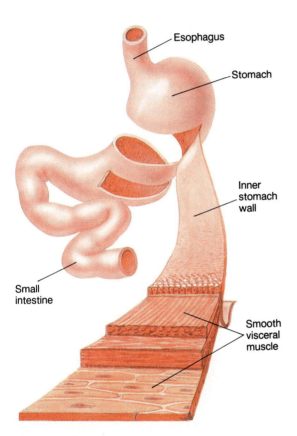

**Figure 17-13**
**Smooth visceral muscle.** Slowly contracting sheets of uninucleated, nonstriated muscle cells under involuntary control.

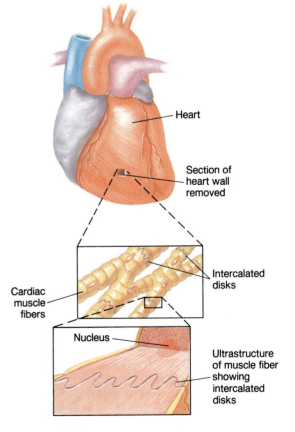

**Figure 17-14**
**Cardiac muscle:** Striated, involuntary, uninucleated cells interconnected by membranes of intercalated disks.

## THE VERSATILE EPITHELIUM

An animal's form and function cannot be fully appreciated without considering the structural and protective roles of its epithelial tissue (or simply, epithelium). The *epithelium* is one or more cellular layers that form the internal and external surfaces of an animal. It therefore delineates the organism, by providing the border between organism and external environment. Anything entering or leaving an animal's body must pass through at least one layer of epithelial cells. Although the most apparent epithelial layers are on the outer surface of an animal, most of a vertebrate's epithelial surfaces lie hidden along the digestive and respiratory tracts, or lining blood vessels and body cavities. Many internal organs, the heart, for example, are packaged in at least one layer of epithelia.

One of the most important jobs of the epithelium is to protect the organism and its structures. An animal's external surface takes the most physical punishment. The skin must be especially resistant to tearing while remaining elastic enough to accommodate some stretching. Tight junctions between epithelial cells provide the strength necessary to keep the envelope intact. In addition, protective epithelium is usually stacked in several layers, further fortifying the envelope. These multicellular layers are referred to as *stratified epithelia* (stratum = layer). Many areas of the body, however, are covered by a single layer of cells, called the *simple epithelium.* This thin sheet provides little protection and functions primarily as a filter. For example, regions of the kidney that filter wastes from the blood are lined with simple epithelia, as are the microscopic sacs in the lung across which oxygen enters the bloodstream and carbon dioxide escapes into the lung spaces to be exhaled. In addition to protection and filtration, some epithelial cells specialize in manufacturing substances to be secreted. Such secretory epithelial structures (glands) produce and release hormones, mucus, oils, milk, and digestive enzymes.

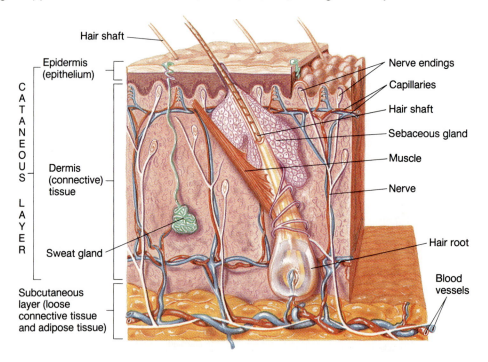

Skin is deceptively complex. It is not simply several layers of epithelial tissue, but a biological cooperative of all four tissue types (see accompanying figure). Thus, skin is an organ, the largest organ of your body. The outer **epidermis** consists of superficial layers of dead cells produced by the underlying living epithelial cells. Below the epidermis lies the *dermis,* a layer in which connective tissue predominates. Embedded in the dermis are blood vessels (a complex of epithelium, muscle, and connective tissue), various glands (epithelial tissue), blood (connective tissue), muscles, nerves, and follicles (epithelial tissue) for constructing hair or feathers. Skin is firmly secured to the underlying layer (called the *subcutaneous layer*) by connective tissue. (An overview of these tissue types is provided in the Connections Box on page 218.) Although it is not part of the skin itself, subcutaneous tissue is considered, along with skin, part of the **integumentary system**, the body's protective external covering.

# Synopsis

## MAIN CONCEPTS

- **Skeletal systems** protect, support, maintain form, and, along with muscles, help move an animal and its parts. Some animals use only internal water pressure to support the animal's body. Others have a rigid exoskeleton external to the animal's living tissues. This may be a calcified shell or a hardened cuticle of chitin. Still other animals have an internal endoskeleton that includes spicules of sponges, calcified plates of sea stars and sea urchins, and bones and cartilage of fishes, frogs, birds, snakes, mammals, and all other vertebrates.

- **Levered movement**. Rigid external or internal skeletons are attached to muscles, which move parts of the skeleton at movable joints. Each bony "lever" is moved by an antagonistic muscle pair. One muscle reverses the action of the other, so that the lever can return to its original position.

- **Architecture of bone**. In solid bone, osteocytes in the calcified matrix are supplied by blood vessels in Haversian canals. Spongy bone is lightweight bone with reinforced hollows that contain red marrow. Osteocytes build new bone by depositing hard mineral salts around flexible strands of collagen. Osteoclasts destroy bone by decalcifying them during calcium deficiencies or remodeling of existing bone.

- **Muscular systems** move an animal's body and propel substances through that body. Muscle contraction generates force by shortening. The effects are movement of levers, constriction of tubes, collapse of a space (forcing out its contents), and expansion of ribcage. The contraction is due to shortening of actomyosin, a complex of two proteins that form a sliding filament. An ATP-driven ratchet mechanism pulls actin filaments over myosin, drawing the end points closer. The movement is triggered by calcium. Actomyosin relaxes when calcium is pumped out of the cell.

## KEY TERM INTEGRATOR

Skeletal Systems

| | |
|---|---|
| **hydrostatic skeleton** | Support system that uses water pressure to maintain body form. |
| **exoskeleton** | External encasement for protection and support. In arthropods, the chitinous veneer is periodically shed and replaced with a larger size during **ecdysis**. |
| **endoskeleton** | Internal support system. In vertebrates it consists of **cartilage**, bones, or both. The **axial skeleton (skull, vertebral column, ribs**, and **sternum**) forms a protective framework to which the **appendicular skeleton (pectoral girdle, scapulae, clavicles**, pelvic girdle, and bones of the extremities) are attached. Adjacent bones are held together at **joints** by **ligaments**. Synovial cavities in mobile joints cushion and lubricate. |
| **ossification** | The building of bone from **collagen** and mineral salts by **osteocytes**. Most bone growth occurs as the **epiphyseal plates**. The osteocyte product is either **compact bone** or the **spongy bone** that contains the **red marrow**. The **medullary cavity** of long bones contains **yellow marrow**. Bone is built in cylindrical **lamella**, concentric cylinders around each **Haversian canal** in compact bone. Osteocytes are linked to each other by **canaliculi**. Bone is disassembled by **osteoclasts**. |

| Muscular System | |
|---|---|
| **skeletal muscle** | Voluntary muscle composed of **muscle fibers** in parallel alignment and usually attached to bone (by **tendons**). **Myofibrils** in each fiber are composed of the the **contractile protein, actomyosin**. The fiber's **striated** appearance reveals the presence of **sarcomeres**—overlapping strands of **actin** and **myosin** that shorten muscle length by sliding toward each other. |
| **motor unit** | A muscle fiber and its motor neuron. An activating neurological impulse sweeps from the **sarcolemma** down the **transverse tubules (T-tubules)** to the **sarcoplasmic reticulum**, which then releases calcium that triggers contraction. **Recruitment** initiates this process in additional muscle fibers to generate stronger force of contraction. |
| **smooth muscle** | Involuntary muscle composed of uninucleate cells in sheets or the circular bands that comprise some **sphinctors**. **Visceral smooth muscle** constitutes the major type, **multiunit muscle** found in blood vessels and other nonvisceral locations. |
| **cardiac muscle** | Fibers of striated heart muscle under involuntary control. Adjacent cells are connected by the branching process and by **intercalated disks** through which impulses are relayed from cell to cell. |

# Review and Synthesis

1. Slugs, octopuses, and thousands of other animals have no rigid skeleton. Explain what keeps these animals from being flattened by gravity.

2. Discuss the role of the following antagonistic relationships: osteocytes and osteoclasts; biceps and triceps; calcium and troponin/tropomyosin.

3. Describe the network that links osteocytes with each other and with Haversian canals. Why is this network necessary to the life of bone?

4. Supply the appropriate term(s) from Figures 17-3 and 17-8 that each of the following describe: one of the major sites of red blood cell formation; allow hand to rotate; 360° range of motion; central opening larger in women; contracting this muscle points front of foot downward; muscle that pulls front of foot upward.

5. Complete this table:

| Muscle Type | Voluntary or Involuntary | Striated or Unstriated | Attached to | In Fibers or Sheets |
|---|---|---|---|---|
| skeletal | ? | ? | ? | ? |
| ? | ? | unstriated | ? | ? |
| ? | involuntary | straited | ? | ? |

6. Describe the events in a muscle fiber that follow the arrival of an excitatory nerve impulse. Carry the discussion through to the point where the fiber returns to its original relaxed state.

## Additional Readings

Huxley, A. 1980. *Reflections on Muscles.* Princeton University Press, Princeton, N.J. (Intermediate.)

Kessel, R., and R. Kardon. 1979. *Tissues and Organs; A Text-Atlas of Scanning Electron Microscopy.* W.H. Freeman, San Francisco. (Introductory.)

Knight, B. 1980. *Discovering the Human Body.* Lipincott and Crowell. (Introductory.)

Morey, E. 1984. "Spaceflight and bone turnover." *Bioscience* 34:168–72. (Intermediate.)

Vander, A., J. Sherman, and D. Luciano. 1986. *Human Physiology.* 4th ed. (Chapter 10.) McGraw-Hill, New York. (Intermediate to advanced.)

Chapter 18

# Processing Food and Wastes — The Digestive and Excretory Systems

*T*he biology student lowered himself into his warm dinner, a bathtub full of chicken soup. Having learned that bacteria and some other microscopic organisms acquire nutrients by directly soaking them up from the surrounding medium, the enterprising student was

attempting to duplicate this digestive feat. In 24 hours, the soup would indeed be teeming with thriving bacteria, but the student would be hungry. If he depended on this activity as his sole source of nutrients, starvation would eventually end his misguided experiment. Even if nutrients could pene-

trate the epithelial layer of his skin, his body was too large for diffusion to deliver them to all his living parts. Furthermore, few of the nutrients would be small enough to cross plasma membranes and enter the cells they did reach. They would therefore be wasted, since the nutritive value of

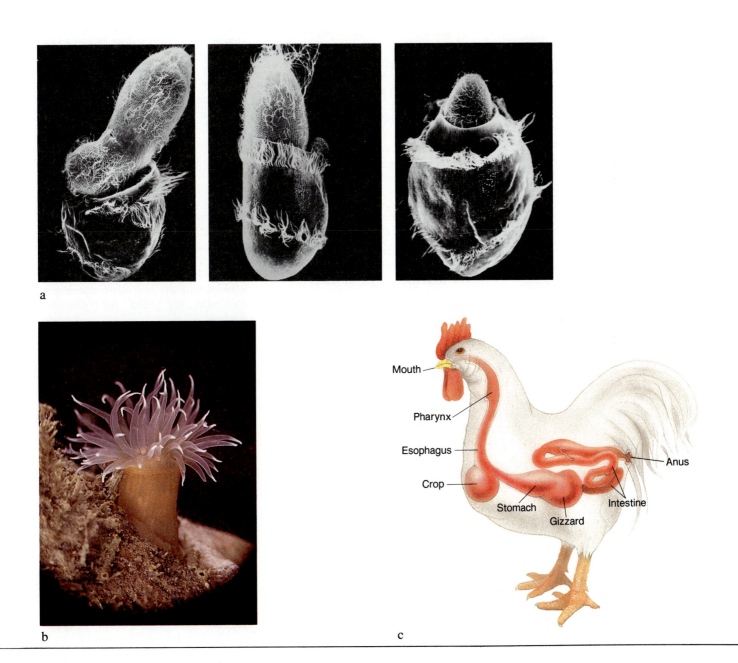

**Figure 18-1**
Digestive diversity. *(a) Intracellular digestion. The protozoon Didinium* engulfs a much larger *Paramecium* for dinner. *(b) Incomplete digestiva tract* of a sea anemone. *(c) Complete digestive tract* of a bird. The crop stores food; swallowed stones in the bird's muscular gizzard grind food.

food is unleashed only inside living cells. Except for the simplest animals, converting nutrients to a form that can be distributed throughout an animal's body and absorbed by all its cells requires the coordinated participation of a system of specialized organs, the **digestive system**.

As cells use nutrients for fuel or molecular building blocks, they generate metabolic wastes. If not evacuated, these compounds may poison the cells. Wastes are discharged from cells and, in most animals, enter a circulatory system. They are then carried to places in the body where they can be *excreted* from the organism, sometimes after being detoxified (rendered harmless) by a specialized organ, such as a liver. These tasks are performed by the **excretory system**. An animal's digestive and excretory systems form a trio with the circulatory system (in our case, the bloodstream) to bathe each cell in a life-sustaining nutrient solution that is free from dangerous levels of toxic wastes.

# The Digestive System

At what point does food enter the body? Most people answer, "as soon as you put it in your mouth" or "when you swallow it." Yet eating does not introduce food into your body. Substances in the stomach or intestines are still outside of you, just as your finger poking through the hole of a doughnut remains outside the pastry. The digestive tract is actually a pouchlike or tubelike continuation of the animal's external surface into or completely through its body. The walls of the tract are like a more absorbent version of skin, forming a barrier between the external environment and the internal tissues of the body. To enter the body, nutrients must be absorbed across the epithelium (see Connections: The Versatile Epithelium, Chapter 17) that lines the digestive tract. Digestion prepares food for just that. **Digestion** is the process of disassembling large particles into molecules that are small enough to be absorbed into the animal's body and eventually to enter the cytoplasm of cells, where the nutritive value of food is harvested. Eating only initiates the digestive process, launching food on a journey through tunnels

and chambers where digestion proceeds to completion.

## Digestive Diversity

The simplest form of digestion is the type that inspired the overzealous student to recline in the soup. Most microscopic single-celled organisms rely on **extracellular digestion**, that is, digestion that occurs outside the cell. These organisms secrete digestive enzymes that hydrolyze complex substances into small, absorbable molecules that can enter the cell. Such a simple strategy works for single-celled organisms such as bacteria, yeast, and a few protozoa. Most protozoa, however, engulf food particles into membrane-bound sacs, called food vacuoles, in which **intracellular digestion** occurs as large particles and molecules are enzymatically hydrolyzed to an absorbable size. (This process, called *phagocytosis*, is discussed in Chapter 7.)

Multicellular animals generally process food by extracellular digestion in specialized cavities or tubes that contain digestive enzymes. Nutrients released by grinding and chemical breakdown of food particles in these digestive areas are absorbed and distributed to the cells of the body. (This process is supplemented by direct engulfment and digestion of nutrients by the cells that line the digestive tract.) In many animals, unabsorbed wastes leave the digestive cavity through the *anus,* an exit located at the far end of the digestive tract. Animals lacking an anus have simple cavities with only one opening through which food enters and wastes escape. They are said to have **incomplete digestive systems**.

More complex animals need the specialization afforded by a **complete digestive system**, which includes a tube with openings at both ends—a mouth for entry and an anus for exit. Food moves sequentially through different compartments of complete digestive systems, each region performing its specialized tasks. Complete digestive tracts are "disassembly lines," that is, assembly lines in reverse. Several digestive strategies are illustrated in Figure 18-1.

An animal's digestive anatomy corresponds to the type of food it regularly consumes. Animals that harvest

microscopic food particles from water, for example, can do so because they have filters or sticky surfaces that screen these tiny nutrient bits from water. Teeth or other devices for crushing large pieces would be a useless adaptation for these *filter feeders.* Sponges, clams, and some other aquatic animals circulate water over mucus-lined surfaces that trap food particles. Some filter feeders extend feathery or highly branched extensions that harvest microscopic bits of food (Figure 18-2).

Many filter feeders devote virtually all their time to feeding. Even their sexual activities don't interrupt their quest for food. The majority of higher animals, however, feed periodically and must interrupt other activities, such as reproduction and caring for young, to eat. Discontinuous feeders are equipped with storage compartments that allow the animal to consume more than can be used immediately, and to harvest these nutrients over a sustained period, freeing time for other pursuits. The *crop* of birds and earthworms and the large stomachs of humans and other mammals eliminate the need for constant feeding.

Some animals obtain nutrients from fibrous plants that they are incapable of digesting. Cattle, antelopes, buffalo, giraffes, and several other *ruminant* animals possess additional stomach chambers, one of which is heavily fortified with cellulose-digesting microorganisms. In this chamber, called the *rumen*, bacteria and protozoa break down the otherwise indigestible cellulose and are themselves digested as the contents of the rumen travel through the rest of the digestive tract. Horses and other grazing animals that lack rumens may still utilize plant cellulose because of an elongated *cecum*, a blind (closed-ended) sac that extends from the intestine. The cecum serves as a fermenting vat in which microbes disassemble cellulose to products that are digestible by the animal.

## Vertebrate Digestive Systems

An examination of the human digestive system provides insight into the digestive strategy of mammals in general. The digestive tract is 30 feet (9 meters) of mouth, esophagus, stom-

a                                                    b

**Figure 18-2**
**Filter feeders.** (*a*) A clam filters microscopic food from the water it pumps through its body. (*b*) Some sea cucumbers collect microbes on their feathery tentacles, which are periodically placed in the mouth.

ach, intestines, and anus, plus accessory organs (Figure 18-3). All digestive structures are secured to the *peritoneum,* the connective tissue lining of the abdominal cavity.

Mechanical and some chemical digestion begins in the mouth, where the food is ground into a softened mass called a *bolus.* The ingested material receives a smoother ride through the digestive tubes because of a lubricating layer of mucus that coats the entire length of the tract. This mucus is secreted by the cells that line the tube, a living layer called the **mucosa.** Underlying the mucosa are two layers of smooth muscle which churn the bolus and propel it through the tract by sequential waves of muscle contractions (Figure 18-4). This movement, known as **peristalsis,** moves solids and liquids alike.

## MOUTH AND ESOPHAGUS
Digestion begins in the **mouth.** Food is cut and ground by the teeth (see Bioline: Teeth), mixed with saliva, and rolled into a pliable ball by the tongue. Saliva is the watery secretion from three pairs of salivary glands that empty their products through ducts into the mouth. Saliva contains starch-hydrolyzing enzymes that initiate chemical digestion in the mouth and may help protect teeth from decay by breaking down and dislodging starchy food particles trapped between teeth. (In spite of this natural protec-

tion, tooth decay remains the most costly disease in the United States.) These enzymes work best when the oral pH is neutral. Saliva contains bicarbonate ions ($HCO_3^-$) to buffer the oral secretions, maintaining optimum conditions for starch digestion. It also contains an enzyme, called *lysozyme,* that destroys many types of dangerous bacteria. *Mucin* in saliva lubricates the food and binds it together into a bolus.

When consistency is suitable, the bolus is forced into the back of the mouth (the *pharynx*) by an upward thrust of the muscular tongue and swallowed. Swallowing is voluntarily initiated, but smooth muscle reflexes complete the complex sequence that directs the bolus down the **esophagus,** keeping it out of the adjacent airway. The *soft palate* elevates and closes off the nasal cavity. The *glottis* (the "wrong hole" in swallowing) is sealed by the *epiglottis* when the *larynx* elevates during swallowing, leaving the esophagus the only open passageway for the bolus. (This explains why you can't breathe while swallowing.) These reflexes, controlled by a neural center in the medulla of the brain, occasionally fail in maintaining the proper sequence, and food or liquids enter the trachea (windpipe) by mistake. The explosive coughing reflex triggered by such misdirection protects the respiratory tract, which gracefully accepts only gases (Chapter 19). Once the bolus enters the esophagus, peristalsis propels it into the stomach.

## STOMACH
The **stomach** continues the job started by teeth, that is, the mechanical breakdown of food solids and limited chemically mediated digestion. The stomach's entrance is kept closed by a ring of muscle, the **cardiac sphincter,** which prevents the caustic contents of the stomach from backing up into the esophagus, which could be injured by stomach acid. (Occasionally, this happens anyway, causing "heartburn" or "acid stomach," which is felt neither in the heart nor in the stomach, but near the heart in the esophagus.) As peristaltic waves reach the stomach opening, the cardiac sphincter briefly relaxes to allow entry of the bolus. As food enters, it forces the collapsed stomach to expand, its muscles remaining relaxed to allow it to fill. Powerful contractions then sweep down the stomach toward the exit opening. The escape route, however, is blocked by a contracted **pyloric sphincter,** and the material is forced back toward the center of the stomach chamber. This churning action thoroughly mixes the contents for several hours, after which the pyloric sphincter opens slightly to allow the mixture to gradually squirt into the small intestine.

The average human stomach can hold and store about a liter (almost a quart) of material, which is churned into a pastelike consistency and mixed with *gastric juice* (gastro = stomach), forming a solution called *chyme.* The contents of gastric juice prepare food

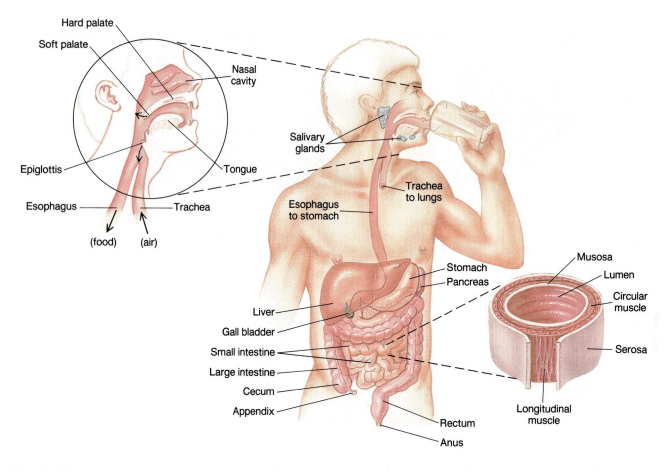

**Figure 18-3**
**The human digestive system.** In addition to the hollow digestive tract, the system includes accessory organs that aid digestion. The mucosa and musculature of the digestive tract are shown in the intestinal cross section.

for chemical digestion in the small intestine. One of the compounds in gastric juice, *hydrochloric acid (HCl)*, lowers the pH of the stomach contents to around 2.0. This extremely acidic environment kills most microbes, including those that could cause illness if allowed to survive in the **gastrointestinal tract** (the stomach and intestines). Stomach acid also dissolves food solids and opens up highly folded protein chains so that they are more easily attacked by proteolytic (protein-digesting) enzymes.

Although most digestive enzymes are in the small intestine, protein digestion begins in the stomach with the action of *pepsin*, an enzyme that attacks peptide bonds between certain amino acids. Pepsin degrades proteins to smaller fragments, but not to the single amino acids that are small enough to be absorbed into the circulation.

Pepsin not only attacks protein in food, but can also digest the proteins that constitute the muscles and tissues of the stomach itself. In other words, the living tissue that lines the stomach is just another piece of meat to the gastric juices. To prevent the stomach from becoming its own next meal, glands secrete a thick layer of mucus (Figure 18-5), forming a barrier between the gastric juices and the vulnerable stomach lining. Producing active pepsin would be particularly hazardous, since the enzyme-producing cells of the stomach would be the immediate targets of their own digestive enzymes. These cells avoid such a suicidal mission by producing *pepsinogen*, an inactive version of pepsin that digests nothing. Only after pepsinogen diffuses through the mucous layer does the hydrochloric acid of the stomach convert it to pepsin, the active protein-digesting enzyme.

Even with these protections, a stomachful of concentrated acid and tissue-destroying enzymes would be dangerous if they were continuously produced, for example, when the stomach was empty and a meal was not imminent. Hydrochloric acid secretion is timed to occur immediately before and during a meal. Endocrine

# B I O L I N E

## TEETH: A MATTER OF LIFE AND DEATH

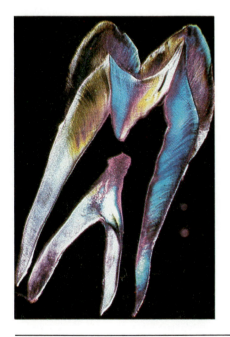

The elephant's heart was in excellent condition. So were its other vital organ systems. The giant was free of infection and disease. Yet she was dying, slipping toward the same end that has claimed her kind for thousands of generations. Although surrounded by food, the elephant was starving to death.

The coarse diet of plants on which an elephant dines steadily grinds away its teeth. During its lifetime, the worn-out molars are periodically replaced so that eating can continue. Only six sets of teeth can be produced, however, and usually by the age of 60 years, the last set has been ground flat. The otherwise healthy elephant can no longer chew its food and starves to death.

Teeth are especially important to digestion by herbivores (plant eaters), since most plants must be mechanically processed before nutrients can be released by chemical degradation. Chewing plant material requires teeth that can crush the tough cell walls to release the intracellular nutrients. The teeth must have broad-ridged surfaces that grind together, as do our molars. The knifelike cutting edges of the front teeth (incisors) are specialized for cutting off bites of food. Dogs, cats, and other carnivores that feed exclusively on meat have little need for teeth with grinding surfaces, since there are no cell walls in their diets to crush. Grinding surfaces have been replaced by sharp cutting edges that rip and slice the bolus into smaller pieces. Four pointed teeth, called canines (or "fangs"), are instrumental in securing and killing prey. The reduction in the length of human canine teeth reflects our different strategies for capturing prey. Primitive humans used their hands (and weapons) rather than their teeth for this purpose.

Animals that swallow their food whole do not use their teeth to assist digestion. Many of these animals have rows of sharp teeth, such as the backward curving teeth of snakes used exclusively for capturing prey (and delivering nasty bites to aggressors). The sharp teeth of sharks provide another fascinating, if not chilling, example of teeth adapted strictly for tearing off chunks of meat to be swallowed without being chewed.

---

cells embedded in the gastrointestinal mucosa produce *gastrin*, a hormone that arouses the HCl-secreting *parietal cells* of the stomach. Nerve impulses generated by chewing, smelling, tasting, or even seeing food stimulate gastrin secretion. Pressure exerted on the internal surface of the stomach as it fills with food activates mechanoreceptors that stimulate parietal cells to secrete acid. Before too much acid accumulates, negative feedback inhibits further secretion, maintaining safe levels. The stomach secretes about two liters of gastric juice per day.

When protective mechanisms fail, the stomach may begin digesting portions of itself, causing painful and dangerous *peptic ulcers*. If the production

of mucus is insufficient or acid secretion is excessive, craters can be digested in the stomach (or small intestine). These open sores can erode through the stomach wall, and gastric contents can escape into the abdominal cavity, causing inflammation or infection of its living lining, the peritoneum. This condition, called *peritonitis*, can be fatal if not rapidly treated. People with ulcers should avoid caffein (coffee, tea, cola, chocolate), alcohol, tobacco, and prolonged stress. All these factors promote gastrin secretion, which in turn acidifies the stomach.

A healthy stomach merely initiates digestion and absorption, processes that await final treatment in the small

intestine. Only a few small molecules, such as aspirin and alcohol, enter the bloodstream through the stomach walls. This explains the rapid onset of their effects. Most other substances are not absorbed until they enter the small intestine, some six hours after ingestion.

## The Small Intestine

After leaving the stomach, chyme spends the next three to six hours in the **small intestine**, which is made up of 21 feet (6.4 meters) of highly coiled muscular tube about one inch (2.6 centimeters) in diameter (Figure 18-6). During this time, virtually all the nutrients are digested to small

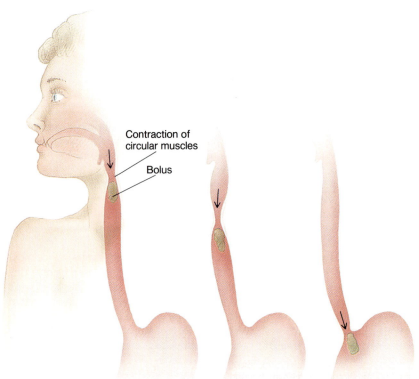

**Figure 18-4**
**Movement by peristalsis.** Each region of smooth muscle surrounding the tube contracts immediately after the one just preceding it, creating waves of contraction that sweep along the tube. These waves force ingested material through the entire length of the digestive tract.

**Figure 18-5**
**Gland in stomach lining discharging mucus.** In addition to secreting mucus, each gastric gland also contains cells that produce two components of gastric juice: hydrochloric acid (HCl) and pepsinogen (which is converted to the active enzyme pepsin in the stomach).

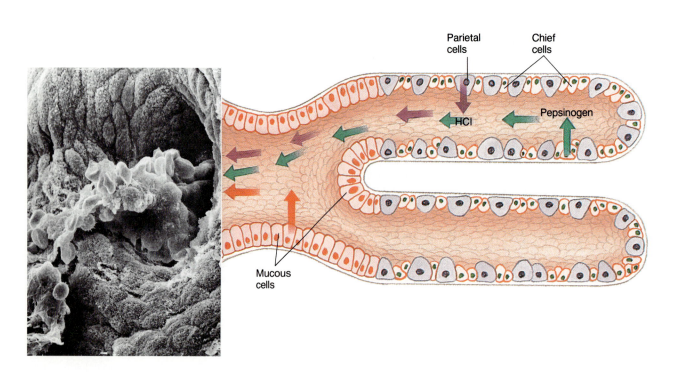

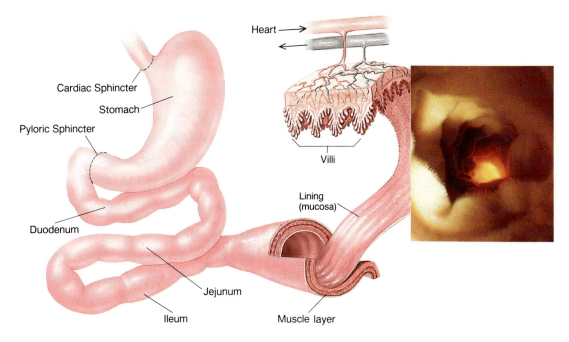

Heart

Cardiac Sphincter

Stomach

Pyloric Sphincter

Villi

Duodenum

Lining
(mucosa)

Jejunum

Ileum

Muscle layer

**Figure 18-6**
**Anatomy of the small intestine.** The surface area is increased by the folding of the lining and by the presence of villi, which are clearly evident in the inset photo. The average person sends about 330 pounds (727 kilograms) of solid food and 640 quarts (680 liters) of ingested liquids through this tunnel every year.

organic molecules (such as simple sugars, amino acids, and nucleotides) and absorbed into the bloodstream. The intestinal muscles mix the contents and move them steadily through the three parts of the intestine. The first 15 inches (38 centimeters) or so forms a C-shaped curve called the *duodenum*. The *jejunum* comprises the next two-fifths of the small intestine, and the *ilium* constitutes the remainder of the tube.

The velvety inner surface of the intestinal wall is a model of efficient absorption. Its texture is provided by fingerlike projections, called **villi**, which protrude from the entire surface. Villi increase the absorptive surface of the intestine in the same fashion that the texture of terry cloth towels enable them to soak up much more water than smooth cloth towels. Each villus (singular) is covered with smaller projections, called **microvilli**, that further increase the surface area of the small intestine to enormous dimensions. Together, intestinal villi and microvilli create an interior surface of about 2700 square feet (300 square meters),

150 times greater than the surface of your skin. Packed inside your lower abdomen is an intestinal surface about half the size of a basketball court.

Villi are laced with a rich network of capillaries. These tiny blood vessels allow the circulatory system to collect absorbable nutrients and distribute them to all the cells in the body (Figure 18-7). The capillary bed in each villus surrounds a central vessel of the lymphatic system, a portion of the circulatory system that carries lymph. *Lymph* is a bloodlike fluid that lacks red blood cells (Chapter 20). This saclike lymphatic vessel, called a *lacteal*, accepts the byproducts of lipid digestion absorbed from the intestinal lumen and transports them to the bloodstream. (The term *lumen* refers to the space within a hollow structure.) Most other substances are pumped by active transport across the epithelium directly into the capillaries.

The small intestine can absorb molecules only after they have been reduced to sufficiently small size by enzymes. These enzymes come from the pancreas (see next section) and from

glands embedded in the intestinal mucosa that produce proteolytic enzymes that clip amino acids from peptide chains. Other intestinal glands release *carbohydrases* that split disaccharides into two simple sugars. For example, intestinal glands secrete *lactase*, an enzyme that severs the disaccharide lactose (milk sugar) into its two simple sugar components, glucose and galactose. Many individuals cannot drink milk or eat milk byproducts without developing severe diarrhea because they lack the ability to produce lactase. Undigested sugar in the large intestine upsets its osmotic balance and draws an excessive amount of water into the lumen.

## THE PANCREAS

The **pancreas** is a gland that manufactures digestive enzymes and discharges them directly into the duodenum. It also releases sodium bicarbonate, which helps neutralize the severe acidity of chyme as it squirts from the stomach through the pyloric sphinctor. Pancreatic bicarbonate brings the contents of the small intestine to a

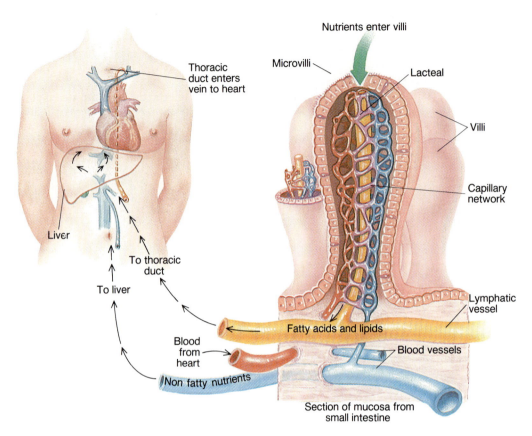

**Figure 18-7**
**Microanatomy of villi.** Nutrients are absorbed by capillaries or a central lacteal and distributed via the bloodstream to the rest of the body. Tiny microvilli further increase the absorbing surface. Substances absorbed from the small intestine pass through the liver (which detoxifies harmful molecules) before entering the general circulation.

slightly alkaline pH of 8.5. This pH is optimum for the hydrolytic activities of the pancreatic enzymes, which digest all four major types of macromolecules.

- *Trypsin, chymotrypsin* and other proteases degrade proteins to amino acids.

- *Nucleases* break down nucleic acids to nucleotides and nitrogenous bases.

- *Carbohydrases* oxidize complex carbohydrates sugars to simple sugars.

- *Lipases* hydrolyze triglyceride lipids to fatty acids and glycerol.

Pancreatic lipases work poorly, however, unless the lipids are emulsified by bile from the liver.

## THE LIVER AND GALL BLADDER

Among the more than 100 specialties performed by the **liver** is the production of an emulsifier called **bile**. This colored liquid, a solution of bile salts, pigments, and cholesterol, is transported either to the pancreatic duct for introduction into the duodenum or to the **gall bladder** for storage until needed following the next meal. Bile contains no enzymes and performs no direct digestion, but it is nonetheless required for efficient digestion of fats. Because of their water-insoluble nature, fats form large globules, exposing few molecules to intestinal lipases. Although churning breaks up these globules, the smaller droplets immediately rejoin each other unless an **emulsifier**,

such as bile salts, prevent them from doing so. The hydrophobic portion of the bile salts attaches to the fat, leaving its hydrophillic end facing out. As the intestine continues to churn, the diameter of each fat droplet decreases to a size that can be thoroughly digested by lipases.

## THE LARGE INTESTINE

Each day, digestion and absorption in the small intestine require about 2 gallons (7.5 liters) of water, about one-tenth of the body's total supply. If this much water were allowed to continue through the digestive tract and escape with the feces, we would quickly dehydrate. Recovering this precious water is the principal job of the **large intestine**.

The diameter of the digestive tract abruptly expands to 3 inches (7.5 centimeters) at the juncture between the small and large intestines. The large intestine is about 5 feet (1.5 meters) in length and is shaped like an inverted "U." The final 7 inches (18 centimeters) is the **rectum**, which joins with the anus (refer back to Figure 18-3). The large intestine, except for the rectum, is called the **colon**.

Chyme is emptied into the large intestine at a point several centimeters from the start of the colon, leaving a short blind pouch, the cecum (Figure 18-8). A small blind tube, the *appendix*, extends from the cecum. The cecum and appendix have no known function in humans. They are vestigial reminders of our evolutionary descent from organisms whose enlarged ceca (plural of cecum) aided digestion of plant material.

By the time it enters the large intestine, chyme is a watery solution from which nutrients have been removed. Remaining are undigested plant material (cellulose and lignin), large numbers of bacteria that normally populate the intestines, bile pigments, mucus, and sloughed cells from the intestinal lining. (The entire epithelial lining renews itself every 36 hours.) Most of the volume of chyme is water. Membrane "pumps" in the colon's epithelium actively transport sodium from the lumen into the cells of the intestinal walls. Water follows passively, responding to the steep osmotic gradient created by relocating sodium.

In this way, the large intestine recovers up to 99 percent of the water that enters it.

As water is absorbed, the solids and wastes are concentrated into a mass called *feces*. Just enough water is retained to keep the feces moist and pliable. Pressure-sensitive neurons in the colon detect when feces accumulate in the rectum and respond by provoking a *defecation reflex*. Impulses from the colon travel to the spinal cord, which responds by relaying a command to contract the muscles of the rectum, forcing the feces out through the anus. Because one of the two anal sphinctors is under voluntary control, we can consciously delay defecation (expulsion of feces) until an appropriate time.

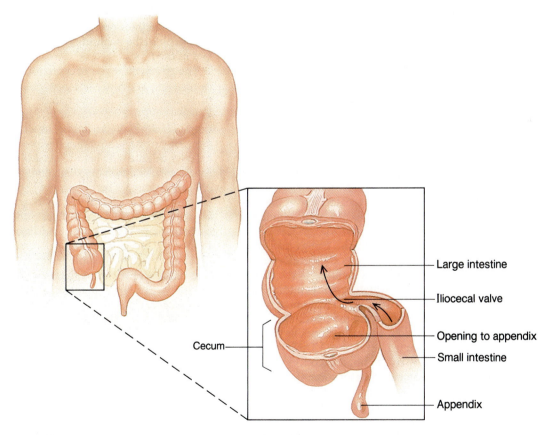

Large intestine

Iliocecal valve

Opening to appendix

Small intestine

Cecum

Appendix

**Figure 18-8**
**The large intestine** receives unabsorbed wastes and water from the small intestine through the ileocecal valve. The reabsorption of water is enhanced by convolution in the interior wall. The short cecum and appendix in humans is all that remains of larger cambers that housed cellulose-digesting microorganisms.

## NORMAL FLORA

Huge numbers of bacteria reside in the healthy human digestive tract. These microbes are collectively called the **normal flora**. Each gram of saliva in the healthy mouth contains millions of oral bacteria. In the large intestine, the number of resident microbes is so large that there are billions of bacteria per gram of feces. In fact, they constitute almost half the dry weight of human feces.

These intestinal bacteria metabolically attack organic substances in chyme and use them as nutrients, often producing unpleasant-smelling byproducts. These organisms are not freeloaders, however, for they manufacture vitamin K, biotin, folic acid, and other nutrients we can absorb and utilize. Although these services are not required when the diet is well balanced (see Bioline: Balanced Information for a Balanced Diet), the normal flora contribute enormously to the well-being of all people by helping protect us from disease-causing microbes. Our indigenous microbes successfully compete with many dangerous microbes for the body's limited space and nutrients.

## HUNGER

Food-seeking behaviors usually follow sensations of hunger, the body's message to your conscious mind that nutrients need to be replenished. During the hours following a meal, the concentration of glucose in the blood is high enough to stimulate neurons, called *glucoreceptors*, in the hypothalamus. Activated glucoreceptors inhibit the brain's *hunger center* (also in the hypothalamus). The sensation of hunger reoccurs when glucose levels fall slightly, signalling glucoreceptors to release their inhibitory grip on the hunger center. Yet glucoreceptors cannot claim all the credit for regulating hunger, since the sensation subsides as soon as our stomachs become full, long before glucose from the meal is absorbed into the bloodstream. Short-term hunger inhibition is probably the direct result of stomach distention, which stimulates pressure-sensitive neurons in the stomach wall. Otherwise, we could not satisfy our hunger for at least half an hour (the minimum time food spends in the stomach before beginning to enter the duodenum), regardless of how much we eat.

# The Excretory System

As important as nutrition is, bathing cells in nutrients would be futile if high concentrations of toxic wastes generated by cellular metabolism were allowed to accumulate. Removal of these wastes from an organism is an essential process called **excretion**. Whereas defecation unloads unabsorbed wastes that never really enter the body, excretion removes undesirable byproducts that form within the organism, wastes that would disrupt the delicate physical and chemical balance required for life to continue. In other words, homeostasis depends on excretion (see Connections: Homeostasis, Chapter 20).

## Organs That Participate in Excretion

Unicellular organisms and very simple animals can export their metabolic "sewage" directly into the surrounding medium to be washed away. But larger, more complex animals require specialized structures that selectively remove the wastes from the body fluids without throwing out the nutrients in the process. One simple waste, carbon dioxide, leaves the body through the gills of aquatic animals or the moist respiratory surfaces of air-breathing animals. Lungs and other respiratory organs are therefore excretory organs as well. So is the skin of humans, which perspires a solution of water, salts, and some nitrogenous wastes. (The major advantage of perspiration, however, is cooling the body rather than excretion.) The large intestine plays a small excretory role, receiving excesses of some types of salts that are passed into the cavity across the intestinal wall. The liver excretes breakdown products of dead red blood cells, which it first converts to bile pigments or to *urochrome*, the yellow pigment that gives urine its color. The liver also converts highly poisonous ammonia ($NH_3$) released during degradation of high-nitrogen compounds (proteins, nucleic acids) to **urea**. This relatively nontoxic nitrogenous molecule is tolerated in the tissue fluids until eliminated.

In many animals, special organs remove these toxic wastes, especially the high-nitrogen compounds. In insects, a series of tubes extend into the body cavity fluids, from which are absorbed wastes. These tubes (called *Malpighian tubules*) collect urine and pour it into the rectum for elimination from the body. Another invertebrate, the earthworm, has a series of more elaborate excretory organs, called *nephridia*. Each nephridium is a tube surrounded by capillaries. Cilia draw body fluid into and through the tube, which serves as a collecting duct for wastes. Wastes accumulate in the fluid that forms in the collecting duct. The waste-containing fluid is called urine. Salts and other nutrients that enter the nephridium are pumped out of the collecting duct and returned to the blood so they will not be lost. The urine leaves the worm's body through the opening (the nephridiopore) at the end of the tube.

The principal organs of excretion in humans are a couple of fist-sized structures attached to the lower spine, the **kidneys**. Without them, or a functional replacement, an animal quickly dies. Surprisingly, a mammal without kidneys will die long before metabolic wastes can accumulate to high enough concentrations to be poisonous. The kidney-deficient animal can no longer regulate the level of *electrolytes* (ions such as sodium, potassium, and chloride) in the blood. As potassium accumulates, it fatally disrupts the rhythmic beating of the heart. Clearly, the kidneys are responsible for more than merely removing wastes. They are better considered organs of homeostasis, since they help maintain a stable internal environment. They stabilize blood pH, regulate the volume of blood, determine the osmotic balance of plasma (the liquid portion of blood) and tissue fluids, and prevent either the accumulation or loss of excess water, depending on the environment in which the organism lives.

Kidneys are biological cleaning stations that remove wastes from blood, forming **urine**, a solution fated for discharge from the body. Human urine consists of urea and other nitrogenous

# B I O L I N E

## BALANCED INFORMATION FOR A BALANCED DIET

Although still embroiled in controversy over such topics as the impact of dietary cholesterol and saturated fats on our total health, the science of human nutrition has established that the healthiest diets are those that balance the three molecular food groups—carbohydrates, triglyceride lipids (fats and oils), and proteins. Foods that provide these three also contain enough energy, organic building blocks, vitamins, and minerals to supply the needs of the average person without supplements from bottled vitamins.

### Energy

The body can obtain its energy needs from any of the three food groups. High-energy molecules, such as carbohydrates and triglyceride lipids, provide the most convenient form of *kilocalories* (1000 calories, the measure of molecular energy discussed in Chapter 3). Although people differ in their daily energy needs, the average has been established as 2700 kilocalories for men and 2000 for women, assuming average weight and moderate level of activity. Rather than restricting caloric intake, nutritionists recommend balancing energy consumption with expenditures. In other words, a person's level of activity should match the calories consumed. Excess calories that are not used for fueling body activities are converted to fat. Even protein can be fattening if its consumption is not balanced with physical activity. Once the protein synthesis needs are satisfied, amino acids are converted to glucose in the liver and may end up as fat. Conversely, a person whose caloric intake has been reduced below the body's energy demands loses weight and may become emaciated.

### Carbohydrates

Although carbohydrates have been indicted as sources of "empty calories," they do provide the most convenient form of glucose. Glucose is the "all-purpose" energy source and the only one that can be used by all brain and nerve cells. When these cells are deprived of energy, the resulting temporary *hypoglycemia* (glucose deficit) may cause unclear thinking, clumsiness, depression, or muscle shaking due to diminished neurological function.

Other cells can harvest energy from fatty acids, but using them instead of glucose tends to acidify the blood, a condition called *acidosis*. (Diabetics who cannot utilize glucose when their insulin levels plunge may lapse into acidosis comas if their blood pH gets too low.) Thus, moderate carbohydrate consumption is part of a balanced diet.

Not all carbohydrates are easily digestible. Cellulose, the polysaccharide of plant cell walls, resists disassembly in the human digestive tract. Yet it promotes health by providing bulk, fiber that assists in the formation and elimination of feces. Low-cellulose diets cause constipation and have been linked to colon cancer.

### Triglyceride Lipids

Fats and oils are rich sources of energy, as well as the preferred form for long-term storage of energy surpluses. (Twice as much energy can be stored in lipid as in an equivalent weight of carbohydrate or protein.) Lipids also supply some lipid-soluble vitamins (A, D, E, and K) and are needed for manufacturing phospholipids for cell membranes. Two fatty acids (linolenic and linoleic acid) required for phospholipid construction cannot be manufactured in the body and must be acquired from a dietary source. If these *essential fatty acids* are not included in the diet, the integrity of cell membranes is compromised. Fat is also an efficient shock absorber and thermal insulator. Nonetheless, excess fat in the diet may contribute to several undesirable conditions, from obesity to cardiovascular disease (Chapter 20).

Although still controversial, high concentrations of cholesterol in the blood are believed to increase the likelihood of heart disease. Yet cholesterol is an important component of cell membranes, and we simply cannot do without it. Recent evidence indicates that diets very low in cholesterol may increase the incidence of cancer. Moderating, but not eliminating, cholesterol and saturated fat consumption appears to be the most prudent course until nutritionists resolve the controversy.

## Proteins

Dietary protein is needed to supply the amino acids from which we assemble our functional proteins (such as enzymes, hemoglobin, some hormones, and antibodies) and the structural proteins that comprise much of the fabric of our tissues. We can manufacture all but eight amino acids metabolically. The absence of even one of these eight *essential amino acids* prevents complete synthesis of proteins. Consequently, the quality of protein in the diet is just as important as the quantity. The quality of protein in meat generally exceeds that obtained by eating plants. Grain and vegetable protein is often deficient in one or more amino acids. Corn, for example, provides no lysine. Corn can be supplemented with another vegetable, such as beans, that contains lysine but lacks one of the amino acids supplied by corn. This complementary combination of vegetables provides a complete source of amino acids. Vegetarian diets should include such a complement of vegetables to avoid protein deficiency.

## Vitamins and Minerals

Vitamins are organic compounds required in tiny amounts. They are also much smaller than the giant macromolecules; rather, they are closer in size to a single glucose molecule. They are not metabolized for energy, and, except for choline, which is incorporated into cell membranes, they do not become structural components of tissue. Most vitamins, notably the B-complex group, are coenzymes that assist essential enzyme-mediated reactions, as discussed in Chapter 5. Others (vitamins C and E) perform protective functions, preventing random oxidation and subsequent damage of cell membranes. Vitamin D promotes bone growth by stimulating calcium absorption. Retinal function depends on vitamin A, which becomes a constituent of rhodopsin, a pigment needed for vision. Blood clotting proceeds properly only in the presence of vitamin K.

Although plants can synthesize all their own vitamins, animals cannot. Humans must acquire 13 vitamins from their diet or suffer vitamin deficiency disorders, some of which can be fatal. A well-balanced diet normally provides all the vitamins needed.

A dietary supply of *minerals* is just as important as vitamins. Without calcium and magnesium, a battery of enzyme-mediated reactions would simply not occur. Calcium is also needed for bone growth and muscle function. Iron forms the molecular core of cytochromes (needed by all cells for respiration) and of hemoglobin, the oxygen-transporting pigment in blood. Phosphorus is needed for phosphorylation and nucleic acid synthesis. Sodium, potassium, and chloride are required for maintaining osmotic balances and transmitting nerve impulses. A number of other minerals (iodine, copper, zinc, cobalt, fluorine, and silicon) are needed in very tiny amounts, so small that they are referred to as *trace elements*.

wastes, as well as a variety of ions (salts) dissolved in water. From the 50 gallons (190 liters) of blood that human kidneys filter each day, a little over a quart of urine is formed. The kidneys of terrestrial animals are remarkably adapted to conserve water by forming highly concentrated urine. These organs are so efficient that some desert animals, notably kangaroo rats, may live their entire lives without taking a drink. These animals get all the water they need from the moisture in food and the water generated when hydrogen combines with oxygen during cellular respiration.

## The Urinary System

The structures that form and export urine constitute the **urinary tract**, exemplified here by the human system (Figure 18-9). From the kidneys, urine is moved by peristalsis down two muscular tubes, the **ureters**, and into a holding tank, the **urinary bladder**. When enough urine accumulates to stretch the bladder wall, the smooth muscles of the organ contract, forcing the contents down a canal that carries the urine out of the body. This canal, called the **urethra**, is ordinarily kept closed by two sphinctor muscles. The involuntary internal sphinctor automatically relaxes when the bladder contracts, but the fluid is retained by the outer sphinctor, which relaxes in

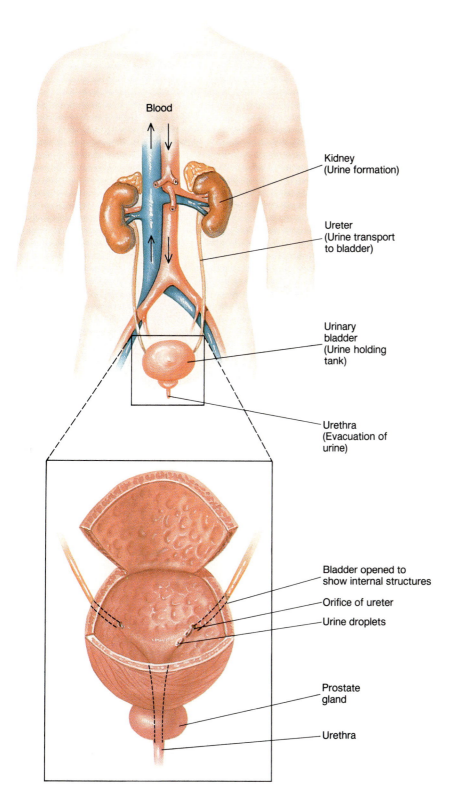

**Figure 18-9**
**The human excretory system.** Wastes are removed from blood by kidneys, forming urine. Ureters provide passageways to the bladder, where urine is held until released from the body through the urethra.

response to conscious commands from the brain. *Urination* can therefore be delayed according to the dictates of either convenience or urgency.

Not all animals employ the same excretory strategy as terrestrial mammals, which must sacrifice some water to dissolve and carry away urea. Birds, reptiles, and insects convert ammonia to *uric acid,* an insoluble waste compound with the consistency of paste. A solid, concentrated waste product such as uric acid wastes little water.

Animals that excrete uric acid develop in eggs outside the parent, where they are literally enclosed with their excretory products until they hatch. Whereas urea would dissolve in the egg's fluids and eventually poison the embryo, uric acid precipitates as a solid residue that settles to the egg's lowest point, providing a "gravity driven self-cleaning" system.

## The Mammalian Kidney

The formation of urine is a complicated process in which the kidney determines what will remain in the body and what will be discarded. This determination influences the composition of the body fluids.

The kidney is structured to allow materials to be exchanged between two sets of tubes—the blood vessels and the *renal tubules* ("renal" = kidney) in which urine forms (Figure 18-10). The blood and contents of the urinary tubes never mix but are separated by the membranes that form the tubes themselves. These membranes do the filtering.

Blood flows into each kidney through the renal artery; cleansed blood flows out through the renal vein. The renal artery branches into smaller vessels that form a network of capillaries. Blood first passes through the **glomerulus,** a capillary bundle embedded in a double-membraned container, the **Bowman's capsule.** The capsule is actually an invagination of the *proximal* (nearest) end of the renal tubule, forming an indentation into which the glomerulus protrudes.

The renal tubule then convolutes among the vascular network, providing a large surface area for exchange of materials between the two systems. The **proximal convoluted tubules** are

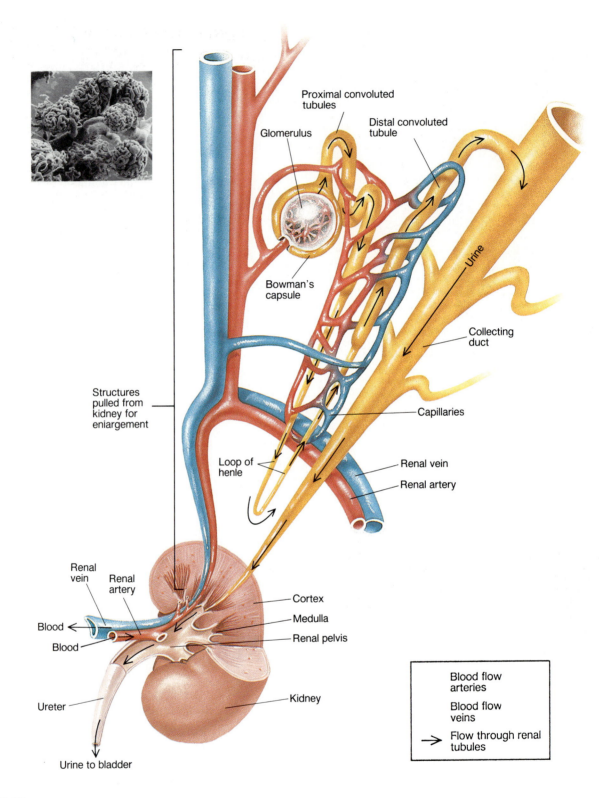

Proximal convoluted
tubules

Distal convoluted
tubule

Glomerulus

Bowman's
capsule

Urine

Structures
pulled from
kidney for
enlargement

Collecting
duct

Capillaries

Loop of
henle

Renal vein

Renal artery

Renal
vein

Renal
artery

Cortex

Medulla

Renal pelvis

Blood

Blood

Kidney

Ureter

Urine to bladder

| | |
|---|---|
| | Blood flow arteries |
| | Blood flow veins |
| → | Flow through renal tubules |

**Figure 18-10**
**Structure of the kidney.** Blood flowing into the kidney through the renal artery is purified by urinary tubes (shown in yellow). Cleansed blood then leaves the kidney through the renal vein. Water, nutrients, and wastes pass from blood into the nephron at the Bowman's capsule. Nutrients and about 95 percent of the water are recovered from the nephron and reabsorbed by capillaries back into the bloodstream. The liquid that remains, urine, moves from the renal tubules into collecting ducts, which empty into the hollow renal pelvis. Urine then travels to the bladder through a ureter. (inset) A scanning electron micrograph of several glomeruli.

separated from the **distal convoluted tubules** (distal = distant) by an elongated section, called the **loop of Henle**, whose shape dips down into the kidney's medulla and then ascends back up to the cortex. Urine forming in the renal tubules pours into collecting ducts that empty into a hollow space, called the renal pelvis. From there, the liquid flows through the ureter to the bladder. The renal tubule, starting with the Bowman's capsule and ending at the collecting duct, forms a **nephron**, the functional unit of the kidney. Each kidney houses over a million nephrons.

## Urine Formation by the Nephron

Extraction of wastes from blood to form urine is a three-part process, fol-lowed by an additional step that delivers urine to the bladder (Table 18-1). A close examination of these three urine-forming processes (pressure filtration, selective reabsorption, and tubular secretion) reveals how the kidney combines excretory functions with an ability to maintain homeostasis.

### PRESSURE FILTRATION

Blood pressure forces fluid out of the glomerulus capillaries into the Bowman's capsule. The one-cell thick membranes act as a sieve that allows water, glucose, salts, urea, and other small molecules across, while retaining blood cells and large molecules (step 1, Figure 18-11). The proteins left behind in the blood create an osmotic gradient that would draw water back into the bloodstream if not for the blood pressure within the glomerular capillaries. This pressure (generated by the heart pumping blood through the capillaries) forces the filtrate to collect continuously in the Bowman's capsule. An animal relying solely on this somewhat unselective process to form urine would have little chance of survival, since it would be throwing out critical resources (water, salts, some hormones, and nutrients) with the trash.

### SELECTIVE REABSORPTION

Although all animals must recover the nutrients from the crude filtrate to survive, humans and other terrestrial animals must also recover most of the water to prevent dehydration. About 50 gallons (190 liters) of the blood's

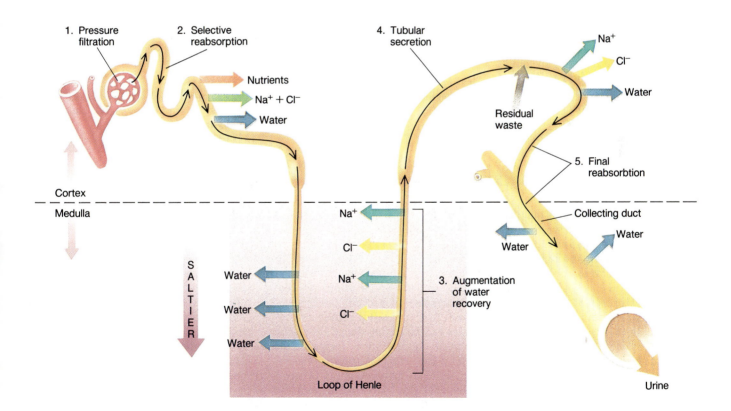

**Figure 18-11**
**Urine formation in the nephron.** Nutrients, salts, and water forced into the nephron by pressure filtration are recovered during selective reabsorption. Water recovery is augmented by the loop of Henle, which dips deep into the medulla, where increasing concentrations of salt produce a steep osmotic gradient. Osmotic pressure then draws water from the urine so that it can be recovered by blood vessels. The concentrated urine then moves into the distal tubules, where tubular secretion removes residual wastes too large to have entered the nephron by pressure filtration. Urine then enters the collecting duct, which passes through the salty medulla. Therefore, even more water is extracted from the urine during final reabsorption.

liquid are filtered by the human kidneys each day. If you excreted it all, your body would lose 10 percent of its water in less than 10 minutes. A half hour of this would be fatal.

As the water-rich, nutrient-laden glomerular filtrate moves along the proximal tubule, active transport exports glucose, salts, and other valuable nutrients out of the renal tubule (step 2 and 3, Figure 18-11). Before these compounds can be reabsorbed back into the bloodstream, they accumulate in the interstitial fluid (the liquid that surrounds cells). The higher solute concentration of the fluid around the proximal tubule creates an osmotic gradient that draws water out of the filtrate. Therefore, active transport recovers nutrients, and water follows along passively.

The efficiency of water recovery is boosted by the loop of Henle. Ionized salt ($Na^+$ and $Cl^-$) is actively pumped from the ascending side of the loop, making the surrounding fluid very salty (shown here as a color gradient). The steep osmotic gradient draws water out of the descending side of the loop. (The ascending tube is impermeable to water.) Thus, urine is very concentrated (a high waste-to-water ratio) by the time it reaches the distal convoluted tubules for tubular secretion.

## TUBULAR SECRETION

The composition of urine is modified by the addition of large waste molecules (such as creatine and urochrome), uric acid, and ions during tubular secretion at the distal tubules (step 4, Figure 18-11). It is here that the kidney exerts its regulatory checks and balances that maintain homeostasis of the tissue fluids. The kidney regulates blood pH by excreting more hydrogen ions and absorbing more bicarbonate ions when the blood becomes slightly acidic. (This process is reversed when blood is becoming too alkaline.) Tubular secretion also provides the kidneys a way to stabilize the "saltiness" of the blood by changing the rate at which sodium is recovered from urine before leaving the kidney. When the sodium concentration in the blood decreases, as it does when the body sweats profusely, the adrenal cortex

**TABLE 18-1**

| A Summary of Renal Function | | |
|---|---|---|
| **Process** | **Function** | **Location** |
| Pressure filtration | Forces large volumes of the blood's rich liquid out of the vessels and into the renal tubule. | Glomerulus and Bowman's capsule |
| Selective reabsorption | Recovers nutrients, salts, and water from the rich filtrate in the renal tubule, transporting them back into the bloodstream through the adjacent capillary network. | Proximal convoluted tubules; loop of Henle |
| Tubular secretion | Excretes waste molecules too large to squeeze into the renal tubule during pressure filtration. Ions ($H^+$, $K^+$, and bicarbonate) are exchanged in amounts needed to keep the blood's pH and osmotic pressure stable. | Distal convoluted tubule; collecting duct |
| Excretion | Urine that accumulates in collecting duct is excreted into the renal pelvis. From here it travels through ureters to the urinary bladder. | Distal collecting duct, renal pelvis, and beyond |

releases the hormone aldosterone (Chapter 16), which stimulates the reabsorption of sodium (and potassium) from the distal convoluted tubules and collecting ducts. When the sodium concentration increases, as after a salty meal, aldosterone release is inhibited, less sodium is pumped back into the bloodstream, and the ionic excess is excreted.

The volume of the blood is a function of its water content, regulated at the collecting ducts. Blood volume decreases when too little water is absorbed from urine back into the bloodstream. This causes the blood to become too concentrated. Osmoreceptors in the hypothalamus detect

the increase in the blood's osmotic strength and order the posterior pituitary to release antidiuretic hormone (ADH), as described in Chapter 16. Because ADH increases the renal tubule's permeability to water, more of it is reabsorbed from urine. This dilutes the blood while concentrating urine (step 5, Figure 18-11). When blood becomes too dilute, ADH release decreases, the duct's permeability to water decreases, and excess water is excreted in a more dilute urine.

The hypothalamus also alerts the consciousness when the water content of blood is low. What you perceive as thirst is your hypothalamus instructing you to consume more water.

# Synopsis

## MAIN CONCEPTS

- **A trio of systems.** The digestive and excretory systems act in concert with the circulatory system to assure that each cell is bathed in a nutrient solution that is free of dangerous levels of toxic wastes.

- **Digestion.** Large, complex food substances are dismantled into molecules small enough to pass through plasma membranes and to enter cells, where their nutritional value is unleashed. Complete digestive systems are continuous tubes with a mouth at one end, an anus at the other. In humans, several accessory organs manufacture and deliver digestive enzymes, a fat emulsifier (bile), and acid-neutralizing bicarbonate. Ingested material is forced through the entire tract by peristalsis. A thick mucus lining provides a barrier between living tissues of the digestive tube and digestive juices. Materials in the tract do not enter the body until they are absorbed through the digestive wall. In vertebrates, virtually all nutrient absorption occurs in the small intestine. The large intestine recovers water, leaving a solid mass of undigested material for elimination as feces.

- **Excretion.** The wastes generated by cellular metabolism are evacuated by the excretory system. Humans eliminate most of their wastes by forming and discarding urine. Kidneys remove from blood the major nitrogenous waste, urea, which is forced into the renal tubules of the nephrons from the glomerulus into the Bowman's capsule. The filterate contains nutrients and water, which are recovered by reabsorption into the capillaries surrounding the renal tubules. Additional wastes are then added to urine, which is further concentrated in collecting ducts.

- **Homeostasis.** The kidneys help stabilize the pH and osmotic strength of blood and regulate blood volume.

## KEY TERM INTEGRATOR

### Digestive Systems

| | |
|---|---|
| digestion | The disassembly of food either outside of cells (**extracellular digestion**) or inside cells (**intracellular digestion**). |
| digestive system | **Incomplete digestive systems** have one opening for entrance and exit of materials, compared to the two openings of **complete digestive systems**. In vertebrates, the **mucosa** that forms the tube connecting the two openings secretes mucus, plus enzymes and acid that disassemble food as it moves along by **peristalsis**. The food enters the tract through the **mouth**; swallowing forces the bolus into the **esophagus**, which carries it to the **cardiac sphinctor** and into the **stomach**. The material is now in the **gastrointestinal tract**. The mixture squirts into the **small intestine** through the **pyloric sphinctor** and is mixed with enzymes from the **pancreas**. The **gall bladder** contributes **bile** from the **liver**. **Villi** and **microvilli** increase the small intestine's surface area. The final leg of the journey is through the **large intestine** (the **colon** plus the **rectum**), the area richest in **normal flora** microbes. |

## Excretory Systems

**excretion**
In vertebrates, the **kidneys** remove nitrogenous wastes, usually in the form of **urea**, a soluble molecule carried away in **urine**. This liquid is formed in **nephrons**, each of which begins as a **Bowman's capsule** wrapped around a **glomerulus**. The capsule narrows into a tube, the **proximal convoluted tubule**, which takes a sharp turn as it becomes the **loop of Henle** embedded in the kidney's **medulla**. The **distal convoluted tubule** then conveys the waste to the **collecting duct**. The **renal pelvis** accepts the finished urine, which is eliminated through the **ureters, urinary bladder**, and **urethra**, the other organs of the **urinary tract**.

## Review and Synthesis

1. Describe the functional interconnections between the systems for digestion, circulation, and excretion.

2. Why is food not considered to be inside the body immediately after it is swallowed? With this consideration in mind, is digestion in humans intracellular or extracellular? At what point do nutrients actually enter the body?

3. Trace the fate of a mouthful of food through the entire digestive tract. Describe the changes that occur in each portion of the tract and discuss the activities of saliva, gastric secretions, pancreatic secretions, bile, and the intestinal normal flora.

4. Explain why each of the following can result in an ulcer of the digestive tract walls: failure of the cardiac sphinctor to prevent backflow of stomach contents; thinning of the mucous layer of the stomach or small intestine; inappropriate gastrin secretion; prolonged and extreme anxiety (stress).

5. Select an item from the right-hand column that would describe the most probable outcome of the condition on the left:

   | | |
   |---|---|
   | premature defecation reflex | protein deficiency |
   | blocked bile duct | white, chalky feces |
   | blocked pancreatic duct | diarrhea |
   | lactase deficiency | |

6. Utilization of absorbed amino acids sometimes requires removal of the amino ($NH_2$) group from the molecule. Trace the fate of the released nitrogen from its toxic form (ammonia) through its elimination from the human body. How does this compare with the way birds dispose of their toxic nitrogen wastes?

7. In addition to cleansing the blood of wastes, describe three other ways your kidneys help maintain homeostasis.

8. Draw and label a typical nephron and relate each part to the three urine-forming processes.

9. What would happen if your renal medulla were no saltier than your renal cortex?

## Additional Readings

Dantzler, W. 1982. "Renal adaptations of desert vertebrates." *Bioscience* 32:108–13. (Intermediate.)

Hamilton, E. 1982. *Nutrition: Concepts and Controversies.* West Publisher, San Francisco. (Introductory.)

Langley, L. 1965. *Homeostasis.* Reinhold, (Advanced.)

Moog, F. 1981. "The lining of the small intestine." *Scientific American* 245:154–76. (Intermediate to advanced.)

Nilsson, L. 1978. *Behold Man.* Little, Brown, Boston. (Introductory.)

*Readings from Scientific American.* 1978. *Human Nutrition.* W. H. Freeman, San Francisco. (Intermediate to advanced.)

Vander, A., J. Sherman, and D. Luciano. 1986. *Human Physiology: The Mechanism of Body Function.* 4th ed. McGraw-Hill, New York. (Intermediate.)

Chapter 19

# Gas Exchange — The Respiratory System

The metabolic "fires" of virtually all animals depend on oxygen to keep burning. During cellular respiration, oxygen unleashes the chemical energy stored in food molecules, generating more than 90 percent of the ATP that drives energy-consuming processes. Animals can survive for weeks without food and for days without water, but only a few minutes without oxygen.

The term "respiration" takes on additional meanings when referring to multicellular animals. The collection of processes that fall under the umbrella of this term are those used by an animal to

- Acquire and absorb oxygen from the environment
- Transport oxygen to the individual cells of the body.
- Generate ATP, using oxygen as the terminal electron acceptor of biological oxidations (also called *cellular respiration*).

You have already become acquainted with the dynamics of cellular respiration (Chapter 10). This chapter focuses on the acquisition of oxygen and its entrance into the circulatory system for distribution to cells, where it is needed for cellular respiration. In this chapter, you will also discover how the same distribution system eliminates the metabolic waste $CO_2$. Because the evacuation of $CO_2$ accompanies the acquisition of $O_2$, the respiratory system is often called the **gas exchange system**.

## Strategies for Acquiring Oxygen

Molecular oxygen ($O_2$) is a gas that comprises 21 percent of the atmosphere at sea level. Its ability to dissolve in water is essential to the lives of all $O_2$-dependent animals, both those that live on dry land and those that live underwater. When $O_2$ molecules strike a dry surface, they merely bounce off; dry membranes cannot capture a gas. A moist surface, however, readily extracts $O_2$ from air because many of the oxygen molecules dissolve in water when they strike it. The gas exchange membranes of all animals must remain moist to avoid suffocation. The moisture you see condensing as you exhale on a cold day testifies to this fact.

The successful respiratory system not only captures and distributes $O_2$, but also discards the respiratory refuse, carbon dioxide ($CO_2$). To enter or leave the body, these gases must cross a barrier of living tissue, the **gas exchange membranes**. In the simplest animals, the gas exchange membrane is the skin or, in unicellular organisms, the plasma membrane. Most multicellular animals, however, have developed more elaborate systems that overcome the limitations of "skin breathing." These limitations are imposed by the relatively small surface area of the skin. The skin simply doesn't provide enough contact with air to harvest the amount of $O_2$ needed for larger, highly active animals.

One way to increase the efficiency of respiration is to expand the gas exchange surface. Another way is to couple the gas exchange membranes to a circulatory system that quickly distributes $O_2$ to remote inner tissues. The gas exchange membranes of these ani-

**Figure 19-1**
**Simple gills.** This nudibranch (sea slug) increases its $O_2$ absorbing efficiency with extensions of its external surface. The abundant supply of circulatory vessels seen beneath the animal's transparent skin distributes the collected $O_2$ to the rest of the body.

mals abound with tiny blood vessels (capillaries) that quickly exchange gases between the circulatory system and the environment. The transaction is aided by the thinness of the two barriers through which gases must pass—usually a one-cell thick gas exchange membrane and a one-cell thick capillary wall. The circulatory system also carries $CO_2$ from tissues to the gas exchange membrane, where it is discarded.

The successful respiratory strategy is one that is tailored to the medium in which the animal lives. Animals that absorb $O_2$ from water must use a different type of respiratory system than animals that extract it from air.

## Extracting Oxygen from Water

The amount of $O_2$ dissolved in water is at best 21 times less than in air, depending on temperature (warm water holds less $O_2$ than does cold water.) As a result, multicellular aquatic animals are either locked into a much slower rate of activity than are air-breathing animals, or they must spend substantial amounts of energy to boost the amount of $O_2$ they can extract from water. In some simple animals such as sponges, anemones, hydras, and flatworms, all cells are close enough to their outer surfaces that no specialized respiratory structures are necessary. Many larger aquatic animals, however, enhance their $O_2$-collecting capacity by using **gills**. These are highly *vascularized* (rich in blood vessels) extensions of the gas exchange membrane. The complexity of an animal's gills reflects its $O_2$ needs. Simple gills are mere flaps or tubes of skin, evaginations that increase the gas exchange surface (Figure 19-1). Such a system is adequate for some nudibranchs (sea slugs), sea stars, and many other slow-moving aquatic animals.

Animals with greater $O_2$ requirements have more complex gills to harvest additional oxygen. Complex gills subdivide into minute fingers that amplify the surface area without increasing the space required to house them. In addition, they are usually part of a more elaborate respiratory system, one that actively and steadily propels water over the gill surfaces, constantly

replenishing the animal's $O_2$ supply and purging $CO_2$ from the blood.

Gill complexity has reached its current pinnacle in fishes. Most fish suck in water through their mouths, direct it over the gills in a single direction, and then discharge it through holes in the sides of their bodies. Such one-way flow is imperative to the success of highly active "water-breathers," for it enormously multiplies the amount of $O_2$ extracted from water. Blood can harvest much more $O_2$ from water if the two liquids flow in opposite directions than in the same direction. This principle is called **countercurrent flow** (Figure 19-2). Because gas exchange follows the dictates of simple diffusion, $O_2$ is absorbed only if the oxygen concentration in the tissue fluids and capillaries is lower than that in the water. Countercurrent flow maintains a diffusion gradient advantageous for both $O_2$ absorption and $CO_2$ elimination.

Some aquatic animals rely on neither gills nor skin-breathing. These animals have descended from air-breathing terrestrial ancestors. Porpoises, whales, sea snakes, and some water spiders still rely on air to supply their $O_2$, even if they have to bring an external supply of it under water with them (Figure 19-3).

## Extracting Oxygen from Air

Although air is richer in $O_2$ than is water, it tends to dry out gas exchange membranes. Terrestrial animals either have to live in extremely moist environments, such as water-saturated soil, or spend much of their precious water keeping their gas exchange surfaces moist. The amount of water lost is minimized by using something like a gill in reverse, that is, gas exchange membranes that extend inside the body, where most of the moisture can be retained. This has been accomplished in several ways, the two most common of which are the use of tracheae by most terrestrial arthropods, such as insects, and the use of lungs by amphibians, reptiles, birds, and mammals (and by a few species of fishes).

**Tracheae** are moist, branching tubes that extend into the interior from a series of holes, called *spiracles*, in the sides of the animal's body. Each

trachea divides and subdivides into smaller tubules, called *tracheoles*, that are so pervasive that all the animal's living cells are within diffusion distance of an $O_2$ source. Therefore, the distribution of $O_2$ needs no assistance from the arthropod circulatory system. In most arthropods, air flow into tracheae is passive, unassisted by any method of forced gas movement. Such a system, however, would be inadequate for larger organisms. Without actively drawing in more air, they would not obtain enough $O_2$ and without help from a circulatory system, $O_2$ could not reach all the tissues. Some insects use muscle activity to enhance air flow into the tracheae, providing enough $O_2$ to grow larger than most other terrestrial arthropods, but even these active breathers cannot achieve sizes greater than a few inches.

The other major terrestrial respiratory adaptation is found principally in vertebrates and is capable of providing the respiratory needs of such giants as whales and elephants, as well as moderately large animals such as yourself. **Lungs** are highly vascularized, invaginated organs that localize gas exchange membranes in one part of the body. Air passes through a tube into the inflatable lungs where $O_2$ enters blood and $CO_2$ leaves it. The $O_2$-poor, $CO_2$-rich air is then forced out, usually through the same tube.

The simple lungs of amphibians supplement other gas exchange membranes—the skin and the vascularized membranes of the mouth. Their lungs are mere hollow sacs with relatively little surface area. Animals that depend exclusively on lungs for respiratory function, however, require larger gas exchange surfaces. Sectioning the interior of each lung into millions of tiny pouches, called **alveoli**, provides an enormous surface area in a relatively small space (Figure 19-4). Alveoli resemble the air pockets in a kitchen sponge. Each alveolus is rich with capillaries for rapid gas exchange between blood and air.

The efficiency of $O_2$ extraction from the lungs is greatly amplified by the presence of **hemoglobin**, a blood protein that temporarily binds $O_2$ and releases it in the tissues. The average person extracts about 10 pounds (4.5

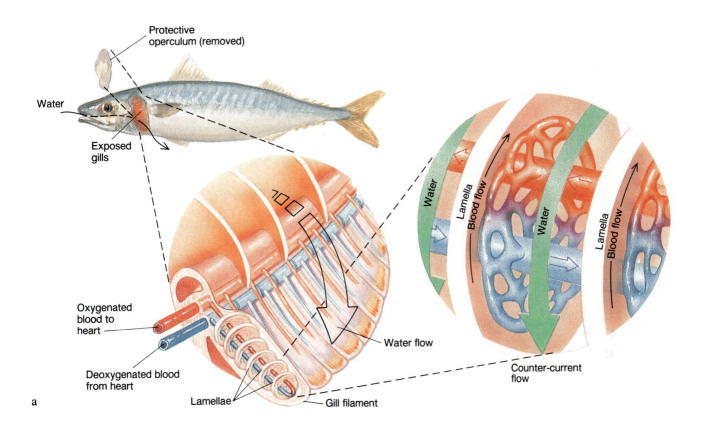

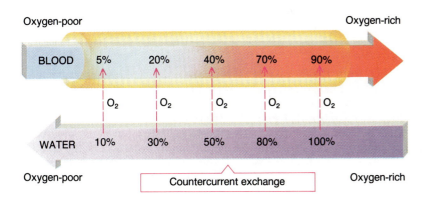

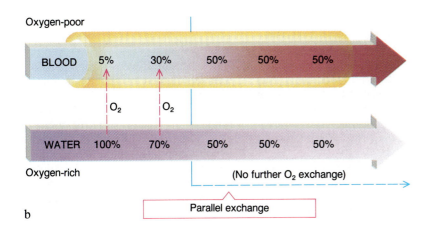

**Figure 19-2**
**Complex gills.** Fish have the most complex gills, with enormous surface area compacted into a small space (*a*). Blood flowing through each filament changes from blue to red as it acquires $O_2$. The process is enhanced by a countercurrent exchange system, in which water is forced past the lamella in a direction opposite that of the blood flow. As blood acquires $O_2$, it moves toward water that is even richer in the gas, so that at every point along the exchange surface, the $O_2$ level in the blood is lower than that of water. (*b*) This favors the diffusion of $O_2$ into the bloodstream at both ends of the lamella. If the flow were in the same direction, absorption would occur only at one end, and much less $O_2$ could be obtained.

**Figure 19-3**
**Carrying more than dinner**, this diving spider also carries its own $O_2$ supply in a glistening bubble of air around its abdomen and breathing pores. The spider is hauling its catch home to the larger submerged air bubble in which it dines, sleeps, and breeds.

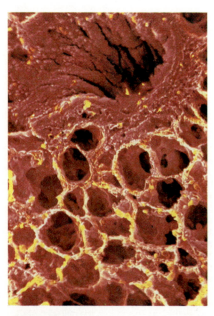

**Figure 19-4**
**Like a living honeycomb**, alveoli fill the interior of the human lung with air pockets. Each lung is literally a sac filled with smaller sacs.

kilograms) of $O_2$ from 13,000 gallons (50,000 liters) of air inhaled each day. Without hemoglobin, this number would plunge to no more that one-fifth of an ounce. This amount of $O_2$ is only slightly more than we could obtain by breathing water (and the results would be equally fatal).

High concentrations of hemoglobin dissolved directly in blood, however, would thicken the blood and impair its flow through the vessels. Yet very high concentrations are needed to carry enough $O_2$. This problem has been solved by encapsulating hemoglobin in **erythrocytes** (red blood cells). The name of these cells refers to the bright red color of hemoglobin when saturated with $O_2$, although in its $O_2$-deficient state the protein is a subdued purple or blue.

Air entering the lungs can be either forced in by increasing external pressure (called **positive pressure breathing**) or, as in humans, sucked in by lowering internal pressure when the sealed cavity that surrounds the lungs expands. Air flows toward the decreased pressure inside the lungs; this process is called **negative pressure breathing**. Compression of the cavity expels the inhaled air. Both types of breathing are illustrated by the frog depicted in Figure 19-5.

## The Human Respiratory System

The respiratory system of mammals is well represented by that of humans (Figure 19-6). The epithelium of the airways leading to the lungs houses millions of mucous-secreting **goblet cells**. The blanket of mucous secreted by these cells keeps the surfaces of the airways moist, so that even the driest air is humidified by the time it reaches the gas exchange membranes in the lung. The sticky mucous layer also

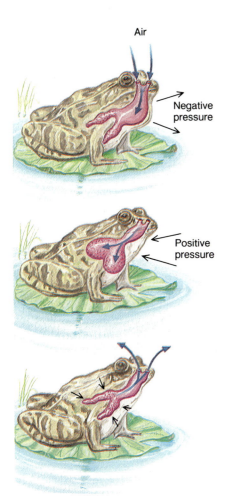

Air

Negative
pressure

Positive
pressure

**Figure 19-5**
**Breathing by positive and negative pressure**. Frogs use a combination of both to inhale air. Expanding the mouth creates negative pressure that sucks air through the nostrils. Then, with closed nostrils, the mouth cavity collapses, generating positive pressure that forces the air into the lungs. The frog exhales simply by opening its nostrils again, letting the pressurized air escape.

traps microbes and other dangerous air-borne particles in the upper regions of the respiratory tract, before they can enter the lungs and cause serious pulmonary (lung) injury or infection, pneumonia, for example. Particles trapped in the mucus are moved steadily toward the mouth as hairlike cilia of the underlying epithelial cells beat back and forth, propelling the entire mucous layer along at about 1 inch per minute. Trapped microbes end up in the throat, where they are swallowed and then killed by the stomach acid. The importance of the **ciliated mucosa**, as this layer of protective cells is called, becomes uncomfortably apparent when the mucous membranes dry out, as they tend to do in very dry air. As a result, the impaired defenses are often unable to eliminate microbes that cause respiratory infections. This, coupled with the fact that people crowd indoors during cold winter months (increasing the likelihood of exposure to sick persons), explains the seasonal fluctuations of colds and other mild respiratory infections. These and more serious respiratory disorders are the subject of this chapter's Bioline, Trouble in the Airways.

## Nasal Cavity

Air enters the respiratory system through either the mouth or *nostrils*, the openings into the **nasal cavity**. Air entering through the nostrils passes through a forest of nasal hairs. By tumbling the air, these hairs promote the sticking of inhaled particles to the mucus lining the cavity. This also increases contact between inhaled air and the warm mucous membranes, heating the air and ultimately enhancing gas exchange in the lungs. (Warmth increases diffusion rates.) Air leaving the nasal cavity on its way to the lungs next enters the **pharynx** (the throat) and, from there, the larynx.

## Larynx

A short passageway, called the **larynx**, connects the pharynx with the lower airways. Just before the larynx, the pathway branches in two directions. One is the entrance to the digestive tract. The other, an opening called the **glottis**, leads to the larynx and lower airways. During swallowing, the glottis is covered by the **epiglottis**, a flap that prevents food and liquids from entering the lower respiratory tract by mistake (Chapter 18). Otherwise, the glottis remains open, allowing air to enter the **trachea** ("windpipe"). Air passing through the glottis silently rushes between the wide open vocal cords (inset, Figure 19-6). When vocal cords are pulled together by muscular contraction in the larynx, however, they cover the glottis. Escaping air

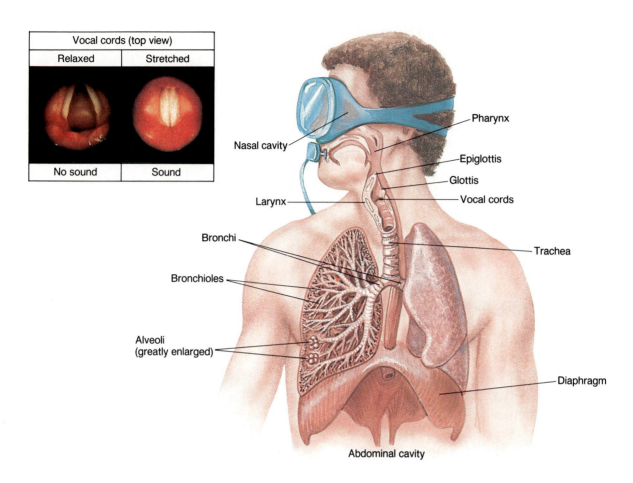

| Vocal cords (top view) | |
|---|---|
| Relaxed | Stretched |
| No sound | Sound |

Nasal cavity

Pharynx

Epiglottis

Glottis

Vocal cords

Larynx

Bronchi

Trachea

Bronchioles

Alveoli (greatly enlarged)

Diaphragm

Abdominal cavity

**Figure 19-6**
**The human respiratory tract**. The volume of the lungs is about three times greater than the maximum volume of air one can inhale in a single breath. In spite of the large amount of "dead air" in each lung, air that is replenished provides enough $O_2$ to satisfy our needs. Photo inset: — the vocal cords.

noisily vibrates the stretched vocal cords, creating the sound of the voice. The sound is amplified by the hollow chamber formed by the laryngeal cartilage (the "Adam's apple" or "voicebox"). Men generally have larger larynxes than do women, as evidenced by their deeper voices and more prominent Adam's apples.

## Trachea and Bronchial Tree

Once air enters the trachea, it is in the *lower respiratory system,* which encompasses all the respiratory structures below the glottis. C-shaped bands of cartilage reinforce the trachea and some other lower airways, pre-

venting the tubes from collapsing and impairing air flow. The trachea divides into two **bronchi**, the tubes through which air enters each of the two lungs. In the lungs, the bronchi branch into smaller tubes, the **bronchioles**, which rebranch into even smaller bronchioles. The abundantly branched complex is referred to as the **bronchial (or pulmonary) tree**.

## Lungs

The ultimate destination of inhaled air is the alveolus, the functional unit of the lung. It is in the alveoli that $O_2$ and $CO_2$ are exchanged. Clustered at the end of the bronchioles, each alveolus is a hollow, thin-walled sac with a moist

inner surface and a highly vascularized outer surface (Figure 19-7). Gases readily pass through the two membranes, so that blood moving through the capillaries can quickly pick up a fresh supply of $O_2$ and unload $CO_2$. The oxygenated blood travels from the alveolus directly to the heart, which pumps it to the tissues of the body. The "used" air is forced out of each alveolus and back into the bronchioles to be expelled from the airways during exhalation.

If your lungs were mere hollow bags, like those of a frog, the gas exchange surface would be less than 5 square feet (about half a square meter), too small to absorb enough $O_2$ to keep you alive.

# BIOLINE

## TROUBLE IN THE AIRWAYS

At one time, most human illnesses were believed to be caused by some imagined property of air. Long before it was known that nocturnal mosquitoes transmitted malaria, for example, the disease was thought to have been acquired by breathing mysterious vapors residing in night air. (The word malaria means "bad air.") People were simply afraid of air. And their fears were not entirely unwarranted, for the respiratory tract is indeed the most common route through which disease-causing microorganisms enter the body. In addition, a multitude of other disorders can arise from within the respiratory tract. Trouble in the airways is still the leading reason why people consult their physicians.

### Invaders of the Airways

The lungs are especially susceptible to microbial invasion because their warm, moist surfaces provide a hospitable environment for growing bacteria and other microbes. The problem is compounded by the amount of exposure between these vulnerable surfaces and microbe-laden air. The surface area of your lungs is 20 times greater than that of your skin, and in spite of many mechanical defenses along the route (ciliated mucosa, nasal hairs, etc.), eight microbes enter the lungs every breath or so (about 200,000 per day). Although most are either exhaled or destroyed, some may become entrenched and cause infectious disease. Here are a few diseases caused by bacteria or viruses that enter the body through the respiratory tract:

- Common cold — a mild virus infection localized in the mucosa of the nose, pharynx, and sometimes eyes and ears.

- Influenza (the "flu") — a virus infection that starts in the respiratory tract and spreads to other parts of the body, causing fever, general aches, extreme fatigue, and discomfort.

- Common childhood diseases (mumps, measles, and chickenpox) — infections of the respiratory tract that spread to other parts of the body.

- Serious systemic diseases — meningitis, encephalitis (infections of the brain and brain lining), scarlet fever, diphtheria, smallpox, and tuberculosis — potentially fatal diseases that begin in the airways and spread to nonrespiratory organs.

- Croup syndrome — viral infection of the epiglottis that, in children, may cause the airway to swell shut and asphyxiate the victim.

- Pneumonia (inflammation of the lung) — includes bacterial and viral pneumonia as well as other conditions for which lung infection is a primary complication, for example, Legionnaire's disease and the *Pneumocystis* lung infections associated with Acquired Immunodeficiency Syndrome (AIDS — see Connections: Fighting AIDS, Chapter 31)

### Obstructions of the Airways

When an inhaled object becomes stuck in the glottis, it can be forced out by creating high pressure in the lungs. Wrapping your arms around the victim from behind and applying vigorous pressure just below the ribcage can often blow the obstruction out of the glottis. This is called the *Heimlich maneuver*.

Diseases that may obstruct the airways include the previously mentioned croup syndrome, asthma, emphysema, bronchitis, and cancer. Asthma is an allergic condition that causes the lower airways to swell following exposure to a substance to which the victim is allergic. Swollen airways can trap air in the lungs, making the victim unable to exhale. Emphysema also traps air in the lungs, but only after years of bronchial destruction have caused airways to collapse. Long before this outcome, however, infiltration of fibrous tissue into the alveoli reduces gas exchange efficiency to the point where a source of pure oxygen is required. Bronchitis, inflammation of the bronchi, may also block the lower airways but can usually be successfully treated with medication.

Lung cancer is the proliferation of malignant tumor cells in the lungs. The tumor can obstruct the passageways to the lung and may spread to other areas of the body, seeding the formation of new tumors. Even with surgical removal of the cancerous lung, the disease is commonly fatal.

## Dying for a Cigarette?

Although its prevalence has declined to its lowest point in 40 years, cigarette smoking remains the greatest cause of preventable death in the United States. It is responsible for 300,000 premature deaths each year. Smokers account for three of every four lung cancer deaths, and they are more susceptible than nonsmokers to cancer of the esophagus, larynx, mouth, pancreas, and bladder. Hardening of arteries (atherosclerosis), cardiovascular disease, and peptic ulcer strike smokers with greater frequency than nonsmokers. Emphysema and bronchitis are 20 times more prevalent among smokers. Cigarette smoke also impairs the protective efficiency of the ciliated mucosa by stimulating an overproduction of mucus and by paralyzing the action of the cilia. Even if a smoker's cilia were intact, trapped microbes would descend into the lungs from the sheer weight of the additional mucus, making pneumonia one of the many health risks of smoking.

Smokers also endanger other people. Each year they are responsible for the deaths of thousands of "innocent bystanders," nonsmokers who share the same air with smokers. The risks of passive (involuntary) smoking are indisputable; second-hand smoke can make you seriously ill. Being married to a smoker is especially hazardous, increasing the risk of lung cancer in nonsmokers by as much as 34 percent. Children of smokers have double the frequency of respiratory infections as children who are not exposed to tobacco smoke in the home. Each year 12,000 nonsmokers die of lung cancer in the United States. According to the National Academy of Sciences (1987), 20 percent of these deaths are attributable to inhaling other people's tobacco smoke.

Another "innocent bystander" is a fetus developing in the uterus of a woman who smokes. Smoking increases the incidence of miscarriage and stillbirth, and decreases the birth weight of the infant. Once born, these babies suffer twice as many respiratory infections as babies of nonsmoking mothers.

---

The spongy interior of your lungs houses more than 300 million alveoli, providing about 750 square feet (70 square meters) of $O_2$-collecting, $CO_2$-discharging surface (about the size of a badminton court). Ironically, the lung tissue harvests little of the absorbed $O_2$ directly for its own needs. Like the rest of the body, lung tissue obtains its $O_2$ from oxygenated blood pumped there by the heart (Chapter 20).

The lungs are suspended in the **thoracic cavity** (the chest). Cradled in the ribcage, they are sealed in a waterproof, airtight double-membraned sac called the **pleura**. A single sheet of muscle, the **diaphragm**, separates the thoracic cavity from the abdominal cavity. This enclosed configuration not only protects the lungs, but also provides a means of *ventilation,* that is, a means of breathing.

## Breathing

Mammals are negative pressure breathers. We do not breathe with our lungs, but with our diaphragm and the muscles of our chest cavity. Contracting the dome-shaped diaphragm flattens it downward, pushing out the ribs and expanding the volume of the ribcage (Figure 19-8). Contracting the *intercostal muscles* pulls adjacent ribs closer together, forcing the breastbone upward and outward, further enlarging the space within the thoracic cavity. These activities create a partial vacuum that is immediately filled by environmental air rushing through the airways into the alveolar spaces (similar to the way opening a bellows draws

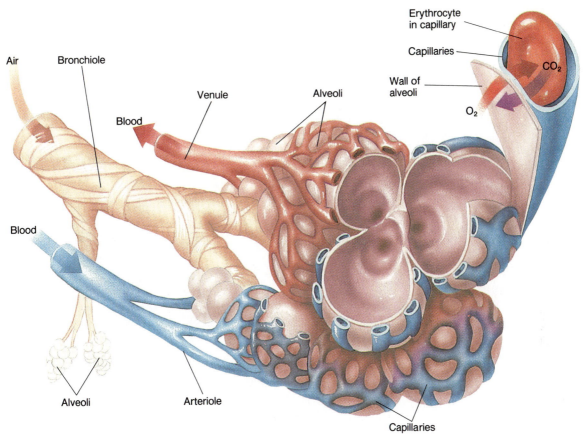

Air

Bronchiole

Blood

Venule

Alveoli

Erythrocyte
in capillary

Capillaries

Wall of
alveoli

CO₂

O₂

Blood

Alveoli

Arteriole

Capillaries

**Figure 19-7**
**Alveoli, the lung's trade centers**. Each alveolus is a thin-walled bubblelike chamber (the size of a pin point), surrounded by capillaries. Gases quickly move in the directions indicated by the arrows.

in air). When inhalation is complete, the intercostal and diaphragm muscles relax, and the ribcage returns to its smaller configuration. This compresses the lungs, forcing air out. Exhalation is therefore a passive endeavor, although it can be assisted by contracting another set of intercostal muscles that are antagonistic to those muscles responsible for inhalation. Contracting abdominal muscles forces the diaphragm upward into the thoracic cavity, forcing even more air out. Blowing up a balloon and the puffing associated with vigorous exercise are examples of active exhalation.

Although you may hold your breath, it is impossible to voluntarily

stop breathing to the point of serious O₂ deprivation. As soon as consciousness is lost, involuntary control mechanisms restart automatic breathing. These same control mechanisms keep you breathing while you sleep or when your attention is on other matters. The breathing control center (called the **respiratory center**) is located in the medulla oblongata, a portion of the brainstem that regulates automatic activities (Chapter 15). Since the cerebral cortex is not required for breathing, once voluntary control over breathing ceases, the respiratory center in the medulla automatically steps into the management position. In this way you continue to breathe as needed. But

how does the respiratory center know when you need to take a breath and how deep that breath should be?

Chemoreceptors in several areas of the body constantly monitor the concentration of CO₂ in the blood and in the brain's internal fluids. Between breaths, the level of dissolved O₂ decreases and CO₂ accumulates. When CO₂ concentrations exceed a threshold, the respiratory center in the medulla fires excitatory signals to the diaphragm and intercostal muscles. This prompts the person to automatically inhale. As the lungs inflate, stretch receptors send inhibitory signals to the respiratory center, which then stops exciting the respiratory muscles, and

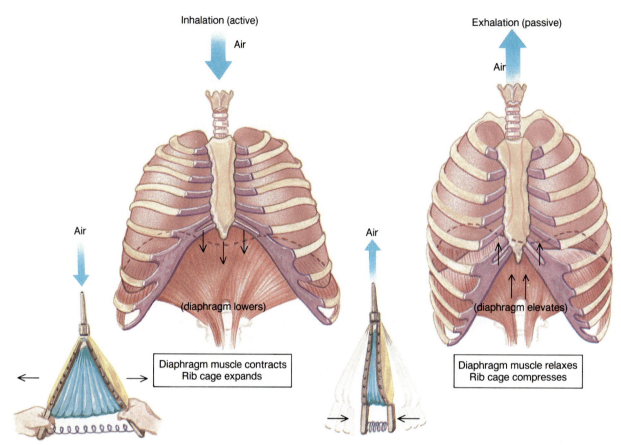

Inhalation (active)

Air

Air

(diaphragm lowers)

Diaphragm muscle contracts
Rib cage expands

Exhalation (passive)

Air

Air

(diaphragm elevates)

Diaphragm muscle relaxes
Rib cage compresses

**Figure 19-8**
**The body's bellows**. Contracting the diaphragm and intercostal muscles sucks air into lungs by
expanding the volume of the thoracic cavity. The inhaled air is expelled when these muscles relax.

the person exhales. This complete breathing cycle expels $CO_2$, reducing its level in the blood. The respiratory center remains inactive until $CO_2$ concentrations again accumulate to the threshold concentration, at which time another cycle is initiated.

When the body is active, cellular respiration increases, generating more $CO_2$, which in turn boosts the breathing rate and supplies more $O_2$ to meet the higher energy demands. Only rarely does the level of $O_2$ in the blood influence breathing rates. When severe $O_2$ deficiencies are detected by chemoreceptors in the carotid arteries, however, breathing can no longer be ignored. Every breath becomes a matter of great urgency.

The respiratory center also responds to changes in blood pH. Carbon dioxide forms a mild acid (carbonic acid, $H_2CO_3$) when it dissolves in water, so the blood tends to become slightly more acidic as $CO_2$ concentration increases and more alkaline as $CO_2$ is eliminated. Even a slight elevation in the blood's acidity triggers the medulla to increase the breathing rate. When enough $CO_2$ is exhausted to return blood pH to 7.4, the medulla reinstates a normal ventilation rate. Thus, breathing is another homeostatic mechanism for maintaining constancy in the body. It not only regulates $O_2$ and $CO_2$ levels, but also helps stabilize blood pH (see Connections: Homeostasis, Chapter 20).

## Gas Exchanges

Two types of gas exchanges are constantly occurring in your body—one at the alveolar membranes and the other in the rest of the body's tissues. The principle underlying the exchange of gases is the same in both locations —passive diffusion along a concentration gradient (Chapter 7).

### In the Lungs
Inhaled air is rich in $O_2$ (about 21 percent) and poor in $CO_2$ (about 0.4 percent). Blood flowing to an alveolus carries the opposite complement of gases, that is, a higher $CO_2$ concentration and a reduced $O_2$ cargo. These two concentration gradients drive $O_2$

into the bloodstream and $CO_2$ into the alveolus airspace, an exchange completed in a quarter of a second. The two layers of cells that separate the air in the alveolus from the blood in the adjacent capillaries offer little resistance to the passage of either gas.

The exchange is greatly enhanced by the presence of hemoglobin (Figure 19-9). As soon as $O_2$ enters blood, it complexes with hemoglobin (forming **oxyhemoglobin**) in the erythrocytes, which reduces the concentration of free $O_2$ in the bloodstream. This keeps the free $O_2$ concentration in blood low compared to that in the airspace, maintaining a steep concentration gradient even after substantial $O_2$ has entered the blood. In this way, blood becomes "supercharged" with about 70 times more $O_2$ than could be absorbed without the participation of hemoglobin.

## In the Tissues

Blood leaves the lungs carrying high concentrations of $O_2$ (bound to hemoglobin) and a low $CO_2$ load. Cellular respiration in tissues, on the other hand, depletes $O_2$ while generating $CO_2$. When oxygenated blood reaches needy tissues, a concentration gradient again drives the exchange. Oxygen diffuses into the tissues as $CO_2$ enters the blood, both moving passively from areas of higher concentration to lower concentration. Such a system helps deliver $O_2$ where it is needed most. In well-oxygenated tissues, the concentration gradient is not as steep, and so, little $O_2$ leaves the bloodstream at these sites. This conserves the blood's $O_2$ cargo for use by more active tissues in which the steeper concentration gradients favor the transfer of the gas.

The $O_2$-hemoglobin linkage is very loose. This is a necessary situation, since $O_2$ would do little good if it couldn't escape the bloodstream and enter tissues. At a lower $O_2$ concentration, such as is found in metabolically active tissues, oxyhemoglobin releases $O_2$, which then diffuses toward the de-

ficient areas. Hemoglobin's ability to do the right thing at the right place is assisted by two additional properties of the protein.

- It tends to pick up $O_2$ at cooler temperatures (as in the air-cooled lungs) and release $O_2$ in warmer temperatures (as in tissues).

- Hemoglobin's grip on $O_2$ weakens as the concentration of $CO_2$ increases. This phenomenon is known as the *Bohr effect*. Hemoglobin therefore releases most of $O_2$ where it is needed most, that is, in metabolically active tissues where $CO_2$ accumulates to higher concentrations.

Hemoglobin doesn't return to the lungs empty-handed. It picks up some of the $CO_2$ from the tissues, forming a loosely bound complex called **carbaminohemoglobin** ("carb" referring to $CO_2$; "amino" referring to the amino acid of hemoglobin to which $CO_2$ attaches). Yet only about 11 percent of the blood's $CO_2$ is transported by he-

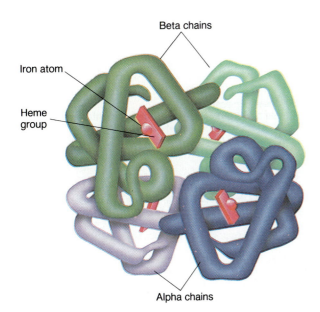

Beta chains

Iron atom

Heme group

Alpha chains

**Figure 19-9**
**Hemoglobin**, one of the body's largest proteins (composed of more than 10,000 atoms), carries four $O_2$ molecules when fully laden. The four protein subunits house iron-rich *heme* groups that act like $O_2$ magnets. Each red blood cell contains about 300 million hemoglobin molecules, the pigment that gives blood its color.

moglobin. Most of the $CO_2$ that enters erythrocytes dissolves in the cytoplasm. It then complexes with water molecules to form carbonic acid ($H_2CO_3$), which immediately disassociates into $H^+$ and a **bicarbonate** ion ($HCO_3^-$). The velocity of this reaction is accelerated some 250 times by the enzyme **carbonic anhydrase** housed in erythrocytes. Most $CO_2$ is therefore converted to bicarbonate ions as soon as it diffuses into a red blood cell, keeping the $CO_2$ concentration low compared to that in the interstitial fluid. This sharply steepens the $CO_2$ concentration gradient, boosting diffusion efficiency so that the gas can be adequately evacuated from the tissues.

## BACK IN THE LUNGS

This cascade of events is reversed at the alveolus. Carbaminohemoglobin releases $CO_2$ in response to the low $CO_2$ concentration in the airspace, reverting to "$O_2$-hungry" hemoglobin. In addition, under these conditions the carbonic anhydrase-mediated reaction reverses its direction, converting bicarbonate to $CO_2$ and water. The total effect is to maintain a $CO_2$ concentration gradient steep enough to quickly drive the gas into the alveolus airspace for exhalation from the body.

Bicarbonate ions provide more than a means of transporting $CO_2$. They also buffer the blood against fluctuations in pH, maintaining it at a pH of 7.4. As $CO_2$ dissolves in water, the $H^+$ released from carbonic acid dissociation is immediately absorbed by hemoglobin. This leaves the bicarbonate ion in the plasma, where it is neutralized by a sodium ion, forming sodium bicarbonate. Sodium bicarbonate is such an effective buffer that "heartburn" sufferers frequently use it to neutralize stomach acid. In its pure powdered form, it is known as baking soda.

The transport and exchange of gases is summarized in Figure 19-10.

**Figure 19-10**
**Transport and exchange of gases.**
*In Tissues*
1. Oxygen leaves hemoglobin and diffuses into tissues.
2. As $CO_2$ diffuses from tissues into red blood cells, some of it complexes with hemoglobin.
3. Most of the $CO_2$, however, reacts with water to form bicarbonate ($HCO_3^-$).
4. This releases a hydrogen ion, which is quickly neutralized by hemoglobin.
5. Some bicarbonate diffuses from the cell and is carried in the liquid portion of the blood as sodium bicarbonate ($NaHCO_3$).

*In Lungs*
6. Sodium bicarbonate disassociates, and the bicarbonate ion diffuses into red blood cells.
7. Carbonic anhydrase quickly releases $CO_2$ from bicarbonate, and the gas diffuses into the alveolus. The reaction requires the $H^+$ carried by hemoglobin to combine with the remaining oxygen atom from $HCO_3^-$).
8. As $O_2$ diffuses into red blood cells, it complexes with hemoglobin, so a steep concentration gradient is maintained.
9. Carbaminohemoglobin releases $CO_2$, driving up the intracellular concentration of $CO_2$, which then quickly diffuses into the alveolus in response to the steep concentration gradient. The free hemoglobin picks up $O_2$ (step 3).

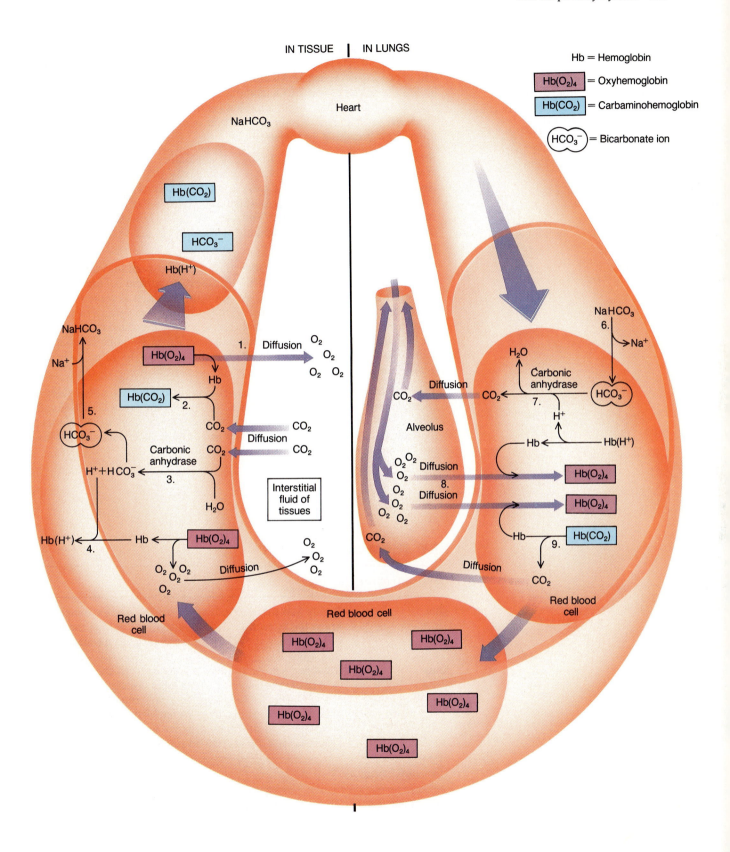

IN TISSUE | IN LUNGS

Hb = Hemoglobin

Hb(O$_2$)$_4$ = Oxyhemoglobin

Hb(CO$_2$) = Carbaminohemoglobin

HCO$_3^-$ = Bicarbonate ion

Heart

NaHCO$_3$

Hb(CO$_2$)

HCO$_3^-$

Hb(H$^+$)

NaHCO$_3$
6.
Na$^+$

NaHCO$_3$

Na$^+$

5.

HCO$_3^-$

Hb(O$_2$)$_4$  1.  Diffusion  O$_2$
O$_2$
O$_2$  O$_2$

Hb

Hb(CO$_2$)  2.

CO$_2$

HCO$_3^-$

CO$_2$  CO$_2$
Diffusion
CO$_2$  CO$_2$

Carbonic
anhydrase
3.
H$^+$+HCO$_3^-$

H$_2$O

Interstitial
fluid of
tissues

H$_2$O

Carbonic
anhydrase

CO$_2$  CO$_2$
Diffusion

Alveolus

H$^+$

Hb  ←  Hb(H$^+$)

Hb(O$_2$)$_4$

7.

HCO$_3^-$

Hb(H$^+$)  ←  Hb  ←  Hb(O$_2$)$_4$
4.

O$_2$  O$_2$
O$_2$
O$_2$  Diffusion
O$_2$
O$_2$
O$_2$  O$_2$

8.
Diffusion

Hb(O$_2$)$_4$

Hb(O$_2$)$_4$

O$_2$
O$_2$
O$_2$

Diffusion

CO$_2$

Hb  9.

Hb(CO$_2$)

CO$_2$

Diffusion

Red blood
cell

Red blood
cell

Red blood cell

Red blood
cell

Hb(O$_2$)$_4$

Hb(O$_2$)$_4$

Hb(O$_2$)$_4$

Hb(O$_2$)$_4$

Hb(O$_2$)$_4$

Hb(O$_2$)$_4$

Hb(O$_2$)$_4$

## Synopsis

### MAIN CONCEPTS

- **The need for gas exchange.** All animals must obtain $O_2$ for cellular respiration and must dispose of waste $CO_2$. Simple aquatic animals exchange gases across their water-bathed skin. More complex animals have specialized gas exchange structures (gills in aquatic animals, tracheae in most terrestrial arthropods, and lungs in air-breathing vertebrates).

- **Strategies.** In most complex animals, a circulatory system connects the gas exchange structure with the body's other tissues. (Exceptions are arthropod circulatory systems which need carry no oxygen, since the pervasive tracheoles reach every living cell.) Highly vascularized gas exchange membranes provide surfaces across which $O_2$ enters and $CO_2$ leaves the bloodstream. The lungs' alveoli, which abound with capillaries, increase the gas exchange surface to enormous dimensions.

- **Gas exchange.** Oxygen diffuses from the alveolar airspace into the circulation and complex with hemoglobin. This keeps the concentration of free $O_2$ lower in the blood than in air; as a result, diffusion favors entry of more $O_2$ into the bloodstream. The resultant oxyhemoglobin unloads its oxygen in tissues where the concentration of $CO_2$ is high and $O_2$ is low. $CO_2$ generated in tissues enters the bloodstream and travels to the lungs in several forms — as bicarbonate ions, sodium bicarbonate, or complexed with hemoglobin to form carbamino-hemoglobin.

- **The urge to breathe.** When sensors in the blood vessels and brain detect elevated $CO_2$ concentration, they alert the brain's medulla, which then launch involuntary instructions to take a breath. Each breath disposes of enough $CO_2$ that the urge to breathe momentarily subsides. When $CO_2$ again accumulates, the involuntary breathing cycle is repeated.

### KEY TERM INTEGRATOR

#### Oxygen-Acquiring Strategies

| | |
|---|---|
| **gills** | Richly vascularized extensions of the **gas exchange membrane** for extracting $O_2$ from water and ridding the body of $CO_2$. In fishes, the efficiency is enhanced by a **countercurrent flow** system. |
| **lungs** | Inflatable sacs intimately associated with the bloodstream (where oxygen-carrying **hemoglobin** circulates). The gas exchange surface of lungs can be greatly increased by the presence of **alveoli**. Some lungs are inflated by **positive pressure breathing**, and others by **negative pressure breathing**. Most terrestrial arthropods rely on **tracheae** instead of lungs. |

#### Human Respiratory System

| | |
|---|---|
| **ciliated mucosa** | The mucous-secreting lining of the respiratory tract, equipped with **goblet cells** that secrete the mucus, and cilia that move the mucous blanket and any trapped particles toward the mouth. It lines the **nasal cavity**, **pharynx**, the **trachea**, and the two **bronchi**. |
| **larynx** | The cartilage-reinforced area containing the **glottis** and vocal cords. Its proximity to the **epiglottis** is critical to keeping food and water out of the lower airways during swallowing. |

| | |
|---|---|
| **lungs** | Contain the **bronchioles** and alveoli. Sealed airtight in the **pleura**, lungs inflate when the **thoracic cavity** enlarges due to contraction of the **diaphragm** and intercostal muscles. This occurs automatically in response to impulses from the brain's **respiratory center**. |

Gas Exchanges

| | |
|---|---|
| **oxyhemoglobin** | Hemoglobin carrying $O_2$ molecules. When loaded with $CO_2$, it is called **carbaminohemoglobin**. Most $CO_2$ is dissolved in cytoplasm or plasma as **bicarbonate**, the formation of which is accelerated by **carbonic anhydrase**. |

## Review and Synthesis

1. Distinguish between cellular respiration and breathing.

2. Compare the respiratory strategies of the following: a sponge, a lobster (an aquatic arthropod with gills), a cockroach, a fish, an elephant.

3. Describe the functional interconnection between the human circulatory system and the respiratory system.

4. Why would simple hollow lungs equipped with blood vessels along their walls not meet the respiratory demands of mammals?

5. Lung tissue floats if a piece of it is dropped into water. Explain why. Would a section of a frog's lungs float?

6. The term "cyan" (as in "cyanide") means blue colored. From what you know about hemoglobin, explain why people become cyanotic (blue) when deprived of oxygen.

7. Why do we not bleed blue blood when injury opens our veins?

8. Describe two mechanisms that help match the amount of $O_2$ released with a tissue's need for $O_2$.

9. How does each of the following increase the concentration gradient between gases in air and blood? Hemoglobin; carbonic anhydrase; erythrocytes.

## Additional Readings

American Cancer Society. 1980. *The Dangers of Smoking; Benefits of Quitting.* New York. (Introductory.)

Comroe, J. 1966. "The lung." *Scientific American* 202:56. (Intermediate.)

Gordon, M., A. Bartholomew, C. Grinnell, and F. White. 1982. *Animal Function: Principles and Adaptations.* (Chapter 5.) Macmillan, New York. (Advanced.)

McKane, L., and J. Kandel. 1985. *Microbiology—Essentials and Applications.* (Chapter 20) McGraw-Hill, New York. (Intermediate to advanced.)

Vander, A., J. Sherman, and D. Luciano. 1986. *Human Physiology: The Mechanisms of Body Function.* 4th ed. McGraw-Hill, New York. (Intermediate to advanced.)

Chapter 20

# Internal Transport and Defense — The Circulatory and Immune Systems

Before complex animals could declare their independence from the sea, they had to develop an internal fluid that mimicked the properties of seawater. They also needed a system that circulated this fluid throughout the body, bathing all cells in a life-sustaining solution, even while living on dry land. In other words, they needed a portable ocean.

Some simple aquatic animals use a circulatory system of enormous dimensions—the ocean itself. Diffusion from seawater, however, fails to supply nutrients to cells buried in the remote recesses of larger, more complex animals. In contrast, a *circulatory system* delivers oxygen and nutrients to all cells of a complex animal and carries away their wastes. The liquid portion of the system, the seawater substitute, is called *blood.*

In vertebrates, these internal highways are patrolled by blood cells that guard the body and mobilize an elaborate defense system when the animal's security is threatened by alien intruders. The unrelenting assault by potentially dangerous microbes is blunted by this system, which quickly neutralizes minor breaches of security and launches a full-scale *immune response* against more serious invasions. In higher animals, therefore, the circulatory system does considerably more than just transport nutrients, wastes, and respiratory gases. It also

- Helps protect the organism.
- Channels hormones to their target cells and organs.

- Helps maintain a stable pH throughout the body.

- Helps maintain a constant body temperature in some animals.

- Causes certain structures to change appearance and function when they become engorged with blood (the reproductive structures of many vertebrates, for example).

Although the circulatory system of higher animals still functions as a portable ocean, such a description ignores its many other life-supporting contributions.

## Circulatory Strategies

The most primitive circulatory systems in multicellular animals rely on facilitated movement of seawater

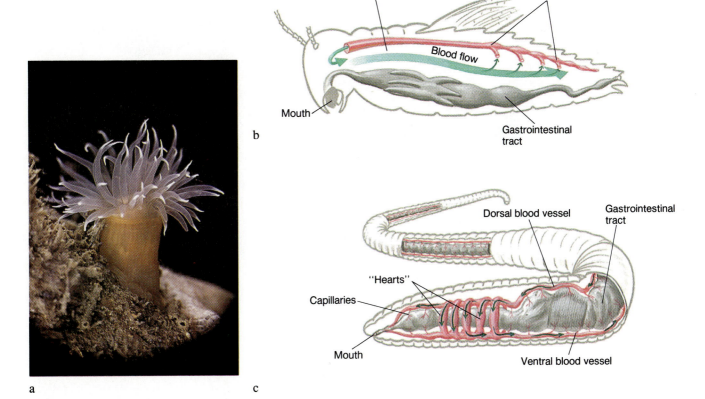

a

b

c

**Figure 20-1**
**Circulatory strategies** *(a)* Gastrovascular cavity of an anemone. *(b)* Open circulatory system of a grasshopper. *(c)* Closed circulatory system of an earthworm.

through open channels and cavities. The flow is maintained by the beating of cilia that line the cavities and often is further enhanced by the movement of the body itself. Many of these animals have combined their circulatory systems and digestive systems into a single *gastrovascular cavity* that digests food in the same cavity in which gases are exchanged (Figure 20-1a). More complex animals have replaced seawater with blood, which is pumped through the body by a heart, recirculating the same fluid over and over. Some of these animals, such as snails, insects, and lobsters, have **open circulatory systems** that contain open-ended vessels to move blood from one part of the animal to the other. Blood empties from the tube directly into tissue spaces and percolates through the body cavity to reenter the tube (Figure 20-1b).

In **closed circulatory systems**, such as the one you have, blood surges throughout the body in a continuous network of closed tubes (vessels). Thus, no whole blood escapes into the tissues (Figure 20-1c). One advantage of a closed system is the ability to concentrate blood where it is needed most (in highly active muscle tissue, for example) by changing the diameters of the vessels in a particular region. Yet in a closed system, blood never comes in contact with most of the cells that are nourished by it. How then can blood unload its life-sustaining cargo where it is needed and pick up the metabolic trash? To answer this question, we must look at the anatomy of the vessels that comprise the closed circulatory system.

## The Cardiovascular System

An animal's **cardiovascular system** consists of (1) the vessels through which blood flows and (2) the heart,

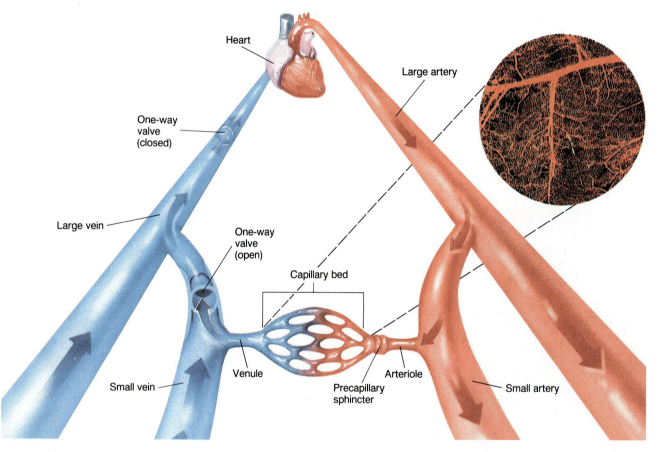

**Figure 20-2**
**The voyage of blood.** In the closed human system, blood travels in the following circular pathway:

heart ⟶ arteries ⟶ arterioles ⟶ capillaries ⟶ venules ⟶ veins ⟶ (back to heart)

The delicate latticework of a capillary bed is shown in the inset. Precapillary sphinctors regulate the amount of blood entering each capillary bed. (Capillary beds are discussed on page 320.)

Labels in figure: Heart, One-way valve (closed), Large vein, Large artery, One-way valve (open), Capillary bed, Venule, Arteriole, Small vein, Precapillary sphincter, Small artery

the muscular pump that propels the blood (cardiac = heart, vascular = vessel). The heart forces blood through large vessels into progressively smaller vessels that ultimately permeate every tissue. Blood then collects in other tiny vessels that merge as they direct the fluid back to the heart. Each of the five types of vessels in the vertebrate circulatory system (Figure 20-2) varies in complexity according to its function (Figure 20-3).

## Blood Vessels
### ARTERIES AND ARTERIOLES
Blood leaves the heart through impermeable **arteries**—large elastic vessels endowed with varying amounts of smooth muscle. Large arteries near the heart are practically devoid of muscle.

However, their abundant elastic tissue allows them to assist blood movement by expanding with each surge of blood pumped from the heart, and then to snap back to their original diameter, forcing the contents onward. As arteries get further from the heart, they become smaller and more muscular, finally branching into the smallest arteries called **arterioles**. Layers of smooth muscle allow arteries and arterioles to change diameter according to the dictates of the sympathetic nervous system. In this way, the amount of blood flowing through the vessels can be increased or diminished according to the changing needs of local tissues. Such **vasoconstriction** (reduction in arterial diameter) and **vasodilation** (increase in arterial diameter) also help regulate blood pressure. Con-

stricting vessel diameter in the arterioles elevates blood pressure in the arteries.

### CAPILLARIES
The tiniest vessels, **capillaries**, are highly permeable tubes consisting of a single layer of flattened cells through which many small molecules pass freely. The lumen of these minute vessels is just large enough for red blood cells to move along in single file (Figure 20-4). Because of their small size, a thousand miles of capillaries infiltrate every few cubic inches of tissue. (There are about 25,000 miles of capillaries in your body, enough to circumscribe the earth at its equator.) This creates an enormous surface area for exchange of materials between blood and the interstitial fluid in tissue.

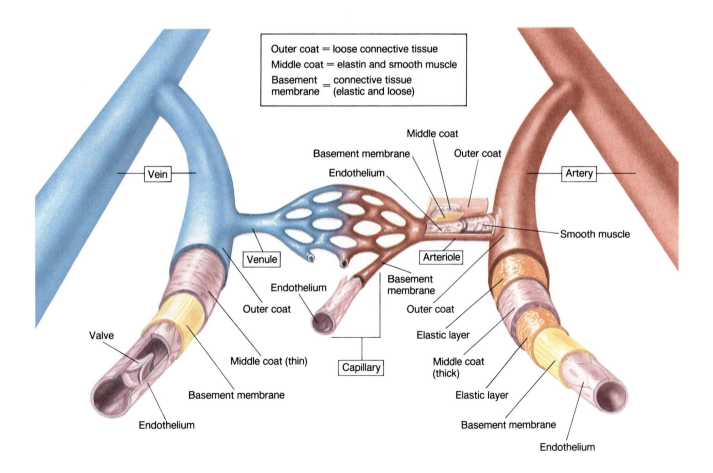

Outer coat = loose connective tissue
Middle coat = elastin and smooth muscle
Basement membrane = connective tissue (elastic and loose)

**Figure 20-3**
**Anatomy of blood vessels.**

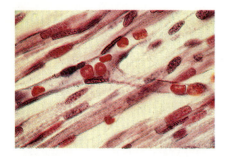

**Figure 20-4**
**Red blood cells moving single file through a capillary.** The thinness of the capillary walls is clearly evident.

Each tissue in the body is infiltrated with a latticework of capillaries called a *capillary bed*. The amount of blood entering the capillary bed of a particular tissue is influenced by contractions of **precapillary sphinctors**. These muscular gates can be narrowed to decrease blood flow to less needy tissues, and opened to allow more flow to areas that momentarily require it. Fewer than one in every ten capillary beds remains open during periods of rest. This number increases sharply during muscle exertion, eating, mental activity, or other enterprises that demand greater blood flow. Precapillary sphinctors respond to involuntary neural and hormonal messages; blood flow is therefore under the command of the autonomic nervous system. (Try willing yourself to stop blushing during your next embarrassing moment.) Automatic regulation of blood distribution also helps "warm-blooded" animals to maintain a constant body temperature in the face of external thermal fluctuations (see Connections: Homeostasis).

All gas, nutrient, and waste exchanges occur across the thin walls of capillaries. Some molecules, such as water, carbon dioxide, and oxygen, passively move across the vessel walls by diffusion or in response to a pressure gradient (Figure 20-5). Other substances are exchanged by active transport, or are moved across the cells of capillary walls in vesicles formed by endocytosis and released on the other side by exocytosis. Barring injury, whole blood never leaves the capillaries to enter the surrounding tissues.

CONNECTIONS
Characteristics of Life p. 2
Excretory System p. 290
pH and Gas Exchange p. 310

## HOMEOSTASIS: THE WISDOM OF THE BODY

The overriding function of blood and the circulatory system is to maintain *homeostasis* (Chapter 1). Homeostatic mechanisms are constantly correcting deviations from a biological norm, maintaining an internal stability that must persist if life is to continue. We have already discussed several homeostatic mechanisms — feedback regulation, osmoregulation, and excretion. Some animals are also capable of very exact **thermoregulation**, maintaining a constant internal body temperature in spite of fluctuations in external temperatures. Such animals are referred to as **endotherms** because their body temperatures are internally regulated. "Warm-blooded" animals, as endotherms are sometimes called, can succeed in environments that are biologically inaccessible to **ectotherms** ("cold-blooded" animals). Living on ice in the frozen polar regions are numerous endotherms (polar bears, penguins, seals, and humans, for example) but virtually no ectothermic animals. Because the internal temperatures of ectotherms are largely dictated by environmental temperatures, their metabolism, and therefore their mobility, slow as temperatures fall. Even in the cool early mornings of moderate climates, snakes, frogs, turtles, and many other ectotherms must bask in direct sunlight to warm their bodies enough to forage for food.

Thermoregulation by endotherms depends on heat generated during metabolism and distributed by blood. When environmental temperatures exceed the animal's stable internal (core) temperature, excess heat must be dissipated into the surroundings. In cold environments, heat must be conserved to avoid *hypothermia* (abnormally low body temperature). In mammals and birds, the "thermostat" that determines which of these options will be exercised is a portion of the brain, the *hypothalamus*. When heat-sensing neurons (thermoreceptors) alert the hypothalamus that external temperatures exceed the body's core temperature, it commands blood vessels in the skin to dilate. As a result, blood, carrying heat from the body, is shunted to the surface, where the heat is radiated to the air. In some mammals, the success of this cooling strategy is enhanced by sweating or panting. Water evaporating from skin or oral membranes carries away heat in the vaporized water molecules, thereby cooling the body.

When external conditions drop below the body's core temperature, the hypothalamus reverses these commands. As a result, surface vessels are constricted, driving blood deeper into the body, away from the cold surface. In many animals, heat loss at the surface is diminished by reflex contractions of the muscles at the base of feather or hair shafts. The elevated shafts trap air that helps insulate the body of animals covered with fur or feathers. (The formation of "goose bumps" on your skin is another vestigial reminder of our common ancestry with these animals.) The thyroid, under orders transferred through the pituitary from the hypothalamus, steps up its release of thyroxin, a metabolism-accelerating hormone. This produces more heat by burning more fuel. Shivering, produced by the reflex contraction of skeletal muscles, generates additional heat.

## VENULES AND VEINS

Blood from capillaries collects in larger vessels called **venules**, which empty their contents into larger **veins** for return to the heart. Blood pressure in veins is substantially lower than that in arteries, having been dissipated by distance and by the resistance within narrow capillary passageways. Inasmuch as most veins in humans direct blood flow against the force of gravity,

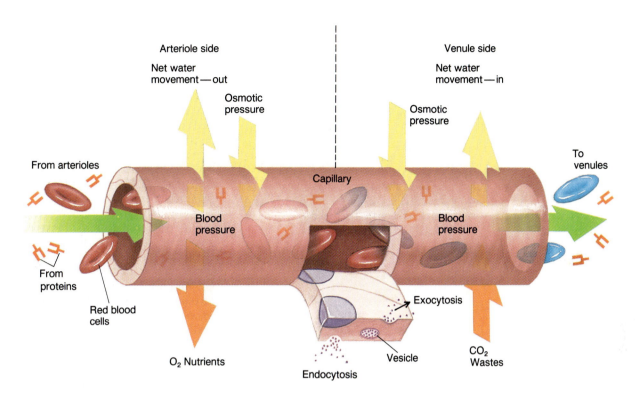

**Figure 20-5**
**Action at the capillaries:** Exchange of materials in a closed system. Substances migrate passively through capillary walls in response to pressure and concentration gradients, and actively by endocytosis-exocytosis and, although not shown, active transport.

it may seem that blood would collect in these low-pressure vessels and go nowhere. A review of their anatomy (Figure 20-2), however, reveals that veins are well suited for their task. Unlike arteries, they have little smooth muscle. Instead, they are thin-walled and stretchable, allowing some pooling of blood during periods of inactivity. Skeletal muscle activity and the pressure exerted during inhalation, however, squeeze veins and force their contents to travel toward the heart. This direction is determined by *one-way valves* (Figure 20-6). Without one-way valves, blood would simply squirt back and forth in the veins.

In humans, the veins in the legs receive the greatest weight of blood—sometimes so great that the valves fail and the blood permanently pools. This causes veins to *varicose* (enlarge). This condition can be deterred by exercising the leg muscles to force the blood upward, relieving the strain on the valves.

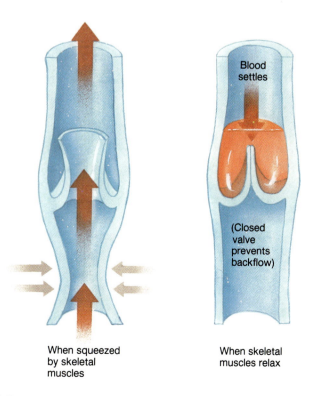

**Figure 20-6**
**No return.** To keep blood flowing through veins, one-way valves open when pressure forces blood toward the heart, and snap shut when the flow is reversed.

## Circulatory Routes of Vertebrates

In some lower vertebrates, such as fishes, blood travels a single loop. The heart forces blood to the gills for oxygenation; from there it goes directly to the tissue capillaries and returns to the heart.

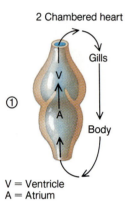

2 Chambered heart

Gills

Body

V = Ventricle
A = Atrium

This simple loop requires a heart with only two chambers—an **atrium** and a **ventricle**. Contraction of the atrium forces blood into the ventricle, which provides the power surge that sends blood through the vessels. Blood then passes through the narrow gas exchange capillaries and continues on to the rest of the body without any additional boost from the heart. As a consequence, the flow of oxygenated blood is more sluggish than if there were a second pump downstream from the respiratory capillaries to restore the pressure.

Amphibians have an additional atrium in their hearts that helps overcome this limitation.

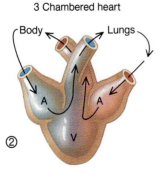

3 Chambered heart

Body          Lungs

One of the two atria delivers deoxygenated blood to one side of the ventricle, while the other fills the ventricle from the opposite side with oxygen-rich blood. With this improved system, the ventricle functions as both an upstream pump and downstream "booster." Although the three-chambered heart provides better flow, it allows a slight amount of blood to mix in the ventricle.

The four-chambered hearts of mammals and birds completely isolates oxygenated from deoxygenated blood.

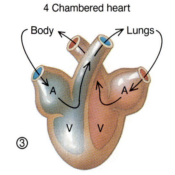

4 Chambered heart

Body          Lungs

The result is two distinct circulatory loops. (1) The **pulmonary circulation** channels blood to the lungs for oxygenation. (2) The **systemic circulation** delivers that oxygenated blood to the tissues and routes deoxygenated blood back to the heart. The integration of these two systems in humans is depicted in Figure 20-7.

### PULMONARY CIRCULATION

In the pulmonary circuit, deoxygenated blood is looped directly from the heart to the lung capillaries through large pulmonary arteries and then back to the heart through the pulmonary veins. These veins return the blood to the other side of the heart, which pumps blood into the systemic circulation with renewed impetus. (Unlike the arteries and veins in the systemic circulation, pulmonary arteries carry *deoxygenated* blood from the heart, and pulmonary veins transport oxygen-laden blood back to the heart.) For such a system to work, the heart is divided into left and right sides, each of which functions as a separate pump.

### SYSTEMIC CIRCULATION

Blood enters the systemic circulation through the largest vessel in the body, the **aorta**, which branches shortly after it emerges from the heart. One branch routes freshly oxygenated blood to the head and arms. The other plunges into the trunk of the body, giving rise to several circuits before splitting into two large iliac arteries that service the legs (Figure 20-8). With one exception, each of these circuits begins with an artery and ends with a vein that empties deoxygenated blood into the **vena cava**, which returns it to the heart. For example, the *renal circuit* carries blood from the aorta to the kidneys for purification and then back to the vena cava. The one exception is the *hepatic portal system,* which begins as capillary beds in the intestine and ends as capillaries in the liver. The large portal veins carry nutrient-rich blood to the liver, where substances are detoxified before entering the vena cava.

One especially important circuit is the one that delivers blood to the tissues of the heart. Blood does not nourish the heart as it travels through the chambers; the heart needs its own set of capillaries. Large **coronary arteries** branch from the aorta immediately after it emerges from the heart, providing cardiac muscle priority access to oxygen-rich blood. (These large arteries are evident on the computer-enhanced view of the heart shown in Figure 20-9.) Deoxygenated blood then collects in coronary veins, which return the blood directly to the heart. Any obstruction in the short *cardiac circuit* can deprive the heart of the oxygen needed to keep it alive. Such obstructions cause *heart attacks* (also referred to as *coronaries*), which not only kill heart tissue, but also cause heart muscle to contract in uncontrolled spasms that fail to pump sufficient blood. Although heart attacks are *not* caused by obstructions in the flow of blood through the chambers of the heart, they ultimately interfere with this flow. Sensations of pain may occur in the chest or be felt in the left arm, shoulder, or back. Pain in these areas may suggest defective coronary arteries and an impending heart attack. If a substantial portion of the heart is deprived of oxygen for enough

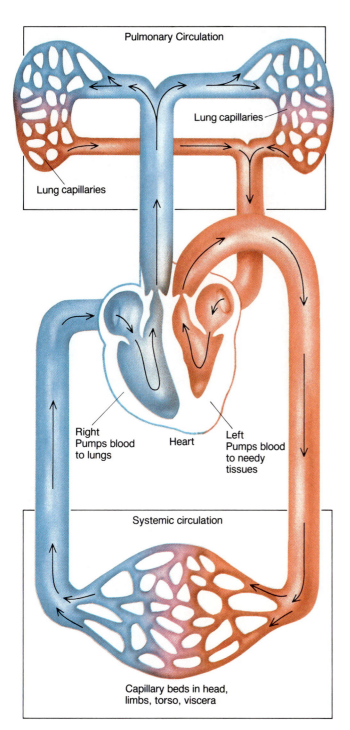

**Pulmonary Circulation**

Lung capillaries

Lung capillaries

Right
Pumps blood
to lungs

Heart

Left
Pumps blood
to needy
tissues

Systemic circulation

Capillary beds in head,
limbs, torso, viscera

**Figure 20-7**
**Our dual circulatory system.** The right side of the heart drives the pulmonary circulation, pumping deoxygenated blood through the lungs and returning oxygenated blood to the left side of the heart, which then propels it through the systemic circulation. Notice that the left ventricle is more muscular than the right.

time, the heart muscle, and the person wrapped around it, will die.

## The Human Heart: A Marathon Performer

The 11 ounces (312 grams) of hollow muscle that lay beneath your sternum is a fist-sized pump that delivers an astonishing performance. The average person's heart works continually for more than 70 years without a breather. During this time it will beat 2.5 billion times and pump more than 50 million gallons (about 200 million liters) of blood, 80 gallons (300 liters) every hour. It recirculates the entire volume of blood in the body every minute, pumping about 10 pints of blood through 60,000 miles (almost 100,000 kilometers) of vessels.

Your heart is a two-sided four-chambered model. Each side has an atrium and a larger ventricle (Figure 20-10). The right atrium receives deoxygenated venous blood from the body and pumps it into the right ventricle, which forces it toward the lungs. The left atrium receives the oxygenated blood returning from the lungs and pumps it into the left ventricle, whose powerful contractions propel the blood through the systemic circulation.

Each heart chamber has two openings, one for entry of blood and another for exit. One-way valves located at the two openings in each ventricle keep the blood flowing in the proper direction, preventing it from exiting through an entrance. The valves operate passively in response to pressure exerted by the contracting chambers. Even with valves, however, blood would flow nowhere if all four chambers contracted simultaneously or randomly. The chambers must contract in a sequence that allows the ventricles to rest while the atria contract and vice versa. This sequence is called the **cardiac cycle**.

### GUIDING BLOOD THROUGH THE HEART

The "signature" of the cardiac cycle is the heart beat: a double sound ("lub-dub") that accompanies a complete contraction sequence, followed by a relaxation interval. During the relaxation interval, positive pressure within

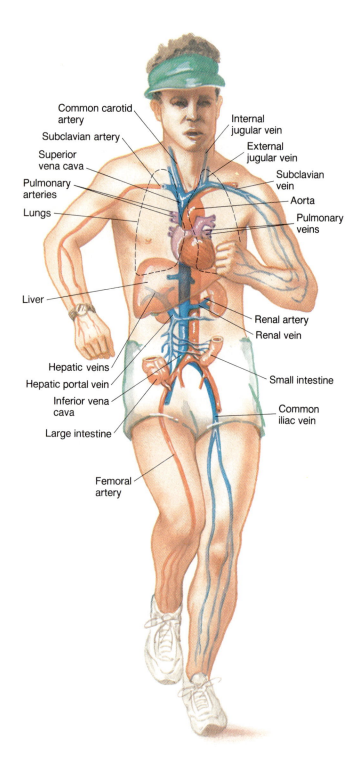

Common carotid artery

Subclavian artery

Superior vena cava

Pulmonary arteries

Lungs

Liver

Hepatic veins

Hepatic portal vein

Inferior vena cava

Large intestine

Femoral artery

Internal jugular vein

External jugular vein

Subclavian vein

Aorta

Pulmonary veins

Renal artery

Renal vein

Small intestine

Common iliac vein

**Figure 20-8**
**The major arteries and veins in the systemic circulation.**

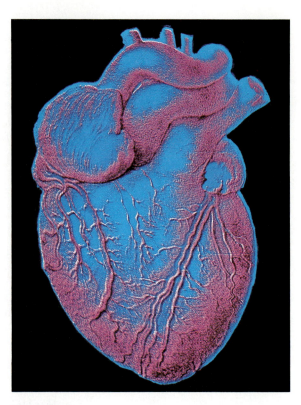

**Figure 20-9**
**Coronary arteries** are the first to emerge from the aorta, giving the heart muscles first claim to freshly oxygenated blood.

the vena cava and pulmonary veins forces the two atria to fill with blood (Figure 20-11). The cycle begins with atrial **systole** (a term meaning "contraction") which squeezes the two atria, forcing their contents through the *tricuspid valve* into the right ventricle and through the *bicuspid (mitral) valve* into the left ventricle. The atria now undergo **diastole**; that is, they relax. Atrial diastole is immediately followed by ventricular systole. The filled ventricles contract, generating back pressure against the tricuspid and bicuspid valves, causing them to snap shut. The contents of the ventricles can escape only through the *semilunar valves,* which pop open in response to the increased pressure. Blood rushes into the aorta and pulmonary artery. The ventricles then

relax, and the semilunar valves close as blood surges back against them, preventing backflow. The cardiac cycle is complete. The double throb that accompanies each cycle is not the sound of the heart contracting, but the clapping of valves as they snap shut.

Valve defects can prevent evacuation of a chamber and cause blood to collect in the heart and back up into the veins. Some valve malfunctions may be repaired; others require that the valve be replaced with a mechanical substitute.

## REGULATING THE HEART BEAT
The human heart normally beats about 70 times every minute. This rate changes according to the body's oxy-

gen demands, emotional state, and other factors that affect the heart rate center in the medulla oblongata of the brain (Chapter 15). Regulation by the autonomic nervous system is called *extrinsic control,* because the determining factor (the sympathetic and parasympathetic autonomic systems) lies outside the heart. Activating the sympathetic nervous system during times of stress bombards cardiac tissue with the neurotransmitter *norepinephrine,* which accelerates heart rate. Once the parasympathetic system regains control, it slows the rate down to its normal rhythm by releasing the neurotransmitter, *acetylcholine.* Some hormones, notably *epinephrine* (adrenaline), also exert an extrinsic influence on heart rate, elevating it during times of stress.

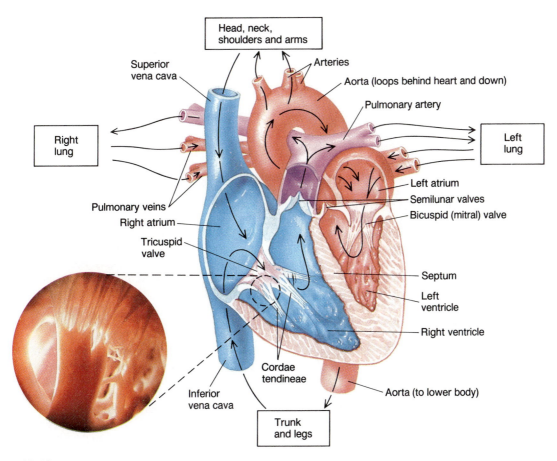

**Figure 20-10**
**The four-chambered heart of mammals and birds** is typified here by a human heart. The muscular thickness of the left ventricle is well suited for propelling blood through the systemic circulation. The **cordae tendineae** secure the valves, preventing them from inverting (prolapsing) under the enormous back pressure generated during ventricular contraction.

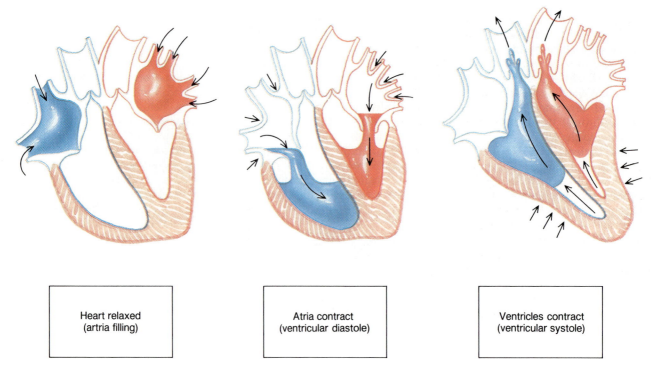

| Heart relaxed (artria filling) | Atria contract (ventricular diastole) | Ventricles contract (ventricular systole) |
| :---: | :---: | :---: |

**Figure 20-11**
**Steering blood through the heart.** The heart doesn't contract all at once; rather, the atria contract first, immediately followed by ventricular contraction. The average duration for the cycle is 0.85 seconds.

Yet heart muscle contracts even when not excited by neurons. Isolated cardiac muscle fibers suspended in a balanced salt solution continue to rhythmically contract about 70 to 80 times per minute, with no external source of excitation. For the heart to function properly, however, its millions of cardiac muscle fibers must contract in synchrony to coordinate atrial and ventricular contractions. In other words, a **pacemaker** is needed to initiate each contraction cycle and to establish the heart's own *intrinsic rhythm.* In the right atrium, a collection of cells called the **sinoatrial (SA) node** serve as the heart's pacemaker. Every 0.85 seconds, this node generates an action potential (much like a neurological impulse). The impulse triggers action potentials in neighboring cardiac muscle fibers, which not only contract, but also relay the impulse to their neighbors by way of the

intercalated discs that connect adjacent fibers. (See "Cardiac Muscle," Chapter 17.) The action potential sweeps through the two atria, causing simultaneous contraction of their fibers (Figure 20-12).

Just as the atria begin to relax, the impulse reaches a second neurological center, the **atrioventricular (AV) node**, located at the top of the ventricles. The AV node responds by launching a new action potential, which races down the septum and up the ventricle walls. The resulting wave of contraction rushes upward, forcing blood out through the semilunar valves, completing the cardiac cycle. Without a pacemaker, the heart would continue to contract but in a chaotic fashion that would pump no blood. Although the intrinsic rhythm can be temporarily overridden by extrinsic control from the medulla or by the influence of epinephrine, the pacemaker regains its intrinsic author-

ity as soon as the extrinsic instructions are relaxed.

Before the development of electronic pacemakers, which can be surgically implanted in the chest, a faulty SA node meant the end of life. Artificial pacemakers deliver an electric shock at intervals that approximate the intrinsic cardiac rhythm.

## Blood Pressure

The function of the heart is to generate enough positive pressure within the cardiovascular system to propel blood through the vessels. But blood pressure can be altered by factors other than cardiac output. For example, vasodilation of all the arteries and arterioles in the body causes blood pressure to plunge because the expanded vessel diameters no longer present enough resistance to maintain pressure. As pressure drops, blood flows more slowly, even if the heart continues to pump at a furious pace. This is what occurs during "shock" following an injury. When circulatory pressure temporarily fails to deliver adequate amounts of oxygenated blood to the brain, the victim faints, even though sympathetic excitation may have accelerated the heart rate. For many people, however, the opposite situation constitutes a more common danger, that is, high blood pressure, or *hypertension.* Although hypertension cannot be felt, if prolonged it increases the risk of heart attacks and strokes (obstructions in the vessels leading to the brain). The condition can be detected by measuring blood pressure with a *sphygmomanometer* (Figure 20-13).

Blood pressure depends not only on resistance, but also on cardiac output, the volume of blood in the system, and the location within the circuit. Pressure is highest in the arteries close to the heart and lowest in the veins. By convention, blood pressure is measured in the large artery of the upper arm, close to the heart, where pressures approach that of the aorta. Blood pressure is reported as two figures, the first representing the systolic pressure (during contraction of the ventricles) and the second the diastolic pressure (while the heart is relaxed). For an average resting adult, systolic pressure is great enough to raise a column of mercury 120 millimeters; the pressure during diastole would elevate the same mercury column 80 millimeters. The conventional way of expressing blood pressure is to combine these two terms as a fraction, in this instance, 120/80. Exertion, emotional upheaval—indeed, any factor that alters heart rate, vessel diameter, or total blood volume—may temporarily change a person's blood pressure.

## Composition of the Blood

The blood of vertebrates is a complex tissue composed of blood cells and cell-like components suspended in a clear, straw-colored liquid called **plasma.** Dissolved in plasma are an array of molecules—antibodies, hormones, and chemicals that aid in the transport of various substances and oppose the escape of blood from injured vessels (Table 20-1).

### Plasma

The average human contains about 10 pints (4.7 liters) of blood, about 55 percent of which is plasma. About nine-tenths of plasma is water, the blood's solvent. The rest is composed of various dissolved substances, predominantly three plasma proteins:

- *Albumin,* which helps maintain osmotic conditions that favor recovery of water that has been forced out of capillaries into the surrounding tissues.

- *Globulins,* which perform a variety of tasks. Some assist in the transport of lipids and fat-soluble vitamins, while others, the *gamma globulins,* contribute to the immunological defenses.

- *Fibrinogen,* which provides the protein network necessary for blood clotting.

Salts, vitamins, hormones, dissolved gases, sugars, and other nutrients comprise just over 1 percent of plasma's volume.

### Red Blood Cells

Oxygen delivery to tissues is greatly augmented by the presence of **erythrocytes,** which are often simply called red blood cells. The amount of oxygen absorbed by *hemoglobin* is 99 times greater than the oxygen dissolved in plasma (Chapter 19). This protein constitutes 95 percent of the red cells' dry weight. Not only are red blood cells the most abundant cells in the body (5 billion in each milliliter of blood), but they are also the simplest, a distinction that reflects their commitment to a single task. A mature erythrocyte is a small (about 5 micrometers in diameter), flattened disc that lacks a nucleus, ribosomes, and mitochondria, organelles that are unneeded by these nonreproducing sacks of hemoglobin (Figure 20-14).

The human body produces red cells

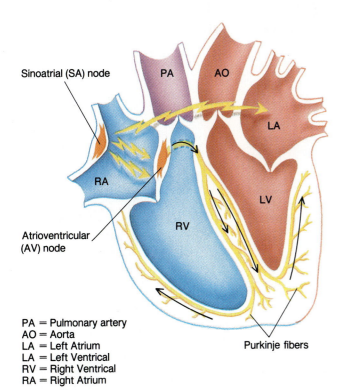

Sinoatrial (SA) node

PA

AO

LA

RA

LV

RV

Atrioventricular (AV) node

Purkinje fibers

PA = Pulmonary artery
AO = Aorta
LA = Left Atrium
LA = Left Ventrical
RV = Right Ventrical
RA = Right Atrium

**Figure 20-12**
**Synchronizing the heart beat.** Impulses generated in the SA node (the pacemaker) sweep through the muscles of the two atria, causing them to contract in unison. When this impulse reaches the AV node, it launches an action potential that is channeled through the ventricular walls by Purkinje fibers, causing the ventricles to contract just as the atria begin to relax.

at the astonishing rate of 2.5 million every second, or about half a ton of them during an average person's lifetime. This massive construction project occurs in the red bone marrow, where undifferentiated precursors, called **stem cells**, proliferate and differentiate into erythrocytes, a process that lasts six days. On day three, the nucleus is expelled, making room for more hemoglobin. Although cell division stops at this point, the cell continues to differentiate until it develops into a mature erythrocyte with a life expectancy of four months. In the end, most retiring erythrocytes are engulfed by scavenger white blood cells that break down the dead cell's remains to components that can be recycled.

Each day the human body manufactures an average of 200 billion mature red blood cells to replace those that have died, or about 1 percent of the total erythrocyte supply. The number produced fluctuates according to demand for and availability of oxygen. For example, persons living at higher elevations produce more red cells, thereby compensating for the reduced amount of oxygen in each breath. The key to this adaptability is *erythropoietin,* a hormone (secreted by the kidney) that initiates the transformation of stem cells into erythrocytes. Erythropoietin production is regulated by a negative feedback mechanism. High oxygen concentration in sensitive tissues reduces the formation of the hormone, slowing the production of erythrocytes. Low oxygen concentration stimulates erythropoietin formation, speeding up erythrocyte production until the oxygen supply in tissues is restored to normal.

## Platelets and Clotting Factors

A hole in an injured vessel can lead to fatal blood loss if the leak is not quickly patched. The vascular system and the blood itself are equipped with three lines of defense that minimize blood loss. The muscles surrounding injured vessels immediately constrict, reducing blood flow to the affected region. Circulating **platelets**, spiny fragments released from special white blood cells, become trapped at the injury site. Their spiny extensions attach to collagen fibers exposed by the tear in the vessel wall. As platelets accumulate, they eventually plug minor leaks.

Platelets also release *serotonin,* a chemical that further constricts vessels to reduce bleeding. If the leak cannot be quickly plugged by these two responses, a third mechanism, **blood clotting**, seals the vessels and provides a framework for repairing the damage.

The resources for clotting circulate in the blood in an inactive form. These materials are plasma proteins called *clotting factors.* One clotting factor, **fibrinogen**, is a rod-shaped plasma protein that, when converted to **fibrin**, generates a tangled net of fibers that binds the wound and stops blood loss until new cells replace the damaged tissue. The command to launch the clotting process must be a discriminating one; random or accidental clotting can clog the vascular system and stop circulation. This command is provided by *thromboplastin,* an enzyme that is activated by the exposed collagen fibers in a damaged vessel wall. Thromboplastin initiates a complex and not yet completely understood series of reactions that eventually lead to fibrin formation and clotting (Figure 20-15).

Once they have helped mend damaged tissue, clots must be disassembled

### TABLE 20-1

| Components of Blood[a] | | |
|---|---|---|
| **Component** | **Percent** | **Function** |
| Plasma | 55 | Suspends blood cells so they flow. Contains substances that stabilize pH and osmotic pressure, promote clotting, and resist foreign invasion. Transports nutrients, wastes, gases, and other substances. |
| White blood cells | <0.1 | Allow phagocytosis of foreign cells and debris. Act as mediators of immune response. |
| Platelets | <0.01 | Seal leaks in blood vessels. |
| Red blood cells | 45 | Transport oxygen and carbon dioxide. |

Plasma

Red blood cells

[a] *Blood settles into three distinct layers when treated with substances that prevent clotting.*

1

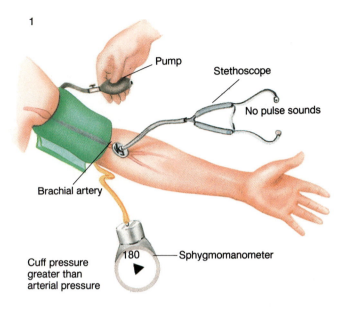

Pump

Stethoscope

No pulse sounds

Brachial artery

180 — Sphygmomanometer

Cuff pressure
greater than
arterial pressure

2

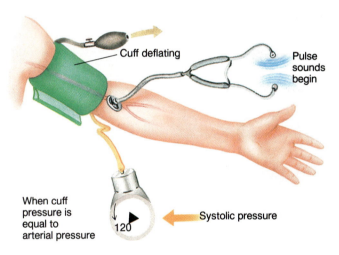

Cuff deflating

Pulse
sounds
begin

When cuff
pressure is
equal to
arterial pressure

120 ← Systolic pressure

3

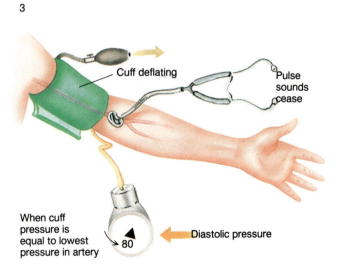

Cuff deflating

Pulse
sounds
cease

When cuff
pressure is
equal to lowest
pressure in artery

80 ← Diastolic pressure

**Figure 20-13**
**Determining blood pressure** using a *sphygmomanometer* and a stethoscope. The blood pressure in this example is 120/80.

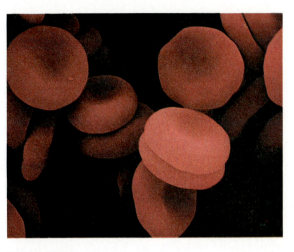

**Figure 20-14**
**Red blood cells: Biconcave sacks of hemoglobin.** The flattened shape of erythrocytes keeps each of the 270 million hemoglobin molecules close to the cell's surface for rapid gas exchange.

to prevent them from accumulating and obstructing circulation. Within each clot is a supply of an inactive protein called *plasminogen.* Activation of plasminogen converts it to *plasmin,* an enzyme that dissolves the clot by dismantling fibrin. Plasminogen-activating chemicals are used therapeutically to dissolve clots that threaten to obstruct the blood supply to the brain or heart, reducing the likelihood of strokes and heart attacks.

## White Blood Cells

Although 1000 times less abundant than erythrocytes, white blood cells, or **leukocytes** (leuko = white, cyte = cell), are necessary for our survival. These cells defend us against invading microorganisms that could overwhelm and destroy our bodies. Leukocytes also function as sanitary engineers, cleaning up dead cells and tissue debris that would otherwise accumulate to obstructive levels. These tasks are accomplished by five types of leukocytes, all of which develop from *stem cells* either in the bone marrow or lymphoid tissue (discussed later in this chapter). Unlike erythrocytes, leukocytes retain all their cellular components; they can therefore divide and produce daughter white blood cells.

Bone marrow normally manufactures about 4 billion leukocytes every hour Iand houses an additional 3 billion in reserve. During most bacterial

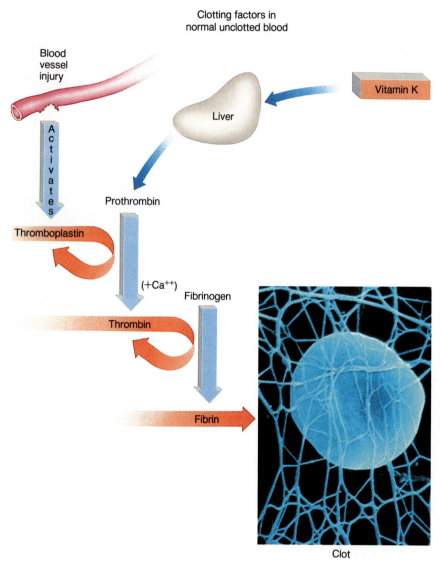

Clotting factors in
normal unclotted blood

Blood
vessel
injury

Vitamin K

Liver

Activates

Prothrombin

Thromboplastin

(+Ca⁺⁺)

Fibrinogen

Thrombin

Fibrin

Clot

**Figure 20-15**
**The cascade of events leading to clot formation.** The complexity of the sequence may reduce
the likelihood of accidental clot formation. Clotting can be prevented by thromboplastin inhibitors,
calcium binders, or vitamin K deficiency. (Inset — A red blood cell ensnared in a fibrin mesh.)

infections, these protective reserves are mobilized, sharply increasing the number of leukocytes to well above their normal values of 5 to 10 million per milliliter of blood. Some types of viruses, on the other hand, cause infections that deplete leukocytes to far below the normal number. Such quantitative fluctuations provide clues that often aid in the diagnosis of disease. This is one reason why your physician may take a "blood count" when you are ill. (Leukocytes can be microscopically counted in a stained sample of a patient's blood.)

Stained blood smears also reveal the different types of leukocytes (Figure 20-16). The most numerous leukocyte is the **neutrophil**, a protective cell that engulfs and destroys foreign intruders. Its cytoplasm is packed with "granules" that give the cell a grainy appearance when stained with certain dyes. The granules are actually *lysosomes*— vesicles filled with digestive enzymes and other chemicals used to destroy particles engulfed by *phagocytosis* (Chapters 6 and 7). Phagocytosis is one of the most potent weapons against invasion by disease-causing microorganisms. Neutrophils are highly mobile cells that are usually the first protective

cells to arrive at the scene of injury or microbial attack.

Another phagocytic leukocyte is the **macrophage**, a cell that develops from another white blood cell, the *monocyte*. Some macrophages wander through the body, engulfing any foreign agents they encounter (Figure 20-17). Macrophages also clean up battlefields littered with the cellular debris of combat between neutrophils and invading microorganisms. Other macrophages are fixed to tissues, such as in the spleen, where they cleanse blood flowing through.

Phagocytosis is not the only mecha-

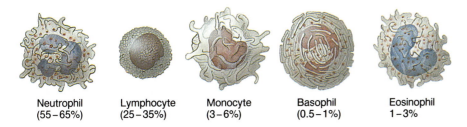

**Figure 20-16**
**Human leukocytes** and their relative abundance in the blood.

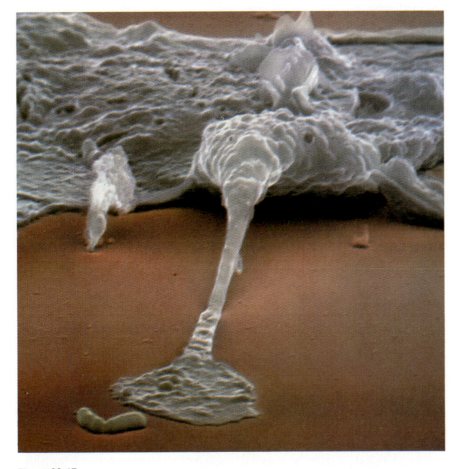

**Figure 20-17**
Phagocytosis, A macrophage about to devour its catch of intruding bacteria.

nism of protection afforded by leukocytes. A group of nonphagocytic white cells, the *lymphocytes,* contribute invaluably to our lines of defense against microbial invasion, as well as combatting cancer and assisting in neutralizing toxic chemicals. Two groups of lymphocytes, called *B-cells* and *T-cells,* are the "masterminds" of the *immune system.* This microbe-specific protective response will be detailed later in this chapter.

Two other nonphagocytic cells, *eosinophils* and *basophils,* are usually present in low concentrations. These cells may act as safeguards against accidentally launching our cellular weapons against our own tissues. Basophils also turn blood into a cellular minefield; when detonated by an alien cell, they spew out chemicals that trigger protective *inflammation.*

## Inflammation

As important as phagocytic cells are to defending the body, they provide little security unless they can be called into the area where protection is needed. The body's overall strategy for attracting these and other protective cells to sites of danger, an infected wound, for example, is called **inflammation**. The effect is to localize infection, destroy dangerous microbes, and help repair tissue injury.

Inflammation is initiated by the release of chemicals from cells following injury or during infection. These chemicals cause local blood vessels to dilate and increase in permeability. This brings additional blood, with its protective cells, into the affected region. These chemicals also attract phagocytic cells, especially neutrophils, to the site of injury, where they escape the blood vessels by squeezing out through the spaces between the cells of the vessel wall. Protective leukocytes thus accumulate in the injured or infected region, often forming a yellowish liquid, called *pus,* that consists of leukocytes and microorganisms suspended in fluid. In some cases, as from an open pimple, pus allows direct drainage of microbes and tissue debris out of the body.

These inflammatory events do not proceed unnoticed. The accumulation of fluid and leukocytes in the inflamed area causes the tissues to swell and hurt due to stimulation of local neurons. The area grows hot and red from the additional blood. In helping the body recover, inflammation creates many of the symptoms we often associate with disease rather than healing. Inflammation has consequently been targeted as a problem to attack rather than an ally. Use of drugs, such as cortisone, to subdue inflammation may provide immediate relief from symptoms, but in doing so, it suppresses this important line of defense. Such anti-inflammatory agents may cripple the inflammatory response, causing existing infections to worsen. Sometimes they allow the normally harmless bacteria that reside in your mouth and throat to cause serious, even fatal invasions of the lower respiratory tract, bloodstream, or central nervous system.

# The Immune System

Lymphocytes are slow and seemingly unarmed, lacking lysosomes and incapable of attacking an invader by phagocytosis. Yet these cells are deadly weapons in the body's defensive arsenal. Without them we would be immunologically crippled and incapable of surviving microbial invasions.

Lymphocytes are the cells of the **immune system**, a two-pronged arsenal that is much more specific in its pro-

tection than phagocytosis and inflammation. This protective response is triggered by the introduction of antigens into the body. **Antigens** are specific foreign materials, usually proteins, each of which is as distinctive as a human fingerprint. The immune system is constantly "fingerprinting" the cells and substances in the body. Most substances pass the test; they are recognized as "self" and are tolerated by the immune system. But when a foreign antigen is detected, lymphocytes mobilize their defenses and usually eliminate the antigen. If the antigen is a toxin or part of a disease-causing microorganism, its elimination disposes of the intruder and provides long-term protection against future attacks by the same microbe or toxin. In other words, it makes the host *immune* to recurrences of the same disease.

The double prongs of immunity are provided by two types of lymphocytes, **B-cells** and **T-cells**. Both types develop from stem cells in the bone marrow, but some are processed by the *thymus* (a gland in the chest) emerging as T-cells. The identity of the B-cell processing organ in humans is still speculative, but in chickens (where it was discovered) it is the *Bursa of Fabricius* (hence, "B" cell). The analogous structure in humans is believed to reside in the bone marrow. Although B-cells and T-cells usually work together to present a united front against disease, a look at how they function individually will be helpful in understanding the combined immune response.

## B-Cell Immunity

When activated by the presence of a foreign antigen, a few B-cells begin to rapidly proliferate. Antigen intrusion stimulates only those cells that are programmed to recognize that particular antigen. The few cells that do react produce a huge population of lymphocytes, all of which are descendants of the antigen-stimulated cells. These progeny cells enlarge and differentiate into **plasma cells** and begin producing **antibodies**, the substances responsible for B-cell immunity. Antibodies are gamma globulins, a group of plasma

proteins commonly called *immunoglobulins.* Most of them are Y-shaped molecules that react specifically with the antigen that stimulated their formation (Figure 20-18). Once bound to the antigen, antibodies can provide a number of protective services.

## PROTECTIVE ACTIVITIES OF ANTIBODIES

Antibodies improve the efficiency of phagocytic defenses. They react with antigens on the surface of intruding cells, coating the intruder with molecules to which phagocytic cells readily attach. Antibody-coated cells are easy targets for engulfment by neutrophils or macrophages. Bacterial cells coated with antibodies also *agglutinate* (clump together). Each phagocytic cell consumes far more bacteria when it engulfs a clump than if it had to engulf them one cell at a time. Some antibodies also combine with a group of blood proteins, called *complement,* following attachment to their antigens. The bound complement punches tiny holes in the plasma membrane of the foreign cell, causing it to burst. In the process, molecular fragments released from complement attract phagocytic cells to the site, recruiting cells that can engulf any bacteria that may have escaped membrane damage.

Not all antigens reside on the surface of bacteria and other cells; intracellular structures, such as ribosomes and mitochondria, have antigens distinct from those on the cell's surface. Some antigens reside on the surface of viruses. Other antigens are soluble molecules such as toxins, produced (and often secreted) by cells. Antibodies that combine with these poisonous substances neutralize their toxicity and render them harmless.

## IMMUNOLOGICAL MEMORY

Although the first exposure to an antigen excites the immune system, it is several days, sometimes weeks, before protective levels of antibodies appear in the blood. During this lag time, called the *primary immune response,* antigen-specific lymphocytes proliferate and the immune system becomes "sensitized" to the antigen. If the antigen persists long enough, or if the body is again challenged with the same antigen, antibody concentrations leap to enormous levels. This *secondary immune response* occurs swiftly enough to provide immediate protection whenever the same antigen intrudes again. Thanks to the secondary response, measles, mumps, chickenpox and many other diseases are suffered only once during a lifetime. The first encounter with the intruder leaves survivors *immune* to subsequent infection.

Immune systems become sensitized to an antigen because some lymphocytes differentiate into **memory cells** rather than antibody-producing plasma cells. Although antibodies and plasma cells wane following removal of the antigenic stimulus, memory cells persist, often for years, providing *immunological recall.* When stimulated by the same antigen, they rapidly proliferate into plasma cells (and more memory cells), generating a secondary immune response in a matter of hours rather than days. Furthermore, the

strength of the reaction is boosted with each exposure to the antigen until an immunological ceiling is reached. Immunity therefore is due to the presence of antigen-specific memory cells that launch a swift secondary immune response whenever the antigen is encountered. This type of protection is called **active immunity**.

Antigen-specific active immunity can be safely induced by using *vaccines,* which are some of our most potent weapons against diseases. Vaccines are modified forms of disease-causing microbes; these variants have lost the ability to cause disease but retain the same antigens as their dangerous counterparts. An immune system exposed to a vaccine (and usually several "booster" doses) builds memory cells without the danger of disease developing during the vulnerable primary response. In this fashion, we can now safely promote active immunity against polio, measles, mumps, diphtheria, tetanus, rabies, and many other dangerous diseases.

## PASSIVE IMMUNITY

Sometimes there is no time to wait for an immune system to become sensitized. An attack of diphtheria or tetanus, for example, or the bite of a poisonous snake, requires immediate immunity to neutralize the life-threatening antigens. In these cases, antibodies that have been previously obtained from the blood of an immune individual can be injected into the person who needs immediate protection, producing a temporary state of **passive immunity** ("passive" because the immune individual does not actively produce the antibodies). In mammals, passive immunity can be naturally acquired by the transfer of antibodies across the placenta from a mother to her developing fetus. The fetus receives immunological protection even before it is capable of producing its own antibodies. Such passive protection is augmented by the presence of antibodies and immune cells in breast milk, providing breast-fed babies more disease resistance than bottle-fed babies.

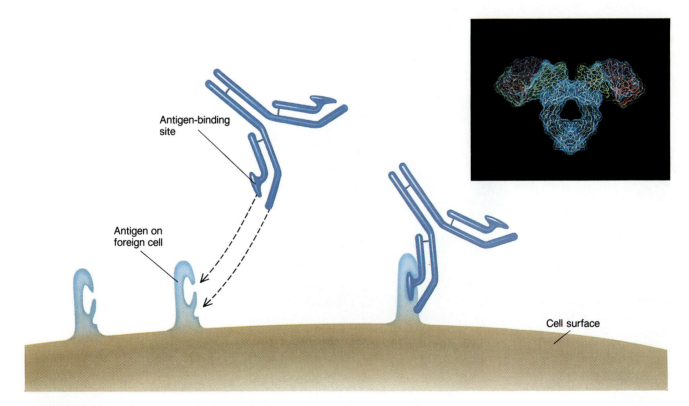

**Figure 20-18**
**Immunological ammunition.** The most common type of antibody is armed with two specific binding sites (one at each tip of the "Y") that react exclusively with a single type of target antigen. (Inset — Computer generated model of antibody.)

## T-Cell Immunity

Although T-cells produce no antibodies, they play a critical role in combating disease. When stimulated by the presence of a foreign antigen, the T-cells recognize that antigen and proliferate into several subpopulations of *activated lymphocytes.* One subpopulation consists of memory cells that, like memory B-cells, provide immunological recall. The presence of T-memory cells shortens the response time from several weeks to one or two days.

The T-cells that provide protection are called *effector cells.* Some of these cells move into contact with the antigenic target and kill it by a still unexplained mechanism. These *killer T-cells* help eliminate virus-infected cells before they can release infectious progeny. They also destroy cancer cells that spontaneously arise from normal cells (Figure 20-19). Other effector cells produce **lymphokins**. These chemicals enhance the phagocytic activities of macrophages, draw phagocytic cells and other protective lymphocytes to the antigen, and directly poison some types of foreign cells. Another lymphokin, called *immune interferon,* helps arrest viral infections by ordering host cells to produce antiviral proteins.

Many antigens elicit the participation of both T-cell and B-cell immune systems in a *combined immune response* (Figure 20-20). For example, *helper T-cells* must cooperate with B-cells to initiate antibody formation against certain antigens, providing a fail-safe system that prevents accidentally triggering antibody production against the host's own tissues. The necessity for helper T-cells is tragically illustrated by the deadly disease AIDS, whose high mortality rate is due to virus-mediated destruction of these cells (see Connections: Fighting AIDS, Chapter 31). With their helper T-cells destroyed, AIDS victims are susceptible to a fatal parasitic lung infection and a rare form of cancer, neither of which occurs in persons with intact immune systems.

*Suppressor T-cells* have the opposite effect as helper cells; they decrease antibody formation, thereby helping avoid exaggerated immune responses that might damage the host. An appropriate combined immune response requires a balance of suppressor and helper cells.

## The Lymphatic System

Mammals are equipped with a **lymphatic system**, a network of fluid-carrying vessels and associated *lymphoid organs* that participate in immunity (Figure 20-21). Lymphatic vessels are a series of one-way channels. Each vessel originates as a bed of *lymphatic capillaries,* which merge into successively larger vessels. Fluid forced out of capillaries by internal pressure would accumulate in tissue spaces if it were not returned to the bloodstream. The lymphatic capillaries do just that—they collect the excess fluid and chan-

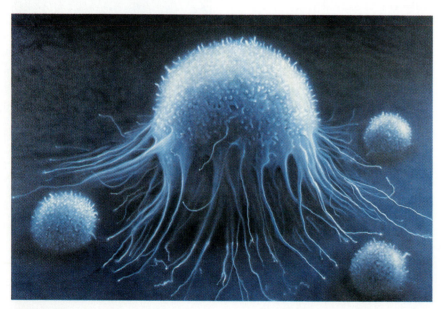

**Figure 20-19**
**Killers close in on a killer.** Killer T-cells approach a large cancer cell before it can produce a life-threatening tumor. Such immune surveillance by T-cells is critical to the prevention of malignancies.

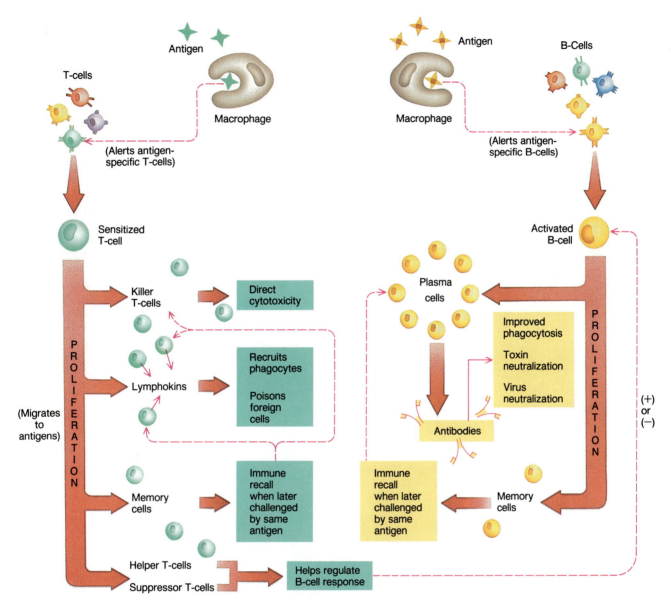

**Figure 20-20**
**The combined immune response.** Macrophages help T and B lymphocytes launch an immune response. The macrophage initiates the response by processing antigens and presenting them to inactive T-cells and B-cells. If the antigen fits the surface receptors on the lymphocytes, that cell proliferates into a population of immunologically activated cells. An immune response against that antigen is launched.

nel it back to the blood vessels. Once inside the lymphatic capillaries, which are similar to blood capillaries in size and thickness, the recovered liquid is called **lymph**. This colorless fluid contains no red blood cells and has a lower protein concentration than blood. The flow of lymph is not powered by the heart but relies on compression exerted by contraction of skeletal muscles during movement and breathing. One-way valves in the lymphatic vessels keep lymph flowing toward veins adjacent to the heart, where the fluid is emptied into the bloodstream. Fats absorbed from the intestinal villi following digestion are also collected and delivered to the bloodstream by lym-

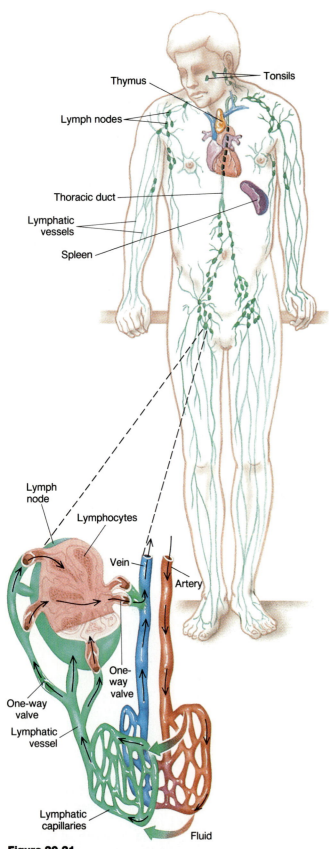

Thymus

Tonsils

Lymph nodes

Thoracic duct

Lymphatic vessels

Spleen

Lymph node

Lymphocytes

Vein

Artery

One-way valve

One-way valve

Lymphatic vessel

Lymphatic capillaries

Fluid

**Figure 20-21**
**The lymphatic system.** Lymphatic vessels capture and return excess fluid leaked from capillaries of the bloodstream. Lymphoid organs include lymph nodes, thymus, spleen and tonsils.

phatic vessels (see Figure 18-7).

Before reentering the bloodstream, lymph passes through a series of **lymphoid organs** that help "purify" it by removing foreign substances and any microorganisms that may be present. Most of these organs are little more than collections of lymphocytes, plasma cells, and macrophages housed in structures called **lymph nodes**. Phagocytosis filters debris and foreign particles from the lymph, whereas immune cells screen the fluid for the presence of specific antigens. During infection, lymph nodes become engorged with immune cells and macrophages recruited to fight the microbial aggressor. "Swollen glands" are therefore common signs of infection. The *spleen* is also a lymphoid organ. It filters blood as well as lymph, while serving as a reservoir of lymphocytes. Another member of the lymphatic system, the *thymus,* processes lymphocytes into committed T-cells.

# Synopsis

## MAIN CONCEPTS

- **Closed circulatory systems** deliver to each of a complex animal's cells a well-oxygenated nutrient solution and carry away the wastes. The system also contributes to homeostasis by stabilizing pH and osmotic balance, delivering hormones, protecting against foreign intruders, and, in some animals, resisting changes in body (core) temperature.

- **The anatomy of blood vessels** is well suited to their location and function. The design of the elastic arteries allows them to snap back after each surge from the heart, squeezing blood onward. To meet the needs of the tissues they service, muscular arterioles either enlarge or reduce vessel diameter, and precapillary sphinctors open or close the entrance to capillary beds. Venules collect the blood leaving capillaries and route it to veins. One-way valves in veins assist the blood's return to the heart.

- **The mammalian heart** is also equipped with one-way valves to keep blood flowing in a single direction. A double pump, the right side of the heart pumps blood to lungs for oxygenation, and the left side powers blood through the general circulation. Although cardiac muscle fibers contract rhythmically without an external stimulus, neurological impulses regulate the contraction sequence by stimulating the sinoatrial node, which initiates contraction of the atria and relays the stimulus to the rest of the heart. This intrinsic regulation may be temporarily overridden in times of stress, fear, or exertion.

- **Blood contains cells and cell-like entities.** Erythrocytes carry oxygen, neutrophils and macrophages phagocytize foreign intruders and tissue debris, platelets aid in clotting, and lymphocytes launch the immune response. Inflammation is a nonspecific response that concentrates phagocytic cells in an area of injury or infection.

- **The antigen-specific line of defense is immunity.** Both branches of the immune system enhance the protective efficiency of phagocytosis. Antibodies that coat foreign cells act as attachment sites for phagocytic cells. Agglutination gathers the intruders into large easy-to-engulf packages. Because lymphokins attract phagocytic cells to the site of antigen stimulation, protective weapons are concentrated at the site of attack. In addition, antibodies and activated T-cells can directly kill or neutralize some types of invaders.

- **The capillaries of the lymphatic system** collect fluid that has escaped from the blood vessels and return it to the bloodstream. Lymph is filtered through lymphatic organs (notably lymph nodes) on its way to the vena cava.

## KEY TERM INTEGRATOR

Cardiovascular System

| | |
|---|---|
| **blood vessels** | The channels through which blood flows. In the **closed circulatory system** of vertebrates, blood travels through **arteries, arterioles,** and **capillaries,** and then back to the heart through **venules** and **veins. Precapillary sphinctors** and muscles in arteries change blood flow to different tissues by **vasodilation** and **vasoconstriction. Endotherms** use this differential blood flow for **thermoregulation.** |
| heart | In birds and mammals, a muscular four-chambered pump with one **atrium** and one **ventricle** on each side. The right side powers the **pulmonary circulation,** while the left side pumps blood through the **systemic circulation.** The largest vessel in the body is the **aorta.** Deoxygenated blood empties into the **vena cava.** Large **coronary arteries** provide cardiac muscle access to oxy- |

gen-rich blood. Each **cardiac cycle** is initiated by the **sinoatrial node** (the **pacemaker**), followed by an impulse from the **atrioventricular node**, so that **systole** alternates harmoniously with **diastole**.

---

Composition of Blood

---

| | |
|---|---|
| **plasma** | Water-based fluid of blood, in which are dissolved nutrients, salts, gases, wastes, and blood proteins such as **fibrinogen**, which, when transformed to **fibrin**, promotes **blood clotting**. |
| blood cells | Cells or cell-like bodies suspended in plasma, all of which arise from **stem cells**. They include **erythrocytes**, **platelets**, and various **leukocytes**. **Neutrophils** and **macrophages** are the phagocytic white cells associated with **inflammation**. |

---

The Immune System

---

| | |
|---|---|
| **antigen** | Foreign material that triggers **lymphocytes** to launch a protective response. |
| **B-cells** | Lymphocytes that respond to antigen stimulation by proliferating into a population of **antibody**-producing **plasma cells** and antigen-specific **memory cells**. Natural antigen exposure or introduction of **vaccines** results in **active immunity**, whereas injection of antibody produced in another individual provides **passive immunity**. |
| **T-cells** | Lymphocytes that respond to antigen stimulation by proliferating into a population of **lymphokin**-producing cells, killer cells, and memory cells. |

---

Lymphatic System

---

| | |
|---|---|
| **lymph** | Fluid lost from the bloodstream and recovered by lymph vessels. Lymph passes through **lymphoid organs** and **lymph nodes** on its way back to the bloodstream. |

---

# Review and Synthesis

1. Name and discuss five essential functions of the mammalian circulatory system.

2. Name three locations where one-way valves are essential to fluid flow. Why are these valves not needed in arteries or arterioles?

3. Explain why a two-chambered heart would be inadequate for humans and other large, highly active mammals.

4. Discuss how, in closed circulatory systems, tissues get what they need from blood without ever coming in contact with it. Include capillary anatomy and function in your answer.

5. Why do people develop varicose veins but not varicose arteries?

6. Metabolism generates heat. How does blood flow prevent heat from building to lethal temperatures? What happens to heat after it is carried away by blood?

7. Physicians sometimes use heparin, a chemical that inactivates thrombin. How would such a chemical affect blood clotting?

**8.** Before the mid-seventeenth century, scientists believed that blood was dumped from arterioles into body spaces and then recovered by venules. In 1661, Marcello Malpighi corrected this misconception by discovering the "missing link" in the circulatory system. What do you think Malpighi found? What instrument must he have used to make his discovery?

## Additional Readings

Edelson, R., and J. Fink. 1985. "Immunological function of the skin." *Scientific American* 252:46. W. H. Freeman, San Francisco. (Intermediate.)

Levi, R., and J. Moskowitzs. 1982. "Cardiovascular research: a decade of promise." *Science* 217:121. (Intermediate.)

McKane, L., and J. Kandel. 1985. *Microbiology: Essentials and Applications,* (Chapters 15 and 16.) McGraw-Hill, New York. (Intermediate.)

Wiggers, C. 1957. "The heart." *Scientific American* 242:84. (Intermediate.)

# 6

# *Reproductive Biology*

Reproduction is a basic characteristic of life. As this painted lady butterfly visits these flowers for food, it transfers the plants' male gametes. The butterflys' striking wings help attract members of the opposite sex.

Chapter 21

# Sexual Reproduction of Flowering Plants

*I*t was quite an experience. Eyewitnesses reported the victim was lured by an overpowering fragrance and then knocked into a pool of cold water. Frantic attempts to climb out were fruitless because the sides of the pool were very high and slimy, making it impossible to get a firm hold. Eventually, the hapless captive discovered a very tight passageway, managed to squeeze through it and escape.

But the ordeal was not yet over. The victim emerged with two sacs, bulging with small spheres, attached to his back. As the victim fled, he became lured by the fragrance of yet another seductress and, just like the first time, was powerless to resist the provocative aroma. The events that followed were the same; he was dunked into a pool of water, made several unsuccessful attempts to climb out, and eventually escaped through a narrow exit. However, when he emerged this time, he was free of the baggage. The sacs had become dislodged while squeezing through the escape passage.

The victim of this real life drama is a male euglossine bee (orchid bee), and the seductress/captor is a bucket orchid flower, *Coryanthes leucocorys* (Figure 21-1). During its ordeal with the cup-shaped flower, the bee inadvertently transferred the orchid's male gametes (housed in the sacs) from one flower to another, a critical phase in the sexual reproduction of the orchid.

Like other flowering plants, *Coryanthes* cannot transfer its own gametes to other individuals. And like most other plants, *Coryanthes'* male gametes are contained in **pollen grains**, the small spheres in the sacs that were attached to the bee's back. The bucket orchid has a marvelously complicated flower that baits and then captures male orchid bees by dunking them into a pool of water. When the bee escapes through a narrow tunnel at the back of the flower, bags of pollen grains are attached or removed from the insect's back, transferring gametes between flowers.

But the bees also benefit from their ordeal. They are attracted to bucket orchids not only because of the flowers' alluring fragrance, but also because *Coryanthes* flowers produce a waxy substance which the male bees need to make their own sexy "perfume" to attract female orchid bees. Thus, the sexual reproduction of the orchid *and* that of the bee are inseparably intertwined, each depending on the other to form a critical link in a chain that leads to sexual reproduction.

## Flower Structure

The center of sexual activity for all flowering plants is the flower. Flowers are the most advanced reproductive structure in the plant kingdom, coming in an array of sizes, shapes, colors, color patterns, and arrangements. This variability is not random, however, for all flowers are well adapted for the success of sexual reproduction. (Details of the reproduction of nonflowering plants—ferns, mosses, conifers, and so on—are covered in Chapter 33.)

### A Typical Flower

A flower is a cluster of modified leaves on a shortened stem called a **pedicel**. In a typical flower, such as the one il-

**Figure 21-1**
**Glamorous and bewitching,** the bucket orchid, *Coryanthes leucocorys,* has two miniature spouts that alternately drip water into the cup, forming a small pool. The flower's fetching perfume lures male euglossine bees to the spouts, and a drop of water knocks the befuddled bee into the pool. As the bee departs, through a narrow groove in the back of the flower, the flower either deposits or picks up bags of pollen grains from the bee's back. By capturing bees and then loading and unloading sacs of pollen, the plant succeeds in transferring its sperm from flower to flower for sexual reproduction.

lustrated in Figure 21-2, the modified leaves occur in four circular groups or **whorls** attached to a **receptacle**, the widened end of the pedicel that forms the base of the flower. The four groups are the calyx, petals, stamens, and the pistil.

The outermost whorl of a flower forms the **calyx**, a collection of modified leaves called **sepals**. Generally, the sepals surround and protect the flower bud during its development. The second whorl is made up of **petals**, collectively forming the **corolla** of the flower. The petals of many plants come in colors and shapes that catch the eye of pollen-transferring animals. Although some petal colors and markings are subdued or invisible to humans, the animals they attract readily see them (Figure 21-3). Petals aren't the only floral structures that attract animals. Sometimes the sepals, entire clumps of flowers, and even leaves are brightly colored or modified to capture the attention of passing animals (Figure 21-4).

It is easy to see how sepals and petals could be modified leaves; they are leaf-shaped and have prominent veins running through them. But the inner whorls of a flower do not appear leaf-like at all. They are the product of extensive evolutionary modifications of spore-producing leaves, modifications that enable them to manufacture and house the plant's gametes. The whorl immediately inside the petals is a collection of **stamens**—modified leaves that form the flower's male reproductive organs. Most stamens have a long, slender stalk, the **filament**, and a swollen end called the **anther**. Pollen grains are produced inside the anther lobes in regions called *pollen sacs*. When pollen is mature, the anthers split open, releasing the pollen.

The central whorl of a flower is made up of **carpels**—modified leaves that form the female reproductive organs. Some flowers have only one carpel, others have more than one carpel, and still others contain a compound unit made of carpels that have fused together. Each separate carpel, or each unit of fused carpels, is called a **pistil**.

The pistil has three parts. At the top of the pistil is (1) a **stigma**, a very sticky area to which pollen adheres. In plants that rely on wind to transfer their pollen, the stigma is enlarged and feathery to glean pollen from the air (Figure 21-5a). In others, the shape of the stigma brings it into contact with pollen-laden areas of an animal's body (Figure 21-5b). For many plants, the pollen must physically and chemically fit the stigma before pollen grains will grow a **pollen tube** to deliver sperm to the egg at the base of the pistil. In this way, the stigma "recognizes" pollen only from members of its own species.

The (2) **style** of the pistil connects the stigma to the (3) **ovary**, the enlarged bottom part of the pistil where eggs are produced. Since most ovaries contain many eggs, an equal number of pollen grains must grow a tube to convey a sperm to each egg. Following the fusion of egg and sperm, the zygote develops into the seed's embryo. At the same time, the ovary (and sometimes other floral parts) matures into a **fruit** that surrounds and protects seeds.

## Flower Diversity

The flower is a product of millions of years of evolutionary refinement. Each type is a floral masterpiece that promotes sexual reproduction.

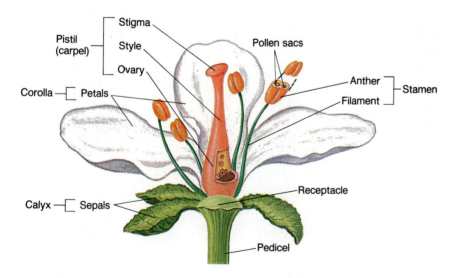

**Figure 21-2**
**A typical flower.** Flowers produce the plant's gametes. The pistil provides a special surface (the stigma) to collect pollen grains, a channel (the style) that directs the growth of a pollen tube containing sperm to the egg in the ovary. Following fertilization, the ovary develops into a fruit that protects and nourishes the developing embryos contained in its seeds.

a          b

**Figure 21-3**
**A bee's-eye view.** Humans and bees see different wavelengths. To us, a meadow cranesbill *(Geranium)* is uniformly lavender *(a)*, but to bees the petals have distinct lines that converge at the center of the flower where nectar is found *(b)*. These "nectar guides" absorb ultraviolet wavelengths, while the remaining parts of the petals reflect them. Nectar guides direct the landing of flying pollinators much like lights along an airport runway.

a          b

**Figure 21-4**
**Flower power.** Color is used to help distinguish the flower from its surroundings and to attract animals, mostly insects. Generally, it is the petals of a flower that are colored, but some plant species lack petals or have inconspicuous flowers. In these cases, other structures may be modified to catch the attention of passing animals. *(a)* The flowers of flannel bush *(Fremontedendron)* have brilliantly colored sepals to attract animals. *(b)* An African daisy is not a single flower, but a miniature bouquet of tiny, densely packed flowers forming a *composite*. The outer *ray flowers* have enlarged petals to attract animals. The petals of the inner *disk flowers* are reduced and barely visible. *(c)* In poinsettia, it isn't the flower that does the attracting, but a cluster of colored leaves just below the inconspicuous, dull flowers.

c

a

b

**Figure 21-5**
**Improving the odds.** *(a)* Plants that rely on wind to transfer pollen grains produce tremendous amounts of pollen, sometimes creating pollen clouds. The chances of pollen hitting a stigma are greatly increased by having a large surface, like the feathery stigma of this grass flower. *(b)* The two-lipped monkey flower stigma *(Mimulus)* "licks" pollen from the head of bees. The stigma quickly closes to exclude pollen from its own stamens.

Flowers differ in

- Number of parts in a whorl.
- Color of petals.
- Symmetry.
- Whether all four whorls of modified leaves are present, or one or more whorls is absent.
- Whether parts are fused or separate.
- Whether the ovary is sunken into a concave receptacle, is surrounded by a cup-shaped receptacle, or sits on top of a convex or conical receptacle.
- Whether flowers occur singly, or in groups called *inflorescences.*

A flower containing all four whorls of modified leaves — sepals, petals, stamens, and carpels (pistil) — is called a **complete flower**. Flowers lacking one or more whorls are **incomplete flowers**.

All complete flowers are **perfect flowers** because they contain both stamens and carpels. **Imperfect flowers** contain either stamens or carpels, making them male or female flowers, respectively. In some species that produce imperfect flowers (walnut trees and corn, for example) male and female flowers are produced on the same plant. Such plants are called **monoecious** plants (mono = one, oecious = house) (Figure 21-6*a*). Others, such as willows and asparagus, produce imperfect male and female

flowers on separate individuals. These are called **dioecious** plants (di = two) because the plants themselves are either male or female (Figure 21-6*b*).

Many flowers have nectaries — glands that secrete sugary nectar. These flowers attract animals and "reward" them for transferring pollen grains with a stream of nutritious nectar. In many cases, flowers provide a two-course banquet of nectar and high-protein pollen.

## Formation of Gametes

As with animals, the formation of gametes in plants begins with meiosis. (To review the life cycle of a flowering plant, refer to Figure 12-1*b*.) Special diploid *mother cells* inside the anthers and ovaries of flowers divide by meiosis to form haploid *spores*. Mother cells in the anthers are called **microspore mother cells**, and the diploid mother cells inside ovaries are called **megaspore mother cells** (micro- = small and mega- = large). (These prefixes do not refer to the size of the spores, but to the size of the gametes they ultimately produce; the male sperm is smaller than the female egg.) **Microspores** produced within anthers divide by mitosis to form pollen grains, the male gametophytes that produce the plant's sperm. In the ovary, **megaspores** divide by mitosis to form female gametophytes that produce the eggs. When gametes fuse dur-

ing fertilization, a zygote is formed and a new sporophyte generation begins.

## Production of Female Gametes

Female gametes are formed within **ovules**, which are round masses of cells attached by a thin stalk to the inner surfaces of the ovary. Each ovule contains one enlarged megaspore mother cell, surrounded by cells making up the *nucellus*. Some ovaries contain only one ovule, whereas others have tens, hundreds, or even millions of ovules. The outer cells of the ovule form one or two protective layers called the *integuments*. The integuments do not completely encircle the ovule, but leave a small opening, the *micropyle*, through which the pollen tube grows to deliver sperm.

The production of an egg, illustrated in Figure 21-7, begins when the megaspore mother cell undergoes meiosis. Three of the resulting megaspores degenerate. The nucleus of the surviving megaspore divides three times by mitosis, manufacturing a female gametophyte composed of eight haploid nuclei. (The first mitotic division produces two nuclei, the second four, and the third eight.) The eight nuclei are then partitioned into seven cells, three cells are positioned at opposite ends of the female gametophyte, and one cell is in the middle. The three

**Figure 21-6**
**Inperfect (uniserval) flowers** on one plant *(a)* are called monecious (the male flowers are pendulent and the female flower is at the tip of the stem). Meadow rue *(Thalictricum)* is diaecious, male *(b)* and female *(c)* flowers are produced on separate plants.

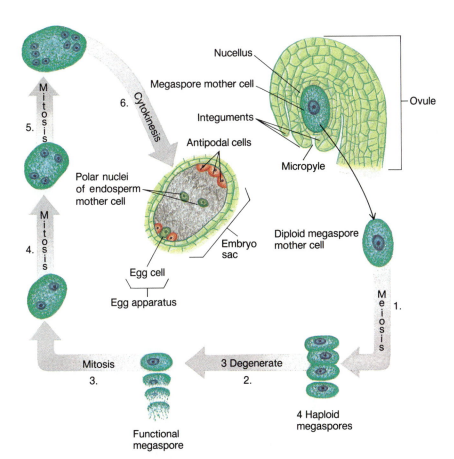

**Figure 21-7**
**Development of the female gametophyte** is a six-step process that takes place inside an ovule. (1) A megaspore mother cell undergoes meiosis to produce four haploid megaspores. (2) Three megaspores degenerate, leaving a single functional megaspore. The nucleus of the functional megaspore divides three times (3, 4, and 5), and (6) then cytokinesis partitions the nuclei and cytoplasm into seven cells, producing the mature embryo sac.

cells near the micropyle form the *egg apparatus*, and the cell in the center is the **egg gamete**. The three cells farthest from the micropyle are called **antipodal cells**.

The cell in the middle of the female gametophyte contains two haploid nuclei. This *bi*nucleate cell is the **endosperm mother cell** which, following fertilization, proliferates into a tissue that nourishes the embryo and the growing seedling until it can photosynthesize on its own. Together, the egg apparatus, antipodal cells, and the endosperm mother cell form a fully developed female gametophyte, or **embryo sac**.

## Production of Male Gametes

As embryo sacs develop in the ovary, pollen grains form in the anthers of stamens (Figure 21-8). The centers of each anther lobe, the pollen sacs, are filled with microspore mother cells. Each diploid microspore mother cell undergoes meiosis to produce four haploid microspores. However, unlike the female megaspores, all four microspores develop into male gametophytes.

Microspores divide once by mitosis to produce a pollen grain, which is a two-celled male gametophyte made up of a *generative cell* enclosed within a *tube cell*. When pollen lands on a matching stigma, the generative cell undergoes mitosis to produce two sperm nuclei, as the tube cell grows a pollen tube to an ovule.

Pollen grains are surrounded by a very tough wall that protects the male gametophyte while being transported to a stigma. Sculpturing of the walls of some pollen grains helps them to adhere to animals (Figure 21-9*a*) or to remain airborne (Figure 21-9*b*). The pollen of different species vary in size, shape, and surface pattern. The characteristics of the pollen grain wall are so unique that botanists use them as "botanical fingerprints" to identify plant species. Ancient plants can also be identified by analyzing the characteristic walls of fossilized pollen.

## Pollination and Fertilization

The transference of pollen grains from an anther to a stigma is called **pollination**. Many plants, like *Coryanthes* described in the chapter introduction, use insects for pollination; others depend on wind, water, or other types of animals, such as birds and a few mammals. Most of the plants that enlist the services of animal pollinators offer food to attract the animal. A few, how-

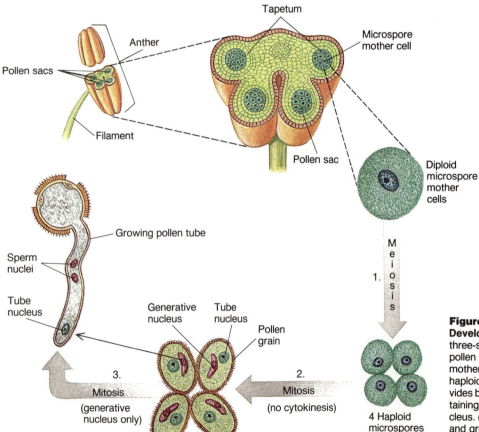

**Figure 21-8**
**Development of the male gametophyte** is a three-step process that takes place inside the pollen sacs of an anther. (1) A microspore mother cell divides by meiosis to produce four haploid microspores. (2) Each microspore divides by mitosis to produce pollen grains containing a tube nucleus and generative nucleus. (3) The pollen is deposited on a stigma and grows a pollen tube, and the generative nucleus divides to form two sperm nuclei.

Figure labels: Tapetum; Anther; Microspore mother cell; Pollen sacs; Filament; Pollen sac; Diploid microspore mother cells; Meiosis; 1.; Growing pollen tube; Sperm nuclei; Tube nucleus; Generative nucleus; Tube nucleus; Pollen grain; 3. Mitosis (generative nucleus only); 2. Mitosis (no cytokinesis); 4 Haploid microspores

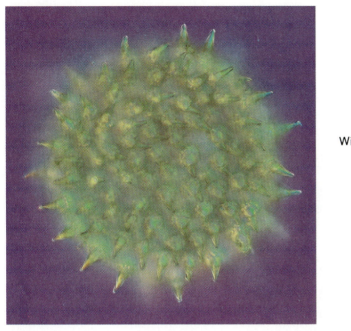

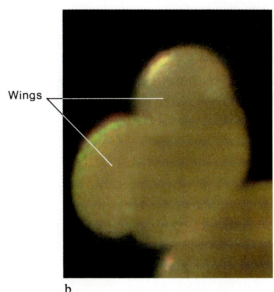

Wings

**Figure 21-9**
**Pollen structure and function.** *(a)* Wall sculpturing of cotton pollen helps them adhere to insects.
*(b)* Bladderlike wings of pine pollen *(Pinus nigra)* help keep them airborne.

ever, employ some unusual tactics to accomplish pollination (see Bioline: The Odd Couples).

Even after pollination is completed, plant gametes are still physically separated. The sperm nuclei are in the pollen grain lying on top of the stigma, and the egg is buried deep within the ovary. But when pollen and stigma match, a pollen tube grows down through the stigma, style, and ovary tissues (Figure 21-10), eventually worming its way through the micropyle of an ovule, then penetrating the embryo sac and discharging the sperm nuclei. One sperm nucleus fuses with the egg (fertilization), producing a diploid zygote. The other sperm nucleus fuses with the diploid endosperm mother cell to form a triploid (3N) **primary endosperm cell**.

## Development After Fertilization

Fertilization sparks dramatic transformations in the flower.

- The zygote develops into the plant embryo.
- The primary endosperm cell produces endosperm tissue.
- The ovule is converted into a seed.
- The ovary develops into a fruit.

The endosperm, seed, and fruit nourish and protect the embryo during its development and during its journey to a new habitat.

### Embryo Development

The first mitotic division of a newly formed zygote produces two different-sized daughter cells (Figure 21-11*b*). The smaller, terminal cell divides by mitosis to form the plant embryo. The larger, basal cell produces the *suspensor*, a column of cells that conducts nutrients from the endosperm to the growing embryo (Figure 21-11*c*).

The initial divisions of terminal cells form a globular mass of embryonic cells. But at this point differences begin to emerge between developing

monocot and dicot embryos. (Recall from Chapter 13 that monocots and dicots are the two major groups of flowering plants.)

In dicots, two centers of division form two cotyledons, making the young dicot embryo heart-shaped (Figure 21-11*d*). The mature dicot embryo consists of an epicotyl, two cotyledons, and a hypocotyl. The **epicotyl** is the portion of the embryo above the cotyledons (epi = above, cotyl = cotyledon). At the tip of the epicotyl is the **plumule**, the shoot tip with several tiny, immature leaves. The cotyledons, or seed leaves, house food reserves. The **hypocotyl** is the portion of the embryo below the cotyledons (hypo = below). At the tip of the hypocotyl is the **radicle**, the part of the embryo that eventually develops most or all of the plant's root system.

In contrast to dicots, only one center of division develops at the apex of a growing monocot embryo, producing a single cotyledon called a **scutellum** (Figure 21-12). A special structure, the **coleoptile**, surrounds the shoot tip,

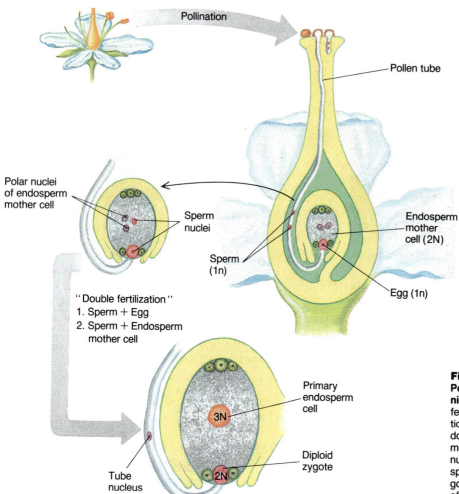

Pollination

Pollen tube

Polar nuclei
of endosperm
mother cell

Sperm
nuclei

Sperm
(1n)

Endosperm
mother
cell (2N)

Egg (1n)

"Double fertilization"
1. Sperm + Egg
2. Sperm + Endosperm
   mother cell

Primary
endosperm
cell

3N

Tube
nucleus

2N

Diploid
zygote

**Figure 21-10**
**Pollination and fertilization: The beginnings of a new generation.** Pollen is transferred from anthers to a stigma during pollination. The pollen grain then grows a pollen tube down the style and ovary and through the micropyle. At the same time, the generative nucleus divides to produce two sperm. One sperm nucleus fertilizes the egg to form a zygote, and the other fuses with the two nuclei of the endosperm mother cell to form a triploid primary endosperm cell.

protecting the young stem and leaves as they emerge from the soil. Another protective structure, the **coleorhiza**, sheaths and protects the root tip as it penetrates the soil.

## Seed Development

Following fertilization, the ovule develops into a seed that consists of the embryo, endosperm, and a seed coat (Figure 21-13). In some plant seeds, the endosperm surrounds the embryo and is part of the seed; in others, the endosperm is stored in the cotyledons and is part of the embryo.

The seed coat is formed from the integuments of the ovule. It may become tough and hard, protecting the embryo from abrasion or other dangers while being dispersed to a new

habitat. Some seeds have projections from the seed coat that help absorb water, attach the seed to the fur of an animal for dispersal, or protect the embryo from hungry herbivores. In nearly all seeds, the micropyle persists and can be seen as a tiny opening in the seed coat. Next to the micropyle is the **hilum**, a scar on the seed coat that marks the point where the stalk from the ovary was attached to the ovule.

## Fruit Development

With sexual reproduction completed, the flower parts that were needed for pollination (sepals, petals, and stamens) wither away. As embryos and seeds develop, the ovary of the flower matures into a fruit. Like many animal species, humans rely on fruits as a

primary source of nutrition. In fact, fruits have had a profound effect on human life, perhaps sparking the development of human civilization (see Bioline: The Fruits of Civilization).

Whether a fruit is green or brightly colored, juicy or dry, soft or hard, all or part of it is formed from a ripened ovary. (Some so-called vegetables, such as cucumbers, tomatoes, squash, string beans, and grains are really fruits because they developed from ovaries.) Some fruits develop from more than just the ovary; other flower parts, an entire flower, the receptacle, and even clusters of flowers may produce the fruit. The basis for dividing fruits into three categories—simple, aggregate, and multiple—depends on whether a fruit develops from the ovary of one pistil (**simple fruits**), from

# B I O L I N E

## THE ODD COUPLES: BIZARRE FLOWERS AND THEIR POLLINATORS

Many plants press into service an army of pollinating "middlemen" of various sizes, shapes, and behavior. Usually, the plant entices pollinators by providing "rewards" such as nectar, pollen, oils, or sometimes food bodies. But some plants seem to resort to nefarious means to reach this end, as in the following cases:

• Case 1: The Imposter

The dull-red, wrinkled flowers of the South African milkweed, *Stapelia gigantea,* look and smell like putrifying flesh, deceiving female blowflies into laying her eggs there. In doing so, the duped female blowfly carries pollen from flower to flower. But the floral hoax has grave consequences for the blowfly. All of the fly's maggots perish from lack of proper food, for, no matter how foul, blowfly maggots cannot live on flower tissue, only animal flesh.

• Case 2: Open House

The upward pointing petal of the Dutchman's pipe *(Aristolochia clematitis)* is covered with millions of tiny grains of wax. Gaining a footing on such a surface would be similar to trying to walk down a mountain covered with greased ball bearings. Even with hooks and suckers, the gnats and flies attracted to the flower cannot gain a foothold on this slick surface. As soon as an insect lands, it slips down the flower tube past a forest of downward-pointing hairs, finally crashing into a cluster of stigmas at the bottom of the chamber. The hairy forest prevents any "visitors" from leaving for at least two or three days. During that time the stamens mature and shower the insects with pollen. Then the flower tilts over and the retaining hairs shrivel. The insects walk out, usually to be quickly lured into captivity by yet another Dutchman's pipe flower, completing the pollination cycle.

• Case 3: The Case of Mistaken Identity

The flowers of the orchid *Ophrys speculum* take advantage of the powerful sex drive and energetic nature of male wasps. To a male wasp, the orchid looks and smells like a female of his own species. The illusion is perfect; as the male wasp "copulates" with a number of orchids, it transports pollen between them. Given a choice between the orchid and a female wasp; the flower is just as successful in arousing the amorous attention of the male wasp as the genuine female of his own species!

• Case 4: RIP—Murder Among the Lilies

The South African water lily attracts hundreds of eager hoverflies, bees, and beetles to its flowers, using an abundance of nutritious pollen as bait. The insects soon learn that water lilies are a rich source of food. Lily flowers go through a female stage first when no pollen is offered and then go into the male stage. But by the time pollen is offered, the mortal deed has already been done, for hidden below the bountiful stamens float the remains of previous visitors.

The deceased visitors happened upon the lily flower on the first day it opened, a day in which the flower was in its female stage. In their feeding frenzy, the insects carrying pollen from other lily flowers in their male stage failed to notice there was no pollen in this flower. The immature stamens form a ring of slick columns surrounding a watery pool. Once they land and start crawling over the stamens in search of pollen, it is too late. Unable to get a grip, they helplessly slid down the slippery stamens into the pool. A wetting agent in the water guarantees that even the lightest fly will sink to the bottom of the pool and drown. At night the flower closes, and pollen collected from visits to other lily flowers washes off the corpses and settles to the bottom of the pool where the stigma is located. Pollination is completed. By morning, the murky burial pool is hidden by a blanket of mature stamens, the flower now luring insects with a bounty of pollen. The lily flower in its male stage is safe, but insects cannot tell the difference between flowers that offer pollen and those that offer certain death, a fact on which the lily has staked its sexual reproduction.

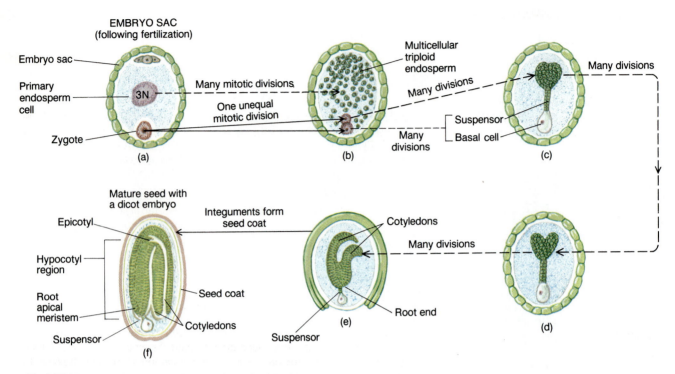

**Figure 21-11**
**Development of a dicot embryo, *Capsella*.** *(a)* A zygote and a primary endosperm cell are formed following fertilization. The remaining cells of the embryo sac (two cells of the egg apparatus and three antipodal cells) degenerate. *(b)* The triploid (3N) primary endosperm cell divides by mitosis to produce nutritive endosperm. The first division of the zygote forms a smaller terminal cell and a larger basal cell. *(c)* Mitotic divisions of the terminal cell forms a heart-shaped proembryo. Divisions of the basal cell forms the suspensor. *(d)–(f)* Continued divisions lead to the formation of a mature dicot embryo. The integuments form the seed coat.

**Figure 21-12**
**Corn (*Zea mays*) fruit (kernel) with mature embryo.** The ovary and integuments of corn fuse to form the outer "fruit + seed" coat. Nutritious endosperm lies outside the corn embryo. The single cotyledon absorbs digested starch and protein from the endosperm.

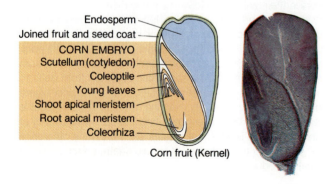

# B I O L I N E

## THE FRUITS OF CIVILIZATION

Our existence depends on fruits and seeds, but our dependency goes beyond the obvious apple, orange, or peanut butter sandwich. Flour for making bread is ground grain, the fruit of grasses. We feed domestic livestock fruits and grains for meat production. In fact, most of our food comes directly or indirectly from fruits and seeds.

The same was true for our early ancestors. Early humans traveled in nomadic tribes, forced to follow seasonal changes in plant growth in order to harvest enough food to feed its members. The perpetual search for food occupied most of the time and energy of every member of the tribe. There was no time for contemplation, no time or energy to develop art or science, or even to begin the first scratchings of a written language. So how did these intellectual and civilized pursuits begin? Human civilization may owe its existence to one of the simplest kinds of fruits, grains.

Historians have identified three geographic regions where human civilization flourished and then radiated out to other areas of the world. Each center of civilization had at least one naturally growing grain that likely provided the nutritional foundation for the development of civilization. The Near East Center had wheat and barley, the Far East Center had rice, and the Middle America Center had corn. Because of the abundance of these grasses and their high nutritional content, a few members of the tribe could gather, tend, and cultivate enough grain to feed the entire group. In addition, grass grains could be stored for long periods without spoiling, eliminating the need to travel from place to place in order to have a fresh supply of food. Some anthropologists speculate that as more and more members of the tribe were freed from the oppression of an endless, exhausting search for food, they had time to contemplate their surroundings and to develop the innovations that allowed civilization to progress. The development of plant and animal husbandry practices freed even more people to specialize in other activities. Eventually, language, arts, and sciences began to flourish, leading to the social and technological advances we enjoy today.

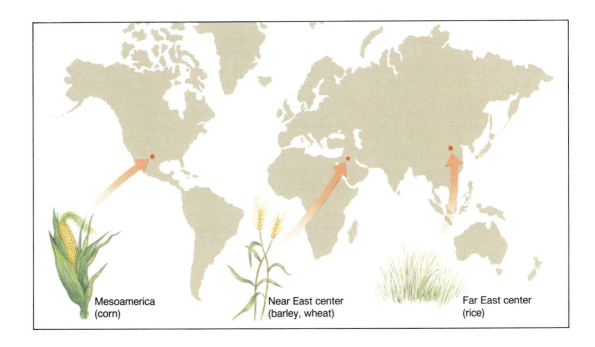

Mesoamerica
(corn)

Near East center
(barley, wheat)

Far East center
(rice)

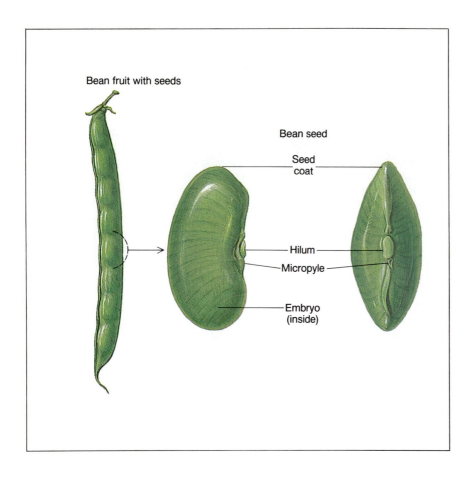

Bean fruit with seeds

Bean seed

Seed coat

Hilum

Micropyle

Embryo (inside)

**Figure 21-13**
**Seeds.** The seed is a life-sustaining capsule that houses, nourishes, and often protects the embryo after separation from its parent.

many pistils in a single flower (**aggregate fruits**), or from the pistils of separate flowers (**multiple fruits**) (Figure 21-14).

Typically, fertilization triggers fruit development. In a few cases, however, fruits develop without fertilization through a process called **parthenocarpy**. Parthenocarpic fruits contain no embryos and therefore no seeds. Some cultivated plants (bananas, pineapples, seedless grapes, seedless watermelons, and seedless cucumbers, for example) lack seeds because their fruits developed parthenocarpically, often aided by sprayed artificial plant hormones that induce the ovary to mature. (Parthenocarpy refers only to fruit development without fertilization; it differs from parthenogenesis in which an embryo, and subsequently a new organism, develops from an unfertilized egg.)

## Seed Dispersal

In one flowering season, a single plant can produce hundreds, thousands, and even millions of seeds, each of which contains an embryo capable of growing into a new plant. Imagine what might happen if a parent plant simply dropped all of its seeds onto the ground directly beneath its stems and branches. As the seeds germinated and grew, they would compete with each other for space, light, water, and nutrients. Intense competition for resources among the developing plants (the **seedlings**) would lead to massive deaths and widespread malnourishment. There simply are not enough resources to support so many individuals concentrated in one area. Seed dispersal solves this problem by reducing competition among members of the same species. Dispersal also scatters seeds to new areas where conditions may be even more favorable for growth and reproduction.

Because of the advantages of dispersal, many fruits have become fashioned through evolution to enhance seed dispersal. The fruits of wild cucumber, witch hazel, and many legumes literally explode, flinging seeds in various directions (Figure 21-15). The majority of plants do not disperse their own seeds however, but enlist some agent—either wind, water, or animals—to do the job for them (Figure 21-16).

Plants recruit a broad range of animals, from worms to bears, to disperse seeds. When the juicy fruits of plants are eaten by animals, their seeds pass undamaged through the animal's digestive system and are scattered with the animal's feces. Rodents gather and stockpile nuts and seeds in caches; the

a. Simple fruits

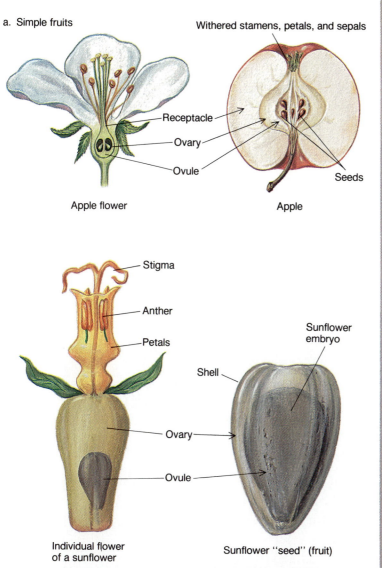

Withered stamens, petals, and sepals

Receptacle

Ovary

Ovule

Apple flower

Seeds

Apple

Stigma

Anther

Petals

Shell

Sunflower embryo

Ovary

Ovule

Individual flower of a sunflower

Sunflower "seed" (fruit)

b. Aggregate fruit

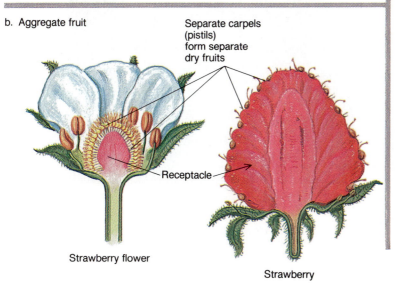

Separate carpels (pistils) form separate dry fruits

Receptacle

Strawberry flower

Strawberry

c. Multiple fruits

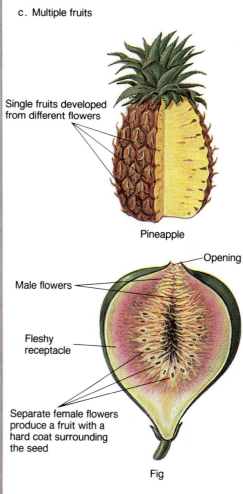

Single fruits developed from different flowers

Pineapple

Opening

Male flowers

Fleshy receptacle

Separate female flowers produce a fruit with a hard coat surrounding the seed

Fig

**Figure 21-14**
**Fruit types.** *(a)* Simple fruits develop from one pistil. An apple is an example of a *fleshy simple fruit,* whereas a sunflower "seed" is an example of a *dry simple fruit.* *(b)* Aggregate fruits, such as a strawberry, are formed from many separate pistils in a single flower. *(c)* Multiple fruits form from several flowers; the matured ovaries of densely packed flowers either form a single fruit (such as a pineapple) or are joined into a single unit by other flower tissues. For example, a fig is an enlarged, urn-shaped receptacle filled with many tiny, separate simple fruits.

**Figure 21-15**
**Exploding fruits** disperse their own seeds. The common vetch *(Vicia sativa)* produces a power-packed legume that explodes, flinging seeds 3 meters (10 feet) from the plant.

a

b

**Figure 21-16**
Airborne seeds and fruits have wings or plumes that help them remain aloft. The broader the wings or the fluffier the plumes, the longer they remain airborne, and the further they travel. *(a)* Plumes of milkweed. *(b)* Wings of swamp maple.

uneaten seeds grow into new plants. Many seeds can be dispersed over very long distances, across oceans to other continents and to distant islands, in mud adhering to the feet of migrating birds. Some animal-dispersed seeds are equipped with barbs, hooks, spines, or sticky surfaces that temporarily fasten them to the animal's fur, aiding dispersal to distant locations (Figure 21-17).

# Germination and Seedling Development

When environmental conditions are favorable, a seed germinates and a new generation of plants begin to grow and develop. During **germination**, the radicle of the embryo thrusts through the seed coat, forming a root that anchors the new plant.

Favorable conditions for germination include proper temperatures, sufficient water, adequate oxygen, and the presence or absence of light. Yet some seeds need more than these conditions to germinate. Some must be exposed to freezing temperatures for exact periods. Others must travel through the digestive system of an animal, be battered and ground against rocks, or even be scorched by fire before they will germinate. (The fire or digestive enzymes of an animal break down tough seed coats, enabling the radicle to break through.) Such adaptations help time germination to coincide with the most favorable conditions. Many adaptations for timing germination are controlled by plant hormones and are discussed in the next chapter.

Of all environmental factors, adequate moisture is often the most critical for seed germination and seedling development. Seeds are very dry, sometimes containing as little as 2 percent water, and, as a result, must become hydrated before germination can begin. As moisture enters the seed, a process called **imbibition**, the tissues soak up water, the seed swells, metabolism speeds up, and cellular division and growth begins in the embryo.

## Seed Viability and Seed Dormancy

The embryos of different plant species remain viable (capable of growing and developing) for different periods of time. The *seed viability* of some species, such as willows, is only a few days, whereas the seeds of other plants remain viable for months or years. Oriental lotus seeds recovered from ancient tombs, for example, were capable of germinating after 3000 years.

There are two reasons why a seed would fail to germinate and grow: either (1) the embryo is dead (that is, the seed is not viable) or (2) the embryo is "resting" (**seed dormancy**). Technically, all seeds are dormant until activated by water. However, some seeds remain dormant even when water is plentiful. They are unable to grow be-

**Figure 21-17**
**Hitchhiking fruits** with hooks, spines, or barbs "bum" a ride to new habitats by attaching to the skin or fur of animals. (Photo is an arphagophyto burr from Madagascar.)

cause of an undeveloped embryo, hormone blocks, or the embryo's inability to break through a tough seed coat. This delays germination until conditions are more favorable.

## Seedling Development

A new seedling's battle for survival begins the instant it is released from the parent plant. Producing a root system to obtain water is more important than breaking through the surface for photosynthesis (nutrients stored in the seed last for several days). Without moisture, the seedling will quickly die of dehydration. This is also the time that most plants establish mycorrhizae associations with soil fungi (Chapter 14), improving their ability to absorb water and minerals.

Seedling development is different for monocots and dicots. To illustrate these differences, we discuss a representative example of each, a dicot bean seedling *(Phaseolus)* and a monocot corn seedling *(Zea)*.

## BEAN SEEDLING DEVELOPMENT

Like all dicot embryos, the bean has an epicotyl, a hypocotyl and two cotyledons. During germination, the bean radicle emerges and grows downward, producing many lateral roots from a single main root (Figure 21-18). The pattern of cell growth causes the bean hypocotyl to bend, forming a hook that forces its way up through the soil, carving a channel so that the delicate cotyledons, young leaves, and apical meristem can be drawn along without injury. Once it breaks through the soil surface, the hypocotyl straightens, exposing the cotyledons and epicotyl to sunlight.

Up to this point, the seedling has depended on the nutrients stored in the endosperm of its cotyledons. But once the cotyledons and young leaves are exposed to light, they manufacture chlorophyll and other pigments, the white seedling turns green, and photosynthesis begins. As food reserves are used up, the cotyledons eventually wither and fall.

## CORN SEEDLING DEVELOPMENT

Unlike dicots, food reserves are usually not stored in the cotyledon of monocots. Instead, the single cotyledon absorbs food from surrounding endosperm, which fuels germination and seedling growth (Figure 21-19).

When a corn kernel germinates, the root tip breaks through the seed coat and protective coleorhiza. At the same time, the coleoptile forces its way up through the soil. The growing shoot tears through the protective coleoptile. Once young leaves are exposed to sunlight, photosynthesis begins.

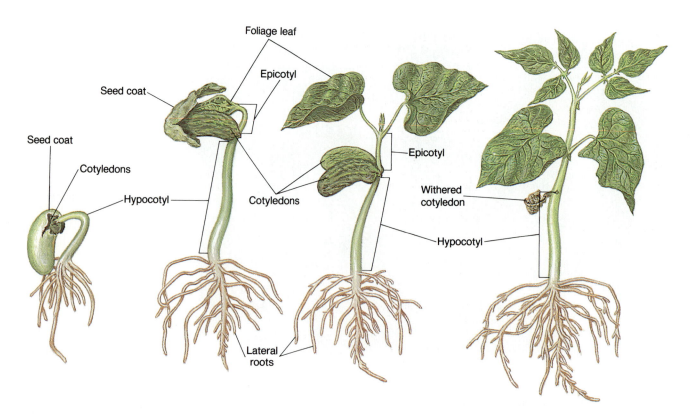

**Figure 21-18**
**Germination and development of a bean (*Phaseolus vulgaris*).**

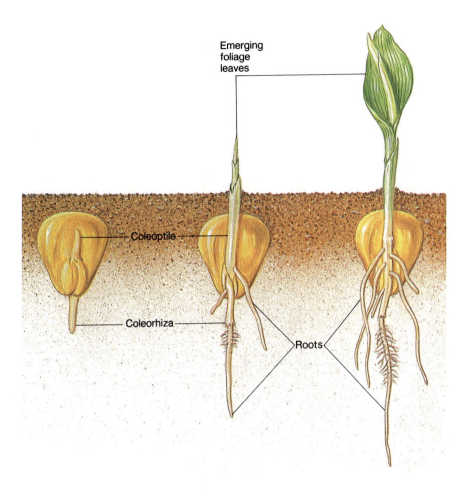

Emerging
foliage
leaves

Coleoptile

Coleorhiza

Roots

**Figure 21-19**
**Germination and development of corn *(Zea mays).***

## Seedling Survival and Growth to Maturity

A seedling's life does not depend solely on its ability to absorb water and minerals, and to photosynthesize enough food to support its growth and development into a mature plant. It must also be able to outcompete other plants for nutrients and light, as well as to avoid the many hazards of life, including continuous onslaughts of plant-eating herbivores, attacks by hundreds of disease-causing microbes, fluctuations in temperature, slashing winds, and a host of other life-threatening environmental conditions. The few seedlings that do survive establish a new generation of plants. The ability of a plant to survive depends on a number of growth and development strategies, and behaviors that are regulated by plant hormones. We discuss plant hormones, and their effect on plant growth and development, in the next chapter.

## Synopsis

### MAIN CONCEPTS

- **Plants rely on wind, water, and animals** to transfer pollen grains for sexual reproduction and to disperse seeds to new habitats. Many plants attract animal vectors by offering food (nectar, pollen, and edible tissues of flowers and fruits).

- **In most plants, sexual reproduction and fruit and seed formation takes place in flowers.** The designs of various flowers promote sexual reproduction by disseminating and collecting pollen from wind, water, or pollen-carrying animal vectors. Brightly colored petals, clusters of flowers, and sometimes colored leaves capture the attention of animals.

- **Meiosis leads to the production of gametes** for sexual reproduction. In plants, diploid mother cells in the anthers of stamens and in the ovary of the pistil divide by meiosis to form spores. The spores then develop into the male gametophyte (pollen grain) containing two sperm gametes or into the female gametophyte containing a single egg.

- **Following the transfer of pollen during pollination,** fertilization occurs in the ovary of the pistil. The resulting zygote grows into the plant embryo; the ovule develops into the seed; and the ovary (and sometimes other flower parts) matures into the fruit. The triploid primary endosperm cell formed by the fusion of one sperm nucleus with the endosperm mother cell produces nutrient endosperm.

- **Adaptations of fruits and seeds** help to disperse seeds, reducing competition between the offspring and parent plant, and among offspring, and distributing seeds to new habitats.

- **Favorable environmental conditions** (adequate water, oxygen, and temperature) break seed dormancy and cause germination. Various plant hormones regulate the timing of seed germination, so that young seedlings do not grow when conditions are unfavorable.

### KEY TERM INTEGRATOR

#### Flower Structure

| | |
|---|---|
| **complete flower** | Has four **whorls**: (1) a **calyx** of **sepals**, (2) a **corolla** of **petals**, (3) **stamens** with **anthers** and **filaments**, and (4) a **pistil** of one or more **carpels**. Each pistil has a **stigma**, **style**, and **ovary.** |
| **incomplete flower** | Lacks one or more of the four basic flower parts. In both complete and incomplete flowers, flower parts are attached to the **receptacle** of a **pedicel.** |
| **imperfect flowers** | Have either stamens or pistils. **Monoecious** plants produce male and female flowers on the same individual, whereas **dioecious** plants have male and female flowers on different individuals. |

#### Formation of Gametes for Sexual Reproduction

| | |
|---|---|
| **microspore mother cells** | Located in the **pollen sacs** of anthers and divide by meiosis to form four **microspores**. Each microspore develops into a **pollen grain** with a **tube cell** and a **generative cell**, which divides to produce two **sperm nuclei.** |
| **megaspore mother cells** | Located in **ovules** and divide by meiosis to form **megaspores**. One megaspore divides three times by mitosis to produce an **embryo sac** containing the **egg apparatus** with **egg gamete**, **antipodal cells**, and an **endosperm mother cell.** |

Pollination and Fertilization

| | |
|---|---|
| **pollination** | The transfer of pollen from anther to stigma. The tube cell grows a **pollen tube** down the style to an embryo sac in the ovary where one sperm nucleus fertilizes the egg; the other sperm fuses with the endosperm mother cell to form a **primary endosperm cell**. |

Embryo, Seed, and Fruit Development

| | |
|---|---|
| embryo | The zygote develops into an embryo having one or two cotyledons (in monocots, the **scutellum**), an **epicotyl** with **plumule**, and a **hypocotyl** with **radicle**. In monocots, the **coleoptile** sheaths the epicotyl, and the **coleorhiza** sheaths the hypocotyl. |
| seed | Develops from an ovule. The **seed coat** forms from the surrounding **integuments**. The **micropyle** and **hilum** scar are visible on a seed. |
| fruit | Formed when the ovary matures, the number of ovaries determines whether the fruit is a **simple**, an **aggregate**, or a **multiple fruit**. During **parthenocarpy**, fruits develop without fertilization. |

Seed Dispersal, Germination, and Seedling Growth

| | |
|---|---|
| **seed dispersal** | Following **seed dormancy** and **imbibition** of water, **germination** occurs and a new **seedling** begins to develop. |

## Review and Synthesis

1. Pick a flower and describe it using the terms introduced in this chapter. Based on its shape, color, smell, markings, size, and so forth, what type of pollinator do you think this plant relies on to transfer pollen? Test your hypothesis by observing the plant for a while. Were you right?

2. Using labeled sketches, illustrate the similarities and differences in sperm and egg production in plants.

3. How does pollination differ from fertilization?

4. Examine an apple, orange, acorn, strawberry, or any other type of fruit. Try to deduce how this fruit functions for the plant. Is it adapted to protect and/or disperse the seeds it contains? How?

5. Compare monocot and dicot seedling development. Make a list of five things they have in common and five ways they differ.

6. Explain why all complete flowers are perfect flowers, yet all incomplete flowers are not necessarily imperfect.

## Additional Readings

Meeuse, B., and S. Morris. 1984. *The Sex Life of Flowers*. Facts on File Publications, New York. (Introductory, fun-to-read.)

Proctor, M., and P. Yeo. 1972. *The Pollination of Flowers*. Taplinger Publishing Co., New York. (Intermediate.)

Rost, T., M. Barbour, R. Thornton, T. Weier, and C. Stocking. 1984. *Botany: A Brief Introduction to Plant Biology*. 2nd ed. John Wiley and Sons, New York. (Introductory.)

Weier, T., C. Stocking, and M. Barbour. 1974. *Botany: An Introduction to Plant Biology*. John Wiley and Sons, New York. (Intermediate.)

Chapter 22

# Plant Growth and Development

Without plants, we would quickly die. Like all other animals, humans depend on plants for the food and oxygen they produce during photosynthesis. But our dependency on plants even goes beyond these basic necessities.

- We use plants to medicate ourselves against disease. (More than half our medicines come directly from plants.)

- We shelter ourselves with plants (lumber).

- We clothe ourselves in plant fibers (cotton and linen).

- We brighten our world with plant extracts (coloring and dyes).

- We spice up our food and our lives with plants (vanilla, paprika, ginger, turmeric, coffee, cocoa, and teas).

- We scent our bodies with plant oils (perfumes and colognes).

- On special occasions, we adorn ourselves with plants (a boutonniere, a corsage).

Our dependency on plants puts us in a vulnerable position. With soaring world population, widespread habitat destruction, rapidly diminishing natural resources, and an alarming extinction rate for the world's plant species, we are rapidly depleting and destroying the very things that supply us with so many necessities and pleasures. More now than ever, our understanding of plants, and how they grow and develop, determines not only the quality of our lives, but even the number of people that can live on earth.

## Control of Growth and Development

How does a tiny saguaro embryo the size of a grain of sand grow into the world's tallest cactus? What makes a root grow down, while stems and branches always reach up toward light? Why do some plants bloom in the spring and others wait for summer, fall, or even winter? And why, when farmers sow the best quality seed, do crops sometimes fail entirely or yield only a fraction of their potential?

Questions such as these do not have simple answers, for the way a plant grows (gets larger) and develops (becomes more complex) results from complicated interactions between three levels of control.

- *Intracellular level* (within a cell)— genetic control over growth and development where certain genes in a cell operate at precise times to code for specific activities and characteristics. (How genes control activities and characteristics is discussed in Chapters 27 and 28.)

- *Intercellular level*—control of growth and development by hormones, potent chemicals produced and transported between plant tissues. Hormones coordinate growth and development in different parts of an organism by triggering cellular reactions in target cells, including determining which genes are expressed.

a

b

**Figure 22-1**
**Nature versus nurture.** The environment profoundly influences how a plant grows. *(a)* Persistent ocean winds have produced this picturesque-shaped Monterey cypress growing near the shoreline. *(b)* At high elevations, continuously strong and frigid winds kill axillary buds on the windward side of stems so that branches grow only from leeward axillary buds, giving a "flag-form" appearance to this alpine fir.

• *Environmental level*—control of growth and development by the environment (includes many factors, such as the direction or intensity of light, amount of moisture, acidity of rain, extremes in temperature, availability of minerals, and so on). The environment plays a critical role in molding plant growth and development (Figure 22-1).

# Plant Hormones — Coordinating Growth and Development

To date, five groups of plant hormones have been discovered—auxins, gibberellins, ethylene gas, cytokinins, and abscisic acid. Not all plant hormones fit the general definition for a hormone: a chemical synthesized in one part of an organism that stimulates or inhibits a specific response in target tissue elsewhere in the organism.

Unlike most animal hormones that trigger highly specific responses in specific target tissues (Chapter 16), plant hormones work in more general, and often more complex, ways. For instance,

• One plant hormone may initiate a response in one part of the plant and an entirely different response in another part.

• The effects of plant hormones often change as the concentration of the hormone increases or decreases.

• The effects of a plant hormone may vary, depending on the time of year or the developmental stage of the plant.

• Since many plant hormones work together to trigger growth and development changes, combinations of hormones evoke responses that are different from what each hormone would produce alone.

These variables make working with plant hormones one of the most challenging research areas in botany. They are also the reasons why researchers have not yet answered all of the questions concerning the role of plant hormones in plant growth and development.

## Auxin

As a plant grows, its increase in size is the result of cell elongation and enlargement, and the production of new cells by mitosis. **Auxins** are a group of growth hormones that promote cell elongation by softening cell walls. This allows internal turgor pressure to stretch the wall, thereby enlarging the cell.

Auxins influence plant growth and development in a number of ways, including

• Causing stems to bend. Unequal distributions of auxins in a stem produce unequal cell elongation. Stem bending enables plants to become favorably oriented to their environment. For example, stem bending allows newly germinated horizontal seedlings to grow up through the soil (Figure 22-2) and enables shaded plants to grow toward light (Figure 22-3).

• Stimulating lateral and adventitious root development for increased water and mineral absorption.

• Stimulating divisions of vascular cambium cells for increased production of secondary vascular tissues.

• Inducing cells to differentiate into xylem to increase water and mineral transport, and to add reinforcement for greater support.

• Promoting the development of flowers and fruits.

Several chemicals are classified as auxins. One, **indoleacetic acid (IAA)**, was the first plant hormone to be isolated (see Discovery: Plant Hormones). It took more than 9000 kilograms (20,000 pounds) of oat meristems (a center of IAA production) to obtain a single gram (0.035 ounces) of the hormone, less than the weight of a paper clip. These results testify to the potency of hormones; they are active in extremely small amounts.

IAA is responsible for a familiar phenomenon known as **apical dominance**, a growth pattern in which shoot tips prevent axillary buds from sprouting (Figure 22-4). The IAA produced in the apical meristem moves down the stem and inhibits axillary buds from growing into new leafy stems. As horticulturists and gardeners will tell you, pinching off the tips of stems stimulates a plant to produce more leafy stems and become bushier. In other words, cutting off apical meristems removes the IAA production center, allowing axillary buds to begin development.

Synthetic auxins manufactured in laboratories are called *growth regulators*. Although chemically different, growth regulators have the same effect as the natural auxins produced by the plant. The active ingredient in Agent Orange, the defoliant used in Vietnam, is a potent synthetic auxin that quickly kills plants by disrupting normal growth. Unfortunately, Agent Orange also contained the contaminant Dioxin, which is now known to be one of the most toxic substances for humans.

## Gibberellins

**Gibberellins** include more than 50 compounds. All promote growth by stimulating both cell elongation and cell division (Figure 22-5). When gibberellins are applied to dwarf plants, normal growth is restored; this suggests that dwarfs lack the gene needed to manufacture their own supply of gibberellins. Normal stem growth, however, is not regulated by gibberellins alone; auxins and cytokinins (discussed below) are also involved.

In addition to affecting stem growth, gibberellins are critical to embryo and seedling development. In barley, gibberellins mobilize food reserves stored in the seed's endosperm by stimulating the **aleurone layer** (the outer row of endosperm cells) to synthesize a starch-hydrolyzing enzyme (Figure 22-6). Other functions of gibberellins include inhibiting seed formation, stimulating pollen tube growth, increasing fruit size, initiating flowering, and breaking dormancy in seeds and axillary buds.

## Ethylene Gas

Some enterprising entrepreneurs have made a great deal of money from an

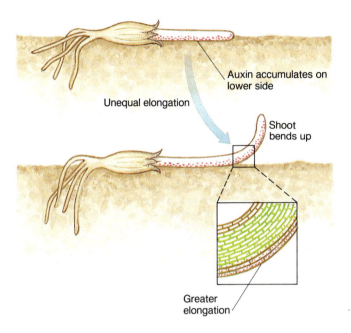

**Figure 22-2**
**Which way is up?** When a seedling is on its side, auxin accumulates on the lower side of the shoot, increasing cell elongation there and causing the shoot to grow up.

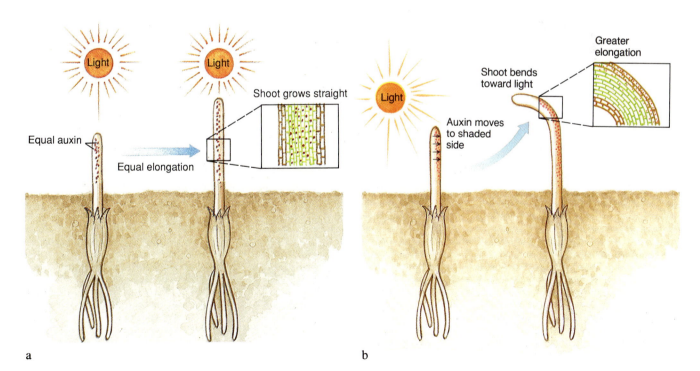

a

b

**Figure 22-3**
**Light and stem growth.** *(a)* Shoots grow straight when light comes from above because auxin is evenly distributed. *(b)* Shoots grow toward light because auxin is transported to the shaded side, increasing cell elongation. In this way, a plant can grow out from under the shade of another plant.

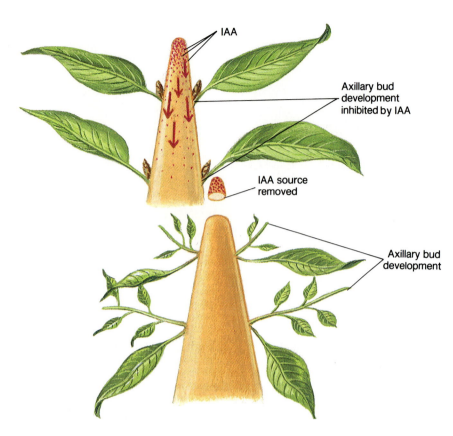

**Figure 22-4**
**Apical dominance.** *(a)* IAA, produced by apical meristems, moves down the stem and inhibits axillary bud growth. *(b)* When the stem is clipped, the source of IAA is removed and axillary buds grow into new stems.

IAA

Axillary bud development inhibited by IAA

IAA source removed

Axillary bud development

**Figure 22-5**
**Gibberellin-enhanced growth.** Applications of gibberellins to a cabbage plant cause skyrocketing vertical growth due to rapid enlargement of internode cells.

# D I S C O V E R Y

## PLANTS HAVE HORMONES

A little more than 100 years ago Charles Darwin performed some simple experiments that suggested plants may produce growth regulators, or hormones. The following outline briefly charts the history of major discoveries in plant hormone research.

| Investigator | Experiment | Conclusions |
|---|---|---|
| 1880<br>C. Darwin | | Unidirectional light causes grass shoots to bend toward the light.<br><br>Covering or removing the shoot tip stops bending; suggests the factor that causes bending is in shoot tip. |
| 1910<br>P. Boysen-Jensen | | When a block of gelatin is inserted between a severed tip and stump, the shoot still bends; suggests factor is a chemical that passes through gelatin.<br><br>Mica inserted on the dark side stops bending, but shoot bends normally when mica is inserted on lighted side; suggests factor promotes elongation on dark side, but does not inhibit elongation on light side. |
| 1918<br>A. Paal | IN DARK | Displaced tips cause bending in the opposite direction; suggests factor in tip affects cells directly below. |
| 1926<br>F. Went | a b c<br>d e | METHODS<br>  a. Removal of tip stops shoot growth.<br>  b. Plain agar block on excised tip does not cause growth.<br>  c. If agar block contains juice from crushed tip, shoot elongates.<br>d.–e. Off-set agar blocks + juice cause bending. Bending is proportional to amount of juice.<br><br>CONCLUSIONS: chemical produced in tip causes shoot to grow and bend in proportion to its concentration. Went names the chemical *auxin*. Auxin moves away from light, explaining bending toward unidirectional light. |
| 1926<br>E. Kurosawa | Infected rice | DISCOVERY: GIBBERELLINS. Rice infected with fungus, *Gibberella fujikuroi,* grows taller than noninfected plants. Purified extract from fungus stimulates cell elongation and division. |
| Up to 1930<br>Various | | DISCOVERY: ETHYLENE. Many observations that insense, smoke, and gas for lighting caused fruit to ripen faster. By 1930, ethylene gas component was determined to cause this effect. |
| 1941<br>J. van Overbeek | cork formation around cut | DISCOVERY: CYTOKININS. Coconut milk stimulates growth of embryos on plain agar. Injuries stimulate new cell growth around wounds. Active ingredient for both responses is the same. Since factor causes cell division, it is named cytokinin. |
| 1965 | dormant bud contains ABA | DISCOVERY: ABSCISIC ACID. Factor isolated from dormant buds and seeds causes growing tissues to stop growing. Many tests confirm that abscisic acid almost always inhibits growth. |

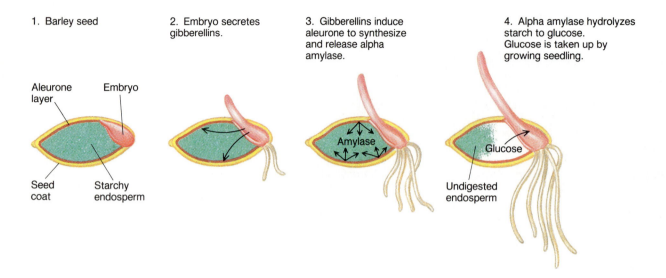

1. Barley seed

2. Embryo secretes gibberellins.

3. Gibberellins induce aleurone to synthesize and release alpha amylase.

4. Alpha amylase hydrolyzes starch to glucose. Glucose is taken up by growing seedling.

Aleurone layer

Embryo

Seed coat

Starchy endosperm

Amylase

Undigested endosperm

Glucose

**Figure 22-6**
**Getting food out of storage.** Gibberellins released during seed germination mobilize the endosperm's food reserves, providing nutrients to the seedling until its shoot emerges from the soil and leaves begin to photosynthesize.

unusual plant hormone, **ethylene gas**. Their advertisement claims that when green fruits are placed in a "specially designed" bowl, they ripen two or three times faster than normal. This is true. However, you can accomplish the same thing by putting the fruit in any closed container, including an ordinary paper bag. The covered bowl (or bag) merely traps ethylene gas released from the ripening fruit. Ethylene gas hastens fruit ripening, and as fruits ripen, they produce even more ethylene gas. Enclosing fruits, then, sets up a positive feedback system that causes them to ripen faster and faster. It is production of ethylene gas by overripe fruit that explains why "one rotten apple can spoil the whole barrel."

The discovery that ethylene gas stimulates fruit ripening has dramatically lowered the cost of shipping fruits to markets. Tomatoes, bananas, apples, oranges, and many other fruits are usually picked green (reducing the chance of spoilage), transported in ventilated crates (to prevent ethylene gas buildup), and then gased with synthetic ethylene at distribution centers for last-minute ripening.

In addition to its role in fruit ripening, ethylene gas causes stems to grow thicker in response to mechanical stress (Figure 22-7). Trees that have been staked while growing, and thus protected from bending stresses caused by wind, usually have weak trunks and stems. On the other hand, plants allowed to bend in the wind develop much sturdier stems.

Ethylene gas interacts with other plant hormones to elicit specific growth responses. For example, ethylene gas plus abscisic acid (discussed below) controls leaf fall; ethylene gas plus auxin initiates flowering in mangoes and bromeliads (including pineapples); and ethylene gas plus gibberellins controls the ratio of male and female flowers on some monoecious plants.

## Cytokinins

**Cytokinins**, like auxins and gibberellins, are growth-promoting hormones. Cytokinins stimulate rapid cell division.

In combination with auxins and gibberellins, cytokinins help orchestrate cell division in meristematic re-

gions so that stems and roots grow normally. Depending on the species, cytokinins also promote cell enlargement in young leaves, eliminate apical dominance, control flowering and fruiting, and help control shoot and root development in tissue masses used to clone plants by tissue culture (see Chapter 39, The Biotechnology Revolution).

Finally, cytokinins delay **senescence**—the aging and eventual death of an organism, organ, or tissue. Cytokinins delay senescence of leaves, flowers, and fruits by increasing nutrient transport to them.

## Abscisic Acid

Not all plant hormones promote growth. **Abscisic acid** (abbreviated **ABA**) almost always inhibits growth by either reducing the rate of cell division and elongation, or halting them altogether.

Suspended growth (dormancy) can be a life-saving response to severe environmental conditions. With repressed growth, dormant plants are better able to withstand the stresses of dry, hot periods, or freezing temperatures.

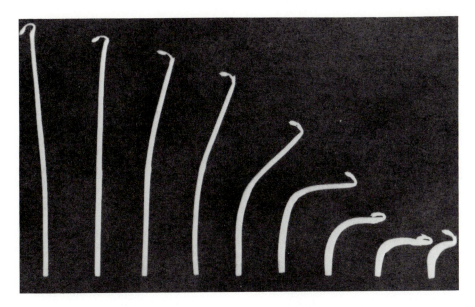

**Figure 22-7**
**Ethylene gas and the emergence of seedlings.** While pushing through soil, stresses stimulate ethylene gas production in young seedlings. The bigger the obstacle, the greater the stress and the more ethylene gas produced. As shown here, increased levels of ethylene gas cause the stems of these pea seedlings to thicken and curve. Because of this response, a seedling is better able to force its way through the soil and emerge.

During fall and winter, levels of abscisic acid remain high in the axillary buds of deciduous trees, plants that drop their leaves during unfavorable periods. High abscisic acid levels counteract growth-promoting auxins and gibberellins, preventing cell division and elongation at a time that would expose delicate growing tissues to lethal temperatures. ABA concentrations fall at the onset of favorable growing conditions, releasing buds from their dormancy and causing plants to burst out with new growth.

When abscisic acid accumulates in guard cells during periods of water shortage, it causes stomates to close. This enables the plant to help recover its water balance.

Abscisic acid was given its name because early investigators believed it caused flowers, fruits, and leaves to fall. This process is called **abscission**. The hormone retains its name, even though there is uncertainty over whether ABA is involved in abscission at all.

## Other Hormones?

Investigators have been alert for evidence of other plant hormones. For example, experiments strongly suggest that a substance produced in leaves stimulates development of flower buds. Although this elusive flowering chemical has yet to be isolated, it has been named **florigen**.

In addition to the possibility of a flowering hormone, some investigations suggest the existence of a root growth hormone. However, like the flowering hormone, the root-forming hormone has not yet been isolated.

Recently, scientists have discovered that some plants produce chemicals that act like insect growth hormones. These plant chemicals mimic the insect's hormones so closely that they disrupt the insect's normal growth and development, preventing them from reaching reproductive maturity (see Bioline: Biochemical Warfare).

## Timing Growth and Development

In most parts of the world, spring triggers a lavish surge of plant growth and development. Seeds are jolted out of dormancy, ghostly deciduous trees erupt with new leaves, and branches become crowded with masses of blossoms. But not all plants bloom in early spring. Some wait until late spring, summer, or even fall before producing flowers, or before their seeds germi-

nate. Clearly, plants are able to synchronize their growth and development with the changing seasons, some with great precision. Indian tribes in Arizona based their yearly calendar on the punctual flowering and fruiting of the saguaro cactus; a new year was marked by the appearance of the first saguaro fruits.

In addition to yearly (seasonal) cycles, plants also have daily cycles. For example, some flowers open only at certain times of the day—morning glory (*Ipomoea purpurea*) at 4 o'clock in the morning, fire weed (*Epilobium angustifolium*) at 6 o'clock in late afternoon, nightshades (*Datura*) at 7 o'clock in the evening, and the queen of the night (*Selenicereus grandiflora*) at 9 o'clock at night. Carl von Linné (Linnaeus), the Swedish scientist who introduced the binomial system for naming organisms (Chapter 1), established a "flower clock" in the Botanical Gardens of Uppsala, using plants that bloomed throughout the day and night to indicate the time of the day.

But how do plants "know" the time of day or season of year? Part of the answer lies in the ability of plants to detect and respond to environmental changes, using temperature and the length of daylight, or **photoperiod**, as the indicator of a season. The other part of the answer lies inside the plant itself, where a *biological clock* measures the passage of time. Together, temperature, photoperiod, and biological clocks regulate cell metabolism and hormonal production, so that plant growth and development remain synchronized with changes in the environment.

## Temperature and Growth

Plants grow in nearly all parts of the world, including Death Valley, California, where the highest air temperature ever recorded reached 57.8° C (136° F), and Siberia, USSR, where the world's lowest recorded temperature was −87° C (−126° F).

Each plant species has a particular range of temperatures it can tolerate, including an optimum temperature at which it grows best, and maximum and minimum temperatures it can tolerate for limited periods of time. Where a plant grows, then, is partly determined by the prevailing temperatures and the length of temperature extremes.

Because plant responses are so sensitive to temperature, wide temperature fluctuations or prolonged exposure to extreme temperature can threaten a plant's life. However, a dormant plant is able to withstand greater temperature extremes for longer periods. On the days the highest and lowest air temperatures were recorded in Death Valley and Siberia, most of the plants were dormant. Had they been actively growing, vulnerable meristems and primary tissues would have been killed, jeopardizing the life of the entire plant.

Plants don't instantly become dormant in response to adverse conditions. Dormancy requires preparations, including building stores of food, water, and other nutrients, and, in some cases, growing protective structures (such as bud scales) to shield delicate tissues. To complete preparations in time, plants must be able to detect the approach of unfavorable growing conditions. Dormant plants must be able to respond when conditions are again favorable, so they may return to active growth and development. But how are plants able to "anticipate" these changes in weather?

Because temperature is closely tied to other weather factors, gradual temperature changes act as a predictor for upcoming weather. By responding to temperature trends, plants synchronize growth and development to upcoming weather.

Some plants must be exposed to cold temperatures before they will flower or before their seeds will germinate. The promotion of flowering or germination by cold treatment is called *vernalization.* Pitted fruit trees (peaches, apricots, plums, etc.) and apple trees fail to produce flowers and fruits unless they are exposed to cold temperatures for precise numbers of hours (ranging from 250 to more than 1200 hours, depending on the species). Similarly, many seeds germinate only after they are exposed to near-freezing temperatures for several weeks. Under natural conditions, required periods of cold indicate that bleak winter conditions have passed.

## Light and Growth
### PHOTOPERIODISM

In addition to responding to temperature cues, plants "anticipate" changes in growing conditions by responding to variations in photoperiod, the relative length of light during 24 hours. Except at the equator, all areas of the earth experience yearly changes in photoperiod. In the Northern Hemisphere, the period of daylight gradually lengthens until late June and then gradually shortens until early December. Changes in photoperiod are so reliable that on any given day of the year, the length of daylight and night is nearly constant from one year to the next in a particular region.

Plants monitor photoperiod so that the plant's state of growth and development is appropriate for the season. The ability of organisms to respond to changes in photoperiod is known as **photoperiodism.** Photoperiodic responses in plants include initiating dormancy, flowering, germination, and stem and root development.

### Phytochrome

Plant photoperiodic responses are controlled by a light-absorbing pigment named **phytochrome** (phyto = plant, chrome = coloring agent). Phytochrome exists in two forms, each form absorbing different wavelengths of light. One form of phytochrome absorbs red light at wavelengths of 660 nanometers and is called $P_r$ (*Phytochrome r*ed). The other form absorbs longer, far-red light at wavelengths of 730 nanometers and is called $P_{fr}$ (*Phytochrome f*ar-red). $P_{fr}$ is the active form of phytochrome; it promotes or inhibits a photoperiodic response.

Phytochrome flips back and forth between its two forms (Figure 22-8): when $P_r$ absorbs red light it immediately converts to $P_{fr}$, and when $P_{fr}$ absorbs far-red it immediately converts to $P_r$. To make things even more complicated, $P_{fr}$ is destroyed by enzymes or converted into $P_r$ in the dark.

The conversion of phytochrome enables plants to measure the length of daylight and night during a 24-hour period. At sunrise and throughout the day, red light predominates, so phytochrome exists as $P_{fr}$. The first appearance of $P_{fr}$, then, marks the beginning of a new day. At sunset, far-red light predominates, so phytochrome exists as $P_r$; the appearance of $P_r$ denotes the end of daylight. Since $P_{fr}$ accumulates in the plant during daylight, and $P_r$ accumulates in the dark, the proportion of $P_{fr}$ to $P_r$ enables plants to chemically measure the relative length of daylight and darkness.

How phytochrome stimulates or blocks photoperiodic responses is still not known. Some studies suggest that phytochrome is embedded in cell membranes and controls the passage of hormones or other molecules that alter cell activity.

### Photoperiod and Flowering

It is crucial for plants to time flowering, so that it coincides with favorable conditions that last long enough for flowers, fruits, and seeds to mature before severe weather returns. Flowering must also coincide with the activity of pollinators.

Plants are classified into three groups, depending on how changes in daylength affect when they flower:

• **Short-day plants** flower in late summer or fall when the length of day-

# B I O L I N E

## BIOCHEMICAL WARFARE: PLANT CHEMICALS THAT MIMIC ANIMAL HORMONES

How do plants protect themselves from the voracious appetites of insects and other herbivores? Some produce or store toxic chemicals that quickly kill the hungry beasts. Others form tangles of spines and thorns that, even though they may not kill the diners outright, certainly make it difficult or very uncomfortable for them to graze casually. But at least 125 species have evolved another, more subtle way of deterring foragers. These plants actually tamper with the growth hormones of certain insects that try to eat them, disrupting the insects' own pattern of growth and development.

Certain firs, spruces, yews, and a few other plants successfully duplicate some of the insect's own hormones that control its metamorphosis. Insects that undergo metamorphosis change from one form to another, essential changes that produce a reproductively competent adult. (A caterpillar changes into a butterfly, for example.) But when insects eat these plants, the "mimicking hormone" disrupts its metamorphosis and the insect fails to change into a reproductive form. Since the insects cannot reproduce, the population falls, and the herbivore problem disappears. Some scientists are investigating whether these hormone "lookalikes" can be used as insecticides for controlling agricultural pests.

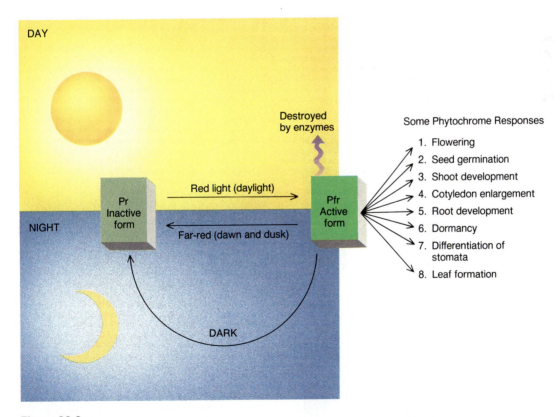

**Figure 22-8**
**Phytochrome and development.** The active form of phytochrome, $P_{fr}$, is formed in daylight, when $P_r$ absorbs red light. $P_{fr}$ reverts to inactive $P_r$ at dawn, dusk, and in the dark, or is destroyed by cell enzymes. The length of the dark period largely determines how much active $P_{fr}$ remains to stimulate or inhibit various plant responses.

light becomes shorter than a specific, critical period (Figure 22-9a). Short-day plants include strawberries, violets, sages, dahlias, asters, and chrysanthemums.

- **Long-day plants** flower in the spring when the length of daylight becomes longer than a critical period (Figure 22-9b). Larkspurs, spinach, wheat, lettuce, and potatoes are examples of long-day plants.

- **Day-neutral plants** flower independently of daylength. Although other factors such as temperature and water supply may affect flowering, most day-neutral plants flower only when mature enough to do so. Day-neutral plants include carnations, roses, snapdragons, sunflowers, and, as most gardeners know, many common weeds.

The difference between short- and long-day plants depends on whether the critical period of daylight represents a maximum length (short-day plants) or a minimum (long-day plants). For example, hibiscus *(Hibiscus syriacus)* and barley *(Hordeum vulgare)* are both long-day plants. Their critical daylength period is 12 hours; days longer than this *minimum* daylength promote flowering. On the other hand, rice *(Oryza sativa)* is a short-day plant. Even though rice has the identical critical daylength as hibiscus and barley (12 hours), only a daylight period shorter than 12 hours initiates flowering. In other words, 12 hours is the *maximum* daylength for rice.

Thus, the actual number of hours of critical daylength does not determine whether a plant is a short-day or long-day plant. Neither does the critical number of hours mean that short-day plants flower only when days are short, and long-day plants only when days are long. For example, chrysanthemums are short-day plants with a critical daylength of 15 hours, whereas oats *(Avena sativa)* are long-day plants with a critical daylength of only 9 hours.

Phytochrome is the key to detecting changes in daylength; the formation and disappearance of $P_r$ and $P_{fr}$ signal the beginning and end of daylight and darkness. These signals help to set a natural timing system within plant cells. After a great deal of research, investigators learned that this timer actually runs at night. Thus, instead of daylength being the critical factor as the terms "long-day" and "short-day" plants" suggest, the critical period is actually the length of darkness or night.

The cell's "dark timer" works like a stopwatch. The timer starts running when $P_{fr}$ begins to disappear at sunset and is stopped when $P_{fr}$ begins to be formed at sunrise. If the dark timer has run for the critical length of time, then a message (florigen?) is sent from the leaves to the shoot apex, stimulating axillary buds to develop into flowers (Figure 22-10). Long-day and short-day plants differ in whether the timer triggers or inhibits the formation of this flowering message.

## Phytochrome and Shoot Development

Light affects the development of a shoot from the moment a seedling nudges its way through the soil surface. Actually, even before a seedling emerges from the soil, the absence of light has already had an impact on

a. Poinsettia, a short-day plant

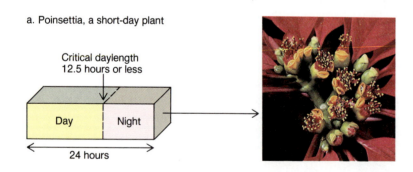

b. Hibiscus, a long-day plant

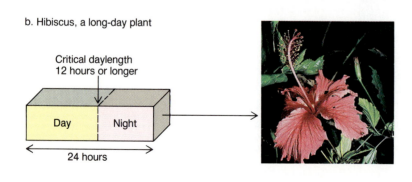

**Figure 22-9**
**Photoperiod and flowering.** Short-day and long-day plants flower in response to changes in photoperiod. *(a)* The poinsettia is a short-day plant, flowering when the critical daylength becomes *less* than 12.5 hours. *(b)* The hibiscus is a long-day plant, flowering when the critical daylength becomes *greater* than 12 hours.

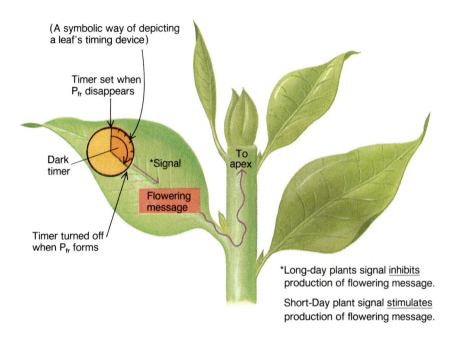

(A symbolic way of depicting a leaf's timing device)

Timer set when $P_{fr}$ disappears

Dark timer

Timer turned off when $P_{fr}$ forms

*Signal

Flowering message

To apex

*Long-day plants signal <u>inhibits</u> production of flowering message.

Short-Day plant signal <u>stimulates</u> production of flowering message.

**Figure 22-10**
**Regulating flowering in long-day and short-day plants.** Plant leaf cells have a "dark timer" that measures the length of darkness. When the night reaches a critical length, the signal either inhibits (long-day plants) or stimulates (short-day plants) the production of one or more hormones that convey a flowering message to the stem, promoting flower bud development.

shoot development. Underground, the shoot elongates rapidly, undeveloped leaves stay small, the shoot hook (if present) remains bent, and the plant remains colorless, devoid of chlorophyll. These features refer to a condition called **etiolation**.

The light-induced appearance of $P_{fr}$ signals the emergence of the shoot from the soil. $P_{fr}$ stimulates hormonal changes that slow down stem elongation, cause the shoot to straighten, stimulate leaf enlargement, and begin chlorophyll synthesis.

Shade affects stem growth by changing the production of $P_{fr}$. A plant growing in the shade receives mostly far-red light, red light having been absorbed by the leaves of taller plants. Far-red light reduces the amount of $P_{fr}$. (When $P_{fr}$ absorbs far-red light, it converts to $P_r$.) Low $P_{fr}$ concentrations boost stem elongation, increasing a plant's chances of growing out from under the shade.

## PHYTOCHROME AND SEED GERMINATION

Some seeds will not germinate in the dark, which means, oddly enough, that they will not germinate under the ground either. However, when these seeds are exposed to light they germinate. (You must continually pull weeds from your garden because the seeds of many weedy plants germinate only after exposure to light, which, ironically, is what they get when you turn the soil to dig them up.) Experiments show that a short flash of red light (but not far-red light) also induces light-requiring seeds to germinate. This is a clue that phytochrome is involved. The red light causes a shift from $P_r$ to $P_{fr}$, removing the germination block.

## Biological Clocks

Plant growth and development is controlled not only by changes in temperature and light, but also by internal biological clocks. One type of biological clock, the "dark timer" mentioned earlier, measures the length of night, launching flowering in some plants.

Other plant activities are also controlled by biological clocks. For example, the leaves of some plants are oriented horizontally during daylight, but fall into a more compact vertical

"sleeping" position at night, presumably keeping the plant warmer. Even when these plants are kept in continuous light and at constant temperature, the leaves continue to fall into the "sleeping" position at exactly the same time each day. This rhythm is said to be **endogenous** (controlled internally), rather than **exogenous** (controlled by environmental rhythms outside the plant). The "sleep" movements of leaves follow a **circadian** (circa = about, diem = day) cycle, recurring at approximately 24-hour intervals. Other endogenous circadian plant activities include flower opening and closing, and nectar and flower perfume production.

## Plant Tropisms

Plants change their direction of growth in response to several environmental stimuli. These responses, called **tropisms**, occur when environmental factors induce changes in the rate of production or distribution of growth-regulating hormones. In addition to tropisms stimulated by light, gravity, or contact with objects (discussed below), other environmental stimuli produce tropic responses such as chemicals (chemotropism), temperature (thermotropism), wounding (traumotropism), and water (hydrotropism).

## Phototropism

A familiar tropism occurs in many houseplants. If left in one spot, they gradually turn their leaf surfaces toward a major source of light, usually a nearby window. Since light is the environmental stimulus, the response is called a **phototropism**. *Flavoprotein,* a blue and violet light-absorbing pigment, is believed to be responsible for phototropic responses. It promotes lateral auxin movement away from the light source, which in turn increases cell elongation in the cells on the shaded side (see Figure 22-3).

## Gravitropism

Changes in growth caused by gravity are called **gravitropisms** or **geotropisms**. For plants, gravity indicates which way is down and which way is up, enabling them to direct the growth of roots and shoots in the right direction. Shoot growth away from the gravitational force is an example of **negative geotropism**, whereas root growth toward the gravitational force is **positive geotropism**.

Investigators believe that plant cells detect the direction of gravity when small particles called *statoliths* are pulled to the bottom of the cell by gravity. The change in position of the statoliths redirects the transport of growth hormones, generating the gravitropic response. In one experiment, a growing shoot was placed on its side. Within 15 minutes, the lower side of the shoot had twice as much auxin than the upper side. This unequal distribution of auxin caused the stem to bend upward (see Figure 22-2). Once upright, statoliths settle into their original position inside the cell, and auxin concentration becomes equally distributed on all sides of the stem. Therefore, the shoot grows straight up.

In roots, the root cap plays an important role in controlling the direction and rate of root growth. When the root cap is experimentally removed,

the root grows faster than normal, but not downward. This suggests that the root cap produces a growth inhibitor that slows the growth of certain root tip cells so that growth is always downward. When a root is placed horizontally, the inhibitor accumulates in the cells on the lower side, reducing their growth relative to cells on the top side and producing downward bending (Figure 22-11). Unlike the situation in stems, root bending is probably not controlled by auxin. Although abscisic acid has been suggested as the growth inhibitor in roots, experiments have yet to confirm this role.

## Thigmotropism

A change in growth stimulated by contact with another object is called a **thigmotropism**. Vines and other plants that climb or cling to objects either have tendrils that wrap around the supporting structure (Figure 22-12a), or pads at the tips of branches that fasten a plant to a surface (Figure 22-12b). Most climbing plants have weak stems and branches that are unable to support their own weight. They rely on another plant, a wall, a fence post, or some other solid object for support.

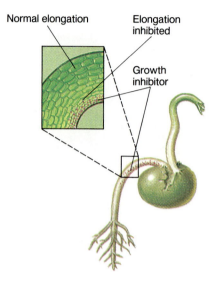

**Figure 22-11**
**Which way is down?** Gravity causes an accumulation of a growth inhibitor on the lower side of a root, slowing cell elongation relative to cells on the upper side, causing the root to turn downward.

a                                         b

**Figure 22-12**
**"Getting by with a little help from friends."** Plants with very weak stems rely on a sturdier plant or other solid object for support. The tendril of a pea plant grows around another plant for support *(a)*, while expanded pads of a virginia creeper fasten growing stems to a wall *(b)*.

# Synopsis

## MAIN CONCEPTS

- **Plant growth and development** is fixed by intracellular genetic controls, which, in turn, are regulated by intercellular hormonal controls. In plants, the environment greatly influences growth and development by affecting the production and distribution of hormones.

- **A plant hormone has more than one effect** on growth and development. These effects vary with hormone concentration, target tissue, time of year, and developmental stage of the plant. Many growth and development responses are the result of combinations of plant hormones.

- **Four of the five groups of plant hormones promote growth** (auxins, gibberellins, cytokinins, and ethylene gas), whereas one inhibits growth (abscisic acid). Although some research suggests the presence of a flowering hormone and a root growth hormone, neither has been isolated.

- **Many plants synchronize dormancy and periods of active growth** with changing seasons by responding to variations in temperature and daylength, and by using an internal biological clock. Dormancy is critical to surviving unfavorable growing conditions.

- **Interconvertible forms of light-absorbing phytochrome** enable plants to monitor changes in photoperiod, thereby synchronizing flowering, seed germination, and other photoperiodic responses with appropriate environmental conditions.

## KEY TERM INTEGRATOR

Coordinating Growth and Development

| Hormone | Action | Function |
| --- | --- | --- |
| **auxins,** includes **indoleacetic acid (IAA)** | Promotes cell elongation | Promote **apical dominance**, root development, formation and differentiation of secondary vascular tissues, flower and fruit development |
| **gibberellins** | Promotes cell division and elongation | Restore normal growth in dwarfs, stimulates **aleurone layer** to produce starch-digesting enzyme |
| **cytokinins** | Promotes cell division | Remove apical dominance, flower and fruit development, leaf enlargement, root and shoot development, delays **senescence** |
| **ethylene gas** | Increases metabolism (?) | Promotes fruit ripening, increases thickness of stems |
| **abscisic acid (ABA)** | Inhibits cell elongation and division | Promotes dormancy, stomate closure. In combination with ethylene gas causes leaf **abscission**. |
| **florigen (?)** | Stimulates flowering | *Proposed* flowering hormone |
| **root-forming hormone (?)** | Causes roots to grow down | *Proposed* hormone |

---

### Timing Growth and Development

| | |
|---|---|
| **photoperiodism** | A response of organisms to changes in **photoperiod**. Plants measure photoperiod by the presence, absence, and relative amounts of two forms of **phytochrome**. |
| daylength | **Short-day** and **long-day plants** flower in response to changes in photoperiod, whereas flowering in **day-neutral plants** is independent of photoperiod. |
| **etiolation** | Characterizes young seedlings until their shoots emerge from the soil. |
| biological clocks | Include the "dark timer" that measures the length of darkness. Twenty-four hour **circadian** sleep movements of leaves are **endogenous** and are not controlled by **exogenous** rhythms. |

### Tropisms

| | |
|---|---|
| **tropisms** | Responses of plants to environmental stimuli. Light stimulates **phototropisms**; gravity stimulates **gravitropisms**, including **positive** and **negative geotropisms**; and contact with objects stimulates **thigmotropisms**. |

---

## Review and Synthesis

1. List five reasons why researchers have so much difficulty working with and determining the functions of plant hormones.

2. Match the hormone with its function. (Multiple matches are possible.)

   ___auxins              A. fruit ripening
   ___cytokinins          B. apical dominance
   ___giberellins         C. senescence
   ___ethylene gas        D. leaf abscission
   ___abscisic acid       E. flowering

3. The seeds of some plants do not germinate in the spring, even though temperatures are warm and water is plentiful. What would prevent these from germinating? Is there any advantage to delay germination past spring?

4. How do plants time growth and development to coincide with favorable weather?

5. How do the characteristics of etiolated seedlings improve their chances of emerging from the soil?

6. List the ways phytochrome influences plant growth and development.

7. Making connections: Relate plant hormones to animal hormones discussed in Chapter 16. How are they similar, and how do they differ? What advantages and disadvantages are there for plant growth and development to be strongly influenced by the environment? Why do you think the environment has less control over animal growth and development?

## Additional Readings

Butler, W., and R. Downs. 1960. "Light and plant development." *Scientific American* 203:55–63. (Introductory.)

Hillman, W. 1979. *Photoperiodism in Plants and Animals.* Oxford Biology Readers, No. 107. Burlington, N.C. (Intermediate.)

Jensen, W., and F. Salisbury. 1984. *Botany.* 2nd ed. Wadsworth Publishing Co., Belmont, Calif. (Intermediate.)

Nickell, L. 1982. *Plant Growth Regulators—Agricultural Uses.* Springer-Verlag, New York. (Intermediate to advanced.)

Northington, D., and J. Goodin. 1984. *The Botanical World.* Times Mirror/Mosby College Publishing, St. Louis, Mo. (Introductory.)

Van Overbeek, J. 1968. "The control of plant growths. *Scientific American* 219:75–81.

Chapter 23

# Animal Reproduction

*A*ll species on earth, from bacteria to humans, follow the same formula for success. Individuals of the species perpetuate their kind by *reproducing,* leaving behind offspring that pass on their genetic characteristics from generation to generation.

## Plant Versus Animal Reproduction

In Chapter 21 you became acquainted with several variations of the reproductive formula as practiced by plants. The manner in which animals produce new individuals is, in many ways, identical to the reproductive biology of plants. But animal reproduction also differs from that of plants in several important features.

- Although some animals may reproduce asexually, most rely exclusively on sexual processes.

- Alternation of generations that characterizes plants is not part of any animal's reproductive cycle.

- Although animal species rarely interbreed, many plant species commonly interbreed with other closely related species. This fact has led to differences in the way some botanists and zoologists define a species (see Connections: The Species Concept).

- Successful reproduction in many higher animals requires that parents live long enough to care for their offspring until they can survive on their own. Only after the offspring have attained independence has the parent's reproductive "mission" been successfully completed.

## Animal Reproductive Strategies

### Asexual Reproduction

Asexual reproduction is common among the simplest animals. It generates progeny without the greater investment of energy and resources associated with gamete production, mate-seeking, and fertilization, processes essential for sexual reproduction. Even these asexually capable animals, however, periodically bear the additional expense of sexual reproduction for its one overwhelming advantage—it increases variation within the species much more rapidly than does asexual reproduction. (See Connections: Asexual versus Sexual Reproduction, Chapter 12.)

Asexual processes (Figure 23-1) produce **clones**—offspring that are identical to the parent. For example, two identical individuals are produced when an organism divides into equal halves by the process of **fission**. Although fission is usually associated with cell division of unicellular organisms, anemones and flatworms also display this reproductive talent. Some animals asexually reproduce by **budding**, a process by which offspring develop as an outgrowth of a parent. The developing animal may emerge from an external surface or from an internal surface. Some sponges form internal buds that are released when the parent disintegrates. (In contrast, offspring that develop from external buds break away without injury to parent or progeny.) Some sea stars and flatworms can *regenerate* a new individual from body fragments. Some worms self-amputate pieces of their bodies, each piece growing into a complete organism. This phenomenon is known as *epitoky.*

One type of reproduction escapes classification as strictly asexual or sexual. Although gametes are produced, **parthenogenesis** yields offspring with-

a          b          c

**Figure 23-1**
**Types of asexual reproduction in animals.** *(a) Fission.* Although usually a unicellular process, some multicellular animals, such as this anemone, reproduce by fission. *(b) Budding.* A budding hydra supporting two attached offspring. *(c) Parthenogenesis.* This brood of aphids developed inside the larger female from unfertilized eggs.

out fertilization. Through this process a haploid (1N) egg develops into a complete adult (The unfertilized egg may also spontaneously double the number of its chromosomes, forming a diploid—2N—adult.) Parthenogenesis commonly occurs among some insects, and less commonly in reptiles, amphibians, and fishes. One advantage of parthenogenesis is that it provides a method for quick proliferation during favorable conditions, a strategy that is later replaced by sexual reproduction when environmental factors are less favorable.

## Sexual Reproduction

Sexual reproduction increases biological variability within a species, thereby increasing the likelihood of producing some offspring that will survive when environmental conditions change. By combining in a single organism traits from two genetically distinct parents, each of the offspring acquires a unique genetic mix. This array of traits distinguishes it from other organisms of the same species, even from others produced by the same two parents. Variation is encouraged by crossing over during meiosis (Chapter 12) and gene mixing during *fertilization*, the fusion of two haploid nuclei. Fertilization in animals combines the genetic material from a *sperm* (the male gamete) and an *ovum* (the female gamete, the egg) to form a *zygote*, the diploid cell from which the new individual will develop. Each parent therefore contributes only half of the offspring's genes.

Sex requires animals to possess some additional equipment that is not needed for asexual reproduction, for example, gamete-producing structures called *gonads*. In many animals, specialized organs are needed for bringing the gametes together. In addition, many animals must devote additional energy and nutrients to assure that at least a few of their progeny survive to reproductive age. Rarely do asexually produced progeny receive, or need, any parental resources other than those spent for manufacturing the new organisms. Although sexual reproduction is biologically expensive, organisms not only make these expenditures, but may also devote a major portion of their resources to it. In some

invertebrates, for example, the gonads comprise more than half the body weight (Figure 23-2). Sex is undoubtedly important to the survival of species, even in animals that retain the option to reproduce asexually.

## FERTILIZATION STRATEGIES
The animal kingdom is characterized by a rich variety of sexual methods.

---

**CONNECTIONS** {
Morphological Species (page 11)
Defining Moneran Species (page 559)
Ecological Species (page 694)
}

### THE SPECIES CONCEPT: ANIMAL VERSUS PLANT SPECIES
Some biologists define a species solely on the basis of whether individuals can successfully reproduce fertile offspring. This approach, often called the *biological species concept,* defines a species as a group of individuals that can interbreed and produce living, fertile offspring. In other words, each species remains *reproductively isolated* from other species. Members of one species cannot perpetuate their own kind by interbreeding with individuals of another species.

The biological species concept works well for defining animal species. Interbreeding between animals of different species is often impossible for the following reasons:

- *structural differences* prevent copulation (for example, male and female genitals do not fit);
- *differences in courtship or mating rituals* prevent identification of a mate;
- *differences in mating season* prevent contact between individuals during reproductive periods;
- *gametes can not fuse* so there is no fertilization; or;
- *production of sterile offspring,* preventing hybrid progeny from reproducing.

Because of these reproductive barriers, members of different animal species cannot successfully interbred, even if they live in the same area.

Unlike animals, however, many plant species are not reproductively isolated, so members of one species commonly hybridize with members of another closely-related plant species. Although the chances of producing fertile offspring may not always be 100%, two morphologically distinct species may still be 75%, 50%, or even 10% interfertile, raising the question "at what percent should taxonomists draw the line between different plant species?" With interbreeding and varying degrees of interfertility, the biological species concept does not work well for defining many plant species. As a result, botanists generally rely more heavily on genetically-controlled morphological traits and evolutionary relationships to distinguish one plant species from another.

The contrast in number of species in the plant and animal kingdoms (450,000 versus 1,300,000, respectively), then, may simply reflect differences in how plant and animal species are defined.

---

Each method has the same purpose: to increase the likelihood that sperm and egg will meet. The simplest strategy, one employed by some jellyfish, sea star, marine worms, and oysters, is to discharge sperm and eggs into the surrounding water, where **external fertilization** occurs whenever the two encounter each other. The likelihood of fertilization is therefore a matter of

Gonads

**Figure 23-2**
**Huge reproductive organs** dominate the an[a]tomy of this jellyfish, testifying to the importan[c]e of sexual reproduction.

chance, a probability that increases as more gametes are produced and as gamete release is synchronized. External fertilization therefore requires the production of enormous numbers of gametes. A single oyster, for example, releases more than 100 million eggs each season. Only a tiny fraction of these will be fertilized and only a tiny fraction of the fertilized eggs will become progeny oysters. If each of these eggs were fertilized and produced an equally prolific adult, in five generations the population of oysters would occupy eight times the volume of the earth.

Nonetheless, resources can be conserved by increasing the efficiency of external fertilization, thereby reducing the number of gametes needed to assure reproductive success. Animals that synchronize the discharge of sperm and eggs concentrate their gametes in the same place at the same time. The precise moment of gamete release is synchronized by chemical or behavioral cues that trigger the coordinated discharge of gametes (see **Animal Behavior II: Sexual Rituals**).

Fertilization occurs only in watery environments, through which sperm must swim to reach and fertilize an ovum. Therefore, animals that mate on dry land must rely on **internal fertilization**, which occurs inside a chamber in the body where a local aquatic environment can be maintained. Internal fertilization has an-

other important advantage that makes it useful even for animals that live in water: it improves the odds of gametes encountering each other, thereby reducing the need for producing large numbers of gametes that will be swept away by currents. This reduces the number of eggs needed from millions to a few. In many internal fertilizers, sperm is transferred using an intrusive structure called a **penis**, which releases male gametes into the female's sex receptacle, a **vagina**. Other internal fertilizers lack these copulatory organs. The male octopus, for example, stores sperm in one of its eight arms. When inserted into the female, part of the arm breaks off and eventually ruptures, showering the eggs with sperm. Afterwards the male regenerates the lost appendage.

As with synchronized external fertilization, timing is important for the success of internal fertilization, since sperm must be present when mature ova are available. In addition to courtship behaviors, fertilization is also encouraged by physical attributes that promote sexual recognition and arousal of reproductive behavior in a potential mate. These physical properties, called **secondary sex characteristics**, advertise an animal's gender to another of the same species but of the opposite sex (Figure 23-3). Secondary sex characteristics are induced by sex hormones.

In many invertebrates, one animal

may possess the **primary sex charac**teristics (external genitals and gonads) of both male and female, making it a **hermaphrodite**. Although hermaphrodites can fertilize themselves, most can only *cross fertilize,* that is, transfer sperm from one individual to another. The exchange is typically reciprocal, with each individual in the pair fertilizing the eggs of the other. Examples of these sexual variations are illustrated in Figure 23-4.

Regardless of strategy, normal fertilization requires a single sperm to release its haploid set of chromosomes into the egg. After sperm cells bind to the jelly coat surrounding the egg, they enzymatically digest through this material until they reach the *vitelline membrane* (the ovum's plasma membrane). Although many sperm attach to sperm-specific receptor sites on the membrane, only one sperm cell penetrates the membrane and enters the cytoplasm through the egg's plasma membrane. Immediately after penetration, the vitelline membrane is transformed into a *fertilization membrane* that prevents entrance of other sperm attached to the jelly coat (Figure 23-5).

## AFTER FERTILIZATION
Although internal fertilization reduces the number of gametes needed, more parental resources are required to satisfy the needs of the developing embryo. The embryos of most external

**Figure 23-3**
**Gorgeous exhibitionist.** This male kingfisher's feathered display is a secondary sex characteristic used to inform a potential mate of its species, gender, and sexual intent.

**Figure 23-4**
**Sexual variety.** *(a) External fertilization.* Male and female salmon discharge sperm and eggs simultaneously and in the same location, enhancing the probability of encounters between opposite gametes in the open water. *(b) Hermaphroditic internal fertilization.* Entwined, these two slugs exchange sperm discharged from the blue penis that extends from each partner's head. This is cross fertilization between individuals having both male and female sex organs. *(c) Internal cross fertilization* between male and female.

a

c

b

# ANIMAL BEHAVIOR II

## SEXUAL RITUALS

"Timing is the thing; it's true. It was timing that brought me to you."

These words from the popular song, "Timing," are far more universal than the "boy-meets-girl" situation that the songwriter was describing. Sexual reproduction among virtually all animal species requires some form of timing. Timing is needed to

- Coordinate the sexual advances of one partner with the receptivity of another.

- Synchronize the release of gametes (for external fertilizers), so that sperm and eggs are discharged at approximately the same time in the same area.

- Promote ripening of the female's ova with the onset of reproductive behavior in males.

- Help mates locate and recognize each other.

- Assure that offspring arrive during the season when conditions are most favorable for their survival.

Reproductive timing coordinates the courtship behavior of many animals. The spotted newts in the accompanying photograph, for example, are engaging in courtship behavior that culminates in both partners releasing their gametes simultaneously. Such synchronized gamete release increases the likelihood of fertilization. Male sticklebacks display an especially elaborate courtship routine to elicit a sexual response from a female. He builds a nest and then performs a zig-zag dance that attracts a female. After she enters the nest, the male strokes her back; she responds to his touch by laying her eggs and leaving the nest. He then enters the nest, discharges his sperm over the eggs, and spends the next few weeks tending the fertilized eggs until they hatch.

The male stickleback carried out all these activities without being taught, nor was he copying the behavior of another fish. He inherited this pattern of courtship behavior. This emphasizes an important evolutionary principle: not only must animals possess the biological equipment needed for sex, but most must also inherit an ingrained set of behavioral instructions for proper use of the equipment. The stunning tail feathers of the peacock would have no reproductive consequence if he didn't know how and when to spread these feathers. The behavior is automatic. When a peahen is nearby he spreads and shakes his tail feathers. This triggers the female to behave according to her inherited instructions, culminating in a reproductive encounter and fertilization of her ova.

The songs of crickets and cicadas, the blinking of fireflies, and the nocturnal croaking of frogs are all inherited behaviors that help mates identify each other as members of the same species and potential sex partners. This helps prevent nonreproductive accidents, such as mating between members of different species or between members of the same sex. Accuracy in mate selection is often reinforced by the presence of visual labels that distinguish the genders. Only the male cardinal, for example, displays the brilliant red coloration that we usually associate with this species. The drab female responds to this gender display with sexual recognition.

Such visual displays are often accompanied by chemical signals that not only serve to help mates recognize each other, but act as powerful sex attractants as well. Some **pheromones** (chemicals that, when released by an animal, elicit a particular behavior in other animals) have these reproductively valuable effects. They are released into air or water, producing a gradient of smell that, in some species, can be detected by a potential mate from a distance of several miles (see the moth on page 240, Chapter 15).

The right smell or the correct approach not only promotes fertilization, but can also protect a mate from being mistaken for a potential meal. Male spiders, which tend to be smaller than the female, could be consumed by his potential mate before fertilizing her eggs. He must approach not only cautiously, but also with some behavior that identifies him as a mate, not a meal. Some spiders vibrate the female's web in a rhythm that the female recognizes as an approaching male. Some male spiders placate their potentially dangerous mates by providing a gift, usually a captured meal wrapped in silk. She usually accepts the gift and allows him to copulate with her. Males occasionally try to trick a female into accepting an empty package. When their deceit is discovered, the "empty-handed" male becomes the gift—he is immediately eaten.

The seasonal nature of reproductive behaviors is striking; for many species, they occur at the same time every year. The onset of these behaviors is often coordinated by some specific environmental change, one that informs the animal's nervous system that the mating season has arrived. For many animals this environmental factor is *photoperiod,* the appropriate length of daylight (or darkness) that triggers involuntary sexual impulses.

One effect of photoperiod is to trigger hormonal surges that, in many female mammals, bring about the onset of *estrus* (sexual "heat," a period of sexual receptivity) during the mating season. Females in estrus produce a sexual attractant, the odor of which stimulates males to begin courtship and other sexual behavior, encouraging mating at precisely the time the female's eggs mature. Such timing not only promotes fertilization, but also restricts the arrival of offspring to the season most hospitable to their survival.

Reproductive behavior in humans is not limited by such restraints. Women do not experience estrus, and sexual arousability in people follows no seasonal pattern. The songwriter mentioned earlier might find it ironic that the human mating behavior to which he referred is one of the few behaviors not governed by biologically determined timing mechanisms.

---

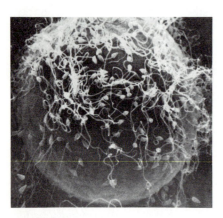

**Figure 23-5**
**The moment of fertilization.** Only one of these sperm cells will enter and fertilize this egg, producing a diploid zygote.

fertilizers rapidly develop into self-feeding individuals. The egg therefore needs little *yolk* (a deposit of lipids, proteins, and other nutrients that nourishes the developing embryo). Many internal fertilizers, notably birds and reptiles, produce eggs with a large supply of yolk, which represents a substantial investment of maternal energy and materials.

Although their eggs contain little yolk, mammals devote more resources to their progeny than any other group of animals. With few exceptions, mammalian mothers supply their developing offspring direct access to the nutrients in the mother's bloodstream. This is accomplished by use of a *placenta* that functionally connects the mother's blood with that of the progeny developing inside her. (Details of the placenta are provided later in this chapter.)

## Mammalian Reproduction

Reproduction is similar for all mammals. They are internal fertilizers, employing a penis to introduce sperm into the female's vagina during sexual coupling, called **coitus**. All produce sperm in **testes** (the male gonads) and eggs in **ovaries** (the female gonads). The human reproductive system typifies the sexual anatomy and physiology of mammals in general.

### The Human Reproductive System—Male
Mature sexual structures in men generate sperm cells (a process known as **spermatogenesis**), transfer sperm to the female's vagina, and produce sex hormones that regulate sexual development, spermatogenesis, secondary sex characteristics, and, to some extent, *libido* (sexual desire). These tasks, as summarized in Table 23-1, are accomplished by a combination of the *external genitals* (the visible sexual structures) and the *internal sex organs* within the body. Together, these components constitute the male reproductive system (Figure 23-6).

### EXTERNAL GENITALS
A boy or man's external genitals consist of the penis and a sac, called the **scrotum**, that contains the testes. Compared to body size, humans have the largest penises of any primates, although they are relatively tiny compared to those of some invertebrates and certainly much smaller than many other mammals. (The penis that emerges from the side of the slug's head is about half the length of its body.) The penis is part of the *urogenital system,* so called because of its dual urinary and reproductive functions. A tube, the **urethra**, that runs through the length of the penis conveys both sperm and urine, although not simultaneously.

To deposit sperm in a vagina, the penis must have penetrating capacity. The erect penis is rigid enough to readily enter a vagina. Unlike dogs, raccoons, and a few other mammals, humans have no bone in the penis for this purpose. Its interior is largely occupied by three long cylinders of spongy *erectile tissue*. Blood is pumped into these spaces in response to sexual stimulation. When engorged with blood, the erectile tissue expands, causing the two changes associated with erection: (1) the penis enlarges, and (2) it becomes rigid. The rigidity of the erect penis is due to the presence of a nonexpandable sheath that surrounds the paired **corpora cavernosa**. The penis can swell no larger than the size of this sheath. As a result, further influx of blood increases internal fluid pressure rather than size, causing the erect penis to become rigid enough for vaginal penetration. The lower erectile body, the **corpus spongiosum**, lies outside the nonexpandable sheath. If it were to become rigid, the great pressure would collapse the urethra that runs through it and prevent discharge of sperm cells. Although the corpus spongiosum engorges with blood during erection, it remains spongy.

The surface of the penis is richly endowed with sensory neurons that increase tactile sensitivity. The most sensitive part of the penis is the **glans**, which contains the urethral opening through which **semen**, the sperm-containing fluid, is expelled during **ejaculation**. In the flaccid (soft) penis, the glans is surrounded by the **prepuce**, the "foreskin" which is sometimes removed shortly after birth by **circumcision**. Circumcision is usually performed for reasons of religion or hygiene, as well as for treating a condition (called phimosis) in which the prepuce cannot retract over the glans. During erection of the uncircumcised penis, the prepuce normally retracts, fully exposing the glans.

**TABLE 23-1**

| Functions of the Human Male's Reproductive System | |
|---|---|
| **Function** | **Accomplished by** |
| Spermatogenesis | Seminiferous tubules |
| Male hormone production: secrete testosterone secrete gonadotropins (FSH & LH) | Interstitial cells of the testes Anterior pituitary |
| Transport of sperm from testes to urethra | Epididymis, vas deferens, ejaculatory ducts |
| Production of seminal fluid | Seminal vesicles, prostate |
| Deposit of sperm in female reproductive system | Penis and muscles of prostate and urethra |

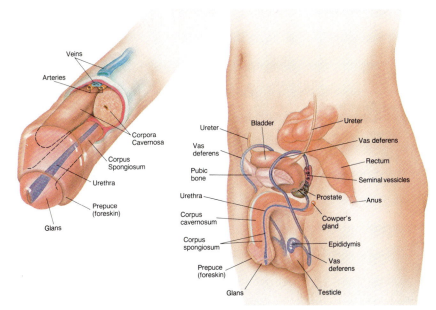

**Figure 23-6**
**Human reproductive system — Male.**
    The blue structures trace the route sperm travels from the testes to the penis. When blood flow to the penis exceeds that leaving the penis, the corpora (plural of corpus) engorge, and the penis becomes erect.

The scrotum is located at the base of the penis. It is lined with a layer of muscle, called the *tunica dartos,* that reduces the size of the sac in response to cold, sexual arousal, and fear. Inside the scrotum, a *septum* between the testes keeps each gonad on its own side. This prevents tangling of the tubes that conduct sperm cells from the testes to the urethra.

## INTERNAL SEXUAL ANATOMY

The internal male reproductive structures consist of a pair of testes, about 50 feet (15 meters) of ducts for transporting sperm to the penis, and glands for manufacturing the fluid portion of semen (review Figure 23-6). The testes have the dual function of producing

sperm and *testosterone,* an important male sex hormone. This hormone is produced by the **interstitial cells** of the testes, whereas spermatogenesis occurs in the highly coiled and compacted **seminiferous tubules** (Figure 23-6). These tubules are lined with a self-perpetuating layer of *spermatogonia,* cells that differentiate into *spermatocytes* and then into sperm cells. Each day as many as a billion sperm finish two and a half months of development in the testes and enter the **epididymis** where they finish maturing.

When sperm leave the testes, they are still incapable of swimming. The sperm's chances of reaching an egg depend on its having enough stored energy to swim a distance roughly equivalent to a six-foot-tall man swimming 18 miles. To conserve stored energy,

sperm are moved passively along their route to the penis, propelled by *peristalsis,* in much the same way that substances are moved through the digestive tract. Eventually, the sperm acquires the ability to swim. Its tail is powered by the rich supply of mitochondria that form a sheath in the sperm's *middle piece.* The nucleus comprises the major mass of the sperm's head. The leading tip is capped with an *acrosome* which contains the enzymes needed for the sperm to penetrate and fertilize the ovum.

Sperm finish maturing and acquire motility in the epididymus. Each epididymis is a tube, approximately 20 feet (6.1 meters) long, coiled into a compact mass attached to each testis. Sperm spend several weeks in the epi-

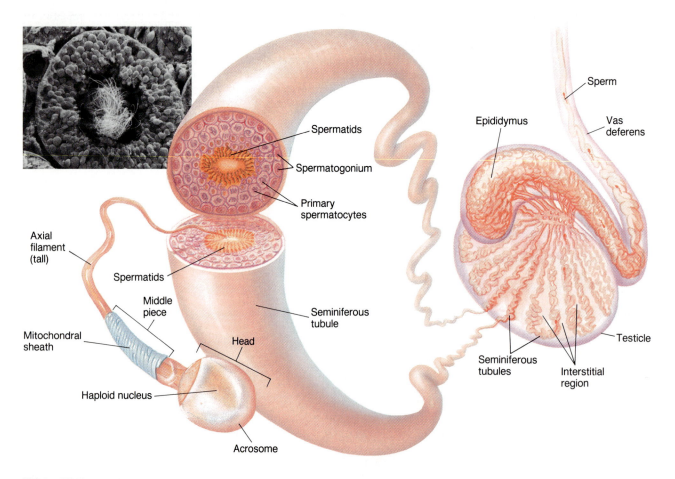

**Figure 23-7**
**Spermatogenesis** occurs in the cells that line the seminiferous tubules. Interstitial cells between the tubules produce the male sex hormone, testosterone. The sperm is shown as it appears after maturation in the epididymis.

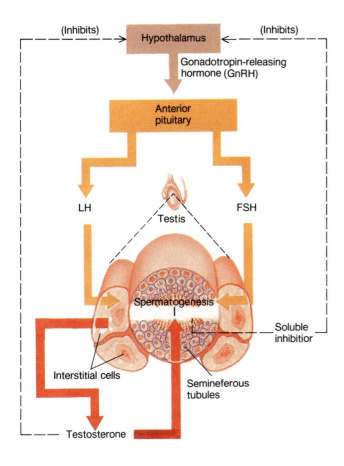

**Figure 23-8**
**Production and regulation of male sex hormones.** Increasing testosterone concentration in the bloodstream inhibits the release of GnRH, reducing further production of LH until the concentration of testosterone begins to drop. This is classic negative feedback. Sperm production is also believed to be regulated by a negative feedback mechanism. But because sperm cannot get into the bloodstream, it has been speculated that some soluble substance (which has been called "inhibin") is produced by seminiferous tubules during spermatogenesis, providing information to the hypothalamus about sperm concentration in the testes. Notice that both testosterone and FSH are needed for spermatogenesis.

didymis and then move into the **vas deferens**, the next 18 inches (46 centimeters) of the pathway.

Sperm can retain viability for three months in the vas deferens. From there they enter the urethra through the **ejaculatory ducts**. Shortly before ejaculation, the gametes are bathed in secretions from the **seminal vesicles**, which produce most of the ejaculatory fluid. These secretions are rich in fructose, a sugar that is metabolized to produce ATP in the sperm's mitochondria, providing energy for the long journey that may await them. Much of the remaining semen volume is comprised of fluids from a muscular gland, the **prostate**. During sexual arousal, the prostate's alkaline secretions accumulate in the urethra, where they mix with sperm and the fluid from the seminal vesicles. (Sperm cells comprise only a small percentage of the semen's final volume. They are so small that 5

billion of them, enough to repopulate the world with humans, would fit into a space the size of an aspirin tablet.)

The complete semen is ejaculated from the penis during **orgasm**, a brief period of peak excitement that constitutes the sexual climax. The ejaculatory fluid nourishes and protects the sperm, provides a liquid medium needed for motility, and temporarily neutralizes vaginal acidity that would otherwise kill sperm cells.

Before ejaculation, two pea-size **bulbourethral (Cowper's) glands** secrete a few drops of clear fluid into the urethra. This fluid lubricates the tube and helps wash away residual urine (which is toxic to sperm). Autonomic instructions during ejaculation contract the urinary bladder sphincter. This keeps urine out of the ejaculate and prevents *retrograde ejaculation,* which is a "backfiring" of semen into the bladder.

## MALE SEX HORMONES

Spermatogenesis falls under hormonal regulation, using negative feedback (as discussed in Chapter 16) to prevent overproduction of gametes. Male sex hormones, especially **testosterone**, have a number of other functions as well. Testosterone stimulates the development of secondary sex characteristics, such as beard growth and hair on the chest, enlargement of the larynx that deepens the voice, and, in some men, male pattern baldness. Testosterone also stimulates the maturation of the penis, testes, and accessory sex structures during *puberty,* a period in late childhood when a dramatic increase in the concentrations of sex hormones instructs the adolescent body to change into its sexually mature form. Testosterone influences sexual behavior by increasing libido and helps regulate spermatogenesis.

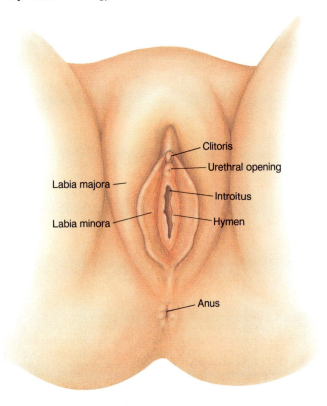

Clitoris

Urethral opening

Labia majora —

Introitus

Labia minora —

Hymen

Anus

**Figure 23-9**
**The vulva.** External genitals of the human female.

Two sex hormones are produced by both sexes but have different effects in men than in women. These are the two **gonadotropins** (gonad-stimulating hormones) produced by the anterior pituitary. One gonadotropin, **luteinizing hormone (LH)**, stimulates testosterone production by interstitial cells of the testes. (In males, LH is sometimes called *interstitial cell-stimulating hormone* or ICSH.) The other is **follicle-stimulating hormone (FSH)**, which, together with testosterone, stimulates spermatogenesis in the seminiferous tubules. Both of these gonadotropins are produced by the anterior pituitary in amounts dictated by the quantity of **gonadotropin-releasing hormone (GnRH)** secreted by the hypothalamus. Testosterone production and spermatogenesis are therefore regulated by the hypothalamus, using the negative feedback mechanism illustrated in Figure 23-8.

## The Human Reproductive System—Female

A mature woman's sexual anatomy performs many tasks essential to fertility and reproductive success. The summary of these tasks, provided in Table 23-2, reveals the greater complexity of the female's reproductive system com-

pared to that of the male. In addition to gamete production, she provides an environment for fertilization to occur, and then nourishes, houses, and protects the developing embryo. Even after birth, her body continues to supply nourishment as breast milk. The hormonal task of regulating fertility and pregnancy is considerably more complex than the simple switching on and off of testosterone and sperm production in males.

### EXTERNAL GENITALS

The female's external genitals are collectively known as the **vulva**. The vulva consists of the structures illustrated in Figure 23-9. The hair-covered **mons veneris** overlies the pubic bone. The other prominent feature of the vulva is a pair of major lips, the **labia majora**. The **labia minora** are smaller lips that, unlike the labia majora, are free of hair.

The **clitoris** protrudes from the point where the labia minora merge just below the mons veneris. Rich in sensory neurons and erectile tissue, the clitoris resembles the glans of the penis in its sexual sensitivity and erectile capacity. In fact, the clitoris is derived from the same embryonic tissue that gives rise to the male's glans. The sole function of the clitoris is to receive

sexual stimulation and transmit it to the central nervous system. (The clitoris is the only human structure dedicated exclusively to the enhancement of sexual pleasure.) Unlike the glans of the penis, the clitoris has no urinary or ejaculatory function. The urethra opens below the clitoris, between the two labia minora.

### INTERNAL REPRODUCTIVE STRUCTURES

A woman's internal reproductive anatomy is illustrated in (Figure 23-10). The vagina is a six-inch (15 centimeter) long collapsed tube that connects the vulva with the uterus. It also serves as a muscular container that receives sperm ejaculated from the penis, and forms the birth canal through which an infant leaves its mother's body during childbirth. The entrance to the vagina may be partially covered by a fold of connective tissue called the *hymen.* (Although often touted as an indicator of a woman's sexual history, the hymen is easily disrupted by nonsexual activity independent of coitus.) When coitis does occur, penile penetration is facilitated by a lubricant that collects along the walls of the vagina during sexual excitement.The lubricant is interstitial fluid forced through the vaginal walls by the increased blood pressure that occurs in sexually aroused genital tissues. The lubricant is not produced by glands.

At the opposite end of the reproductive system lie the two ovaries. The ovaries are suspended in the abdominal cavity, into which they release mature ova. Usually, one mature ovum is discharged every 28 days or so. The egg is quickly swept into one of the **oviducts** (also called *fallopian tubes*) in currents created by cilia that line the tubes. Fertilization occurs in the oviducts. Sperm that were deposited in the vagina get to the egg by first entering the **uterus** through a hole in the *cervix* (the lower tip of the uterus). From the uterus, sperm cells swim into

TABLE 23-2

## Functions of the Human Female Reproductive System

| Function | Accomplished by | |
| | Major Organ | Substructure Of Major Organ |
| --- | --- | --- |
| Produces eggs | Ovaries (female gonads) | Follicle |
| Receives sperm | Vagina | — |
| Provides an environment for fertilization | Oviducts (fallopian tubes) | — |
| Supports the developing embryo; Expells fully developed fetus | Uterus (womb) | Endometrium Myometrium |
| Produces hormones that regulate sexual functions and promote development of secondary sex characteristics | Ovaries, pituitary, hypothalamus | Corpus luteum; anterior lobe of pituitary |
| Nourishes the offspring after birth | Lactating breasts | Mammary glands |

the oviducts, where one of the sperm may fertilize the egg. The probability of fertilization may be enhanced by the presence of *prostaglandins* in semen. These hormones, produced by a man's prostate, induce uterine contractions that may help force sperm into the oviducts.

Following fertilization, the zygote begins dividing in the oviduct as it is swept into the uterus, where it implants in the inner uterine layer (the *endometrium*). Here it develops for the next nine months. An unfertilized egg fails to implant and soon disintegrates.

Like the testes, ovaries have the dual task of gamete and hormone production. Each ovum develops from a haploid cell, called a *primary oocyte,* in the outer layer of the ovary. Here the developing oocyte is housed in a chamber of cells called a **follicle**. A woman is born with all the primary oocytes she will ever have, about 400,000 of them. This is more than enough to last her a reproductive life-

time, since normally only one will mature into an ovum each month. Oocyte maturation is accompanied by the development of a single layer of cells that surrounds the egg. The mature oocyte and its surrounding cell layer constitute a mature follicle.

About two weeks elapse between the onset of oocyte maturation and **ovulation**, the release of the ovum from the follicle (Figure 23-11). The follicle then develops into the **corpus luteum**, which secretes hormones that prepare the uterine endometrium to receive a developing embryo. The preparative changes in the uterus must be coordinated with the monthly cycle of ovum production to assure that the endometrium is receptive when the embryo arrives (about eight days after fertilization). The cycle of ovum production, called the **ovarian cycle**, is therefore synchronized with the **uterine (menstrual) cycle**, the repetitive monthly changes in the uterus that prepare the endometrium for receiving and sup-

porting an embryo. The two cycles are coordinated and co-regulated by female sex hormones.

## FEMALE SEX HORMONES

As in men, the primary sex hormones of women are produced by the gonads in response to release of *gonadotropin-releasing hormone (GnRH)* from the hypothalamus and subsequent release of gonadotropins from the pituitary. Although the gonadotropins (LH and FSH) are identical to those found in men, they exert distinctly different influences in women. FSH prepares for ovulation by stimulating a follicle to ripen. LH triggers ovulation and the subsequent transformation of the follicle into the corpus luteum. Two additional reproductive hormones are distinctively female. These are the **estrogens** (which will be referred to as "estrogen" in the rest of this discussion) and **progesterone**. Estrogen is the primary female hormone, regulating fertility and the development of secondary sex characteristics, such as enlarged breasts and fat distribution that is characteristically female. Its production is greatest during the first two weeks of the 28-day reproductive cycle. Progesterone, which is produced by the corpus luteum, dominates during the last two weeks of the cycle. It prepares and maintains the uterus for pregnancy (Figure 23-12), participates in milk production, and helps prevent the ovary from releasing additional eggs late in the cycle or during pregnancy. All these hormones are needed for synchronizing the ovarian and uterine cycles (Figure 23-13).

## HORMONAL CHANGES DURING PREGNANCY

If implantation occurs, the last half of the cycle is altered to maintain the endometrium and to prevent ovulation for the duration of pregnancy. The signal that alters the cycle comes from the embryo itself. By the time it implants, the developing embryo is no longer a single-celled zygote, but an actively dividing collection of cells. Some of these cells begin secreting **human chorionic gonadotropin (HCG)**, a hormone that prevents the corpus luteum from degenerating. (Tests for detecting HCG in a

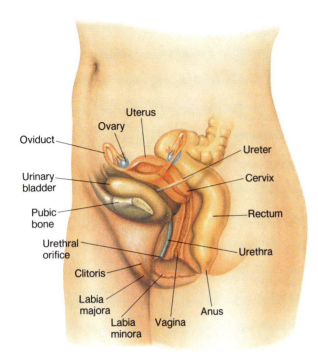

**Figure 23-10**
**Human reproductive system—Female.**

woman's blood or urine provide a reliable means of determining pregnancy.) The sustained corpus luteum continues to produce estrogen and progesterone, which maintain the endometrium. Since these hormones also inhibit the release of FSH, no follicles can mature during pregnancy. It is therefore impossible to ovulate while pregnant. Birth control pills prevent ovulation by artificially increasing the concentrations of these hormones (see Bioline: Preventing Unwanted Pregnancy).

The corpus luteum eventually deteriorates several months into pregnancy, but before it does, secretion of estrogen and progesterone shifts to the placenta. The **placenta** develops from embryonic tissues and part of the endometrium, providing a link for exchanging materials between fetal and maternal blood. Development of the placenta, which is characteristic of nearly all mammals, is discussed in the next chapter.

## ADDITIONAL HORMONAL EFFECTS

Another mammalian property is the one for which the group was named. The breasts of mammalian mothers contain **mammary glands** that produce breast milk, a nutritionally complete liquid that provides nourishment for offspring after they are born. A hormone (**prolactin**) from the anterior pituitary stimulates *lactation,* milk production by the mammary glands. Prolactin is produced throughout pregnancy but doesn't take effect until immediately after childbirth. Nipple stimulation provides positive feedback for additional prolactin secretion.

Nipple stimulation also triggers release of another hormone, **oxytocin**, from the posterior pituitary. Oxytocin instructs the muscle fibers in the breast to contract, forcing milk out of the storage spaces into the milk ducts where it can be sucked out. This phenomenon, called "letdown," occurs within a minute of nipple stimulation. Suckling also blocks the release of gonadotropins, creating a temporary state of infertility. This fact comes with a precautionary note, however. Relying on nursing as a birth control technique commonly results in unexpected pregnancy, because more than half of nursing mothers regain their ability to ovulate before the baby is weaned.

As with men, secondary sex characteristics in women are the product of hormonal influence, notably estrogen. Gonadotropin and estrogen levels increase during puberty (at years 10 through 14 in most girls) inducing the changes characteristic of mature women. Puberty also marks the onset of menstruation. Estrogen halts the growth of long bones, which, along with testosterone's growth-stimulating properties, accounts for the greater average height of men than women. (The growth of boys ceases later in life, when testosterone accumulates to growth-inhibiting concentrations.) During puberty, girls and boys develop many of the same characteristics, for example, pubic and armpit hair, thickening of skin, and an increased lipid content in secretions of the skin's sebaceous glands.

## The Biology of Human Sexual Response

Sexual excitement in men and women is accompanied by two biological phenomena—**vasocongestion** (a local increase in blood flow) and **myotonia** (muscle tension). Vasocongestion in the pelvic region produces penile and clitoral erection, and swelling of the testicles and labia. Vasocongestion also lubricates the vagina by forcing interstitial fluid through its walls during sexual arousal. The increased pressure in the genitals stimulates sensory neurons, providing neurological input that is perceived as pleasurable. Myotonia is most notable during orgasm, the sexual climax. During orgasm, involuntary muscle contraction occurs at 0.8-second intervals. In men these contractions move sperm into the urethra and launch the semen from the penis during ejaculation. Although women don't ejaculate, contractions of the uterus during her orgasm may help draw sperm in through the cervical opening. Following orgasm, the body returns to the nonaroused condition during a period called resolution. Rearousal immediately after orgasm is impossible for most men and many women. The duration of this unresponsive *refractory period* following orgasm is highly variable, lasting from a few minutes to days.

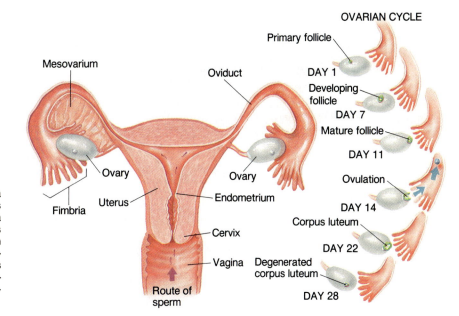

**Figure 23-11**
**28 days in the life of an ovary.** Each month, a primary oocyte in one of the two ovaries (shown here in frontal view) develops into a mature ovum (egg). On day 14, the egg erupts from its follicle. The ovum is swept into an oviduct by cilia. Fertilization occurs in the oviduct. Following ovulation, the follicle develops into the corpus luteum. If a developing embryo fails to implant, the corpus luteum degenerates.

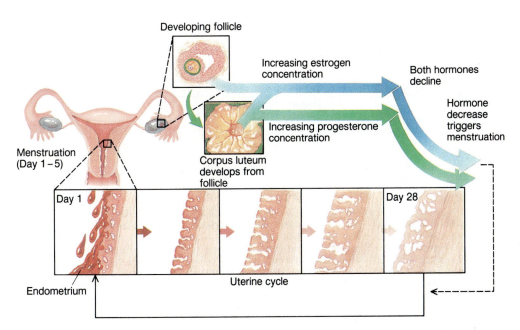

**Figure 23-12**
**Hormonal regulation of the uterine cycle.** Both estrogen and progesterone are produced by the ovaries as part of the ovarian cycle. First estrogen is produced by the developing follicle and initiates endometrium buildup. Then progesterone and estrogen released from the corpus luteum maintains the lining for another 14 days or so.

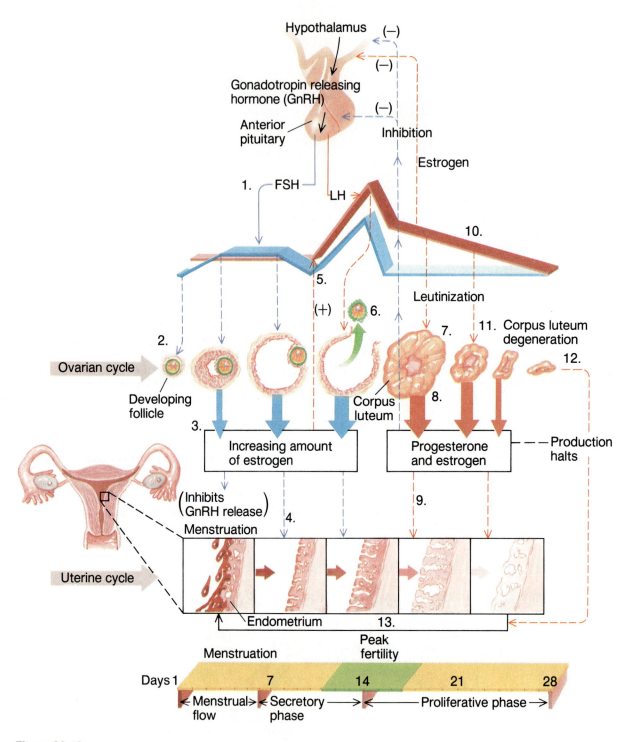

**Figure 23-13**
**Synchronizing the ovarian and uterine (menstrual) cycles** (the numbers correspond to the steps in the figure). (1) FSH is released from the anterior pituitary. (2) FSH promotes development of a follicle and the oocyte it contains. (3) The ripening follicle secretes estrogen which (4) stimulates endometrium development in the uterus. (5) Presumably, when estrogen reaches a critical concentration, it stimulates the hypothalamus and anterior pituitary to flood the body with LH. (6) This surge of LH triggers ovulation. (7) LH also "leutinizes" the follicle, transforming it into a corpus luteum. (8) Progesterone (and some estrogen) produced by the corpus luteum maintain the enriched endometrium, so that it is receptive to implantation. (9) Progesterone from the corpus luteum also inhibits GnRH release, so LH secretion stops, and (10) the gonadotropin declines over the next 14 days. (11) In the absence of implantation, the corpus luteum deteriorates (due to declining LH concentrations) until (12) progesterone and estrogen production suddenly halts. Without these two hormones, the extra endometrial tissue cannot be maintained and is sloughed as menstrual fluid. Menstruation signals the beginning of a new cycle. With the disappearance of progesterone, inhibition of GnRH is relaxed and FSH is again released (1) and another follicle begins to ripen (2).

# B I O L I N E

## PREVENTING PREGNANCY

Throughout history, people have devised many ways to prevent pregnancy. *Contraceptive* (birth control) methods have ranged from inserting elephant and crocodile manure into the vagina in order to block the entrance of sperm into the uterus to "fumigating" the vagina over a charcoal burner after intercourse to kill the sperm. Though primitive, these methods employed the same strategies as do some of the most effective modern contraceptive approaches, that is, physically blocking the uterine entrance or killing the sperm in the vagina. Other modern methods of avoiding pregnancy prevent ovulation, halt spermatogenesis, trap gametes in the oviducts or vas deferens, or interfere with implantation of the fertilized ovum in the uterus.

### Chemical Intervention

*Oral Contraceptives* (effectiveness = 95 to 98 percent)

Artificial estrogens and progesterones inhibit the release of FSH and LH which in turn prevent ovulation. These two hormones, either alone or in combination, constitute *birth control pills,* the most effective temporary contraceptive method known today. The exact formulation for each woman is determined by a physician to minimize adverse reactions, such as nausea, headaches, weight gain, and, after several years, hypertension and, rarely, formation of blood clots. To date oral contraceptives have been limited to women. Use of hormones in men to switch off sperm production also inhibits testosterone production. So far, experimental male "pills" tend to effeminize the user and reduce libido, although the isolation of the hormone "inhibin" in 1987 may be the long-awaited breakthrough in finding an oral contraceptive that shuts off spermatogenesis without influencing testosterone production. (Inhibin is the soluble inhibitor shown in Figure 23-8.)

*Spermicides* (effectiveness = 75 percent)

Spermicides are sperm-killing chemicals inserted into the vagina before intercourse. When used alone they are not very reliable. Spermicides are reliable only when used with a diaphragm or condom (see below).

### Mechanical Intervention

*Intrauterine Device (IUD)* (effectiveness = 98 percent)

The IUD is a plastic or metal wire that is inserted by a physician into the uterus, where it prevents implantation of a developing embryo. In spite of its effectiveness and convenience, once inserted the IUD may cause two serious side effects that negate its attractiveness for some women. Uterine puncture, a potentially fatal complication, occurs in about one in every 1000 women. The IUD may also aggravate existing low-grade infections of the uterus, and lead to a serious and painful condition called *pelvic inflammatory disease.* The most common side effects are painful cramping and irregular bleeding, especially immediately after the device is inserted. Although very infrequent, these complications created enough concern to cause IUDs to be taken off the market in 1987. A few types of IUDs are again being made available.

*Diaphragm with Spermicide* (effectiveness = 98 percent)

The famous lover Casanova was said to have used half of a hollowed out lemon to cover the cervix of a lover in an attempt to reduce the number of new little Casanovas. The modern diaphragm, a thin rubber dome, works in the same way, although with considerably more success because it is precisely fitted to the cervix and is used in conjunction with a spermicide. It is inserted just before intercourse, and must remain in place for 8 hours following intercourse to be fully effective.

*Condoms* (effectiveness = 85 to 95 percent, when used with a spermicide, close to 100 percent)

These thin sheaths, worn over the penis, trap sperm and have the additional advantage of preventing sexually transmitted diseases. (For this reason, a condom is also called a "prophylactic," a word meaning "disease preventing".) Condoms also are important lines of defense in the war against acquired immunodeficiency syndrome (AIDS). Inconvenience and diminished sensations are two drawbacks.

*Vaginal Sponge* (effectiveness = 98 percent, preliminary data)

A relatively new device, the spermicide-saturated, cup-shaped sponge fits over the cervix prior to intercourse, much like a diaphragm, but without the need for physician fitting. This advantage may be offset by a serious problem—an increase in the incidence of *toxic shock syndrome (TSS).* This potentially fatal disease is usually due to excessive growth of a particular strain of toxin-producing bacteria in the vagina. (Most cases of TSS, however, have been associated with the use of superabsorbent tampons by menstruating women.)

*Douche* (effectiveness = poor)

In this age-old method, the vagina is flushed with water or a chemical douche to wash away sperm before they penetrate the cervix. But sperm quickly enter the uterus, and these cannot be removed by this unreliable approach to contraception.

## Permanent Sterilization

*Vasectomy* (effectiveness = 99.85 percent)

In this surgical procedure for men, a portion of each vas deferens is removed and the cut ends of the tubes are tied. Although spermatogenesis continues, sperm are blocked from reaching the penis, so the man's ejaculate contains no sperm cells. Otherwise, the volume of the semen and the sensations of orgasm are unaffected by vasectomy.

*Tubal Ligation* (effectiveness = 100 percent)

Surgically cutting and sealing a woman's oviducts prevents an egg from reaching the uterus or even coming in contact with sperm. Ovulation continues. In most cases the operation is irreversible.

## Natural Methods

*Calendar Rhythm* (effectiveness—variable)

This method requires abstinence from intercourse during a woman's fertile period, about 12 hours before and 48 hours after ovulation. The problem with the rhythm method is predicting when ovulation occurs, especially in women with irregular ovarian cycles. Even when practiced diligently, failure rates range from 15 to 35 percent. Effectiveness can also be improved by keeping a daily record of the woman's body temperature, which rises about half a degree at the onset of her peak fertility days. Another clue is the change in consistency of cervical mucus that occurs just before ovulation.

*Coitus Interruptus* (effectiveness = 70 percent, if practiced)

In this method, also known as withdrawal, the man removes his penis from the vagina before ejaculation. Because it requires willpower and reduces gratification, failures are very likely.

# Synopsis

## MAIN CONCEPTS

- **Asexual reproduction is a biologically economical way to generate identical copies of oneself.** Although asexual processes conserve energy and nutrients, the diversifying advantages of sexual reproduction are so important that most animal species depend on this biologically expensive way to produce progeny.

- **Animals employ several strategies that increase the likelihood of fertilization.** Animals that depend on external fertilization release sperm and eggs at about the same time in the same location. These external fertilizers use pheromones and specific sexual behaviors to help coordinate gamete release; many internal fertilizers use them to locate and identify mates and to coordinate copulatory activity with gamete maturity.

- **Gametes are produced by gonads.** In humans, sperm cells are generated by the testes, and ova are produced by ovaries. Sperm are mixed with fluids from the seminal vesicles and prostate, and the resulting semen is ejaculated from the penis. If deposited in a woman's vagina during coitis, sperm swim through the uterus and into the oviducts, where one sperm cell may fertilize an ovum. The resulting zygote divides to form an embryo that implants in the endometrium for development.

- **Sex hormones govern sexual development and maturation of primary and secondary sex characteristics.** In men, three hormones are needed for gamete production—testosterone and the two gonadotropins, LH and FSH. Gamete production and ovulation in women are stimulated by LH and FSH. Estrogen and progesterone maintain the rich endometrium in preparation for implantation of an embryo. If no embryo implants, the corpus luteum deteriorates, estrogen and progesterone decline, and the endometrium is sloughed off as menstrual fluid.

- **HCG produced by the developing embryo prevents corpus luteum degeneration**, and the estrogen/progesterone concentrations remain high enough to retain the embryo-supporting endometrium. Because these two ovarian hormones also prevent ovulation during pregnancy, a second pregnancy cannot be superimposed on the first. HCG alters normal cycles to maintain pregnancy.

## KEY TERM INTEGRATOR

### Principles of Animal Reproduction

| | |
|---|---|
| **asexual reproduction** | Reproduction without fertilization. Fission, budding, and a nonsexual variation of sexual reproduction called **parthenogenesis** produce offspring that are **clones** of the parent. |
| **sexual reproduction** | Production of progeny following fertilization. **External fertilization** occurs only in aquatic habitats. Animals that rely on **internal fertilization** use an organ, such as a **penis**, to deposit sperm into the female's receptacle, such as a **vagina**. Sexual anatomy consists of **primary sex characteristics** that are responsible for fertility and **secondary sex characteristics** that aid in mate location and gender differentiation. "Double-gendered" animals are **hermaphrodites**. |

### Mammalian Reproduction — Human Male

| | |
|---|---|
| **spermatogenesis** | The making of sperm cells by the **seminiferous tubules** of the **testes** (the male gonads seated in the scrotum). The process is regulated by the pituitary **gonadotropins**, **LH** and **FSH**. Sperm are moved to the **urethra** through the **epididymis**, **vas deferens**, and **ejaculatory ducts**. |

| | |
|---|---|
| ejaculation | The emission of **semen** from the penis during **orgasm**. Semen is a mixture of fluids (produced by the **seminal vesicles** and **prostate**) and sperm cells. Ejaculation is preceded by penile **erection**, the engorgement of the paired **corpora cavernosa** and the **corpus spongiosum** during sexual arousal. The phenomenon is enhanced by the tactile sensitivity of the penis, especially the **glans**. |

### Mammalian Reproduction — Human Female

| | |
|---|---|
| **ovulation** | Discharge of an ovum from a ripe **follicle** of the **ovaries** (the female gonads). The egg is swept into the **oviducts** (the site of fertilization), then into the **uterus**, where it implants if fertilized. Sperm swim to the egg from the **vagina** (where sperm cells are deposited during **coitis**) through the uterus and then into the oviducts. The **ovarian cycle** and **uterine (menstrual) cycle** are aligned by pituitary hormones (LH and FSH) and two ovarian hormones, **estrogen** and **progesterone**, produced by the follicle after it changes into the **corpus luteum**. |
| **human chorionic gonadotropin (HCG)** | Hormone produced by an embryo that alerts a woman's body that she is pregnant. The embryo also helps produce the **placenta**. Another hormone, **prolactin** promotes milk production by **mammary glands**. |

## Review and Synthesis

1. Explain why the following statement is biologically accurate: "The biological advantage of sex is **not** reproduction, but diversification."

2. Compare the advantages and disadvantages of external fertilization with those of internal fertilization.

3. Discuss how behavior, photoperiod, and pheromones help insure that an animal's gametes will be fertilized. Consider internal and external fertilizers.

4. Describe three differences in gamete production by men and women.

5. Trace the "odyssey" of a sperm cell from spermatogenesis to fertilization.

6. The United States has one of the highest teenage pregnancy rates in the world. At the current rate, 40 percent of today's 12-year-old girls will be pregnant by age 20. What do you think could be done to reduce the number of unwanted pregnancies among teens? (There is no right or wrong answer to this question.)

7. Describe one similarity and one difference in each of the following paired items: vas deferens/oviducts; penis/clitoris; penile erection/vaginal lubrication; testes/ovaries; testosterone/estrogen; prolactin/oxytocin.

8. Explain why LH and FSH have such different effects in human males and females. Describe these effects.

9. Explain why a woman who is taking birth control pills must stop them for several days every month to allow menstruation to occur. (In your answer, describe the normal hormonal regulation of the ovarian and uterine cycles and how this is altered by oral contraceptives.)

## Additional Readings

Epel, D. 1977. "The program of fertilization." *Scientific American* 237:128. (Intermediate.)

Langone, J. 1982. "The quest for the male pill." *Discover* 10. (Introductory.)

Witters, W., and P. Witters. 1980. *Human Sexuality, A Biological Perspective.* D. Van Nostrand, New York. (Introductory to intermediate.)

# Chapter 24

# Animal Growth and Development

The newborn humpback whale being pushed by his mother to the ocean's surface for his first breath of air bears little resemblance to the way it looked 12 months ago. At that time he was a single, microscopic cell—a zygote—more potential than whale. Yet, packaged in these few micrograms of cytoplasm were all the blueprints and metabolic machinery needed to direct the zygote's transformation into one of the largest animals on earth.

Although the growth and development of a zygote to mature animal is bewilderingly complex, the mystery is being steadily stripped away through a field of study known as *developmental biology. Growth,* an increase in the size of an organism, is the simplest part of the puzzle. The formation of a mass of identical cells can readily be explained by mitosis alone. Indeed, all the somatic (body) cells in a 100-ton adult whale are genetically identical, having descended by mitosis from the same zygote.

Yet the whale's *development* is far more complicated than this. His body is an intricate composite of nerve cells, bone cells, muscle fibers, and hundreds of other cell types. Every cell has selectively read and executed the instructions found in different sections of the same genetic blueprint, and they have done so at precisely the right time. Some modern theories of gene regulation that determine which portions of a cell's genetic instructions will be obeyed are discussed in Chapter 28. Here we provide an overview of an animal's development from the moment of fertilization to adulthood.

# Embryonic Development

At the moment of fertilization, the zygote begins the changes that transform it into an **embryo**—an organism in the early stages of development, beginning with the first division of the zygote. As soon as a sperm cell penetrates an ovum, the membrane changes its electrical polarity, much like the voltage disturbance of an action potential in nerve cells (Chapter 15). The electrical disturbance spreads over the egg's surface. Less than a second after fusion of the gametes, the membrane has changed in a manner that keeps additional sperm from penetrating, preventing the zygote from acquiring extra sets of chromosomes. The voltage disturbance also stimulates the formation of a fertilization membrane (Chapter 23) around the zygote's cytoplasm that forms a long-term barrier to sperm penetration, even after the membrane voltage has returned to normal.

The voltage disturbance does more than prevent multiple sperm penetration. It also triggers the release of calcium ions into the cell's cytoplasm, ions that initiate the developmental process. Animals then progress through remarkably similar stages in embryonic development. These stages are of particular interest because they

- Provide clues to understanding **differentiation**, the development of distinct cell types with different forms and specialties, even though they all possess the same genes.

- Reveal how early embryonic changes influence the form and structure of a fully developed animal.

- Furnish evidence of evolutionary relatedness among animals.

Figure 24-1 provides an overview of these stages and an introduction to some important terminology.

## Cleavage and Blastulation

Although a zygote is spherical, for most animals it functionally has two distinct "ends." The *animal pole* that gives rise to the embryo lies opposite the *vegetal pole* that will eventually house a nutrient-filled container, the yolk sac. The cytoplasmic components at the two poles differ. The first mitotic division (called **cleavage**) partitions the zygote's cytoplasm, so the cells that develop at the animal pole contain different cytoplasmic components than do the cells at the vegetal pole. The cytoplasmic constituents that each new cell gets helps determine the developmental fate of that cell. Even though they are genetically identical, the cells in the early embryo quickly diverge from each other in form and specialization, ultimately giving rise to very different tissues.

During successive cleavages, cells divide mitotically (Figure 24-2a). Cleavage continues until it produces a solid ball of cells, called a **morula**, which is about the same size as the original zygote. The morula fills with fluid, forming a hollow space, called the *blastocoel*. In many animals, the blastocoel is large, and the cells that enclose it form a hollow ball called a **blastula** (Figure 24-2b). The relative size of the cavity differs from animal to animal, depending on the amount of yolk provided the embryo. Some embryos need little nutrient reserves. Sea urchin embryos, for example, develop outside the parent in water and rapidly become **larvae**, a self-feeding immature form of an animal. The blastocoel of these embryos is large, with virtually none of the cells in the blastula committed to yolk storage. The embryos of amphibians (such as a frog) also develop in water but take longer than sea urchins to achieve the feeding stage, and, as a consequence, require more yolk. The larger yolk leaves less room in the blastocoel. Reptiles and birds, animals whose embryos develop on dry land, must supply their developing offspring with enough food and water to last until they hatch from the shells that enclose them. The enormous yolk and albumen (egg white) reserves that provide these resources occupy most of the egg space. Thus, the growth of early embryonic cells is restricted to a small area on the surface, where it forms an embryonic disk or a **blastodisk**.

## Gastrulation

As soon as blastulation is completed, the embryo begins to change shape. The blastocoel disappears as cells along one side of the blastula migrate toward the opposite side, collapsing the blastocoel space and forming two layers that lie adjacent to each other. The **gastrula**, as the embryo is now called, temporarily consists of two layers: the outer **ectoderm** and the inner **endoderm** as shown in Figure 24-3. The hollow core of the gastrula, the **archenteron**, eventually becomes the animal's digestive tract. The **blastopore** (the opening into the archenteron) is the embryonic forerunner of

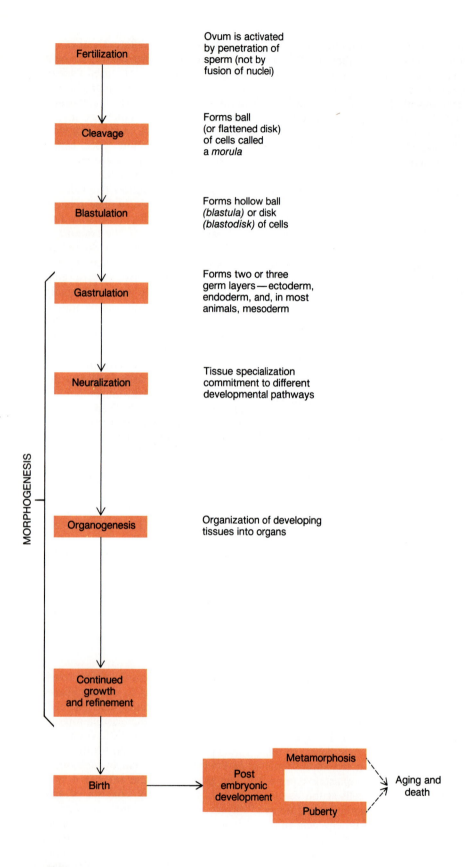

**Figure 24-1**
**Stages of growth and development in animals — an overview.**

the anus in vertebrates and a few other animals (sea urchins and sea stars). Later, another opening develops at the opposite end of the archenteron, an opening destined to become the mouth in these animals.

Gastrulation (gastrula formation) is not complete until formation of the **mesoderm**, the middle embryonic cell layer. In the sea urchin, the mesoderm cells are produced by pouches of the archenteron lining that protrude into space between the endoderm and ectoderm, where the mesoderm cells accumulate. In developing frog and chick embryos, gastrulation is restricted to a smaller space to accommodate the larger yolk deposits (Figure 24-3). The chick blastodisk produces mesoderm by inward migration of the ectodermal cell layer toward the endoderm.

The endoderm, ectoderm, and mesoderm are known as the embryonic **germ layers** of the gastrula. All structures in the mature animal develop from these three layers (Figure 24-4).

## Morphogenesis

The inward migration of cells during gastrulation is the first sign of embryonic **morphogenesis**, the development of form (morpho = form or shape). The developing embryo is shaped by the growth of new cells and by the movement of existing tissues. Each of the morphogenic activities diagrammed in Figure 24-5 participates in the formation of specialized tissues and organs. Morphogenic movement of embryonic cell layers determines the fundamental pattern of an embryo. These fundamental patterns are refined by cell differentiation.

## Differentiation and Tissue Specialization

Cells differentiate in form and function as the embryo acquires its specialized tissues. Exploring how cells differentiate is a challenge at the forefront of biological research. Experiments reveal that, even after cells are committed to their specialty, they still retain all the genes needed for giving rise to every type of cell in the animal (see Connections: Totipotency and Clones, Chapter 28).

Differentiation must therefore be a matter of gene expression; that is,

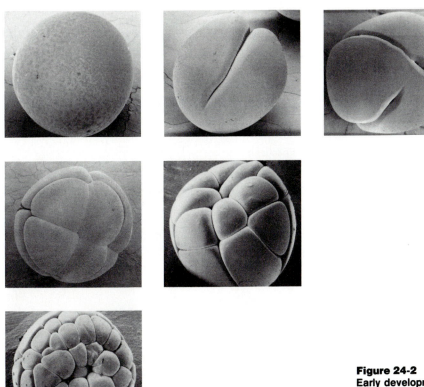

a

**Figure 24-2**
**Early developmental stages.** *(a) Cleavage.* Although cell division increases cell number, the overall size of the morula is about the same as that of the ovum. *(b) Blastulation.* Differences in the size and location of the hollow blastocoel depend on the amount of space (if any) devoted to yolk storage.

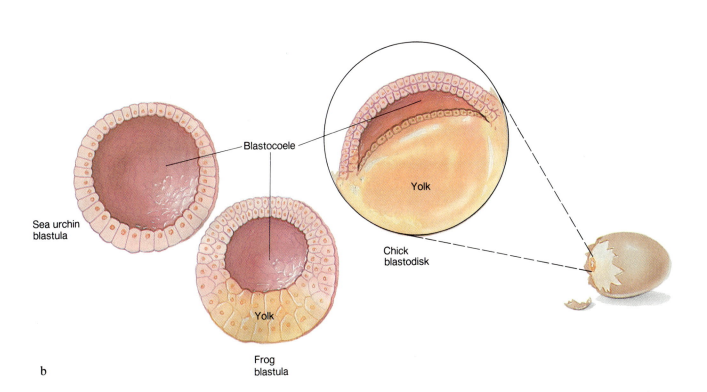

Sea urchin
blastula

Blastocoele

Yolk

Frog
blastula

Chick
blastodisk

Yolk

b

a cell's specialty is determined by which genes are switched on or off as it develops. Cytoplasmic constituents clearly play a major role in determining which genes will be expressed. Recall that the distribution of materials in the zygote is unequal, so some cells get cytoplasmic components that others lack. The effects of unequal distribution can be experimentally demonstrated by stimulating frog zygotes to divide along abnormal planes and then separating the two daughter cells (Figure 24-6). When the first division plane is normal, each cell develops into two identical tadpoles. This occurs because both cells acquire a full complement of cytoplasmic constituents that channel each through the same developmental pathways. When the first division is perpendicular to the normal plane, however, only one cell (the one containing the animal pole constituents) develops into a complete tadpole. The other cell simply forms an undifferentiated ball of cells.

## Embryonic Induction

As important as cytoplasmic constituents are in determining which genes are expressed or repressed, factors outside the cytoplasm also influence development. One of the most important of these comes from neighboring cells in the embryo, one body part *inducing* differentiation of an adjacent part. This phenomenon is known as **embryonic induction**.

A classical example of embryonic induction occurs in the frog blastula. The gastrula's *dorsal lip mesoderm* (the mesoderm cells that develop above the blastopore) has a profound effect on the overlying ectoderm, inducing it to fold into the *neural tube* that eventually forms the spinal cord and brain (Figure 24-7). If a bit of dorsal lip mesoderm is experimentally transplanted from one gastrula to the future belly region of another gastrula of the same species, it induces the ectoderm in this location to develop into a second neural tube. The embryo often continues to develop into two tadpoles attached at the abdomen. This is because the neural tube, once formed, becomes an inducer, instructing the formation of other tissues that in turn

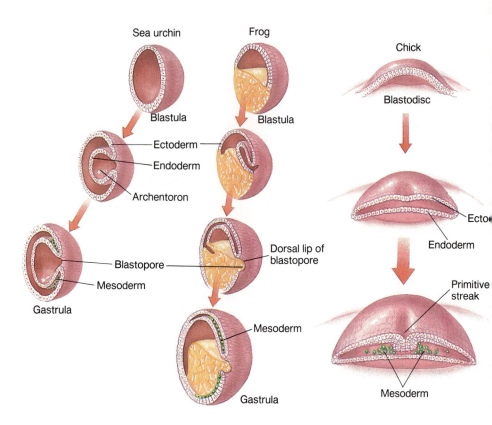

**Figure 24-3**
**Gastrulation and formation of the three embryonic germ layers.**

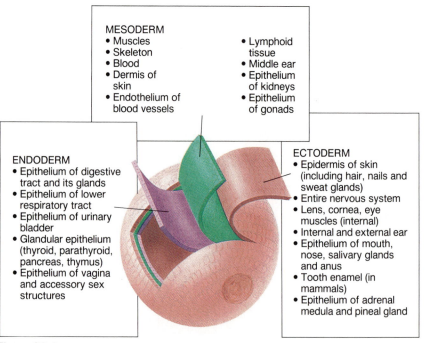

**Figure 24-4**
**The fate of the germ layers in vertebrates.**

MESODERM
• Muscles
• Skeleton
• Blood
• Dermis of skin
• Endothelium of blood vessels
• Lymphoid tissue
• Middle ear
• Epithelium of kidneys
• Epithelium of gonads

ENDODERM
• Epithelium of digestive tract and its glands
• Epithelium of lower respiratory tract
• Epithelium of urinary bladder
• Glandular epithelium (thyroid, parathyroid, pancreas, thymus)
• Epithelium of vagina and accessory sex structures

ECTODERM
• Epidermis of skin (including hair, nails and sweat glands)
• Entire nervous system
• Lens, cornea, eye muscles (internal)
• Internal and external ear
• Epithelium of mouth, nose, salivary glands and anus
• Tooth enamel (in mammals)
• Epithelium of adrenal medulla and pineal gland

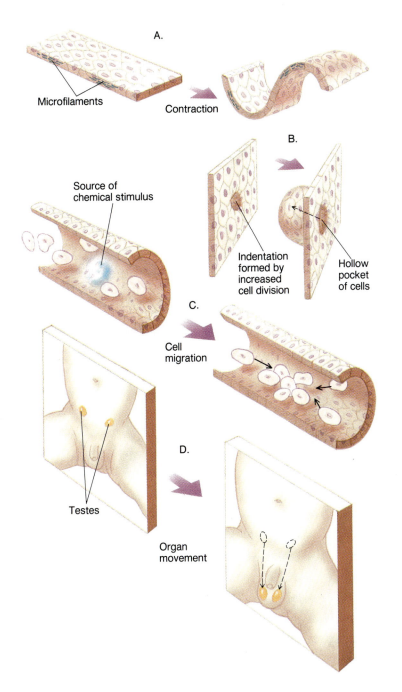

**Figure 24-5**
**Morphogenesis. Shaping the embryo by moving masses of cells.** *(a) Contraction of microfilaments* pulls cell sheet around shortened side. *(b) Invagination and evagination* causes cell layers to form pockets. *(c) Chemotaxis (movement in response to a chemical)* draws cells to the source of the chemical. *(d) Movement of whole organs* translocates them to their functional locations.

induce still other changes until development is complete. In normal embryos, this step-by-step process directs development along the proper sequence, each tissue emerging only after the appearance of its antecedent tissue. Proper sequence is essential for normal development of an organism,

in much the same way that the foundation, walls, and roof of a building must be assembled in the correct sequence.

The development of the eye in vertebrates exemplifies how induction brings about sequential development. Perfect vision depends on precise posi-

tioning of the lens so that it can focus incoming light on the retina. The embryonic eye begins its development as an outpocket of the rudimentary forebrain, which forms an optic vesicle advancing toward the overlying epidermal layer (Figure 24-8). When the optic vesicle contacts epidermal cells, it induces them to invaginate and form the lens. At the same time, the optic vesicle forms the cup that will become the retina. Postponing lens development assures proper positioning of the lens in the developing eyeball.

The instructions for induction are chemical. In fact, induction still proceeds even when tissues are experimentally separated by membrane filters that prevent contact, but allow the migration of large molecules. Most inducer chemicals have yet to be discovered, but some have been identified as hormones. As inducers, hormones usually participate in the differential development of tissues already committed to a generalized developmental pathway, one that has two or more alternate routes. For example, pregenital tissue is identical in

male and female human embryos. The presence of male hormone channels this tissue into the developmental pathway that produces a penis and scrotum. In the absence of the hormone, the same tissue forms the vagina and vulva.

Induction continues throughout development, even throughout life, largely through the influence of hormones.

## Organogenesis

Organ formation (**organogenesis**) is a complex process in which two or more specialized tissues develop in precise relationship with one another. We have already discussed the role of induction in lens formation and neural tube development, both of which are also examples of organogenesis. Cell

growth, cell division, morphogenesis, and induction clearly participate in organogenesis, as does another phenomenon, the *systematic death of cells*. This removes embryonic cells from strategic locations, creating openings and hollows, and sculpting the form of a structure. Human fingers, for example, are "carved" from a paddlelike structure by the programmed cell death of the web tissue between the digits. (A preview of Figure 24-12 reveals these changes in the embryo's hand.)

Tubes and chambers are hollowed out from solid pieces by the death of interior cells. Eyelids form when a line of cell death cuts through the continuous piece of skin that covers the eyeball, producing a slit that allows the upper and lower lids to separate. Pro-

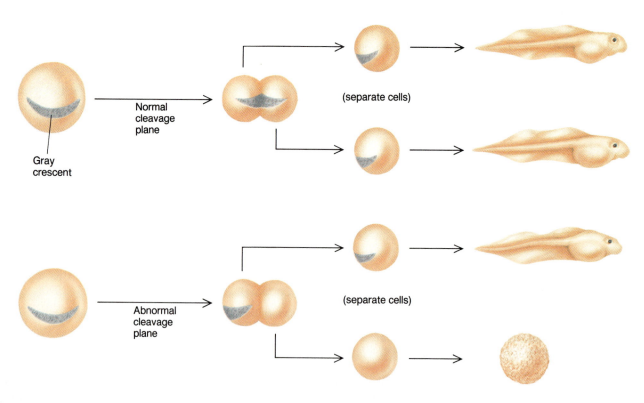

**Figure 24-6**
**The gray crescent — cytoplasmic director of cell destiny.** This experiment demonstrates that the cytoplasm in the crescent-shaped gray area of a fertilized frog zygote helps direct embryonic development. Normal cleavage divides the zygote through the gray crescent, so each daughter cell receives an equal share of the components in the cytoplasm. If these two cells are experimentally separated, each grows into a normal tadpole. However, if the first division is altered so it doesn't bisect the gray crescent, only the cell containing the gray crescent develops normally.

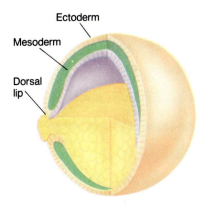

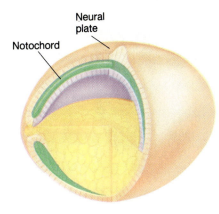

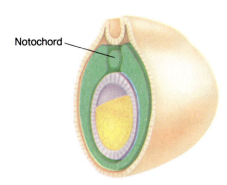

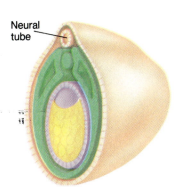

**Figure 24-7**
**Neural tube development.** Mesoderm in the dorsal lip of the gastrula induces notochord formation, which in turn induces neural plate development. The neural plate then folds into the neural tube, the embryonic forerunner of the brain and spinal cord.

grammed cell death continues to participate in development even after the animal has been born. The disappearance of a tadpole's tail is a familiar example of strategic cell death during postembryonic development.

# Postembryonic Development

Unlike plants, which grow and development throughout life (such *open growth* is discussed in Chapter 22), most animals produce no new structures in the course of normal development after maturity. This is particularly true for animals that have acquired roughly their adult form by the time they are born or hatch. In these organisms, postembryonic change generally constitutes an increase in size, although not all parts grow at equal rates. At birth, for example, a human's head is disproportionately larger than other body parts. Thereafter the arms, legs, and trunk grow much faster than the head, so that the body eventually loses its neonatal (newborn) look and acquires the proportions characteristic of the adolescent and adult. (If you make a circle with your arms touching at the fingertips above your head, you can see how large your head would be if it had grown at the same rate as the rest of your body.)

The development of reproductive capacity is another example of a postembryonic developmental stage, one that is often accompanied by the appearance of secondary sex characteristics (see Chapter 23). In humans, this change is called **puberty**.

## Metamorphosis

Postembryonic changes are most striking for animals that go through a larval stage before developing into an adult. Most types of larvae require no further assistance from their parents. Larvae eventually acquire sexual maturity when they change into the adult form, a process called **metamorphosis**. Metamorphic transformations can alter the form of an organism so greatly, that, in some animals, larva and adult scarcely resemble each other (Figure 24-9). During metamorphosis,

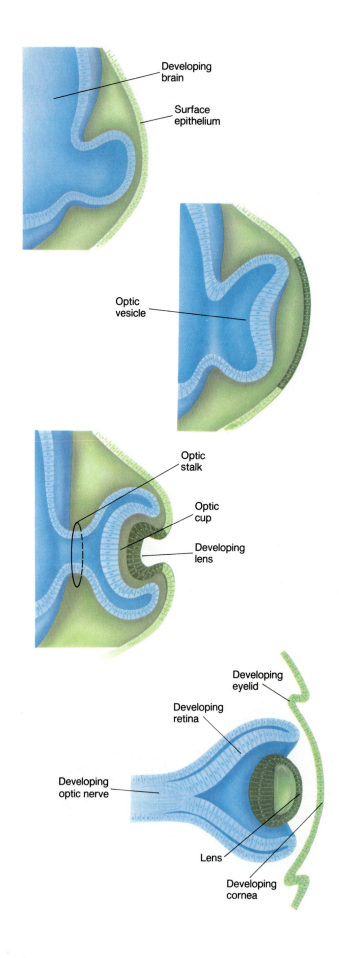

**Figure 24-8**
**Induction of lens formation.** The optic cup "instructs"
the adjacent epidermal cells to invaginate
and develop into the lens of the eye.

reactivation of the same processes that produced the larva—morphogenesis, cell growth and division, and systematic cell death—reorganize the larva's tissue into those of the adult. For example, in butterfly and moth larvae (caterpillars), slow-growing patches of cells called *imaginal disks* are tucked away in the caterpillar. After the larva encases itself and enters the inactive **pupa** stage, programmed cell death sweeps through the larval tissue. As larval tissue dies, cells in the imaginal disks begin to rapidly divide and differentiate, replacing the destroyed tissue with cells that form the adult. The mature insect that emerges is specialized for reproduction and dispersal, unlike the ravenous but slow-moving larva.

Not all insects pass through a pupal stage during metamorphosis. In some insects, like grasshoppers and cockroaches, the larvae look like miniature versions of the adult. These larvae, called **nymphs**, go through several **molts**, in which they shed their hard outer layer (the *cuticle*) as they grow and secrete a new layer over the larger body. With each molt, their appearance is slightly altered. The last of these molts, the final step in metamorphosis, produces the sexually mature adult.

Insects are not the only animals that undergo metamorphosis. The familiar transformation of a tadpole into a frog is a continuous series of events that gradually produce an air-breathing adult from the water-bound larva.

Metamorphosis is associated with genetically programmed changes in behavior as well as anatomy, for an animal must "know" how to use its new structures for them to be of any advantage (see **Animal Behavior III**). A butterfly would not succeed if it continued to act like a caterpiller.

## Regeneration

Animals possess reserves of undifferentiated *stem cells* that are used to replace tissue that has either died or been injured. But replacement of lost parts (**regeneration**) may depend less on these stem cells than on *dedifferentiation* of cells that have already specialized. During dedifferentiation, cells reverse their commitment to their current specialty and revert to a less specialized form. Following amputation of a limb, for example, an adult salamander first repairs the damage with scar tissue formed from stem cells. The scar is soon replaced by a bud of dedifferentiated tissue, which then progresses along a developmental pathway that parallels the embryonic development of the organ until the missing limb has been replaced with a fully functioning arm.

Regenerative capacity is greatest during embryonic development, yet

**Figure 24-9**
**Metamorphosis.** This adult frog has changed from its larval form (a tadpole) to its sexually mature adult form.

many animals retain this ability into maturity (Figure 24-10). When sponges, sea stars, hydras, flatworms, and earthworms are cut into pieces, fragments that are large enough can regenerate a complete organism. In contrast, regenerative capacity is markedly diminished in birds and mammals, which cannot replace limbs or whole organs (although a damaged liver can regenerate itself if a large enough healthy portion remains). Nonetheless, birds and mammals do retain enough regenerative capacity to heal most types of wounds.

## Human Development

### The Embryo's Lifeline

The development of a human embryo begins with fertilization and, six to eight days later, implantation in the prepared endometrium (Figure 24-11). The embryo enzymatically burrows into the endometrium, rupturing uterine blood vessels as it penetrates. By the time the entry site reseals itself, the embryo is surrounded by a pool of maternal blood, a pool that is continually replenished by fresh blood from maternal vessels. Branching projections, called *villi*, sprout from the **chorion**, the embryo's outer membrane covering. These chorionic villi provide a large surface for exchange of respiratory gases, nutrients, and wastes between embryo and mother until a placenta is constructed from the embryo's chorion and the mother's uterine lining. Veins and arteries develop in the villi, connecting the embryo's early circulatory system with the blood-filled space, which serves as a temporary life support system. These connecting vessels soon become channeled within an **umbilical cord**, one end of which is attached to the embryo's belly and the other end to the placenta. In the placenta, the embryo's vascularized chorionic villi are bathed in maternal blood. The two blood supplies are separated by only a thin layer of cells permeable to nutrients, gases, and wastes.

The placenta, which also produces hormones essential to continuing normal pregnancy, takes over HCG (human chorionic gonadotropin) production from the developing embryo. Recall from Chapter 23 that HCG stimulates the corpus luteum to secrete estrogen and progesterone, which inhibit ovulation and prevent menstruation during pregnancy. The placenta eventually abandons HCG production and begins direct production of estrogen and progesterone.

### Stages of Development

During the first two months following fertilization, all the major organs appear, many in a functioning capacity. By the end of the eighth week, the embryo has acquired a distinctively human form. Up to this point, the developing offspring is still called an embryo. During the last seven months in the uterus, however, it is referred to as a **fetus**. In the fetal stage, organ refinement accompanies overall growth as the fetus develops the ability to survive outside the uterus.

The 266-days between conception and birth are traditionally grouped into three stages, each called a *trimester*. The most dramatic changes occur during the first of these periods.

### THE FIRST TRIMESTER

During the first three months of development, the embryo is transformed

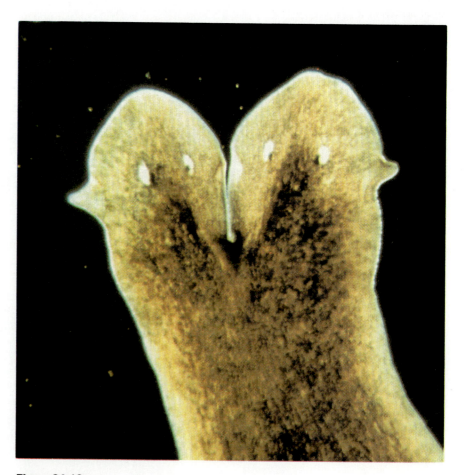

**Figure 24-10**
**Each as good as the original.** When this flatworm's head is split, each half regenerates its "lost" half, producing two complete heads.

# ANIMAL BEHAVIOR III

## THE DEVELOPMENT OF BEHAVIOR

Behavior is a biological process, as much a part of an animal as are its anatomical features and physiology. An animal's behaviors appear just as do its physical structures—during embryonic and post embryonic development. In mammals, for example, suckling movements begin during embryonic development in the uterus, even before the behavior is needed. That is also the time when reflex grasping behavior develops. Other behaviors develop later in life as part of post-embryonic development. Singing a birdsong, spinning a cocoon, weaving a spider's web, or learning to walk, run, fly, or speak are all behaviors that develop some time after birth or hatching. Some of these are purely innate (instinctive) behaviors with highly predictable stereotyped patterns that are executed automatically at a programmed time. Others are *learned behaviors*, those that are influenced by an animal's experience. Both types of behaviors, instinctive and learned, are considered part of an animal's normal development.

Instinctive patterns dictate virtually all the behavior of small animals with short life spans. Most invertebrates, for example, perform stereotyped activities that cannot be modified by experience. Small animals have neither large enough brains to house the elaborate neural circuits needed for complex behavior and learning, nor time for learning to occur. Large animals live longer and have brains with a larger *associative cortex*. This is the portion of the brain responsible for the development of **associative learning**, the linking, through experience, of distinct and otherwise unrelated objects or phenomena. Examples of associative learning are everywhere. Here are just a few.

- The swarms of sea gulls that follow large boats have learned to associate food with these vessels (pieces of dead fish are frequently found in the vessel's wake).

- Baby birds learn to open their mouths at the sound of their mother.

- Some wild animals that used to avoid humans now seek out campsites, having learned that campers are sources of food. Raccoons and bears have even learned how to successfully beg for handouts from people. They have made an association between three otherwise unrelated entities—the presence of humans, their own behavior (begging), and food.

(Unfortunately, the association has increased the danger of animal attacks on people.)

- You are doing associative learning right now as you read this passage and extract its meaning.

Thinking is perhaps the most complex form of associative learning. It is fundamental to post-embryonic development in humans. Perhaps the simplest form of associative learning is **imprinting**, an innate response by newborn birds, one that allows the chick to immediately identify its parent. The first moving object the newly hatched bird sees and hears is adopted as its parent, and the chick follows it automatically. When investigators substituted a ticking clock for the parent bird, the new chicks imprinted on the clock, adopting it as "mother." When the clock was moved, the chicks followed.

Somewhere between imprinting and thinking is **imitation**, a somewhat complex way by which genetic programs are modified by learning the behavior displayed by another individual. The hunting skills of cheetahs, lions, and other cats may seem "instinctive," but young cats learn to hunt by copying the behavior of adult cats. Cats raised in isolation don't learn the behavior. The way some birds learn to sing their particular mating song provides another example of imitation. Immature birds come equipped with the genetic ability to sing—the ability to produce all the right notes. But in some species the bird must receive "lessons" to create the proper song. Only when it hears another bird of the same species sing does it learn how to put the notes together in the correct sequence. A deaf bird can sing, but the song is garbled. Instinct provides the basic foundation, but learning refines or modifies the basic program.

As with other large animals, human behavior is a composite of instinct and learning. Our behavior is so thoroughly modified by learning, however, that it is difficult to determine the degree to which it is innate. For example, most behavioral biologists agree that humans, like other vertebrates, inherit a biological propensity for aggressive behavior and a drive to defend territorial boundaries. But our innate territoriality and aggression is modified by the associative cortex; otherwise we would be engaged in perpetual war. Our actual behavior is usually the result of choices made available to us by experience. This too is part of our behavioral development.

from a single-celled zygote, into an ensemble of organ systems that are only slightly less complex than a full-term baby. Cleavage, blastulation, and implantation occupy the first week or so following fertilization. At the time of implantation, the embryo is called a **blastocyst**, a mass of embryonic cells enclosed in a hollow ball of cells, the **trophoblast**. (These and other structures discussed in the next two paragraphs are depicted in Figure 24-11.) The trophoblast secretes the enzymes that allow the blastocyst to embed in the uterine wall, after which the trophoblast differentiates into the chorion, the outermost of four *extraembryonic membranes*. In addition to its previously discussed roles, the chorion secretes the menstruation-preventing hormone HCG. The innermost membrane, the **amnion**, envelops the young embryo and encloses the *amniotic fluid* that suspends and cushions the developing body throughout its duration in the uterus. A third membrane forms the **yolk sac**, which, in humans, encloses a clear fluid rather than a supply of nutrients. The yolk sac manufactures blood cells for the early embryo (until the liver temporarily takes over the job) and later helps form the umbilical cord. The fourth membrane, the **allantois**, is rich in blood vessels and eventually helps form the vascular connections between mother and fetus. (These same four extraembryonic membranes are also present in the eggs of birds and reptiles, a similarity that reveals evolutionary relatedness. Other embryonic similarities among animals are discussed in Connections: Embryonic Development and Evolution page 414.)

A few days after implantation, the embryo beings to gastrulate. The cell mass in the blastocyst becomes a double layer of cells called an *embryonic disk,* separated from the chorion by a newly formed space, the *amniotic cavity.* The two layers in the embryonic disk are the endoderm and ectoderm. An elongated indentation creases the disk's ectodermal surface. The middle germ layer, the mesoderm, is formed by cells migrating inward along this indentation, the so-called "primitive streak". Toward the end of neurulation, the embryo folds around part of

its yolk sac, bringing its "edges" together to form a hollow "tube within a tube," as illustrated below.

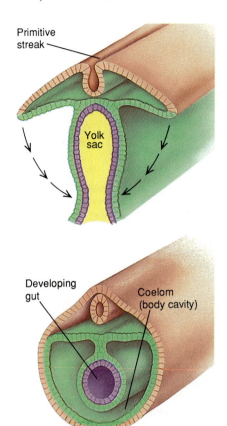

The enclosed space becomes the body cavity and the yolk sac gives rise to the digestive tract. The endoderm forms the inner lining of the embryo, and the ectoderm covers the outer surface. Much of the mesoderm is sandwiched between these two layers, although some mesodermal cells remain in extraembryonic spaces. The bending of the embryo disk into a tube pulls the overlying amniotic cavity with it until it completely surrounds the embryo in a bubble of amniotic fluid.

Development proceeds in a *cephalocaudal* order, that is, the head end develops faster than the tail end (cephalo = head, caudal = tail). The young embryo's head therefore acquires enormous proportions compared to the rest of the body, which doesn't "catch up" until after puberty.

The embryo begins forming its first major organ system during its third week. At this time ectodermal ridges arise along the primitive streak, the first indications of the neural tube

(Figure 24-12). Blocks of tissue, called **somites**, form in pairs along the developing spinal column. From these paired sections the vertebrae, part of the ribs, some muscles, and surrounding tissue develop. The fetal heart begins to form almost as quickly, when two tubes of mesodermal tissue fuse to form a hollow pump that, early in the embryo's third week, begins the first of its 2.5 billion or so heart beats. Mesodermal tissue in the yolk sac and chorion begin manufacturing blood as early as the fifteenth day. By the end of the first month the embryo, though still less than half a centimeter (3/16 inch) long, has begun forming its lungs, liver, and several other internal organs. The heart is now a four-chambered pump. The first signs of eyes and a nose appear, and four buds protrude from the side of the C-shaped embryo. By the end of the following week the buds will be vaguely recognizable as developing arms and legs.

During the second month, the embryo begins developing the structures that will specialize in digestion. The liver takes over its temporary job as the main blood-producing organ, while an intricate cartilage skeleton begins to ossify. The brain develops cerebral hemispheres, and spinal nerves branch from the vertebral column. The face acquires a distinctively human look, with slate-colored eyes and the beginnings of eyelids. Muscles are forming and assuming their permanent relationships. Fingers and toes become unmistakably evident. The embryo becomes a fetus.

The third month is devoted primarily to growth and development of existing structures. Formation of genitals during the third month, however, reveals the sex of the fetus. The limbs are clearly recognizable, with well-sculpted fingers and toes, complete with nails. The head becomes erect as the neck becomes better defined. The fetus begins reflex movements that go undetected by the mother.

The first trimester is unquestionably the most dangerous time for the developing individual. The magnitude of the changes that occur during this time renders the embryo particularly vulnerable to disease and agents that would be relatively harmless to an

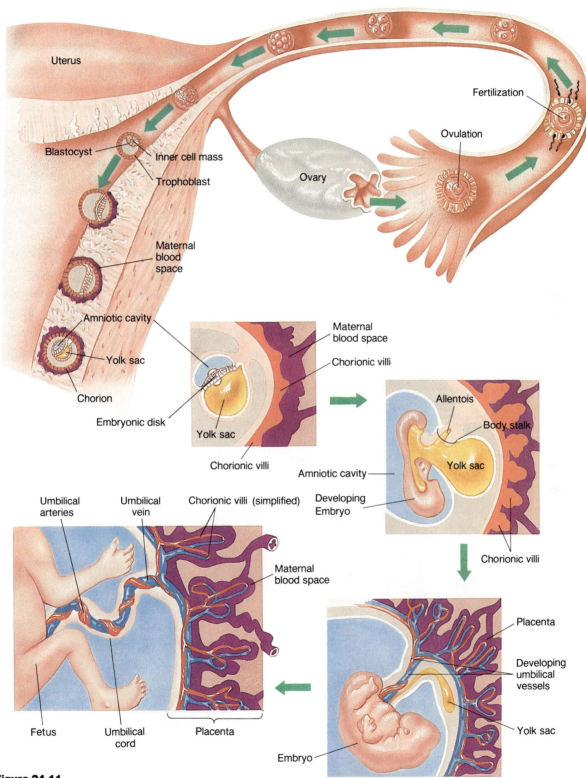

**Figure 24-11**

**Implantation and placenta formation.** Following fertilization, cleavage and blastulation produce a ball of cells called a blastocyst. By the time of implantation in the uterine wall, gastrulation produces the three-layered embryonic disk (in orange) that gives rise to the embryo (the mesoderm is not shown). The amnion contains the liquid in which the embryo floats as it develops. The chorion and allantois help form the placenta and umbilical cord. Notice that fetal and maternal blood do not mix, but are separated by the selectively permeable membranes of the chorion and capillaries.

older fetus or a newborn human (see Bioline: The Dangerous World of a Fetus).

## THE SECOND TRIMESTER

As the 7.5-centimeter (3-inch) fetus enters its fourth month, it begins sucking and swallowing reflexes, and soon starts kicking. The mother does not feel these movements, which are often referred to as *quickening,* until the fifth month. The ossifying skeleton is clearly distinguishable by X-ray examination, and the bone marrow begins blood production. The fetus acquires a downy coat of soft body hair, called *lanugo,* that covers all the skin. Lanugo is believed to help fat droplets secreted by fetal sebaceous glands stay on the skin. These droplets form a greasy protective coating, the *vernix caseosa,* that conditions the fetal skin during its aquatic tenure in the amniotic fluid. Midway through the pregnancy, two heart beats become discernible in the mother — her own (72 beats per minute) and that of the fetus (up to 150 beats per minute). Convolutions appear in the cerebral cortex of the brain, and sense organs begin supplying the fetus with limited information about its environment.

Although most of the organ systems are at least partially functional by the end of the second trimester, the fetus has little chance of survival if born before the seventh month. Even with expert medical care, a 23-week-old fetus has only a 1 in 10 chance of surviving outside the uterus. If born a month later its chances of survival jump to 50 percent.

## THE THIRD TRIMESTER

During the last three months of pregnancy, the fetus increases its body weight by 500 to 600 percent — from one and a quarter pounds (about half a kilogram) to about 7 pounds (3.2 kilograms). This growth spurt slows to about a 30 percent increase during the final month. The brain and peripheral nervous system enlarge and mature at an especially rapid rate during the final trimester. Neurological performance later in life, including intelligence, depends on providing the proteins needed for fetal brain development. (Women who suffer protein deficiency during this time have babies that are mentally slower throughout life. This condition cannot be reversed by providing the child with a protein-rich diet.) By the end of the third trimester,

**3 weeks —**
Organogenesis begins with neural tube formation, seen here as enlarged crests that delineate the neural groove from which the spinal cord and brain develop.
LENGTH — 0.25 cm (1/10 in).

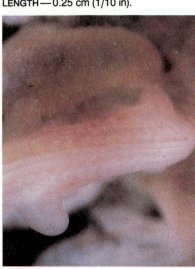

b

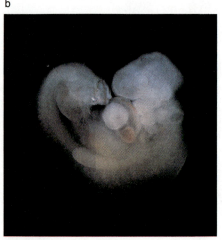

**4 weeks —**
A pumping heart, somites, developing eye, arm and leg buds are evident. The embryo has a long tail, gill arches, and a relatively enormous head.
LENGTH — 0.7 cm (1/3 in).

**5 weeks —**
Internal organ development is well underway. Fingers are faintly suggested. Fetal circulatory system is evident and body stem is now an umbilical cord.
LENGTH — 1.2 cm (1/2 in).

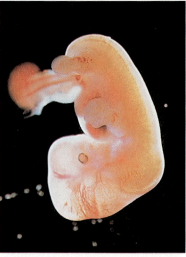

a

c

**Figure 24-12**
**Some stages in human embryonic and fetal development.**

the fetus is capable of regulating its temperature and can control breathing. This ability to control breathing, together with the degree of lung maturation, generally determines a premature infant's chances of survival.

The formerly lean fetus changes in appearance as fat deposits form under the skin, imparting the rounded, chubby shape characteristic of babies. Lanugo falls away before birth, although the cheesy vernix caseosa still covers part of the skin. In males, the testes descend into the scrotum. Most fetuses change position in the uterus, becoming aligned for a head-first plunge through the birth canal.

One system that does not mature by the time of birth is the immune system —human babies are born "immunologically incompetent." During the final fetal month, however, antibodies from the mother's blood cross the placenta and temporarily fortify the baby's defenses against infectious diseases until its own immune arsenal becomes competent. In breast-fed babies, this immunological gift is supplemented by the maternal antibodies and immune cells in breast milk.

## PARTURITION (BIRTH)
After nine months of development, a human fetus is six billion times larger than the original zygote. Yet this astonishing size increase probably plays no direct role in inducing birth. In fact, the fetus is believed to have little to do with initiating the uterine contractions of **parturition** (birth). Although the birth-triggering factor(s) remains a mystery, in most women the onset of labor is preceded by a drop in the concentration of progesterone, the hormone that inhibits contractions of the smooth uterine muscles during pregnancy. Once labor begins, a cascade of events promote uterine muscle contractions of increasing strength and frequency. The posterior pituitary releases oxytocin, which stimulates muscle contraction in the uterus. This response then triggers a reflex release of even more oxytocin, establishing a positive feedback loop that intensifies contractions. (Oxytocin also plays a role in lactation, as discussed in Chapter 23.) Prostaglandins (Chapter 16) are also believed to participate in labor; this contention is supported by the ability of aspirin, a prostaglandin antagonist, to delay parturition. Both

d

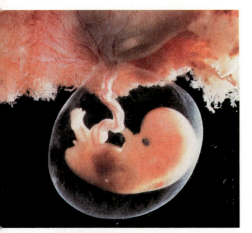

**2.5 months—**
The fetus, seen here floating in the amniotic cavity, now has all major organ systems. The umbilical blood vessels and placenta are well defined. Tail and gill arches have disappeared.
LENGTH—3 cm (1 1/2 in).

**5 months—**
The fetus looks very much as she will at birth. Bone marrow is assuming more of the blood-producing duties. Mother begins to feel fetal movements, even hiccupping.
LENGTH—25 cm (10 in).

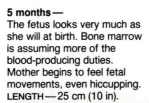

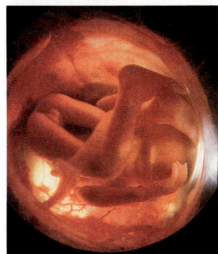

e

f

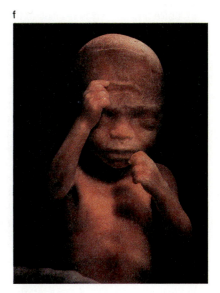

**5.5 months—**
The vernix caseosa is accentuated by a groove the fetus has scraped away with its thumb. Internal organs now occupy their permanent positions (except testes in males).
LENGTH—30 cm (12 in).

CONNECTIONS —

Classification and Phylogeny (page 11)
Evidence for Evolution (page 498)
Animal Diversity (page 605)

## EMBRYONIC DEVELOPMENT AND EVOLUTION

Comparing the physical characteristics of different animals often reveals striking similarities. Many of these similarities show **homology** between two structures, that is, different versions of the same anatomical design. Although they perform different functions, a bat's wing and a human hand possess a structural similarity that both species inherited from a common ancestor far back in evolutionary time. Not all physical similarities, however, are due to common ancestry. The wings of insects and birds, for example, are not homologous, even though both are used for flying. One way to distinguish true homology from structures that evolved independently of each other is to look at the complexity of the similar features. The multi-bone composites that form the skulls of chimpanzees and humans are virtual matches of each other, a similarity that would not exist had they evolved independently. These structural similarities furnish clear evidence that chimps and humans descended from a common ancestor.

Another way of detecting homology is to examine animal embryos for similarities that cannot be detected in the fully developed animals. Embryonic similarities among vertebrates are especially striking. For example, the resemblance between the four extraembryonic membranes of mammals and those in reptile and bird eggs testifies to the common ancestry of all three groups. Even when the original function of a membrane is no longer needed by a particular species, the membrane remains as a vestigial reminder of the organism's origins. (Although humans and other mammals have no yolk in the embryo, they still have a yolk sac left over from an early "yolked" ancestor.)

But the similarities among the embryos themselves provide the most startling evidence of evolutionary relationships (see accompanying figure). In every vertebrate embryo, the head end develops faster than the tail end. Early events in the formation of the central nervous system are similar for all vertebrates, starting with neural tube development on the dorsal ("back") side of the embryo. All vertebrate embryos, including the embryos of humans and other air-breathing mammals, develop embryonic gill slits or pouches that either become functional gills (as in fishes) or become modified into different structures (such as the eustachian tubes that drain fluid from your middle ears into your mouth). Human embryos, like all vertebrate embryos, possess an elongated tail, although only briefly. Gill pouches and tails are more vestiges of our primitive ancestry, as is the "head kidney," the first kidney to appear in the vertebrate embryo. Although in reptiles, birds, and mammals it never produces urine, the head kidney is nonetheless retained. Apparently, it is an *organizer* that directs (or induces) the later development of the functional kidneys.

These are just a few of the similarities that reveal our evolutionary ties to other animals.

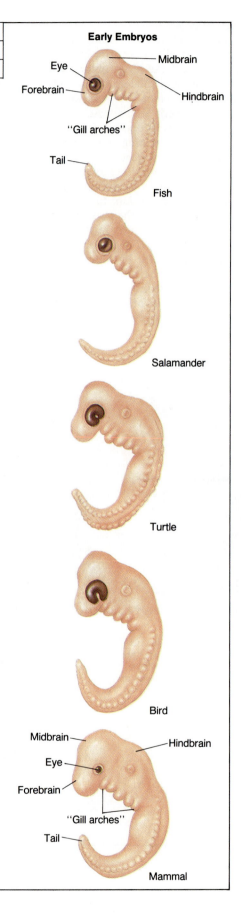

Similarities in early embryos of vertebrates.

# B I O L I N E

## THE DANGEROUS WORLD OF A FETUS

"The womb" is often used as a metaphor for warmth, protection, and security, a world free of fear and danger. Yet a fetus may be exposed to a great many threats while residing in its mother's womb (the uterus), including nutrient deficiency, infectious microbes, radiation, alcoholic beverages, and the chemicals in drugs, cigarettes, and environmental pollutants. The placenta screens out most harmful substances, but some dangerous agents pass through the placental barrier and enter the fetal circulation. Radiation passes through *all* living barriers, bombarding the fetus with its *teratogenic* (embryo-deforming) potential. Large doses of X-rays during the first trimester of pregnancy, for example, may cause mental retardation, skeletal malformations, small head size, and predispose the person for later development of leukemia. When a cell is impaired during the first trimester, the millions of cells that develop from the affected cell inherit the impairment. The effect is therefore amplified. These cells fail to differentiate normally and produce organs that are malformed, some so severely that the fetus or newborn child has little chance of survival.

The virus that causes *rubella,* or "German measles," illustrates the vulnerability of the embryo during its first trimester. The disease is so mild that a pregnant woman with rubella may be unaware that she has the disease. But the virus passes through the placenta and infects the emerging embryonic organs, interfering with their normal development. The result is physical malformations, mental retardation, deafness, or death in about half of the babies exposed during their first six weeks of development. Yet second- and third-trimester fetuses exposed to rubella suffer no impairment. Routine vaccination against rubella has dramatically reduced these tragic occurrences.

Another microbe that can reach the fetus is the bacterium that causes syphilis, a killer that is not so selective about the trimester. A mother with syphilis can transmit the disease to her developing baby during any trimester. One-third will die in the uterus, and, if born alive, the infant may suffer permanent neurological damage and other symptoms of advanced syphilis. Fortunately, most microbes cannot cross the placenta, so the fetus usually remains healthy while the mother is combating a cold, the flu, or most other microbial onslaughts.

Many chemicals to which the mother may be intentionally or accidentally exposed can compromise fetal development. Some of these chemicals may be found in medications, such as the mild tranquilizer, thalidomide. In the 1950s thousands of women who took this drug during early pregnancy delivered babies whose arms and legs had failed to develop. Although stricter drug regulations resulted from this episode, many pharmaceutical products that are potentially dangerous to fetuses are still available. These include vitamins in large doses (especially D, K, and C), cortisone, antibiotics, birth control pills, tranquilizers, anticoagulants, and thyroid drugs. Drugs such as LSD, marijuana, and alcohol expose the fetus to proven toxins or teratogenic agents. Alcohol consumption by a pregnant woman during a critical period of fetal development can lead to *fetal alcohol syndrome.* This collection of deformities includes mental retardation, small head size, facial irregularities, and later learning disabilities.

Tobacco smoke also constitutes a serious danger to the developing fetus and contributes to infant mortality. Heavy cigarette smoking during pregnancy impairs fetal growth and reduces resistance to respiratory infections, such as bronchitis and pneumonia, during the first year of life. Carbon monoxide in tobacco smoke competes with oxygen for hemoglobin binding sites, reducing the oxygen available to the fetus. Nicotine constricts blood vessels, further reducing the blood supply. Tobacco smoke may also contain teratogens and may contribute to development of heart and brain defects, cleft palate, and sudden infant death syndrome (SIDS).

oxytocin and prostaglandins are administered by injection to induce labor when its delay threatens the health of the mother or fetus.

Parturition occurs in three distinct stages (Figure 24-13). During the first stage, the mucous plug that blocks the cervical canal and prevents microbial invasion of the intrauterine environment is expelled. The amniotic sac ("bag of waters") ruptures during this stage, and its fluid is discharged through the vagina. Contractions during this stage force the cervical canal to dilate until its diameter enlarges to about 10 cm (3.9 inches), marking the onset of the second stage of labor, expulsion of the fetus. Powerful contractions move the fetus out of the uterus and through the vagina, usually head first. Fewer than 4 percent of all deliveries are "breech," in which the buttocks or legs come first. Contractions during the final stage expel the detached placenta, called the *afterbirth,* from the uterus. Additional contractions help stop maternal bleeding by closing vessels that were severed when the placenta detached.

A newborn infant quickly acclimates to its terrestrial environment. With the umbilical cord clamped and severed, the baby rapidly depletes its source of maternal oxygen, and, still unable to breathe, respiratory waste ($CO_2$) accumulates in the blood. The brain's respiratory center responds to this increase by activating the breathing process, and the lungs inflate with air. But before the lungs can oxygenate the blood, a circulatory modification is required, since, in the uterus, the unneeded fetal lungs were bypassed by the bloodstream. The septum between the right and left atria of the fetal heart is perforated by a large hole, called an "oval window," that allows blood to flow directly from one side of the heart to the other rather than traveling through the pulmonary circulation. At birth, the oval window is immediately closed by a hinged flap, forcing the blood to travel through the lungs to get to the other side of the heart. At the same time, blood vessels to and from the now useless umbilical cord are sealed by muscular contractions, as are vessels that bypassed the fetal lungs and liver.

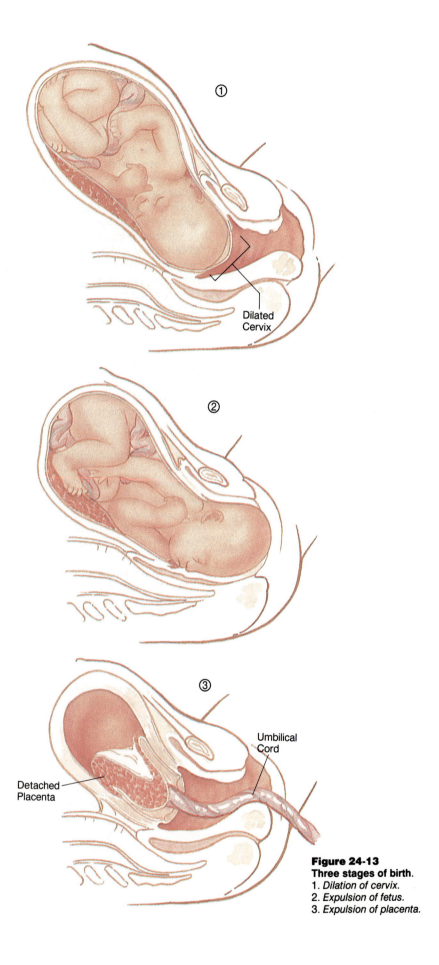

**Figure 24-13**
**Three stages of birth**.
1. *Dilation of cervix*.
2. *Expulsion of fetus*.
3. *Expulsion of placenta*.

## Synopsis

### MAIN CONCEPTS

• **Following fertilization, a zygote gives rise to a new individual** that contains cells differentiated for dramatically different specialties. Early differentiation is determined by what cytoplasmic components a cell acquires following cleavage of the zygote. These components apparently determine which genes will be expressed in that cell.

• **Embryonic induction determines the developmental fate of embryonic tissue and the sequence of differentiation.** Each structure is formed only after the structure that induces its formation is in place. In vertebrates, embryonic induction initiates organogenesis of the neural tube, which then induces the formation of additional structures. Morphogenesis moves tissues and organs to form the embryo's characteristic pattern. Programmed cell death modifies the structure of organs and eliminates unneeded tissue.

• **Animal embryos typically pass through a triple-layered gastrula (or gastrula-like) stage** that consists of endoderm, ectoderm, and a middle layer of mesoderm. All subsequent tissues in the adult animal develop from these three germ layers.

• **All vertebrate embryos look remarkably similar in early development.** They all develop gill pouches, somites, and an elongated tail, and all begin organogenesis with neural tube formation. In addition, reptiles, birds, and mammal embryos possess the same four extraembryonic membranes. These similarities reflect the common evolutionary origin of all vertebrates.

• **Human development occurs in the uterus.**
   —The blastocyst implants in the endometrium.
   —Gastrulation forms a triple-layered embryonic disk.
   —The outer membrane (the chorion) taps maternal blood (and later develops into the embryonic portion of the placenta).
   —The embryonic disk bends into a tube (the archenteron, which is destined to become the digestive tract).
   —Organogenesis forms the neural tube and, by the end of the second month, the embryonic beginnings of virtually all the organs.

   During the remaining seven months, the fetus refines these structures and grows in size until uterine contractions expel it (and minutes later the placenta) through the vagina.

### KEY TERM INTEGRATOR

---

Embryonic Development

---

| | |
|---|---|
| **embryo** | An organism in the early stages of development. **Cleavages** begin transforming the zygote into a multicelled **morula**. This ball (or disk) of cells then becomes a hollow structure, such as a **blastula** (a **blastocyst** in humans). Invagination produces a double-layered **gastrula** which acquires a middle layer of **mesoderm** sandwiched between the **ectoderm** and **endoderm**. These three **germ layers** of the embryo **differentiate** to form all the adult's structures. |
| **embryonic induction** | The development of new specialized tissue in response to signals from adjacent embryonic cells. The process assures that **differentiation**, **morphogenesis**, and **organogenesis** produce parts that fit together for proper form and function. |

Postembryonic Development

| | |
|---|---|
| **metamorphosis** | Transformation of a **larva** or **nymph** into an adult. The change may occur during a **pupa** stage or more gradually. Some insects go through several **molts**. |
| **regeneration** | Development of replacement parts when tissue is lost or destroyed. |
| **puberty** | The attainment of reproductive capacity in humans. |

Human Development

| | |
|---|---|
| **fetus** | The human embryo after two months of development until **parturition**. The embryo's lifeline is the **umbilical cord**, which connects the embryo to the placenta. The **trophoblast** gives rise to the **chorion**, the earliest connection to the maternal blood supply. Blood vessels in chorionic villi, along with the **allantois**, help form the umbilical cord. Throughout development, the embryo and fetus float in fluid enclosed by the **amnion**. |

## Review and Synthesis

1. Arrange these terms in the order of development: neural tube formation; parturition; blastulation; fertilization; gastrulation; gametogenesis (meiosis).

2. What role does cleavage of the zygote have in determining the developmental fate of each cell in the blastula?

3. How does embryonic induction help determine the sequence of organogenesis? Why is sequence so important in embryonic development?

4. Distinguish between morphogenesis and cell differentiation. Why is the movement of each cell into its proper location as important as gaining its particular specialty through differentiation?

5. Describe four movements associated with morphogenesis. Which movements participate in the following? neural tube formation; formation of the eye by embryonic induction; final placement of testicles; gastrulation.

6. Three types of postembryonic development (metamorphosis, regeneration, and puberty) were discussed in this chapter. In which group(s) of animals would each of these types be most important?

7. Why would a single mutation in a cell of a four-week old embryo be much more detrimental than the same mutation in a cell of a five-month old fetus?

8. Although the yolk sac and allantois are vestigial structures in human embryos, both contribute in an important way to human development. How? Also discuss the role of the other two extraembryonic membranes in the development of a human. What is the evolutionary significance of all four membranes?

## Additional Readings

Hayflick, L. 1980. "The cell biology of human aging." *Scientific American* 242:58–65. (intermediate.)

Hopper, A. and N. Hart. 1985. *Foundations of Animal Development.* 2nd ed. Oxford University Press, London. (Intermediate to advanced.)

Human Body Series. 1984. *Reproduction, The Cycle of Life.* Chapters 3, 4. U.S. News Books. (Introductory.)

Wolpert, L. 1978. "Pattern formation in biological development." *Scientific American* 239:154–64. (Intermediate.)

# 7

# *Genetics and Inheritance*

# Chapter 25

# *On the Trail of Heredity*

- Experts in genetics counsel potential parents who have a high probability of bearing children with genetic deficiency diseases.

- Forty years ago, PKU victims would have suffered permanent mental retardation. Thanks to geneticists persons born with PKU (phenylketonuria) can now lead normal, healthy lives. PKU is one of 250 known "inborn errors of metabolism," inherited genetic defects that disable critical enzymes. Many of these, including PKU, are successfully treated by regulating diet, so that the victims can lead healthy lives.

- For decades, a working knowledge of modern genetics has been used to create new varieties of plants and animals with traits considered advantageous by breeders. Now scientists are directly changing the genes themselves, redesigning organisms to improve crops, milk production, and meat supplies, and to create inexpensive high-grade sources of medicines and other valuable products (Figure 25-1).

- Soon we may be able to cure hemophilia, cystic fibrosis, or some of more than 2500 other genetic disorders. The tool is *gene therapy*—splicing into human chromosomes functionally perfect genes to replace the faulty ones. Gene therapy may require treating a patient before that person is even born.

Our discovery of the fundamental principles of heredity and genetics has launched a scientific revolution, one that has vastly enriched our understanding of life. The principles of genetics help account for the enormous diversity of organisms on earth, the existence of millions of species instead of just one very successful one. They explain a driving force underlying evolution and illustrate how a species becomes adapted to its environment.

They have taught us the importance of sex, how it strengthens a species and improves the long-term survival of species by providing a wealth of genetic "opportunities" compared to those obtainable from asexual reproduction.

The principles of modern genetics have also launched a new age of applicability, often called "genetic engineering" (Chapter 39). But let's shine the spotlight back in time and examine how this revolutionary age in genetics began.

It all started quietly and with little notice. Modern genetics was born in a garden, where one man patiently pursuing his hobby discovered the founding principles, the "laws" of inheritance. The man was a nineteenth-century monk named Gregor Mendel (see Discovery: The Premier Genetic Detective, page 431).

## Mendel's Discoveries

Mendel's results provided the principles on which an entire branch of genetic studies, called *Mendelian* genetics, is based. Briefly stated using modern terms, he found that

- Hereditary traits are determined by discrete units of inheritance known as **genes** (Mendel called genes "factors").

- A gene for a particular trait may have more than one form of expression. (Each alternative form of the gene is called an **allele**.)

- A (diploid) organism has two copies of each gene, so one individual may possess two identical or two nonidentical alleles.

- The two alleles for a particular trait separate from each other during gamete formation, each going to a different gamete.

- The distribution of a pair of alleles for one trait does not influence how other pairs of alleles (for different traits) are distributed. Hence, alleles for different genes are assorted independently of one another (see Chapter 12), producing a random combination of characteristics in offspring.

**Figure 25-1**
**Redesigning Organisms.** Scientists can now genetically alter some organisms to improve their value to us. The scientist in this photo holds a flask containing nucleotides, the building blocks from which genes are constructed. Such modern technology is the product of one and one half centuries of investigations into the nature of heredity.

The significance of Mendel's work is difficult to extract from these condensed versions of his findings. Let's therefore examine his conclusions and see how he arrived at them. Later we'll consider some exceptions to these principles.

## Organisms Possess Two Alleles for Each Gene

At the core of Mendel's studies lay his conclusion that every organism possesses physical units of heredity (his "factors") that determine its genetic characteristics, and that each of these factors (genes) was responsible for determining a specific trait. He also proposed that there were two alternate forms for a determining factor, so a single trait may be expressed with different characteristics. In garden peas, for example, there are two forms of the factor that specifies flower color—one allele for white and another for purple. Thus, Mendel's pea plants produced either white or purple flowers, depending on which alleles the plant inherited from its parents.

We now know that, in diploid organisms, the two alleles for each trait occupy the same position on two separate homologous chromosomes, as depicted in Figure 25-2. Each genetic location, called a **locus**, corresponds to a gene for a particular trait. If an organism has two identical alleles, it is referred to as **homozygous** for that particular trait. (We typically use capital and lower case letters to symbolize the two forms of the same allele. The two homozygous states can be represented as *AA* or *aa*.) A homozygous individual *breeds true* for that trait because all its offspring receive the same allele from this parent. **Heterozygous** (*Aa*) organisms are those that possess two different alleles for a trait. (Notice that these terms describe traits rather than the whole organism; a single individual can be heterozygous for some traits and homozygous for others.) In the heterozygous pea plants studied by Mendel, however, only one of the alleles, the **dominant** form (*A*), is expressed, masking the presence of the other allele (*a*), which is thus called **recessive**. All seven of the traits Mendel studied had two forms of expression, one dominant and one recessive (Table 25-1).

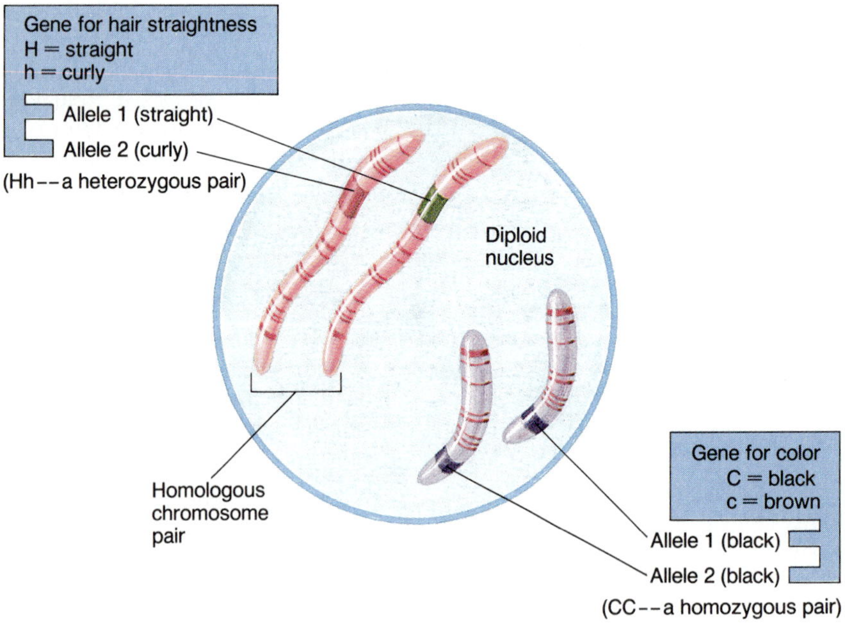

Mendel soon discovered that even if two plants looked identical for a particular trait, they would sometimes produce progeny that resembled neither parent for that characteristic. Two purple-flowered pea plants may produce some offspring with white flowers (just as two dark-haired people can produce a blond child). Even following self-fertilization, nonparent-like traits sometimes emerged, a tall pea plant producing both tall and dwarf offspring. As Mendel thus discovered, it is not always possible to deduce which alleles an organism possesses simply by observing its outward appearance. Apparently, organisms have alleles whose effects remain "hidden" when masked by the presence of certain other alleles. In other words, an individual's **genotype** (genetic makeup) cannot be determined solely from its **phenotype** (the observable characteristics that are the expression of its genotype).

Mendel demonstrated how the genotype for a trait could be tracked by following the phenotypes of large numbers of carefully bred individuals and their offspring. He would begin his experiments by crossing two strains that bred true for differing forms of a

**Figure 25-2**
**Stylized representation of alleles on homologous chromosomes in a diploid nucleus.** Alleles are identified by using letters—capital letters for dominant alleles, the same letter in lower case for the recessive counterpart. ("Dominant" and "recessive" are discussed in the text.) Although each chromosome would actually contain thousands of genes at different loci, only one gene (two alleles) per pair of chromosomes is identified in this illustration. Mendel died before the discovery of chromosomes and diploidy, yet his findings predicted the existence of this type of genetic arrangement.

Gene for hair straightness
H = straight
h = curly

Allele 1 (straight)
Allele 2 (curly)
(Hh – – a heterozygous pair)

Diploid nucleus

Homologous chromosome pair

Gene for color
C = black
c = brown

Allele 1 (black)
Allele 2 (black)
(CC – – a homozygous pair)

TABLE 25-1

| Seven Traits of Mendel's Pea Plants | | |
| --- | --- | --- |
| **Trait** | **Dominant Allele** | **Recessive Allele** |
| Height | Tall | Dwarf |
| Seed color | Yellow | Green |
| Seed shape | Round | Angular (wrinkled) |
| Flower color | Purple | White |
| Flower position | Along stem | At stem tips |
| Pod color | Green | Yellow |
| Pod shape | Inflated | Constricted |

trait. In other words, he would obtain a **hybrid** offspring by crossing two plants homozygous for the opposite forms of a trait. Often he limited his observations to a single trait. (These single-trait experiments are called **monohybrid crosses**.) In other experiments he tracked the fate of two traits through several breedings, called **dihybrid crosses**. Monohybrid and dihybrid crosses provided results that enabled Mendel to formulate the principles of allele segregation and independent assortment.

## Segregation of Paired Alleles

Mendel's simplest experiments consisted of two consecutive generations of monohybrid crosses. For example, he cross pollinated true breeding tall plants with true breeding dwarf plants. (These "parents" are referred to as the $P_1$ generation.) In the first generation of offspring, called the $F_1$ (*first filial*) generation, the recessive trait disappeared altogether; there were no dwarf plants regardless of how many peas or progeny plants he examined. When these tall $F_1$ plants were allowed to self-fertilize, however, the recessive trait (dwarfism) reappeared in a few of their offspring (the $F_2$, or second filial, generation), in a ratio of approxi-

mately 3 tall to 1 dwarf. Each of the seven traits selected by Mendel showed this same pattern of inheritance.

Mendel sought to understand what happened to the disappearing traits during the $F_1$ generation, and how they could reemerge in the $F_2$ offspring. This question led to his proposal of the existence of paired genetic factors, and to his discovery of dominant and recessive characteristics. He concluded that, although the $F_1$ offspring looked identical to the parent ($P_1$) with the dominant characteristic, they differed from this parent in the types of factors they possessed. The true breeding parents were homozygous ($AA$ or $aa$), whereas the hybrid progeny were heterozygous ($Aa$), the recessive allele masked by the dominant allele. The recessive trait was expressed only in individuals that lacked the dominant allele—in other words, plants that are homozygous for the recessive allele ($aa$). (Although the terms "allele," "homozygous," and "heterozygous" were introduced after Mendel's death, we use them here to clarify his explanations.)

Mendel concluded that the only way a recessive trait could reappear in such a constant 3 to 1 ratio was if, during sex cell formation, the paired alleles

separated from one another, each allele going to different sex cells. The $F_1$ generation would all be heterozygous ($Aa$), since each received one dominant allele ($A$) and one recessive allele ($a$) from their homozygous parents ($AA$ and $aa$) (Figure 25-3). Alleles again segregate during gamete formation by the $F_1$ heterozygous plants. An "$Aa$" parent makes two genotypes of gametes, one carrying "$A$" and the other carrying "$a$". If an $Aa$ individual self-fertilizes, each of its offspring has an equal chance of receiving either allele. This creates three possible genotypes in the $F_2$ generation—homozygous dominant ($AA$), heterozygous ($Aa$), and homozygous recessive ($aa$).

One way to visualize the results of such crosses is to use the **Punnett square method**, as illustrated in Figure 25-4. Punnett squares predict the possible genotypes and their expected ratios, when there is an equal probability of acquiring either of the two alleles. Mendel's results revealed a ratio of dominant to recessive phenotypes of approximately 3 to 1. This is just what the Punnett squares predict it should be, supporting Mendel's contention that alleles separate from each other during gamete formation. This phenomenon is now recognized as the Law of Segregation ("Mendel's First Law"). The ultimate validation of Mendel's first law came years after his death with the discovery of homologous chromosomes and their separation during meiosis (Chapter 12), the mechanism that explains the segregation of alleles during gamete formation (Figure 25-5).

Since the principles of dominance and segregation of alleles apply to all diploid organisms, not just to garden peas, Mendel's findings can help you understand how you inherited some of your characteristics. For example, a "widow's peak" along the forehead, the ability to roll the tongue into a tube, nearsightedness, and the absence of fingerprints are dominant traits. Anyone with one of these dominant alleles will develop the corresponding trait. A child of parents who are both heterozygous for one of these traits has only a 1 in 4 chance of being homozygous recessive, acquiring a phenotype

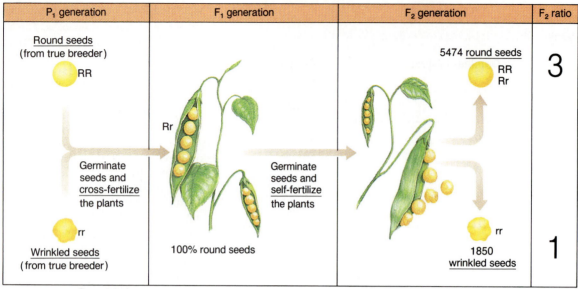

| P₁ generation | F₁ generation | F₂ generation | F₂ ratio |
|---|---|---|---|

R = allele for round seeds
r = allele for wrinkled seeds

**Figure 25-3**
**Mendel conducted monohybrid crosses** like this to track the course of a single trait through two generations. For each of seven plant traits he found a similar pattern of inheritance. From this pattern, he concluded that there are paired genetic factors, one for each variation of the trait (shown here as *R* and *r*), that segregate during gamete formation. The single factor pairs up again with another factor during fertilization. The values given are those actually obtained by Mendel.

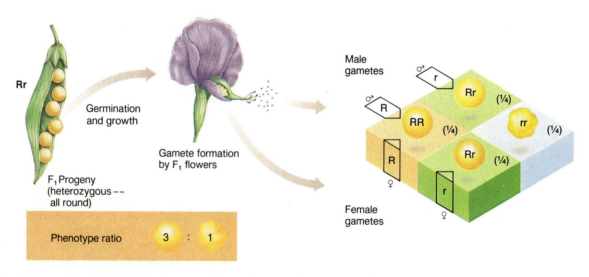

**Figure 25-4**
**The Punnett square method of determining probability ratios.** This visual representation is used here to predict seed shape in offspring following self-fertilization of heterozygous pea plants. Each F₁ gamete will contain either an *R* (smooth) or *r* (wrinkled) allele. All possible male gametes are listed along the top and all female gametes along the side. The possible combinations of alleles following fertilization are shown in the boxes, each representing the genotype created by the union of those two alleles. The *genotype* ratio is 1 : 2 : 1 (*RR : Rr : rr*); there are twice as many *Rr* boxes as either *RR* or *rr*. The 3 : 1 *phenotype* ratio explains why recessive traits reappeared in one-fourth of Mendel's F₂ progeny after disappearing from the F₁ generation.

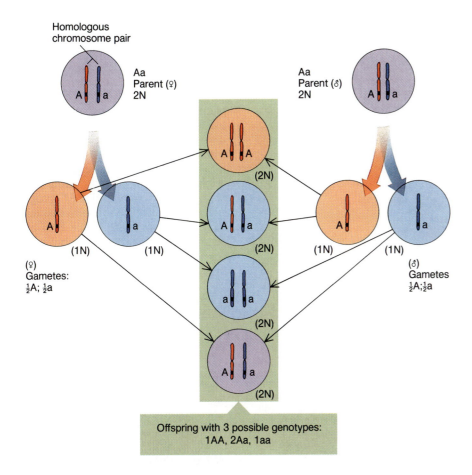

**Figure 25-5**
**Meiosis explains Mendel's findings.** Because they occur on separate homologous chromosomes, alleles (*Aa*) segregate from each other during meiosis to form gametes. Each gamete has an equal probability of receiving an "*A*" or "*a*". Fertilization randomly reestablishes the paired state, so each allele has an equal chance of joining with either of the others. Each offspring will therefore acquire one of the indicated genotypes. It will be twice as likely to be *Aa* than either *AA* or *aa*, since there are two chances of acquiring this heterozygous combination.

the individual is heterozygous (Figure 25-6). A homozygous parent showing the dominant trait can only donate the dominant allele, producing offspring with only the dominant phenotype. If any of the progeny show the recessive trait, the "test parent" must be heterozygous. The results of testcrosses supported Mendel's belief in paired factors for each trait and the segregation of those factors during gamete formation.

## A PROBLEM OF PROBABILITY
Predicting the outcome of genetic crosses is like that of any randomly occurring event: given enough occurrences, the results are statistically predictable. Each sperm (or egg) of a heterozygous individual has an equal chance of acquiring either of the alleles for a particular trait. During self-fertilization (or fertilization with a gamete from another heterozygous individual), each allele has an equal chance of being teamed up with the same type or the alternative type of allele. Such random distribution produces the combinations and ratios predicted by Punnett squares, as long as the number of offspring is large. For example, if you flipped a coin three times and happened to get "heads" all three times, the results could be attributed to chance variation. Three flips are not enough to conclude that the probability of obtaining heads is 100 percent. But if you flipped the coin 100 times, the likelihood of getting heads every time would be greatly reduced; the frequency of obtaining heads would be much closer to 50 percent. In other words, with 100 flips you would be more likely get a heads-to-tails ratio that approached 1:1 than after three flips. Therefore, the larger the number of offspring examined, the closer the results are to the predicted ratio.

The results of genetic crosses can be predicted by using simple multiplication instead of Punnett squares, since *the probability of two independent events occurring together is the product of the probability of each occurring alone.* With garden peas, the chances of a gamete acquiring the allele for white flowers from a heterozygous purple parent is one in two, or 1/2. During fertilization, the probability of

that contrasts with that of the parents. This is just as Mendel's principles predicted. But does this mean that persons with widow's peaks are three times more numerous than persons with straight hairlines? Not at all. In fact, for some traits, the recessive phenotype is far more common than the dominant phenotype. After all, most of us *do* have fingerprints, even though it is a recessive condition. In this case, the dominant allele is so rare that nearly everyone is homozygous recessive and therefore has a complete set of fingerprints.

## THE TESTCROSS

As stated earlier, outward appearance cannot reveal whether an individual who shows a dominant trait is homozygous or heterozygous for that characteristic. (A yellow pea may be homozygous or heterozygous for seed color.) If Mendel's principle of segregation is accurate, then it should be possible to determine genotype by crossing the individual in question with a mate that shows the recessive trait. This strategy, known as a **testcross**, reveals the presence of a "hidden" recessive allele if

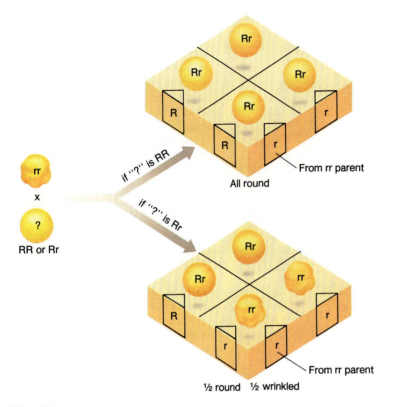

**Figure 25-6**
**The testcross** allows us to determine the genotype (*RR* or *Rr*) of an organism that shows the dominant trait. When crossed with an organism that shows the recessive trait (*rr*), a homozygous dominant (*RR*) will yield only progeny with the dominant characteristic, since all the offspring will have a dominant allele. If the test organism is a heterozygous individual (*Rr*), about half the progeny will possess the recessive phenotype. (Only two squares need to be shown, since the *rr* parent can donate only one type of allele—an "*r*".)

the zygote receiving both alleles for white flowers is 1/4, since each gamete has a probability of 1/2 for having this allele $(1/2 \times 1/2 = 1/4)$. In other words, one of every four offspring of the heterozygous parent will be homozygous recessive (white-flowered); the other three will show the dominant phenotype (purple flowers). This is another way of saying the ratio of dominant to recessive phenotypes is 3 to 1, which is the approximate ratio Mendel consistently obtained from thousands of genetic crosses.

You could also express the possible gamete types from a heterozygous individual as "$P + p$" ($P$ is the dominant allele for purple flowers). Crossing heterozygotes, you would obtain

$$(P + p) \times (P + p) = PP + 2Pp + pp.$$

The result is one homozygous domi-

nant (*PP*), to 2 heterozygous (*Pp*), to one homozygous recessive (pp) offspring. This is the same as that obtained with Punnett squares—a *genotype* ratio of 1:2:1 and a *phenotype* ratio of 3:1.

## Alleles on Nonhomologous Chromosomes Assort Independently of One Another

Mendel made another important finding when he switched to dihybrid crosses, studying the inheritance of two characteristics at a time. He found that the distribution of the two alleles at one genetic locus had no bearing on the distribution of the other pair of alleles. Receiving the allele for purple flowers, for example, did not influence the likelihood of inheriting the allele for yellow seeds or for constricted

pods. In every dihybrid cross Mendel performed, the characteristics assorted independently of each other. From these results grew Mendel's *Principle of Independent Assortment*. It applies to genes on separate, nonhomologous chromosomes.

When he crossed a true breeding plant that possessed two dominant traits [for example, round (*R*) and yellow (*Y*) seeds—genotype *RRYY*] with another true breeder showing the recessive alternatives of the same characteristics [wrinkled (*r*), green (*y*) seeds—genotype rryy], all the F$_1$ progeny possessed the dominant characteristics, as predicted by the Law of Segregation. When these doubly heterozygous F$_1$ plants (*RrYy*) were allowed to self-fertilize, the F$_2$ generation displayed four phenotypes in an approximate ratio of 9:3:3:1.

Mendel explained these findings by proposing that alleles for different genes assorted independently. Therefore, inheriting one of the alleles for seed shape (*R* or *r*) had no influence on which seed color allele (*Y* or *y*) was inherited. The probability of acquiring any phenotypic combination of the two traits can be calculated by multiplying the probability of inheriting each phenotype alone.

(round + wrinkled) × (yellow + green)

(**3** + **1**) × (**3** + **1**)

**9** + **3** + **3** + **1**

round yellow   round green   wrinkled yellow   wrinkled green

For many people, this is easier to visualize by using the Punnett square method. If alleles at different locations assorted independently of each other, an allele for seed shape could team up with either allele for seed color. This would produce gametes with four possible combinations (*RY, Ry, rY, ry*). Figure 25-7 reveals all possible F$_2$ combinations that can arise when these gametes fertilize each other. The genotype-phenotype relationships shown in the figure are summarized in Table 25-2. When Mendel examined the results of 556 F$_2$ fertilizations, he found four phenotypes in the 9:3:3:1 ratio predicted

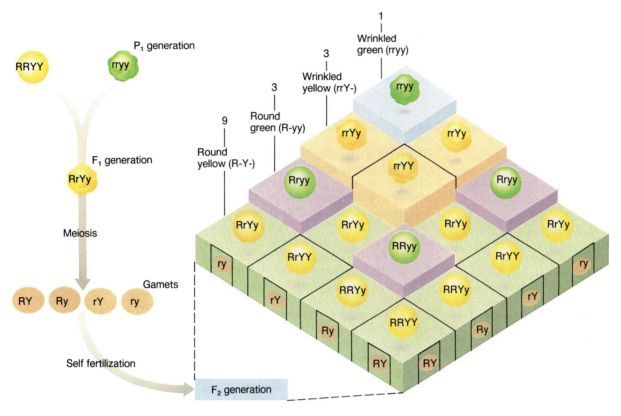

**Figure 25-7**
**Mendel's dihybrid crosses** led to his principle of independent assortment. When plants that breed true for round (*RR*) yellow (*YY*) seeds are crossed with those producing wrinkled (*rr*) green (*yy*) seeds, the F$_1$ progeny are heterozygous for seed shape and color. Independent assortment allows either *R* or *r* to team with either *Y* or *y*, generating gametes with four genotypes, Self-fertilization of an F$_1$, plant produces an F$_2$ generation with the nine genotypes and four phenotypes shown by the Punnett square. When Mendel performed this experiment with hundreds of peas, he obtained ratios very close to the 9:3:3:1 phenotype pattern predicted here. Mendel also determined the genotypes of the F$_2$ progeny by allowing them to self-fertilize, then examining their offspring. The F$_2$ generation consisted of all nine genotypes in a ratio that approximated the predicted 1:2:1:2:4:2:1:2:1 ratio.

TABLE 25-2

| | **Dihybrid cross between heterozygotes.** | |
|---|---|---|
| **Genotype** | **Number Showing the Same Genotype** | **Number Showing the Same Phenotype** |
| *RRYY* | 1 | |
| *RRYy* | 2 | |
| *RrYY* | 2 | 9 round/yellow |
| *RrYy* | 4 | |
| *RRyy* | 1 | |
| *Rryy* | 2 | 3 round/green |
| *rrYY* | 1 | |
| *rrYy* | 2 | 3 wrinkled/yellow |
| *rryy* | 1 | 1 wrinkled/green |
| total | 16 | 16 |

by probability. These traits did indeed assort independently.

But each chromosome houses a multitude of genes. Genes that reside on the same chromosome are physically linked to each other and don't assort independently. (This phenomenon, called "linkage," is discussed with chromosomes in the next chapter.) This is only one of several exceptions to Mendel's "laws." Although Mendelian genetics forms the basis for understanding the fundamental patterns of inheritance, several exceptions to Mendel's principles were discovered as the science of genetics grew. Here are a few findings that are variations of the patterns explained by Mendelian genetics.

# Mendel Amended

## Variations in Dominance Patterns

Some traits don't have two alternative forms of appearance or don't display the simple dominant-recessive relationship. These variations, as well as Mendel's classical patterns, are easier to understand if we discuss them in terms of "gene products." The gene product for the purple flower allele in Mendel's pea plants, for example, is the pigment that imparts the color (or, more accurately, an enzyme that catalyzes the formation of the pigment). The allele for white flowers is a variant of this gene that produces no pigment at all, leaving the flower white. This explains why purple is dominant over white in garden pea flowers; as long as there is one purple flower allele in the heterozygous pair, pigment will be produced and the flower will be purple. Only in the homozygous recessive configuration does the white flower appear, somewhat by default. Many genes control their traits in such a fashion, the recessive allele simply being an inoperative form of the dominant allele; it has lost the dominant allele's ability to produce an active gene product.

With this in mind, let's consider some traits that depart from the classical dominant–recessive relationship.

### INCOMPLETE DOMINANCE AND CODOMINANCE

Some traits at first seem to follow the notion of "blending" rather than Mendelian inheritance. For example, when red-flowered snapdragons are crossed with white-flowered snapdragons, all of the progeny are pink, as though the traits blended. Self-pollination of pink snapdragons produces an interesting mix of offspring—50 percent pink, 25 percent red, and 25 percent white—revealing that blending has not occurred. (Blending would continue to produce all pink flowers generation after generation.) Although heterozygous offspring are pink, the homozygous progeny regain the color of their red or white "grandparents." In this case neither allele is dominant

over the other. The effect of each gene is modified by the presence of the other, so that together they form a third phenotype. This phenomenon is known as **incomplete** (or **partial**) **dominance**; it leads to heterozygous individuals that are phenotypically distinguishable from either homozygous type.

The same is true of **codominance**, in which both alleles in a heterozygous individual are *fully expressed*. Unlike incomplete dominance with its "diluted" intermediate phenotype, both expressions of a codominant trait exist simultaneously and unmodified in a heterozygous individual. A classic example of codominance is the ABO blood group antigens used to determine blood type in humans (Chapter 20). The codominant alleles ($I^A$ and $I^B$) direct the blood cell to manufacture the A antigen (producing type A blood) and the B antigen (producing type B blood). Heterozygous persons produce both A and B antigens (type AB blood). What about type O blood? Read on.

### MULTIPLE ALLELES

The human ABO blood group system can be used to illustrate another departure from classical Mendelian genetics —the existence of more than two possible alleles for a single trait. Although a diploid individual can possess only two, the existence, of additional alleles increases the number of possible genotypes. In the ABO system three types of alleles exist: $I^A$, $I^B$, and $I^O$ ($I$ = the genetic locus; the superscripted letter = the allele at locus I). The $I^O$ allele (sometimes written "$i$" to show that it is recessive) is an inoperative variant that, like the allele for white flowers in pea plants, dictates no gene product. A homozygous person with two $I^O$ alleles would lack both A and B antigens. These individuals are blood type O. An inoperative allele such as $I^O$ is recessive to both of the alternative alleles. Therefore, an $I^A/I^O$ heterozygous pair would produce the same phenotype as the homozygous $I^A/I^A$, that is, a person with type A blood. (These atypical symbols were coined by a physician—Carl Landsteiner— whose representations differ from the standard used by geneticists.)

Mendel discussed inheritance patterns for only those traits with two versions of the gene, yielding three possible genotypes. Had he selected a locus that had three possible alleles, the number of genotype combinations that are possible would jump from 3 to six. If we add a fourth allele, there are 10 possible genotypes. Had he selected traits with **multiple alleles** that showed more than two phenotypes, it is doubtful that Gregor Mendel could have understood the implications of his observations.

## Interactions Between Genes at Different Loci

Not only did Mendel's seven pairs of alleles segregate independently, but they also *functioned* independently of each other. In many cases, however, the expression of a pair of alleles is influenced by the genotype at other gene sites. For example, some brown-eyed people have parents who are both blue-eyed. This could be quite disturbing to these people when they realize that the allele for brown eyes ($B$) is dominant over the one for blue eyes ($b$). Since both parents are apparently $bb$, who donated the dominant allele ($B$) needed to provide your darker eye color?

But before any of you confront your parents, consider this possibility: either parent could be carrying a "silent" $B$ allele. The expression of the genotype $Bb$ may be modified by other genes that mask the dominant allele, so that brown pigment would not be produced. Because of such gene interaction, about 1 in every 50 persons carrying a single $B$ allele will be blue-eyed.

One form of gene interaction is called **epistasis**, in which a particular genotype blocks the expression of a gene pair at another locus. Such epistatic genes are responsible for one type of *albinism* in animals (the absence of pigment in skin, hair, and eyes). Coloration is normally determined by genotypes at the loci that dictate pigment formation. A homozygous recessive configuration at the locus for albinism, however, suppresses the expression of these other genes regardless of their genotype and

# D I S C O V E R Y

## GREGOR MENDEL: THE PREMIER GENETIC DETECTIVE

In 1865 an obscure monk stood before a group of scientists and presented them the key that would unlock the door to one of life's great puzzles. He had begun to solve the mystery of *heredity*, the passage of traits from one generation to the next. The group of scientists had the rare privilege of being the first to be presented with one of the great triumphs of the human mind. Yet this esteemed audience responded not with questions or other signs of interest, but with polite applause. They simply had no way of knowing that they had just witnessed the birth of modern genetics.

The monk was Gregor Mendel, a scientist who conducted his experiments in the unlikely world of an Austrian monastery. Using common pea plants and a keen mind, Mendel formu-

lated principles that would eventually change the course of biological science in a direction that ultimately led to an understanding of the mechanism that underlies the genetic control of all living things.

During Mendel's time, attempts to make sense of heredity were laden with myths and misconceptions, and even the most widely accepted theory of heredity simply didn't fit the facts. His contemporaries supported the "blending" view of inheritance, which contrasted sharply with Mendel's conclusions. According to the blending explanation, parents produce hereditary fluids that mix together to form progeny possessing blended characteristics. (The term "blood relative" is a holdover of this misconception.) Blending would always produce offspring that were homogenized versions of their parents, intermediate forms halfway between each parent in every characteristic. If blending were the pattern of heredity, sexual reproduction would not contribute to the rich diversity of life. Instead, organisms would tend to become more and more alike until they eventually became identical. Such a concept is inconsistent with the existence of almost 2 million known species on earth, for blending would produce just one completely blended species. Blending also failed to explain why children are sometimes "chips off the old block," closely resembling one parent but not the other, or why traits may skip a generation and reappear in grandchildren. Mendel's work had placed a "dead end" sign on the path followed by proponents of genetic blending. The sign was ignored for more than 30 years.

### Mendel's Methods

Mendel mated pea plants and determined the pattern by which certain characteristics were inherited by the offspring. His success was largely due to sound scientific methodology, as described in Chapter 2. Other investigators had approached the problem of inheritance by examining the multitude of traits in whole organisms. Such an unmanageable number of variables prevented any possibility of establishing scientific controls on an experiment. Mendel recognized the value of *controlled* experimentation.

He first established a reliable system for tracking inheritance patterns. Mendel spent years developing a collection of garden pea plants that, when bred with themselves, always produced offspring that were identical to the parent in the specified *genetic trait* (an inheritable property). In other words, he used plants that *breed true* for each trait he examined. His choice of the pea plant was also strategic (see figure on page 432) for the plant can either fertilize itself or cross fertilize with other pea plants. Mendel could therefore allow his plants to self-fertilize, or he could prevent self pollination by amputating the pollen-producing anthers. These pollenless flowers were then fertilized by brushing on pollen from another plant (pollination and fertilization in plants is discussed in Chapter 21). In this way, he controlled which plants mated with each other. With accidental matings eliminated, Mendel was able to correlate parental traits with those of the progeny.

Mendel was brilliant in his ability to simplify experimental procedure, to make it manageable. He selected only seven traits to examine, and he looked at only one or two of those traits in each experiment. This aspect of controlled experimentation reduces the number of variables, enables researchers to conclude that the observed effect is the result of a defined cause (the experimental variable). Equally important, in each experiment Mendel examined hundreds, often thousands, of progeny, minimizing the variation that is due to chance alone. The strength of his findings lay in his mathematical analysis of his results, an analysis that revealed the basic pattern of inheritance.

This pattern gave us our first glimpse of the way genetic traits are transmitted from generation to generation. But during his lifetime, Mendel received little recognition for his achievements. He never knew that, years after his death, his elegant experiments would not only be recognized as accurate, but would also shatter the mystery that shrouded heredity and replace it with scientific principles. His triumph was a lonely one.

Mendel's work emerged before its time, preceding the discovery of chromosomes, genes, diploidy, and meiosis, all of which would have provided a physical basis for understanding his principles. It was these discoveries that prompted scientists to finally realize the significance of Mendel's work. After 35 years of obscurity, his results and insightful conclusions were rediscovered by geneticists around the turn of the century. Having emerged from the darkness, Mendel's work still stands as a supreme example of scientific investigation.

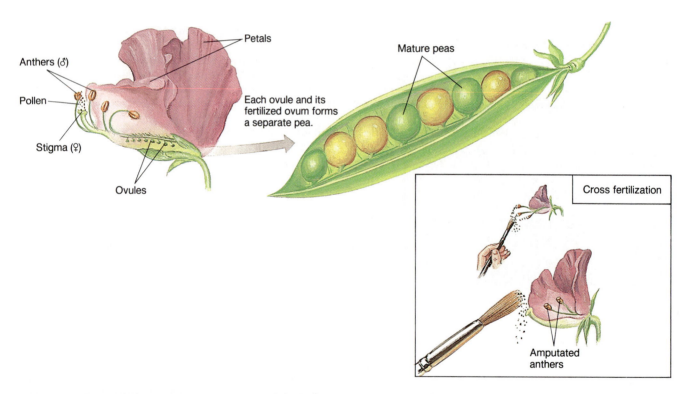

**Reproduction by Mendel's pea plants.** The flower may self pollinate with pollen falling from its anthers onto the stigma. Amputation of the anthers prevents self-pollination, so the source of sperm can be controlled during cross-pollination experiments (shown in inset). Each pea is a seed, the product of a separate fertilization event. Seeds develop only following fertilization, so results are not complicated by "fatherless" offspring. These properties contributed to Mendel's scientific success.

blocks pigment formation (Figure 25-8). The epistatic gene masks the genotype of the genes for pigmentation, producing an albino individual with white skin and hair, and eyes tinted pink by blood. The recessive albinism allele is uncommon. Although some animals are intentionally bred for this trait (as are white rats and rabbits), the remote likelihood of inheriting two such rare recessive alleles makes naturally occurring albinism a rare phenomenon.

## Pleiotropy

The gene product of a single gene can have a far-reaching effect on many other characteristics. This phenomenon is called **pleiotropy**. Unlike epistasis, a pleiotrophic gene doesn't affect other genes; it simply produces a wide variety of effects. For example, the disease *cystic fibrosis* occurs in persons who inherit two copies of a recessive allele specifying the production of an abnormal glycoprotein that thickens mucus. A multitude of phenotypic ab-

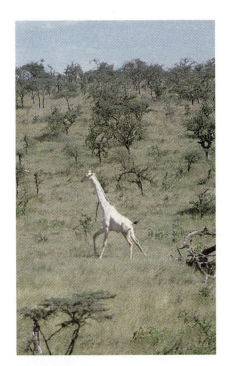

**Figure 25-8**
**A rare albino giraffe.**

normalities are expressed in these persons, all related to the thickness of mucus. The mucus impairs the functioning of digestive glands, sweat glands, and respiratory mucous glands. As a result, digestive enzymes are poorly secreted, and so food is improperly digested and absorbed. Sweat contains abnormally high levels of salt. The mucus that lines the airways is too thick to be propelled out of the lower respiratory tract by the ciliated mucosa (Chapter 19) and, laden with trapped microbes, settles into the lungs. As a result, pneumonia and other lung infections are the most frequent causes of death in persons with cystic fibrosis.

Not all pleiotropic effects are detrimental. Many geneticists believe that pleiotropy is an almost universal property of genes and that the far-reaching secondary effects of most genes are simply not yet recognized.

## Polygenic Inheritance

If height were determined by simple dominance at a single locus, people would come in two sizes, tall and dwarf. This type of variation is called **discontinuous** because all possible variants fall into distinct categories. But in humans and most other organisms, there is **continuous variation** in height, that is, a graded change between the two extremes in phenotype. Skin color in people is another familiar example of continuous variation.

Traits that show continuous variation are not determined by a single gene, but by a large number of genes at different loci. This phenomenon is called **polygenic inheritance** (Figure 25-9). The overall expression of a polygenic trait is therefore quantitative; it is the sum of the influences of many genes. Two persons who differ in just one pair of alleles show only slight color differences compared to two individuals who differ in all the polygenic determinants.

Most of the traits currently being studied by geneticists are polygenic in nature. These include animal and plant traits that are being selected by breeders for improving livestock and crops, as well as such elusive human

characteristics as intelligence. Polygenic inheritance poses challenges to researchers that single gene inheritance doesn't. Namely, the results are additive. Therefore, the outcome of an experimental cross is the sum of many genotypic variables. Interpreting these quantitative results is further complicated by the influence of environmental factors, which can interact with an organism's genotype to alter the phenotype. A genetically tall person, for example, will not achieve his maximum height potential if deprived of adequate nutrition. Another example is familiar to most of you. A light-skinned person who spends time in the sun will be darker than a genetically similar individual sheltered from sunlight.

## Environmental Influences

Nearly all phenotypes are subject to modification by the environment. Two organisms with the same genotype may have different phenotypes in different environments. For example, plants with inferior genotypes that have been grown under ideal conditions may develop into hardier individuals than plants with superior genotypes grown in suboptimal conditions. In humans, height, weight, skin color, intelligence, and countless other traits are modified by environmental influences.

Descriptions of environmental alteration of phenotype within a single genotype number in the hundreds. An excellent example is the two-color pattern of siamese cats. The cat has one gene for hair color, and the genotype at this locus is the same in all the cat's cells, those producing light fur as well as those making dark fur. The gene directs the synthesis of a temperature-sensitive enzyme that manufactures the dark pigment. This enzyme is operative only on the cooler areas of the body (feet, snout, tip of the tail, and ears). The rest of the cat's body is so warm that the enzyme is inactivated and the fur remains a light color. The cat is two-colored because of differences in environmental conditions (temperature) in various parts of the animal's body.

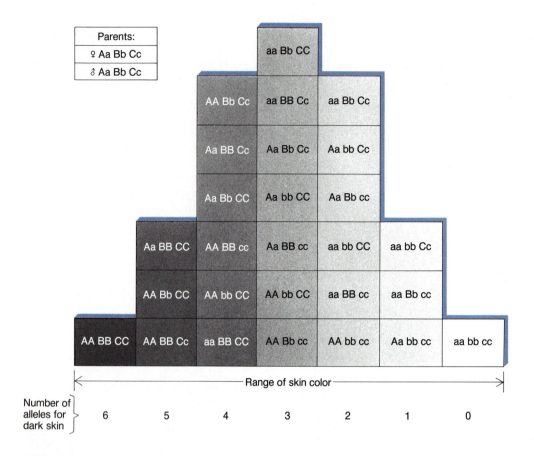

**Figure 25-9**
**Polygenic inheritance and skin color.** People don't come in two contrasting colors, black and white. Most of us are somewhere between these extremes. There are at least three genes (at three different genetic loci) that determine the amount of pigment in human skin. At each locus there are two versions of the gene, one allele for maximum pigment (the capital letter) and one allele for no pigment (the lower case letter). A heterozygous genotype shows incomplete dominance, so there are three possible phenotypes at each locus. This figure shows the 27 genotypes and 7 phenotypes that could be produced by two people who are heterozygous for all three genes (*AaBbCc*). The darkness of skin is determined by the total number of alleles for dark pigment in the genotype.

## Mutation

Allele segregation and independent assortment increase genotype diversity by promoting new combinations of genes. But such gene shuffling doesn't explain how there came to be such a vast diversity of life: if all organisms descended from a common ancestor, with its relatively small complement of genes, where did all the genes present in today's millions of species come from?

New genes arise from existing genes by **mutation**, a permanent change in the genetic message that may alter or destroy the original trait. The mutated allele is as stable as was the *wild type,* the term used to distinguish the original allele from the new altered form (called the *mutant* allele). Because they are stable, mutations are inheritable by the daughter cells of the mutant cell. The new trait is also inheritable by multicellular offspring if the mutation occurred during gamete formation, so

that sperm or eggs contain the mutant allele. Mutation thus produces new genes, which are then exposed to the selective environment. Most mutant alleles are detrimental. For example, a mutation might inactivate a gene product needed for a critical life function. Yet occasionally these stable genetic changes create neutral or advantageous characteristics. In this way, mutation provides the raw material for evolution and the diversification of life on earth (Chapter 29).

# Synopsis

## MAIN CONCEPTS

- **The enduring nature of genes.** Although a particular (recessive) trait might disappear for a generation or more, it can eventually reappear because the allele for that trait is retained intact. Alleles for such "disappearing" traits are recessive to dominant alleles and are masked by the contrasting dominant allele in heterozygous individuals (*Aa*). These recessive traits are expressed when the homozygous recessive (*aa*) condition is reestablished.

- **Mendel's "Law of Segregation."** Paired alleles segregate from each other when the homologous chromosomes on which they reside separate during meiosis. Each resulting gamete receives one of the two alleles.

- **Mendel's Principle of Independent Assortment.** Alleles of genes that reside on different chromosomes segregate independently of each other. Dihybrid crosses between organisms that are heterozygous for two traits (*AaBb*) yield the approximate $9:3:3:1$ phenotype ratio predicted by probability equations (or by the Punnett square method). This is the ratio expected when two events do not influence each other.

- **Genes that don't "obey" Mendel's laws.** (1.) Genes that reside close together on the same chromosome are physically linked and do not segregate independently of each other. (2.) Some alleles show incomplete dominance, the pair of alleles "diluting" each other's effect in heterozygous individuals. (3.) Codominant alleles are fully expressed in heterozygous individuals. (4.) Multiple (more than two) alleles for a particular trait can combine to form more than three genotypes. (5.) Genes also interact with each other to modify the phenotype, such as epistatic genes suppressing the expression of the alleles at another locus. (6.) A single gene may be pleiotropic, having a multitude of effects. (7.) One trait may be due to multiple genes at several loci (polygenic inheritance). (8.) Tracking the fate of alleles from generation to generation is also complicated by the effect of environment on phenotype. (9.) New genes can emerge (and old ones disappear) by mutations, changes in the genetic message of a gene. Such changes have produced the rich diversity of genes in the biosphere.

## KEY TERM INTEGRATOR

Mendelian Genetics

| | |
|---|---|
| **genes** | Discrete units of inheritance that occupy a specific **locus** on the chromosome and dictate the formation of (usually) one particular **trait**. There are usually at least two **alleles** for a gene, one of which may be **dominant** over the expression of the **recessive** allele. |
| genetic crosses | A diploid individual may be either **homozygous** or **heterozygous** in its **genotype**, but its **phenotype** doesn't always tell you which. In individuals showing the dominant trait, a **testcross** will reveal the genotype. In **monohybrid** or **dihybrid crosses**, the probability ratios of offspring can be predicted by using simple multiplication or the **Punnett square method**. |

---

Amended Mendelian Genetics

---

| | |
|---|---|
| **incomplete dominance** | Production of an intermediate phenotype when contrasting alleles are present in a heterozygous individual. It is distinct from **codominance**, in which two alternative alleles are both fully expressed and unmodified by each other. Both phenomena increase the number of phenotypes possible, as does the existence of **multiple alleles** for a particular gene. |
| **epistasis** | Modification of gene expression in which a particular gene at one locus masks the expression of at least one other gene at a different chromosomal location. |
| **polygenic inheritance** | The interaction of many genes to determine a particular trait. It results in **continuous variation** rather than **discontinuous variation**. |
| **pleiotropy** | A single gene product having multiple phenotypic effects. |
| **mutation** | A stable change in genotype. |

---

# Review and Synthesis

1. How did Mendel's work predict the existence of diploidy, chromosomes, genes, and meiosis?

2. Although the expected phenotype ratio following a monohybrid cross is 3:1, it tells us nothing about the relative frequencies of these individuals in the general population. Explain why.

3. Why can't the Principle of Independent Assortment be discussed without an initial understanding of the Law of Segregation?

4. Without doing a testcross, which of the following could you definitely say would or would not breed true for the trait in question? a hybrid for hairy fingers; a purple-flowered pea plant; a red-flowered snapdragon; a pink-flowered snapdragon; a person with type O blood; a person with AB blood; a tall pea plant; a white-flowered pea plant.

5. The formula for predicting phenotype results of a dihybrid cross (page 428) can also be used for predicting the expected genotypes (and ratios) when a *RrYy* plant self-fertilizes. Complete the following formula and compare your results with those obtained using a Punnett square. Notice that instead of a 3:1 *phenotype* probability per trait, you are using a 1:2:1 *genotype* probability.

$$(1RR + 2Rr + 1rr) \times (1YY + 2Yy + 1yy) = ?$$

6. Design a mating experiment which would demonstrate that intermediate dominance is not evidence of genetic blending. Explain how your proposed experiment proves that genes remain intact, even in organisms with the intermediate phenotype.

7. How does each of the following increase genetic diversity?
   meiosis
   fertilization
   mutation

8. For simple dominant traits that show Mendelian inheritance patterns, why is it redundant to say "heterozygous dominant" or "homozygous recessive"?

## Additional Readings

Baskin, Y. 1984. *The Gene Doctors: Medical Genetics at the Frontier.* Morrow, New York. (Introductory.)

Braun, P. (ed.). 1988. *Biology, Annual Editions.* 5th ed. (Unit 2, Genetics.) Dushkin Publishing Co., (Introductory.)

Harrison, D. 1970. *Problems in Genetics with Notes and Examples.* Addison-Wesley, Reading, Mass. (Intermediate.)

Mendel, G. 1865. "Experiments in plant hybridization." Reprinted in *Classic Papers in Genetics,* ed. J. Peters, pp. 1–20. Prentice-Hall, Englewood Cliffs, N.J. (Introductory.)

Sturtevant, A. 1965. *A History of Genetics.* Harper & Row, New York. (Introductory to intermediate.)

Chapter 26

# Genes and Chromosomes

They have been called biological computers and living strings of pearls. Such comparisons, however, diminish the importance of their two-pronged biological mission, which is absolutely essential to life. They control the molecular activity in cells so that biological order prevails, and "remember" genetic traits so that inheritable characteristics can be passed on to progeny. They reside deep inside a cell's interior, dictating which inheritable features to manufacture and, in doing so, instructing an organism to develop into itself. In stained cells, these structures stand out as thick, dark strands, inspiring biologists to call them by a name that means "colored bodies."

The name stuck; today these structures are still referred to as **chromosomes**. They are the bearers of the genes (Figure 26-1).

In the years after chromosomes were discovered in the nuclei of salamander cells, a flurry of interest in heredity spread throughout the scientific community. Improvements in microscopy led to the discovery of mitosis, meiosis, haploidy, and diploidy. Chromosomes were proposed as the likely physical location for hereditary information and the means for distributing this information to progeny. According to the chromosomal theory of inheritance, a cell simply duplicated the strands and donated one complete copy to each daughter cell during division. Researchers investigating the theory's validity inevitably bumped into a long ignored, 35-year-old paper written by an Austrian monk. Gregor Mendel's scientific "ghost" had awakened.

Fresh information about chromosomes revealed the significance of Mendel's conclusion—his identification of dominance, genetic "factors" (genes), and inheritance ratios. The discovery that each chromosome had a partner, a homologous mate, supported Mendel's concept of paired factors. His Law of Segregation suddenly had a physical explanation—during sex cell formation, the two alleles of a pair separated as meiosis reduced the diploid number in half by separating

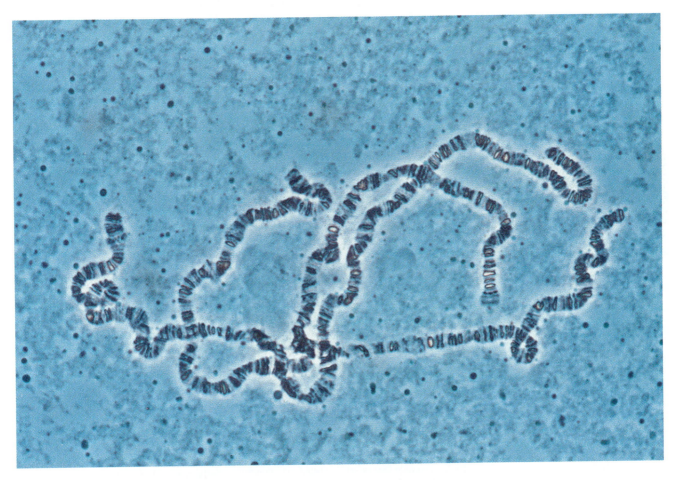

**Figure 26-1**
**Genes on display?** Giant polytene chromosomes from the salivary gland of larval fruit flies show thousands of distinct bands. Some of the dark bands may represent individual genes.

homologous chromosomes into different gametes. In a single year (1900), Mendel's inheritance ratios were rediscovered independently by three geneticists. All realized that they had been "scooped" by Mendel 35 years earlier, even before improved microscopy put chromosomes on display. Yet, as important as the new microscopes were in revealing the hereditary apparatus in cells, a living tool attracted the genetic limelight in those early days of hereditary research. It was the common fruit fly, *Drosophila melanogaster,* and it surrendered a wealth of information about the role of chromosomes in inheritance.

## Lessons From Fruit Flies

Genetic investigations can be enormously time consuming. Experimental matings yield results only when offspring have appeared, often a wait of a year or more. Subsequent generations would require even more years to reveal information. Geneticists therefore needed an easy-to-maintain organism that produced multiple generations in a year. They found it swarming around rotting fruit. Fruit flies could be easily raised by the thousands, completing their entire reproductive cycle in about two weeks (Figure 26-2). Geneticists couldn't have picked a better partner for their work. Once streamlined by *Drosophila,* genetic research blossomed into what many consider the "golden age of genetics," a period that has yet to end.

Here are some early breakthroughs that were forged from the fertile association between geneticist and fruit fly.

### Genetic Markers

Inheritance in fruit flies, pea plants, or any other organism can be charted only if there is detectable evidence of genetic differences. At first, all *Drosophila* looked alike. After culturing the flies for many generations, however, the process of random mutation introduced variant characteristics. Flies with mutant white eyes instead of red eyes appeared; others had smaller wings; still others had bar-shaped eyes rather than the normally large round eyes (Figure 26-3). Mutant flies were isolated and allowed to mate. Offspring with the same mutant characteristic were mated with each other. In this way, many strains that bred true for a mutant trait were obtained. Flies with these genetic differences were then mated with wild-type (nonmutant) flies, and inheritance patterns were charted by examining the $F_1$ and $F_2$ phenotypes. Mutations that show up as observable traits act as **genetic markers** that allow investigators to track an allele from generation to generation.

### Linkage

As an exception to Mendel's Principle of Independent Assortment, traits whose genes reside on the same chromosome generally do not assort independently. The existence of nonindependently assorting traits disturbed the apparent tidiness of Mendel's findings, which were quickly being elevated to the status of "guiding principles." Let's reconsider the $9:3:3:1$ ratio from the $F_2$ dihybrid cross on page 429 ($RrYy \times RrYy$). Had the genes for seed color and shape lay next to each other on the same chromosome, the two dominant alleles would have stayed together during gamete formation, as would the two recessive alleles, reducing the possible gamete phenotypes from four (*RY, Ry, rY,* and *ry*) to two (*RY* and *ry*). Such crosses yield progeny showing 2 possible phenotypes instead of four. Investigators quickly discovered that several *Drosophila* traits, body color and wing shape, for example, tended to stay together rather than assort independently. This phenomenon became known as **linkage**, with the traits segregating as though the genes were physically connected to each other. In fact, they are linked together, residing on the same chromosome.

This fact became apparent as lists of

**Figure 26-2**
*Drosophila melanogaster,* **the geneticist's ally.** These petite flies reproduce rapidly in bottles with simple yeast growing on a medium. The flies dine on the yeast. Each female produces hundreds of offspring, each of which can reproduce before they are two weeks old. Results from one year of *Drosophila* experimentation would require 30 years to obtain using an organism that reproduced only once a year.

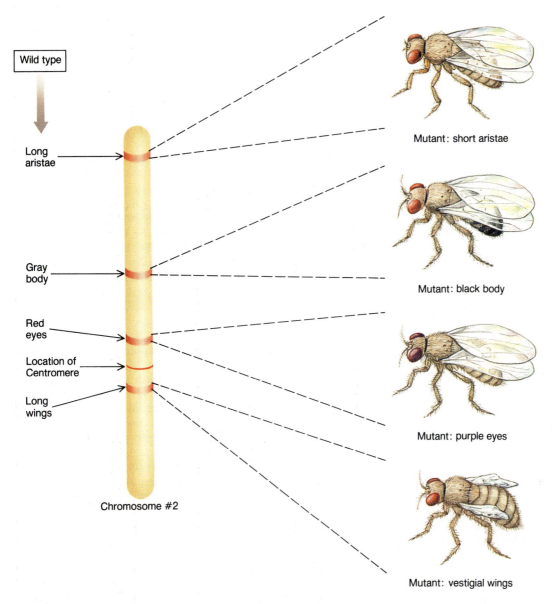

Wild type

Long aristae

Gray body

Red eyes

Location of Centromere

Long wings

Chromosome #2

Mutant: short aristae

Mutant: black body

Mutant: purple eyes

Mutant: vestigial wings

**Figure 26-3**
**A gallery of mutants.** The mutation that produced each variation pictured here occurs at a particular location on one of the chromosomes. The normal (wild-type) characteristic that corresponds to each mutant trait is printed next to its location on the chromosome.

linked traits grew longer. *Drosophila* testcrosses revealed four **linkage groups**. The traits in each group were linked with each other but not to the traits in the other three linkage groups. This corresponds exactly to the four chromosomes in the haploid gametes of fruit flies. In virtually all eukaryotic organisms, the number of linkage groups is identical to the haploid num-

ber of chromosomes. This is potent evidence that genes reside on chromosomes.

## Crossing Over — The Gene Exchange

But *Drosophila*'s investigators also discovered that traits in a linkage group (on the same chromosome) occasion-

ally segregated independently, as though on different chromosomes, ending up in separate gametes. The genes for body color and wing length, for example, reside on the same chromosome. Therefore, crossing a heterozygous gray-bodied, long-winged fly (*BbWw*) with a black-bodied, short-winged mate (*bbww*) should yield only two phenotypes, the same as the parents. Here are the results one would expect if linkage were 100 percent:

(BbWw) × (bbww)

Yet contrary to these predictions, in about one-sixth of the progeny a pair of alleles appeared to have traded places with each other on homologous chromosomes. These offspring showed neither of the predicted phenotypes. Instead, they inherited a combination that would be possible only if the genes were no longer connected (a *Bbww* fly, for example, gray-bodied and short-winged). Perplexed investigators realized that these alleles did not always remain linked. Another concept emerged — **crossing over**, the exchange of genetic segments between homologous chromosomes.

Recall from Chapter 12 that, during prophase I of meiosis, each chromosome (consisting of two sister chromatids joined by their centromere) aligns with its homologue to form a four-chromatid complex called a *tetrad*. While closely grouped, homologous chromatids entertwine, and portions of the strands *cross over* to the other strand, resulting in an exchange of chromosome segments. This trading of strands happens at intersections (chiasmata) where homologous chromatids overlap. Each of the two strands physically breaks and rejoins with the severed portion of the other chromosome (Figure 26-4). Although

such "breakage and reunion" of chromosomes may seem haphazard and dangerous, it is a biological maneuver that is both orchestrated and advantageous. A complex of 50 or more enzymes devoted strictly to supervising the events of crossing over assures that chromosomal segments are traded frequently and in an orderly fashion. Even the overlap process is directed, so that equal lengths of chromosome are normally exchanged. Otherwise, one chromosome would lose some genes in the trade, and the other would acquire some unneeded spares.

Crossing over has tremendous evolutionary advantages. This tightly governed process rearranges gene combinations during meiosis. As a result, gametes, and therefore progeny, receive chromosomes containing a different assortment of alleles than the parent. In humans, an average of 40 to 50 crossing-over events occur during the formation of each ovum or sperm, creating new combinations of potential traits. In virtually all eukaryotic species, crossing over boosts the biological advantages of sexual reproduction. It increases the diversity of individuals within a population by reshuffling existing alleles.

Crossing over also solves a problem created by the placement of many genes on the same chromosome. If the assortment were rigidly fixed, then mutation would be the only force that could change the chromosome. The formation of a lethal mutation would cancel out all favorable alleles sharing that chromosome, since any offspring inheriting the advantageous alleles would also get the lethal allele and die before reproducing. Crossing over shuffles alleles between chromosomes, randomly separating favorable and unfavorable alleles. Chromosomes with the most favorable gene combinations give rise to organisms with a greater likelihood of surviving. Hence, crossing over provides a means of eliminating an unfavorable allele without bringing down all the other genes on the same chromosome.

## Mapping the Chromosome

Studies on fruit flies also revealed that genes are arranged in a linear sequence, much like towns along a highway. This gene configuration was

proven by doing with genes what one does with towns — create a map that shows not only the sequence of genes, but also the relative distances between them. But genetic mappers cannot drive along a chromosome and record the genes and their in-between distances as they pass by them. Investigators had to use an indirect way of determining gene location. Crossing over supplied them with this indirect approach.

The mapping technique is based on the *frequency of crossing over* between two genes on the same chromosome. The frequency is determined by counting the relative number of progeny with a "mixed phenotype," that is, possessing two traits that would not be found in the same individual if the alleles remained linked to each other on the same chromosome. In *Drosophila,* for example, the linked genes for body color ($B$ = gray, $b$ = black) and wing length ($W$ = normal, $w$ = short) became observably unattached from each other in about 180 of every 1000 offspring produced by the heterozygote parent *(BbWw).* In other words, the frequency of crossing over between *b* and *w* is 18 percent. (A total of 820 flies were *BbWw* — linkage retained; 180 were *Bbww* or *bbWw* — linkage broken by crossing over.) In this way, crossing over frequencies between genes for any two observable traits can be determined, as long as both genes reside on the same linkage group, that is, on the same chromosome.

Genetic markers at opposite ends of a long chromosome are more likely to be separated by crossing over than are two markers that lie adjacent to each other. This means that the crossing-over frequency between two genes reflects their relative distance from one another — the greater the distance, the greater the frequency. This technique, called **linkage mapping** (or simply "chromosome mapping"), is used to construct genetic maps that show relative positions and distances between genes on a chromosome. By convention, a 1 percent frequency is equated to one *map unit*. The alleles w and b are thus 18 units apart ($180/1000 \times 100 = 18\%$).

Mapping not only provides relative distances, but also delineates the se-

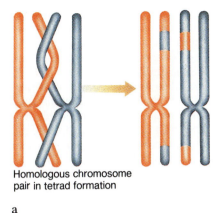

Homologous chromosome pair in tetrad formation

a

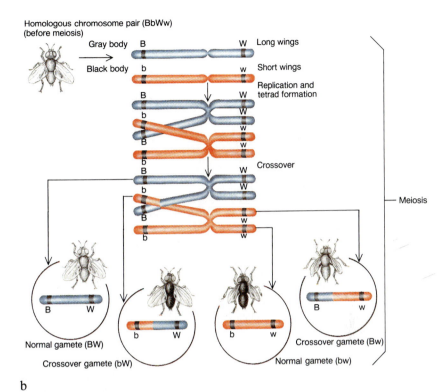

b

**Figure 26-4**
**Crossing over—the gene exchange.** *(a)* Tetrad formation during synapsis, showing three possible crossover intersections (chiasmata). *(b)* Simplified representation of a single crossover in a *Drosophila* heterozygote (BbWw) at chromosome #2 and the resulting gametes. ("↓" indicates the location of each centromere.)

quence of genes on a chromosome. Let's consider a third genetic marker ("*Frd*," the allele for freckled body) that shares the same chromosome as *b* and *w*. Mating experiments show that *w* is 32 units from Frd, which is 50 units from *b*. Only one sequence of these three genetic markers is consistent with the distances indicated by the crossover data. The sequence is shown in the following genetic map:

| b | 18 | w | 32 | Frd |
|---|----|---|----|-----|

50

Through this technique, hundreds of gene loci have been mapped on this one fruit fly chromosome alone.

Genetic mapping not only demonstrates that genes are arranged in a fixed linear order along a chromosome, but also provides geneticists with valuable research tools. The availability of accurate genetic maps of human chromosomes helps in the quest to locate human genes, notably those responsible for genetically acquired diseases such as diabetes and

other metabolic disorders. Mapping human chromosomes may hasten the development of *gene therapy,* that is, replacement of faulty genes with functional substitutes.

## Giant Chromosomes

*Drosophila* also gave investigators access to the visual details of chromosomes. Cells from the salivary gland of *Drosophila* larva contain chromosomes that are not only longer, but about 100 times thicker than chromosomes in other *Drosophila* cells. At one stage of larval development, salivary gland cells temporarily cease dividing, yet the cells keep growing in size, and DNA replication continues, providing the additional genetic material to accommodate the needs of these enormous cells. The extra DNA strands remain attached to each other in perfect side-by-side alignment, producing giant chromosomes with as much as 1054 times more DNA than normal chromosomes. These unusual *polytene chromosomes,* when stained and examined microscopically, are

rich with visual detail. Thousands of dark bands (about 6000 of them) transect the strands; each band is separated from its neighbor by a lighter *interband* region (see Figure 26-1). Visible changes in a particular chromosomal band often correlate with a change in a specific trait. In this way, geneticists can visually locate and map many *Drosophila* genes.

## Sex and Inheritance

Humans are a lot like flies. Most of the genetic breakthroughs revealed by *Drosophila* apply to us and to most other organisms as well. We even resemble fruit flies in the fundamentals of sex. An animal's gender, for example, is determined by chromosomes (although in some exceptional instances, gender may change in response to environment—see Bioline: The Fish That Changes Its Sex).

## Gender Determination

The chromosomes of males and females of the same animal species look identical, except for one chromosomal pair. This pair acquired the label **sex chromosomes** to distinguish them from the other chromosomes, the **autosomes**, those that look the same in both genders. In some insects, such as grasshoppers, there is only one type of sex chromosome, called an **X chromosome**. Females have two copies of this chromosome (XX), whereas males have only one (X-). Therefore, the otherwise diploid males are "haploid" with respect to their sex chromosomes.

In other insects and most higher animals, the chromosomal number is the same in both genders. (Both sexes in humans, for example, possess 46 chromosomes in their somatic cells and 23 in their gametes.) But one of the sex chromosomes looks different in males. Whereas the cells of females harbor a pair of X chromosomes, males possess one X chromosome paired with a smaller **Y chromosome** (illustrated for *Drosophila* in Figure 26-5). About half the sperm produced by males will contain the Y chromosome, and the other half will possess the X. Since normal ova contain only an X chromosome, the sex of a progeny individual depends on whether the ovum is fertilized by an X-bearing sperm (forming an XX pair that programs the formation of a female) or a Y-bearing sperm (producing the XY male combination). Gender in most animals is thus determined by the male. King Henry VIII of England, who disposed of many wives because they failed to bear him a son, should have pointed his lethal finger squarely at himself. Whether due to chance or to some defect in his Y-bearing sperm, it was *his* gametes that determined the gender of his royal progeny.

Although some plants possess sex chromosomes and gender distinctions between individuals, most have only autosomes, each individual producing male and female parts.

## Sex Linkage

For fruit flies and people alike, there are hundreds of genes on the X chromosome that have no counterpart on the smaller Y chromosome. Most of these genes have nothing to do with determining gender, but their expression usually depends on gender. For

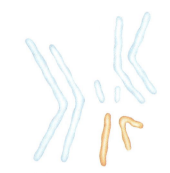

**Figure 26-5**
**Dissimilar homologues** characterize the sex chromosomes in this male fruit fly. Males possess an X and a Y sex chromosome, whereas females possess two X chromosomes and no Y. In the three pairs of autosomes, each chromosome is indistinguishable from its homologue.

example, in females, a recessive allele on the X chromosome will be masked if a dominant counterpart resides on the other X, and thus will not be expressed. In males a recessive allele on his X chromosome *would* be expressed, since there is no counterpart on the Y chromosome. (The male is essentially haploid for the allele.) These inheritable characteristics that reside on X chromosomes are called **sex-linked**, or sometimes "X-linked"

(Figure 26-6). Thus, for sex-linked traits, a female is either heterozygous or homozygous, but a male is normally **hemizygous**, a condition in which an individual possesses only a single allele for a gene instead of the usual two copies. (In birds and a few other animals, the XY individuals are female; as a result, the sex-linked hemizygous condition is found in females instead of males.)

So far, some 200 human sex-linked traits have been described. Many of these traits are expressions of mutations that result in male-associated genetic disorders. These include a type of heart valve defect, a form of mental retardation, several optical and hearing impairments, three types of muscular dystrophy, and faulty color vision.

One sex-linked deficiency has altered the course of history, tormenting some of Europe's royal dynasties. The disease is hemophilia, or "bleeder's disease"—the inability to produce a clotting factor needed to quickly halt blood flow following injury. Nearly all hemophiliacs are males. (Although females can inherit two recessive alleles for hemophilia, it is extremely rare.) In general, women who have acquired the rare defective gene are heterozygous *carriers* for the trait. The phenotype of a carrier discloses no suggestion that she harbors the recessive

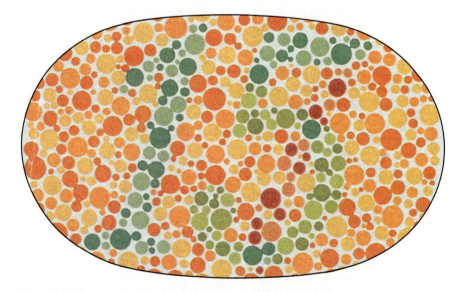

**Figure 26-6**
**Distinctively male.** There are hundreds of sex-linked traits, all associated with the XY configuration. They appear in males but rarely in females. Red-green colorblindness is such a trait. People with this defect in vision cannot see the number "15" in the above pattern.

# B I O L I N E

## THE FISH THAT CHANGES ITS SEX

In vertebrates, gender is generally a biologically inflexible commitment: one develops into either a male or a female as dictated by the sex chromosomes acquired from parents. Yet even among vertebrates, there are organisms that can reverse their sexual commitment. The Australian cleaner fish, a small animal that sets up "cleaning stations" to which larger fishes come for parasite removal, can change its gender in response to environmental demands. Most male cleaner fish travel alone rather than with a school. Except for a single male, schools of cleaner fish are comprised entirely of females. Although it might seem logical to conclude that maleness engenders solo travel, it is actually the other way around. Being alone fosters maleness. A cleaner fish that develops away from a school becomes a male, whereas the same fish developing in a school would have become a female.

But what of the one male in the school, the one with the harem? He may have developed as a solo and then found a school in need of his spermatogenic services. But there is another way a school may acquire a male. If the male in a school dies, one of the females, the one at the top of a behavioral hierarchy that exists in each school, becomes uncharacteristically aggressive and takes over the behavioral role of the missing male. She begins to develop male gonads, and within a few weeks, the female becomes a reproductively competent male, indistinguishable from other males. Furthermore, the sex change is reversible. If a fully developed male enters the school during the sexual transition, the almost-male fish developmentally backpedals, once again assuming the biological and behavioral role of a female.

---

allele. The functional allele on the other X chromosome produces enough blood clotting factor to assure a normal phenotype. But the Y chromosome has no allele for producing clotting factor. Therefore, boys who inherit the defective allele (from mothers who are carriers) develop hemophilia.

The mutant gene can be traced in the royal family of England back to Queen Victoria. She transmitted the recessive gene to three of her children, a male who eventually died of hemophilia and two daughters who were heterozygous carriers. The disease plagued several royal families who married Victoria's daughters and their descendants, in some cases killing the heir to the throne. The son of Nicholas, the last Russian czar, had hemophilia, acquired from his mother, Alexandra, a granddaughter of Queen Victoria. Their desperation to save their son contributed to the overthrow of the czar.

Pedigrees reveal that sex-linked traits tend to skip a generation (Figure 26-7). When a male with an X-linked characteristic fathers offspring, none of them shows evidence of the trait (unless his mate also carries the allele). All male progeny receive a normal gene from the mother. All daughters will be carriers. If carriers conceive offspring with a normal mate, the chance that a male will inherit the sex-linked characteristic is 50 percent (versus a 0 percent chance in females).

Sex linkage should not be confused with another type of inheritance, called **Y-linked inheritance**. Y-linked traits are dictated by those few genes carried only on the Y chromosome. In mammals, male reproductive anatomy is Y-linked, the Y chromosome carrying genetic information that instructs the embryonic gonads to develop into testes. The testes then produce the hormone testosterone which in turn directs the embryo to develop other male reproductive structures. (In the absence of Y, females develop, even if there is not a second X chromosome.) The Y chromosome also carries a gene needed for sperm production.

## Balancing the Gene Dosage

The amount of gene product manufactured by a cell is influenced by the "gene dosage," the number of copies of the allele that directs its synthesis. Increasing gene dosage usually intensifies the expression of the trait. It would seem, then, that a mammalian female who is homozygous for a trait on the X chromosome would express that trait with twice the intensity as a hemizygous male, since she would have two alleles versus the male's one. Yet this does not happen; in both genders, most of these traits are expressed more or less equally. This puzzling observation is explained by the presence of a darkly staining aggregate of chromatin, called a **Barr body**, which is normally found in the interphase cell nucleus of females but never in males (Figure 26-8*a*). Early in the normal embryonic development of a female, one of the two X chromosomes is retired, and the Barr body represents its inactivated remains. Thus, like males, each cell of females has only one functional copy of each gene on one X chromosome.

This may appear to you to contradict what we know about sex-linked traits. If the cells of females have only one active X chromosome, why aren't they hemizygous like males? Wouldn't sex-linked traits appear as frequently in females as males, since there is no

active homologous allele to mask recessive alleles on the X chromosome? This would be the case if the same X chromosome were inactivated in each of the female's cells. But the X chromosome "chosen" for retirement is randomly determined early in embryonic development, when the embryo consists of about 4000 cells. In about half the cells, one X remains active while in the other cells the homologous X prevails, creating two genetically distinct types of cells scattered randomly throughout the embryo. The millions of cells that descend from each of these cells form a patch of tissue that may be genetically and phenotypically different from adjacent tissue patches in which the other X chromosome directs the action. As a result, every female is a mosaic of genetically different tissue patches. This is why her phenotype doesn't reveal her hemizygosity. She contains enough cell patches with the dominant allele

to compensate for the lack of gene products in patches containing the sex-linked recessive allele.

Sometimes, however, the genetic patchwork can be seen, as in calico cats and tortoise-shell mice (Figure 26-8*b*). The allele for black coat color is on one X, and that for yellow is on the other X. When the X is inactivated in the embryo, all the hair follicles that arise from that particular cell will be hemizygous for one or the other color, generating the patchwork color pattern of these animals. This explains why male calico cats are virtually nonexistent, since males are hemizygous for one color allele from the moment of fertilization. (Such color patches occur only in animals in which pigmentation is determined by genes on the X chromosome, a condition that does not exist in humans, hence the absence of "calico women.")

## Sex Influences Determined by Autosomal Genes

Many of the traits that distinguish males and females within the same species are determined by genes that reside on autosomes rather than on sex chromosomes. Sex hormone differences between the two genders affect the expression of these autosomal genes. Milk production by mammals is the exclusive domain of females, even though males possess genes for manufacturing milk. Because these genes require the presence of female sex hormones for their expression, the phenotype normally develops in only one sex. This condition is called **sex-limited gene expression**. Beard growth in men and other secondary sex characteristics fall into this category of gender-directed phenotype.

Sex hormones can have another type of influence on phenotype called **sex-influenced dominance**. Both male and female sheep can produce horns, but horns are three times more common among males, even though the allele for horn growth is on an autosome. In sheep that are heterozygous for horn growth, the allele that orders horns to develop acts as a dominant in males (so the horns will grow) but as a recessive in females (so horns fail to appear). A more familiar example is

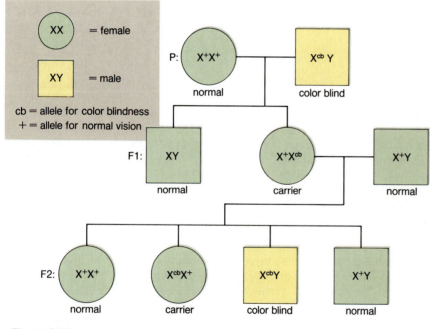

**Figure 26-7**
**The "generation gap."** Pedigrees show the tendency of sex-linked traits (in this case, color blindness) to skip a generation, producing a phenotypic gap—a generation of normal men and normal carrier women, but no persons showing the trait. Since the Y chromosome has no gene for color vision, the trait is hemizygous in males, making them far more likely candidates to express the defect than females. Hemophilia and other sex-linked traits follow this basic inheritance pattern.

the greater tendency of men to lose their coveted scalp hair. In the heterozygous genotype, the allele for pattern baldness behaves as a dominant in the presence of testosterone (a male sex hormone) but as a recessive in its absence. Most heterozygous women are therefore exempt from having to parade their genotypes around on the tops of their heads.

# Aberrant Chromosomes

In addition to mutations that alter or destroy the genetic meaning of a single gene, pieces of a chromosome may be lost, extra segments may appear, different-sized pieces may be exchanged between nonhomologous chromosomes, or a part of the chromosome may flip over so it reads backwards. Unorthodox arrangements such as these may doom the organism from

the moment the chromosome is altered, but, like single-gene mutations, they occasionally confer advantageous traits. These structural modifications in chromosomes produce large-scale genetic changes that may propel evolution forward in giant steps.

In all these structural alterations, the chromosome breaks and rejoins with another segment, as in crossing over, but differing in the ways described in Table 26-1. Since each of the structural aberrations shown in the table follows chromosomal breakage and reassembly, their incidence is increased by exposure to agents that damage DNA—virus infection, irradiation with X-rays or ultraviolet light, or exposure to chemicals that can break the DNA backbone. Some chromosomal aberrations have been linked to cancer, and many of these DNA-damaging agents are potential *carcinogens* (cancer causers).

Each aberration presented in Table 26-1 is described in more detail in the discussion below.

## Deletions

Forfeiting a portion of a chromosome usually results in a loss of critical genes and subsequent death of the organism. These aberrations, called **deletions**, commonly follow breakage of DNA. Broken DNA strands often have "sticky ends" that tend to adhere to the free sticky ends of other broken DNA pieces. Usually, the broken segments immediately reattach to their former positions, enzymes seal the nicks in the strands, and the damage is repaired. Sometimes, however, the broken strands drift to other positions and reconnect with another broken area of the same chromosome, deleting the genes between the two points of injury. Deletion shortens the chromo-

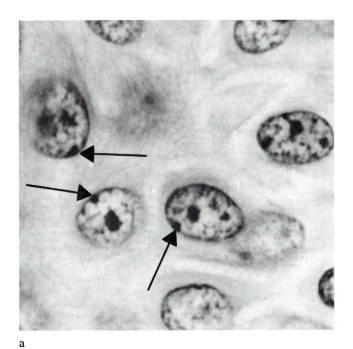

a

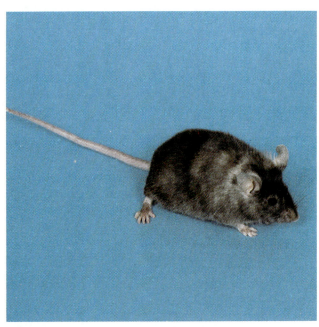

b

**Figure 26-8**
**Chromosome in retirement — the Barr body.** *(a)* The inactivated X chromosome in the nucleus of a woman's cells appears as a darkly staining Barr body. *(b)* Random inactivation of either X chromosome in different cells during early embryonic development creates a mosaic of tissue patches that descend from each cell after inactivation. These patches are visually evident in the tortoise-shell mouse shown here, as well as calico cats.

TABLE 26-1

## Modifications in Chromosome Structures

| Type of Alteration | Example of How Change May Occur | Some Possible Effects (🟢 Favorable   🟠 Harmful) |
|---|---|---|
| DELETION | a  b  c  d  e — Damage — (Deleted segment discarded) c d — Faulty repair — a b e | 🟢 Rarely favorable; perhaps elimination of detrimental genes |
| | | 🟠 Loss of critical genes is lethal; disrupts chromosome separation during meiosis |
| DUPLICATION | Breakage  a b c d  →  a b c d  a d (Deletion)  a b c c b Duplication | 🟢 Provides backup genes to cover needs if mutation inactivates a gene; increases gene dosage to boost production of some products |
| | | 🟠 May interfere with chromosome separation; may disrupt gene function if duplication occurs within a gene |
| INVERSION | a b c d  Inversion  a c b d | 🟢 Increases genetic diversity by changing gene positions |
| | | 🟠 Reduced fertility; loss of control of gene expression |
| TRANSLOCATION | | 🟢 Enormous genetic changes may generate rapid evolutionary advances |
| | | 🟠 May activate genes that cause cancer; reduce fertility; may result in gain or loss of whole chromosome |

some, so that when it pairs with its unaltered homologue during synapsis, part of the longer strand has no homologous region for alignment. This part forms a loop extending from the paired homologues. The loop can be seen microscopically in stained preparations, a flag that a deletion has occurred at that location in the shorter chromosome. Matching deletions with the disappearance of a particular trait in the mutant helps geneticists identify the location of specific genes.

This technique is called deletion mapping.

## Duplications

Occasionally, a segment of a chromosome is repeated. Such **duplications** often arise following breakage of two chromosomes and misaligned reunion during repair, so that one member loses a section (a deletion) and the other member gains an extra copy of the region. Duplications may have advantages, such as providing backup

genes that protect the organism in case mutation inactivates a critical allele. It also increases gene dosage, intensifying production of gene products needed in large quantities, for example, hemoglobin in vertebrates. But duplications may also be lethal or fatal, especially when they occur in the middle of a gene and disrupt its genetic message.

## Inversions

Sometimes the segment between two points of breakage is resealed into the chromosome in reverse order. This may occur when the chromosome forms a loop, breaks in two points, and then rejoins with itself, as depicted in Table 26-1. Although the chromosome's length and number of genes remain unaltered, pairing and separation of homologous chromosomes during cell division may be disrupted. Moreover, inversions can result in partial sterility if crossing over occurs within the inverted region and the normal homologous chromosome, since the resulting gametes may have deleted and duplicated segments:

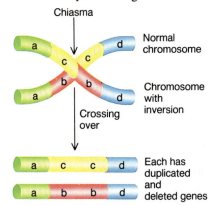

By changing the gene positions, inversions may change gene expression. For example, a gene that is affected by an adjacent gene may no longer be expressed, or may produce a different phenotype after being separated from its neighbor by inversion.

## Translocation

A broken chromosomal segment may also be joined to a *nonhomologous* chromosome, resulting in a **translocation**. (Translocation should not be confused with crossing over, since crossing over involves the equal exchange of chromosome segments between *homologous* chromosomes.) In most translocations, each of the two broken chromatids acquires the genes lost by the other, as shown in Table 26-1. When one piece is significantly shorter than the other, the abnormality is visible by inspecting enlarged photographs of the chromosomes. (See "Karyotyping" later in this chapter.) Most translocations either are fatal or reduce the number of gametes that can result in the production of viable offspring. Furthermore, as with inversions, repositioning genes may alter their expression.

Translocations increase the incidence of cancer in humans. Genes that cause cancer apparently can be activated when nonhomologous chromosomes trade DNA segments. One of the first indications of this was the discovery of the *Philadelphia chromosome*. This unusual chromosome is an aberrant version of human chromosome number 22 with one of the chromatids notably shortened. The Philadelphia chromosome may be responsible for one form of leukemia, an often fatal malignancy of the white blood cells. For years it was believed that the missing segment represented a simple deletion, but with improved techniques for observing chromosomes, the missing genetic piece was found attached to one of the chromatids of chromosome number 9, apparently moved there by translocation. Although the leukemia cells still have all the same genes as normal cells, the gene positions are different. Rearrangement may disrupt normal gene controls that ordinarily prevent overgrowth of cells. Loss of these genetic controls may be responsible for the rampant growth of cancer cells, which eventually outcompete the normal cells in the body. As an encouraging note, this form of cancer is not inheritable. The translocation apparently strikes a single cell while undergoing mitosis during embryonic development, and affects only those blood-producing cells that descend from this altered cell. In addition to this form of leukemia, several other types of cancer are associated with translocations.

Translocations also play an important role in evolution, generating large-scale changes that may be pivotal in the branching of separate evolutionary lines from a common ancestor. Such a genetic incident probably happened during our own recent evolutionary history. A comparison of the 23 pairs of chromosomes in human cells with the 24 pairs in the cells of chimpanzees, gorillas, and orangutans reveals a striking similarity. Six chromosomes are identical, and all but two of the others would be identical except for a series of traceable inversions and translocations. But how did humans "lose" the twenty-fourth chromosome? A close examination of human chromosome number 2 reveals that we didn't really lose it at all. In the ape species, if the two chromosomes that don't match any human chromosomes are fused together to create a single chromosome, they form a perfect match, band for band, with human chromosome number 2 (Figure 26-9). Fusing two chromosomes into one reduces the haploid chromosomal number from 24 to 23.

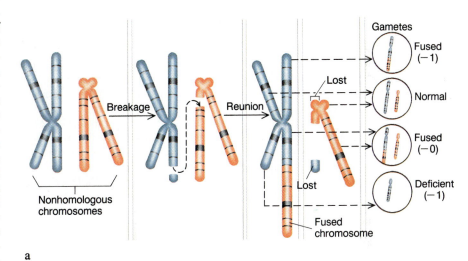

a

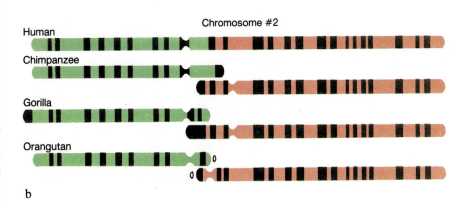

b

**Figure 26-9**

**Chromosomal fusion by translocation.** *(a)* Simultaneous breakage of terminal segments from nonhomologous chromosomes may translocate the entire length of one onto the other. Such an event can create two types of gametes that contain the fused chromosomes—one with the original number of chromosomes (−0) and one in which the chromosomal number has been reduced by one (−1). *(b)* Translocation may have played an important role in the evolution of humans from their apelike ancestors. If the only two ape chromosomes that have no counterpart in humans are fused, they match human chromosome number 2 band for band.

# Changes in Chromosomal Number

## Addition or Loss of One Chromosome

During mitosis or meiosis, chromosomes normally separate into their respective daughter cells, each of which gets an equal number. (Normal chromosomal separation is referred to as *disjunction*.) But sometimes the separation process goes wrong, and **nondisjunction** is the result. Nondisjunctions may send an extra chromosome to one daughter cell, while the other cell gets one less. This mistake generates a condition called **aneuploidy** which nearly always has serious consequences. Loss of a chromosome usually creates fatal genetic deficiencies. Normal fertilization of the other gamete, the one with the extra chromosome, results in a *trisomic* zygote, one with three homologues of that chromosome instead of the normal two. When trisomy shows up in human chromosome 21, the zygote develops into a child with *Down syndrome.* Down syndrome is characterized by mental impairment, alteration in some body configurations, frequent circulatory problems, and lowered resistance to infectious diseases. Down syndrome was the first disorder in humans shown to be due to a chromosomal aberration.

Nondisjunction occasionally alters the number of sex chromosomes of an individual. A zygote with only one X (and no Y) develops into a girl with *Turner syndrome* in which secondary sex characteristics fail to develop, genital development is arrested in the juvenile state, and body structure is slightly abnormal. The opposite situation, the presence of an extra X chromosome, results in an XXX "superfemale." Usually, the extra X chromosome has no observable phenotypic effect, although subnormal intelligence is frequently associated with the condition. Males with an extra X chromosome (XXY) are less fortunate. These individuals develop *Klinefelter syndrome,* which is characterized by mental deficiency, under-

development of genitalia, and the presence of feminine physical characteristics (such as breast enlargement). Men with an extra Y chromosome (XYY "supermales") appear normal except for a tendency to be taller than average. In general, however, their IQ scores are lower than normal. Although still a contentious issue, XYY men may have a tendency toward impulsive behavior that could account for the fact that the number of XYY men relative to XY men is much greater in our prisons than in the general population.

## Additional Sets of Chromosomes

Abnormal chromosomal separation during meiosis can also affect whole sets of chromosomes in a resulting zygote. Increasing the number of chromosomal sets produces a situation called **polyploidy** (a 3N or greater number of chromosomes). Polyploids can be generated by any of several atypical events—by the failure of duplicated chromosomes to separate so they remain in the same nucleus; by fertilization when one or both gametes has more than the haploid number of chromosomes; or by the fusion of three or more gametes during fertilization.

In most higher animal species, polyploidy is lethal, or at least interferes with fertility, because of the extra sex chromosomes. The absence of sex chromosomes in most plant species is one of the factors that makes them more tolerant of polyploidy. Half of all known plant species are polyploid (Chapter 33) or contain some polyploid tissue. As long as there is an even number of chromosomal sets (4N, 6N, etc.), gametes receive equal shares and the plant retains fertility. Plants that have odd-numbered sets (such as triploid—3N—plants) tend to be sterile, but the species may still survive through asexual reproduction.

Polyploidy in plants may confer advantages over their diploid counterparts, especially when the extra set of chromosomes is the result of hybridization of two closely related plant spe-

cies. The polyploid is often hardier than either of the diploid species since it combines the advantages of both (for example, disease resistance of one and drought resistance of the other).

In addition to enhanced hardiness, polyploid plants have several other advantages over their diploid counterparts, such as increased yields and production of fruit with special qualities. Many of our crop plants, including wheat, strawberries, and potatoes, are polyploid. Some seedless fruits are produced by triploid plants. Although these plants are usually sterile (hence their inability to make seeds) they can be propagated vegetatively (by cuttings or grafts). Polyploids of pears, apples, and grapes produce giant fruit that show commercial promise.

# Karyotyping

Detecting nondisjunction and chromosomal aberrations would be impossible without a means of physically viewing the chromosomes so that they could be counted and classified. This is now routinely done by preparing a **karyotype**, a visual display of an individual's chromosomes. Human karyotypes are usually prepared from white cells in blood. (An example can be found in Chapter 6, Figure 6-10.) After treatment with a chemical that halts mitosis in metaphase, cells are mounted on a microscope slide, stained, magnified, and photographed. The chromosomes are cut from the photograph and pasted on an identification template according to decreasing length. The banded appearance of the stained chromosomes distinguishes between chromosomes of similar length. The technique (called *G-banding* for the Giemsa stain used to enhance the bands) makes it easier to detect structural changes in chromosomes. In addition to revealing the location of chromosomal aberrations associated with genetic disease, G-banding allows a more precise comparison of chromosomes from different types of organisms, revealing similarities that help us trace the evolutionary origins of different species.

# Synopsis

## MAIN CONCEPTS

- **Genes reside on chromosomes in fixed linear orders**. The physically connected set of alleles on the same chromosome constitutes a linkage group, a group of alleles that do not assort independently.

- **Linked genes don't always stay linked**. Crossing over shuffles alleles between homologous chromosomes. This increases genetic diversity and allows beneficial alleles to escape certain elimination whenever a lethal mutation occurs on the same chromosome.

- **Linkage maps of alleles can be constructed for chromosomes**. The further apart two genes are from one another on a chromosome, the more likely they will be separated by crossing over between the two. By determining the frequency with which alleles become "unlinked" during meiosis we can construct genetic maps of chromosomes.

- **Gender is determined by sex chromosomes** in most animals and a few plants. In humans two X chromosomes (XX) program the individual to become a female, and XY a male. Sex-linked traits are normally expressed when males inherit a recessive allele on the X chromosome.

- **Chromosomes may rejoin incorrectly after breakage**. The rejoined strands may lose a segment (a deletion), acquire extra genes (a duplication), or reconnect in a reverse order (an inversion). Segments may also be transferred to nonhomologous chromosomes (a translocation).

- **The number of chromosomes can change**, such as the addition or deletion of a single chromosome (for example, trisomy), or an increase in their number of chromosome *sets* (resulting in polyploidy).

## KEY TERM INTEGRATOR

### Genes on the Same Chromosome

| | |
|---|---|
| **linkage** | The tendency of two alleles at different locations on the same **chromosome** to be inherited together. Tightly linked **genetic markers** usually remain together when passed on from parent to progeny. Genes that reside on the same chromosome constitute a **linkage group**. |
| **crossing over** | The mutual exchange of chromosomal segments between homologous chromosomes. The resulting frequency of crossing over between known linked genes is used to construct **linkage maps** of chromosomes. |

### Chromosomes and Gender

| | |
|---|---|
| **sex chromosomes** | Chromosomes that determine an organism's gender. Unlike **autosomes**, mammalian sex chromosomes have different homologues, an **X chromosome** and a **Y chromosome**. |
| **sex-linked traits** | Traits whose genes reside on the X chromosome. Males (XY) are **hemizygous** for these genes, so any recessive allele on the X chromosome will be expressed. In females (XX) the same gene would likely be masked by a dominant allele on the other X chromosome, even though one of the female's X chromosomes is inactivated, forming a **Barr body**. Unlike sex linkage, **Y-linked inheritance** primarily governs development of testes and maleness. |

---

Atypical Chromosome Structure and Distribution

| | |
|---|---|
| **deletion** | Loss of a chromosomal segment. |
| **duplication** | Gaining at least one extra copy of a chromosomal segment. |
| **translocation** | Transfer of chromosomal segments between nonhomologous chromosomes. |
| **nondisjunction** | The failure of chromosomes to separate during meiosis, leading to **aneuploidy** in cells that receive an extra chromosome, or **polyploidy** in cells that get extra chromosomal sets. |
| **karyotype** | Stained, microscopically enhanced chromosomes arranged according to decreasing size. |

---

# Review and Synthesis

1. The traits of one well-studied organism fall into seven distinct linkage groups. How many chromosomes are in the *somatic* cells of this organism?

2. Why wouldn't crossovers be detected in offspring of the following cross: EEFFGG × eeffgg.

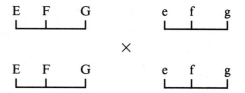

Draw the tetrad of each of these parents and do a couple of crossovers on paper. Repeat the above, but this time replace the homozygous dominant parent with a full heterozygote (EeFfGg)

3. Describe two biological advantages of crossing over.

4. Draw a simple genetic map showing the gene order and relative distances using these data:

| *crossover frequency* | *between these genes* |
|---|---|
| 36% | X-Y |
| 10% | Y-Z |
| 26% | X-Z |

5. Alleles on opposite ends of a chromosome are so likely to be separated by crossing over between them that they segregate independently. How would we ever know that these two genes belong to the same linkage group? (Let's say this chromosome has 45 known genes.) Here's a hint: if A is linked to B and B is linked to C, then A is linked to C.

6. Why do sex-linked traits tend to skip a generation? Under what circumstances would this not happen? (For instance, how could a color blind man have a child that is also color blind?)

7. Explain why a YØ sex chromosome deficiency would be fatal in humans, whereas an XØ deficiency would not.

8. If you were offered a calico kitten, why would you not have to ask whether it is a male or female?

9. Describe the similarities or differences between each pair of terms:
genetic marker—mutant allele
sex linkage—sex-influenced dominance
triploidy—trisomy
linkage—independent assortment
heterozygous—hemizygous
translocation—crossing over

**10.** List and describe four types of sex influence on inheritance.

**11.** As an official of the Olympic games, you received a laboratory report stating that stained cells from a female athlete were found to be devoid of Barr bodies. Would you disqualify her from competition? Justify your decision to the eager reporters waiting outside your door.

**12.** The hard, obtrusive seeds of wild bananas render the fruit virtually inedible to humans. The commercial variety of bananas is triploid and delicious. Why would triploidy make such a difference?

## Additional Readings

Baer, A. 1977. *The Genetic Perspective.* Saunders, Philadelphia. (Intermediate.)

Croce, C., and G. Klein 1985. "Chromosomal translocations and human cancer." *Scientific American* 252:54. (Intermediate.)

Harrison, D. 1970. *Problems in Genetics with Notes and Examples.* Addison-Wesley, Reading, Mass. (Introductory to intermediate.)

Lewontin, R. 1982. *Human Diversity.* W. H. Freeman, San Francisco. (Intermediate.)

McKusick, V. 1971. "The mapping of human chromosomes." *Scientific American* 244:104–13. (Intermediate to advanced.)

Marx, J. 1985. "Putting the human genome on the map." *Science* 229:150–51. (Intermediate.)

Chapter 27

# The Molecular Basis of Genetics

To many people, heredity seems magical. For thousands of years, observers of heredity, unable to construct a realistic explanation of how parents pass on traits from generation to generation, ascribed the phenomenon to mystical forces. It may in fact still seem magical to you. After all, how can the remarkable feats of a gene be explained in physical and chemical terms alone? How can a physical structure like a chromosome "memorize" the genetic information needed to direct the construction of a particular organism, with its thousands of genetically controlled characteristics, and then order the organism to develop those characteristics? And how is genetic information preserved throughout a multitude of generations? Although an organism dies, its inheritable characteristics are transmitted to its progeny, and to generations of subsequent offspring. The result is the virtual immortality of many inheritable traits. (The use of phospholipid bilayers for cell membranes, for example, has probably been preserved throughout the 3.5 billion year history of life on earth.) Yet today the answers to these questions are becoming increasingly clear. There is no "magic"—just a fascinating and understandable explanation of gene function. Although important questions have yet to be answered (and some have yet to be asked), scientists have demystified the gene.

To do this, investigators had to hunt for clues on the most subtle level of life—the realm of molecular interactions. As the quest for answers drew scientists into the molecular domain, it created a new field of life science, called **molecular biology**. The goal of molecular biologists is to discover the physical and chemical principles that make cells operate. Like other geneticists, these scientists experimented on organisms. But instead of *Drosophila*, they chose less complex organisms whose simpler anatomy posed fewer distractions. These scientists chose bacteria and the viruses that infect bacteria.

## Bacteria — Leading the Way

As models for investigating molecular genetics, bacteria have several advantages over eukaryotic organisms, mostly because of their relative simplicity.

- Bacteria are much easier to understand than eukaryotic organisms. The most commonly used bacterium, *Escherichia coli* (or more often *E. coli*), has about 2000 genes, compared to the 5000 genes of *Drosophila* or the estimated 40,000 or more genes of humans.

- Bacteria have a single chromosome* rather than a diploid set. Dominant traits cannot hide the presence of recessive alleles, making the organism's genotype much easier to deduce from its phenotype.

- Bacteria are easy to propagate rapidly, more so than even *Drosophila,* requiring thousands of times less laboratory space to study the same number of eukaryotic organisms.

- Billions of bacteria can be propagated asexually from a single cell, producing a genetically homogeneous population.

---

* *Bacterial geneticists consider the DNA of bacteria to be a true chromosome even though it lacks histones and doesn't condense during cell division.*

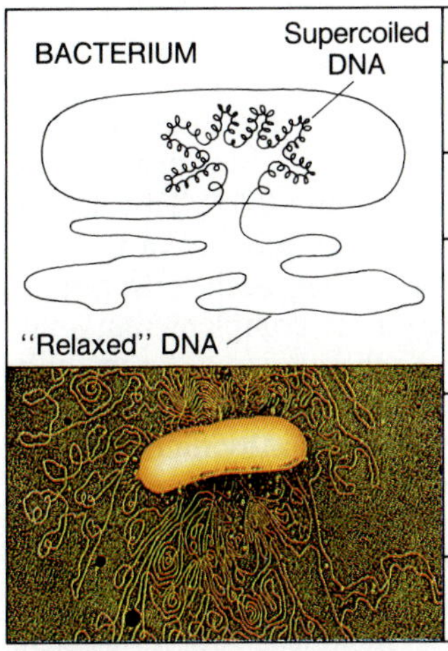

| CHARACTERISTIC | PROKARYOTIC | EUKARYOTIC |
|---|---|---|
| Configuration of DNA | Circular | Linear (open-ended) |
| Length (average) | 1000 $\mu$m | 1.8 meters (in humans) |
| Number of chromosomes per cell | 1 | at least 2, up to 768 |
| Associated proteins | DNA-binding proteins; Gyrases | Histones; Nonhistone chromosome protein |
| DNA housed in nucleus | No | Yes |

**Figure 27-1**
**The prokaryotic chromosome,** a model of complexity and simplicity. It is a single strand with "ends" fused to form a circle. The length of the DNA in the circle is about 1000 times the diameter of the cell. To conserve space, it is packed in the bacterium as "supercoiled" DNA, held together by DNA-binding proteins. *Gyrase* is an enzyme that coils and uncoils the chromosome. Each cell possesses only one copy until just before division, when the chromosome duplicates. The adjacent table compares the chromosomes of prokaryotic and eukaryotic cells. The photo is of *Escherichia coli* broken open and spilling out its chromosome.

Relatively simple organisms like bacteria are nonetheless enormously complex, reflected by the huge amounts of information stored in their single chromosomes. Yet it is a mere fraction of the information stored in the chromosomes of eukaryotic cells. Each human cell, for example, contains so much genetic information that it would require over a million of these pages to express it in written form. Yet the molecule that accomplishes the task is, like bacteria, relatively simple. It is DNA (see Chapter 4), a linear polymer of gigantic length. (In spite of its length, it is extremely narrow — less than one-trillionth of an inch thin.) A single circular strand of DNA constitutes the simple bacterial chromosome (Figure 27-1).

Since the mechanism of gene function was largely elucidated by studying bacteria, these organisms are used here to illustrate several genetic mechanisms in action. Much of what has been learned from bacteria is applicable to all organisms, from viruses to humans. In spite of the greater complexity of eukaryotic chromosomes (discussed in Chapter 28), their DNA accomplishes its tasks in fundamentally the same way as does the DNA of bacteria and viruses, another testimony to the unity of life.

# DNA — Life's Molecular Supervisor

The way DNA exerts its molecular authority over an organism's genetic destiny is a threefold process:

1. **Storage of genetic information.** DNA is the molecular "blueprint," a record of precise instructions that determine what inheritable characteristics a cell can manufacture.

2. **Inheritance.** Genetic instructions are "written" in a form that can be copied and passed on to (and "read" by) an organism's descendants, without the parent sacrificing its own set of blueprints needed for continuing its life.

3. **Expression of the genetic message.** Cells translate the encoded genetic instructions into real characteris-

tics by using a gene's information to direct the synthesis of a specific protein. The protein then participates in the development of a particular trait. One protein might help form the structure of an organism; another might be an enzyme that catalyzes the formation of a trait. Or the protein might execute a specific biological task (such as oxygen transport), or regulate the expression of other genes. Each gene is a blueprint for the formation of a particular protein.*

In a sense, the genetic basis of life is a matter of "writing, reading, and following instructions" on a molecular scale. Just as a reader of a sentence deciphers a code by translating a linear string of letters into a meaningful thought, cells use a language, a *genetic code,* that they translate into genetic traits. The cell transcribes ("writes") a copy of the genes' encoded instructions to send to the cytoplasmic "workers" (ribosomes). These instructions direct their activities, telling them what protein to build (Figure 27-2). The cytoplasmic workers must be able to "read" the instructions, that is, *translate* them into the exact gene product ordered for construction. Periodically, new copies of the instructions must be produced, so each progeny cell will receive all the information needed, to organize their growth and activities. Without genetic instructions, the only thing created is chaos.

## Operation 1 — Storage of Genetic Information

DNA is a linear polymer composed of repeating **nucleotides**, which are monomers analogous to chemical "letters." When arranged in a line, *nucleotides encode information by virtue of their sequence,* in the same way a sentence contains information by virtue of its arrangement of letters. The genetic "alphabet," however, consists of only four letters: the four types of nucleotides in DNA. But four "let-

ters" is more than enough to "write" an unlimited variety of genetic messages.

Each nucleotide contains the three components depicted in Figure 27-3 — a 5-carbon sugar, a phosphate group, and a *nitrogenous (nitrogen-containing) base.* In DNA the sugar and phosphate groups are identical in all nucleotides. The identity of a nucleotide is determined by which of the four nitrogenous bases it contains (Figure 27-4). Take a moment to memorize the identity of these bases, as well as which ones are purines (A and G) and pyrimidines (T and C), two important categories of nitrogenous bases. These distinctions are needed to help you understand the mechanisms for encoding, replicating, and translating genetic information.

### DNA's MOLECULAR BACKBONE

The nucleotide phosphate group and sugar are used to link nucleotides together to form a single linear nucleic acid strand. The phosphate of one nucleotide covalently bonds to a free ribose of another, forming a chain that can continue to grow as long as there are free phosphate and ribose groups. These sugar-phosphate linkages form the molecular backbone of the DNA molecule, with the nitrogenous bases protruding to the side (Figure 27-5).

### BASE PAIRING

The typical DNA molecule contains two parallel strands, not just the one depicted in Figure 27-5. This double strandedness is a product of the specific affinity that each nitrogenous base has for its complementary nucleotide. Each guanine (G) pairs only with a cytosine (C), and each adenine (A) pairs only with thymine (T). The sequence of nucleotides in one strand is therefore *complementary* to the order in the parallel strand (Figure 27-6). The two strands are held together by hydrogen bonds between complementary nucleotides. The nucleotide pairs (A-T and G-C) form cross links that resemble the steps of a ladder, with the ladder's "rails" formed by the parallel sugar-phosphate backbones. The molecule actually resembles a spiral staircase more

---

* Two notable exceptions to this statement are the genes that direct the formation of tRNA and ribosomal RNA, two forms of nucleic acid that are discussed in a later section of this chapter.

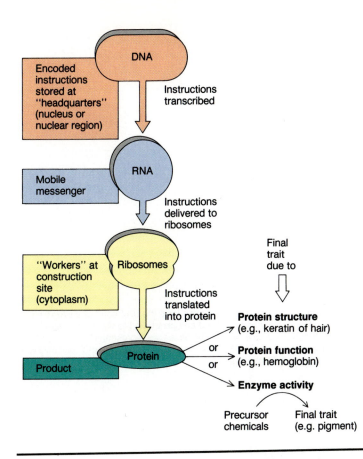

**Figure 27-2**
**The flow of genetic information** from DNA to expressed traits is often compared to following construction specifications encoded in a blueprint. The analogy works better if we compare it to a set of written instructions instead of a blueprint, because cells "read" genetic messages that are transcribed in linear sequences, more like sentences than pictures. The gene product is usually a protein, which may form a specific structure or perform a particular function. If the protein is an enzyme, it catalyzes formation of the trait for which the gene codes, for example, brown pigment for eye color. Without the gene, there would be no enzyme and the eyes would be another color.

**Figure 27-4**
**Nitrogenous bases in DNA.** Four bases, two purines (adenine and guanine) and two pyrimidines (thymine and cytosine), determine the identity of the nucleotides in DNA. Purines and pyrimidines in nucleotides associate with each other by forming hydrogen bonds. The association is specific—guanine (G) and cytosine (C) are held together by *three* H-bonds, whereas adenine (A) and thymine (T) form *two* H-bonds. This specific "base pairing" is symbolized diagrammatically next to each molecular representation. These schematic shapes are used to show base pairing in subsequent illustrations.

**Figure 27-3**
**Molecular skeleton of a nucleotide.** Nucleic acids (DNA and RNA) are composed of repeating nucleotide units. Each nucleotide consists of the three parts shown here. The sugar determines whether the molecule is DNA or RNA. In DNA (*deoxyribo*nucleic acid), the sugar is *deoxyribose*; in RNA (*ribo*nucleic acid) the sugar is *ribose*. Each of the three parts contributes to nucleic acid function. Different nitrogenous bases distinguish each nucleotide type from the other three so that messages can be written using four different nucleotide "letters."

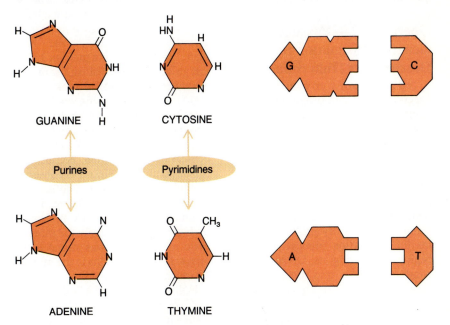

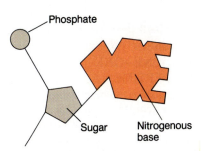

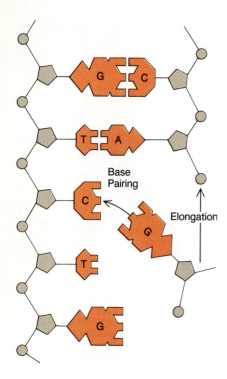

Base Pairing

Elongation

**Figure 27-5**
**Backbones and base pairing.** Sugar-phosphate linkages form the backbone of a simple linear strand of DNA, leaving the nitrogenous base exposed and protruding to the side, where it can form hydrogen bonds with other nitrogenous bases. The bases are represented in their symbolic form to help show the specificity of base pairing in DNA; adenine (A) and thymine (T) pair only with each other, as do guanine (G) and cytosine (C).

than a ladder. Each consecutive base is slightly rotated to the right of the one just below it, twisting the molecule into a double-stranded helix.

DNA's double strandedness is critical to its role in heredity and the ability of cells to translate genetic information into expressed characteristics.

## Operation 2 — Passage of Genetic Information to Descendants (Inheritance)

A cell provides its progeny with the necessary genetic information by producing exact copies of its DNA (a process called **replication**), and then giving each daughter cell a copy. Specific base pairing, as described above, is essential to DNA replication. The cell also maintains a reservoir of free, unpaired nucleotides available for constructing new DNA, as well as all enzymes and accessory proteins needed for the replication process. The cell begins duplicating its DNA by unwinding the helix and separating the strands (Figure 27-7). The strands separate, exposing the nitrogenous bases of the individual strands. The enzyme

**DNA polymerase** assembles free nucleotides, aligning them with the complementary nucleotides in the unpaired region of each single strand, restoring double strandedness. In this way, a new complementary strand is constructed along each single strand of the original double helix. The sequence of the new strand is dictated by the order of nucleotides in the existing strand, which serves as the "template." (A template is a mold or a physical guide.) When finished, the process generates two identical molecules of DNA, each containing one strand from the original DNA molecule and one newly synthesized strand. This form of DNA synthesis is called **semiconservative replication**, because half the original DNA strand is conserved in each new molecule. The result of semiconservative replication is the production of two DNA molecules, each containing precisely the same genetic message that was stored in the original molecule. This is the molecular mechanism of heredity.

DNA is often called a "self-replicating" molecule. This description is somewhat inaccurate, however, for

DNA cannot replicate by itself. It requires cellular machinery, such as specific enzymes. Enzymes are needed to unwind the helix, to separate ("unzip") the strands, and to polymerize a new complementary strand along the exposed bases of each single strand. (This explains why free nucleotides don't pair with each other. DNA polymerases, the nucleotide-assembling enzymes, match free nucleotides only with those that are already part of an existing DNA strand.)

Although more complex than their prokaryotic counterparts, eukaryotic chromosomes also replicate in a semiconservative manner. Overall differences between the two are illustrated in Figure 27-8.

## Operation 3 — Expression of the Genetic Message
### ONE GENE – ONE POLYPEPTIDE

An organism is the manifestation of its particular constellation of proteins. A gene typically dictates the formation of a single type of protein, a relationship often referred to as the "one

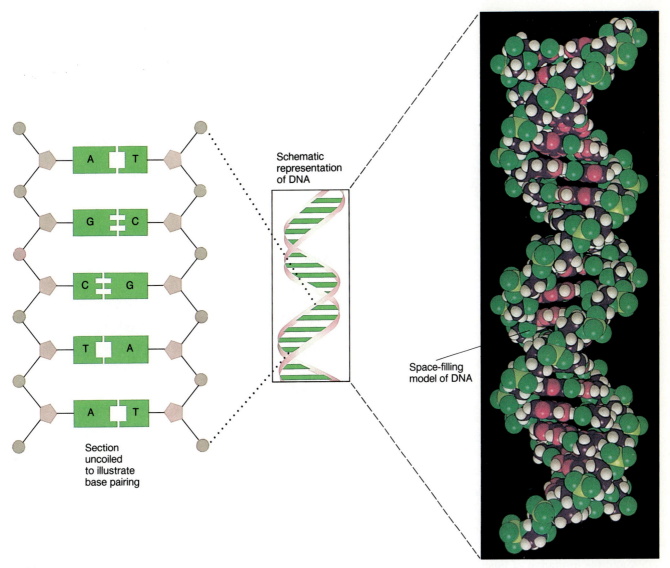

Schematic
representation
of DNA

Space-filling
model of DNA

Section
uncoiled
to illustrate
base pairing

**Figure 27-6**
**The double helix.** Base pairing creates a double-stranded DNA molecule that twists into a helix, resembling a spiral staircase. the nucleotide sequence in the two strands is complementary, so if you were given the sequence in one strand, you could predict the nucleotide order in the other strand. The computer-generated image (inset) better represents the actual appearance of DNA.

gene–one protein" hypothesis of genetic control. A gene specifies the formation of one type of polypeptide chain, the string of amino acids that folds into a functional protein. In proteins composed of more than a single type of polypeptide, different genes specify the formation of each chain. Since these proteins require two or more genes to direct their synthesis, the one gene–one protein relationship is more accurately "one gene–one

polypeptide." Either way, the order of nucleotides in DNA determines the sequence of amino acids in protein, which determines the structure and ultimately the activity of that protein. (The role of amino acid sequence in determining protein shape and activity is discussed in Chapter 4.)

## PROTEIN SYNTHESIS
To proceed further, we must first answer two questions. How does genetic

information stored in DNA get to the cytoplasm? And once the information is there, how is that message "read" and used to construct the corresponding protein? These tasks require the participation of the second type of nucleic acid, RNA, which is very much like DNA. RNA is a linear polymer with fixed sequences of nucleotides. It is distinguished from DNA by three properties that enable it to bridge the gap between encoded information in

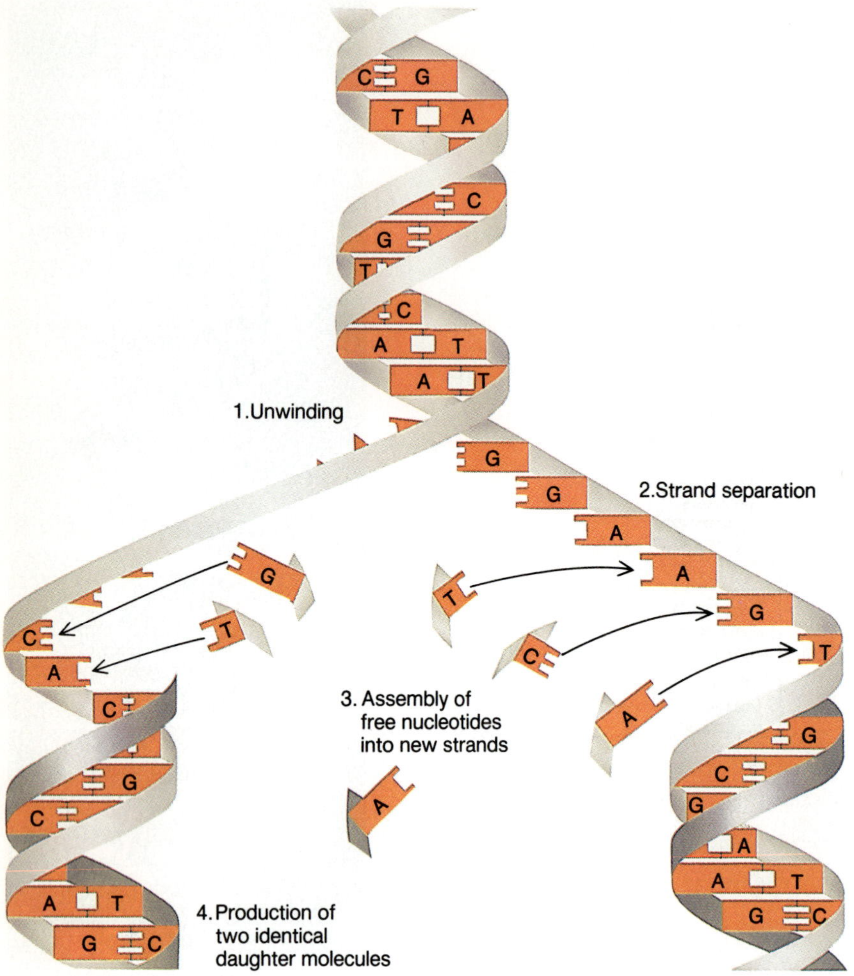

1. Unwinding

2. Strand separation

3. Assembly of free nucleotides into new strands

4. Production of two identical daughter molecules

**Figure 27-7**
**Replication of DNA** produces two identical copies of a chromosome by the mechanism shown here. During cell division, each daughter cell receives one of the copies.

DNA and the formation of functioning proteins:

- The nucleotides in RNA contain *ribose,* a sugar that differs from DNA's deoxyribose only in the presence of one more oxygen atom. The difference is an important one, however; it allows enzymes to distinguish between the two types of nucleotides.

- As in DNA, there are four distinct nucleotides, but one of them, *uracil,* is unique to RNA. It replaces thymine as the nucleotide that pairs with adenine. (There is no thymine in RNA.)

- RNA is a single-stranded molecule, so its nitrogenous bases are exposed and available for base pairing.

Transcribing the Message

The RNA that carries genetic information to the ribosomes in the cytoplasm is aptly called **messenger RNA (mRNA)**. It is constructed in a manner that in some ways resembles the assembly of a new DNA strand during replication. The double helix temporarily separates, and a strand of mRNA assembles along one of the single DNA strands in the separated region. This process is called **transcription** (Figure 27-9). **RNA polymerase**, the enzyme that directs the process, distinguishes between DNA and RNA nucleotides, and assembles only RNA nucleotides in the growing chain. It also distinguishes between the two strands of the DNA molecule. It selects one of the DNA strands, the "sense strand," as the template that determines the nucleotide sequence in the mRNA molecule.

In this way, genetic information stored in a gene is encoded in a mRNA molecule by virtue of its nucleotide sequence. The instructions in a gene are thus "transcribed" (copied) into a form that can be carried to ribosomes by a mobile messenger molecule. This is why mRNA synthesis is called "transcription." At the ribosomes, the mRNA directs the synthesis of a specific protein. In other words, the transcribed instructions are *translated* into a particular sequence of amino acids. In eukaryotic cells, mRNA leaves the double-membrane nucleus through the nuclear pores.

Translating the Message

As you translate the line of characters in this sentence, you do so by recognizing groups of letters, words that have specific meanings. In constructing proteins, the linear array of nucleo-

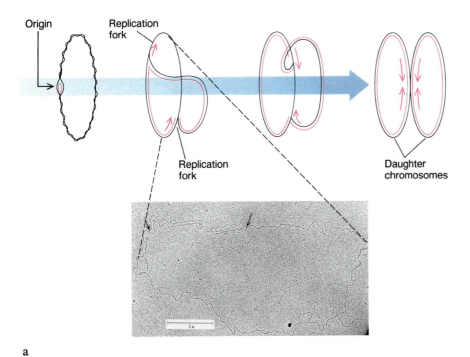

a

b

**Figure 27-8**

**Comparison of bacterial and eukaryotic chromosomal replication.** In both, replication occurs at "forks" that travel away from each other in opposite directions. Replication is therefore bidirectional. The circular chromosome of bacteria has only two replication forks *(a)*. Chromosomal duplication is completed when the forks meet each other halfway around the circle. The chromosome in the photo has completed about one-sixth the process. In eukaryotic chromosomes *(b)*, replication begins at several points, each with two replication forks that travel away from each other until they either meet another fork or come to the end of the molecule. Five distinct replication sites are apparent in the photo of a single DNA molecule from a mammalian cell. The dark lines are areas with newly replicated DNA.

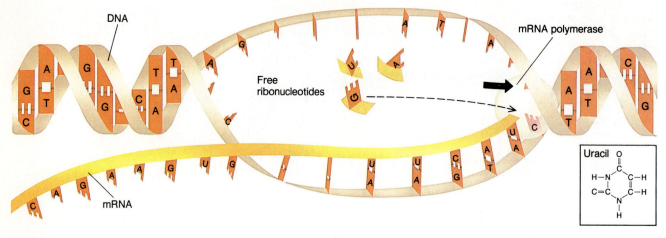

a

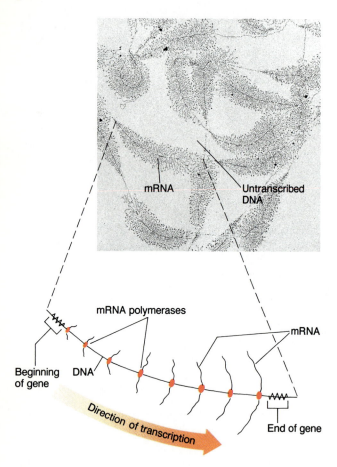

b

**Figure 27-9**

**Transcription — dispatching the molecular messenger.** *(a)* Only one strand of DNA (called the "sense strand") is used to encode information in the message sent to the ribosomes. This template strand determines the nucleotide sequence in the mRNA molecule assembled during transcription, which proceeds in the direction that mRNA polymerase is moving. The region is only temporarily separated, and the two strands immediately rejoin following transcription. The uracil (see inset) that replaces thymine in RNA differs from thymine in the single chemical group shaded for emphasis. *(b)* Photo of transcription. Closely spaced polymerases speeding along DNA produce hundreds of mRNA molecules, seen here protruding from the sides of several genes. The longer mRNA strands are the oldest ones; newly emerging strands are short.

tides in mRNA is also translated in groups, molecular "words" called **codons**, each of which represents only one specific amino acid. Each codon is three nucleotides long. With a four-letter genetic alphabet (A, G, C, and U), 64 triplet combinations are possible, more than enough to assign each of the 20 amino acids at least one unique codon. The presence of a particular codon triplet in mRNA orders the insertion of the corresponding amino acid in the growing protein chain. The codon CAC, for example, would specify the insertion of histidine at that point in the protein being synthesized. The codon CUA always specifies leucine. In this way, the entire message is read codon-by-codon until the protein is complete.

The mRNA-directed synthesis of a protein is called **translation**, a process that requires mRNA, amino acids, several enzymes, ribosomes, and ATP energy. It also requires another type of RNA that decodes mRNA's encoded message (written in codons) and translates it into the language of proteins (amino acids). This "molecular decoder" is called **transfer RNA (tRNA)**. It works like any decoding device: whenever a particular set of symbols appears in the encoded message, it associates the corresponding "word" with it. Each tRNA is a relatively small length of RNA that folds into a cloverleaf shape. This configuration orients its codon-recognition site opposite its amino acid site on the folded structure (Figure 27-10). The recognition site, called the **anticodon**, is a triplet nucleotide sequence that base pairs with a particular codon in mRNA. Each tRNA carries a particular amino acid, which is always the same one for a given anticodon. Each type of tRNA has a unique anticodon that can base pair with one and only one codon in mRNA. Base pairing between the tRNA's anticodon and the mRNA's codon brings the corresponding amino acid into position for incorporation into the protein being synthesized. For example, the codon UCU in mRNA binds only with the anticodon AGA of the tRNA carrying the amino acid serine. Therefore, UCU in mRNA instructs the cell to insert serine at that point in the newly forming protein. In this way, tRNA bridges the language gap between codon and amino acid. Codon by codon, the cell uses its tRNAs to translate the genetic code (Figure 27-11).

Each amino acid (except two) has several codons that order its insertion. These are genetic "synonyms," a backup system that reduces the danger of lethal mutations disrupting the cell. A change in a single nucleotide in the codon's third position, for example, would change the amino acid inserted unless the new codon were synonymous with the original (which happens frequently). In that case the same amino acid would be inserted, and the mutation would have no effect.

Four codons warrant special mention. Three (UGA, UAG, and UAA) have no corresponding amino acid. These triplets, called *nonsense codons*, spell "stop!" when mRNA is being translated into protein. They are used to terminate synthesis at the comple-

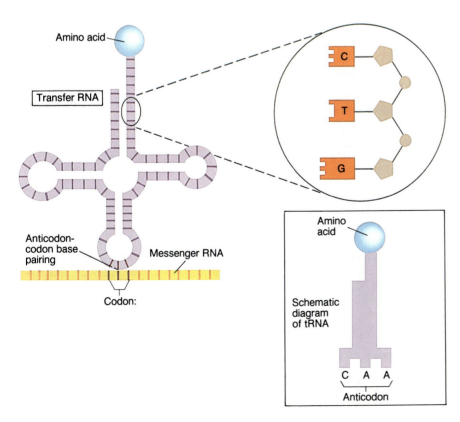

**Figure 27-10**
**Molecular decoder**, tRNA, adapts one "language" (a 3-nucleotide codon "word") into another (a specific amino acid) by using a specific recognition site. This site, the anticodon, recognizes a single codon in mRNA. Each tRNA carries a specific amino acid. In this way a particular tRNA associates the correct amino acid with its corresponding codon in mRNA. The "cloverleaf" structure is shown positioned on the mRNA to illustrate how the anticodon base pairs with its complementary codon. Although folding distorts the cloverleaf into a more complex three-dimensional structure, the simplified schematic version shown in the box will be used from here on to represent tRNA. Energy for "charging" an empty tRNA with the appropriate amino acid comes from the hydrolysis of ATP.

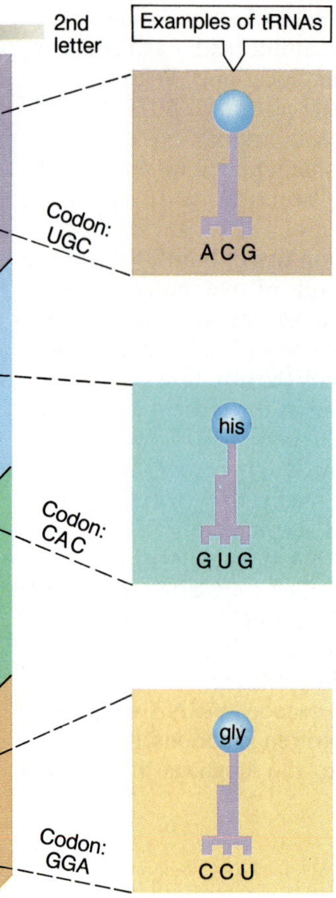

**Figure 27-11**
**The genetic code** is a universal biological language. The correlation between codon and amino acid indicated in this decoder chart is the same in virtually all organisms. To use the chart to translate the codon UGC, for example, find the first letter (U) in the indicated row. Follow that row right until you reach the second letter (G); then find the amino acid that matches the third letter (C). UGC specifies the insertion of cysteine.

Cells have no such chart to help them decode the message in mRNA; tRNAs do the job, a few of which are shown here. Remember that codons are triplet sequences in mRNA; they are complementary to the anticodon on the tRNA.

tion of a protein. Another codon, AUG (specifying the amino acid methionine), is the "start" codon. It always appears at the beginning of an mRNA and initiates the synthesis of the corresponding protein.

Translation occurs at ribosomes (Figure 27-12). In the cytoplasm, one large and one small ribosomal subunit begin translation by assembling into a single unit at the start of a mRNA molecule. The assembled ribosome

has a groove through which mRNA travels. It also has sites that position two tRNA molecules so that their amino acids lie adjacent to each other on the large subunit. The ribosome then moves along the mRNA strand, and amino acids are incorporated into the growing polypeptide chain in the order specified by the mRNA. A simplified version of these events is depicted in Figure 27-13. (Many of the enzymes and accessory molecules

have been omitted for clarity.) The steps in the figure are numbered to correspond to the following description.

Step 1. *Initiation.* Protein synthesis starts with the initiator codon, AUG. This fixes the "reading frame," assuring that translation begins with the correct nucleotide. If the message were initiated one or two nucleotides over, all the remaining triplets would be in-

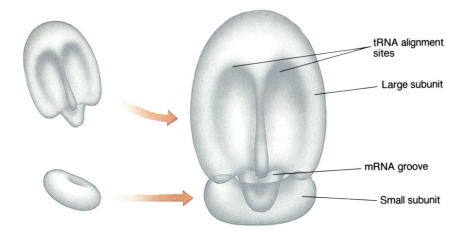

**Figure 27-12**
**Assembly of a functional ribosome.** Two subunits, one large and one small, fit together to form an orientation groove for mRNA and create sites for accepting two tRNAs at a time. An enzyme (not shown) on the large subunit catalyzes the formation of a peptide bond that connects the two aligned amino acids. Thus, ribosomes are more than mere "workbenches" on which proteins are synthesized. They are more like "workers" that help assemble the protein.

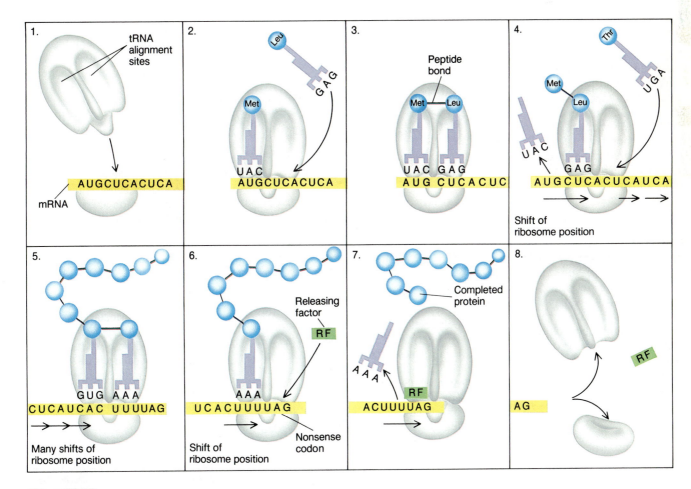

**Figure 27-13**
**General steps in translation** occur in the cell's cytoplasm. The mRNA in the figure is only a short portion of the entire molecule (which for most proteins would exceed 600 nucleotides in length). Each step is discussed in the text.

correctly read and the wrong amino acids inserted. In the example below, the upper brackets indicate the correct reading frame. The brackets below show the incorrect codons that would be produced if initiation began two nucleotides off.

```
correct ⟶      met  leu  his  pro
               ⎡‾‾⎤⎡‾‾⎤⎡‾‾⎤⎡‾‾⎤
mRNA ⟶  — A U G C U G C A U C C A —
               ⎣__⎦⎣__⎦⎣__⎦
incorrect ⟶     ala  ala  ser
```

Since AUG codes for methionine, using this initiator codon for all proteins means that every protein must begin with that amino acid. Yet for most proteins, a methionine at the beginning of the chain would disrupt its function. Cells avoid this limitation by using an enzyme to clip away the initial methionine shortly after a protein's construction is underway. In this way methionine remains as the first amino acid only in those proteins that need it there for proper function.

*Steps 2–5. Chain Elongation.* With the initiator tRNA in place, there remains a second free site on the ribosome for another tRNA to align with the next codon in line (which is CUC in this illustration). CUC pairs with the anticodon GAG of the tRNA carrying the amino acid leucine. The two amino acids are aligned next to each other on the large ribosome subunit and are enzymatically joined by a peptide bond (Chapter 4). The ribosome shifts *(translocates)* to the right by three nucleotides. The first tRNA departs, leaving its amino acid. Translocation brings the third codon into position on the ribosome. The complementary tRNA aligns and orients its amino acid next to the previous one, and another peptide bond forms between these. The growing protein chain is now three amino acids long. Translocation releases the second tRNA and brings the fourth codon into position. This process continues adding amino acids in the proper sequence until the entire polypeptide chain is synthesized.

*Steps 6–8. Chain Termination.* The completion of a polypeptide chain is signalled with a nonsense (stop) codon in the mRNA strand. Since these triplets have no associated amino acids,

their presence produces a region on the mRNA to which no tRNA can bind. Therefore, the amino acid inserted just before the nonsense codon becomes the terminal member of the chain. Although nonsense codons have no corresponding tRNA, they are recognized by proteins called *releasing factors* that help dislodge the finished polypeptide chain from the ribosome, after which the ribosome disassociates into its two subunits.

Protein synthesis occurs very rapidly, and efficiency is increased by simultaneous translation of each mRNA molecule by many ribosomes. As soon as one ribosome has translated the first few codons, the mRNA site is again available for another ribosome to assemble. As a result, ribosomes are often found in chains held together by mRNA, a complex called a **polysome** (Figure 27-14). Since a protein strand is assembled at each ribosome of the polysome, a single mRNA may be simultaneously generating dozens of identical polypeptide chains.

The flow of genetic information from its stored form in DNA to its expression as a specific protein is summarized in Figure 27-15.

## A Reexamination of the Gene

This molecular explanation of genetics provides us with a fresh look at the gene, which can now be redescribed as "a segment of DNA with a unique nucleotide sequence that encodes information specifying the construction of a particular protein (or polypeptide chain)." Yet this cannot be interpreted as *the* definition, for some genes direct the formation of tRNA or ribosomal RNA rather than proteins. Furthermore, recent discoveries, such as "silent regions" (untranslated genetic segments) in the middle of genes and genes that overlap with each other, reveal that genetic expression is often more complicated than the verbatim linear conversion of information from DNA to protein. Nonetheless, while new findings continue to force geneticists to modify their view of the gene, the core of the

definition remains the same as that described by Mendel over 100 years ago. Genes are physical units of inheritance (Mendel's "factors") that direct an organism to form particular traits.

Perhaps no other characteristic illustrates the unity of life as strikingly as the gene. The language that your cells use to write their genetic messages is identical to that of bacteria, plants, fungi and even viruses, which are little more than "portable genes" (see Bioline: Viruses). The genetic code is universal: all cells "speak the same language" and use the same genetic vocabulary.* The codon CCC, for example, specifies the insertion of proline, regardless of the organism. When the human gene for insulin production is experimentally introduced into the bacterium *Escherichia coli*, the nucleotide sequence is translated into the same protein as in a human cell, and the bacterium begins manufacturing insulin. In addition to the code, the mechanics for inheritance and gene expression are virtually universal. All cells even use similar shaped tRNA.

These universals have important biological implications. The likelihood of even two species developing all these universal properties by coincidence is astronomically small. Such universality among the earth's nearly two million species supplies powerful evidence of a common ancestor early in evolution that passed on this system to billions of years of descendants (Chapter 29).

## Mutation

Discovering the molecular basis for gene action has provided us with a clearer understanding of mutation, a phenomenon that occurs in all species. Changes in DNA nucleotide sequence show up in mRNA as altered codons, which translate into different amino

---

* *Two minor variations exist in five species of microorganisms. In Paramecium, for example, UAG = glutamine rather than "stop." Mitochondria, which have their own protein-synthesizing machinery, also have some minor differences in codon recognition. In spite of these few unique "words," they use the same genetic language as do all organisms.*

# B I O L I N E

## VIRUSES: "PORTABLE" GENES

A few biologists view organisms as very elaborate mechanisms for "preserving their own genetic identity" (their unique nucleotide sequence) from generation to generation. Are individuals nothing more than vessels for transmitting their genes through time? The answer to this philosophical question is an unconditional "yes" for one group of organisms, the **viruses**. These parasitic agents perpetuate themselves by infecting host cells, and, once inside the cell, produce progeny viruses.

The structure of a typical virus is a masterpiece of biological simplicity, consisting of nothing more than a strand of nucleic acid containing viral genes, surrounded by a protective protein coat.

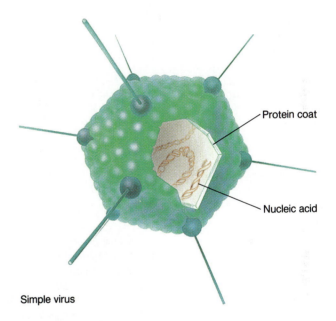

Protein coat

Nucleic acid

Simple virus

They have no organelles, no cytoplasm—in fact, none of the structures that characterize cells. These noncellular entities can still be considered organisms, however, because they reproduce and show heredity. Progeny viruses are "carbon copies" of their parent virus. All it takes for a virus to succeed is a set of genes with the information needed for producing more viruses, and a way of getting those genes into a host cell (another function of the protein coat). Once inside, the virus abandons its coat and exists as a single strand of nucleic acid. During this stage in the life cycle, the "chromosome" *is* the organism: The genes are in a new "vessel" (the host cell), but the organism is still a virus, with a distinct identity from the cell that contains it.

Inside the host cell, the viral genes are transcribed and translated by the host cell's enzymes and ribosomes. The protein synthesis machinery doesn't distinguish between the viral and cell's nucleic acid; it simply follows the dictates of the new genetic instructions. In this way, the virus commandeers the cell's metabolic machinery, redirecting it so that instead of producing new cell material, the cell becomes a virus-producing factory, as described in the illustration of bacterial infection by a virus (on the following page).

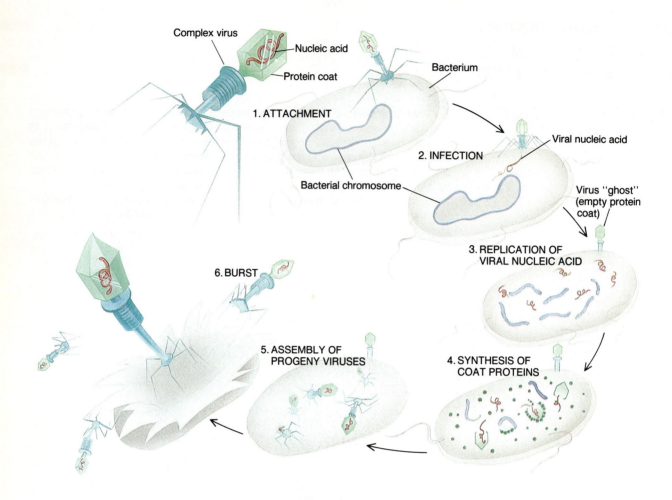

Host cell polymerases replicate the viral nucleic acid, sometimes generating several hundred copies of the viral genes. These genes also instruct the cell to produce huge quantities of coat proteins that automatically aggregate around the new strands of viral nucleic acid, assembling hundreds of progeny viruses. The cell eventually bursts, liberating viral offspring, each with the capacity to infect another cell.

So here is an organism that changes vessels, yet still remains a virus. For viruses, the essence of the organism is its set of genes, not the accessory equipment used to preserve that particular information.

acids inserted into the corresponding protein. Since amino acid sequence determines the activity of the protein, the substitution of even a single amino acid for another can alter the phenotype, producing a mutant trait. In the previous chapter we discussed some *chromosomal mutations,* alterations of the physical structure of the chromosome. The following discussion describes various *gene mutations,* changes in the nucleotide sequence in DNA.

## Types of Gene Mutations

**Point mutations** are changes that occur at one point within a gene, often involving one nucleotide. It may be a *base substitution,* in which one "letter" is replaced by another, or it may be the addition or deletion of one or more nucleotides from a gene. Some point mutations fail to affect phenotype. For example, if the point mutation (in DNA) generates a codon (in mRNA) synonymous with the ousted one, the protein remains unchanged.

Or, if a new amino acid is inserted, but is chemically similar to the one it replaces, the protein's activity remains the same.

But point mutations can be lethal if they impair the function of a critical protein. For example, a single nucleotide change in DNA converts CAG (the codon for glutamine) into the nonsense codon *UAG,* prematurely terminating the chain and preventing a complete protein from being produced. Even point mutations that

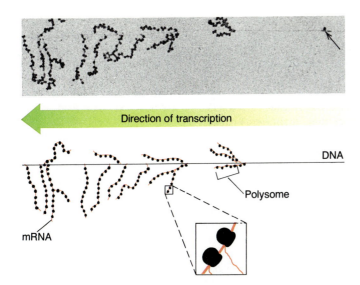

**Figure 27-14**
Ganging up on a messenger. In bacteria, as soon as part of an mRNA molecule is transcribed, ribosomes attach to it and begin translation. The result is an aggregate of ribosomes that follow each other along the mRNA. The complex of ribosomes held together by an mRNA molecule is called a polysome. The artists have added proteins to the polysome in the box to emphasize that each ribosome is synthesizing a protein strand.

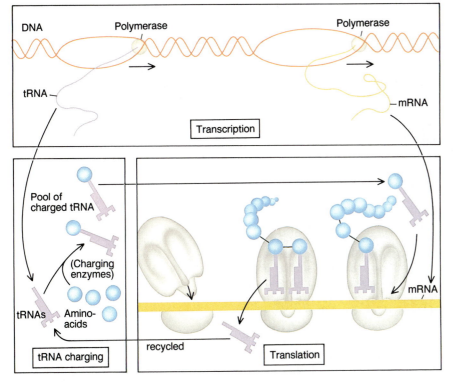

**Figure 27-15**
**Overview of protein synthesis.** All the processes diagrammed here occur in a single cell. The steps are compartmentalized for clarity. The genetic information in DNA ultimately dictates the amino acid sequence in protein, explaining why cells are biologically "obedient" to their genes.

change a single amino acid may create serious consequences. Recall from Chapter 4 the tragedy of changing a single amino acid (out of more than 300) in hemoglobin. The result was sickling of red blood cells, an often fatal condition. The genetic difference between healthy persons and those with sickle cell anemia is *one nucleotide base*. This single point mutation in a hemoglobin gene changes the codon from GAA, which specifies glutamic acid, to *GUA*, which inserts valine in its place.

A point mutation may also generate a more sweeping type of genetic change, a **frameshift mutation**. Inserting or deleting one or two nucleotides in a gene throws off the reading frame of the rest of the mRNA molecule, changing all the codons "downstream" from the point of change. The message is read in the wrong sequence, generating a useless protein that contains a string of incorrect amino acids beyond the shift point. Frameshift mutations may also generate nonsense codons that prematurely terminate the chain. To illustrate, delete the second nucleotide from "—CCU—AUA—GUC—," which codes for proline, isoleucine, and valine, and you produce "—CUA—*UAG*—UC . . . ," which codes for leucine followed by "stop."

## Mutagens

The probability of a gene being spontaneously altered by mutation is about one in a million. It is often the result of base mispairing during DNA replication, the polymerase mistakenly inserting a noncomplementary nucleotide. The low frequency of spontaneous mutation may be increased, however, by exposure to **mutagens**, chemical or physical agents that induce genetic changes. Some mutagens directly modify nucleotides, chemically changing them to another base. (The removal of $HNO_3$ from cytosine, for example, changes it to uracil.) Other mutagens add or delete single nucleotides during DNA replication, generating frameshift mutations.

Some mutagens don't change nucleotide sequence directly, but damage the DNA in a way that mobilizes the cell's system of **repair enzymes**. The powerful mutagenic potential of ultra-violet (UV) light, for example, is due to its ability to damage DNA rather than any capacity to directly change nucleotide sequence. Here's how:

UV-induced injury occurs in adjacent thymines, which become chemically linked to form a single duplex (called a "dimer") that prevents normal DNA replication and transcription.

UV damaged DNA
(contains T-T dimer)

a  — C-C-T-A-T-T-A-G-C-A —
       ⋮ ⋮ ⋮ ⋮ ⋮ ⋮ ⋮ ⋮ ⋮ ⋮
   — G-G-A-T-A-A-T-C-G-T —

The cell mobilizes its repair enzymes. First, a DNA disassembling enzyme (a DNase) removes the portion of the single DNA strand that contains the lesion.

DNase

b  — C-C-                 G-C-A —
       ⋮ ⋮                 ⋮ ⋮ ⋮
   — G-G-A-T-A-A-T-C-G-T —

DNA polymerase

— C-C-T-A-T-T          G-C-A —
    ⋮ ⋮ ⋮ ⋮ ⋮ ⋮          ⋮ ⋮ ⋮
— G-G-A-T-A-A-T-C-G-T —

A ploymerase then replaces the discarded portion of the strand, using the remaining single strand as the template.

ligase seals "nick"

c  — C-C-T-A-T-T-A-G-C-A —
       ⋮ ⋮ ⋮ ⋮ ⋮ ⋮ ⋮ ⋮ ⋮ ⋮
   — G-G-A-T-A-A-T-C-G-T —

Once the nicks in the strand are sealed by the enzyme *ligase*, the repaired DNA is indistinguishable from the pre-irradiated DNA—unless a mistake in base pairing occurs during replication, generating a mutation.

oops!   (point mutation)

d  — C-C-T-G-T-T-A-G-C-A —
       ⋮ ⋮ ⋮ ⋮ ⋮ ⋮ ⋮ ⋮ ⋮ ⋮
   — G-G-A-T-A-A-T-C-G-T —

Agents that mobilize repair enzymes are indirectly mutagenic. By increasing the amount of DNA replication in a cell, they increase the likelihood that mistakes will occur in the nucleotide sequence.

Overexposure of human skin to UV light (such as that in sunlight) causes mutations in skin cells; many of these mutations induce cancer. In fact, about 90 percent of all known *carcinogens* (cancer-causing agents) are also mutagens. They cause the genetic changes that result in the key property of all cancer cells—the breakdown of controls that normally prevent excessive cell proliferation. The fact that any mutagen is a potential carcinogen has provided us with an efficient and inexpensive test for screening large numbers of materials to detect those that might cause cancer. We simply expose a culture of bacteria to the chemical in question and count the number of detectable mutant cells produced. This number is compared to the number of mutants in a control culture treated in the same way but without exposure to the chemical being tested. An increased mutation rate in the chemically treated bacteria flags the substance as a potential human carcinogen. This procedure, called the *Ames test,* has detected cancer-causing potential in many substances to which people are frequently exposed, including materials in hair dyes, cured meats, and cigarette smoke.

# DNA Recombination

During chromosomal crossing over (Chapter 12 and 26), chromosomes exchange segments. It is actually the strands of DNA that are severed and rejoined during this and similar processes. This breakage of DNA and its subsequent reattachment to a different DNA strand (or to a different place in the same strand) is called **DNA recombination**.

DNA genetically recombines by breakage and reunion of the sugar-phosphate backbone, forming new covalent bonds at a site different from that of the original break. Crossing over is just one of the ways recombination reorganizes DNA in a cell. Genes can move to new locations on chromosomes through a controlled recombination process. (These "jumping genes" are discussed in Chapter 28.) The introduction of "foreign" DNA from another type of cell, or even from viruses, may also lead to recombination.

Our understanding of recombination has been greatly enhanced by studying gene transfer from one bacteria cell to another (Figure 27-16). Much of what we know about the molecular events of recombination is a byproduct of studying these one-way gene transfers in bacteria, which are generally followed by recombination between the transferred genes and the recipient's DNA. One of these genetic transfer mechanisms (transformation) led to the discovery that DNA is the genetic substance. This finding launched one of the most remarkable periods in the history of scientific achievement (see Discovery: The DNA Detectives). Similar techniques are used to intentionally introduce new genes into organisms in an attempt to genetically engineer the gene recipients with new properties that are advantageous. In this capacity, manipulating DNA recombination has produced new gene combinations, and therefore new organisms, the likes of which have never been created by nature (Chapter 39).

# D I S C O V E R Y

## THE DNA DETECTIVES

For almost 100 years the genetic substance was hiding in "plain sight." DNA was discovered in 1868 by the German biochemist, Friedrich Miescher (who called it "nuclein"), yet no one knew at that time that they had discovered the genetic substance of life. DNA was generally dismissed as a candidate for this ultimate position of molecular authority.

Yet the field of genetics grew anyway, as researchers amassed more information on the workings of heredity. Thomas H. Morgan demonstrated that genes reside on chromosomes. Sir Archibald Garrod's studies of metabolic disorders showed that mutations in human genes caused disease by inactivating enzymes needed for critical metabolic functions. George Beadle and Edward Tatum later theorized that genes must control enzymes, and then conducted experiments that supported their "one gene – one enzyme" hypothesis. Yet DNA remained in the shadows, too simple chemically to be considered the "designer" of such complex molecules as protein. Then, in 1944, three scientists led by Oswald Avery performed experiments that pointed the finger directly at DNA as the genetic substance.

They used two genetically stable strains of bacteria called pneumococci. One strain was dangerous, producing capsules that surround the bacteria and enable them to evade the human defense system and cause pneumonia. The other strain was identical to this except it produced no capsule. As a result, it was quickly destroyed by the host defenses before it could cause pneumonia. Avery grew a batch of the virulent (disease-causing) strain and heated it until all the cells were dead. Theoretically these dead cells should have been harmless. Yet Avery knew they weren't entirely harmless for he was familiar with the 16-year old work of a scientist named Fred Griffith. Griffith discovered that if mixed with a culture of living, nonvirulent pneumococci, somehow the heat-killed cells transferred their genes for capsule formation to living, nonencapsulated cells. The "harmless" cells that received the donated genes suddenly began killing the mice into which they were injected. Griffith had discovered genetic transformation in bacteria, but he had no idea what the "transforming principle" was. Avery reasoned that the bacterial substance responsible for this phenomenon contained the gene that specified capsule formation, and he set out to determine the identity of this genetic substance. He found that purified DNA from encapsulated bacteria transformed naked nonvirulent cells to the encapsulated state. He concluded that DNA was the genetic substance. Avery repeated the experiment after treating the transforming principle with an enzyme that specifically destroys DNA. No gene transfer occurred, verifying that DNA was the master molecule of heredity in bacteria. The modern era of molecular genetics had begun.

Eight years later, Alfred Hershey and Martha Chase employed an ingenious system to confirm DNA's genetic role. They propagated two batches of bacterial viruses in the presence of radioactive isotopes. In one batch the viral protein was labeled with radioactive sulfur; in the other batch the DNA contained radioactive phosphorus. Shortly after infecting bacteria with the labeled viruses, the investigators used radiation detectors to track the fates of protein (which contains sulfur but no phosphorus) and DNA (which contains phosphorus but no sulfur). They waited just long enough for viruses to attach to the bacterial surfaces and inject their genetic material, then put the preparation in a kitchen blender which broke the virus coats from the bacteria. Radioactivity detectors revealed that the phosphorus (DNA) stayed with the bacteria while the sulfur (protein) was in the medium and could be washed away. Even in the virtual absence of viral protein, the virus could still propagate in their bacterial hosts. DNA's role in determining heredity had been confirmed.

Yet DNA was still far from being a preoccupation of the scientific community. In fact for a while in 1952, only one person, Rosalind Franklin, was working full time on DNA. She was firing X-rays through crystals of the molecule, trying to get a better picture of what DNA looked like. The beams scattered as they traveled through the crystallized DNA before hitting (and exposing) a piece of photographic film. The resulting *X-ray diffraction* picture provided an indirect look at the orientation of the atoms that make up DNA. Although she didn't interpret her pictures correctly, she gave them to a couple of scientists who eventually did. By 1953, the stage was set for the central event in the DNA story, a crowning achievement by two young scientists whose names were about to become household words—James Watson and Francis Crick. Revealed in the seemingly shapeless smudges of Franklin's X-ray diffraction pictures was the very essence of the DNA molecule—a double helix. Watson and Crick knew from Erwin Chargaff's chemical analysis of the polymer that it consisted of four nucleotides, that the amount of adenine always equaled that of thymine, and that the amount of guanine always equaled that of cytosine. Using cardboard cutouts of the nucleotides, Watson and Crick constructed a model of the DNA molecule that not only fit all the available physical and chemical data, but also provided a mechanism for "self replication" and encoding information. Almost 40 years later, their model still remains the centerpiece of molecular genetics.

Watson and Crick not only discovered DNA's structure, but they also suggested ways it could direct protein synthesis. (Crick even proposed that RNA is copied from DNA and carries the genetic message to the cytoplasm, where some kind of "adapter molecule"—later discovered to be tRNA—matches each amino acid with a particular nucleotide sequence.) Not until 1960 was the flow of genetic information from DNA to mRNA to protein finally elucidated, and Crick's insightful predictions validated.

By 1961, one great task remained—breaking the genetic code itself. What is the "language" that cells use to instruct ribosomes to insert the correct amino acids in the proper sequence in a developing protein? Most experts believed the scientists could not even begin to crack the code in less than five or ten years. But the codebreakers received a terrific boost from H. G. Khorana, who worked out a technique for constructing synthetic messenger RNA (with a known nucleotide sequence) in a test tube. Researchers started with an artificial messenger (pioneered by Marshal Nirenberg) that contained one type of nucleotide—uracil. When they placed these linear strands of uracils (called "poly-U") in a test tube containing the protein synthesis machinery extracted from bacteria, and added all 20 amino acids, the system followed the artificial messenger's instructions and manufactured a protein. This was all that the codebreakers needed. Once again, they set up the experiment, except this time 20 duplicate test tubes were set up. Each contained all the protein synthesis machinery, but only one type of amino acid instead of all 20 (a different amino acid in each tube). The poly-U messenger failed to stimulate protein synthesis in 19 of the tubes because they lacked the amino acid that the UUU codon incorporates. Only in the tube containing the amino acid phenylalanine did the ribosomes build a protein (a chain of phenylalanines). In genetic code, the codon UUU spells "phenylalanine." Over the next four years, synthetic mRNAs were used to test all 64 possible codons to determine which amino acids would be incorporated. The result was the universal decoder chart shown on page 464. The genetic code was cracked.

In just 20 years scientists worked out the molecular basis of life and, in doing so, delivered a deeply satisfying message: the universe inside every living organism is within our cognitive reach. Life is explainable in physical and chemical terms, with concepts that can be understood without evoking superstition or mysticism. They proved that life on its most subtle level is governed by rational principles rather than by mystical forces that defy explanation.

In condensing this history, many who contributed to our understanding of molecular genetics could not be mentioned. These people include Linus Pauling, Maurice Wilkins, M. S. Meselson and F. W. Stahl, scientists who helped mold the field of modern genetics.

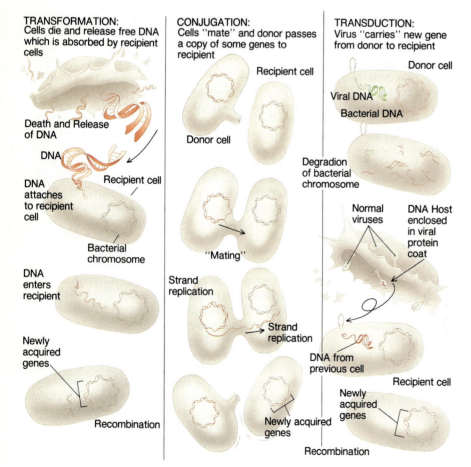

**TRANSFORMATION:**
Cells die and release free DNA which is absorbed by recipient cells

Death and Release of DNA

DNA

DNA attaches to recipient cell

Recipient cell

Bacterial chromosome

DNA enters recipient

Newly acquired genes

Recombination

**CONJUGATION:**
Cells "mate" and donor passes a copy of some genes to recipient

Recipient cell

Donor cell

"Mating"

Strand replication

Strand replication

Newly acquired genes

Recombination

**TRANSDUCTION:**
Virus "carries" new gene from donor to recipient

Donor cell

Viral DNA

Bacterial DNA

Degradion of bacterial chromosome

Normal viruses

DNA Host enclosed in viral protein coat

DNA from previous cell

Recipient cell

Newly acquired genes

Recombination

**Figure 27-16**
**Genetic transfer in bacteria** serves the same purpose as does sexual reproduction and crossing over in eukaryotes: it increases genetic variation by allowing recombination to produce new combinations of genes. All three mechanisms are one-way transfers of DNA from donor to recipient. In each case, recombination incorporates the newly acquired genes into the bacterial chromosome. The new genes replace the recipient's homologous genes, which are discarded.

# Synopsis

## MAIN CONCEPTS

- **The chemistry of the gene.** DNA is a helical molecule of two parallel chains of nucleotides. The sequence of nucleotides in the two strands is complementary to each other because of specific A–T and G–C base pairing between the opposite nucleotides on the two strands. Each gene's genetic "meaning" is encoded in its unique nucleotide sequence.

- **DNA replication—preserving the gene's message from one generation to the next.** The doublestranded molecule separates into two single strands. DNA polymerase assembles free nucleotides along the single-stranded molecules, adhering to the dictates of base pairing. (Each T is paired with A; each C is paired with G.) The result is two complete DNA molecules, each of which is identical to the original and is destined for different daughter cells.

- **Genes direct protein synthesis.** One gene determines the formation of one specific type of protein (or polypeptide chain). The protein may itself be the trait (a structural protein), it may be an enzyme that catalyzes the formation of the trait, or it may perform another task such as oxygen transport.

- **The flow of genetic information.** Genetic instructions flow from DNA to RNA to protein.
  The information spelled out by the gene's nucleotide sequence is encoded in a molecule of mRNA during transcription.
  The nucleotide sequence of mRNA is then translated into a specific amino acid sequence as the mRNA directs protein synthesis at a ribosome. Translating mRNA messages requires tRNAs, which insert the correct amino acid for each codon in the messenger. In this way, translation assembles amino acids in the precise order dictated by DNA.
  The protein folds into its final shape according to the dictates of its amino acid sequence, its shape providing the ability to perform its particular task. The gene's encoded information is now expressed as a specified trait.

- **Mutation** is a change in the genetic message. Gene mutations may be at one point, often leading to the insertion of a wrong amino acid or generating a chain-terminating nonsense codon. Because addition or deletion of one or more nucleotides (except in multiples of three) throws off the reading frame, incorrect amino acids are inserted "downstream" from the point of mutation. Exposure to mutagens increases the rate of mutation.

- **DNA recombination** is the transfer of DNA segments to other strands or to different places in the same strand.

## KEY TERM INTEGRATOR

Genetic Messages—Coding and Decoding

| | |
|---|---|
| **nucleotides** | Molecules containing a 5-carbon sugar (ribose or deoxyribose), a phosphate, and one of five nitrogenous groups. They are the molecular subunits (monomers) from which DNA and RNA are constructed. |
| **DNA replication** | Creating duplicate copies of DNA, each of the two copies having the same nucleotide sequence as the original molecule. The process is mediated by the enzyme **DNA polymerase**, which assembles free nucleotides along the two separated strands. The molecule is thus duplicated by **semiconservative replication**. |

| | |
|---|---|
| **messenger RNA** | A single-stranded polymer assembled during **transcription**. Its nucleotide sequence is complementary to the DNA strand used as a template along which **mRNA polymerase** assembles the ribonucleotides. mRNA's complementary sequence determines the order of amino acids in protein synthesized during **translation**. Each **codon** in mRNA calls for the insertion of a particular amino acid, carried into position by the **tRNA** that bears the complementary **anticodon**. Peptide bonds link the amino acids together as the protein is synthesized. The message in DNA has been decoded. |

### Changes in Genetic Messages

| | |
|---|---|
| **mutations** | Inheritable changes in DNA. Gene mutations occur within a gene (as opposed to chromosomal mutations). Substituting one nucleotide for another results in a **point mutation**. Addition or loss of nucleotides can result in **frameshift mutations**. Often mutations occur when **repair enzymes** are replacing damaged DNA. Many **mutagens** are agents that stimulate repair enzymes, enhancing the likelihood of mutations. |
| **DNA recombination** | The rejoining of DNA pieces with those of a different strand or with the same strand at a point different from where the break occurred. |

## Review and Synthesis

1. Describe how semiconservative DNA replication accounts for inheritance.

2. You have about 46 trillion double strands of DNA in your body. At most, how many of these contain the original strands donated by your parents?

3. DNA, RNA, and protein are "colinear" molecules. Why is this critical to the flow of genetic information?

4. What is the importance of specific nucleotide base pairing to inheritance and to expression of genetic messages?

5. Describe at least two similarities and two differences between the processes of DNA replication and transcription.

6. Arrange these in their order of participation in gene expression: protein folding, mRNA, tRNA, mRNA polymerase, translocation, releasing factor, trait development, DNA strand separation, assembly of ribosome subunits.

7. Explain the importance of protein synthesis to the development of genetic traits.

8. Why doesn't a cell run out of tRNA? (Give two reasons.)

9. Using the genetic code chart on p. 464, construct the short polypeptide chain coded for in this sense strand of DNA: TACGGATCGCCTACG— (remember to include the mRNA sequence.)

10. The more copies of a document a person must type, the more likely there are to be typographical errors. How does this statement help explain why agents that stimulate DNA repair enzymes are mutagens?

11. Reconsider dominant and recessive alleles. With your knowledge of mutation, propose a mechanism that explains the emergence of a recessive trait

that is really nothing more than the absence of the dominant allele's gene product. Relate this mechanism to the sudden appearance of the allele for hemophilia in Queen Victoria and her descendants.

**12.** What is the evolutionary significance of the following statements?
"All organisms store genetic information in DNA."
"In all organisms the flow of genetic information is identical."
"All organisms use the same genetic code."
Add two more universals of life that support your answer. (These can be from previous chapters.)

## Additional Readings

Braun, P. 1988. *Annual Editions — Biology.* 5th ed. Section II (Genetics). Dushkin. (Introductory.)

Dickerson, R. 1983. "The DNA Helix and how it is read." *Scientific American* 249:294. (Advanced.)

Hoagland, M. 1981. *Discovery — The Search for DNA's Secrets.* Van Nostrand Reinhold, New York. (Introductory.)

Lewin, B. 1985. *Genes II.* John Wiley and Sons, New York. (Intermediate to advanced.)

Readings from *Scientific American. Molecules to Living Cells.* 1980. W. H. Freeman, San Francisco. (Intermediate.)

Watson, J. 1968. *The Double Helix.* Atheneum, New York. (Introductory.)

Chapter 28

# Orchestrating Gene Expression

I f it happened to you, it would be a disaster. But to a desert lizard, losing a leg is a relatively small sacrifice to escape becoming a road runner's dinner. The lizard soon replaces the amputated limb by performing a feat of genetic "alchemy," turning cells into different types of cells. Fully differentiated cells reverse their developmental commitment, become unspecialized, and then redifferentiate into the complex of specialized cells that comprised the replacement organ. Muscle cells, for example, become cartilage cells, and cartilage cells give rise to new bone cells. Such transformations are possible because of an ability to regulate which genes are expressed in each cell.

## Why Regulate Gene Expression?

Regeneration is a natural demonstration of **totipotency**—the genetic potential of one type of cell from a multicellular organism to give rise to any of the organism's cell types, even generate a whole organism. (A single parenchyma cell in the phloem of a carrot, for example, can proliferate and develop into a fully mature plant.) Totipotency is a cellular property of virtually all multicellular eukaryotic organisms (see Connections: Totipotency and Clones), a property confirmed by hundreds of experiments.

In spite of their differences in form and function, the various cells in a multicellular organism have the same genes, having all descended from the same totipotent zygote. The distinct characteristics of each cell depend on which genes are expressed and which remain "silent." In other words, an organism is not just the product of its genetic endowment—it is also the product of molecular "decisions" that determine the expression of that potential (Figure 28-1).

Selective gene expression presents one of the most compelling puzzles in biology: "how are different genes turned on and off in response to internal or external signals?" Yet embryonic development and cell differentiation, as discussed in Chapters 21–24, is only one facet of gene regulation.

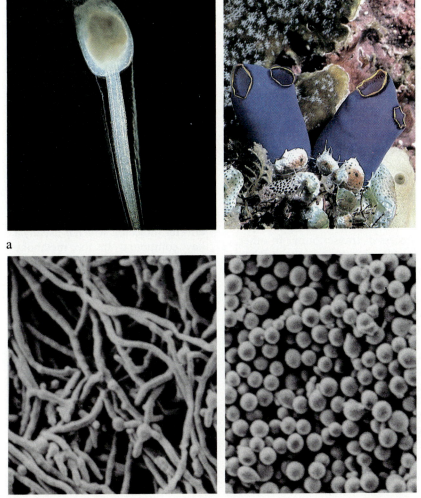

a

b

**Figure 28-1**
**Two examples of gene orchestration.** *(a) During normal development*. The free-swimming sea squirt larva looks more like a tadpole than the adult sea squirt, yet larva and adult are genetically identical. Their striking distinctions in form and function are due to selection of different genes for expression as growth and development proceed. *(b) In response to environmental change*. Many fungi can turn off genes for growth in the mold form (fluffy and filamentous) and switch on a different block of genes, those for unicellular growth (the yeast phase). Moisture and temperature determine which block of genes is expressed.

Selective gene expression is also essential for the homeostatic functioning of all organisms. Here are just a few of the phenomena that depend on controlled gene expression:

- *Response to changes in the chemical environment.* Organisms conserve energy and materials by leaving most genes inactive until their gene products are needed. For example, some digestive enzymes are produced only when the substrate of that enzyme is present. Variations in the chemical environment may also trigger developmental changes.

- *Response to changes in the physical environment.* Fluctuations in temperature, moisture, pH, and photoperiod (daylength) can trigger changes in gene expression. For example, many seasonal changes in the structure and behavior of plants and animals are likely due to photoperiod, different genes being expressed in the long days of late spring than in short winter days. Although a particular plant has the genetic capacity to produce flowers, the genes remain unexpressed until the correct season.

- *Mechanism of hormonal control.* In target cells, specific genes are switched on or off in the presence of the corresponding hormone (see Chapters 16 and 22).

- *Muscle and bone development in response to use or disuse.* Physical activity unleashes the expression of otherwise silent genes for the production of contractile protein (actomyosin) by muscle cells or for the proliferation of bone-building osteocytes.

The picture of genes we have painted so far is still incomplete. DNA segments that direct the formation of a gene product—an enzyme or structural protein—are certainly genes. But these **structural genes** are supplemented by others that code for no enzyme, structural protein, or any observable physical trait. The sole function of these **regulatory genes** is to control the expression of structural genes.

# Gene Regulation in Prokaryotes

Although the mechanics of gene regulation in eukaryotes remains unsolved, several mechanisms in prokaryotes have been well documented. The best understood of these is a regulatory unit called the **operon**.

## The Bacterial Operon

Genes for enzymes of a complete metabolic pathway tend to be regulated as a functional unit in bacteria. This co-regulation is assisted by clustering functionally related genes together. The three genes that code for the trio of lactose-utilizing enzymes, for example, lie adjacent to each other on the bacterial chromosome. When lactose is present, the three enzymes work in concert to transport the sugar into the cell and begin its catabolism for energy and carbon. All three enzymes are needed for the process to proceed to completion. The absence of any one eliminates the need for the other two, since lactose utilization is blocked at the point of the missing enzyme. It is therefore useful to regulate the three genes as a unit. Clustering genes that participate in the same metabolic pathway, making them "next door neighbors," facilitates cooperative regulation.

These clustered genes represent only part of the operon. The rest consists of regulatory genes and a **promoter**, a short segment of DNA to which mRNA polymerase attaches to begin the process of transcription. In other words, the promoter is the mRNA polymerase binding site. Controlling the ability of mRNA polymerase to engage at the promoter provides the basis for regulating expression of the operon's genes. If the polymerase cannot attach to the promoter, transcription does not occur and the gene products cannot be produced. The regulatory genes provide a mechanism for allowing polymerase attachment only when the gene products are needed. For example, when no lactose is present, mRNA polymerase cannot attach to the operon's promoter, the enzyme-producing genes are not transcribed, and none of the three sugar-

digesting enzymes is produced. In this way, the cell conserves its resources. Here is how the system works.

## Components of the Operon

A typical bacterial operon, depicted in Figure 28-2, consists of these components:

- The genes that house the information for producing the enzymes of a single metabolic pathway. These genes usually lie adjacent to one another and are regulated as a unit. Turning on one gene turns on all the structural (enzyme-producing) genes of an operon.

- A regulator region that is composed of a promoter and an **operator**. The operator is a short DNA segment that separates the promoter from the operon's structural genes that code for the enzymes. The promoter and operator may overlap by several nucleotides.

- An **R (regulator) gene** that directs the synthesis of a *repressor* protein. This repressor can attach to the operator of the corresponding operon, partially covering up the promoter. This prevents mRNA polymerase attachment, which in turn prohibits transcription of the operon's enzyme-producing structural genes. In other words, with the repressor locked onto the operator, the genes of that operon are switched off. Equally important, the repressor can bind with another molecule, a small **effector** molecule, which is usually a key compound in the metabolic pathway being regulated. The ability of the repressor to fasten to the operator depends on whether it carries the effector molecule. In this way, the presence of the effector molecule determines whether the structural genes of the operon are turned on or switched off. Consequently, *gene expression in an operon is ultimately responsive to the concentration of the effector molecule*, the cellular "barometer" that indicates the need for the operon's enzymes.

Turning on enzyme synthesis in response to high substrate concentrations is called *induction* of the operon.

| CONNECTIONS | Plant Development (p 363) |
| | Animal Development (p 399) |
| | Biotechnology (p 722) |

## TOTIPOTENCY AND CLONES: LITTLE CAUSE FOR FEAR

*Clones*—just the word tends to evoke images of a "brave new world" in which all people are identical copies of the ideal human. Yet, often the very people stirring up the fear have only a vague idea of what a clone really is and how long people have been intentionally cloning organisms. **Clones** are asexually produced offspring that, except for mutations that occur during the growth process, are genotypically identical to their one and only genetic parent. Although science reporters and fiction writers generally present cloning as a product of new technology, people have been cloning organisms for thousands of years (and nature has been doing it for billions of years). Every time someone takes a "cutting" from a plant, removing a slip for vegetative propagation, he or she is cloning an organism. The resulting plant is genetically identical to the original from which it was cut, reproduced by mitotic cell divisions only. Mitosis diligently preserves the organism's genetic repertoire rather than scrambling it as do meiosis and fertilization. In the business of raising oranges, apples, avocados, bananas, and dozens of other types of fruits, cutting techniques long ago replaced the practice of growing plants from seeds.

But some cloning techniques are new, in fact, revolutionary. The revolution is based on the totipotency of cells taken from multicellular organisms, especially plant cells. If undifferentiated cells from fully developed plants are removed and placed in an appropriate growth medium, they proliferate into a tumorlike growth called a callus. The calluses grow into fully developed plants, each one an identical clone of the plant that donated the original cells. This technique forms the basis of **tissue culture**—the growing of plant and animal cells in laboratory glassware containing a nutrient medium (see accompanying figure 1).

Tissue culture not only provides a powerful research tool for studying differentiation, but also is a valuable means of propagating clones of crop plants that show advantageous traits, such as increased resistance to disease or drought. Tissue culture is

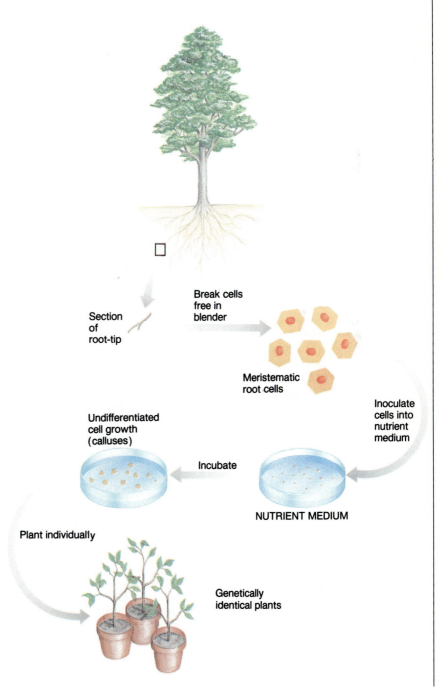

Section of root-tip

Break cells free in blender

Meristematic root cells

Inoculate cells into nutrient medium

NUTRIENT MEDIUM

Incubate

Undifferentiated cell growth (calluses)

Plant individually

Genetically identical plants

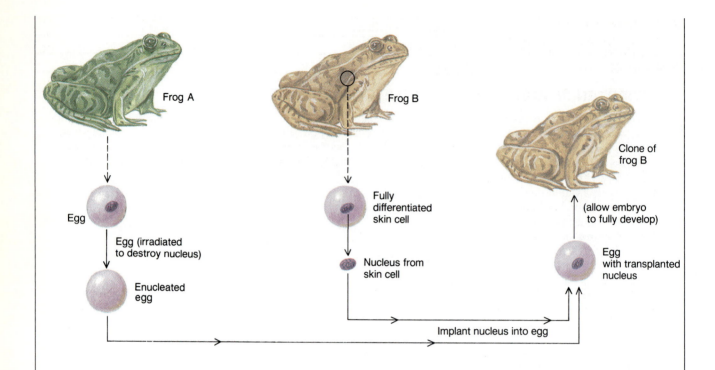

being used to produce improved varieties of grains, ornamental flowering plants, and trees used for reforestation projects. As recombinant DNA techniques develop (Chapter 39), tissue culture should soon allow us to combine into a single organism many advantageous genes from different individuals.

Culturing cells into whole multicellular organisms is more successful for plants than for animals, mainly because most differentiated animal cells don't readily reverse development and give rise to a new multicellular organisms. Yet some animal cells do retain totipotency, as illustrated by the ability of differentiated cells to regenerate lost organs in lizards. Totipotency is also demonstrated by experimentally transplanting a nucleus from a fully differentiated cell into the cytoplasm of an egg cell that has no nucleus. (Its nucleus is previously destroyed by ultraviolet light irradiation.)

The results of one such experiment are shown in Figure 2. This experiment shows that the nucleus of a fully differentiated skin cell still contains all the genes necessary to produce a complete frog. Genes are neither discarded nor permanently inactivated during differentiation and development.

The frog that develops from this experiment is a true clone, identical to the frog that donated the nucleus. Nuclear transplants have produced true clones like this in many other animals, including at least one mammal (a mouse). Many researchers believe that the cells of practically all animals, including mammals, retain full genetic capacity and totipotency. If they are correct, we may soon be using these nuclear transplant techniques to improve livestock animals, cloning thousands of copies of an especially hardy animal or a prolific milk producer.

Many types of animal cells can also be grown in tissue culture. Some of these are cancer cells, which resemble embryonic cells in their growth potential. They have acquired cellular immortality, dividing indefinitely in a suitable nutrient medium. (Because these cells don't differentiate to form multicellular tissues or whole organisms as do the cells in plant tissue culture, the term "cell culture" is preferred over "tissue culture" to describe these techniques.) The cultured animal cells have many scientific and practical uses, such as providing host cells for virus propagation in the laboratory (viruses proliferate only inside living host cells) and for studying the characteristics of cancer cells. Most types of animal cells can be cultured in the laboratory, although they die after a predetermined number of generations unless some genetic change transforms them into immortal cancerlike cells.

## Inducible Operons

Catabolic (degradative) pathways are regulated differently than anabolic (biosynthetic) pathways (Chapter 8). In catabolic pathways, such as the lactose-utilizing reactions, the substrate to be dismantled (lactose in this example) serves as the effector, the key to whether the enzymes are needed. When present, the substrate of a catabolic pathway *induces* transcription, the catabolic enzymes are synthesized, and the substrate is then utilized (Figure 28-3 — left side). This is control of gene expression by the process of enzyme (or gene) **induction**. The effector that turns on the genes of an inducible operon is called the *inducer*. It is usually the substrate of the pathway.

In inducible operons, the repressor protein is *active* in the absence of the inducer. Therefore, when no substrate is available for use, the repressor locks onto the operator and blocks transcription of the genes, and the enzymes are not synthesized (Figure 28-3 — lower left side). The presence of the substrate (the inducer), however, induces enzyme synthesis by binding to the repressor protein and preventing it from attaching to the operator. In this state mRNA is transcribed, enzymes are synthesized, and the substrate is consumed. The pathway is induced.

## Repressible Operons

In biosynthetic pathways, the need for the end product determines the activity of the genes that code for the pathway's enzymes. The cell *turns off* enzyme synthesis when there is an ample supply of the end product (the effector in regulating biosynthetic pathways). Like induction, this activity also conserves energy and materials, since the pathway's enzymes are not needed when there is a surplus of the product. Unlike inducible operons, however, the repressor protein is *inactive* in the absence of the effector. When the effector attaches to the repressor, the repressor is activated. The repressor–effector complex binds to the operator and represses gene transcription. In this way, the elevated concentration of the end product turns off gene expression. The operon is called *repressible* because the presence of the effector produces a state of enzyme (or gene) **repression** (Figure 28-3 — right side). The absence of the effector *derepresses* (allows activation of) the operon's structural genes, which are then transcribed and expressed.

In both inducible and repressible systems, gene repression is reversible, so the enzymes are again synthesized as soon as the effector's concentration changes to permissive levels. This is a negative feedback system (Chapter 8) that operates at the level of enzyme synthesis.

# Gene Regulation in Eukaryotes

Gene regulation in eukaryotic cells cannot be explained by operon-type mechanisms; operons are not even found in eukaryotic cells. Although eukaryotes and prokaryotes employ the same fundamental strategies for storing, transmitting, and expressing genetic information, there are many essential differences, notably in the structure of the chromosome. The structural peculiarities of eukaryotic chromosomes have provided some of the keys to understanding how these genes are regulated.

## Eukaryotic Chromosomes

Before exploring how chromosome structure affects gene expression, let's examine another problem of structure — packaging enormous lengths of DNA into a space as tiny as a cell nucleus. In every nucleated cell in your body, about 2 meters (6 feet) of DNA are packed into the nucleus, a sphere 100,000 times smaller than the dot on this "i." The nucleic acid is not randomly stuffed into a nucleus, but is organized in an arrangement that allows DNA to replicate without tangling and to periodically condense into chromosomes in preparation for cell division. In some chromosomes, the packing arrangement allows segments of DNA to extend from the condensed mass for transcription during mitosis and meiosis.

Eukaryotic chromosomes are associated with a rich supply of intranuclear proteins, which comprise more than half their mass. Most of these are

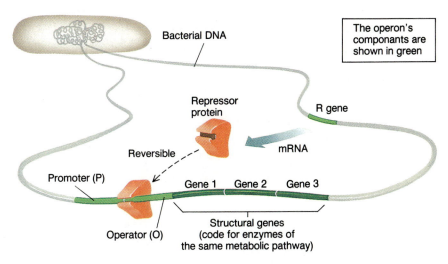

**Figure 28-2**
**The bacterial operon — one model of how gene expression is regulated**. Clustered structural genes (1, 2, and 3) lie "downstream" from a promoter (the mRNA polymerase attachment site) and an operator. A regulator gene (R) produces a repressor protein that determines whether the structural genes are expressed. With the repressor attached to the operator, the polymerase cannot join with the promoter and initiate transcription of the operon's structural genes. As a result, no enzymes are produced. When the repressor is not attached, genes are transcribed. In this way, the repressor protein acts as the molecular "switch" that enables cells to turn genes on and off.

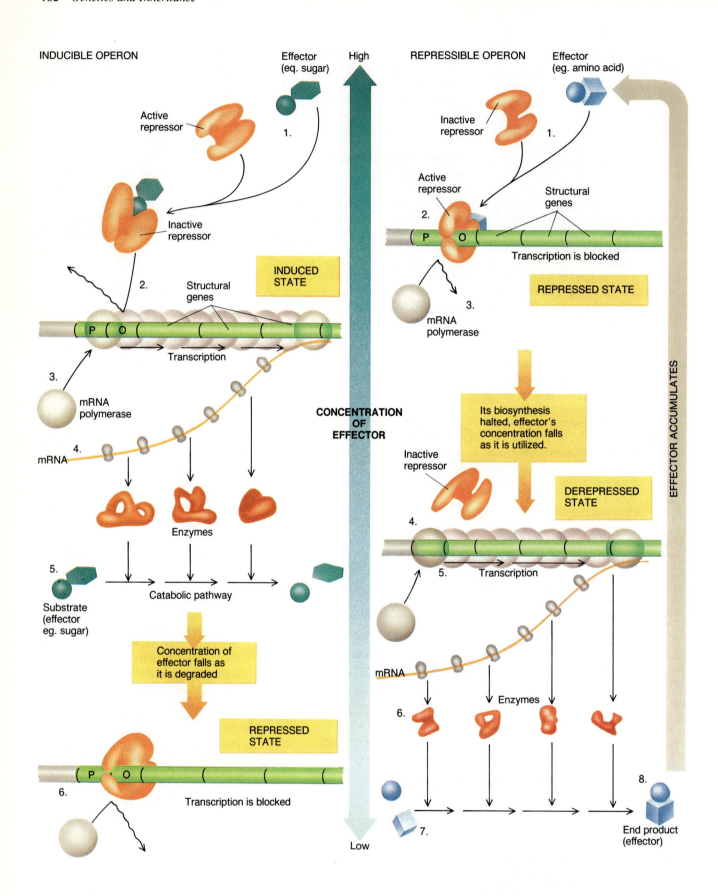

INDUCIBLE OPERON

Effector (eq. sugar)

High

REPRESSIBLE OPERON

Effector (eg. amino acid)

Active repressor

1.

Inactive repressor

1.

Inactive repressor

Active repressor

Structural genes

2.

P   O

Transcription is blocked

INDUCED STATE

REPRESSED STATE

2.

Structural genes

mRNA polymerase

3.

P   O

Transcription

CONCENTRATION OF EFFECTOR

3.

mRNA polymerase

Its biosynthesis halted, effector's concentration falls as it is utilized.

mRNA

4.

Inactive repressor

DEREPRESSED STATE

Enzymes

4.

5.

Substrate (effector eg. sugar)

Catabolic pathway

5.

Transcription

Concentration of effector falls as it is degraded

mRNA

Enzymes

REPRESSED STATE

6.

P   O

6.

Transcription is blocked

7.

8.

End product (effector)

Low

EFFECTOR ACCUMULATES

small alkaline proteins called **histones**. They help neutralize DNA's acidity and electrical charges, so that adjacent strands packed together into the tiny nucleus don't repel each other. (Otherwise, packing would be impossible.) Histones are associated with DNA in units called **nucleosomes**, each of which consists of a central cluster of eight histones (two each of four different types) around which the DNA double helix circles twice (Figure 28-4 —boxed area). A fifth type of histone ($H_1$) locks the nucleosome complex together, so that the DNA cannot unwind from the histone core. Additional coiling and folding are needed to complete the packaging process, since nucleosome formation shortens DNA to only one-sixth its fully extended length.

The current model of coiling needed for complete chromosomal condensation into its metaphase configuration is shown in the rest of Figure 28-4. This highly coiled DNA apparently condenses around a "scaffold," composed of nonhistone protein, that can be seen with an electron microscope in preparations treated to remove histones (Figure 28-5). Although the scaffold resembles the shape of metaphase chromosomes, its role in the architecture of the chromosomes remains unknown.

It is also not known if histones participate in the regulation of gene expression in eukaryotes, although preliminary evidence suggests they may. Whatever their regulatory role, it is probably universal for all eukaryotes, from single-celled protozoa to humans. This view is supported by the similarity of histones among these organisms. All of them have virtually identical histones, as well as the same number of DNA base pairs (146) wrapped around each eight-histone unit. The nucleosome has worked well as a DNA-organizing unit for more than a billion years. Once again the unity of life supplies evidence for the evolutionary kinship of organisms, descendants of the same early ancestor.

## RELATIONSHIP OF CHROMOSOMAL STRUCTURE TO GENE EXPRESSION

Evidence from comparisons of gene activity of chromatin (uncondensed DNA) and chromosomal (condensed) DNA suggests that genes are not expressed when DNA is condensed into chromosomes. Yet metabolic activity continues during mitosis and meiosis, when DNA is condensed, posing a paradoxical question: how are the enzymes that direct metabolism during these periods synthesized? Several possibilities exist. Transcription and translation during interphase (when DNA exists as chromatin) may generate enough enzymes to last through cell division. Another possibility is the production of stable mRNA that remains in the cytoplasm, directing protein synthesis long after transcription of those genes has ceased during cell division.

A third possibility was revealed by microscopic studies of giant polytene

**Figure 28-3**
**Gene regulation by operons.** Inducible and repressible operons work on the same principle—if the repressor is active, genes are turned off; if the repressor is inactive, genes are expressed.
INDUCIBLE OPERON
  1. In high concentration, the effector (in this hypothetical case, a specific sugar to be consumed) binds with the repressor protein and
  2. prevents its attachment to the operator (O).
  3. Without the repressor in the way, mRNA polymerase attaches to the promoter (P) and
  4. transcribes the structural genes. Thus, *when the effector concentration is high, the operon is induced*, and the needed sugar-digesting enzymes are manufactured.
  5. Sugar is utilized. If the sugar is not replenished, its concentration dwindles until the enzymes are no longer needed.
  6. Now there is not enough effector present to combine with the repressor, which then regains its ability to attach to the operator and prevent transcription. *When effector concentration is low, the operon is repressed (turned off)*, preventing synthesis of unneeded enzymes.
REPRESSIBLE OPERONS
  1. In high concentration, the effector (in this case, an amino acid, the end product of a biosynthetic pathway) binds with the *inactive* repressor and
  2. changes its configuration so it can attach to the operator,
  3. preventing transcription of the structural genes. Thus, *when effector concentration is high, the repressible operon is repressed*, preventing overproduction of the end product.
  4. When the effector concentration is low, most repressor remains unbound by the effector and therefore fails to attach to the operator. Transcription proceeds, and
  5. genes are transcribed. *When effector concentration is low, a repressible operon is derepressed* (activated).
  6. Enzymes are synthesized, and
  7. the needed end product is manufactured.
  8. As biosynthesis proceeds, the effector accumulates to high enough concentration to again repress the operon (step 1).

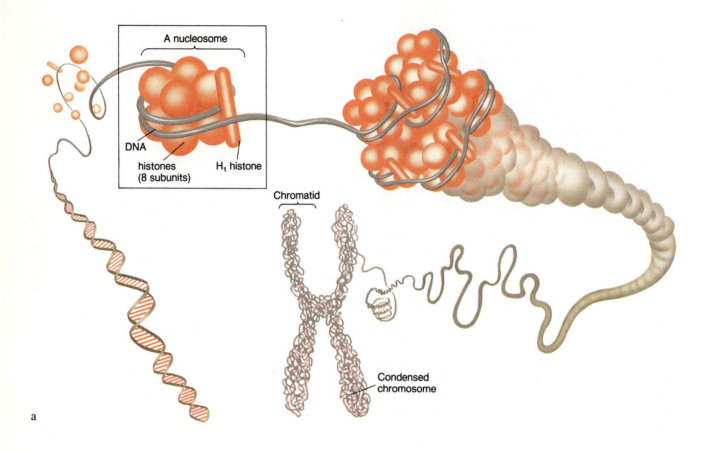

a

**Figure 28-4**
**Packing with nucleosomes**. *(a)* Current model depicting packaging of DNA into condensed chromosome. The nucleosome (boxed area) is the fundamental packing unit. Each nucleosome consists of eight histones encircled by two turns of DNA. The complex is locked into place by another histone (H$_1$). The nucleosomes are coiled into thicker fibers that bend into "looped domains." These thickened strands coil again, forming the arms of the condensed chromosomes. *(b)* Nucleosomes are visible in this electron micrograph of uncondensed chromatin. The nucleosomes are the "beads" on the DNA "string."

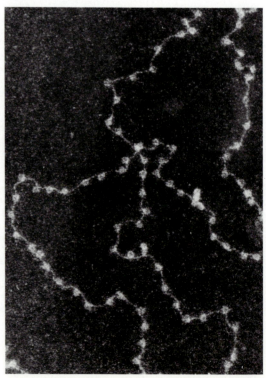

b

chromosomes of insect larvae (Chapter 26), which display a characteristic "puffing" at various points (Figure 28-6). Several lines of evidence demonstrate that these **chromosomal puffs** are sites where DNA unravels from its condensed state so that it can be transcribed during cell division. The puffs are rich in RNA (unlike the condensed portions of the chromosome), evidence of transcription in these extended regions. In other investigations, larvae grown in the presence of radioactive uracil concentrate the radioactivity in these puffed regions, indicating that transcription (uracil incorporation into mRNA) occurs only along these extended stretches of DNA. Further evidence is provided by differences in the puffing pattern of a particular chromosome according to the type of tissue. Even though the tissues were from the same larval insect, different points on the chromosome are puffed in the cells of different tissues, perhaps pinpointing the different genes that are active in distinct cell types.

Another microscopically visible phenomenon supports the hypothesis that genes can be expressed only when DNA is uncondensed. During meiosis, the condensed, synapsed chromosomes continue transcribing mRNAs from loops of DNA that extend from the main axis of these **lampbrush chromosomes**, so named because of the appearance of these loops (Figure 28-7). Unlike chromosomal puffs, hundreds of loops protrude along the entire axis of the chromosome, reflecting the enormous production of proteins needed to prepare for early embryonic development. Inactive genes remain condensed along the axis until they are called into action for later steps in development.

## SILENT DNA

One structural property of eukaryotic chromosomes seems especially puzzling. Most of the DNA in a cell is silent, that is, untranslated and thus not expressed. One would expect there to be *some* silent DNA. Genes that are expressed in some of an organism's cells are silent in other cells that don't need these gene products for development, maintenance, or growth. Since

**Figure 28-5**
**DNA is freed from its protein scaffold** (which still retains the shape of a metaphase chromosome) by selectively removing histones. The huge "cloud" of DNA released is about two million times longer than its condensed length.

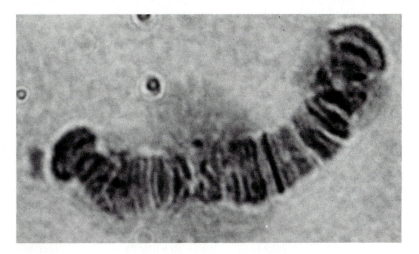

**Figure 28-6**
**Giant puff** extends from a larval fly's polytene chromosome (two smaller puffs are also evident). The puffs reveal areas of active transcription. With the DNA extended, genes in these areas are accessible to mRNA polymerase.

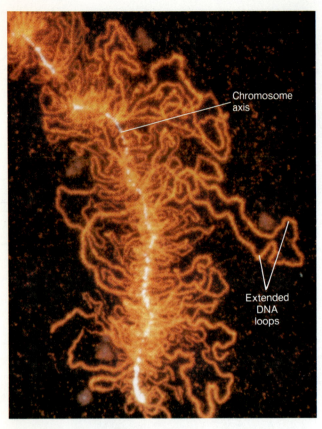

**Figure 28-7**
**Lampbrush chromosomes** get their name from the lateral loops of DNA that extend from the chromosome's scaffold along its main axis, giving it a brushlike appearance.

all cells in a multicellular organism arise from the same fertilized zygote, every cell carries around additional "genetic baggage." Your skin cells, for example, never use their genes for becoming blood cells or for manufacturing hemoglobin even though they acquired these genes from the totipotent zygote that produced you.

But most of the silent DNA appears to be unexpressible in *any* cell, regardless of its developmental fate, constituting neither structural nor regulatory genes. The role of this extra DNA is still unexplained. (Indeed, it may not even have a role, but simply be genetic garbage that poses no real disadvantage to an organism.) On the other hand, the extra DNA may have an important function. Some investi-gators suggest that it enhances the advantages of sexual reproduction by lengthening chromosomes, thereby increasing the frequency of crossing over. Others believe that the extra DNA is part of a eukaryotic regulatory mechanism, as discussed later in this chapter.

## DISCONTINUOUS GENES — EXONS SEPARATED BY INTRONS

Much of the excess DNA in eukary-otes resides between genes, forming "spacers" that separate genes from each other. But some of this silent DNA has been found to lie smack in the middle of a gene, separating it into parts. In some cases, several interven-ing sequences split a single gene into many separate sections. (The record for intervening sequences within a gene coding for a single protein is more than 50.) The discovery of split genes in 1977 shattered the classical model, which depicted the gene as an uninter-rupted strip of genetic information transcribed into a molecule of mRNA that is the same length as the gene.

Investigators soon named the inter-rupting pieces **introns**, to suggest "in-tervening sequences." The segments of DNA that are transcribed and trans-lated into amino acid chains are called **exons**, for "expressed sequences." Fol-lowing transcription, exon regions in the complementary RNA molecule are fused together into a final mRNA that directs synthesis of the protein (Figure 28-8). Such assembly requires that the transcribed introns be pre-cisely snipped out of the pre-messen-ger RNA (called the *primary tran-script*), an "editing" process that occurs in the nucleus. The edited mRNA then leaves the nucleus and is translated by ribosomes into the corre-sponding protein.

Split genes may help regulate gene expression. This possibility is consid-ered in the next section, together with several other proposed regulatory mechanisms.

## Eukaryotic Gene-Regulatory Mechanisms

Two fundamental categories of ge-netic regulatory mechanisms are be-lieved to exist in eukaryotes.

- Alteration of the physical structure of a gene or chromosome (example —duplicating a gene whose product is needed in high quantities).

- Modulation of gene expression (example—turning off transcrip-tion when the gene product is not needed).

Within these two categories, several specific mechanisms are possible (Table 28-1). The next two sections in this chapter expand on the informa-tion in the table. As you will see, some alterations in structure (category I) have been shown to influence gene ex-pression (category II).

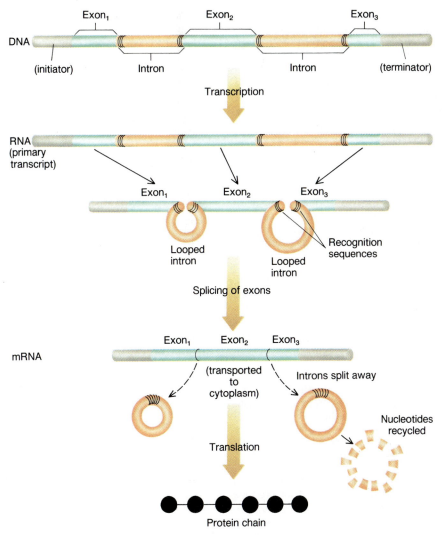

**Figure 28-8**
**Editing a gene's message**. Split genes are expressed by splicing the exons in the primary mRNA into an edited molecule that is colinear with the final protein. This occurs by cutting out the intervening spacer sequences (introns) with absolute precision to prevent nucleotides from being accidentally removed from exons. Specific nucleotide sequences identify the beginning and end of each intron for editing. These signals pinpoint the location where the RNA is to be cut.

## REGULATION BY ALTERING GENE OR CHROMOSOME STRUCTURE

### Changing the Number of Genes.

*Gene amplification* increases the number of copies of a gene whose product is needed in especially high concentrations. Hemoglobin production, for example, is directed by multiple gene copies in red blood cells (before they expel their nuclei). Yet only a single copy of each hemoglobin gene exists (unexpressed) in the chromosomes of other cells. The opposite situation is also possible: genes no longer needed for a particular developmental pathway could be discarded. This extremely rare process is called *diminution*.

## Gene Rearrangement.

Genes that move to new sites on the same or different chromosome are called **transposable elements**. These "jumping genes" were first discovered in corn more than 40 years ago, but their significance to gene expression has only recently been realized. Regimented changes in gene location may be one of the mechanisms that control tissue differentiation during embryonic development. For example, a block of inactive genes could be turned on by relocating an activator segment to a position next to them. Assembling several incomplete exons into a complete structural gene can also change the developmental fate of the cell. In addition to the regulatory effects of changing position, transposable elements often duplicate themselves before migrating to a new location. Therefore, jumping genes can increase the number of gene copies in a cell. This may explain one mechanism of gene amplification.

The discovery of exons and jumping genes offers clues to solving a nagging theoretical problem of immunology and genetics. Chapter 20 emphasized that the immune systems of mammals can distinguish between at least 1 million antigens and make antibodies, each specifically tailored to react with one type of antigen. If each type of antigen-specific antibody requires a different gene to direct its synthesis, then a totipotent zygote must contain, in addition to the genes for directing normal tissue development, an extra million genes for antibody production. This would mean that every cell in the body is burdened with an impossible genetic load. The space required for housing so many extra genes is simply not available in a nucleus.

We now realize, however, that it is not necessary for the zygote to house this huge variety of antibody genes if an antigen-specific composite gene could be assembled from various "minigene" segments (exons). In mice (and probably all other antibody producers as well), this is exactly what happens. Instead of containing a million antibody genes, the zygote has a library of several hundred gene seg-

TABLE 28-1

| | Gene Expression in Eukaryotes — Possible Mechanisms of Control | |
|---|---|---|
| **Category** | **Level of Control** | **Description of Mechanism** |
| I. Alter gene or chromosomal structure | Gene amplification | Increases number of copies of a gene, boosting the concentration of the gene product. |
| | Diminution | Discard genes that are no longer needed. |
| | Rearrangement | Genes "change address," move to new locations on chromosome, perhaps assembling an intact gene from several exons or changing control of an adjacent gene. |
| | Modification | Chemically alters a portion of DNA, perhaps to inactivate the modified gene. |
| II. Modulate expression of functional genes | Transcriptional controls | Differential gene activation increases or decreases rate of mRNA synthesis from a gene. |
| | Post-transcriptional controls | Modify mRNA to activate or inactivate it; prevent mRNA transport to cytoplasm; change rate of RNA processing, so amount of mRNA ready for transcription depends on the cell's need for the corresponding protein. |
| | Translational controls | Either increase mRNA stability (when gene product is needed) or produce highly unstable mRNA that rapidly deteriorates before much translation can occur (when gene product is abundant); stockpile stable, "masked" mRNA which is activated (unmasked) when needed. |

ments that can be selectively called on to participate in the formation of a particular antibody. In the embryonic mouse lymphyocyte, which is not yet capable of producing antibodies, the various genetic regions for antibody formation are broken into shorter segments by many stretches of silent DNA. But by the time each lymphocyte matures into a cell capable of antibody production, selected exons have "jumped" together to form a ge-netic unit that generates one and only one type of antibody. This mechanism of gene rearrangement allows the construction of 6.4 million different types of fully operative antibody genes from just a few hundred minigenes devoted to antibody production.

### Modification of DNA.
Some DNA is chemically modified, presumably to inactivate genes that either have not yet been called into ser-vice or are no longer needed. For example, some genes are chemically silenced by the addition of a methyl ($-CH_3$) group to the genes' cytosines. (The resulting methylated DNA is theoretically untranscribable.) Many studies have reported a direct correspondence between the degree of methylation and the level of gene expression. When methylation is chemically inhibited in eukaryotic cells, genes that are normally silent become expressed. Furthermore, DNA methylation is tissue specific; different genes are methylated in distinct tissue types of the same organism.

Such gene inactivation is reversible. In fact, turning genes on may be accomplished by reversing chemical inactivation. In a human sperm cell, for example, the gene for hemoglobin, a protein it neither needs nor produces, is methylated. But once the developing fetus reaches the stage in which "adult-type" hemoglobin is required, these genes become "undermethylated." This event corresponds with hemoglobin gene expression in developing red blood cells. In brain tissue, which doesn't produce hemoglobin, these genes remain methylated and turned off.

## REGULATION BY MODULATING THE EXPRESSION OF FUNCTIONAL GENES
Modulating the expression of intact genes is a more common way of controlling eukaryotic gene expression than altering gene structure, especially for genes that are repeatedly turned on and off in response to fluctuations in the organism's external or internal environments. Mounting evidence suggests that eukaryotes regulate gene expression by temporarily turning on or silencing a gene's mRNA production. In chickens, for example, when ovalbumin (egg-white protein) synthesis is induced by a hormone, the number of copies of mRNA from the ovalbumin gene increases from none to hundreds. But the primary question remains unanswered: if no operonlike system exists in eukaryotes, what determines whether a particular gene is transcribed?

Several hypothetical models have been proposed to answer this question. One relatively recent discovery about the structure of DNA bears mentioning here, since it may provide a "switch" for activating or silencing transcription. This is the discovery that DNA, classically depicted as being a "right-handed" molecule (its spirals turn to the right), also exists in the "left-handed" configuration. In this alternative state, the molecule is referred to as **Z-DNA**, as opposed to the classical **B-DNA** (Figure 28-9). The Z configuration was long thought to be unstable in living cells, but is now believed to exist in the cell's chromosomes. The chromosomal interbands (light-staining bands of less condensed, genetically active material) may consist of Z-DNA, whereas the

dark-staining, dense, inactive bands of the same chromosomes are comprised of classical B-DNA. It may be that when a whole gene is in its B-type configuration, DNA is untranscribable. Only when at least part of it changes to the Z form is it copied into mRNA. Evidence supporting this view is provided by the fact that at least one gene-activating protein binds to B-DNA and changes it to Z-DNA. Perhaps this activator-induced transition from right- to left-handed DNA unwinds the promoter, making it accessible to mRNA polymerase. Thus, transcription of the adjacent gene(s) would occur only when the activator changes the local DNA structure to the Z configuration.

Once a gene is transcribed, its expression may still be controlled by en-

hancing or interfering with the ability of mRNA to reach the ribosomes. One possible *post-transcriptional regulatory mechanism* has already been introduced in this chapter—the split gene phenomenon. Long mRNA molecules that still contain the transcribed introns cannot leave the nucleus. Thus, they have no access to cytoplasmic translational equipment until they are edited into the final mRNA molecule containing only exons. Some mechanism may postpone editing until the corresponding protein is needed. In addition, other proteins interact with mRNA from the moment of its transcription, perhaps escorting it or preventing its transport to the cytoplasm, or somehow influencing its recognition by ribosomes once it is there.

Once the mRNA arrives in the cytoplasm, *translational regulation* can take over modulating the amount of protein synthesized from that messenger. Protecting mRNA from deterioration increases the number of proteins that can be produced. The combination of efficient utilization and stabilization of mRNA is referred to as "translational amplification." By increasing the number of active mRNA molecules, it creates the same result as gene amplification. Prolactin, the hormone that triggers milk production by mammals, operates on this principle. In the presence of prolactin, mRNA carrying the coded amino acid sequence for casein (milk protein) is highly stable, so its translation is augmented. Withdrawal of prolactin decreases the stability of casein mRNA, and milk protein production stops almost immediately.

Another variation of translational regulation combines mRNA stabilization with a means of temporarily "masking" the mRNA, modifying it so it is not translated. In this way huge amounts of mRNA for certain genes can accumulate in dormant cells (such as a "resting" oocyte) until some triggering event activates development by "unmasking" the messengers, at which time synthesis of the corresponding proteins can begin. The phenomenon is well documented in unfertilized sea urchin eggs, which store

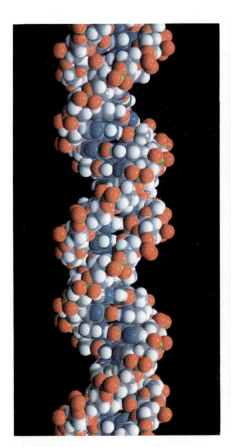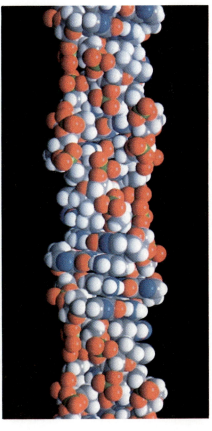

**Figure 28-9**
**Two forms of DNA**, the classical B (on the left) form and the newly discovered Z form, which may be instrumental in gene expression.

untranslated mRNA for months. Moments after fertilization, proteins encoded by these masked transcripts begin to be produced. The mechanism by which mRNA is unmasked, and thus activated, remains unknown.

*Post-translational control* was not included in Table 28-1 because it does not regulate gene expression (even though it clearly participates in regulation of metabolic activity and developmental events). Instead, post-translational mechanisms operate by controlling the *activity* of the proteins already synthesized. One such mechanism depends on the sensitivity of *allosteric enzymes* to the concentration of an effector (Chapter 5). Another mechanism is the controlled survival time of proteins—some are rapidly degraded, whereas others last indefinitely. It is clearly more efficient to prevent the synthesis of unneeded enzymes than to invest energy and resources in proteins that will only be turned off or destroyed. However, the ability to regulate enzyme activity increases the sensitivity of the system. Proteins can be instantly activated or inhibited by changes in effector concentration, avoiding the delay inherent in regulating gene expression. (It takes time for new proteins to be synthesized when genes are turned on and for existing proteins to deteriorate following inhibition of gene expression.) Most gene products are subjected to regulation of both gene expression and enzyme activity.

The quest to understand the mechanisms of gene control in eukaryotes is at the leading edge of biological investigation. It not only contributes to our understanding of life on one of its most fundamental levels, but it also helps us learn why and how these controls malfunction in people, leading to human disease. This understanding enhances our ability to treat or abolish disease. Cancer, for example, is undoubtedly the result of something going wrong with gene-regulatory mechanisms. An understanding of normal regulatory mechanisms and the defective controls responsible for cancer go hand in hand. Progress on one front leads to breakthroughs on the other (see Connections: Cancer).

# CONNECTIONS

| The Cell Cycle, p 157 |
| Chromosomal Aberrations, p 447 |
| Mutagens and Cancer, p 470 |

## CANCER: THE LOSS OF GENETIC CONTROL

Cancer is a "catch-all" name used for a diverse group of diseases that have at least one thing in common—the loss of genetic control over cell proliferation. Multiple factors may trigger this loss of genetic control. Cancer-causing agents (carcinogens), mutations, reduced immunity, and viral infection may all contribute to the development of cancer, some types probably requiring the participation of all these "causes."

Cancer cells have been compared to embryonic cells—undifferentiated cells that rapidly proliferate until normal controls regulate growth and differentiation. Proliferation of cancer cells, however, is not turned off by normal controls. In normal dividing cells, receptor molecules on the plasma membrane detect contact with adjacent cells and relay an inhibitory signal to the chromosomes. In response, the cells stop dividing. This regulatory phenomenon, called **contact inhibition**, prevents cell overgrowth. Loss of contact inhibition by cancer cells is likely due to their inability to turn off the production of a growth factor.

Researchers investigating the factors that cause this loss of control have found that some cancers may be due to a genetic mutation, perhaps in a regulatory gene. Bladder cancer cells, for example, differ from normal cells by a single nucleotide—a point mutation in a gene. The fact that 90 percent of all known carcinogens are also mutagens supports the connection between mutation and cancer.

In addition to mutational changes, many cancerous cells have been discovered to contain an **oncogene** (onco = tumor), the cancer-causing gene. How oncogenes lead to uncontrolled cell growth is still unanswered, but it is clear that gene segments similar to oncogenes reside in the DNA of normal, noncancerous cells. This "proto-oncogene" probably contributes a useful function, perhaps during embryonic development, and is then either inactivated or tightly controlled so that, in normal cells, growth inhibition continues. In cancer cells, the gene may be activated by mutation or by changing its position, associating it with new regulatory segments that either activate it or fail to modulate its expression.

Some types of viruses may even bring new oncogenes into a human cell, as they have been shown to do in other animals. The nucleic acid of such a virus inserts itself into the host cell's chromatin, so that a new copy of the viral genes is produced each time the cell's DNA replicates. In some cells, the virus DNA leaves the host's DNA, eventually propagating and releasing progeny viruses. Sometimes a normal host gene that resides next to the viral nucleic acid is picked up by the virus DNA as it splits from the host's DNA and becomes packaged in the virus protein coat. The progeny viruses may now carry a human oncogene, one that is activated when it is reintegrated in the DNA of newly infected cells. Although most cancers in humans are not associated with virus infection, these tiny infectious particles have provided researchers with new understanding of the process. Oncogenes were discovered in viruses, and much of what we believe about the mechanism of triggering cancer comes from research on oncogenic (cancer-causing) viruses. In addition, viruses are being used to help "sniff out" other genetic segments that may be possible oncogenes.

Perhaps once we understand the abnormal proliferation of cancer cells, we will finally be able to explain the genetic regulatory mechanisms that control the growth of normal cells.

# Synopsis

## MAIN CONCEPTS

- **Selective expression of the genetic endowment**. Every cell has far more genes than it uses at any particular time. As the needs of the organism change, so will the genes that are expressed. As an organism develops, different genes are turned on and off. In this way, one totipotent zygote gives rise to hundreds of different cell types, all genetically identical but differing sharply in which genes are expressed.

- **The operon**. In prokaryotes, the fundamental regulatory mechanism is the operon, which regulates synthesis of the enzymes catalyzing the sequential steps of a metabolic pathway. Each operon's effector is a key metabolic compound whose concentration determines whether or not the pathway is needed at that time. In catabolic pathways, the *substrate* is the effector. In biosynthetic pathways the *end product* is the effector.

- **Regulation of eukaryotic gene expression**. Histones, nucleosome complexes, and large amounts of silent DNA that breaks up genes into introns and exons characterize the eukaryotic chromosome. RNA transcribed from split genes cannot leave the nucleus for translation until the intron regions have been edited out. Such editing fuses exons into a messenger molecule that can be translated into protein. Deferring editing of RNA can delay gene expression until needed.

- **Regulation by altering the structure of genes or chromosomes**.
  - *Gene amplification* boosts the amount of a particular product.
  - *Diminution* disposes of genes no longer needed, such as a gene required in a stage in embryonic development.
  - *Gene rearrangement* can switch on gene expression by moving an inactive gene next to an activator sequence or by moving several exons together to form a complete functional gene. (The reverse of these processes could be used to inactivate genes no longer needed.)
  - *Chemical modification of DNA* renders a gene untranscribable until the chemical group is removed.

- **Regulation of protein synthesis**.
  - By changing the configuration of the promoter DNA from the Z form to the B form.
  - By preventing mRNA from leaving the nucleus until the gene product is needed.
  - By modifying mRNA stability (the less stable the RNA, the fewer times it can be translated into protein before it deteriorates).
  - By masking the mRNA to "disguise" a stable messenger, so that it won't be translated until there is a need for the corresponding protein, at which time the mRNA is unmasked.

## KEY TERM INTEGRATOR

### Gene Regulation — Some Generalizations

| | |
|---|---|
| **totipotency** | The ability of a cell to give rise to any type cell in a multicellular organism. Single totipotent cells can be grown in **tissue culture**, producing **clones** of the organism that donated the cells. |

### Gene Regulation in Prokaryotes

| | |
|---|---|
| **operon** | Cluster of **regulator (R) gene**, regulatory sequences, and a block of **structural genes** that code for enzymes of a single metabolic pathway, transcribing them into one long mole- |

cule of mRNA. Transcription is blocked when the operon's repressor protein is fastened to the **operator**, blocking the access of mRNA polymerase to the **promoter**. Repressor–operator attachment is determined by the presence of the **effector**, a key metabolic compound. The effector binds with the repressor protein, leading to operon **induction** if it prevents repressor–operator attachment, or **repression** if it promotes such attachment.

## Gene Regulation in Eukaryotes

| | |
|---|---|
| **nucleosomes** | Complex consisting of eukaryotic DNA and **histone** proteins, forming the fundamental packaging unit in the nucleus. Histones may assist in gene regulation. Nucleosomes are present in condensed chromosomes and in noncondensed strands, such as chromatin and the unpacked transcribable strands found in **chromosomal puffs** and **lampbrush chromosomes**. |
| **exons** | "Split genes," DNA segments that represent portions of a functional gene that has been divided by spacers, called **introns**. Removing these spacers from the **primary transcript** makes the mRNA transcribable. Some exons may be **transposable elements** that can be moved together to create an intact gene when it is needed or to place a regulatory sequence next to a structural gene. |
| **Z-DNA** | Structural variation of DNA in which the sugar-phosphate backbone shows a zig-zag pattern rather than the smooth, double-helical configuration of classical **B-DNA**. It may have a regulatory role, perhaps serving as an attachment site for gene-activating chemicals. |
| **contact inhibition** | A regulatory mechanism that prevents overgrowth of cells by halting mitosis of cells when they come in contact with neighbor cells. Loss of such regulatory restrictions, a characteristic of cancer cells, is likely due to the influence of an **oncogene**. |

# Review and Synthesis

Study the following graphs before answering questions 1 through 4.

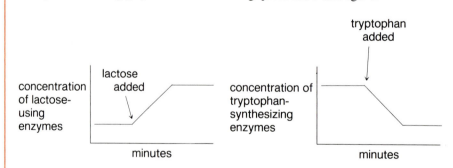

1. In each graph, (a) identify the effector, (b) determine whether it is the starting substrate or the end product of the regulated pathway, and (c) identify which pathway is catabolic and which is biosynthetic.
2. Which of the following does each graph represent? induction; repression; feedback inhibition of enzyme activity; or nonregulated enzyme synthesis.
3. Describe the cascade of events that are responsible for the sudden changes following addition of the sugar lactose or the amino acid tryptophan.

4. Redraw each graph for each of the following situations:
   (a) An organism with a defective R gene.
   (b) An organism with its operator deleted by mutation.
   (c) An organism with a repressor protein that cannot bind with the effector.
5. What is the advantage of clustering structural genes so that all the enzymes for a metabolic pathway are regulated together rather than independently?
6. Describe how the expression of the following genetic segment may be regulated:

   **exon#1**/intron/**exon#2**/intron/**exon#3**

7. List and provide examples of seven levels of controlling gene expression (four that alter genetic structure, and three that modulate the expression of fully functional genes).
8. What is the relationship between gene regulation and cancer?
9. How do "mini genes" enable you to immunologically respond to a million antigens without burdening each of your cells with a million extra genes, one for each antigen?
10. Hormones dramatically alter gene expression (the distinctions between men and women are fundamentally the result of sex hormones selecting different genes for activation). Create a model by which a hormone might temporarily activate a gene and another hormone might turn off that gene later in development. Your model can use any of the mechanisms discussed in this chapter.

## Additional Readings

Brown, D. 1981. "Gene expression in eukaryotes." *Science* 211:667. (Intermediate to advanced.)

Chambon, P. 1981. "Split genes." *Scientific American* 244:60. (Intermediate.)

Darnell, J. 1983. "The processing of RNA." *Scientific American* 249:90. (Intermediate to advanced.)

Jacob, F. 1973. *The Logic of Life: A History of Heredity*. Pantheon Books, New York. (Introductory to intermediate.)

Slamon, D., J. deKernion, , M. Verma, and M. Cline. 1984. "Expression of cellular oncogenes in human malignancies." *Science* 226. (Advanced.)

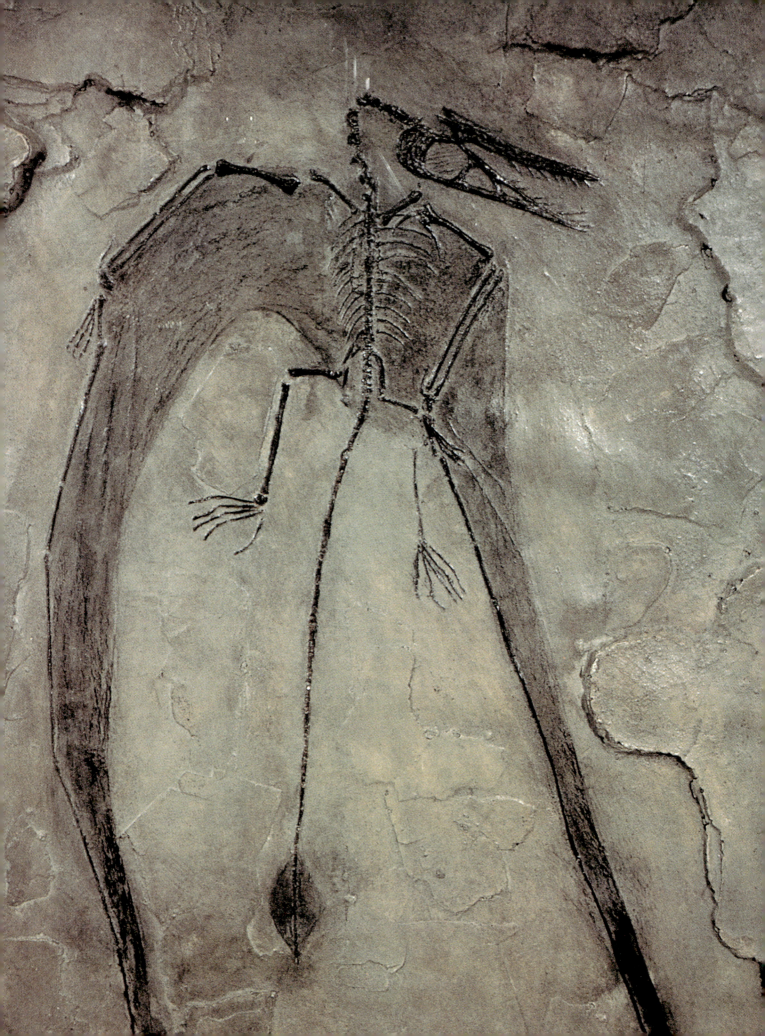

# 8

## *Evolution*

Chapter 29

# Mechanisms of Evolution

The Theory of Evolution

The Evolutionary Process: An Overview

Evidence for Evolution

The Basis of Evolution
   DISCOVERY Charles Darwin: Exploring Life on the Voyage of the HMS *Beagle.*
   Changes in the Gene Pool: The Basic Indicator of Evolutionary Change
   Genetic Variability: The Foundation for Change

The Forces of Evolutionary Change
   Mutation: The Source of New Genes
   Genetic Drift
      Bottlenecks
      Founder Effect
   Gene Flow: Exchanges Between Populations
   Natural Selection
   DISCOVERY Discovering How Evolution Works
   BIOLINE A Gallery of Remarkable Adaptations
      Types of Natural Selection
         Stabilizing Selection
         Directional Selection
         Diversifying Selection
      Balanced Polymorphism
      Sexual Selection

The Origin of Species
   Isolating Mechanisms
   Mechanisms of Speciation

Patterns of Evolution
   Phyletic Evolution
   Punctuated Equilibrium
   Divergent Evolution and Adaptive Radiation
   Convergent Evolution
   Coevolution
   Extinction

Synopsis
   Main Concepts
   Key Term Integrator

Review and Synthesis

Additional Readings

"A fish out of water" can't survive very long, or so we might think. Yet the mudskipper is a fish that not only survives long treks into the world of air, but also walks across mud flats and even climbs trees (Figure 29-1). In water, the mudskipper's fins and gills work exactly like those of a typical fish. Its fins propel and steer the mudskipper through the water, and its gills extract dissolved oxygen (Chapter 19).

But how does the mudskipper remain alive out of water with a body and breathing machinery that are so unmistakably adapted for life in water?

Unlike other fishes, the mudskipper's gills and fins have slight modifications that enable it to survive and move on land. In bulging gill pouches, the mudskipper takes along a supply of water to "breathe," a scuba tank in reverse. For getting around, reinforced forefins serve as stubby arms to crawl across the mud or to shimmy up tree trunks in search of snails for a meal.

A hobbling mudskipper with water-engorged pouches illustrates how structures originally styled for one way of life can become refashioned for new functions. The mudskipper's makeshift legs and water bags are modifications of structures designed strictly for aquatic life. The possession of these structures is evidence that the mudskipper's ancestors lived strictly underwater, and those ancestors were not mudskippers.

Every species has a history of ancestors that possessed other features and behaviors. To trace the history of change in an organism's ancestors is to follow its course of evolution.

## The Theory of Evolution

How did there come to be so many kinds of organisms, each with unique characteristics that enable it to survive in one (or more) of the earth's many diverse habitats? How can we explain the fact that there were organisms living in the past with characteristics very different from those living today? Why were the earth's first very ancient organisms much simpler than many of those living today? And why are there degrees of similarities and differences between organisms that are so definite that scientists are able to organize life into a classification hierarchy? The theory of evolution provides the best scientific explanation for questions about the patterns of life on earth.

The principal conclusions of the theory of evolution are:

- Evolution occurs over very long periods of time. (It is often difficult for us to perceive evolution because it generally requires millions of years.)

- Evolution is the result of many small, random inheritable changes. These changes eventually add up to form new species.

- In general, the more similar species are in appearance, the more closely they are related.

- Most species become extinct (die out). Once extinct, a species never reappears.

- Evolution continues today.

**Figure 29-1**
**The mudskipper** is a fish that can live for hours out of water. Strengthened forefins and modified gill chambers are evidence that the mudskipper evolved from strictly aquatic ancestors.

# The Evolutionary Process: An Overview

**Evolution** is a process whereby the characteristics of organisms change over time, eventually leading to the formation of new species. At the foundation of the evolutionary process is the genetic variability that naturally exists within a species. Each individual has a unique genetic makeup, which, in turn, produces slight differences in the characteristics among the members of the species. Some of these inherited differences improve an individual's chances of surviving and reproducing offspring with similar characteristics. These more favorable characteristics gradually accumulate over time, producing species that are suited to their environment.

In addition to providing the basis for **adaptation** (characteristics that improve an individual's chances of survival or reproduction), the variability among members of a species may enable some individuals to survive changes in the environment or allow individuals to occupy new habitats that require slightly different adaptations than did the previous habitat. Under new environmental conditions, individuals with characteristics better suited for survival and reproduction will eventually produce a new species with characteristics distinct from the original parental species. In this way, one species may change into another species over time as the environment changes, or one species may produce one or more new species in new habitats. Either way, new kinds of organisms evolve.

This, in brief, is how evolution works. But what is the evidence that this process really happens in nature?

# Evidence for Evolution

Scientists from many disciplines have amassed a great deal of evidence that supports the theory of evolution. This evidence includes:

- **Intermediate anatomical modifications** — alterations in the anatomy of structures that can be traced back to an ancestral form. The mudskip-per's water pouches and stubby fin-legs are clear examples of re-fashioned gills and fins of an ancestral fish.

- **Fossil record**. The 3.5-billion-year-old record of **fossils**, the remains of organisms preserved in rock, displays an unmistakable progression from very simple cells to complex life forms. Sometimes the entire history of evolutionary change is clearly documented in a series of fossils. For example, the evolution of the modern horse *(Equus)* can be traced through a succession of fossils that extend back over 60 million years to a 25 cm (about 10 inches) tall "dawn horse" called *Eohippus* (Figure 29-2).

- **Radioisotope dating**. Because isotopes decay at a characteristic rate, scientists are able to estimate the age of a fossil or stratum of rock in which the fossil occurs (Chapter 3). By studying fossils and pinpointing their age, scientists can verify that life forms existed in the distant past that were totally unlike any alive today. In addition, radioisotope dating helps scientists organize the history of life on earth in a chronological sequence, enabling them to see progressive changes in the characteristics of species over time.

- **Embryology and vestigial organs**. Organisms carry remnants of their ancestors. Some remnants of the past are only displayed during embryonic development (see Connections: Embryonic development, Chapter 24), while others, called **vestigial organs**, remain as useless relics of once useful ancestral structures. The underdeveloped pelvis and leg bones in snakes, the diminished toe bones in horses, and the appendix in humans are examples of vestigial organs (Figure 29-3).

- **Biochemical similarities**. All organisms are constructed of the same complex molecules (proteins, lipids, carbohydrates, and nucleic acids); all use ATP to power metabolic reactions; and all use the identical process and coding system for storing and expressing genetic information.

These similarities in chemical makeup and biochemical processes suggest organisms evolved from a common ancestor.

- **Homologous structures**. The presence of homologous structures in a variety of species that otherwise bear little resemblance indicates that they shared a common ancestor in the past. **Homologous structures** develop in the same way and are anatomically identical, although they may work in different ways. For example, the forelimbs of horses, bats, whales, and humans illustrated in Figure 29-4 all have the same framework of bone structure, yet function differently.

- **Molecular homology**. All organisms store hereditary information in nucleic acids, either DNA or RNA. When comparing species, the more similar the nucleic acids, the more closely the species are related. For example, the DNA of humans and chimpanzees is 98 percent similar, indicating that humans are more closely related to chimpanzees than any other type of living organism. This great similarity strongly suggests that humans and chimpanzees evolved from the same common ancestor.

Together, anatomical modifications, the fossil record, radioisotope dating, similarity in embryo development and biochemistry, vestigial organs, and homologous chemistry and structures supply clear evidence that species do not remain forever fixed; they all evolve. Evidence for evolution comes from many different sciences (biology, geology, physics, chemistry, archaeology, paleontology, and others), unifying and strengthening the theory of evolution. Recall from Chapter 2 that in science the word "theory" does not mean "guess" but refers to an explanation that has been confirmed again and again in many different studies. This is certainly the case with the theory of evolution.

Further verification that species are capable of changing comes from the selective breeding practices of agricul-

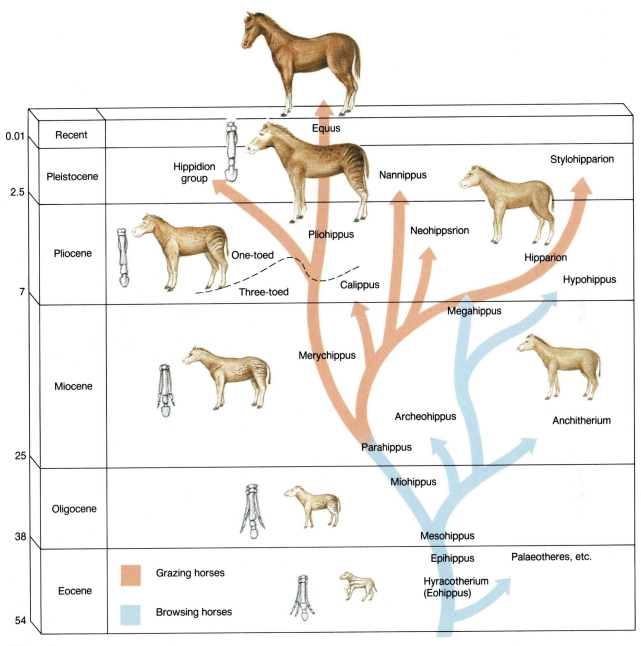

**Figure 29-2**
**The evolution of the modern horse (Equus).** A progression of fossils can be traced back over 60 million years to the "dawn horse" called Eohippus. This small ancestor had four toes on its front feet and three on its hind feet. The evolutionary pattern that led to the modern horse included a reduction in the number of toes, the development of more complex teeth, and an increase in size. All species became extinct except the ancestral line that culminated with *Equus*.

A. Snake

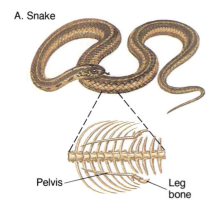

Pelvis

Leg bone

B. Horse

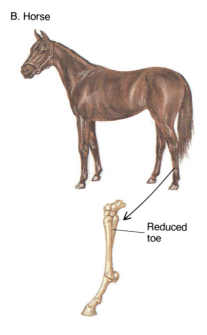

Reduced toe

C. Human

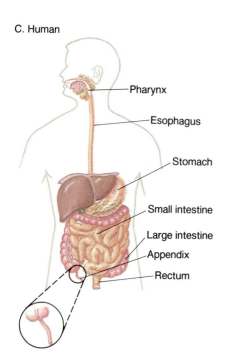

Pharynx

Esophagus

Stomach

Small intestine

Large intestine

Appendix

Rectum

turists and animal breeders. By only breeding individuals with desired characteristics, selective breeding has increased crop and livestock productivity, improved their resistance to disease, developed dwarf plants and animals (dwarf peach trees and the Shetland pony, for example), and produced a great variety of dogs (from a tiny Chihuahua to a hulking St. Bernard) from a single ancestral species. The ability of breeders to purposefully select changes in characteristics demonstrates that the features of organisms can change over succeeding generations and that species have the capacity to evolve. However, breeders are only able to select changes that occur naturally. They do not cause the changes, although advances in biotechnology may make this a possibility in the near future (Chapter 39).

The key to understanding evolution, then, is to discover what *natural* forces cause species to change over time. Uncovering these forces has occupied the time and energy of many scientists over the years, but one in particular, Charles Darwin, stands out as the champion of evolutionary theory. Although Darwin investigated a variety of biology topics, from the pollination of orchids to the taxonomy of beetles, his name will always be linked with the study of evolution because he was the first to present a clear explanation of how evolution could take place. Darwin's explanation not only revolutionized thinking about biology, but also changed forever the way humans viewed themselves in the natural world (see Discovery: Charles Darwin: Exploring Life).

All living species are only temporary products of the still ongoing process of evolution. Some species may remain relatively unchanged for millions of years (a mere blink of an eye in geologic time), but because the environment constantly changes, all will eventually change or become extinct because they did not change, or because their changes were not suitable for survival.

## The Basis of Evolution

Evolution is the great unifying concept in biology. Evolution explains how there came to be so many different types of species on earth, why they look and function the way they do, and why many species have become extinct. It also explains the natural division lines along which the great diversity of species can be classified into genera, families, orders, classes, phyla or divisions, and kingdoms (see Chapter 1).

### Changes in the Gene Pool: The Basic Indicator of Evolutionary Change

Scientists estimate that between 10 and 20 million species (possibly more) have evolved since the beginning of life on earth. Each species is composed of individuals with the potential to interbreed and share features that set them apart from individuals of other species. Some species are composed of a single group of individuals living in just one area, a single population. Many other species are made up of more than one population, each occupying a different locality.

But whether a single population or many, the inherited characteristics of all individuals of the species account for the distinctiveness of a species. These inherited characteristics are controlled by genes, the sum of which in all individuals in a population form a **gene pool**—a reservoir of the total genetic information available for exchange during sexual reproduction. Any single individual contains only a portion of the genes in a gene pool. In species with more than one popula-

**Figure 29-3**
**Vestigial organs: A legacy of ancestors.** *(a)* Snakes retain degenerated leg and pelvic bones inherited from four-legged ancestors. *(b)* Horses still possess degenerated toe bones left over from their three- and four-toed ancestors. *(c)* The human appendix is a degenerated cecum, a chamber used by our vegetarian ancestors for housing cellulose-digesting microorganisms.

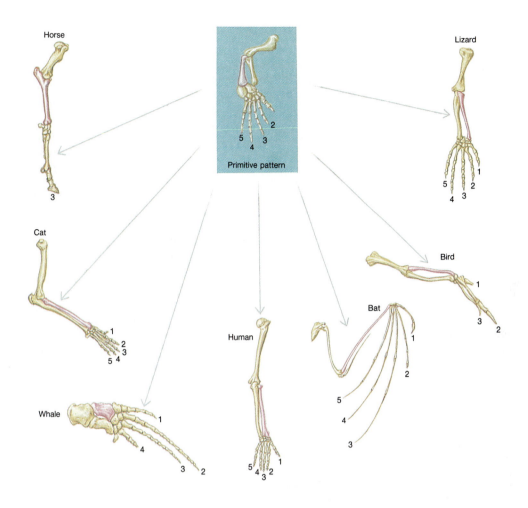

**Figure 29-4**
**Structural homology.** In spite of different uses, the forelimbs of a number of very different vertebrates all have the same framework of bone structure, modifications of a primitive skeletal pattern in an ancient ancestor from which they all descended.

tion, each population usually forms a discrete group of interbreeding individuals and has a slightly different gene pool from the other populations of the same species.

Although a species is composed of populations of individual organisms, individuals themselves do not evolve. Individuals can grow, develop, reproduce, and die, but they cannot change into a different kind of organism. It is the gene pool that evolves as new genes and variants of genes (alleles) are added or removed, and as the prevalence of genes and alleles become modified with time. In other words, populations evolve. Because of the central role of the gene pool, evolution

is sometimes defined as a change in the makeup of a gene pool over time. The variability of genes and alleles in a gene pool, then, is the raw material for evolutionary change.

## Genetic Variability: The Foundation for Change

In any population, the characteristics (phenotypes) of individuals vary slightly. Not all individuals are identical in appearance; for evidence, simply look at the great variety of faces in a large crowd of people. The phenotype of any individual is the result of an in-

terplay between its genetic makeup and the effects of the environment on the expression of traits (Chapters 25 and 28). Since all members of a population usually have the same gene loci, variation results when individuals carry different alleles for a particular locus.

**Allele frequency** is the relative occurrence of a certain allele in the individuals of a population. For example, if three-quarters of the people in a crowd have attached ear lobes, the allele frequency for attached ear lobes is 75 percent and the frequency for free ear lobes is 25 percent. Any change in allele frequency in a gene pool will change the occurrence of genetically determined characteristics in the following generations. In other words, a change in allele frequency means a population is evolving.

Under certain conditions, allele frequencies do not change, and there is no evolution. (This is explained by the Hardy-Weinberg Theorem discussed in Appendix II.) A number of conditions combine to prevent evolution.

- No mutation occurs.
- There is an extremely large population size.
- No new members enter the population from another area and no existing members exit the population.
- All individuals have the equal chance of reproducing; in other words, no phenotypes increase survival or are favored over others for reproduction.

But in nature, these conditions are rarely, if ever met; mutations occur, population size varies, individuals often move between populations, and some individuals do have greater success in surviving and reproducing. Together, these are the main forces that cause changes in allele frequencies, making evolution the normal pattern in nature.

## The Forces of Evolutionary Change

Under natural conditions, a gene pool is continually changing. Changes in the frequency of alleles in a gene pool are caused by

- **Mutation**: small, randomly produced inheritable changes in DNA that introduce new genes and alleles into a gene pool.

- **Genetic drift**: random changes in allele frequency that occur by chance alone.

- **Gene flow**: the addition or removal of alleles when individuals exit or enter a population from another locality.

- **Natural selection**: differential survival and reproduction of organisms with phenotypes that are better suited to the environment. Natural selection increases the frequency of favorable alleles while decreasing the frequency of less favorable ones.

These four forces, alone and in combination, determine the course and rate of evolutionary change (Figure 29-5). But unlike genetic drift, gene flow, and natural selection, only mutations actually change alleles; thus, mutation is the ultimate source of the raw material for evolution.

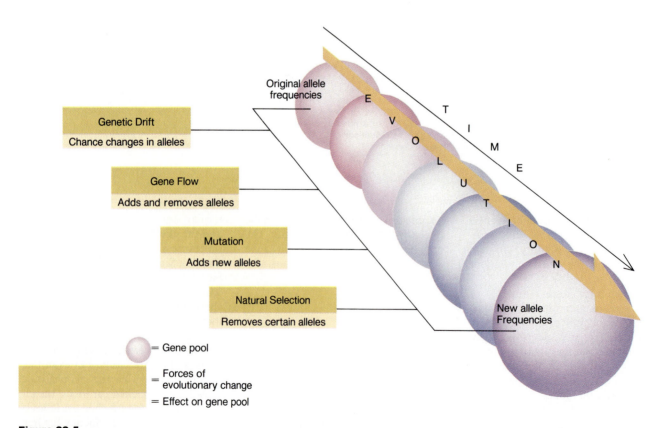

**Figure 29-5**
**Forces of change**. During evolution, the frequency of alleles in a gene pool changes with time. The forces that cause allele frequency changes are mutation, gene flow, genetic drift, and natural selection.

# D I S C O V E R Y

## CHARLES DARWIN: EXPLORING LIFE ON THE VOYAGE OF THE HMS *BEAGLE*

How lucky can you get? What if someone offered you a free trip around the world and all they asked in return is that you pursue your favorite hobby during the trip? Such incredible good fortune is exactly what happened to 22 year-old Charles Darwin when one of his professors, Dr. John Henslow, recommended that he be appointed "naturalist" for a round-the-world scientific expedition.

It was 1831 and Charles Darwin had just graduated from the University of Cambridge. Although Darwin was now a college graduate, he was still unsure of his lifelong profession. Before Cambridge, he had dropped out of medical school where he would have followed in the footsteps of his father, Dr. Robert Darwin, and his well-known grandfather, Dr. Erasmus Darwin. But with the crude surgical methods available at that time, Darwin could not bear the thought of causing pain to a fellow human. With options running out, he had finally decided to study theology and become a clergyman in the Church of England when this "once in a lifetime" offer was proposed to him. An enthusiastic Darwin accepted the commission on the *HMS Beagle,* a small (25 meter or 75 feet long), 3-masted surveying ship that was to map the coastlines and harbors of South America (Figure a).

Charles Darwin had been an avid insect and plant collector since he was a young boy. While studying at Cambridge, Dr. Henslow had taught Darwin techniques in preserving speci-mens, and the value of making careful observations and taking meticulous notes when studying the complexities of nature. Armed with this experience, Darwin was confident that he was up to the task of collecting and organizing the many thousands of specimens he would gather from around the world. But he had no way of anticipating that the rich diversity of life he would encounter on the voyage would spark a revolutionary idea, a mechanism for evolution based on verfiable observation.

The *HMS Beagle* set sail from Devonport, England on December 27, 1831. For the next 5 years, the *HMS Beagle* followed a course mapped in figure b. The often stormy voyage was grueling for the sea-sick prone Darwin. But the young, energetic naturalist never missed an opportunity to collect and document the unexpected variety of organisms and fossils he saw when the *HMS Beagle* set anchor.

The many long periods between ports-of-call gave Darwin the opportunity to read, reread, and analyze the books that were on board—precious time to assimilate new knowledge and connect together the many exciting observations he made on the voyage. Among these books was *Principles of Geology,* a textbook written by Charles Lyell (1797–1875). Lyell's book meticulously presented evidence that the earth was much older than previously believed, and that the earth itself gradually changed over long periods by such natural forces as mountain building, erosion, and flooding. Darwin had seen for

a

himself many examples of such changes clearly etched on the earth in the places he visited during the voyage. He wondered, if something as seemingly unalterable as the earth could change over time, then perhaps the organisms that live on the earth change as well? That would explain why many of the fossils he had collected did not look identical to any of the organisms living today. But, what natural forces could cause organisms to change?

After returning to England in 1836, Darwin set out to find an answer to that question. He read an essay by Thomas Malthus on "the principles of populations" which argued that although organisms have the potential to produce large numbers of offspring, the sizes of populations were held in check by natural factors, for example limited food supply. This idea, together with the knowledge and experience he had gained during the voyage of the *HMS Beagle,* helped Darwin piece together a mechanism for the way species could change; in other words, an explanation for how evolution works. It took Darwin another 23 years of research, analysis, and painstaking documentation before he felt confident enough to present his *theory of evolution by natural selection* to the world. And when he did, the way people viewed the living world would never be the same again.

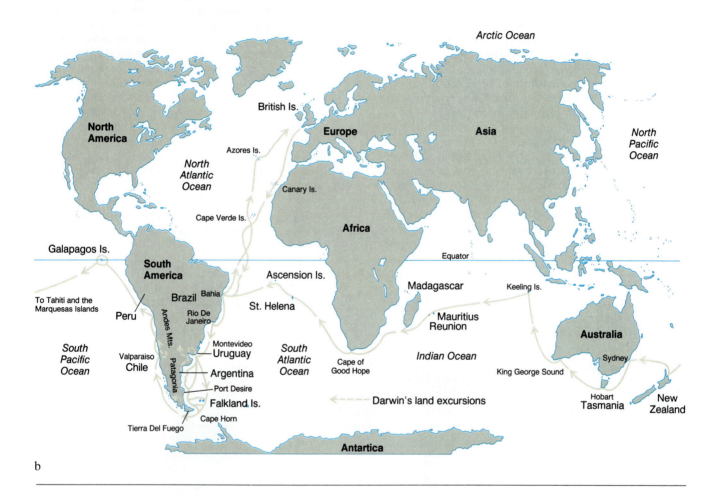

b

## Mutation: The Source of New Genes

A mutation is a random change in the structure of DNA (Chapter 27). Mutations add new genes or alleles to the gene pool and thus supply the genetic foundation on which the other evolutionary forces operate. Because mutations are random, evolution is neither predictable nor goal-oriented toward the production of perfect species.

Thanks to a few Hollywood and Japanese horror films, the mention of "mutation" usually produces images of grossly deformed creatures— multiheaded monsters or 12-story tall grasshoppers. But in nature, a single mutation rarely produces a radical change in the appearance of an organism. Over hundreds of generations, however, the accumulation of many mutations may add up to large scale changes.

Some mutations are beneficial, some others detrimental, and still others are neutral, having no immediate effects. Many harmful mutations are immediately removed from the gene pool because they produce a change so serious that the owner dies during development. (As many as 30 to 50 percent of all human zygotes, for example, fail to develop because of lethal mutations.) Many other harmful mutations are masked by a dominant allele. For example, each of us is believed to carry 7 or 8 lethal recessive genes; the fact that we are alive testifies to the role of dominant alleles.

Mutations occur spontaneously in all cells of the body, but only mutations in reproductive cells contribute to evolutionary change because only they can be passed on to the next generation. That is, they are inheritable. Although mutations are the ultimate source of *new* genes and alleles, recombination during meiosis and sexual reproduction provide new combinations of existing genes and alleles (Chapter 12).

**Figure 29-6**
**Male western Grebes** scurrying across the water.

## Genetic Drift

Genetic drift is a change in gene frequency that results from some chance event. Genetic drift occurs in populations of all sizes, but the effects of genetic drift are often more pronounced in small populations.

To illustrate how genetic drift can affect evolution, let's briefly consider two populations of western Grebes (a species of birds). We will examine one very tiny population and a large population with many individuals. Grebes have the unusual ability to run atop water, propelled by rapidly-beating, lobed feet (Figure 29-6). This behavior is used during courtship and also helps Grebes escape danger. Up to a point, the larger the foot size, the greater a Grebe's ability to skitter across water. Therefore, alleles for large feet are more favorable for survival and reproduction than those for small feet. Let's assume that within both Grebe populations, foot size varies continuously from small to large. While migrating to their breeding ground, a late snow storm kills 99 percent of all the Grebes in both populations. In the small population, by chance all of the large-footed Grebes are killed, removing the more favorable alleles from the gene pool. The next generation of Grebes from this population will all have small feet. A chance event (the snow

storm) increased the frequency of a characteristic that has little adaptive value. Genetic drift has occurred.

In the more numerous population, the probability that all large-footed Grebes will be accidentally killed by the storm is less simply because of the larger number of individuals, so the favorable genes for large feet are more likely to get passed on to the next generation. Thus, given the same circumstances, genetic drift is less likely to occur in the larger population.

This example illustrates how small populations are more susceptible to genetic drift. Because of this, genetic drift can cause dramatic changes in allele frequencies when a few individuals colonize a new habitat (called a **founder population**), or when the population size becomes drastically reduced during a population crash. Population crashes produce a phenomenon called a bottleneck.

### BOTTLENECKS
The size of populations naturally fluctuates as the environment changes during the year or as conditions change from one year to the next. During extreme periods, many individuals in a population may be killed, greatly reducing the population size. Fewer

individuals in the population reduces the diversity of available alleles in the gene pool. This produces a genetic **bottleneck** that retards the population's ability to reestablish its former richness.

Species that are near extinction have very small populations, creating a bottleneck that often prevents the species from reversing its path to extinction.

### FOUNDER EFFECT
When one or a few individuals split off from the main population and establish a new population in a different area, the founding population has a more limited gene pool than the larger parental population. Often genetic drift in the founding population changes allele frequency so dramatically that phenotypes quickly become different from the parental population, sometimes forming a new species. This is called the **founder effect**.

Many examples of the founder effect are found in isolated locations, particularly on oceanic islands. Striking differences between species of land snails (all in the same genus, *Cerion*) on hundreds of Bahama islands are likely the result of the founder effect (Figure 29-7). Because *Cerion* land snails are hermaphroditic, a single

**Figure 29-7**
**The founder effect and *Cerion* snails**. Genetic drift in founding populations of land snails led to remarkably different *Cerion* shells on different islands in the West Indies islands.

snail can found a new population on an island, restricting the initial gene pool to the alleles of a single individual. Since most plants are hermaphroditic *and* are capable of asexual reproduction, the founder effect has been prominent in plant evolution as well.

## Gene Flow: Exchanges Between Populations

In species with more than one population, it is common for individuals to migrate from one population to another, or for gametes, pollen, or spores to disperse between populations. These movements transfer alleles from gene pool to gene pool. The transfer of alleles between populations is called **gene flow**. Gene flow decreases the differences between gene pools, reducing the distinctions between separated populations (Figure 29-8*a*).

When gene flow is prevented, the gene pools of the isolated populations will diverge over time because of differences produced by mutations, genetic drift, and natural selection within each population. Diverging gene pools eventually produce indi-

viduals with such unique characteristics that the populations are recognized as different species (Figure 29-8*b* and *c*).

## Natural Selection

The most powerful driving force in evolution is natural selection. Only natural selection leads to changes in allele frequencies that favor adaptation. It is true that mutation, genetic drift, and gene flow produce changes in allele frequencies in gene pools, but these changes do not necessarily improve the chances of survival or reproduction.

The theory of natural selection, proposed simultaneously by Charles Darwin and Alfred Wallace (see Discovery: Discovering How Evolution Works), states that individuals with characteristics that are better suited to their surroundings have a greater chance of surviving and reproducing than individuals that are less well suited. In other words, better adapted individuals are *naturally selected* by the environment.

Natural selection can be explained in the following steps:

1. Each population of organisms has the potential to reproduce large numbers of offspring. But since environmental resources (food, space, nutrients, and so on) are generally limited, members of a population often produce more offspring than can be supported by the environment.

2. The scarcity of environmental resources leads to competition among individuals, each struggling to gather the resources needed for survival. As a result of competition, not all individuals survive long enough to reproduce or are as successful in reproduction.

3. The individuals in a population usually vary in small ways in structure and behavior. Many of these variations are inheritable; they can be passed on from one generation to the next.

4. Some inherited traits are more adaptive than others; that is, they improve an individual's chances of securing environmental resources reproducing.

5. Better adapted individuals generally survive longer and reproduce more offspring. This is called *differential reproduction*. Through differential reproduction, adaptive alleles are *selected for* (passed on) and, as a result, increase in frequency over many generations. Alleles that are less adaptive are *selected against* (not passed on) and decrease in frequency over many generations.

Because better equipped individuals win the "struggle for survival," natural selection is sometimes referred to as "survival of the fittest." Since the phrase "survival of the fittest" may incorrectly suggest that only the strongest survive or that natural selection is goal directed to produce perfect organisms, many biologists prefer not to use this phrase. It is more correct to say that naturally selected traits accumulate over generations, creating species with remarkable structural, behavioral, and reproductive adaptations that fit the demands of their environment (see Bioline: A Gallery of Remarkable Adaptations).

A well-documented example of natural selection was the change in color-

# D I S C O V E R Y

## DISCOVERING HOW EVOLUTION WORKS

Charles Darwin (1809–1882) ignited one of the most heated debates in the history of science. In the end, he had completely revolutionized how scientists interpreted the way nature works.

The furor began in 1859 when Darwin proposed that generations of interbreeding individuals slowly accumulated minute differences that gradually led to the formation of new species. Although Darwin was not the first to propose that species evolve, he was the first to offer a logical, well-documented explanation of how living things could change from one form into another over long periods of time.

The nonscientific community was outraged by Darwin's theory, for it went against religious teachings that species were created exactly the way they are now and would stay that way forever. But the combination of Darwin's careful documentation and the fact that Darwin's theory clearly supported many observations on the nature of life convinced many scientists that Darwin's theory of "descent with modification by natural selection" was a major driving force in evolution. Darwin's theory opened new ways of studying and interpreting nature. The Darwinian revolution had begun.

Since Darwin first published his work, volumes of research have added new insights into how evolution works, stirring debate among scientists that continues today. These disagreements sometimes give the false impression that many scientists are questioning evolution itself. The controversy, however, is not over whether evolution has taken place (the evidence is overwhelmingly clear that it has), but over how evolution works. Let's briefly follow the turbulent and fascinating history of how scientists discovered the main driving forces of evolution.

### Pre-Darwinian Thought

At one time, each species was thought to be a distinct, unchanging entity. Discoveries by early scientists, including Copernicus and Galileo, disclosed that the universe and the earth itself changed significantly over vast periods of time. These findings raised questions whether all natural things, including species, remained fixed through time. The unearthing of fossils of extinct species, as well as fossils that revealed changes in the characteristics and complexity of organisms, added fuel to the thought that living things naturally change with time. To explain these changes, some philosophers proposed that there were periodic God-made catastrophes followed by the recreation of new organisms. This explanation became known as *catastrophism.*

But catastrophism and other religious interpretations do not help to explain why there are so many similarities in groups of species (even between fossilized and living species), or why there is an unmistakably clear progression from simple to more complex life forms. These are better explained if all species evolved from a common ancestor. If organisms do indeed evolve from a common ancestor, then there must be natural forces that cause species to change into new species.

Jean Baptiste de Lamarck (1744–1829) was a dedicated scientist who strongly supported the theory of evolution. In the early 1800s, he proposed a mechanism for evolutionary change called *the theory of inheritance of acquired characteristics.* According to Lamarck's theory, new or improved organs and structures developed at will to meet an organism's needs *during its lifetime.* These better suited structures, along with those that had become more efficient with use (amassing strong muscles or growing longer necks in the case of giraffes, for example), could be passed from parent to offspring. Structures that were not used would wither away and disappear. In other words, the intent or will of an organism brings about change. (If Lamarck's theory were correct, you could at will increase your IQ to help you pass exams, or will yourself a new organ, an extra heart, perhaps, to improve circulation.) As you might expect, scientists found no supportive evidence for Lamarck's theory. But the theory still had an impact on evolutionary thought, for it generated a great deal of discussion and sparked new ideas about how evolution works. One new idea was proposed simultaneously by Charles Darwin and Alfred Wallace. Both Darwin and Wallace agreed with Lamarck that changes did occur in populations over time, but they maintained that these changes were not caused by the will or needs of an organism. Although they both came up with the same explanation at the same time, Darwin is remembered more because he provided the most extensive and carefully organized documentation for this revolutionary theory.

## The Darwinian Revolution

Darwin called his theory *"the origin of species by means of natural selection."* Much of the knowledge and insight for formulating the theory of natural selection came from a five-year voyage aboard the *HMS Beagle* (see Discovery: Charles Darwin Exploring Life). As naturalist on the *Beagle,* Darwin collected and studied many thousands of South American plants and animals. He observed at first hand the remarkable diversity of life in many habitats.

Upon returning to England, Darwin studied an essay by Thomas Malthus (1766–1834) on the principles of plant and animal population growth. Malthus showed that populations increased geometrically, soon outgrowing limited environmental resources. But in Darwin's travels and studies, he had observed that most populations remain more or less the same size from year to year. How could both be true?

Darwin reasoned that individuals making up a population must compete for limited resources. As a result of this struggle for existence, only a fraction of the individuals, those with more favorable traits that helped them secure the resources, would survive long enough to reproduce. Since Darwin knew favorable traits could be passed on to offspring (through selective breeding of livestock and crops), so too could favorable traits of the more fit individuals be naturally passed on to their off-

spring. In other words, favorable traits were *naturally selected* by the environment. This was the theory of natural selection.

In 1842 Darwin began to gather and organize evidence for natural selection. After 17 years, he published *On the Origin of Species by Means of Natural Selection.* The response was thunderous. Discussions, protests, and personal attacks followed. But Darwin's 30 years of careful observation, study, and thinking provided compelling evidence. The theory of evolution by natural selection remains the explanation for the most powerful force that drives evolution.

## Neo-Darwinism and the Modern Synthesis

Since natural selection was proposed, advances in genetics, biochemistry, ecology, and paleontology have enabled scientists to identify mutation, genetic drift, and gene flow as other natural forces of evolutionary change. The pioneering work of S. Chetverikov, T. Dobzhansky, E. Mayr, F. Sumner, G. G. Simpson, and many others led to what became known as **the modern synthesis** or **Neo-Darwinism,** which emphasizes the role of genetics in explaining how evolution works.

## Punctuated Equilibria: Beyond the Modern Synthesis?

In general, the modern synthesis holds that evolution is gradual; generations change slowly over long time periods. But Niles Eldredge and Stephen Jay Gould have recently proposed that the evolution of new life forms may also occur in spurts followed by long periods without change. In other words, long periods of stability in species are punctuated by episodes of evolution of new species. The theory of punctuated equilibrium is supported by some examples from the fossil record and is gaining support as one pattern of evolution.

Evolution continues, and so do the investigations into how evolution works. Undoubtedly, new theories on the mechanism of evolution will be proposed and tested. Questioning and testing—this is how science advances.

---

ation of peppered moths *(Biston betularia)* that occurred during the industrial revolution in England. There are two forms of peppered moths: (1) a white and black speckled form and (2) a solid dark form. Before the industrial revolution, speckled peppered moths were much more numerous than the dark form. The speckled form was camouflaged from hungry birds because they blended in perfectly with the lichen-covered tree trunks and rocks on which they rested (Figure 29-9*a*). On the other hand, the occasional dark form was easily spotted and quickly eaten by birds scouting

for a tasty peppered moth. In this way, the gene for producing dark moths was naturally selected against (by birds) and occurred in very low frequencies in the gene pool. That was the way things were until the industrial revolution changed the color of the English countryside.

In the 1890s, booming industrial cities released tons of black soot into the air, blackening nearby tree trunks and rocks, and reversing natural selection for the peppered moth. Light colored speckled moths now stood out against the dark sooty background, while dark moths blended in (Figure

29-9*b*). Dark moths survived longer and produced more offspring, increasing the frequency of the gene for producing dark pigment. The percentage of dark moths quickly jumped to 98 percent by 1898 near one industrial city (Figure 29-10).

The effects of industrial soot on the frequency of speckled and dark peppered moths, known as *industrial melanism,* were verified in a study completed in the 1950s. Speckled and pigmented peppered moths were marked for recapture and then released into polluted and pollution-free areas. When survivors were recap-

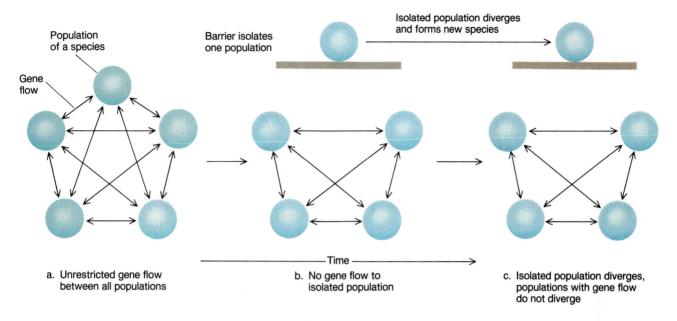

**Figure 29-8**
**Gene flow** maintains the similarity of gene pools and prevents divergence in species composed of more than one population. However, if a barrier isolates populations, separated gene pools diverge, eventually producing separate species.

**Figure 29-9**
**The peppered moth**. *(a)* The dark form is rare in nonpolluted areas of England where it stands out on light, lichen-covered tree trunks, while the speckled form remains camouflaged from hungry birds. *(b)* Near industrial areas where pollution has killed lichens and darkened tree trunks, the situation is reversed and the survival of the dark form is favored.

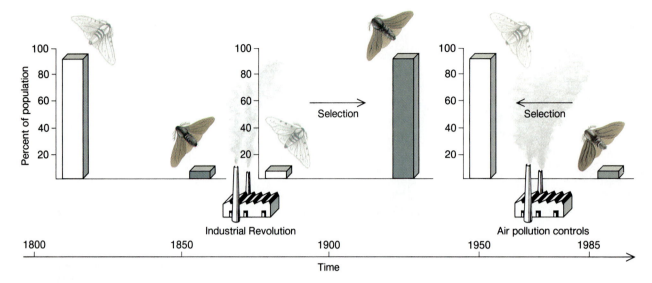

**Figure 29-10**
**Natural selection of the peppered moth in England.** Before the industrial revolution speckled moths were more abundant than dark moths. During the industrial revolution, dark moths became more prevalent because they blended in with soot-covered trees. Following recent pollution controls, speckled are again more numerous than dark moths.

tured, the results confirmed that speckled moths were favored in pollution-free areas and dark moths were favored in polluted areas (Table 29-1). But the story of how natural selection changes the peppered moth isn't over yet, for recent air pollution controls have reversed the trend of natural selection again. Lichens have begun to grow once again on tree trunks and rocks, and the frequency of speckled moths is increasing in areas near industrial centers with air pollution controls.

Although natural selection has changed the abundance of dark and speckled moths back and forth with time, the changes did not produce a new species of peppered moth. This illustrates that natural selection does not always lead to the formation of new species. However, the traits of better adapted individuals often do accumulate through natural selection to the extent that a new species is created.

## TYPES OF NATURAL SELECTION

Natural selection changes the range and frequency of certain phenotypes in a population. There are three possible outcomes, each of which is considered a type of selection (Figure 29-11).

- **Stabilizing selection**—the distribution of phenotypes remains more or less the same from generation to generation because individuals with extreme characteristics die off. Stabilizing selection produces populations of individuals with uniform characteristics.

- **Directional selection**—phenotypes steadily shift in one direction because individuals at one extreme die off.

- **Diversifying selection**—extreme phenotypes become more frequent from generation to generation because individuals with the average, or mean, phenotype die off. Diversifying selection results in populations with individuals that have two or more distinctive phenotypes.

All three types of selection may affect evolution because different traits in a single species may be modified by different types of selection. In Grebes, for example, selection of individuals with large feet leads to directional selection for foot size, while selection of individuals with medium-size bodies results in stabilizing selection for body size.

### Stabilizing Selection

One phenotype is preserved over time during stabilizing selection because the same alleles are selected for, caus-

TABLE 29-1

| Comparison of Number of Released and Recaptured Speckled Versus Dark Peppered Moths in Polluted and Pollution-free Areas | | | | |
|---|---|---|---|---|
| | **Number Released** | | **Number Recaptured*** | |
| **Location** | **Speckled** | **Dark** | **Speckled** | **Dark** |
| Birmingham (Polluted) | 64 | 154 | 16(25.0%) | 82(53.2%) |
| Dorset (Pollution-free) | 393 | 406 | 54(13.7%) | 19(4.7%) |

* Percent recaptured is in parentheses.

# B I O L I N E

## A GALLERY OF REMARKABLE ADAPTATIONS

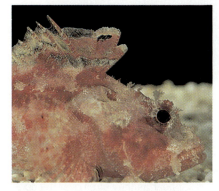

a

b

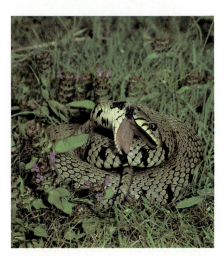

c

Natural selection has resulted in some truly remarkable adaptations. Some adaptations are morphological (for example, hard teeth, sharp claws, horns, and trichomes). But as the following examples illustrate, adaptations can also be behavioral, or those that improve chances of reproduction.

a. **A fishy lure — morphological adaptation.** The rare scorpion decoy fish *(Iracundus signifer)* has a dorsal fin that resembles a smaller fish, complete with its own "dorsal fin," "eye," and "mouth." The scorpion fish swishes its dorsal fin back and forth, luring smaller predators. Within a tenth of a second, the hopeful diner quickly becomes dinner, fatally fooled by a very artful angler.

b. **Disguises — morphological adaptations.** The tiger swallowtail butterfly *(Papilio glaucus)* progresses through a series of larval and pupal stages, each with its own deceptive morphological adaptations. The first larval stage resembles bird droppings. (What predator would eat that?) Three stages later (photo), the green larval caterpillar blends in with the leaves it eats. The caterpillar also has false "eyes" that frighten away predators. The pupal stage resembles a broken twig on a tree trunk, camouflaged from hungry predators.

c. **Playing dead — behavioral adaptation.** When threatened, the ringed snake *(Natrix natrix)* feigns death by dropping its head, dangling its tongue out, and lying completely motionless. Most predators avoid dead organisms.

d

e

f

d. **Safe as the ground you walk on—behavioral adaptation**. The camouflaged horned toad *(Ceratophrys ornata)* buries its body in mud, leaving only its eyes and large jaws protruding. Unwary prey quickly disappear as they stroll over the concealed head.

e. **Torpedo seeds—reproductive adaptation**. The seeds of red mangrove trees germinate while on the tree. When released, the streamlined radicle slices through the water, planting the seedling upright. Seedlings that don't reach bottom are able to float for months until they run aground in shallow water and take root.

f. **Two in one—reproductive adaptation**. Life as an individual is over almost as soon as it begins for the male deep-sea angler fish *(Edriolychunus schmidti)*. To ensure that sexes find each other in the blackness of the deep sea, the newly hatched male angler fish permanently attaches itself to the female by sinking its jaws into her body. The female's skin grows over the male's body and their blood systems connect. The male becomes incorporated into the female body, reduced to nothing more than a small sperm factory.

ing the allele frequency to remain fixed. For example, until recent medical intervention, mortality was greater in human babies with high and low birth weights, but very low for babies of intermediate weights. Since only the alleles for intermediate body weight were selected for, baby size remained the same from generation to generation.

Many scientists believe that the maidenhair tree, chambered nautilus, and horseshoe crab are examples of long-term stabilizing selection. All have been called "living fossils" because they have remained relatively unchanged for millions of years. Present-day individuals look almost identical to multimillion-year-old fossils of their ancestors (Figure 29-12). (On the other hand, some scientists argue that the lack of change may be not due to stabilizing selection, but rather to restricted mutation rates, or to the unchanging environments of these organisms.)

## Directional Selection

When environmental conditions change, as they often do, the most common phenotype in the population may lose its selective advantage. Under new conditions, phenotypes at either one extreme or the other may be better suited, increasing survival and reproduction of individuals with extreme phenotypes (Figure 29-11*b*). When a phenotype at one extreme is repeatedly selected by the environ-

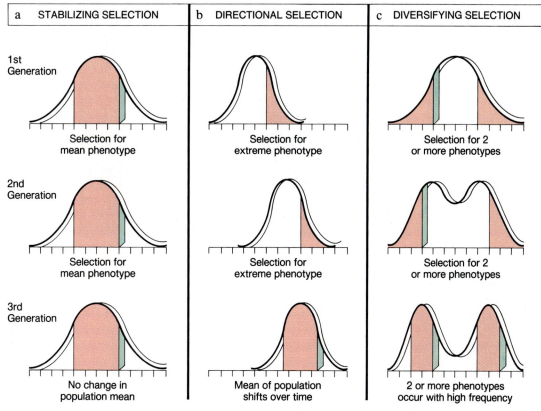

**Figure 29-11**
**Three patterns of natural selection.** The shaded areas represent the ranges of phenotypes for a particular trait that are favored by natural selection. A selection of the mean phenotype (a.) produces no change over three generations during stabilizing selection, whereas selection of extreme phenotypes leads to shifts in the characteristics of a population during directional (b.) and diversifying selection (c.)

ment, phenotypes shift steadily in one direction. The shift in frequency from the light peppered moth to the dark form during the industrial revolution is an example of directional selection. Another example is found in the evolution of the horse, illustrated in Figure 29-2. Here, directional selection led to increased height of the animal and increased hardness of its tooth enamel.

Changing environmental conditions are not the only factors that promote directional selection. When new favorable alleles are added to a gene pool (by mutation or gene flow perhaps), the more favorable alleles may gradually replace less favorable ones by directional selection. Directional selection may also occur in founding

populations where different environmental conditions select phenotypes other than those selected for in the former site.

Diversifying Selection

Diversifying selection occurs when two or more uncommon phenotypes increase in frequency, while once-favored intermediate forms decrease (Figure 29-11c). Diversifying selection promotes **dimorphism** (two forms of a trait) or sometimes **polymorphism** (many forms of a trait) in the population.

An example of polymorphism is found in female African swallowtail butterflies *(Papilio dardanus)* (Figure 29-13). Although they are all members

of the same species, female swallowtail butterflies have strikingly different appearances (phenotypes). All females would make a tasty meal for birds, but each female phenotype survives and reproduces because it accurately mimics a local population of untasty butterflies, one that clearly advertises its rank flavor to bird predators.

BALANCED POLYMORPHISM

When two or more alleles for a single trait are maintained at fairly high frequencies from generation to generation, the population shows **balanced polymorphism**. The diversity in phenotypes of female African swallowtail butterflies is an example of balanced

a

b

c

**Figure 29-12**
"Living fossils"—organisms that have remained virtually unchanged after millions of years of stabilizing selection. *(a)* Maidenhair tree *(Ginkgo biloba)*—unchanged for 100 million years. *(b)* Chambered nautilus *(Nautilus)*—*unchanged for 180 million years.* *(c)* Horseshoe crab *(Limulus polyphemus)*—unchanged for 360 million years. *(Inset*—145-million-year-old fossil from Germany.)*

polymorphism because more than one phenotype is naturally selected and therefore persists over time.

Balanced polymorphism also results when heterozygotes are naturally selected over homozygotes. This is the case with sickle-cell anemia, a genetic disease that deforms human red blood cells. In tropical zones of Africa, heterozygotes (*Aa*) have a greater chance of surviving than either homozygotic pair (*AA* or *aa*) because the recessive

allele increases resistance to malaria. Selection for heterozygotes maintains the recessive allele at a high level, even though the homozygous recessive (*aa*) combination is fatal.

## SEXUAL SELECTION
Not all naturally selected adaptations improve an individual's chances of survival. The spectacular tail feathers of a peacock and the cumbersome antlers of male deer can actually

hinder the individual (Figure 29-14). But since these adaptations improve the chances of mating and reproducing, they are naturally selected. This form of natural selection is called **sexual selection**.

Sexual selection often leads to **sexual dimorphism**, visibly different males and females of the same species (Figure 29-15). Sexual dimorphism promotes mating by advertising gender to the opposite sex.

## The Origin of Species

Mutation, genetic drift, gene flow, and the various types of natural selection cause evolution to take place by changing the frequency of alleles in a population over time. When the characteristics of a population change sufficiently, a new species evolves.

New species evolve only through the modification of old species. There are two ways this can happen.

1. A new species evolves when the characteristics of an ancestral species change so much that it becomes a different species. (One species turns into another).

2. New species evolve as an offshoot of an ancestral species (one species becomes two or more species), a process called **speciation**. Of these two, only speciation increases the diversity of organisms (the number of species) and accounts for the millions of species that have populated the earth.

## Isolating Mechanisms

Generally, the members of different species cannot interbreed; in other words, they are **reproductively isolated**. But how do the members of the same species ever become reproductively isolated to the point of diverging into two or more new species?

Before speciation can occur, some members of a population must become isolated from other members, forming two separate gene pools. Isolation must be complete so that there is no gene flow between the separated gene pools. Once gene pools are isolated, they diverge because of differences in mutation, genetic drift, and natural selection. Over time, the isolated populations amass morphological and behavioral differences that eventually prevent them from interbreeding, so that even if reunited, the populations remain reproductively isolated. When populations are morphologically distinct and reproductively isolated, they are different species.

Barriers that prevent the exchange of alleles (gene flow) are called **isolat-**

Inedible butterfly model: *Danaus chrysippus*; *Amauris niavius*

Swallowtail female mimic: Female mimic (*P. dardanus*); Female mimic (*P. dardanus*)

Nonmimicking female

**Figure 29-13**
**Altered states.** Female African swallowtail butterflies mimic the appearance of local foul-tasting butterfly species (the models), creating strikingly different female phenotypes even though they all belong to the same species. These different forms of females are examples of diversifying selection and balanced polymorphism.

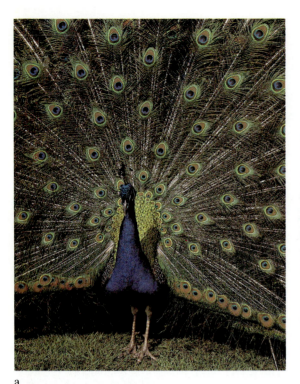

a                                        b

**Figure 29-14**
**Show and tell**. Although they may actually hinder the individual's mobility, brilliant displays of feathers *(a)* or racks of antlers *(b)* are products of sexual selection for characteristics that increase chances of mating.

**Figure 29-15**
**Sexual dimorphism** in sealions and mallard ducks.

**ing mechanisms**. Isolating mechanisms are often grouped into three categories based on whether isolation occurs before mating (Pre-mating Isolating Mechanisms), prevents mating (Sexual Isolating Mechanisms), or prevents fertilization, embryo development, or the growth of fertile offspring (Post-mating Isolating Mechanisms). Examples of each group are presented in Table 29-2.

## Mechanisms of Speciation

There are three types of speciation depending on how close the isolated populations are located to each other.

- **Allopatric speciation** (allo = other, patri = habitat) results when a physical barrier (geographic isolation) separates parts of a population, cutting off allele exchange. Most species are produced by allopatric speciation.

- **Parapatric speciation** (para = beside) occurs in populations that lie adjacent to each other. Gene pools diverge either because gene flow is reduced (possibly due to seasonal, mechanical, or gamete isolation), or because the environment varies sufficiently that different traits are selected in each isolated population.

- **Sympatric speciation** (sym = same) occurs in populations with overlapping distributions. *Polyploidy,* the multiplying of chromosome number, is one way that sympatric speciation may occur. The only way polyploid individuals can sexually reproduce is if they find another comparable polyploid (an extremely low probability), or if they can self-fertilize or reproduce asexually. Since many plants are self-fertile *and* are capable of asexual reproduction, polyploidy has been very important in plant evolution and speciation. More than 40 percent of flowering plant species living today are polyploids. The formation of viable hybrids (offspring produced from matings between parents of different species) is another mechanism for sympatric speciation. Many species of California manzanitas are products of hybridization.

TABLE 29-2

| Isolating Mechanisms |
|---|

I. Pre-mating Isolating Mechanisms
- *Ecological Isolation*—Different habitat requirements separate groups, even though they occur in the same location. Example: Head and body lice are morphologically very similar, yet live in different "habitats" on a single human body. Head lice live and lay eggs in the hair on the head of a human, whereas body lice live and lay their eggs in clothing. Both suck blood for nutrition.
- *Geographical Isolation*—Emerging mountains, islands, rivers, lakes, oceans, moving glaciers, or other geographic barriers keep groups isolated. Example: Different tortoises are found on different Galapagos Islands; surrounding oceans keep tortoise populations isolated.
- *Seasonal Isolation*—Differences in breeding seasons prevent gene flow, even when populations occur in the same area. Example: Two populations of bigberry manzanitas grow close together in the mountains of southern California, yet do not interbreed because one completes blooming two weeks before the other begins to bloom.

II. Sexual Isolating Mechanisms
- *Mechanical Isolation*—Physical incompatibility of genitalia. Example: Genital structures differ in shape for alpine butterfly species, even though these butterflies look nearly identical in all other ways.
- *Behavioral Isolation*—Differences in mating behavior prevent reproduction. Example: Many animals have evolved complicated courtship activities before breeding. Some species of fruit flies *(Drosophila) are indistinguishable to our eyes, yet do not reproduce with each other because of differences in courtship behavior.*

III. Post-mating Isolating Mechanisms
- *Gamete Isolation*—Sperm and egg are incompatible. Gamete isolation is a common isolating mechanism in many plant and animal species.
- *Hybrid Inviability*—Zygotes or embryos fail to reach reproductive maturity. Example: Hybrid embryos formed between two species of fruit flies fail to develop.
- *Hybrid Sterility*—Fertilization is successful between two species, but hybrid progeny are sterile. Example: A mule is a sterile hybrid from a mating between a horse and donkey.

## Patterns of Evolution

There is more than one pattern of evolutionary change. Each pattern can be illustrated by a graph in which morphological changes are plotted against time (Figure 29-16). The resulting graph represents an ancestral *lineage* or "evolutionary tree."

### Phyletic Evolution

The gradual transformation of one species into another is called **phyletic evolution**. Phyletic evolution occurs when the morphology of a species gradually shifts (in small steps) to the point at which new species are formed.

Although only a few complete examples of phyletic evolution can be found in the fossil record (see Figure 29-2), many biologists believe that phyletic evolution produced the majority of species. The scarcity of fossil evidence for phyletic evolution may simply be due to the incompleteness of the fossil record itself, which only records forms that can become fossilized, a small fraction of the earth's species. In addition, it has been estimated that the fossil record usually represents only one step in a thousand (or more)—certainly not enough to

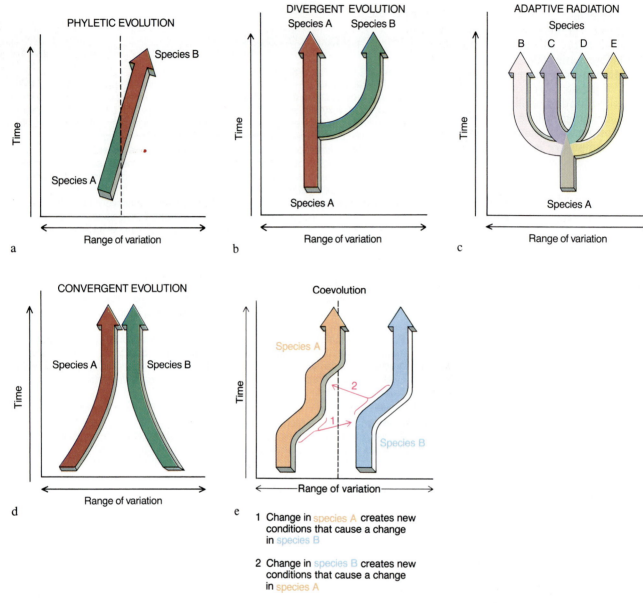

**Figure 29-16**
**Patterns of evolutionary change.** *(a)* Phyletic evolution—one species changes into another species. *(b)* Dizergent evolution—one species produces one or more new species. *(c)* Adaptive radiation—one species changes into many new species. *(d)* Convergent evolution—different species develop similar characteristics. *(e)* Coevolution—recitrocal changes.

document all of the gradual changes that might occur during phyletic evolution.

Some evolutionists, however, question whether phyletic evolution is adequate to fully explain all evolutionary changes. They argue that the abrupt changes in the fossil record may indeed depict evolution accurately. This has led them to propose another pattern of evolutionary change known as the punctuated equilibrium theory.

## Punctuated Equilibrium

The relatively sudden appearance of new species followed by long periods of little or no change, both of which are seen in the fossil record, may actually reflect how some evolutionary changes occur. The **punctuated equilibrium** theory for evolution has been proposed to explain this pattern in the fossil record.

According to this theory, episodes of speciation are followed by long periods (3 million years, on an average) when species do not change, a period of *species stasis* (Figure 29-17). In contrast to phyletic evolution, no intermediate forms would be produced. But what could spark these speciation episodes?

Spurts of speciation could occur when a dominant group of organisms became extinct, thereby opening new opportunities for other groups to

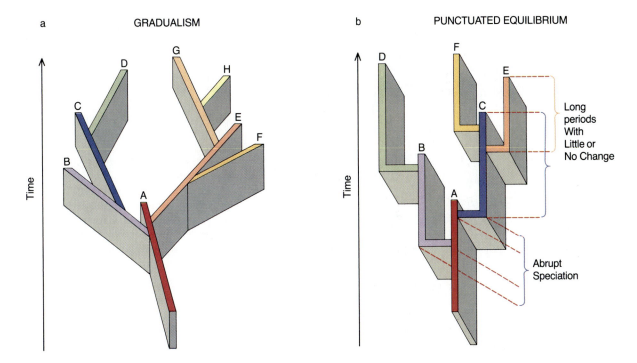

a  GRADUALISM

b  PUNCTUATED EQUILIBRIUM

**Figure 29-17**
**Gradualism vs Punctuated Equilibrium.** a. During gradualism, species arise through gradual accumulation of small changes. b. During punctuated equilibrium, species arise abruptly and remain unchanged for long time periods.

flourish. This is what happened about 60 million years ago when the dinosaurs became extinct and the surviving mammals flourished.

Spurts of speciation could also be caused by an abrupt environmental change. Individuals in a population with characteristics that were fortuitously preadapted to the change would survive and reproduce. These survivors might act as ancestors for a number of new species with slight variations, and yet still retain the adaptive trait(s). Speciation episodes might also occur when a founding population colonizes a new site rich with environmental opportunities. A diversity of environments would select for a diversity of phenotypes, eventually leading to the formation of many new species. But perhaps more frequently than any other cause, an abrupt episode of speciation would result when a *key characteristic* appeared in a population.

This characteristic would provide a new degree of success so that the owners would quickly spur a whole cluster of new species. Key characteristics include the evolution of flowers the modification of gill arches into bony jaws, the clustering of sensory structures into a head and brain, and the evolution of a shelled egg. Such key characteristics may have been responsible for producing completely new groups of organisms, or what taxonomists call higher taxa (genera, families, orders, phyla or divisions, and kingdoms). For example, the evolution of the flower spawned an entire new division of plants, the Anthophyta, and the evolution of a shelled egg led to the evolution of reptiles from amphibians. Major key characters and their role in evolution are discussed further in the next chapter.

Although gradual phyletic evolution and punctuated equilibrium are

alternative explanations for how new species evolve, one does not necessarily exclude the other as an evolutionary trend. The question now being debated among many biologists is whether one occurs more frequently than the other, and whether one better explains the enormous diversity of life forms that have existed since the beginning of life 3.5 billion years ago.

## Divergent Evolution and Adaptive Radiation

**Divergent evolution** occurs when one or more new species emerge as branches from a single ancestral lineage (Figure 29-16b). When many species diverge from a single ancestral line, the pattern is called **adaptive radiation** (Figure 29-16c).

There are many examples of adaptive radiation. One well-documented example occurred on the Galapagos Islands (Figure 29-18). When the is-

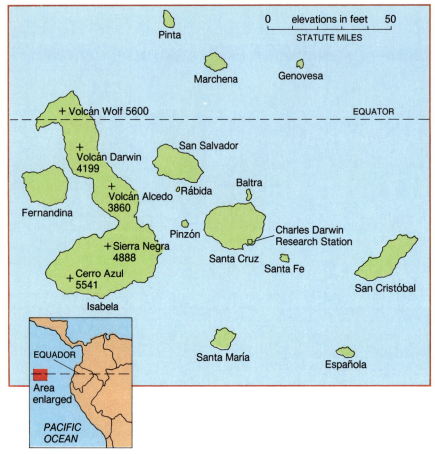

**GALAPAGOS ISLANDS**

**Figure 29-18**
**The Galapagos Islands**—the site of adaptive radiations that produced a rich variety of species from just a few ancestors that originated on mainland South America.

lands formed about 2 million years ago, new sets of environmental opportunities opened up for organisms that could swim, float, fly, or be blown by wind from the mainland of South America. When Charles Darwin visited the Galapagos Islands during the voyage of the *HMS Beagle,* he was particularly struck by the variety in a small group of birds, now known as "Darwin's finches" (Figure 29-19). The diversity of available habitats on the islands sparked the adaptive radiation of 14 species of finches from a single ancestor.

## Convergent Evolution

Organisms that occupy very similar kinds of environments frequently resemble one another, even though they are not closely related. Similar environments impose similar selective pressures that can foster the development of similar adaptations. This development of superficial similarities in different lineages is called **convergent evolution** (Figure 29-16*d*).

There are many examples of convergent evolution. The wings of birds and insects are similar adaptations for flight, yet these animals have very different evolutionary origins. Wings are *analogous* in birds, bats, and insects; that is, they are structures that perform similar functions, but did not originate from a common ancestor. (Recall that *homologous structures* are similar because of common ancestry.) The evolution of succulent tissue and the absence of leaves are adaptations to hot, dry environments that have evolved in distantly related plant families.

The similarity between many Australian marsupials (pouched mammals) and many placental mammals on other continents is an example of convergent evolution. Except for the American opossum, marsupials have been restricted to Australia for at least 75 million years. Since the Australian continent has habitats that are similar to those on other continents, marsupials have evolved characteristics that are very similar to distantly related placental mammals. There are, for example, a marsupial "anteater," "wolf," and "flying squirrel" (Figure 29-20).

## Coevolution

Because organisms form part of the natural environment, they too can act as a selective force in the evolution of a species. In nature, species frequently interact so closely that evolutionary changes in one cause evolutionary adjustments in others. This evolutionary interaction between organisms is called **coevolution** (Figure 29-16*e*).

Flowering plants and their insect pollinators have coevolved for millions of years, leading to many finely tuned structural and behavioral relationships between flowers and pollinators (Chapter 21). Another example of coevolution is found in predator–prey interactions, where improvements in the hunting ability of a predator favor the survival of prey with characteristics that increase its ability to escape. (Predator and prey interactions are discussed in Chapter 37.)

## Extinction

It may seem odd to call **extinction**, the loss of a species, an evolutionary trend. But considering that the 2 million different species living today probably represents less than 10 percent of the total number of species that have existed since the beginning of life on earth, extinction is undoubtedly *the* major trend in evolution. It is the ultimate fate of most, if not all, species.

During the history of life on earth, there have been periods of mass extinctions caused by radical environmental changes, widespread disease, collisions with meteors, and changes in magnetic fields, to mention a few.

**Figure 29-19**
**Darwin's finches.** All 14 species of finches inhabiting the Galapagos Islands (and nearby Cocos Island) are thought to be products of adaptive radiation from a single ancestoral finch species. The finches differ primarily in beak size and shape. These distinctions tailor them to different habitats and food sources. They include seed-eating ground finches, nectar-feeding finches with long, curved beaks, insect-eating finches, and one remarkable tool-using tree finch, the woodpecker finch, which uses its beak to bore into wood in search of insect larva and then uses a cactus spine or twig to excavate its prey.

**Figure 29-20**
**Convergent evolution**. Australian marsupials and placental mammals on other continents have similar features because they have adapted to very similar habitats. (Placental mammals are on the left and the Australian marsupial counterpart is on the right.)

But until recently, most species became extinct because of their inability to change and adapt to new conditions. They were genetically "frozen" by the limited variability of their gene pool. The situation is different today, however. Our unbridled destruction of natural habitats has increased the extinction rate from a long-term average of about one species each 1000 years, to hundreds and perhaps thousands of species in a single year. Such an alarming extinction rate not only destroys potentially beneficial species, but will also likely disrupt the balance of nature, causing even more species to become extinct. Many scientists are concerned that the effects of the accelerated extinction rate will soon have detrimental effects on all life on earth, including our own. We will likely witness some of the disturbing effects of extinction during our own lifetime.

# Synopsis

## MAIN CONCEPTS

- **Evolution is the process by which living things become modified over generations.** It is the major unifying concept in biology, for it explains how there came to be so many different life forms, as well as how even very dissimilar organisms (a bacterium and a human, for example) have similar chemical characteristics —illustrating the unity of life.

- **The basis of evolution is a change in the genetic makeup of a gene pool.** Changes in genes and alleles modify the characteristics of individuals over generations.

- **Changes in allele frequencies in a gene pool** result from mutation (adding new alleles or genes), gene flow (transfer of alleles between populations), genetic drift (random shifts in allele frequency), and natural selection. Better suited characteristics accumulate over generations, whereas less suited characteristics decrease.

- **Natural selection is a major guiding force in evolution** that gradually leads to adaptation.

- **New species arise when isolated populations diverge.** Isolated populations eventually develop characteristics or behaviors that prevent individuals from each population from interbreeding.

- **The major evolutionary trends include**
  Gradual changes over long periods of time.
  Abrupt episodes of speciation followed by long periods without change.
  The splitting of an ancestral lineage to form new species.
  The development of similar adaptations in similar environments by distantly related organisms.
  Mutual adjustments in lineages that interact with each other.
  Extinction, the loss of a species.

## KEY TERM INTEGRATOR

Evolution: The Process of Change

| | |
|---|---|
| **evolution** | The change in **allele frequency** in a **gene pool** over generations. Evidence for evolution includes similarities in biochemistry, morphology, and embryology, a record of changes in fossils, **vestigial organs**, **radioisotope dating**, and **homologous structures**. Our use of **selective breeding** proves species can change and favorable traits can become accentuated over generations. |
| **natural selection** | One of the four forces that cause changes in allele frequencies. The other three forces are mutation **mutation**, **genetic drift**, especially during **bottlenecks** and the **founder effect**, and **gene flow**. |

Types of Natural Selection

| | |
|---|---|
| **stabilizing, directional, and diversifying selection** | The three possible outcomes of natural selection. Diversifying selection can lead to **dimorphism** or **balanced polymorphism** in a species. **Sexual selection** for mating characteristics often leads to **sexual dimorphism** between the sexes. |

## The Origin of New Species

| | |
|---|---|
| **speciation** | One species giving rise to two or more **reproductively isolated** new species. **Isolating mechanisms** prevent gene flow between populations and eventually lead to speciation. **Allopatric speciation** between physically separated populations is the most common mode of speciation. **Parapatric speciation** occurs between neighboring populations, and **sympatric speciation** occurs in populations that overlap. |

## Evolutionary Patterns

| | |
|---|---|
| **phyletic evolution** | One species gradually transforming into another. |
| **punctuated equilibrium** | Rapid evolution of clusters of species that remain unchanged for relatively long periods. |
| **divergent evolution** | The formation of one or more species from a single ancestor. One form of divergent evolution is **adaptive radiation.** |
| **convergent evolution** | The development of similarities between unrelated species in response to similar environmental conditions. |
| **coevolution** | Reciprocal evolutionary changes between species. |
| **extinction** | Loss of a species. |

# Review and Synthesis

1. Complete the following graphic summary for evolution by filling in appropriate terms:

| | | |
|---|---|---|
| gene pool | natural selection | allopatric |
| bottleneck | genetic drift | gene flow |
| sympatric | parapatric | mutation |

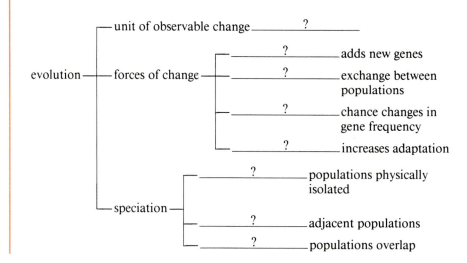

**2.** Match the example with the evolutionary mechanism.

___1. dimor-     A. predator/prey interactions
     phism
___2. coevolution    B. Darwin's finches
___3. gene flow    C. distinctive males and females
___4. stabilizing    D. immigration
     selection
___5. adaptive ra-   E. living fossils
     diation

**3.** Of the four factors that cause allele frequency to change over time, why does only natural selection lead to increased adaptation?

**4.** Why is natural selection based only on inherited variations in characteristics? Could characteristics that are not inherited ever affect the course of evolution? How?

**5.** Although initial pesticide applications kill most of the insects in a population, a few are resistant, survive the pesticide, and pass on the "favorable" resistant gene to their offspring which quickly repopulate the area with pesticide-resistant individuals. To control the pests, farmers must apply stronger doses of pesticides at more frequent intervals. The escalating war against agricultural pests is an example of which type of evolutionary change? Which organism do you suppose will be the ultimate winner of this battle, the farmers or the pests? Why?

**6.** Why must gene flow stop before speciation is possible?

**7.** List the ways in which the fossil record provides evidence for evolutionary change.

**8.** Why are the theories of gradual phyletic evolution and punctuated equilibria not mutually exclusive? What is the basis for the debate between proponents of each?

## Additional Readings

Gould, S. J. 1977. *Ever Since Darwin.* W. W. Norton, New York. (Introductory.)

Lewin, R. *Thread of Life: The Smithsonian Looks at Evolution.* Smithsonian Books, Washington, D.C. (Introductory.)

Mayr, E. 1977. "Darwin and natural selection." *American Scientist* 65:321–27.

Miller, J. and B. Van Loon. 1982. *Darwin for Beginners.* Pantheon Books, New York. (Introductory.)

*Scientific American. Evolution.* 1978. W. H. Freeman, San Francisco. (Introductory.)

Stebbins, G. L. 1982. *Darwin to DNA, Molecules to Humanity.* W. H. Freeman, San Francisco. (Intermediate.)

Chapter 30

# Origin and History of Life

*I*t is one of the greatest detective stories of all time. There were no eyewitnesses, and the detectives were not on the scene until long after the incident had happened. Although a host of clues had been left behind, all the clues had become modified over the years; virtually everything had changed.

What is this baffling case? It is nothing less than the origin and history of life on earth. The detectives in this investigation are scientists from the fields of astronomy, geology, chemistry, physics, and biology. After years of study, the emerging picture of how the earth was formed, how life began, and how life evolved into what it is today is becoming clear. The consensus among scientists is that it all began nearly 4.6 billion years ago with the earth's formation.

## Formation of the Earth

Our solar system was once part of a massive cloud of interstellar gases and cosmic dust (Figure 30-1). As particles of matter pulled together by gravitational attraction, the vast cloud condensed into a gigantic spinning disk. Nearly 90 percent of the matter gravitated to the center, causing temperatures to rise high enough to ignite thermonuclear reactions there. Our sun began to shine.

At the same time the sun was forming, smaller eddies of leftover gases and dust condensed into the nine planets of our solar system. Heavier elements such as iron and nickel sank to form the cores of the planets while light gases floated near the surface. For the planets nearest the sun (Mercury, Venus, Earth, and Mars), most of the lighter gases were blasted away when the sun's thermonuclear reactions began, leaving shrunken, dense planets with virtually no gaseous atmospheres.

On the primitive earth, immense quantities of heat from gravitational contraction, radioactive decay, impacts of meteorites, and solar radiation turned it into a red-hot, molten orb. As time passed, collisions with meteorites became less frequent, and the heat from radioactive decay and gravitation contraction lessened. The earth's surface slowly cooled and a thin crust of crystalline rock formed. Below the crust, the enormous heat of the earth's interior produced massive buildups of hot gases, sparking violent volcanic eruptions that thrust molten rock and gases out through the crust. Repeated eruptions gradually built the

**Figure 30-1**
**Formation of our solar system.**

earth's rugged land masses and filled the once empty atmosphere with clouds of hot gases and steam. The steam condensed and fell back to the earth's surface as rain, first collecting into small ponds, and later forming the earth's oceans, lakes, rivers, and streams. It was in the early ponds that life emerged.

# Origin of Life

Earth is just a little planet, circling a very ordinary star (our sun). Yet today, this seemingly unremarkable planet teems with millions of different life forms. What conditions make a planet suitable for life? To be able to sustain life as we know it,

- A planet must orbit a star at the right distance so that temperatures will not continually exceed 100° C or remain below 0° C. These are the temperature extremes at which water exists as a liquid, unless under very high pressure. Life as we know it requires liquid water.

- The planet's orbit must be nearly circular, maintaining a constant distance from the star and therefore a somewhat stable temperature range within the limits that permits life to occur.

- The planet must be large enough for its gravity to hold a gaseous atmosphere, yet not so large that its gravity creates too dense an atmosphere for sunlight to penetrate.

Within our solar system, only the earth meets all of these criteria.

## Chemical Evolution

Although distance from the sun, orbit, and planet size are prerequisites for the evolution of life as we know it, these conditions did not guarantee that life would eventually appear on earth. The dawn of life had to await **chemimal evolution**—the spontaneous synthesis of increasingly complex organic compounds from simpler molecules.

A variety of organic molecules have been found in samples of interstellar dust, comets, and meteorites, revealing that chemical evolution occurs throughout the universe. The Murchison meteorite that hit Australia in 1969, for example, contained hydrocarbons, fatty acids, and over 100 amino acids. (Recall that life on earth is based on only 20 amino acids.) In addition, scientists are now able to directly observe large numbers of complicated organic molecules in the universe using radiotelescopes, a field of study known as cosmochemistry.

Organic compounds are constructed primarily of four, highly reactive atoms (carbon, hydrogen, oxygen, and nitrogen); a source of energy causes these reactive gases, if present together, to spontaneously assemble into complex organic compounds, regardless of where in the universe they are. But where did the gases that contained these atoms and energy come from for chemical evolution to take place on earth?

In the 1930s two biochemists, J.B.S. Haldane in Scotland and A.P. Oparin in Russia, presented virtually identical hypotheses for how organic compounds could have formed on the primitive earth. Both proposed that the earth's early atmosphere was very different from the one enveloping earth today. Instead of the nitrogen and oxygen gas that comprises most of today's atmosphere, Haldane and Oparin speculated that the earth's early atmosphere was a reducing atmosphere containing a mixture of carbon monoxide (CO), carbon dioxide ($CO_2$), water vapor ($H_2O$), nitrogen gas ($N_2$), and large amounts of hydrogen gas ($H_2$). If the earth's early atmosphere contained oxygen, as does today's atmosphere, atoms and compounds would be oxidized before they could assemble into organic molecules. Life could therefore not emerge in an atmosphere that contained free oxygen.

The earth's early atmosphere would have also contained ammonia ($NH_3$), methane ($CH_4$), and hydrogen sulfide ($H_2S$), all of which are complexes of hydrogen and the other gases. Energy from lightning, ultraviolet rays, and enormous amounts of heat from lava flows and collisions with meteorites ionized these gases briefly, freeing carbon, hydrogen, nitrogen, and oxygen. These ions immediately reacted with each other to form amino acids, nucleotides, hydrocarbons, lipids, and fatty acids, the organic building blocks of life (Figure 30-2). The organic molecules were then washed out of the atmosphere by rain and accumulated in scattered ponds and shallow seas, eventually forming an increasingly rich "organic soup."

Experiments conducted by Stanley Miller and Harold Urey in 1953 reinforced Haldane and Oparin's proposal. In a closed reaction vessel, Miller and Urey repeatedly jolted a mixture of hydrogen gas, ammonia, methane, and water vapor with electricity to simulate lightning (Figure 30-3). After only one week, they found a number of organic compounds in the reaction vessel, including eight amino acids. Among these amino acids were four commonly found in organisms— glycine, alanine, aspartic acid, and glutamic acid.

Since the experiments of Miller and Urey, atmospheric geophysicists have uncovered evidence suggesting that the earth's early atmosphere may have contained a mixture of gases different from that proposed by Haldane and Oparin. Nonetheless, all new proposals still hypothesize an early reducing atmosphere without oxygen. With this new information, geophysicists argue that volcanic eruptions would produce an atmosphere of nitrogen gas, carbon dioxide gas, and water vapor. Nevertheless, experiments verify that organic compounds spontaneously assemble using any of the gas mixtures proposed for the earth's primitive atmosphere. The resulting organic compounds included all 20 amino acids found in organisms, as well as various sugars, vitamins, and nucleotides (the precursors of RNA and DNA). Even ATP, the universal energy carrier of living systems, forms spontaneously if inorganic phosphate is added to the gas mixtures. The fact that many dozens of independent laboratory experiments—many using different mixtures of gases and energy sources—have all been able to produce organic molecules reinforces the hypothesis that the organic molecules necessary for life spontaneously assembled in the earth's primitive atmosphere. (As observed in Chapter 2, repeatability of results is central to the scientific method.)

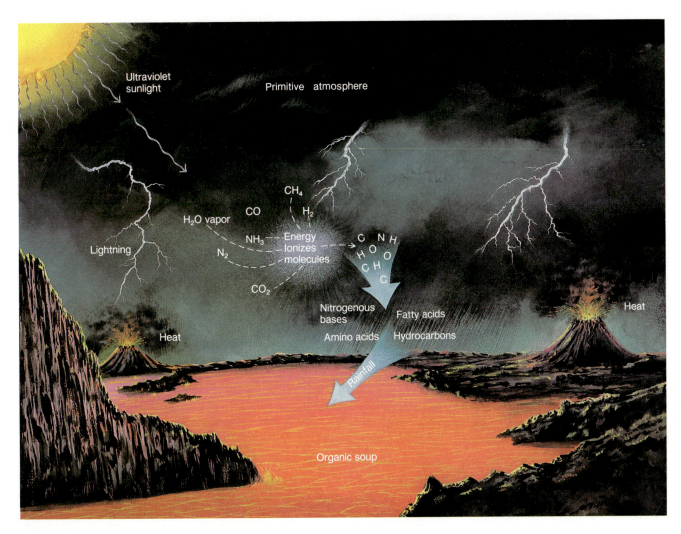

**Figure 30-2**
**Violent beginnings.** Continuous volcanic eruptions poured out lava and gases, building the earth's crust and primitive atmosphere. Plentiful energy from volcanic eruptions, ultraviolet light, and lightning ionized the earth's atmospheric gases. They recombined into a variety of complex "organic" compounds. These compounds were washed out of the atmosphere by rainfall and accumulated in shallow ponds, forming an "organic soup" of the molecular building blocks of life.

Clearly, then, scientists can demonstrate that fundamental organic molecules spontaneously assemble, especially when exposed to the conditions present on the primitive earth. But how were these simple organic molecules brought together to form the more complex polymers of life, such as proteins and nucleic acids, the next step in chemical evolution? Haldane and Oparin suggested that organic compounds accumulated in the organic soup, eventually building to concentrations high enough to permit the synthesis of complex compounds.

Today, most scientists doubt that such high concentrations could have been produced within the soup.

Recent experiments offer new explanations for this stage of chemical evolution as scientists edge closer to restructuring the chemical events that led to the origin of life. In experiments performed in the 1960s, Sydney Fox demonstrated that if amino acids were dry and hot, they joined together to form **proteinoids** (-oid = like, or resembling), complex polymers made up of hundreds of amino acids. (The term "proteinoid" is used because,

like proteins, these macromolecules are constructed of amino acids, but they are unlike any proteins we know today.) On the primitive earth, dry, hot amino acids may have formed along the edges of ponds near hot volcanoes or in hot evaporated pools. The dry, hot amino acids then combined into proteinoids. Scientists also speculate that it was probably also along the edges of ponds or in evaporated pools that spontaneously assembled nucleotides became concentrated, allowing for the formation of nucleic acids (RNA and DNA).

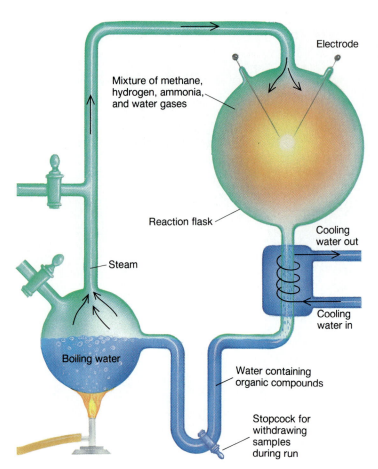

Mixture of methane, hydrogen, ammonia, and water gases

Electrode

Reaction flask

Steam

Cooling water out

Cooling water in

Boiling water

Water containing organic compounds

Stopcock for withdrawing samples during run

**Figure 30-3**
**Reenacting chemical evolution.** In 1953, Stanley Miller and Harold Urey simulated the earth's primitive atmosphere by mixing steam with methane, hydrogen, and ammonia gases in a reaction flask. Discharge from electrodes simulated lightning storms. The products of spontaneous reactions condensed and collected in the U-shaped tube (lower right). When the products were removed for analysis, the researchers found that a number of amino acids and other organic compounds were synthesized under conditions similar to the earth's primitive atmosphere.

## Evolution of Life

It may seem like a giant step to go from a pond with RNA, DNA, proteins, fatty acids, and hydrocarbons to a living cell capable of metabolism and reproduction. However, it may not be as big a step as we once believed. From radioisotope dating we know that the earth is about 4.6 billion years old and that the oldest fossils of organisms are some 3.5 billion years old (Figure 30-4). Yet the complexity of these ancient fossils shows that they were too advanced to be the earth's original organisms. They descended from ancestors that may have evolved within the earth's first 600 million years. The fact that life evolved relatively "soon" after the earth formed has led many biologists to conclude that life may not be, after all, such an unusual phenomenon in the universe (see Bioline: Is There Life on Other Planets?).

Nevertheless, the leap from organic soup to the simplest living thing is a big one, for even the very simplest cell is highly organized and carries out complicated metabolic reactions. Before the first cell could have evolved, one or more types of simpler **precells** (pre = before) would have formed. Precells would be able to concentrate organic molecules, allowing molecules to react more frequently. Within the precell, catalysts would direct the synthesis of specific organic polymers; in other words, the precell would be conducting metabolism. As metabolism produced new organic molecules, the precell would grow. The eventual evolution of a genetic code would enable the precell to pass on naturally selected codes for metabolism, allowing the precell to reproduce itself. At this point, the precell would possess three of the basic characteristics of life — metabolism, growth, and reproduction — crossing the line between precell and living cell.

## LABORATORY-PRODUCED PRECELLS

Scientists are trying to produce a precell in the laboratory to demonstrate how life started on earth, but so far only a few precell candidates have been produced. For instance, polymer-rich droplets called **coacervates** can be created by combining proteins and nucleic acids. These droplets are similar in size to a cell and, when supplied with appropriate enzymes, can "grow" and "reproduce" by carrying out simple forms of metabolims (Figure 30-5a).

Laboratory-produced proteinoids have also been proposed as candidates for precells. Proteinoids have thick protein membranes and are easily produced by heating a mixture of dry amino acids and then plunging the mixture into water (Figure 30-5b). On the primitive earth, proteinoids could have formed as extreme heat from nearby lava dried amino acids and

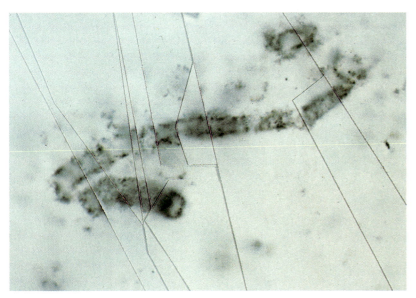

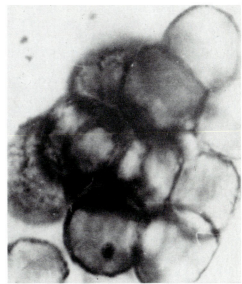

a                                                                              b

**Figure 30-4**
**Imprints of ancient life.** *(a)* The oldest fossil — a 3.5-billion-year-old cast of a filamentous alga from western Australia. *(b)* A 2.5-billion-year-old fossil of an algae colony, perhaps representing the first steps toward the evolution of multicellular organisms.

condensed them into polymers. These polymers would be "plunged into water" when it rained. Proteinoids enlarge (grow) and bud (reproduce) in a manner similar to present-day bacteria.

Other precell candidates have been experimentally produced, but so far none has been formed that possesses all of the characteristics needed to qualify as a primitive cell. But the ease with which so many precells can be produced in the laboratory tells us that the earth's ancient ponds and oceans were probably filled with many kinds

of droplets. Scientists have simply not as yet deduced and combined the right ingredients to form a precell. The precell remains one of the "missing links" in evolution.

## EVOLUTION OF THE GENETIC CODE
The earliest forms of metabolism in a precell required some form of molecular controller so that specific, naturally selected organic polymers could be repeatedly produced. Recent experiments indicate that RNA, not DNA, was the molecule directing synthesis in

these early precells. These experiments reveal that RNA catalyzes many polymerization processes, including the synthesis of short polypeptide sequences. Because of its shape and structure, portions of RNA recognize specific amino acids. As amino acids bond to regions of RNA, they become aligned and then polymerized into unique polypeptide sequences. Because the "information" for the synthesis of specific polypeptides is stored in the RNA molecule itself, this information can be passed on from one generation to the next by making

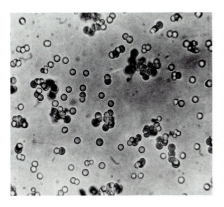

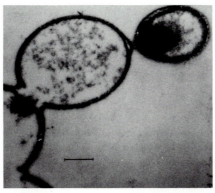

a                                         b

**Figure 30-5**
**Order out of chaos.** One of the first steps toward the evolution of life was the formation of a precell. Experimentally produced coacervates *(a)* and proteinoids *(b)* are analogous to a precell. They partition and concentrate organic compounds, have some catalytic ability, and divide into "daughter" droplets when they grow beyond a stable size.

copies of the RNA molecule and providing each of the progeny with a copy. This was earth's first genetic system.

But where did the RNA come from? Recall that laboratory experiments show that the monomers needed to construct RNA (nucleotides), as well as the other organic molecules needed for life, spontaneously formed and then polymerized in the earth's early atmosphere.

# The History of Life on Earth

The first living cell appeared on earth between 3.5 and 4.6 billion years ago. Life remained unicellular for 3 billion years. For the first 2 billion years, only simple prokaryotes populated the primitive earth. The evolution of single-celled eukaryotes paved the way for the evolution of the first multicellular organisms a billion years later. The multicellular plan was a tremendous success, leading to a proliferation of complex life forms, some of which moved onto the land and eventually took to the air.

Today between 2 to 6 million species of organisms populate the earth. What caused millions of ancient species to change or to die out entirely? In the following sections we describe the events that constitute the history of life on earth.

## The Geologic Time Scale

Geologists divide the earth's 4.6-billion-year life span into four great eras: the Proterozoic (a period that produced few fossils), Paleozoic, Mesozoic, and Cenozoic Eras. With the exception of the Proterozoic Era, remaining eras are divided into periods and epochs on the basis of geological formations and the fossil record. The resulting *geologic time scale* and some of the major events in the history of life are briefly set forth in Figure 30-6.

## Life During the Proterozoic Era

Chemical evolution and the transformation of a precell into a living cell took place during the Proterozoic Era between 0.5 and 4.6 billion years ago. The earliest living organisms were

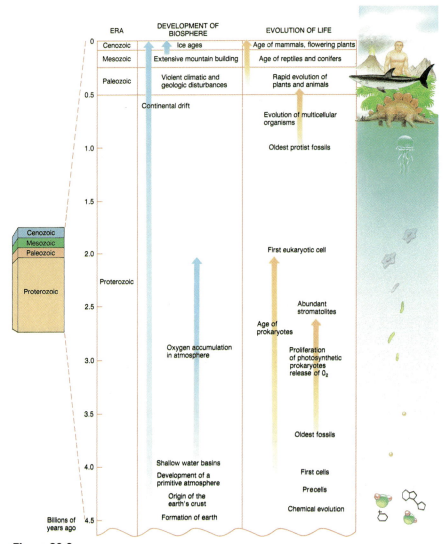

**Figure 30-6**
**The geologic time scale.** The 4.6-billion-year lifespan of the earth is divided into four eras based on changes in the fossil record and rock formations. Major events in the development of the biosphere dramatically affected the evolution of life.

# B I O L I N E

## IS THERE LIFE ON OTHER PLANETS?

**M**any scientists believe that extraterrestrial life is inevitable; indeed, it may even be common in the universe.

The universe is composed of the same kinds of atoms, in essentially the same ratios. The physical and chemical laws that govern their behavior are also the same; atoms react in predictable ways to form molecules, and when supplied with energy, molecules naturally react to form more complex compounds, such as amino acids, sugars, and nucleic acids. Samples of meteorites verify that the organic compounds fundamental to life exist elsewhere in the universe. If conditions for chemical evolution are similar throughout the universe, then why would life have evolved only on earth? Many *exobiologists* (biologists who study the possibility of extraterrestrial life) believe that life is not unique to earth. Some estimate that as many as 1 billion billion ($1 \times 10^{18}$) planets may be suitable for life as we know it.

How do exobiologists arrive at these estimates? Their reasoning goes something like this. There are $10^{20}$ (100 quintillion) known stars, each one a sun. About 20 percent of these stars are essentially identical to our sun in size and brightness. If even 1 percent of these have planets and only 1 percent of those planets are suitable for life as we know it (the correct size and distance from a sun), life might be possible on $10^{18}$ planets. If life exists in forms other than those found on earth, the likelihood of life elsewhere in the universe is boosted even higher. According to 1967 Nobel laureate George Wald, "We find ourselves in a life-breeding universe."

---

simple cells that contained nucleic acids (RNA), enzymes, and ribosomes. Because their genetic information was not complex enough to enable them to manufacture all the organic molecules they required for metabolism and reproduction, these early "chemical heterotrophs" were totally dependent on *abiotic* nutrients, the spontaneously synthesized amino acids, sugars, nucleotides, and ATP that already existed in the surrounding organic soup. With an abundant supply of these nutrients at first, these early heterotrophs proliferated at a rapid rate. But eventually they consumed the organic molecules faster than abiotic synthesis could replace them, causing competition for limited nutrients. Some cells evolved the ability to manufacture their own ATP through fermentation (Chapter 10), giving them a competitive edge over heterotrophs that relied solely on ATP formed in the soup. But even these fermentative heterotrophs depended on spontaneously formed organic compounds. The only organisms that

could be freed from the limitations of abiotically synthesized organic molecules were those that could manufacture their own organic nutrients. These were the earth's first autotrophs.

## EVOLUTION OF AUTOTROPHS

If it were not for the emergence of autotrophs, life might have ended 3.5 billion years ago. The first autotrophs not only supplied themselves with nutrients, but also became food for the heterotrophs. Without autotrophs, life would have remained simple and sparse, limited by the meager supply of abiotically synthesized organic food.

The earth's first autotrophs were probably very much like present-day anaerobic photosynthetic bacteria. They used the energy of sunlight to remove hydrogen atoms from hydrogen gas, hydrogen sulfide, ethanol, or lactic acid. They combined the liberated hydrogens with carbon dioxide to produce organic compounds. These compounds could then be used in glycolysis and fermentation to produce ATP

(Chapter 10).

But the metabolic success of anaerobic photosynthesizers soon gave way to another group of organisms that evolved the ability to extract hydrogen by splitting water, releasing the oxygen as a waste product. These new autotrophs were the first **cyanobacteria**, a 3-billion-year-old group of photosynthetic bacteria that continue to flourish today. Because water was an enormously plentiful resource, these ancient cyanobacteria flourished, tinting the ponds and seas green with their light-capturing photosynthetic pigments and releasing vast amounts of oxygen into the water and air.

Most other organisms were ill equipped to deal with oxygen's tendency to break down organic compounds (see Bioline: The Earth's Deadly Envelope in Chapter 10) and were either driven to extinction or forced into very limited, oxygen-free habitats.

The period between 2.0 and 4.0 billion years ago is referred to as the Age of Prokaryotes. Three evolutionary lines emerged during this time (Figure 30-7). One line, the Urkaryotes, gave rise to a new type of cell — the eukaryotic cell — which would set the stage for the evolution of multicellular life.

## FROM PROKARYOTIC TO EUKARYOTIC CELLS

Eukaryotic cells evolved about 2 billion years ago. With specialized organelles and a nucleus that sequestered genetic material, eukaryotic cells were more organized than their prokaryotic ancestors and were capable of conducting more complex functions. One complex function performed by eukaryotic cells was meiosis. Recall that meiosis leads to the formation of gametes for sexual reproduction (Chapter 12), a process that accelerates the rate of evolution by producing new gene combinations and spreading them quickly through a population. With the ability to reproduce sexually, eukaryotes quickly proliferated.

Early eukaryotes left no fossilized remains. In the absence of fossils, speculations on how eukaryotic cells evolved from simpler prokaryotic ancestors are based on the characteristics and lifestyles of modern unicellular eukaryotes. Here are two modern theories that may explain how this happened.

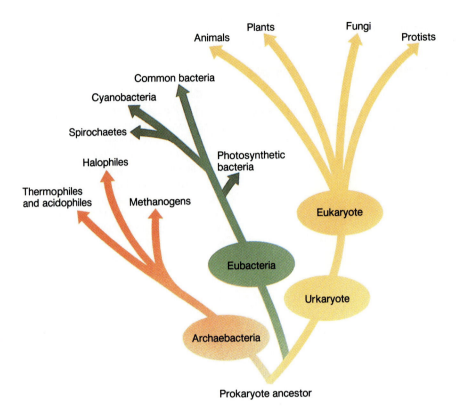

**Figure 30-7**
**The evolutionary tree for all modern groups of organisms.**

### Membrane Invagination Theory.

The **membrane invagination theory** proposes that the plasma membrane of the ancestral prokaryotic cell gradually folded in on itself, forming pockets that isolated specific metabolic reactions (Figure 30-8a). Some of these pockets pinched off, partitioning functions into membrane-enclosed organelles. Over long periods of time, these simple organelles increased in complexity and specialization, eventually forming the complex organelles found in modern eukaryotic cells.

It is easy to see how membrane invagination could have formed the endoplasmic reticulum, Golgi apparatus, and nucleus by the same infolding process. (The nuclear membrane of a eukaryotic cell is often connected to the cell membrane.) The evolution of complex organelles such as chloroplasts and mitochondria, however, is usually explained by another theory, the endosymbiosis theory.

### Endosymbiosis Theory.

The **endosymbiosis theory** proposes that some organelles were once free-living prokaryotic cells that took up residence inside a larger prokaryotic cell, forming a union that was beneficial to all (Figure 30-8b). (The name of this theory uses the word "symbiosis," a term that refers to a close association between different kinds of organisms. Such associations are very common in nature.) The endosymbiosis theory is supported by strong evidence:

- Some bacteria form symbiotic partnerships with eukaryotic cells today. These bacteria resemble mitochondria in structure and carry out the sames steps of aerobic respiration as mitochondria.

- Mitochondria and plastids contain their own nucleic acids, including a circular strand of DNA that closely resembles the chromosomes of prokaryotic cells.

- Mitochondria and plastids can "divide" independent of the cell in which they reside. Because they contain their own genetic information, replication is not solely under the control of the nucleus as it is for other organelles.

- Like an independent cell, a mitochondrion contains its own ribosomes. These ribosomes resemble those of prokaryotic bacteria more than those found in eukaryotic cells.

- Chloroplasts have the identical arrangement of pigments as found in some photosynthetic prokaryotes. Electron transport chains in chloroplasts are identical to those of photosynthetic bacteria.

These and other facts have led many biologists to accept the endosymbiosis theory as the explanation for the origin of possibly all organelles in eukaryotic cells. Even eukaryotic flagella (and possibly even cilia show evidence that

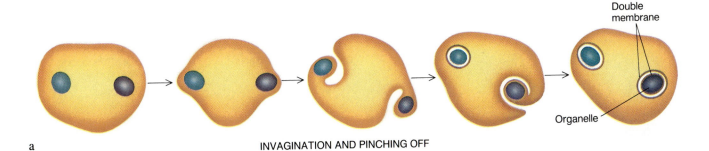

a

INVAGINATION AND PINCHING OFF

Double membrane

Organelle

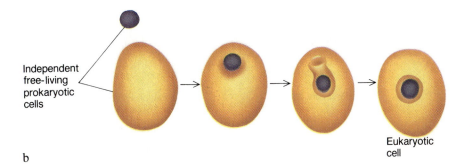

Independent free-living prokaryotic cells

Eukaryotic cell

b

**Figure 30-8**
**Proposals for the origin of eukaryotic cells.**
*(a)* Invagination theory. *(b)* Endosymbiosis theory.

they were once independent prokaryotic organisms.

## FROM UNICELLULAR TO MULTICELLULAR LIFE

Multicellular life began between 0.5 and 1.0 billion years ago. Multicellularity probably began when a number of single-celled eukaryotes aggregated into a colony. Although each cell was capable of surviving on its own if necessary, living in a colony offered advantages over living alone. For instance, it provides increased protection from being eaten and improved ability to locate food. Gradually, cells in the colony became specialized for specific functions. Some cells, for example, moved the colony about, others obtained and processed nutrients, and still others specialized in reproduction. With increased specialization, tasks could be conducted more efficiently. The multicellular plan proved so successful scientists have documented that multicellularity evolved independently more than 17 times.

## Life During the Paleozoic Era

The Paleozoic Era lasted for 345 million years. Conditions on earth changed dramatically on a number of occasions during this long period, causing episodes of mass extinctions, followed by diversification of organisms by adaptive radiation. Many of these changes were the result of shifts in the positions of the earth's continents (see Bioline: Continents Adrift). At times, the drifting continents created broad shallow seas; at other times, land masses became entirely submerged and later rose to form mountain ranges. These shifts in the continents caused changes in the earth's climate as well; as a result, extinction claimed some organisms while new forms evolved. Figure 30-9 illustrates the geologic and climatic changes that occured during the six periods of the Paleozoic Era and the effects these changes had on the evolution of plants and animals.

## Life During the Mesozoic Era

The end of the Paleozoic Era was possibly the most disastrous period in the history of life. Hundreds of thousands of life forms disappeared. Amid the catastrophe two major groups of land organisms (cone-producing gymnosperms and reptiles) survived and emerged to dominate terrestrial life during the Mesozoic Era.

The Mesozoic Era (65 to 240 million years ago) is divided into three periods: the Cretaceous, Jurassic, and Triassic (Figure 30-10). The Mesozoic is sometimes called the Age of Reptiles. Reptiles diversified and spread into all environments, on land, into the air, and even back into the water. On the land, reptiles evolved rapidly into a variety of shapes and sizes as drier environments became more common. A group of small, mammal-like reptiles appeared during the Triassic period. These animals were called the **therapsids**, the ancestors of mammals.

By the end of the Triassic period, another group of reptiles branched off and evolved into the largest land animals ever to roam the earth, the dinosaurs. The dinosaurs diversified in many spectacular ways, ruling the land for the next 125 million years. But the reign of dinosaurs collapsed in "sudden" extinction at the end of the Cretaceous period. How could the dinosaurs fall from dominance to extinction in less than a million years? The answer still remains a mystery (see Bioline: The Fall of the Mighty), but whatever caused the extinction of dinosaurs also caused the extinction of many other forms of life as well, while providing new opportunities for other

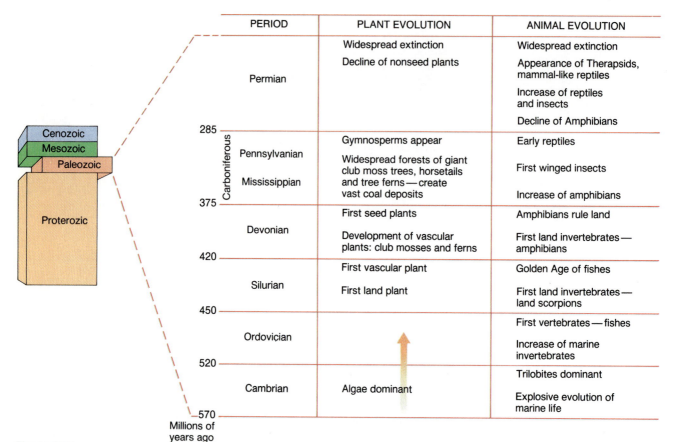

| PERIOD | | PLANT EVOLUTION | ANIMAL EVOLUTION |
|---|---|---|---|
| | Permian | Widespread extinction<br>Decline of nonseed plants | Widespread extinction<br>Appearance of Therapsids, mammal-like reptiles<br>Increase of reptiles and insects<br>Decline of Amphibians |
| 285 | | | |
| Carboniferous | Pennsylvanian | Gymnosperms appear | Early reptiles |
| | Mississippian | Widespread forests of giant club moss trees, horsetails and tree ferns — create vast coal deposits | First winged insects<br>Increase of amphibians |
| 375 | | | |
| | Devonian | First seed plants<br>Development of vascular plants: club mosses and ferns | Amphibians rule land<br>First land invertebrates — amphibians |
| 420 | | | |
| | Silurian | First vascular plant<br>First land plant | Golden Age of fishes<br>First land invertebrates — land scorpions |
| 450 | | | |
| | Ordovician | | First vertebrates — fishes<br>Increase of marine invertebrates |
| 520 | | | |
| | Cambrian | Algae dominant | Trilobites dominant<br>Explosive evolution of marine life |
| 570 | | | |

Millions of years ago

**Figure 30-9**
**The Paleozoic Era**. At the beginning of the Paleozoic Era, the seas were brimming with algae and primitive invertebrates, such as sponges, jellyfish, worms, starfish, and trilobites. Later in the Paleozoic, plants and then animals began to colonize the land. Near the end of the Paleozoic Era, forests of giant tree ferns, club mosses, seed ferns, horsetails, and early conifers blanketed huge expanses of the earth's surface. Amphibians were the dominant land animal. The end of the Paleozoic Era was marked by the greatest episode of mass extinction ever recorded in the fossil record. Among the survivors were two groups of organisms—the conifers and reptiles; they would dominate terrestrial life during the Mesozoic Era.

species to emerge.

Birds, insects, and the first mammals somehow survived whatever caused this mass extinction. Once dinosaurs and other large animals became extinct, mammals quickly filled their vacated habitats, giving rise to a number of groups that would dominate animal life during the Cenozoic Era.

## Life During the Cenozoic Era

The Cenozoic Era is known as the Age of Mammals (Figure 30-11). Mammals possess features that enabled them to edge out the remaining reptiles in most habitats. With insulating hair and thermoregulatory abilities, mammals could be active at times

when "cold-blooded" reptiles were sluggish. They could also inhabit colder climates than reptiles. Cold environments became more prevalent during the Tertiary Period as massive volcanic activity and colliding continents forced the earth's crust to new heights. Intermittent ice ages during the Pleistocene epoch caused mass extinctions and migrations that remolded the distributions of life on earth. Once vastly distributed, tropical vegetation became restricted to the few mild climates. It was during the ice ages that giant mammoths, mastodons, saber-toothed tigers, and other large mammals evolved; all of these mammals are now extinct (Figure 30-12). Following the last ice age,

semiarid and arid areas developed, and the survivors began the most recent period of adaptive radiation.

To date, mammals and flowering plants are the most successful groups of organisms of the Cenozoic Era. Both groups possess a combination of features that enabled them to radiate into practically every habitat on earth. Interestingly, mammals and flowering plants have a common key adaptive characteristic: both protect their young until they are able to survive on their own. Flowering plants protect their embryos in seeds and fruits, whereas mammals give birth to live young and then nourish them with milk until grown.

After 4 billion years of evolution,

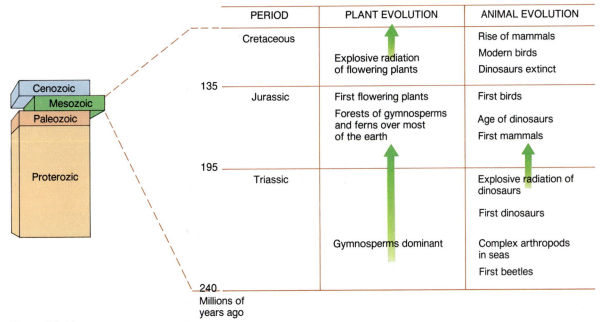

| | PERIOD | PLANT EVOLUTION | ANIMAL EVOLUTION |
|---|---|---|---|
| | Cretaceous | Explosive radiation of flowering plants | Rise of mammals<br>Modern birds<br>Dinosaurs extinct |
| 135 | Jurassic | First flowering plants<br>Forests of gymnosperms and ferns over most of the earth | First birds<br>Age of dinosaurs<br>First mammals |
| 195 | Triassic | | Explosive radiation of dinosaurs<br>First dinosaurs |
| | | Gymnosperms dominant | Complex arthropods in seas<br>First beetles |
| 240 | | | |

Millions of years ago

**Figure 30-10**
**The Mesozoic Era**. The Mesozoic Era is often called the Age of Reptiles. Reptiles evolved into a variety of shapes and sizes, giving rise to the dinosaurs and the ancestors of mammals. The first flowering plants also evolved during the Mesozoic Era. The end of the Mesozoic Era, like the end of the Paleozoic Era, was marked by mass extinctions.

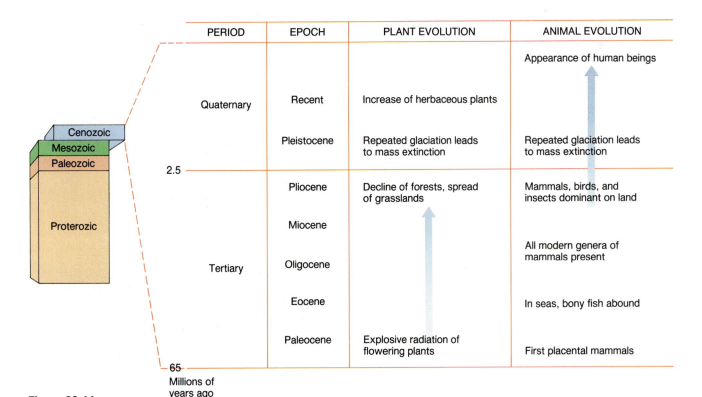

| | PERIOD | EPOCH | PLANT EVOLUTION | ANIMAL EVOLUTION |
|---|---|---|---|---|
| | | | | Appearance of human beings |
| | Quaternary | Recent | Increase of herbaceous plants | |
| | | Pleistocene | Repeated glaciation leads to mass extinction | Repeated glaciation leads to mass extinction |
| 2.5 | | Pliocene | Decline of forests, spread of grasslands | Mammals, birds, and insects dominant on land |
| | | Miocene | | |
| | Tertiary | Oligocene | | All modern genera of mammals present |
| | | Eocene | | In seas, bony fish abound |
| | | Paleocene | Explosive radiation of flowering plants | First placental mammals |
| 65 | | | | |

Millions of years ago

**Figure 30-11**
**The Cenozoic Era**. Mammals and flowering plants evolved rapidly during the Cenozoic Era. Intermittent ice ages continually remolded the distributions of life and caused the extinction of many organisms. Because mammals became the dominant land animal, the Cenozoic Era is also called the Age of Mammals.

**Figure 30-12**
This gallery of Ice Age animals includes a saber-toothed tiger (center foreground),
giant mammoths, bison, and the dire wolf, all of which are now extinct.

# BIOLINE

## CONTINENTS ADRIFT: THE CHANGING FACE OF THE EARTH

• On April 18, 1906, the ground shifted 4.7 m (15.5 ft) along the San Andreas Fault in California and caused the devastating 1906 San Francisco earthquake.

• In 1968 an expedition to Antarctica unearthed a collection of reptile and amphibian fossils similar to fossils found in China and India, thousands of miles away. These ancient vertebrates could not have traveled such long distances, nor could they have survived in the harsh climate of Antarctica. For that matter neither could the swamp plants that formed huge coal deposits there.

• Fossils of the same reptiles, marine animals, and plants have been found in South Africa and in South America; yet there was no way these organisms could have journeyed across the wide Atlantic.

You maybe wondering how a destructive earthquake and strange fossil distributions are related. Both are the result of *continental drift,* the continuous shifting of the earth's land masses explained by the theory of *plate tectonics.*

The theory of plate tectonics maintains that the earth's crust is not a continuous outer shell, but broken into 16 or so plates that move independently of each other (figure a). The enormous heat released by radioactive decay deep within the earth's

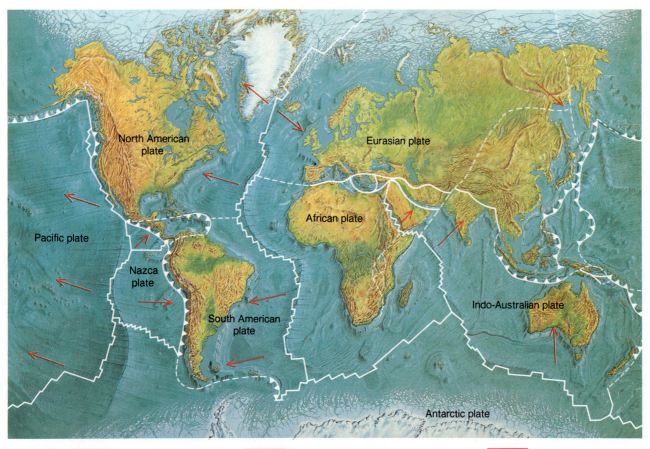

Key:  ▲▲▲ Subduction zone     ▬ Movement of plate     〰 Collision zone

a     ----- Uncertain plate boundary     ┌┐┌┐ Spreading ridge offset by transform faults

interior creates slow-moving convection cells of molten rock in the mantle, upcurrents under the earth's crust on which the plates float and constantly shift positions. Some plates slide against each other as they move. The Pacific and North American Plates, for instance, are sliding toward each other, causing shifts in land and repeated earthquakes along the San Andreas Fault, the line where the two plates "sideswipe" each other. If two plates crumple when they collide, mountain ranges are built (figure b). The

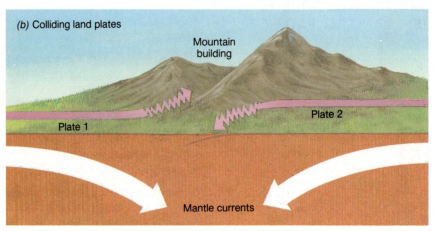

*(b) Colliding land plates*

Mountain building

Plate 1

Plate 2

Mantle currents

b

Andes, Alps, and Himalayas were formed in this way. However, sometimes one of the plates crunches the other beneath it when they collide (figure c). Finally, some plates

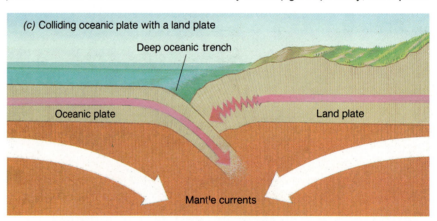

*(c) Colliding oceanic plate with a land plate*

Deep oceanic trench

Oceanic plate

Land plate

Mantle currents

c

are forced apart by upcurrents of molten rock flowing from the earth's interior and hardening into new crust on the surface. This is what is causing the sea floor to spread along the Mid-Atlantic Ridge between the North American Plate and the Eurasian Plate (figure d).

Continental drift occurs very slowly. The plates move 1 to 5 cm (0.5 to 2 inches) per year. But over geological time, the movement has been dramatic, just as have been its effects. Continental drift has caused some or all of the mass extinctions that punctuate the history of life on earth. For example, the extinction of 95 percent of marine invertebrate species and the majority of land plants and animals at the end of the Mesozoic Era coincided with the formation of a supercontinent, Pangaea, as all plates collided. The building of Pangaea about 250 million years ago lowered the sea floor and forced up high mountains, dramatically changing conditions for life. Most organisms did not survive the changes.

Fifty million years after Pangaea formed, it began to fragment, separating into two large continents, Laurasia and Gondwana. Populations of organisms became separated (figure e). Over the next 200 million years, these continents split, and the smaller continents drifted to their present positions, carrying with them fossils of organisms that once flourished half a world away.

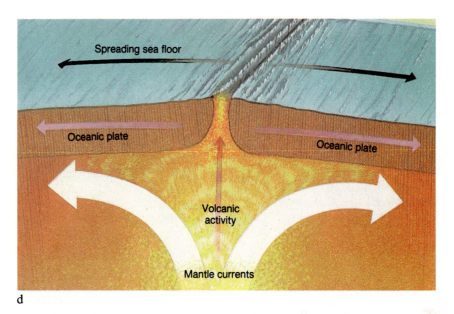

d

Fifty million years from now, biologists will find fossils along the western edge of Canada of organisms that lived in tropical Mexico as continental drift continues to change the size, shape, location, and even the number of continents on the surface of the earth.

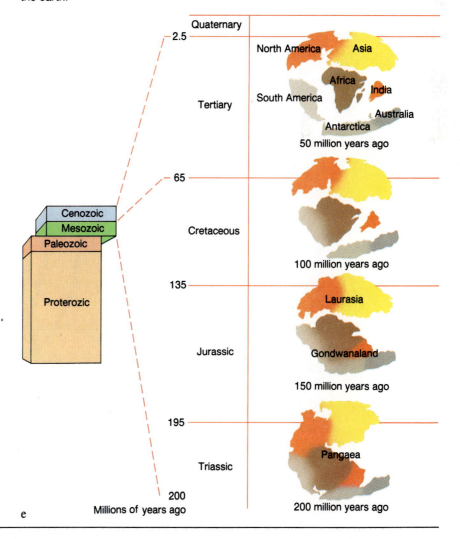

e

the earth is now populated with at least 2 million species and possibly as many as 30 million species. They range from unicellular prokaryotic bacteria similar to the first forms of life on earth, to advanced multicellular mammals and flowering plants. During the last 200,000 years, a tiny fraction in geologic time, a new kind of mammal appeared: one who quickly developed the power to influence the future course of evolution of all forms of life. This new species was *Homo sapiens*, the human species.

## Human Evolution

The mass extinction of dinosaurs and other large animals at the end of the Mesozoic Era may have provided an opportunity for humans to evolve. The extinction of all large vertebrates left enormous numbers of vacated habitats, triggering adaptive radiation of surviving vertebrates, particularly mammals. Among these ancient mammals was a small arboreal (tree-dwelling) primate, the ancestor of monkeys, apes, and humans.

## Our Primate Heritage

Primates evolved more than 65 million years ago. The ancestral line of primates split during the Paleocene Era (60 to 65 million years ago) and eventually gave rise to the two subgroups of modern primates, the **prosimians** (tree shrews, the Loris and Potto, Lemurs, and Tarsiers), and the **anthropoids** which include the New World and Old World monkeys, and the **hominoids**, members of the ape and human families (Figure 30-13).

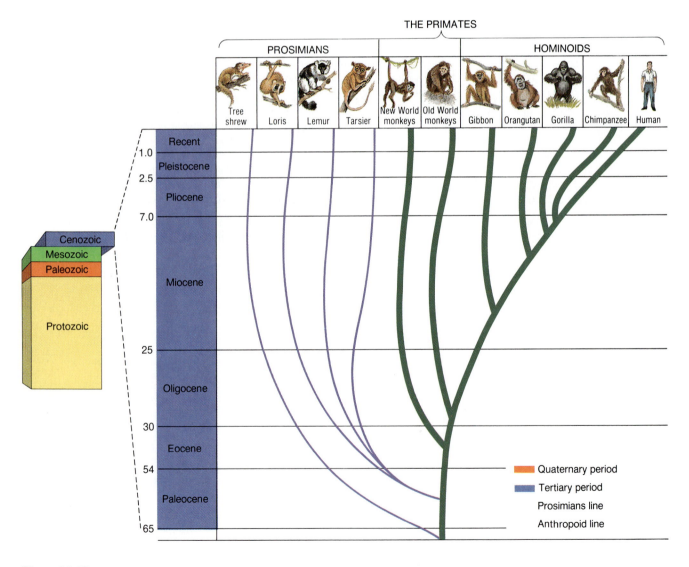

**Figure 30-13**
**The primate evolutionary tree.** Sixty to 70 million years ago a common ancestral line diverged into the branches leading to the anthropoids and all modern prosimians. The closer the divergence points, the more similar groups of organisms are to one another. Thus, gorillas and chimpanzees are more closely related to humans than are monkeys.

# B I O L I N E

## THE FALL OF THE MIGHTY: WHY ARE THE DINOSAURS EXTINCT?

Dinosaurs were the largest animals ever to live on the surface of the earth. The fierce *Tyrannosaurus rex* weighed over 6 tons and stood over 4.5 m (15 ft) tall and was 15 m (50 ft) long. However, *Tyrannosaurus* was dwarfed next to *Brachiosaurus,* which weighed between 70 and 80 tons, and reached 15 m (50 ft) in height and 24 m (80 ft) in length. Each *Brachiosaurus* rib was over 3 m (10 ft) long. But as large as *Brachiosaurus* was, recent discoveries suggest it may not be the largest dinosaur. A team of paleontologists have unearthed a pair of shoulder blades and vertebrae that are at least 20 percent larger than those of *Brachiosaurus.* Tentatively named "Supersaurus," this newly-discovered giant would have weighed about 100 tons, making it 15 times larger than today's largest land animal, the African bull elephant.

But not all dinosaurs were giants; some were as small as a modern pigeon. The term "dinosaur" ("terrible lizard") refers to a diverse group of reptiles that dominated the land, air, and water for the entire Mesozoic Era, a reign exceeding 125 million years. (With a mere 70 million years behind us, mammals have a long way to go to match the staying power of dinosaurs.) But about 65 million years ago all dinosaurs "suddenly" disappeared from the fossil record, creating one of science's most perplexing mysteries.

Many hypotheses have been presented to explain the sudden extinction of the dinosaurs. One proposes there was a drastic change in climate that lowered temperatures and killed off the dinosaurs. Others suggest that epidemic disease was responsible. Or perhaps it was a change in the earth's magnetic field, or mammals eating virtually all dinosaur eggs, or alkaloid poisoning by flowering plants. One recent theory proposes that a collision between earth and a 10-kilometer (6-mile) wide asteroid killed the dinosaurs. The collision filled the atmosphere with dust, cutting off the sun's rays and stopping vital photosynthesis. The evidence to support this theory comes from the discovery of large deposits of iridium at the end of the Mesozoic. (Iridium is rarely found in the earth's crusts, but occurs in great abundance in extraterrestrial materials such as comets, meteors, and asteroids.)

One fact is clear: the dinosaurs did not disappear alone. The list of victims includes:

- 25 percent of the families of shallow marine invertebrates.

- All large animals (no vertebrate larger than a modern house cat survived).

- Hundreds of species of marine plankton.

- A large number of plant species.

- Some primitive mammals.

Any explanation for the extinction of dinosaurs must also explain the disappearance of all these other victims. It may have been a combination, or possibly all, of the proposed disasters that led to the mass extinction of organisms at the end of the Mesozoic Era. For now, we still don't know.

Pteranodon

Anatosaurus

Brachiosaurus

Brachiosaurus

The human family is named **Hominidae**, humans are therefore called **hominids**. (The terms "hominoids" and "hominids" are easily confused; hominoids are apes and humans; hominids are humans only.)

Early primates lived in widespread tropical forests. But changes in the world's climate reduced tropical forests and altered the course of primate evolution. Beginning 40 million years ago, temperatures began to decrease and seasonal cycles of cold and warm, and wet and dry periods became established over many areas of the earth. Tropical forests became reduced, and savannas of scattered trees, open grasslands, and deserts expanded in their place. The reduction in rich tropical forests led to the extinction of many primates. However, some surviving primates left the forests and entered the savannas to forage for food; among these pioneers were the ancient ancestors of apes and humans.

## *Proconsul africanus:* The First Hominoid?

Many paleontologists believe that hominoids evolved from a cat-sized primate named *Aegyptopithecus zeuxis*. A 28-million-year-old fossil of *A. zeuxis* was unearthed in an area of Egypt called the Fayum Depression. The evidence suggests that descendants of *A. zeuxis* migrated throughout Africa. Fossils of one descendant, *Proconsul africanus*, were unearthed in Kenya. Since this 18-million-year-old fossil had a mixture of monkey and ape characteristics, *Proconsul africanus* is considered to be the first hominoid.

## *Australopithecus afarensis:* The First Hominid?

As recently as 1975, most anthropologists maintained that the split in the ancestral line that separated apes and humans occurred at least 14 million years ago, and possibly 20 or even 30 million years ago. The discovery of an 8-million-year-old fossil with advanced human features, however, casts considerable doubt on these estimates. Additional evidence that the split between apes and humans occurred more recently comes from biochemical studies. The close similarity in the protein and DNA structure of humans, chimpanzees, and gorillas

(Table 30-1) suggests that humans and apes diverged as recently as 4 million years ago. (The genetic difference between chimpanzees and humans is only 1.2 percent. The more similar the DNA of two species, the more recent their evolutionary split.)

An archaeological "gold mine" of 200 fossilized bones from 14 individuals was unearthed in Ethiopia between 1974 and 1975. One of these fossils is the famous "Lucy" skeleton (Figure 30-14). The group of 4-million-year-old fossils was nicknamed "the first family." With a humanlike body and apelike head, these were the first true hominids. They were given the name *Australopithecus afarensis*.

*Australopithecus afarensis* became extinct about 3 million years ago. What happened after *A. afarensis* disappeared from the fossil record is not yet clear. Was *A. afarensis* a common ancestor to all subsequent hominids, or only to other species of *Australopithecus*, or was some other yet unearthed ancestor involved? The three proposals for hominid evolution are illustrated in Figure 30-15. Determining which proposal is correct awaits the discovery of more revealing fossils.

What the fossil record makes clear is that three human species flourished simultaneously in Africa about 2 million years ago: *Australopithecus africanus*, *A. robustus*, and *Homo habilis*.

*Australopithecus africanus* and *H. habilis* were both bipedal, about the same height and weight, and both had teeth that suggested they ate the same diet. *A. africanus* was more apelike in appearance, however; its cranium was flatter and its face protruded more than that of *H. habilis*. But the greatest difference between the two was brain size: the brain of *H. habilis* was almost twice the size of *A. africanus*. A larger brain apparently gave *H. habilis* a competitive edge at a time when tropical forests were being replaced by less productive woodlands and savannas. Food became more scarce, and both *A. africanus* and *A. robustus* became extinct. Only *H. habilis*, the "handyman," survived.

## *Homo habilis:* The First Tool Maker

Designating *Homo habilis* as the first member of the genus *Homo* establishes the date of the evolution of the human genus at 2 million years ago, and pinpoints Africa as the location of early human evolution (Figure 30-16). Placing *H. habilis* in the human genus was not based on its 650- to 800-cubic-centimeter brain capacity, for traditionally only fossils with a cranial capacity of at least 1000 cubic centimeters were designated as humans. Rather, it was the ability of *H. habilis*

TABLE 30-1

| Percentage Difference in DNA Nucleotide Sequences between Humans and Other Primates | |
|---|---|
| **Species Compared** | **% DNA Difference** |
| Human–Chimpanzee | 1.2 |
| Human–Gorilla | 1.4 |
| Human–Orangutan | 2.4 |
| Human–Gibbon | 5.1 |
| Human–Old World Monkey (green monkey) | 9.0 |
| Human–New World Monkey (capuchin monkey) | 15.8 |
| Human–Lemur | 42.0 |

to make and use stone tools, together with evidence of a new behavior pattern, gathering and sharing food. Thus, instead of strictly morphological characteristics, *H. habilis* qualified as a human because of its behavior and degree of social interaction.

*Homo habilis* became extinct about 1.6 million years ago and was replaced by a more adept tool maker, a more proficient hunter, and even a more socially complex human, *Homo erectus*.

## *Homo erectus:* Settler of New Continents

*Homo erectus* proved to be a resourceful and adaptable human throughout its long existence, between 1.6 million

and 300,000 years ago. For example, *H. erectus*

- Fashioned tools for specific uses. *Homo erectus* crafted teardrop-shaped hand axes to kill animals for food, and stone flakes and cleavers for removing meat from bones.

- Made routine, cooperative hunts.

- Established home sites for sharing food and raising young.

- Routinely used fire.

- Migrated out of Africa, establishing new human populations in Europe and China. The Peking man of China was a population of *H. erectus* (Figure 30-16).

Although the body of *H. erectus* was similar to that of modern humans (except for being slightly more robust), its skull had a low sloping forehead and prominent brow ridges just above each eye (Figure 30-17). During the 1-million-year span it inhabited earth, the brain capacity of *H. erectus* increased from 850 cubic centimeters to 1100 cubic centimeters. Brain enlargement was a naturally selected trait. The combined advantages of cooperative hunting, food sharing, increased sophistication in making tools, and the possible use of language all improve with greater mental capacity.

## *Homo sapiens:* "Wise Humans"

Fossils with features that are intermediate between *H. erectus* and modern humans have been found in Europe. They were designated as the first known *Homo sapiens*. The one exception to intermediate features was brain size; the brain of these fossils averaged 1450 cubic centimeters, compared to 1400 cubic centimeters for modern humans. Because the brain capacities are different between these fossilized humans and modern humans, each is considered a separate "variety" or type of *Homo sapiens*. The human fossils were named *Homo sapiens neanderthalensis* or neanderthal man; modern humans are named *Homo sapiens sapiens*.

Neanderthal humans were not dull-witted and apish brutes. Except for heavy brows, a slightly protruding face, and a somewhat stocky build, they were very similar in overall appearance to us (Figure 30-17). They lived between 40,000 and 200,000 years ago in a variety of environments (Figure 30-16), including exceptionally cold areas where they lived in caves, built shelters of bones and skins, and used fire to keep warm. Neanderthals crafted about 60 types of tools. In addition to improved tool-building skills, they had a highly developed social system based on a division of labor between members. Neanderthals left evidence of abstract thinking and self-awareness. Many burial sites were found with carefully positioned bodies, food, weapons, and flowers, suggesting a belief in life after death.

Neanderthal fossils disappear from

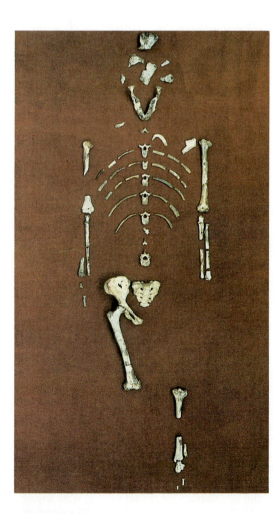

**Figure 30-14**
"**Lucy,**" nearly 40 percent intact, is one of the most complete fossils of *Australopithecus afarensis*. Lucy and other members of "the first family" have been proposed to be the original members of the human family.

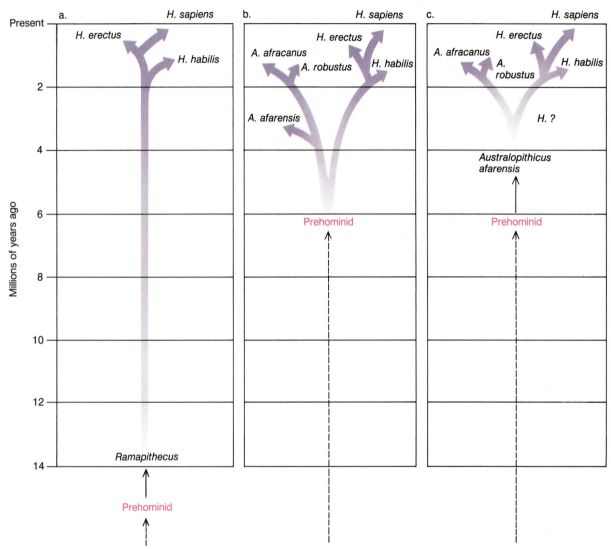

**Figure 30-15**
**Proposed hominid evolutionary trees.** All three proposals agree on the ages of fossils, but disagree on the relationships and dates of divergence. The range in the estimated age of the human family ranges from *(a)* 14 million years, to *(b)* 6 million years, to *(c)* as young as 3.8 million years.

the fossil record between 35,000 and 40,000 years ago. Some scientists suggest that *H. sapiens neanderthalensis* evolved into modern humans by phyletic evolution (Chapter 29), whereas others propose that modern humans diverged from the neanderthal line, coexisted with them for awhile, and eventually displaced them.

Humans that lived 35,000 years ago were virtually indistinguishable from people today. Modern humans appeared in Africa, Europe, and Asia. Each population had characteristics that were slightly different from each other, just as people from these areas differ today.

All early *H. sapiens sapiens* were skillful group hunters, often hunting animals much larger than themselves. Group interactions during hunting and at campsites spurred the further development of language, which in turn, increased socialization and the advancement of culture. By passing knowledge from generation to generation in a process called *cultural evolution*, modern humans soon developed

a complex culture that rapidly advanced technology and art. Cultural evolution rather than biological evolution began to dominate human evolution, leading to the agricultural revolution about 15,000 years ago, the industrial revolution 200 years ago, and the nuclear, space, and computer revolutions that dominate modern human life. What's next? With the accelerated pace of cultural evolution, the future course of human evolution even within the next 50 years defies prediction.

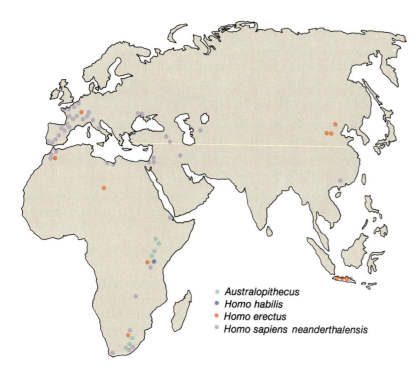

**Figure 30-16**
**Distributions of hominoid fossils.** Apelike *Australopithecus* species are limited to east and south Africa. The first human fossils, *Homo habilis*, were found in the Olduvai Gorge in Tanganyika, Africa. *Homo erectus* migrated out from Africa to inhabit parts of Europe and Asia. Fossils of *Homo sapiens neanderthalensis* are widely distributed throughout Europe and the Mediterranean, with representatives in Southeast Asia and Africa.

- *Australopithecus*
- *Homo habilis*
- *Homo erectus*
- *Homo sapiens neanderthalensis*

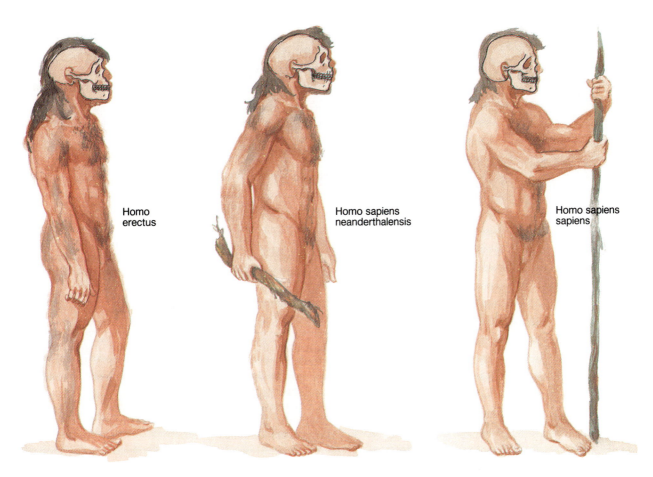

Homo
erectus

Homo sapiens
neanderthalensis

Homo sapiens
sapiens

**Figure 30-17**
Comparision of *Homo erectus, H. sapiens neanderthalensis,* and *H. sapiens sapiens.*

## Synopsis

### MAIN CONCEPTS

- **The earth condensed out of a massive cloud of dust and gases 4.6 billion years ago.** Life evolved in less than 1 billion years, the result of spontaneous chemical evolution.

- **The first life forms were heterotrophic prokaryotes.**

- **The evolution of autotrophs** about 3.0 billion years ago was essential for the continuation of life on earth.

- **Unicellular eukaryotes evolved about 1.0 billion years ago,** paving the way for the evolution of multicellular organisms about 700 million years ago.

- **Repeated changes in the shape, size, and location of the earth's continents** drastically changed the earth's climate and sea level. These changes greatly affected the evolution of organisms.

- **Severe ice ages and expanding dry habitats** caused mass extinctions at the close of the Paleozoic Era.

- **Reptiles and gymnosperms dominated the land during the Mesozoic Era** (65 to 240 million years ago). Adaptive radiation among reptiles produced a diversity of organisms, including the dinosaurs. Flowering plants evolved, as well as insects, mammals, and birds. The Mesozoic Era ended with another episode of mass extinction; all dinosaurs and many other organisms, including 95 percent of marine animals, became extinct.

- **The present Cenozoic Era is called the Age of Mammals.** Flowering plants became the dominant land plants.

- **Monkeys diverged from the primate ancestral line** about 30 million years ago, and apes and humans diverged sometime between 4 and 20 million years ago.

- **The first member of the human family (Hominidae),** *Australopithecus afarensis,* lived 3 million years ago in Africa.

- ***Homo habilis,* the first tool maker,** became extinct 1.6 million years ago. The genus of humans, *Homo,* continued as *H. erectus* for the next million years.

- ***Homo sapiens* has two varieties:** *H. sapiens neanderthalensis* which lived until 40,000 years ago, and *Homo sapiens sapiens* or modern humans which evolved 35,000 to 40,000 years ago.

### KEY TERM INTEGRATOR

Evolution of Life

| | |
|---|---|
| **chemical evolution** | The assembly of simple chemicals into more complex molecules. The first living cell evolved from a **precell**. Possible precells produced in laboratories include **coacervates** and **proteinoids**. |

The History of Life on Earth

| | |
|---|---|
| **cynaobacteria** | Autotrophs that evolved from the first living organisms—unicellular, prokaryotic heterotrophs. |
| evolution of eukaryotes | Explained by the **membrane invagination theory**, the **endocymbiosis theory**, or both. |
| **therapsids** | Ancient reptiles that evolved into mammals. |
| **Hominoids** | Include the ape family, and the human family **(Hominidae)**. Humans are **hominids**. |

# Review and Synthesis

1. Match the following organisms with its classification category:
   - ___ 1. Hominidae
   - ___ 2. Cyanobacteria
   - ___ 3. Hominoid
   - A. first autotrophic heterotroph
   - B. *Australopithecus afarensis*
   - C. *Proconsul africanus*

2. What characteristics qualify a structure as a precell? Why are these characteristics so important?

3. Retrace the sequence of chemical evolution that led to the formation of the first living cell. Identify the steps already reenacted by scientists as well as those not yet duplicated in the laboratory.

4. Why were periods of mass extinctions often followed by rapid diversification of new species through adaptive radiation?

5. Identify the periods of mass extinctions during the history of life on earth. Which episodes correlated with major shifts in the earth's continents?

6. Why was increased brain size naturally selected during the evolution of modern humans?

7. Try to envision the future 10, 50, and 100 years from now. Where do you think cultural evolution will have led humans? How will humans have affected the evolution *and* extinction of other organisms on earth? What priorities should biologists today establish to circumvent any negative changes?

# Additional Readings

Bellamy, D. 1979. *Forces of Life. The Botanic Man.* Crown Publishers I., New York. (Introductory.)

Cech, T. 1986. "RNA as an enzyme." *Scientific American* 255:64–75.

Folsome, C. E. 1979. *The Origin of Life. A Warm Little Pond.* W. H. Freeman, San Francisco. (Introductory.)

"Life. Origin and evolution." 1979. Readings from *Scientific American.* W. H. Freeman, San Francisco. (Introductory.)

McElhinny, M. W., and D. A. Valencio (eds.). 1981. *Paleoreconstruction of the Continents.* American Geophysical Union, Boulder, Colo. (Advanced.)

Margulis, L. 1982. *Early Life.* Science Books International, Boston. (Intermediate.)

Pilbeam, D. 1984. "The descent of hominoids and hominids." *Scientific American* 250:84.

Sagan, C. (ed.). 1973. *Communication With Extraterrestrial Intelligence.* MIT Press, Cambridge, Mass. (Intermediate.)

# 9

## *The Kingdoms of Life: Diversity and Classification*

200 years ago biologists classified all organisms as either plants or animals—the two prominent kingdoms represented in this photo. But a closer examination of the soil, water, and air discloses very different types of organisms, necessitating defining three additional kingdoms of life.

Chapter 31

# *The Monera Kingdom*

ew locations would seem less hospitable to life than a deep thermal rift on the ocean's floor. In addition to total darkness and the enormous pressure exerted by one vertical mile of seawater, superheated water (350° C or 662° F) spews from cracks in the ocean's bottom. If not for the pressure, the seawater would instantly explode into steam. It would seem that life would have little chance here, for such high temperatures would melt lipids, denature proteins, and (at sea level) boil cytoplasm. Yet in 1982, samples revealed that life had indeed invaded this hostile environment. Microscopic organisms were proliferating at temperatures well above that which, in air, would burst your book into flames.

The organisms growing in this prohibitive environment were *bacteria,* members of the kingdom Monera. Although bacteria are often branded as villains because of a few disease-causing species, only a very few species are harmful to humans. Their activities and habitats are remarkable and diverse. Consider the following cases.

- The Red Sea turns reddish orange, the color for which it was named.

- A new life form emerges on the young earth, an organism that uses sun energy to run its biological machinery. Its waste product, oxygen gas, permeates the oceans and atmosphere, allowing for the eventual emergence of plants and animals.

- Deadly diphtheria bacteria fail to gain a fatal foothold in a child's respiratory tract because billions of "friendly" bacteria living in the child's pharynx produce a chemical that inhibits the invader's growth.

- As ripples from the boat's wake sweep across the water's surface, the black ocean suddenly brightens with an eerie phosphorescent glow.

- The snow white polar bears in the San Diego Zoo turned green (Figure 31-1), the color persisting despite vigorous washing.

All these phenomena were caused by bacteria and cyanobacteria, the two groups of organisms that make up the

**Figure 31-1**
*Green* **polar bears?** Normally pure white, these polar bears are green because photosynthetic prokaryotes (cyanobacteria) are living inside the hollow core of the bears' outer hairs, just one of the unusual habitats occupied by members of the Monera kingdom. San Diego zookeepers remedied this cosmetic complication by bathing the bears in salt water.

Monera kingdom. Although they are the earth's simplest cells, they are essential to the evolution and maintenance of the biosphere.

In addition to bacteria and cyanobacteria, we also include a discussion of *viruses* in this chapter. Viruses are not usually represented in the five-kingdom classification scheme because they are noncellular, and as such, considered by some biologists to be nonliving, and not organisms at all. As we will see later, viruses have properties that justify their being assigned to a gray zone between the living and the inanimate.

## Kingdom Monera — The Prokaryotes

All members of the Monera kingdom are single-celled prokaryotes. They are distinguished from the eukaryotes of the other four kingdoms by the presence of a **nucleoid** instead of a true nucleus (Chapter 6), unique ribosomes, and absence of membrane-bound or-

ganelles (such as mitochondria, chloroplasts, and the Golgi apparatus).

Although composed of the same chemical groups as eukaryotes (proteins, nucleic acids, lipids, and carbohydrates), prokaryotes possess several biochemical differences. The most universal of these is **peptidoglycan**, a compound in the prokaryotic cell wall. The polymer stretches around the cell in all directions, surrounding it with a cross-linked matrix that reinforces the cell wall. Because bacteria have no means of evacuating excess water, the peptidoglycan wall provides enough strength to protect the fragile cell from swelling and bursting from the osmotic influx of water. Some bacteria, those surrounded by many layers of peptidoglycan, can withstand internal osmotic pressures 25 times that of the normal atmosphere without bursting.

The rigidity and shape of the cell wall determine the shape of the bacterial cell. Three shapes are most prominent in the kingdom (Figure 31-2), although variations on these are not uncommon. The most recent addition to the catalogue of shapes is a square bacteria discovered in 1980.

a

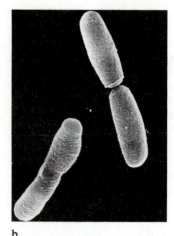

b

c

**Figure 31-2**
**Typical bacterial shapes**—(a) spheres (called *cocci*), (b) rods (called *bacilli*), and (c) spirals. Bacterial cells often remain attached to each other after dividing, producing a characteristic arrangement (the chains formed by these cocci, for example).

## Evolutionary Trends

Of all the organisms alive today, bacteria most resemble our concept of the earth's original life forms, especially anaerobic bacteria such as *Clostridium* (Figure 31-3). A comparison of modern bacteria with ancient fossils reveals that they have not changed much in more than 3.5 billion years.

Modern prokaryotes retain a rich variety of the metabolic strategies "invented" by their primitive ancestors, various ancestral lines "trying different experiments" to best utilize available chemical resources. Many or possibly all of these metabolic strategies developed before the appearance of eukaryotes. Today's eukaryotes possess similar metabolic pathways which are basically variations on the original prokaryotic theme. This suggests that eukaryotes evolved from one or more prokaryote ancestors (Chapter 30).

The primitive nature of modern

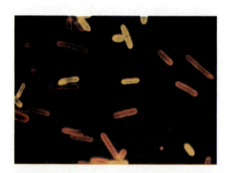

a

b

**Figure 31-3**
**The earth's first life forms** may have been very similar to these modern anaerobic bacteria: (a) Species of *Clostridium* (stained here with fluorescent dye) are common in soil and can grow in oxygen-devoid environments such as injured flesh (causing tetanus and gangrene) and canned foods (the major source of botulism, a fatal form of food poisoning). (b) Species of *Bacteriodes* are among the many kinds of bacteria that inhabit your large intestine. Gas produced in the large intestine is a byproduct of bacterial metabolism rather than human digestion.

bacteria prompts the question, "why did they retain their simplicity when other forms of life were (and continue to be) displaced by complex organisms better suited for local conditions?" The answer is found in the adaptability of bacteria as a group. The cornerstone of this adaptability may be their extremely rapid growth rates that produce enormous numbers of individuals.

## ADAPTIVE ADVANTAGE OF RAPID GROWTH RATES

Most bacteria reproduce by **binary fission**, asexually dividing into two daughter cells of equal size (Chapter 11). These cells in turn divide into four cells, which then produce eight, and so on until millions of new cells appear with each doubling of the population. Since bacteria may divide every few minutes, the rate of growth can be astonishingly fast. The common bacterium *Escherichia coli,* for example, doubles in number every 20 minutes; in less than 48 hours a single *E. coli* cell could theoretically produce a population of progeny equal to the volume of the earth. (As we discuss in Chapter 38, its growth is halted long before such an eventuality, usually as the nutrient supply in its local environment becomes exhausted.) Such rapid growth rates often allow bacteria to outcompete more slow-growing competitors, many of which are simply overgrown and displaced.

## ADAPTIVE ADVANTAGE OF ENORMOUS NUMBERS

Bacteria populate virtually all the earth's habitats, and, in some especially inhospitable places, they are the only inhabitants. One explanation for this adaptability is that wherever bacteria occur, they are usually present in very large numbers. The greater the number of organisms, the greater the likelihood that some members of the species will possess a combination of traits that will enable it to survive a catastrophic change — even if the change is so severe that it kills all other inhabitants. With a rapid growth rate, the survivors can then quickly repopulate the region with their progeny.

## Overview of the Kingdom

Developing a classification scheme for prokaryotes based on evolutionary relationships has been hampered by the microscopic size of the organisms, which makes it difficult to locate fossils. (You would not notice a bacterial fossil as readily as you might a fossilized dinosaur bone.) Although fossilized prokaryotes have been found, the record is not sufficiently detailed to allow us to construct a phylogenetic classification scheme. The current taxonomic system is therefore a practical one that groups living bacteria according to properties that facilitate their identification.

One of the most troublesome problems with classifying bacteria is their structural simplicity. Bacteria lack the complexity needed to categorize them solely on the basis of morphology (form), which is the principal approach used for classifying protists, fungi, plants, and animals. Thus, bacteriologists must supplement morphological observations with information about the organism's biochemistry, metabolic pathways, growth requirements, and several other criteria described in Connections: The Species Concept.

Monerans are divided into two divisions: (1) **Schizophyta** (schizo = split, phyta = plants), the group classically called "bacteria"; and (2) **Cyanophyta** (cyano = blue), a prokaryotic group that was formerly known as "blue-green algae." The current model for categorizing the monerans looks something like the following illustration.

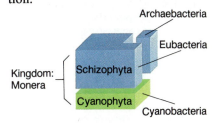

## Division Schizophyta — the Bacteria

The Schizophyta division comprises well over 90 percent of the moneran species, most of which are "typical bacteria," much like the one shown in

Figure 31-4. This figure depicts a **eubacterium**, literally, a "true bacterium." The name is a bit misleading because the other group of schizophytes, the **archaebacteria**, are also true bacteria.

### THE EUBACTERIA

The vast majority of prokaryotes are eubacteria. Examples of eubacteria are literally everywhere — the rod-shaped *E. coli* that inhabits your intestine, the spherical *Streptococcus* that causes "strep throat and scarlet fever," the corkscrew-shaped agent of syphilis, *Treponema pallidum.* Although eubacteria have been termed "typical" bacteria, the tag erroneously minimizes the tremendous variation that exists in this group. This variation includes

- *A diversity of metabolic strategies.* Each bacterium's metabolic pathways are suited for its particular habitat. Bacteria vary tremendously in their tolerance to environmental stresses, including extremes in osmotic pressure, pH, temperature, water availability, and oxygen deprivation. Each of these stresses requires a different metabolic strategy. Some bacteria are *facultative anaerobes,* that is, organisms that use the more efficient oxygen-dependent pathways in aerobic habitats and then switch to fermentation or anaerobic respiration (Chapter 8) when oxygen is depleted.

- *The ability to conduct photosynthesis.* Photosynthetic bacteria use **bacterial chlorophyll** to convert radiant energy into chemical energy for use as food. These photosynthetic pigments absorb longer wavelengths of light than the chlorophyll of eukaryotic photosynthesizers. In water, this ability allows these bacteria to survive at a lower depth than unicellular algae, which grow only near the surface.

- *The formation of endospores.* A few types of eubacteria produce a single spore that enables the species to survive drying, exposure to ultraviolet light and caustic chemicals, and extreme heat. Boiling medical instruments and foods before canning is

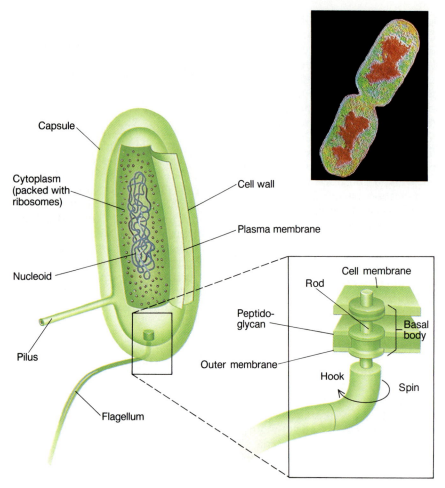

**Figure 31-4**
**Anatomy of a typical bacterium.** Compare the simplicity of this prokaryotic cell with the complexity of eukaryotes in Chapter 6. The blow up details the hook and basal body connection to a flagellum. The shaft of the basal body rotates, causing the flagellum to spin.

- *Motility.* Many bacteria are capable of directed movement, most propelling themselves with flagella. Some spiral-shaped bacteria spin on their axis and swim by "corkscrewing" through their liquid environment. Still others, gliding bacteria, move by an unexplained mechanism along slime tracts they secrete. Even flagellated bacteria, however, are unusual in the way the locomotor appendage works. At the base of the flagellum, just beneath the bacterium's plasma membrane, is a *basal body* that functions as a motor, rotating the flagellum like a helical-shaped propeller that either pushes or pulls the organism (Figure 31-4).

- *Parasitic bacteria.* Some bacteria live in or on host organisms. **Obligate intracellular parasites,** for example, are organisms that can succeed only in the infeated host cells. One group of these, the **rickettsia,** causes several serious diseases, among which is a disease that redirected human history by changing the outcome of several important wars. This disease, epidemic typhus, incapacitated whole armies. Napoleon, for example, lost more soldiers to typhus than to combat-related injuries. Another group of intracellular bacteria, the **chlamydia,** are even smaller than rickettsia and are barely visible using light microscopes. Their small size does not diminish their impact, however; chlamydia causes a number of diseases, such as trachoma, the world's leading cause of human blindness. Chlamydia infections of the genitals also represent one of the most common forms of sexually transmitted diseases in the Western world.

## THE ARCHAEBACTERIA

Archaebacteria (archae = ancient) differ from eubacteria in several properties. Archaebacteria have shapes that are atypical of other bacteria, possess unnusual lipids in their cell membranes, have cell walls that lack peptidoglycan, and even resemble eukaryotes in a few biochemical processes, pigments, and proteins. Some inhabit areas where conditions are too extreme even for eubacteria. For exam-

unreliable because the endospores survive and germinate into actively growing bacteria once conditions have returned to normal. Some of these bacteria cause serious infections or produce dangerous toxins in canned or vacuum-wrapped foods. Instead of boiling, pressure cookers are now used to superheat steam to 121° C, hot enough to kill the most heat-resistant endospores. Modified pressure cookers, called *autoclaves,* are high-efficiency sterilization instruments used in hospitals and microbiological laboratories. Endospores are also resistant to the effects of aging, as illustrated by the unearthing of bacterial endospores at

an archaeological excavation site; the dormat spores proved to be alive after 1300 years of "suspended animation."

- *The formation of a capsule.* A few eubacteria form a capsule around themselves for nutrient storage, protection from drying, or to avoid the protective defenses that would eliminate the bacterium from an infected host. Many types of bacteria that cause human pneumonia, for example, can only do so if they are encapsulated; without capsules these bacteria are quickly engulfed and destroyed by the body's protective white blood cells.

ple, archaebacteria thrive in the Dead Sea's supersaturated salt water, in hot springs, and in superheated marine waters such as those described in the chapter's introduction. They may even be found in strong sulfuric acid solutions, surviving pHs as low as 1.0, which is acidic enough to dissolve metal. Other archaebacteria live side by side with eubacteria in habitats devoid of oxygen, such as mammalian intestinal tracts, raw or partially processed sewage, and the smelly ooze at the bottom of stagnant bodies of water. In these anaerobic habitats, the archaebacteria often generate methane, a flammable organic gas known by many common names, including natural gas or swamp gas.

Because of these unique properties, some bacteriologists have proposed creating a new kingdom for the archaebacteria. However, other bacteriologists argue (and we agree) that because they are prokaryotic they should remain in the Monera kingdom. Archaebacteria were originally viewed as the most primitive type of bacteria that evolved into eubacteria. Today, however, most bacteriologists believe that a common ancestor produced two parallel lines of evolution—one line for archaebacteria and the other for eubacteria (review Figure 30-7).

## Division Cyanophyta—The Cyanobacteria

Members of the Cyanophyta division are called **cyanobacteria**. They were originally called "blue-green algae" because they possess chlorophyll *a* and release oxygen as a waste product of photosynthesis, characteristics similar to algae. Nonetheless, their lack of chloroplasts (and other organelles) and the presence of peptidoglycan in their cell walls clearly places them in the Monera kingdom rather than the protist kingdom with algae. Instead of chloroplasts, cyanobacteria possess a network of membranes in which the photosynthetic pigments are embedded. These internal membrane systems not only mimic the role of chloroplasts, but may also have been the evolutionary forerunners of chloroplasts.

**CONNECTIONS**
Morphological Species (page 11)
Animal versus Plant Species (page 380)
Ecological Species (page 694)

# THE SPECIES CONCEPT: DEFINING MONERAN SPECIES

In the previous boxes in this series on the species concept, morphology was emphasized as the most commonly used criterion for classification; most species are distinguished entirely by differences in their external appearance. But the morphological simplicity of monerans makes it impossible to classify and identify most bacteria and cyanobacteria solely on the basis of appearance. For example, hundreds of rod-shaped bacteria all look identical, yet they possess other features that reveal their differences, leading to their being assigned to different genera and species. The criteria which taxonomists use for classifying bacteria must therefore include not only external appearance, but a number of other more subtle criteria as well. Some of these criteria are given in the accompanying table. Notice that the starting point in characterizing bacteria is, as with other organisms, observing their appearance, both microscopic (Figure 1) and macroscopic (Figure 2).

a

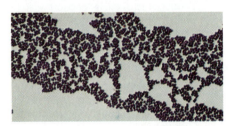

b

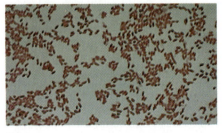

c

**Figure 1**
Colorless before staining, stains help reveal additional characteristics used by taxonomists to classify and identify bacteria. The most important stain, the Gram stain, identifies a bacterial species as either gram positive (*a*) or gram negative (*b*) according to the color of the cells after staining. Flagella (*c*) and other cell structures can be detected using special stains that specifically enhance their appearance.

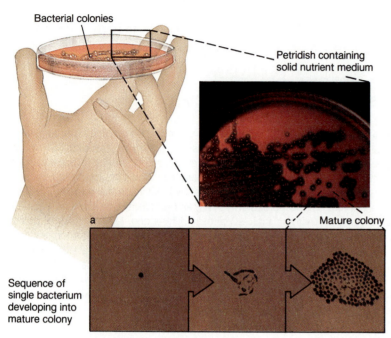

Bacterial colonies

Petridish containing solid nutrient medium

a       b       c    Mature colony

Sequence of single bacterium developing into mature colony

**Figure 2**
**Colony characteristics** are also used by taxonomists to distinguish different species of bacteria from each other. Colonies begin invisibly, as a single cell or a small group of cells, then proliferate into billions of bacteria that form a macroscopic (visible) mass.

| Criteria | Examples |
|---|---|
| Microscopic appearance | |
| Cell shape and arrangement | Cocci in clusters; bacilli in chains |
| Differential staining | Gram-positive or gram-negative |
| Distinguishing structures | Endospores, flagella, capsule |
| Macroscopic characteristics | Colony texture, size, shape, and color |
| Biochemical properties | Unique amino acids in cell wall; composition of capsule. |
| Physiological activities | Motility; oxygen requirements; carbon and energy requirements; ability to use specific sugars; metabolic byproducts; obligate intracellular growth |
| Immunological specificity | Unique antigens of the cell, determined by reacting with antibody preparations of known specificity |
| Genetic analysis | Extraction of DNA, followed by determination of guanine-cytosine (G-C) content and nucleotide sequence |

Based on these criteria, taxonomists have described some 2000 species of monerans.* Although the Monera kingdom is the least diverse of all kingdoms of organisms, they are the most numerous organisms on earth.

---

* *As described in* Bergey's Manual of Determinative Bacteriology *(8th ed.), the authoritative manual on bacterial classification.*

The more than 200 species of cyanobacteria are diversified in form and physiology. Some grow as single cells, although hundreds of cells may sometimes be held together in a gelatinous matrix. Others form long filaments or clusters of cells bound together by cell-to-cell linkages. In some species, an occasional cell is enlarged, forming a **heterocyst** (Figure 31-5). These specialized structures "fix" molecular nitrogen, converting it from its diatomic form ($N_2$) to a reduced state, such as ammonia ($NH_3$) or amino groups ($-NH_2$). In this way, cyanobacteria and nitrogen-fixing eubacteria make nitrogen available to other organisms, a process that is critical to the continuance of life on earth. (The nitrogen cycle is discussed in Chapter 36.)

## Activities of Prokaryotes

In addition to fixing nitrogen, monerans perform other activities that have a profound impact on all organisms in the biosphere (see Bioline: Living with Bacteria). Many of these activities are the exclusive domain of prokaryotes, allowing them to occupy habitats that are uninhabitable by eukaryotes. **Chemosynthetic bacteria**, for example, use $CO_2$ as a carbon source, as do plants, algae, and other photosynthetic autotrophs. Yet they do so in areas devoid of light, the usual energy source of autotrophs. These bacteria extract their energy from inorganic minerals, as described in Chapter 9. (This is roughly analogous to a person trying to exist by eating rocks for energy and depending on inhaled $CO_2$ to supply the carbon needed to build tissues.) The organisms that live near deep, continually dark ocean trenches depend on chemosynthetic bacteria as their ultimate source of food (See DISCOVERY: Living on The Fringe, page 133)

The vast majority of bacteria are heterotrophic, consuming organic molecules for energy and carbon. Most of these are *decomposers,* extracting their energy and carbon from dead organic matter, an activity that disassembles organisms, molecule by molecule, soon after they die.

In addition, bacteria engage in some very novel biological activities. Some have phosphorescent pigments that

# B I O L I N E

## LIVING WITH BACTERIA

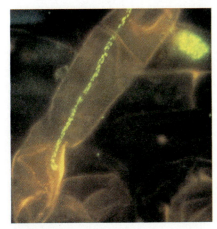

**Photo 1**
**An orderly invasion** of alfalfa root cells by beneficial nitrogen-fixing bacteria.

No organism can avoid sharing its environment with bacteria. We even have to share our bodies with them. However, we are learning how to live with them more successfully than in the past. Compare today's awareness with the state of knowledge in King Henry VIII's day, when it was customary for a person (including the King himself) to bathe only three times—once at birth, a second time upon marriage, and the last time after death. This was also a time when the contents of chamber pots (indoor "toilets") were emptied through windows, often onto sidewalks below or into the streets. No one suspected that typhoid fever, cholera, and many other fatal infections were transmitted by bacteria shed in the feces and urine of infected persons. Flies carried pathogens (disease-causing microorganisms) from open cesspools to people's lips and food. Runoff from flooded cesspools contaminated drinking water. Food decomposed very rapidly, presenting most people with the dilemma of whether to eat rotted food or starve. Rats patrolled the streets, harboring such notorious killers as the bubonic plague bacterium, the agent of the "Black Death." The average person's lifespan was about half today's average, a grim testimony to their total lack of microbial awareness.

Bacteria thrive in every habitat that supports life. A single gram of fertile garden soil harbors more than 2 billion bacteria. Each time you inhale, millions of bacteria enter your respiratory tract. Ten times more bacteria reside in your large intestine than there are cells in your body. Billions more occupy your skin. These bacteria constitute your body's *normal flora*. You benefit from their presence every time they successfully outcompete the disease-causing bacteria that could otherwise infect your body.

Listed below are a few more ways bacteria benefit not just us, but the entire biosphere.

- Decomposers form an essential link in biogeochemical cycles, recycling the earth's limited nutrients (Chapter 36).

- Nitrogen-fixing bacteria and cyanobacteria provide usable nitrogen to the biosphere. Some of these bacteria live in root nodules of legumes (see photo 1). Alfalfa and soybeans are two such plants whose resident bacteria free them from the need for nitrogen fertilizer. Planting a crop of legumes improves the fertility of agricultural fields that have been exhausted by other crops.

- Cattle and sheep are unable to digest cellulose, the chief constituent of their grassy diet. Bacteria (and protozoa) living in the rumen, a special stomach chamber, do the job for them.

- Bacteria are used to produce yogurt, some cheeses, vinegar, and several other food products.

- Modern waste water treatment plants employ bacteria to process raw sewage to a safe substance free of pathogens.

- The newest applications use genetically engineered bacteria to produce antibiotics and other compounds of medical and industrial importance. Chapter 39 ("The Biotechnology Revolution") discusses many of the modern strategies for employing the microbial "workforce."

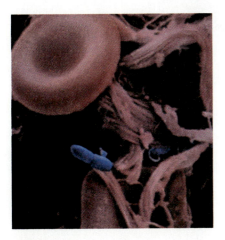

**Photo 2**
**Deadly opportunist.** The presence of pathogens, such as these bacteria (blue rods) that have colonized a burn wound, poses a constant threat of infection. With our modern microbial awareness, we can combat such threats with a battery of control measures.

In spite of these benefits, that small minority of bacteria that causes disease (see photo 2), spoils our food, fouls our water, and deteriorates useful products still represent a formidable foe, one that requires sustained efforts to minimize their detrimental impact. Considering their benefits and the many critical roles they play in the biosphere, however, bacteria more than compensate for the inconvenience and suffering. Life on earth could not continue without them.

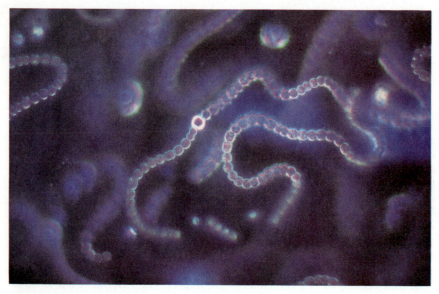

**Figure 31-5**
**Swollen heterocysts** enable chains of *Anabaena*, a cyanobacterium, to "fix" molecular nitrogen (N₂), that is, convert it to a biologically usable form.

tween these two seemingly paradoxical states.

Formulating a concise definition for a virus is difficult at best. In general, viruses have five characteristic properties (and many possess a sixth).

- *They are obligate intracellular parasites,* requiring a host cell to perform virtually all biological functions that lead to the production of new viruses.

- *They are incapable of independent metabolism.*

- *They are smaller than the tiniest bacteria,* so small they can be seen only with an electron microscope.

- *They possess only one type of nucleic acid,* either DNA or RNA, but never both. The DNA or RNA of the virus directs the host cell to produce a crop

glow when energized by respiration (Chapter 10). One unusual bacterium carries a magnetic "compass" to help it home in on a nutrient-rich, oxygen-free habitat (Figure 31-6). Another bacterium, *Hyphomicrobium vulgare,* puzzled scientists by its ability to grow on the surface of purified water, with no apparent source of nutrients or energy. Scientists eventually discovered its source of energy and nutrients—formaldehyde and other small volatile organic molecules that give the air of some laboratories their peculiar smell. Its ability to concentrate such a dilute source of nutrients from air is unmatched by any other heterotroph.

## Viruses

Although the cell is the fundamental unit of life, there are a group of "organisms," the **viruses,** that are not cells at all. A fully assembled virus is as inanimate as a crystal of organic chemicals, but once inside a host cell, the virus's biological apparatus "awakens." There it adopts its "living mode," showing two characteristics of life—reproduction and heredity. A review of the typical viral reproductive cycle on page 468 (Chapter 27) demonstrates how a virus alternates be-

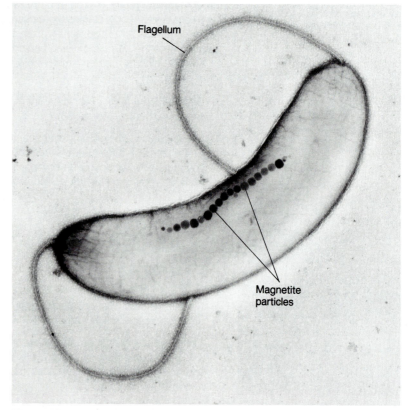

Flagellum

Magnetite particles

**Figure 31-6**
**The "compass bacterium"** (*Aquaspirillum magnetotacium*) lives in swampy waters. Particles of magnetite (a component of iron ore) align in response to geomagnetic forces, allowing the bacterium to distinguish up from down. The bacterium rotates like a compass needle, pointing down toward the nutrient-rich sediment at the swamp bottom. Homing pigeons and honeybees also contain magnetic crystals, perhaps helping them locate their destinations in a manner similar to this magnetized prokaryote.

of progeny viruses. The host cell obeys the viral genes as faithfully as it would its own.

- *While in a host cell, viruses undergo an "eclipse phase,"* during which they disintegrate into their molecular constituents. It is during eclipse that the virus's nucleic acid commandeers the cell's genetic machinery, instructing it to produce new copies of the virus nucleic acid and the proteins needed for assembling new viruses. Eclipse ends when these components assemble into progeny viruses, sometimes hundreds of them in one cell.

- *Some viruses can be crystallized and still retain their infectivity.* One such crystal is shown in Figure 31-7, a purified crystal of polioviruses. The viruses in the gemlike crystal are still capable of infecting host cells.

These unusual properties of viruses are a function of their structural simplicity. The least complex viruses consist of nothing more than a molecule of nucleic acid (the viral genes) surrounded by a protein coat. This coat protects the genes and allows the virus to attach to and infect its specific host cell. Some viruses possess an additional layer, an outer envelope (Figure 31-8). The envelope is acquired as progeny viruses leave their host cell, during which they are enclosed in a bag made from the host's plasma membrane. Other viruses are rod-shaped helices of nucleic acid on which the protein subunits attach, resembling a tightly coiled spring.

Because they are between the living and inanimate, you may be tempted to view viruses as the "missing link"—the ancient transitional form that evolved into the first cells. However, their absolute requirement for a host cell disqualifies them as candidates for the earth's initial life forms.

None of the earth's 1.5 million known species is free of virus infection. Although not always harmful, some cause serious diseases in plants and animals. Viral diseases in humans range in severity from common colds to fatal cases of smallpox, rabies, and AIDS (Acquired Immunodeficiency Syndrome). All viral diseases have one thing in common: no antibiotics are

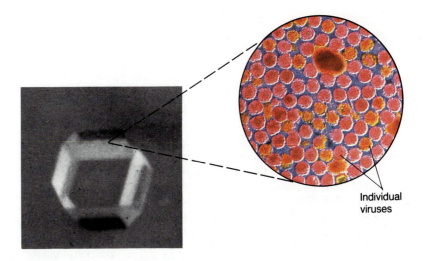

**Figure 31-7**
**"Living gems."** This crystal consists of billions of symmetrically arranged polioviruses. Even when crystallized, the viruses retain their ability to infect cells.

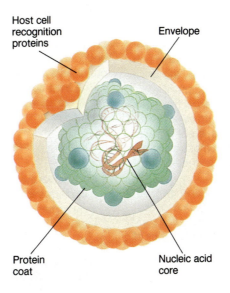

Host cell recognition proteins · Envelope · Protein coat · Nucleic acid core

**Figure 31-8**
**An enveloped virus** contains a nucleic acid core housed in a protein coat and surrounded by an outer phospholipid membrane. This outer layer is acquired as the virus escapes a host cell and takes along a modified portion of the cell's plasma membrane. Membrane modifications include the addition of proteins needed for the virus to attach to its next host cell. Some viruses are even simpler than this, consisting of nucleic acid and protein only, without the envelop.

effective in hastening recovery. Cures for viral diseases still elude medical science, mainly because viruses become a functional part of the host cell. As a result, it is difficult to find a chemical that selectively attacks the virus without similarly destroying the host. The body's natural immunity, however, eliminates most viral infections if the infected person survives long enough to mobilize defenses and control virus proliferation.

In spite of the difficulty in treating virus infections, most of the more seri-

ous ones have been controlled, largely with *vaccines*—inactivated or attenuated forms of pathogenic viruses that specifically stimulate and sensitize the immune system (Chapter 20). Vaccination has reduced or eliminated the threat of polio, smallpox, rabies, yellow fever, measles, rubella, and mumps. Some viral diseases, however, still elude attempts to control them. The most exasperating of these is the common cold; the most frightening is the disease AIDS (see Connections: A Moving Target).

**CONNECTIONS** ⎱
*Pneumocystis* infection (page 307)
T-cell immunity (page 334)
Transcription (page 460)

# FIGHTING AIDS: TRYING TO HIT A "MOVING TARGET"

The major virus that causes Acquired Immunodeficiency Syndrome (AIDS) is proving to be a formidable opponent. Not only has it already killed half the people diagnosed as having AIDS, but it may eventually kill all infected individuals. The virus has several biological "tricks" that have so far thwarted all attempts to control the infection, much less cure it. But several research teams believe they are closing in on **HIV (human immunodeficiency virus)***, as the causative agent is now known (Figure 1). Much of the search for an effective weapon against AIDS focuses on enhancing our understanding of how this virus infects human cells, replicates itself, and paralyzes a person's immune system.

## The Biology of HIV

The virus that causes AIDS attacks and kills T-lymphocytes, notably helper T-cells, a critical component of the immune system (Chapter 20). With immunity crippled by the virus, victims contract fatal cases of otherwise rare forms of infection, usually *Pneumocystis* pneumonia, or cancer (Kaposi's sarcoma). The disease is becoming more prevalent in the general population, although its incidence is still greatest among intravenous drug abusers, male homosexuals, hemophiliacs requiring frequent transfusions, and residents of countries where the AIDS virus is endemic.

The incubation period for the disease is still unknown, but it is believed to be between five and ten years. During most of this time the only thing that remains of the virus is a copy of its genetic information, filed away in the chromosomes of the host cell itself.

HIV belongs to a group of entities called **retroviruses** — RNA viruses that reverse the typical flow of genetic information described in Chapter 27 (DNA to RNA to protein). Retroviruses have an unusual enzyme, **reverse transcriptase**, that transcribes a strand of DNA using viral RNA as the template. The DNA copy, which carries all seven of the known HIV genes, becomes integrated into the T-cell's chromosome and is now called a **provirus**. The integrated provirus is replicated every time the host cell duplicates. As long as the provirus stays dormant in the chromosome, the cell remains undamaged. When viral replication is induced, however, new progeny viral RNA strands are copied from the proviral DNA. The cell fills with numerous copies of viral RNA, some of which serve as mRNA for the production of viral enzymes and proteins for assembling the viral coat. Progeny HIV assemble inside the cell. Their escape kills the host cell, ultimately depriving the immune system of helper T-cells. These events are depicted in Figure 2.

Provirus activation and subsequent development of AIDS may be triggered by a "precipitating" infection, such as in the case with herpes virus. Perhaps the chemical signals that prepare T-cells for immune combat against another infecting agent also activate the silent provirus.

## Launching a Counterattack Against AIDS

The offensive against AIDS consists of a three-pronged attack. Education is the first, and currently most effective, weapon — teaching people how to minimize the risk of contracting AIDS and how to keep from spreading it if infected. The second prong is to search for a cure, a chemical that will kill or control the virus with minimum toxicity to the infected person. At present, the drug AZT (3'-azido-3-deoxythymidine) is the only chemical licensed for use against AIDS. It restores some T-cell immunity in about 30 percent of the patients who receive it, but seriously impairs blood cell production in some recipients. Other drugs currently under study may have fewer side effects, but none so far promises a cure.

The third approach is the development of a vaccine to immunize against the virus, creating an AIDS-resistant population. The promise of immunologically conquering AIDS in the same way we have conquered smallpox and polio is hampered by three serious limitations.

- HIV attacks the very system responsible for immunity, so even in a vaccinated person, the virus may be able to incapacitate immune cells before they can launch an attack.

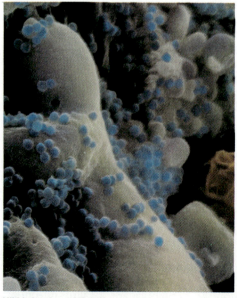

**HIV infection of T-cell**

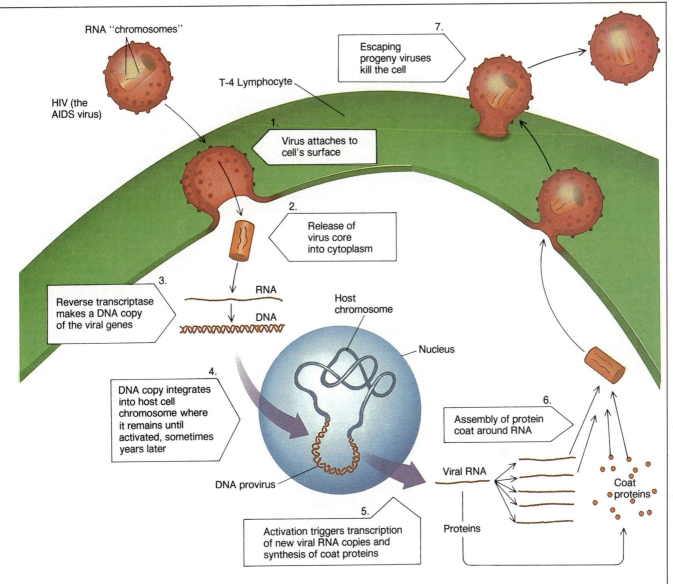

RNA "chromosomes"

HIV (the AIDS virus)

T-4 Lymphocyte

7.
Escaping progeny viruses kill the cell

1.
Virus attaches to cell's surface

2.
Release of virus core into cytoplasm

3.
Reverse transcriptase makes a DNA copy of the viral genes

RNA

DNA

Host chromosome

Nucleus

4.
DNA copy integrates into host cell chromosome where it remains until activated, sometimes years later

DNA provirus

6.
Assembly of protein coat around RNA

Viral RNA

Coat proteins

5.
Activation triggers transcription of new viral RNA copies and synthesis of coat proteins

Proteins

**AIDS virus particles** coating the surface of a white blood cell.

- Since inactivated or attenuated (weakened) virus vaccines contain whole viruses, they may reactivate, and the vaccine itself could cause AIDS.

- HIV is an immunological "moving target," one that disguises itself from the immune system by changing its antigenic identity. It has the ability to mutate so rapidly that it stays one step ahead of the immune system. By the time antibodies are produced against an HIV antigen, those antibodies find no antigen with which to react. The virus has changed that antigen into a new one, which in turn will be changed again by the time the immune system can respond to it. (The effect would be similar to a criminal being able to periodically change his face and fingerprints. He would no longer be recognizable to the police, and would be able to walk among them with no worry of detection.)

In 1987, researchers finally gave the public some good news about AIDS. They found what they called "soft spots" in the virus—a stable antigen that is common to all subtypes of HIV. This raises the hope that a vaccine consisting of this purified protein can stimulate protective immunity against all antigenic strains of the virus. This strategy also comes with a bonus. There are no whole viruses or even viral nucleic acid in such a "protein fraction" vaccine, thus eliminating the possibility of the vaccine itself causing an AIDS-like disease.

---

\* *There are actually two AIDS viruses, HIV-1 and HIV-2. Only one case of HIV-2-caused AIDS had been diagnosed in the United States as of mid-1988. The HIV virus in this essay refers to HIV-1, although much of the information also applies to HIV-2.*

# Synopsis

## MAIN CONCEPTS

- **All monerans are unicellular prokaryotes, the simplest living cells.** Prokaryotic cells are smaller than eukaryotic cells, and lack a nucleus and other membrane-bound organelles.

- **Dividing by binary fission, monerans can produce enormous numbers of offspring within a relatively short period.** Such rapid growth rates virtually assure the production of genetic variants, greatly enhancing the likelihood that some individuals will survive environmental changes. In addition, rapid growth rates may give bacteria the competitive edge over other organisms in certain habitats.

- **The Monera kingdom contains two divisions:** the Schizophyta, which includes eubacteria ("true" bacteria) and archaebacteria ("ancient" bacteria); and Cyanophyta, whose members are called cyanobacteria ("blue-green algae").

- **Monerans are found in virtually all habitats,** including environments too severe for eukaryotes to live.

- **Monerans are essential to the continuance of life on earth.** As decomposers, they recycle essential nutrients. As nitrogen fixers, they transform molecular nitrogen into a form that can be used by all organisms. As chemosynthetic autotrophs, they form the basis for some marine food chains.

- **A few bacteria are notorious for their ability to cause infectious disease, spoil our food, and decompose useful products.**

- **Viruses are noncellular entities** composed of DNA or RNA, protein, and sometimes a surrounding envelope. In a host cell, the viral DNA or RNA commandeers the cell and directs it to make a new crop of viruses.

## KEY TERM INTEGRATOR

### The Monera Kingdom (the prokaryotes)

| | |
|---|---|
| **Schizophyta** | Moneran division consisting of two groups of bacteria. These cells have a **nucleoid** instead of a nucleus, no membrane-bound organelles, and divide by **binary fission.** One group, the **eubacteria**, are "typical" bacteria that have **peptidoglycan** as the structural matrix of their cell walls. This group also includes the **rickettsia** and **chlamydiar**, both of which are **obligate intracellular parasites. Pathogens** are found among all eubacteria, although most of them are harmless or beneficial saprophytes, photosynthesizers, or **chemosynthetic organisms.** The other group, the **archaebacteria**, differ from typical bacteria in their eukaryotic-type lipids, their lack of peptidoglycan, and their ability to thrive in especially severe conditions. |
| **Cyanophyta** | Division of monerans that contains **cyanobacteria**, prokaryotic photosynthesizers that use chlorophyll *a* rather than **bacterial chlorophyll** and that perform nitrogen fixation, often with specialized cells called **heterocysts.** |

### The Viruses

| | |
|---|---|
| **Viruses** | Noncellular particles that infect a living host cell, usually redirecting the cell's biological machinery for producing progeny viruses. Some **retroviruses**, such as **Human Immunodeficiency Virus (HIV)** that causes AIDS, possess **reverse transcriptase** that makes a DNA copy of the virus's RNA genes. This DNA **provirus** integrates into the host's chromosome and silently replicates with the cell's DNA until viral propagation is activated. |

## Review and Synthesis

1. Describe three features that characterize all the members of the Monera kingdom.

2. The simplicity of prokaryotic cells creates classification and identification difficulties. What are six criteria frequently used for classifying and identifying bacteria?

3. Discuss the evolutionary relationship between eubacteria and archaebacteria.

4. Explain how the lysozyme (an enzyme that destroys peptidoglycan) in your tears, saliva, and sweat helps protect you from bacterial infections.

5. Compare the action of eukaryotic and prokaryotic flagella.

6. One serious problem in modern medicine is the overuse of antibiotics. Considering the adaptability of bacteria (rapid growth, huge numbers), why would such a practice threaten to decrease the value of antibiotics as agents to treat infectious disease? (Hint: In an environment flooded with antibiotics, bacteria with certain mutations may survive and repopulate the region.)

7. Other than wiping out infectious disease, what would be three consequences of eliminating all bacteria from the biosphere?

8. On page 562, six properties of viruses were listed. Only three of these are unique to viruses, however. Identify these three and explain why the other three properties are not unique to viruses?

9. In Chapter 20, the antiviral substance interferon was discussed. Why would pure interferon be more effective for treating virus diseases than antibiotics?

## Additional Readings

Gallo, R. 1987. "The AIDS virus." *Scientific American* 256:47–56. (Intermediate.)

McKane, L., and J. Kandel. 1985. *Microbiology, Essentials and Applications.* McGraw-Hill, New York. (Intermediate.)

Simons, K., et al. 1982. "How an animal virus gets into and out of its host cell." *Scientific American* 246:58–66. (Introductory.)

Temin, H. 1972. "RNA-directed DNA synthesis." *Scientific American* 226:24–33. (Intermediate.)

Chapter 32

# The Protist and Fungus Kingdoms

*I*t's an imperfect world, and few people are more aware of it than taxonomists. One hundred and fifty years ago everything was classified into three categories: the nonliving was placed in the mineral class, and all else was assigned to either the Animal or Plant kingdom. Being a taxonomist was simpler in those days, but even then, disturbing exceptions intruded on the tidiness of their system. Mushrooms, for example, are plantlike but, like animals, they have no chlorophyll and cannot photosynthesize. Nonetheless, these heterotrophs were forced into the Plant kingdom. So many "in-between" organisms were discovered that eventually the plant–animal dichotomy had to be declared a flop.

The first crippling blow to the two-kingdom system was delivered just over 125 years ago by a biologist named Ernst Haeckel. Growing weary of trying to "squeeze round objects into square holes," he proposed the creation of another "hole," a third kingdom called "the Protists." This kingdom contained all those organisms that had both plant and animal characteristics. But the invention of the electron microscope some 75 years later revealed that even Haeckel's scheme was not adequate. His Protist kingdom actually contained organisms comprised of two fundamentally different kinds of cells—prokaryotes and eukaryotes—that were too different to belong in the same kingdom.

Today an organism that is neither plant nor animal is assigned to one of three new kingdoms—the monerans, protists, or fungi—which is part of the five-kingdom system described by Robert Whittaker in 1969 (Chapter 1). The kingdom containing the simplest organisms (the monerans) was discussed in the previous chapter. This chapter examines the organisms in the two smallest kingdoms of eukaryotes, the protists and the fungi.

## The Protist Kingdom

Many protists are "taxonomic misfits" that continue to create problems for classifiers. This is not to say they are unfit, however. Actually, they are very well suited organisms, exceeded only by bacteria in the number of environments to which they have adapted, including some of the most inhospitable habitats in the world (Figure 32-1). Nonetheless, protists as a group are still a hodgepodge of eukaryotic organisms that, rather than sharing broad taxonomic similarities, have been assigned to the Protist kingdom more or less by default. They simply don't fit in any of the other four kingdoms. Most protists are unicellular, but many are multicellular. Some are photosynthetic, others are heterotrophic; some are funguslike, others are plantlike, and still others are animal-like. Most are microscopic, yet some are giants, such as the enormous seaweeds that grow in coastal ocean waters.

**Figure 32-1**
**Shocking pink**, this snowfield is tinted by the presence of snow algae, *Chlamydomonas navalis*, an example of the ability of some protists to thrive in conditions prohibitive to most other organisms. Heat from metabolic reactions melts snow and provides the liquid water needed for the microbe's growth.

## Evolutionary Trends

Protists are credited with several evolutionary advances, not the least of which is sexual reproduction. For some protists sex is not a reproductive strategy—they just conjugate and exchange part of their genetic library with each other. This gene mixing increases genetic diversity of the species, but reproduction continues to be asexual. For other protists, however, sex is coupled with reproduction, a process similar to that in plants and animals, at least from a nuclear standpoint. For example, a diploid protist undergoes meiosis, dividing into gametes of opposite mating types, which then fertilize each other. The resulting diploid zygote develops into a progeny individual.

The major evolutionary "breakthrough" of the protists, however, is the development of eukaryotic cellular features, such as a nucleus, mitochondria, chloroplasts, and other double-membraned organelles (Chapter 6). Three varieties of early eukaryotes evolved, the ancestors of what we today call the "higher" kingdoms (Fungus, Plant, and Animal). Modern protists still retain most of the features of these three ancestral types.

## Protozoa — Protists Resembling Animal Cells

**Protozoa** (proto = first, zoa = animals) are unicellular protists, most of which lack cell walls, ingest food particles (or absorb organic molecules), move about, and produce no multicellular spore-bearing structures. These animal-like traits distinguish protozoa from the other two groups of protists (the algae and slime molds). Some protists, however, possess both plant and animal characteristics, such as the *Euglena* species shown in Figure 32-2. These taxonomic fence straddlers possess at least one flagellum that imparts motility, and they lack cell walls. As long as they are confined to darkness, they are animal-like heterotrophs that absorb dissolved nutrients from their medium. When exposed to light, however, these versatile protists become autotrophs. Their chloroplasts proliferate, and the cell turns into a green photosynthetic, plantlike cell.

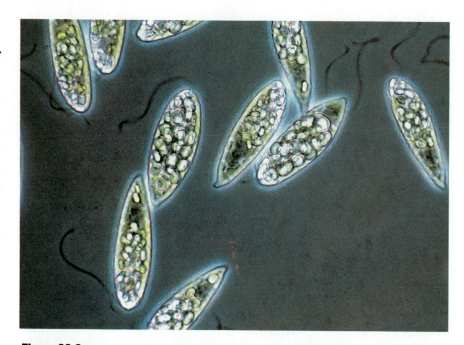

**Figure 32-2**
**Euglena—animal-like and plant-like.** This flagellated protozoon has characteristics of an animal cell, yet is green with chlorophyll like plants. Although it has no cell wall, a tough flexible layer (a pellicle) lies just beneath the plasma membrane. *Euglena* is neither plant nor animal; it is a protist.

Although most protozoa are harmless to humans, like bacteria and viruses they are often feared because of a few pathogenic species (Figure 32-3). More than half the people on earth will contract some type of protozoan infection during their lifetimes. Among the most notorious protozoa are four species of *Plasmodium* that cause malaria. The mosquito-born protozoa kill more people than any other pathogenic microorganism. Another protozoon (*Trypanosoma*) causes deadly "sleeping sickness"—encephalitis (inflammation of the brain)—after being injected into the body by the bite of an infected insect.

Not all protozoan infections are transmitted by insects. Some spread in food and water that have been contaminated with feces. *Entamoeba histolytica,* a gastrointestinal amoeba, is responsible for amoebic dysentery (bloody diarrhea) in more than a million persons in the United States alone. Many infected people die when the amoeba forms cysts in the brain. Another gastrointestinal infection is

often contracted by hikers and campers drinking from a "pure" mountain stream. Streams are often contaminated with the *Giardia lamblia* cysts shed in the feces of infected muskrats and beavers, neither of which are affected by the organism. (Boiling water before drinking kills the cysts of this protozoon and protects against the disease.) This organism may cause more infections than any other protozoon in the United States. A single infected person can spread the disease through a day care center, a school, or a hospital, especially if hygienic practices (such as hand washing after diaper changes) are lax. Another protozoon, *Pneumocystis,* has been thrust into the limelight by current events. Until the early 1980s, this organism and the disease it caused were considered rare. This is no longer so. The immune-crippling disease AIDS renders its victims especially susceptible to fatal pneumonia caused by this protozoon (see Connections: Fighting AIDS, Chapter 31). About 63 percent of fatal AIDS victims have died of

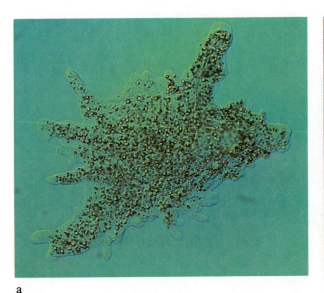

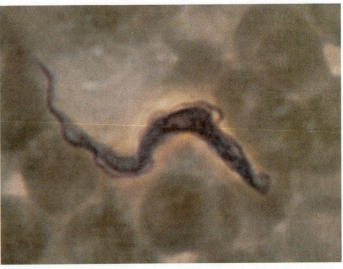

a                                                          b

**Figure 32-3**
**The delicate and the deadly.** This delicate-looking *Amoeba proteus* (*a*) is a harmless resident of stagnant water that preys on other microbes that it engulfs by phagocytosis. The other protozoon shown here (*b*) is one of the most deadly human pathogens. It is *Trypanosoma gambesia,* an insect-borne flagellate that causes African sleeping sickness (a fatal brain infection). Trypanosomes have an "undulating membrane" attached to their flagellum, forming a broad surface that helps them swim through thick body fluids.

*Pneumocystis* infections.

Yet protozoa do far more good than harm. They join with other microorganisms to decompose dead organisms and recycle nutrients. Many of these unicellular heterotrophs constitute part of the **zooplankton**, an important link in aquatic food chains. Feeding on microscopic algae, zooplankton convert the nutrients of algae (some of which, such as cell wall cellulose, are nutritionally inaccessible to animals) into protozoan tissue that is digestible by virtually all consumers. Some herbivores, notably cattle, goats, and other ruminant animals, house their own supply of "zooplankton" in their digestive tracts. These protozoa (along with cellulose-digesting bacteria) enable the animal to utilize grass and other high-cellulose food sources that they would otherwise be unable to digest (see page 53).

Most of the 40,000 species of protozoa live in water, moist soil, or inside other organisms. Dehydration is therefore not a threat. Yet some protozoa actually require temporary exposure

to drying conditions or other stressful environments (such as stomach acid) to trigger the events that complete their life cycles. Most of these protozoa are **polymorphic**; that is, they change form during different stages of their lives. When conditions are ideal, these protozoa exist as **trophozoites**, the actively feeding, growing form of the organism. As conditions become drier, hotter, or otherwise less favorable, the microbes are transformed into protective **cysts**—dehydrated, heavily encased dormant forms of the organism. Cysts withstand adverse conditions that would kill the trophozoite of the same organism (Figure 32-4).

Some polymorphic protozoa go through several hosts during the completion of their life cycles. These protozoa usually change forms as they change hosts, adopting the morphology suited for life inside each animal. (The malaria parasite's form in a human, for example, bears no resemblance to its form in a mosquito.) The existence of many forms of one organism complicates the process of iden-

tifying and classifying these species.

Classifying protozoa is traditionally based on their mode of motility (Table 32-1). These propulsive structures often double as a means of food acquisition. Protozoa move by *cilia, flagella,* or **pseudopodia** (pseudo = false, pod = foot). Pseudopodia are finger-like extensions of cytoplasm that flow forward from the "body" of an amoeba; the rest of the cell then follows. Even amoebas encased in protective shells (Figure 32-5) can use their pseudopodia for locomotion and food gathering by extending them through holes in the encasement.

Pseudopodia, cilia, and flagella are also used for food gathering. Some ciliated protozoa acquire food by using their cilia to create currents that sweep particles into the cytostome (the "mouth"). Flagella can be used as harpoons to spear prey. The role of pseudopodia in phagocytic engulfment of food particles was described in Chapter 7.

The predominant mode of reproduction among protozoa is asexual,

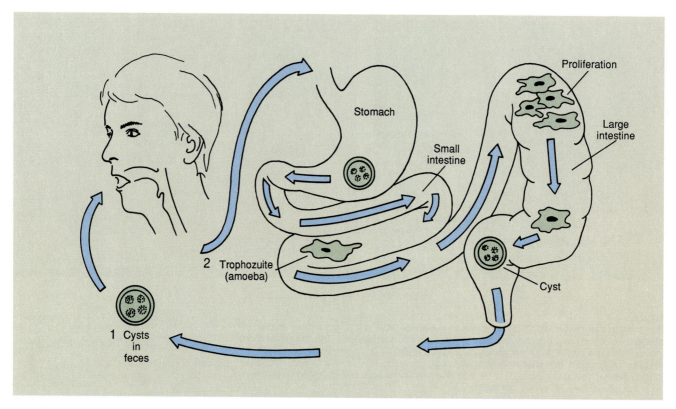

**Figure 32-4**
**Example of polymorphism.** The life cycle of *Entamoeba histolytica* (the gastrointestinal pathogen that causes amoebic dysentery) shows how the organism exists in two forms—(1) a
protective cyst (for surviving outside the host and the trip through the caustic stomach acid) and
(2) an actively growing trophozoite (for proliferation in the hospitable intestine). The trophozoites
are transformed into cysts just before being discharged in the feces.

one cell dividing into two. If the plane
of division is along the length of the
cell, the division is called **longitudinal
fission**; if it is perpendicular to the
cell's length, it is called **transverse fission**. Members of the sporozoa group
(the fourth group in Table 32-1) have
evolved a distinctive variation on this
asexual theme, a variation called **multiple fission**. The cell's nucleus divides
many times without a corresponding
division of cytoplasm (cytokinesis),
producing a multinucleated cell. The
cytoplasm is then partitioned around
each nucleus, forming a large number
of progeny cells that escape when their
cellular "container" bursts.

## Algae — Protists Resembling
## Plants

Taxonomists have trouble with algae.
They have come to little agreement on
their classification and have yet to
create a uniformly acceptable defini-

tion of them. Some biologists propose
doing away with the "algae" label once
and for all, insisting that these organisms have too little in common to belong to the same group. Our definition
of **algae** includes any unicellular or
simple multicelled photosynthetic eu-
karyote, a general description that, like
the others, meets with exceptions.

The problems with classifying algae,
however, do not undermine our recognition of their importance. Every
breath you take depends on them,
their photosynthetic activities generating three-fourths the molecular oxygen on earth. Virtually every animal in
the ocean depends either directly or indirectly on algae for food, and about
half the world's organic material is
produced by algae. In addition, some
species of anemones, corals, clams,
arthropods, and other animals have
gained a competitive edge by forming
mutually beneficial partnerships with

algae. The animal provides the photosynthetic residents protection within
their bodies and $CO_2$ for photosynthesis; the algae supply the animal with
supplemental food and oxygen (Figure
32-6).

Algae have also been recruited as a
modern weapon against human starvation. Using sunlight energy and
human waste, the alga *Chlorella* transforms our refuse into edible food
(called *single cell protein*) that can be
fed to livestock and, when perfected,
directly to billions of people. Furthermore, oceanic kelp beds not only harbor a rich variety of marine life, but the
kelp is also a source of soil fertilizer
and a variety of other commercial
products, such as algin used as a thick-
ener in foods and hand cream.

The size and complexity of algae
ranges from relatively simple single-
celled species to large, complex multicellular forms. But even the most

TABLE 32-1

| Classification of Protozoa According to Motility | | | |
| --- | --- | --- | --- |
| Motility Group | Organelle of Locomotion | Characteristics | Example |
| Mastigophora | Flagella | Heterotrophic or, in some cases, photosynthetic; excess water expelled through contractile vacuole; binary fission along long axis; no sexual reproduction | *Euglena* |
| Ciliata | Cilia | Cilia sweep food particles into mouth; complex cell morphology; many types expel excess water through contractile vacuole; genetic transfer by conjugation; asexual reproduction by fission across long axis | *Paramecium* |
| Sarcodina | Pseudopodia | Amoeba move by extending "fingers" of cytoplasm; feed by phagocytosis; reproduce asexually by binary fission; some species encased in shells of silica or calcium | *Amoeba* |
| Sporozoa | None | Trophozoite stage nonmotile; intracellular parasites; complex life cycles that alternate between sexual and asexual reproductive modes; one species may require two or more different hosts to complete life cycle | *Plasmodium* (agent of malaria) |

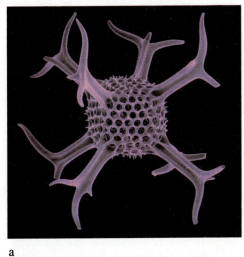

a

b

**Figure 32-5**
**Shelled amoebas.** (a) The scanning electron microscope captures the beautiful intricacy of some radiolaria. (Complexity doesn't require multicellularity.) Pseudopodia protrude through the holes in the shells to gather food particles. The intricate shells are composed of silica; they are literally glass. (b) Foramnifera enclose themselves in "snail-like" shells of calcium carbonate. Geological upheaval raised sediments of enormous numbers of shells deposited millions of years ago when these amoebas abounded in oceans. The result is the White Cliffs of Dover in England.

a                                        b

**Figure 32-6**
**Biological teamwork.** Algae living inside the bodies of animals provide supplemental food and oxygen (as well as their unusual green color). A few such "photosynthetic animals" are shown here. They include some species of *Hydra* (a), clams (b), and anomenes (see Figure 37-1).

complex algae still lack many of the structures found in members of the Plant kingdom. They have no roots, stems, or leaves, and none can produce flowers or seeds. It is their photosynthetic capacity that most closely resembles plants, but some even lack this trait. Nonphotosynthetic algae are heterotrophs, either feeding on particulate organic matter or, in at least one case, parasitizing a living host. One parasitic genus, *Prototheca,* infects people and causes a crippling skin infection of the feet.

Every type of alga falls into one of the six divisions described below. (The first three divisions are unicellular algae, and the last three are multicellular.) All these organisms are composed of eukaryotic cells. In fact, there is no such thing as prokaryotic algae. Organisms once called "blue-green algae" are now called cyanobacteria (Chapter 31).

### Divisions of Algae

• *Chrysophyta*—golden brown and yellow-green algae. Chrysophytes are the major producers of food for all animals living in marine waters.

Although yellow-green algae are included in this group, many taxonomists place these algae in their own division (Xanthophyta). Other algae in this division are the **diatoms**, golden-brown algae distinguished most dramatically by their intricate silica shells (Figure 32-7). Diatoms secrete their delicate "glass" coats of armor, each of which is perforated with thousands of tiny holes to allow contact between the environment and the enclosed photosynthetic cell. Diatom shells are of great economic significance, ideal for use in polishing, filtering, and insulating materials. Commercially, they are referred to as "diatomaceous earth," harvested from enormous geological deposits that resemble fine, white powder. These deposits accumulated for millions of years, the settling diatom shells forming thick layers on the ocean floor. Many of these deposits have been thrust to the surface by geological activity.

The golden color of chrysophytes is provided by accessory pigments that assist chlorophyll in photosynthesis. These pigments are precursors of the fat-soluble vitamins A

and D. These vitamins are concentrated in the liver oils of fishes that dine on chrysophytes. For years people have harvested fish liver extracts as rich sources of these vitamins. (Who can forget the taste of cod liver oil?)

• *Pyrrophyta*—"fire algae." The cells in this group are the **dinoflagellates**, single-celled photosynthesizers that have two flagella. One flagellum moves the alga through its medium, and the other spins the cell on its axis. Dinoflagellates are important to the marine food chain (second only to diatoms). Ironically, periodic dinoflagellate "blooms" are often hazardous to many members of the food chain. **Blooms** are massive growths of algae that occur when conditions are optimal; **red tide** is such a bloom. During a red tide, growth of one red-pigmented dinoflagellate species is so extensive that it tints the coastal waters and inland lakes a distinctive red (Figure 32-8). Some red tide dinoflagellates produce a powerful nerve toxin that kills any fish that eats them. During red tide, the water's surface may be

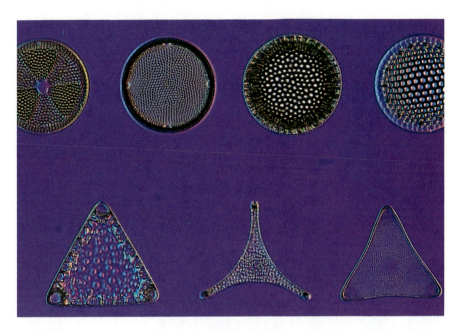

**Figure32-7**
**Gallery of diatoms** reveals the beauty of the silica shells secreted by these golden brown algae. The living cells that occupied them have long since died.

blanketed with dead fish.

In contrast, the nervous systems of shellfish are not susceptible to the toxin, and so to them red tide is a nutritional gold mine. The toxin becomes concentrated in the tissues of oysters, mussels, and clams. Although the shellfish are unharmed, people who eat them may consume enough toxin to cause paralysis, often beginning within an hour of consuming the toxin. Within eight hours, the victim may die of *paralytic shellfish poisoning,* even with medical treatment.

- *Euglenophyta*—This group of unicellular algae (discussed earlier in this chapter) includes *Euglena* and similar protists (called euglenoids) that are both plantlike and animal-like. Exposure to light induces their plantlike property, the production of chlorophyll *a* and *b* (supplemented with carotenoids). The typical euglenoid (Figure 32-2) retains its flagella in its alga form; its movement is directed by its response to light. An "eyespot" (photoreceptor) at the base of the flagellum senses the direction of the light source, enabling

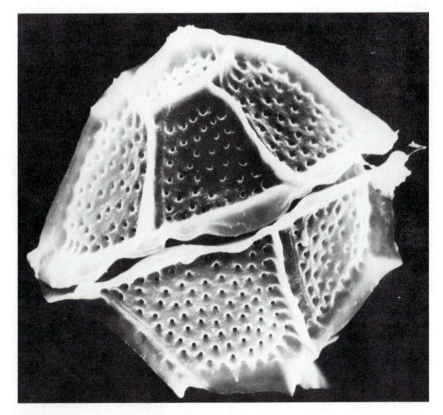

**Figure 32-8**
**Intriguing but deadly.** Some dinoflagellates, such as this *Gonyaufaux,* have killed countless fish and people. Dinoflagellate blooms tint the ocean red (and bioluminescent at night, inspiring their division name, the fire algae).

the organism to swim toward or away from the light to reach the optimal brightness.

Most euglenoids are freshwater protists, although a few are found in the gastrointestinal tracts of animals. Like some ciliated protozoa, euglenoids eliminate excess intracellular water by collecting it in a contractile vacuole and periodically squeezing out its contents.

• *Phaeophyta*—brown algae. This group of multicellular algae, mostly seaweeds, are distinguished by the presence of an accessory pigment (fucoxanthin) found in no other group. The brown pigment helps them gather blue-green light, the wavelengths that penetrate deep into water. Some phaeophytes are the giants of the Protist kingdom. Underwater "forests" of giant kelp contain individuals long enough to firmly attach to the ocean bottom, with their tops reaching the light-rich ocean surface 300 feet above (Figure 32-9a).

Members of the brown algae have developed specialized structures and tissues, such as holdfasts, blades, floats, and conducting vessels. *Holdfasts* anchor the alga to rocks on the ocean bottom; leaflike *blades* provide broad surfaces for light absorption, boosting photosynthetic efficiency; hollow gas-filled *floats* help keep the seaweed upright in the water; and newly synthesized nutrients are shipped from blades down to the holdfasts and lower parts of the organism through con-

a

b

c

**Figure 32-9**
**Multicellular algae.** (*a*) Forests of giant kelp typify the brown algae. The "forest" stays in place by virtue of holdfasts that anchor each organism to the solid ocean floor. Shown here is *Macrocystis*. (*b*) Red algae. (*c*) Green algae form the multicellular link between protists and plants. This *Volvox* colonies consist of hundreds of cells attached together, each spherical colony functioning as a single organism. Daughter colonies inside the mature "adults" are released when the parental colony disintegrates.

ducting tubes reminiscent of phloem in plants.

The brown algae alternate generations between sporophyte and gametophyte stages (Chapters 12 and 21). The large kelp form is the organism's sporophyte stage, the diploid, spore-producing form. Compared to the sporophyte, the haploid gamete-producing stage is usually very tiny, in both size and duration of its existence.

- *Rhodophyta*—red algae. Like brown algae, rhodophytes possess supplemental pigments that absorb deep-penetrating light rays, expanding the range of ocean floor they can inhabitat. These accessory pigments transfer absorbed energy to chlorophyll *a* and *c*. Their accessory pigments also provide color to the 4000 species of these seaweeds, color that ranges from red, blue, or purple to a very dark green.

Except for a few freshwater red algae and free-floating marine forms, rhodophytes attach to the bottom of oceans or estuaries, sometimes hundreds of feet deep if waters are clear enough for sufficient light to penetrate and power photosynthesis. Although much smaller than the giant brown algae, they are easily seen and collected by divers. They are harvested for their many commercial uses, ranging from food (the dark seaweed wrapped around the raw fish of sushi is a red alga) to the production of *agar,* the agent used to solidify nutrient media for growing bacteria, fungi, and other microbes. Rhodophytes are also important members of marine food chains. Some species deposit calcium carbonate that enlarges the structure of coral reefs.

- *Chlorophyta*—green algae. With 7000 species, **chlorophyta** is the largest group of algae. Green algae are the only group found mostly in freshwater habitats (although there are a few marine species), and some are even adapted to terrestrial life, blanketing trees, soil, and porous rocks (or bricks). Most biologists believe that the green algae bridge the evolutionary gap between algae and true plants. Unlike other algae,

chlorophytes share the following characteristics with higher plants.
  —Chlorophyll *a* and *b.*
  —Carotene accessory pigments.
  —Ability to store surplus carbohydrates as starch.
  —Cell walls of cellulose.

Many green algae are unicellular; others are multicellular or colonial (Figure 32-9c). Colonial forms are intermediate between unicellular and multicellular organisms. Individual cells in the colony retain a high degree of functional independence in spite of their physical attachments to other (usually identical) cells in the colony. In true multicellular green algae, cell activities are coordinated and the cells remain tightly bound to adjacent cells, as they are in higher plants. Multicellularity was probably first developed by green algae, an ancestor of higher plants.

One unicellular green alga genus, *Chlorella*, often forms symbiotic associations with some animals discussed earlier (see Figure 32-6). It is also a favorite of scientists in search of new food and oxygen resources for people. *Chlorella* has even been proposed as a food and oxygen generator for astronauts on long space voyages.

Another well-studied single-celled genus is the flagellated *Chlamydomonas* (a green alga that is not always green, as shown in Figure 32-1). Although it can reproduce asexually, this alga also has different *mating types* ("opposite sexes"). Following union of opposite mating types, the resulting zygospore exists for one cell cycle, during which it divides by meiosis to form the haploid nuclei of new mating types.

Many multicellular green algae undergo a parallel reproductive pattern. Like the mating types of *Chlamydomonas,* their gametes are indistinguishable regardless of their mating type. Furthermore, the sporophyte and gametophyte stages of some green algae are also indistinguishable in form. For example, when the flagellated gametes of *Ulva* (sea lettuce) fuse, the zygote proliferates into a green leafy seaweed two cells thick. This diploid sporophyte then forms haploid gametes, each of

which develops into a haploid seaweed that looks identical to the diploid "leaf." This gametophyte then produces gametes, which mate with the opposite type. The resulting zygote develops into a new leafy sporophyte.

In other green algae, the female gametes are large and non-motile, whereas the male gametes are small and motile (analogous to the situation in higher plants and animals). In most of these, indeed in most chlorophytes, the sporophyte is very reduced, the more obvious form of the organism being the haploid gametophyte.

## Slime Molds—Protists Resembling Fungi

Although one would never guess it from their name, **slime molds** are distinctly beautiful. The sexually mature organisms are intricate and attractive, displaying delicate reproductive structures similar to those of molds (discussed in the following section).

Slime molds are either *cellular* or *plasmodial*. **Cellular slime molds** exist as microscopic amoeba-like cells as long as food and water are plentiful. They patrol their habitat for bacteria, spores, and organic debris to engulf and digest, rapidly multiplying as a result. When their food is depleted, the "amoebas" release one of the most developmentally important chemicals known, *cyclic AMP*. This substance is used by virtually all organisms to direct development or govern gene expression (Chapters 16 and 28). Hundreds of amoebas migrate up the cAMP concentration gradient to the point of highest concentration. Here they aggregate into a single unit, a slime-covered "slug," that crawls for a while (which may help disperse the species) and then transforms itself into a funguslike structure. In this form the slime mold produces a **fruiting body**, a spore-producing structure that extends upward in an elevated position. When the spores are released, they are carried on wind or on the surface of animals to new locations. If conditions are suitable, the spores germinate, each forming another amoeba-like cell, and the cycle repeats.

**Plasmodial** (sometimes called

*acellular*) **slime molds** also have a funguslike spore-producing stage in their life cycle. These organisms have no migrating "amoebas," however, instead they produce a **plasmodium**, a huge multinucleated "cell" that feeds on dead organic matter. (A single plasmodium can grow large enough to cover an entire log; see Figure 32-10.) This giant growing mass then forms fruiting bodies, which release spores that germinate into gametes if they reach a location with suitable conditions. Fertilization between gametes yields a zygote that develops into another multinucleated plasmodium.

## The Fungus Kingdom

In the mid-1800s, thousands of Irish people began to notice a terrifying transformation. Potatoes, their major source of nutrition, were turning black. The Irish helplessly watched their potatoes rot. The following spring, the plants themselves were affected, and the food crop of an entire nation failed. The disaster of the potato blight continued for 14 years, and caused more than a million Irish

people to starve to death. Today we know the killer to be a member of the Fungus kingdom.

The Irish potato blight is one of several large-scale disasters inflicted on people by fungi. The vast majority of fungi, however, are beneficial, not just to people, but to the whole biosphere. The following list of fungal activities illustrates a few of their contributions.

- Decomposition of dead organisms and recycling of the nutrients.

- Physical breakdown of solid rock into soil (see lichens below).

- Formation of *mycorrhizae* ("fungal roots"), filaments that grow in intimate association with plant roots, increasing nutrient and water absorption by the plant (Chapters 13 and 14).

- Production of foods, such as cheese, mushrooms, and soy sauce.

- Fermentation of sugar to ethyl alcohol, generating beer, wine, and spirits.

- Synthesis of valuable industrial solvents.

- Production of important therapeutic compounds, notably penicillin and several other antibiotics that save millions of lives that would otherwise be lost to diseases caused by bacteria (Figure 32-11).

## Overview of the Kingdom

Because of their cell walls, fungi were once considered members of the Plant kingdom. But a lack of photosynthesis and a heterotrophic nutritional strategy disqualifies them as plants. Their eukaryotic cellular structure eliminates them as candidates for the Monera kingdom. In 1969, a new and separate kingdom was recognized, a kingdom exclusively for the fungi.

Some members of the Fungus kingdom, notably the **yeasts**, are unicellular, forming colonies similar to those of bacteria. **Filamentous fungi** constitute the multicellular members of the kingdom, organisms comprised mostly of living threads that grow by division of cells at their tips. These filaments, called **hyphae**, are composed of a linear group of cells. The filamentous fungi, most of which are called **molds**,

a                                                        b

**Figure 32-10**
**Plasmodial slime molds,** two dramatically different forms of which are shown here. The protozoalike plasmodium of *Physarium* is an enormous multinucleated "cell" that consumes dead organic matter (*a*). Spore-producing structures (*b*) of slime molds resemble those of fungi.

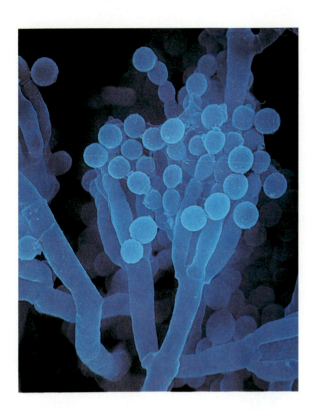

**Figure 32-11**
**Destroyer and savior.** The mold *Penicillium* spoils millions of tons of food each year. (It is the blue-green mold you find on old oranges, bread, and cheddar cheese.) Yet the organism also produces the life-saving antibiotic penicillin. In this photograph, the mold's asexual spores are clearly visible at the tips of special "brushlike" spore-forming structures.

typically grow as fluffy masses. Molds are highly efficient exploiters of available nutrients. During peak growth, a mold colony grows more than half a mile of new hyphae in a single day.

One group of filamentous fungi, however, is not considered a mold, instead, they are called **macrofungi** to signify the large size of their fleshy sexual structures. A mushroom, for example, is so large and elaborate that it is often mistaken for the whole organism. Yet most of the fungus is growing as an unseen filamentous mass under the ground or in the tissues of a tree, with only its large reproductive structures showing.

Fungi therefore range in size from microscopic single-celled yeasts to the large fleshy macrofungi, and there are thousands of species in between, including the fungi that collaborate with algae to form **lichens**. Each member of the lichen contributes to the ability of these "composite organisms" to thrive in a wide range of habitats. In dry environments or those that are poor in organic nutrients, for example, the lichen's photosynthetic alga uses light energy to generate organic nutrients from inorganic compounds, while the fungal filaments gather and conserve what little water is available. Together the alga-fungus team readily grows in conditions that neither the fungus nor the alga could survive alone (Figure 32-12). In some harsh environments, lichens support whole food chains, sustaining even such large consumers as the reindeer caribou of frozen northern Alaska.

## Fungal Nutrition

Fungi are heterotrophs, with most of the kingdom's members obtaining their nutrients by decomposing dead organisms. Yet a few types of fungi don't wait until an organism is dead before they start consuming it. Most of these fungi are pathogenic, causing diseases in both plants and animals. In fact, fungal diseases of people constitute some of the most common and persistent diseases that still challenge modern medicine. It is believed that about 20 percent of the world's population suffers from fungal diseases, ranging from irritating maladies of the skin and mucous membranes (athlete's foot, ringworm, and vaginal yeast infections) to life-threatening systemic infections (histoplasmosis and valley fever).

Plant pathogens can also have disruptive consequences, both to the plant and to organisms that depend on the plants. In addition to the disastrous Irish potato blight, for example, rust and smut fungi continue to reduce human food supplies by attacking agricultural grains. The devastating Dutch elm disease that has downed so many beautiful trees is also caused by a fungus, this one introduced into the tree by the larvae of the Gypsy moth.

## Sex and the Fungi

Except for mushrooms and other macrofungi, most members of this kingdom only infrequently resort to sexual reproduction, relying primarily on asexual reproduction to increase their numbers. Yeast, for example, form small "buds" that enlarge and finally break away from the genetically identical parent. Asexual reproduction is more varied among filamentous fungi. Fragments of hyphae may break away from the parent mold and grow mitotically, forming new individuals. Many

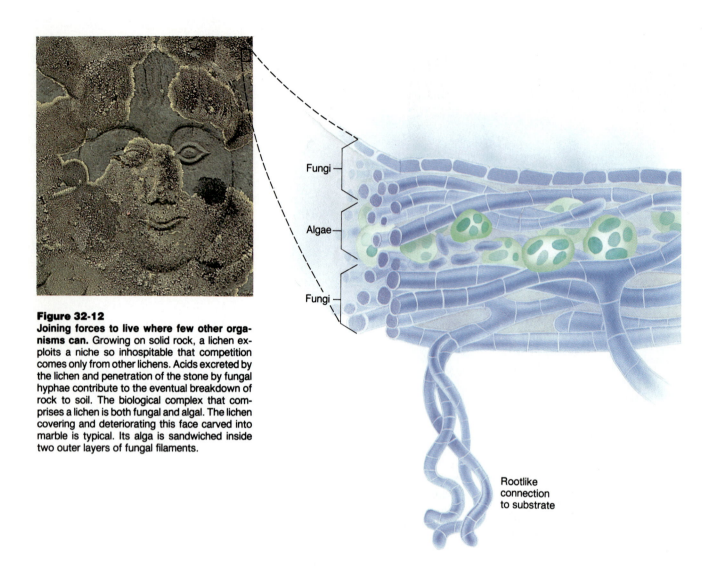

**Figure 32-12**
**Joining forces to live where few other organisms can.** Growing on solid rock, a lichen exploits a niche so inhospitable that competition comes only from other lichens. Acids excreted by the lichen and penetration of the stone by fungal hyphae contribute to the eventual breakdown of rock to soil. The biological complex that comprises a lichen is both fungal and algal. The lichen covering and deteriorating this face carved into marble is typical. Its alga is sandwiched inside two outer layers of fungal filaments.

Fungi

Algae

Fungi

Rootlike connection to substrate

fungi produce asexual spores, which are then released from sporulating structures and germinate into actively growing hyphae.

When fungi sexually reproduce, the diploid stage is short-lived. The zygote produced by fertilization immediately undergoes meiosis to form haploid spores. All the cells in the fungus are therefore haploid, having descended from one haploid spore.

Fungi have a number of sexual strategies, most of them characterized by the formation of a fruiting body that bears the sexual spores. The type of sexual spore produced provides a common criterion for categorizing the fungi. Four of these classes are discussed here (Table 32-2). Another four classes in the fungal division Mastigomycota (the flagellated "lower" fungi) are not included in the table, even though they contain important animal and plant pathogens, including the fungus responsible for the Irish potato blight. The spore types identified in the table are described in the following discussion of each fungal class.

## Classes of Terrestrial Fungi
### ZYGOMYCETES

The black mold growing on a loaf of bread typifies the zygomycetes. These organisms begin as a white cottony mass of *nonseptate* hyphae (see footnote in Table 32-2). One or two days later, the filaments produce sporulating structures that resemble balloons on the end of sticks, held upright by a cluster of *rhizoids* that resemble roots. Each "balloon" is really a sac, a *sporangium,* filled with black asexual **sporangiospores** (Figure 32-13). It is the production of these spores that darkens the mold as it matures. When sporangia rupture, millions of haploid spores are released, some blowing to new locations by the slightest breeze. Spores that settle in favorable conditions germinate into actively growing hyphae and establish new mold colonies.

Although zygomycetes produce no

TABLE 32-2

| The Four Classes of Terrestrial Fungi | | | | |
|---|---|---|---|---|
| **Class** | **Sexual Spores** | **Asexual Spores** | **Hyphae** | **Representative Genera** |
| Zygomycetes | Zygospores | Sporangiospores | Nonseptate* | *Rhizopus* (bread mold) |
| Ascomycetes | Ascospores | Many different types (e.g., conidia and arthrospores) | Septate | *Saccharomyces* (brewer's yeast) *Penicillium* (antibiotic producer) Several genera of morels and truffles |
| Basidiomycetes | Basidiospores | Virtually none | Septate | *Agaricus* (supermarket mushrooms), plus hundreds of other mushroom genera; rusts and smuts (plant pathogens) |
| Deuteromycetes | None | Conidia; arthrospores | Septate | *Candida* (causes vaginitis and other human infections); many genera of pathogenic fungi that cause serious diseases of people |

*\* Nonseptate hyphae are continuous tubes of cytoplasm, whereas septate hyphae have crosswalls that divide the adjacent cells from one another.*

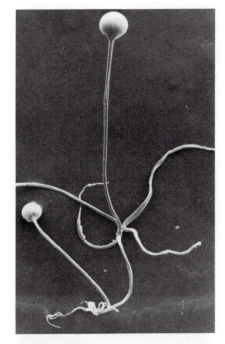

**Figure 32-13**
**Packed with sporangiospores,** the sporangia of this zygomycete (Rhizopus) are ready to burst open and discharge their reproductive cargo. The nonseptate hyphae clearly lack partitions (septa) between adjacent nuclei.

sexual fruiting bodies, they do reproduce sexually. When hyphae of opposite mating types grow in contact with each other, fertilization generates a temporary diploid state. The diploid cell forms a dark warty coat around itself as it develops into a **zygospore**. The zygospore undergoes meiosis (often after months of dormancy), germinates, and develops into a typical sporangium containing haploid sporangiospores.

Zygomycetes not only spoil food, but some also cause serious, often fatal, diseases of people, especially persons with compromised natural defenses. On the brighter side, several human adversaries are also on the fungal "hit list," notably destructive insects and parasites. At least 50 species of fungi, most of them zygomycetes, are quite efficient at trapping or snaring these animals for dinner. Victims of these fungi include small worms (nematodes) that live in soil and destroy the roots of many crop plants and trees. The nematodes are trapped when they crawl through special loop-shaped hyphae of the fungus *Arthrob-*

*otrys.* The loop clamps shut when stimulated, snaring the worm, which is then digested by fungal enzymes and absorbed into the hyphae. Houseflies, grasshoppers, and other insects also fall prey to carnivorous zygomycetes. Scientists are now evaluating the feasibility of using carnivorous fungi as biological insecticides against many types of insect pests—from aphids that destroy millions of dollars of citrus fruit each year to cockroaches in the home.

## ASCOMYCETES

Ascomycetes are often referred to as "sac fungi" because they house their sexual **ascospores** in a sac, called an *ascus* (plural = asci). In unicellular ascomycetes, typified by the common yeast *Saccharomyces cerevisiae* (brewer's and baker's yeast), the cell itself becomes an ascus filled with four ascospores. But most ascomycetes are multicellular and filamentous, some forming very complex fruiting bodies, called *ascocarps,* that contain the spore-filled asci (Figure 32-14). The linear alignment of ascospores in *Neurospora crassa* has allowed geneticists to study genetic crossovers and meiotic distribution of genes. The speed with which the fungus sexually reproduces helped make *Neurospora* a favorite of these researchers, who sometimes refer to it as "the microbiological fruit fly." Investigators merely have to observe the distribution of spores with particular traits (such as color) in an ascus to obtain results of genetic crosses.

To others, the ascomycetes seem more like assassins than assistants. One species causes the blight that has virtually eliminated chestnut trees from the American landscape. Other ascomycetes have downed many other stately creatures, including millions of elms that succumbed to Dutch elm disease. People also fall victim to ascomycetes, which include the agents of dangerous systemic diseases, such as *histoplasmosis* (a systemic infection that starts in the lungs), as well as some forms of common "ringworm" infection.

**Figure 32-14**
**An ascomycete.** The inner surface of the cup fungus is lined with thousands of sacs (asci) that contain the sexual ascospores.

## BASIDIOMYCETES

The members of this class are not only the most familiar fungi (Figure 32-15), but also the most sexually active. The size and complexity of their sexual reproductive structures are paralleled by the importance of sex to the basidiomycetes. Mushrooms and puffballs are "fleshy" structures of compactly intertwined hyphae. The hyphae that form the partitions ("gills") in the cap of a mushroom form millions of club-shaped sporulating structures called *basidia*. Each basidium bears four **basidiospores**, formed by the process shown in Figure 32-16. When mature spores escape the basidiocarp (the fruiting body), they are borne by wind or animals to new locations, thereby dispersing the species.

The fleshiness of basidiocarps is often accompanied by a delicious taste, but a few are also endowed with deadly toxins for which there are no antidotes. Distinguishing between safe and poisonous varieties sometimes requires microscopic examination.

In addition to poisonous mush-

a                                          b

**Figure 32-15**
**Basidiomycetes—fleshy fungi.** These leaf fungi are basidiomycetes that decompose dead organic matter (a). Sexual basidiospores are often so abundant that they form a smokey discharge, as from this puffball (b).

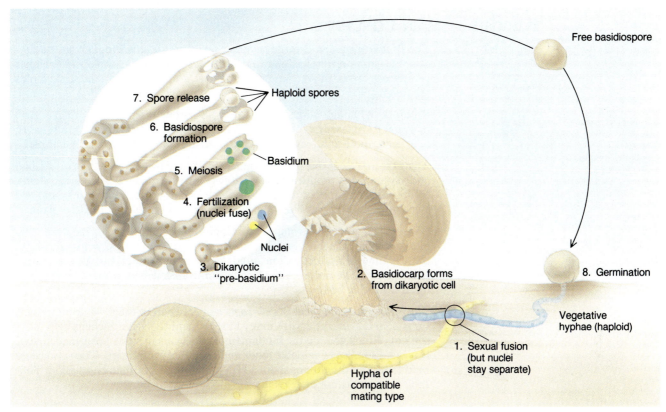

**Figure 32-16**
**The mushroom's sexual cycle.** A mushroom's fruiting bodies (basidiocarps) develop following fusion of sexually compatible hyphae, creating a *dikaryotic* cell (two nuclei per cell). Proliferation of this cell gives rise to a basidiocarp composed of dikaryotic cells. The two nuclei fuse during fertilization in the gills of the basidiocarp to form a single diploid nucleus. The resulting diploid cell (the basidium) immediately undergoes meiosis, forming four haploid nuclei, each of which ends up in a basidiospore. The cycle begins again when hyphae from a germinated basidiospore come in contact with hyphae from another, sexually compatible basidiospore.

rooms, detrimental basidiomycetes include the food-robbing plant pathogens wheat rust and smut, as well as tree-killing bracket fungi.

## DEUTEROMYCETES

The deuteromycetes are sentenced to a life without sex. By definition, these are fungi in which sexual reproduction has not (yet) been discovered. Often referred to as "the imperfect fungi," they lack a *perfect stage,* as the sexual cycle is often called. Some species of deuteromycetes probably lack the genetic mechanism needed for sexual reproduction. Others will remain in this class only until scientists discover their elusive sexual stage. Each year more fungi are removed from the ranks of the deuteromycetes and assigned to the ascomycetes or basidiomycetes according to the type of sexual spore produced. The genus *Penicillium,* for example, is traditionally presented as an important example of a deuteromycete. Yet *P. chrysogenum* (the species that produces the antibiotic penicillin) and *P. roqueforti* (the blue mold that imparts the colorful ribbons that flavor roquefort and blue cheese) were both recently discovered to be ascospore-producing, sexually active fungi.

The most common of all deuteromycetes is *Aspergillus,* perhaps the most ubiquitous organism in the world. It has been found even in the stratosphere, at an altitude of 100 miles. To most people, it is a pest, spoiling food and accelerating the deterioration of useful products. For a few people, it is also deadly. Persons with inadequate immunity often cannot combat this organism. Since exposure to the spores of this fungus is virtually unavoidable, immune-compromised people often die of overwhelming *Aspergillus* infections. *Aspergillus* also compounds the misery of hay fever sufferers, many of whom are allergic to its spores.

# Synopsis

## MAIN CONCEPTS

- **The Protist kingdom.** Protists have few properties in common; all are eukaryotic, but some are unicellular and others are relatively simple multicellular complexes. Protozoa are "animal-like," algae are "plant-like," and slime molds are fungus-like.

- **Protozoa.** Although some protozoa cause serious diseases in humans, most contribute immeasurably to the biosphere. They are critical links in food chains, help decompose and recycle nutrients, and are essential to livestock animals.

- **Algae are indispensable to the biosphere.** They produce most of the world's food and oxygen. They produce the majority of the living material in aquatic systems. Multicellular members include economically important giant kelp and red algae. A green alga was likely the evolutionary forerunner of modern plants.

- **Slime molds resemble protozoa and fungi.** Cellular slime molds have amoeba-like vegetative cells that aggregate and produce a sporulating structure similar to those of fungi. Plasmodial slime molds form a plasmodium, a single giant "cell" with hundreds of nuclei. The plasmodium transforms into spore-filled reproductive fruiting bodies.

- **The Fungus kingdom consists of yeasts and filamentous fungi.** As decomposers, fungi help sustain the earth's many biogeochemical cycles. Lichens help make soil out of rock and are the fundamental producers in some polar food chains. Fungi are sources of valuable foods and resources for humans. They also cause mild to serious diseases in about a fifth of the world's human population. They spoil food and cause diseases in crop plants.

## KEY TERM INTEGRATOR

### Protist Kingdom

| | |
|---|---|
| **protozoa** | Unicellular (or rarely colonial) eukaryotic heterotrophs that lack cell walls. Many are **polymorphic**, alternating between **trophozoite** and **cyst stages**. They are categorized according to the mechanism for motility, which may also double as a food-capturing structure. **Pseudopodia**, for example, propel amoeba and are used to engulf particulate nutrients. They divide by mitosis, leading to **transverse fission**, **longitudinal fission**, or simultaneous release of dozens of progeny cells following **multiple fission**. As **zooplankton**, they constitute a fundamental link in the marine food chain. |
| **algae** | Photosynthetic eukaryotes that grow as single cells, colonial, or simple multicellular organisms. They are fundamental to marine food chains, especially **diatoms** and **dinoflagellates**. (Some of the dinoflagellates are poisonous and, during **blooms** called **red tide**, are frequently associated with fatal poisoning of fishes and people.) |
| **slime molds** | One type of which, the **cellular slime molds**, forms a slimy "slug" when amoeba-like cells aggregate, then differentiate into a spore-producing **fruiting body**. **Plasmodial slime molds** form a huge, multinucleated, amoeboid mass, a **plasmodium**, that transforms into the spore-bearing structures. |

| Fungus Kingdom | |
|---|---|
| **fungus** | Eukaryotic nonphotosynthetic heterotroph with cell walls. Single-celled fungi are **yeasts**; multicellular species are **filamentous fungi**, most of which are **molds** or **macrofungi** (mushrooms, puffballs, bracket fungi, and morels). These filaments, called **hyphae**, may give rise to asexual or sexual sporulating structures. **Sporangiospores** are examples of asexual mold spores, whereas **zygospores**, **ascospores**, and **basidiospores** are sexual fungal spores. |
| **lichen** | Composite organism consisting of fungi and algae. |

## Review and Synthesis

1. Explain why both the two-kingdom and three-kingdom systems were taxonomic flops.

2. Determine which of the three protist groups each of these organisms belongs to: *Euglena; Pneumocystis; Volvox; Trichomonas vaginalis* (a flagellated heterotroph that causes vaginal infections in humans); *Trichonympha* (a flagellated cellulose-digesting heterotroph in the gut of termites).

3. Describe four ways we benefit from algae or their products.

4. Some biologists classify slime molds as fungi. Defend the position of these biologists. What properties justify placing slime molds in the Protist kingdom?

5. Describe at least one beneficial and one detrimental activity of the organisms in each group that follows: Mastigophora; Sarcodina; Rhodophyta; Pyrrophyta; Basidiomycetes; Ascomycetes.

6. How can each of the modes of motility used by protozoa also provide a means of capturing food?

7. Provide four distinctions between true algae and "blue-green algae" (cyanobacteria from Chapter 31).

8. Fertilization in basidiomycetes occurs in two stages. Why, then, are all the cells in a basidiocarp (e.g., a mushroom) dikaryotic (containing two nuclei per cell)? Describe the final fertilization stage and how it produces basidiospores.

9. Discuss the mutual benefits for each organism in the following biological "teams": a "photosynthetic" hydra; mycorrhiza and a pine tree; a lichen; a cow and the cellulose-digesting protozoa in her digestive tract.

## Additional Readings

Baker, D. 1985. "Giardia!" *National Wildlife,* August-September. (Introductory.)

Kessel, R., and C. Shih. 1976. *Scanning Electron Microscopy in Biology: A Student's Atlas on Biological Organization.* Springer-Verlag, New York/Berlin. (Introductory to intermediate.)

McKane, L., and J. Kandel. 1985. *Microbiology: Essentials and Applications.* McGraw-Hill, New York. (Intermediate.)

Margulis, L., and K. Schwartz. 1982. *Five Kingdoms: An Illustrated Guide to the Phyla* W.H. Freeman, San Francisco. (Intermediate.)

Vidal, G. 1984. "The oldest living cells." *Scientific American* 250:48–57. (Intermediate to advanced.)

Chapter 33

# The Plant Kingdom

A fire rages through a forest. Many animals escape the flames, but the plants cannot flee. Anchored to the ground, they must stand against many life-threatening hazards—violent winds, floods, sleet, herbivores, and even fire. Despite what seems to be an enormous handicap, plants not only manage to survive in virtually all habitats, but have become the most prominent form of life on earth.

Of the three most diverse kingdoms of organisms, only members of the Plant kingdom are autotrophs. Plants manufacture their own food, through photosynthesis, using sunlight to combine $CO_2$ from the air and water from the soil to form energy-rich sugars (Chapter 9).

With more than 400,000 species, the Plant kingdom is second only to the Animal kingdom in diversity. Plants are structurally simpler than most animals, but they are more complex than their appearance suggests, for many plant adaptations are biochemical. Plants have evolved a battery of chemicals to defend themselves against assaults by animals, fungi, microbes, and even other plants (Figure 33-1). If taxonomists distinguished plant species on the basis of chemical (rather than structural) differences, the diversity in the Plant kingdom would skyrocket.

## Major Evolutionary Trends

Because ancient plants left so few fossils, it is difficult to document the history of plant evolution. Taxonomists and evolutionary botanists must therefore rely on similarities in biochemistry (mainly photosynthetic pigments), cell structure, growth patterns, and gametes to interpret the course of plant evolution. This evidence suggests that plants evolved from a line of filamentous green algae that invaded land over 400 million years ago (see Figure 30-9). Today only a few plants live in water; most members of the Plant kingdom are terrestrial.

The demands of living on dry land led to a change in the growth pattern of plants, from growing as single rows of cells that form filaments, to forming sheets of cells that form ground-hugging mats. Lying flat against the soggy ground, water diffused into the cells of these early land plants, quickly replacing water used in metabolism or lost to the atmosphere as vapor. With unob-

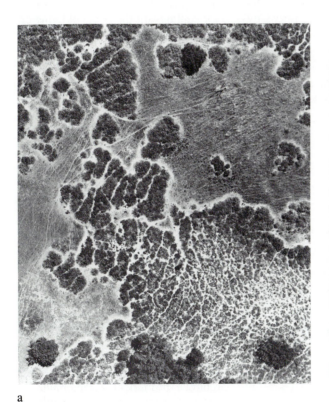

a

b

**Figure 33-1**
**Biochemical warfare.** An arsenal of chemicals not only helps protect some plants from foragers, but also enables them to outcompete rivals for water, space, and sunlight. (*a*)Volatile oils released from the leaves of sage bushes (*Salvia*) help keep the area around the bushes free of competing grasses. These strong-scented oils inhibit seed germination and stunt grass growth as far as 10 meters (30 feet) from established shrubs. (*b*) The roots of some oaks leak chemicals that inhibit the growth of other oak roots, thereby preventing crowding. The pattern of distribution of oaks in this valley is mostly the result of this form of "crowd control."

structed light, abundant $CO_2$, and little or no competition, these early land plants flourished—that is, as long as they remained close to a continuous supply of water. Plants could venture away from the edges of beaches, ponds, or streams only after developing a means of preventing fatal dehydration.

## Conquering Dry Land

Four categories of adaptations enabled plants to make the transition to dry land

- *Controlling water loss.* The above-ground parts of most modern plants are coated with a waxy cuticle that retards evaporation. In addition, outer surfaces are perforated with stomates, tiny pores that open when water is adequate allowing the diffusion of $CO_2$ for photosynthesis. When dehydration threatens, stomates close, reducing water vapor loss (Chapter 13).

- *Vascular tissues.* Plants without vascular tissues must grow close to the ground. The evolution of xylem for transporting water and mineral ions, and phloem for transporting sugars and other organic molecules (Chapter 14), enabled plants to grow taller by providing a means of moving these materials higher. In addition, thick-walled xylem cells help support large numbers of leaves and heavy stems as plants grow.

- *Resistant spores.* Some plants manufacture **spores**, lightweight cells that are specialized for dispersal and for survival in the face of adverse conditions. With thick walls to impede water loss and virtually no metabolism to consume water, dormant spores can survive long periods without additional moisture. When water becomes available, the spores regain activity and grow into new plants.

- *Protective packaging for gametes and embryos.* In addition to protecting the plant body itself, gametes and embryos also need a defense against drying out. This protection has been achieved through the evolution of

    1. *Multicellular gametangia* (gamete-producing structures) which help prevent dehydration of reproductive cells and developing embryos by surrounding them with water-trapping layers of cells.

    2. *Pollen,* water-tight capsules for male gametes, which frees more advanced plants from the need to use water for transferring sperm to the egg for fertilization.

    3. *Seeds,* protective, drought-resistant enclosures for plant embryos, enabling the offspring of seed-producing plants to be dispersed to new localities by water, wind, and animals.

    4. *Fruits* which further clothe the seeds of flowering plants in additional protective layers, enhancing embryo survival and dispersal.

As plants successfully colonized the land, they led the way for animals to follow. Although plants and animals have very different lifestyles, both have evolved some similar adaptations for coping with the rigors of living on dry land (Chapter 34).

## From Prominent Gametophyte to Prominent Sporophyte

During sexual reproduction, all members of the Plant kingdom alternate between a diploid spore-producing generation (the sporophyte) and a haploid gamete-producing generation (the gametophyte)(Chapters 12 and 21). Biologists refer to the sequential change from one generation to the next as **alternation of generations** (Figure 33-2).

A major trend in the evolution of plants has been toward a prolonged sporophyte and a reduced gametophyte generation (Figure 33-3). Complex, advanced plants, such as pines and flowering plants, have a prominent sporophyte generation, whereas more primitive plants, like mosses, retain a prominent gametophyte as did their aquatic ancestors. Diploidy offers an important advantage over the haploid state for complex organisms. Since haploid cells contain only one allele for each gene, the effects of a mutant allele cannot be masked by a dominant allele as in a diploid organism. Diploidy not only allows an

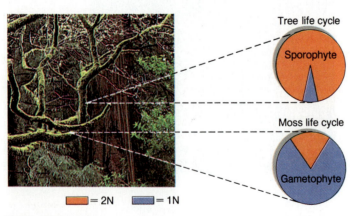

**Figure 33-2**
**Two common residents of many forests are mosses and trees.** Both are classified in the Plant kingdom, yet the moss is composed entirely of haploid (1N) cells whereas the tree is made up of diploid (2N) cells. Trees and mosses, like all plants, alternate between a diploid sporophyte and a haploid gametophyte generation during their sexual life cycle. Some plants, like the tree, have a conspicuous sporophyte, whereas other plants, like this moss, have a conspicuous gametophyte generation.

Tree life cycle

Sporophyte

Moss life cycle

Gametophyte

☐ = 2N    ☐ = 1N

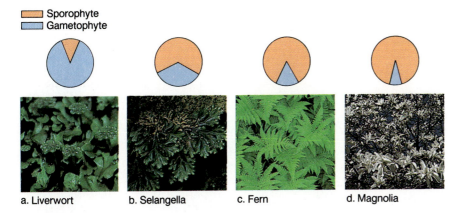

Sporophyte
Gametophyte

a. Liverwort    b. Selangella    c. Fern    d. Magnolia

**Figure 33-3**
**From prominent gametophyte to prominent sporophyte.** Comparing a range of life cycles from primitive to advanced plants reveals a trend toward a prominent diploid sporophyte.

organism to mask lethal mutations without suffering deleterious effects, but also enables an organism to harbor neutral mutations that may later prove to be an advantage.

## Overview of the Plant Kingdom

All multicellular, tissue-forming eukaryotic photosynthesizers are classified within the Plant kingdom. Those few "plants" that do not photosynthesize are still included in the Plant kingdom because they are direct descendants of photosynthesizing plants (see Bioline: The Exceptions). In addition, some taxonomists include three divisions of algae (the red, brown, and green algae) in the Plant kingdom. But since these algae do not form multicellular gametangia and distinctive tissues, many taxonomists place them in the Protist kingdom, as we have done. (All groups of algae are discussed in Chapter 32.)

The first plants to colonize the land formed a **thallus**, a flat, ground-hugging plant body that lacks roots, stems, leaves, and vascular tissues. Once plants moved onto the land, the ancestral line branched, producing two major groups of plants, the **bryophytes** (nonvascular plants) and the **tracheophytes** (vascular plants) (Figure 33-4). Today, there are ten divisions of plants, nine of which are vascular plants and only one of which contains nonvascular plants (see Table 33-1).

## Nonvascular Plants: The Bryophytes

The 24,000 species of nonvascular bryophytes include nearly 15,000 mosses, 9,000 liverworts, and 100 hornworts (Figure 33-5). Because their sperm must swim through water to reach and fertilize an egg, bryophytes are not completely adapted to living on dry land. In addition, bryophytes lack "true" roots for absorbing and conducting water. Instead, water is absorbed directly into outer cells and then slowly diffuses throughout the thallus. Bryophytes produce **rhizoids**, slender cells that resemble roots but function only to anchor the plant, and not to absorb water and minerals. Without specialized conducting and supporting tissues, nonvascular bryophytes can grow no taller than a few centimeters.

The gametophyte is the prominent generation in bryophytes. During the sexual life cycle of mosses, illustrated in Figure 33-6, the multicellular gametophyte produces either male or female reproductive structures or, in some species, both. The roundish male gametangium (gamete-producing structure) is called an *antheridium;* the flask-shaped female gametangium is an *archegonium.*

Sperm produced in the antheridia have two flagella, enabling them to swim to an archegonium to fertilize the egg. In unisexual mosses, the sperm must swim to another plant to encounter an archegonium. This journey is usually impossible unless assisted by rain. As a raindrop strikes a moss stalk, it breaks up into small droplets. Sperm are released into the droplets which then splatter and carry the sperm to other moss gametophytes.

The formation of a zygote following fertilization begins the sporophyte generation. The zygote divides by mitosis many times to form a multicellular sporophyte, which remains attached to the gametophyte for access to food and water. The sporangium of the sporophyte contains cells that undergo meiosis to produce haploid spores, beginning a new gametophyte phase. The spores are dispersed by

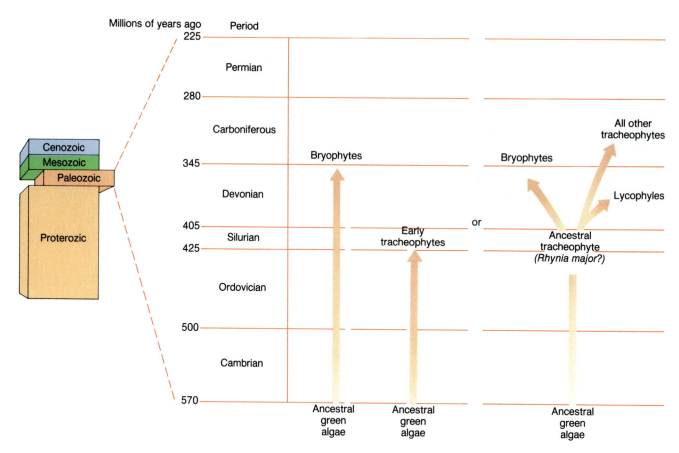

**Figure 33-4**
**Plant origins.** Two hypotheses have been proposed to explain the origin of plants. Although both agree that plants evolved from an ancestral filamentous green alga, one hypothesis contends that bryophytes and tracheophytes arose from different ancestors, whereas the other asserts that both bryophytes and tracheophytes arose from the same ancient tracheophyte (*Rhynia major*, perhaps).

wind and grow into a new multicellular gametophyte with archegonia and antheridia, thus another reproductive cycle begins.

## Vascular Plants: The Tracheophytes

The oldest plant fossils are from an extinct group, the **rhyniophytes** (Figure 33-7). These ancient plants had vascular tissues and thrived in marshy areas during the Silurian period, 420 to 450 million years ago. Curiously, the earliest nonvascular fossils did not appear until 100 million years after the rhyniophytes. This finding has led some botanists to speculate that rhyniophytes gave rise not only to tracheophytes with xylem- and phloem-con-

ducting tissues, but to nonvascular bryophytes as well (Figure 33-4).

Modern tracheophytes are often divided into two general subgroups: (1) the *lower* vascular plants which have more primitive characteristics and (2) the *higher* vascular plants with advanced characteristics (see Table 33-1). All lower vascular plants produce separate sporophyte and gametophyte generations, whereas higher vascular plants have a shortened gametophyte phase that depends on the sporophyte for food and water.

All plants (bryophytes included) produce spores during their sexual life cycle. Lower vascular plants are mostly **homosporous**, that is, they manufacture only one type of spore. These spores grow into a gametophyte

that produces both antheridia and archegonia, a situation that often leads to self-fertilization because male and female gametangia are so close together. On the other hand, higher vascular plants are **heterosporous**, producing two types of spores: a megaspore that grows into a female gametophyte and a microspore that grows into a male gametophyte. By separating gametangia, heterospory increases the chances of cross fertilization, fostering genetic variability.

### Lower Vascular Plants

Lower vascular plants include four divisions: the Psilophyta, Lycophyta, Sphenophyta, and Pterophyta. Between 300 and 430 million years ago, members of Psilophyta, Lycophyta,

TABLE 33-1

| | Non-vascular | Lower Vascular Plants | | | | Higher Vascular Plants | | | | |
|---|---|---|---|---|---|---|---|---|---|---|
| **Characteristics** | **Bryo-phyta (24,000)**[a] | **Psilo-phyta (13)** | **Lyco-phyta (1000)** | **Spheno-phyta (15)** | **Ptero-phyta (12,000)** | **Cycado-phyta (100)** | **Ginkgo-phyta (1)** | **Gneto-phyta (71)** | **Conifero-phyta (700)** | **Antho-phyta (375,000)** |
| Prominent gametophyte | | | | | | | | | | |
| Prominent sporophyte | | | | | | | | | | |
| Water needed for fertilization | | | | | | | | | | |
| Dispersal by spores | | | | | | | | | | |
| Dispersal by seeds | | | | | | ←———————— naked ————————→ | | | | ←enclosed→ |
| True roots, stems, leaves | | | | | | | | | | |

*Characteristics of the Plant Divisions*

[a] *(Number of described species)*

a.      b.      c.

**Figure 33-5**
**The bryophytes**. The division Bryophyta is divided into three classes — a. the Musci (mosses), b. the Hepaticae (liverworts), and c. the Anthocerotae (hornworts).

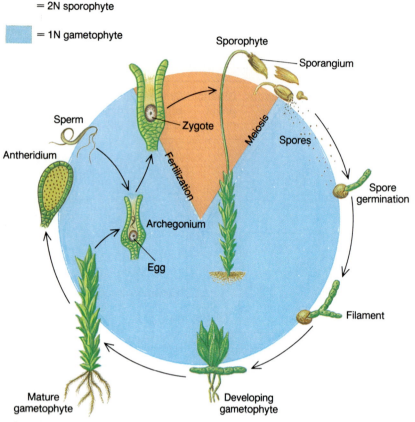

**Figure 33-6**
**Moss life cycle.** Although some mosses produce either male or female gametangia, this moss gametophyte forms both antheridia and archegonia. Following fertilization, the zygote grows into an erect sporophyte, which remains permanently attached to the gametophyte body. The single sporangium at the tip of the sporophyte releases haploid spores that grow into new gametophytes.

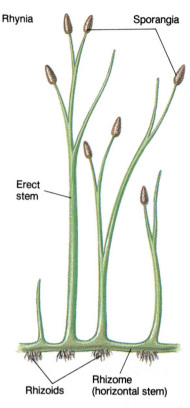

**Figure 33-7**
*Rhynia major,* one of the earliest vascular plants. All rhyniophytes lived 365 to 400 million years ago. The sporophyte of *Rhynia major* illustrated here had erect photosynthetic stems that terminated with a sporangium. It lacked leaves and roots, attaching itself to the ground by rhizoids.

and Sphenophyta were the dominant type of vegetation on earth. Today only about 1000 species belong to these three divisions. Even after adding the 12,000 species of ferns (division Pterophyta), the four divisions of lower vascular plants account for less than 3 percent of all known plant species.

## PSILOPHYTA: WHISK FERNS AND RELATIVES
Some 300 million years ago, lush green forests were filled with species of psilophytes. Today, only one family (containing 2 genera and 13 species) remains of this once diverse division of plants.

*Psilotum* is the most widespread genus of living psilophytes, extending from tropical and subtropical areas into a few temperate zones. The most common species of *Psilotum,* and the one frequently displayed in introductory biology laboratories, is the naked-looking sporophyte of *Psilotum nudum* (Figure 33-8). *Psilotum nudum* looks "naked" because its stems lack leaves; instead, they bear small, widely spaced scales. The axils of some scales contain three-chambered sporangia that manufacture and disperse spores. The spores grow into tiny, underground gametophytes with antheridia and archegonia. The flagellated sperm produced in the antheridia swim

through a film of water to fertilize the egg. The resulting zygote grows and develops into another multicellular sporophyte.

## LYCOPHYTA: CLUB MOSSES AND RELATIVES
Like the Psilophyta, the division Lycophyta has seen more glorious days. About 350 million years ago, members of the genus *Lepidodendron* grew as tall as 50 meters (165 feet) and dominated the forests of the Carboniferous period (Figure 33-9). Today all lycopods are small, herbaceous plants.

The Lycophyta includes three families, six genera, and about 1000 spe-

# B I O L I N E

## THE EXCEPTIONS: PLANTS THAT DON'T PHOTOSYNTHESIZE

They look like a plant, grow like a plant, produce flowers, are multicellular and eukaryotic, but don't photosynthesize. According to the five-kingdom classification system, these organisms wouldn't fit into any kingdom. But these organisms are the inevitable exceptions to any classification system. Although they lack photosynthetic pigments, they are still classified as plants because they evolved from plant ancestors and they still retain plantlike anatomy in spite of their photosynthetic deficiency.

These nonphotosynthetic plants acquire food and nutrients the same way other heterotrophs do, by consuming organic compounds produced by other organisms. Nonphotosynthetic plants are either saprophytes or parasites, and some are both.

The following is a sampling of some of these atypical members of the Plant kingdom.

a *Corallorhiza trifida* (coral-root) is an orchid that is both a parasite and a saprophyte, gathering its nutrients from the roots of other plants or from dead or decaying organic matter. It grows in the beech and fir forests of Europe, Siberia, and North America.

b *Sarcodes sanguinea* (snow plant) is a striking, red, fleshy saprophyte. The heat of the growing inflorescence melts the snow immediately surrounding it. The snow plant grows in the forests of California and Oregon.

c *Orobanchaceae* is an entire family of plants that parasitize the roots of other plants. *Orobanche,* or broom-rape, is the most widely distributed genus and is found in Europe, Asia, Africa, and North America.

d *Cuscuta* is a parasite that entangles its plant host with orange, twisting stems. Outgrowths penetrate the host's vascular tissues and tap its fluids. *Cuscuta* grows all over the world. Common names for *Cuscuta* include dodder, silkweed plants, boxthorn, and matrimony vine.

a

b

c

d

**Figure 33-8**
**A "living fossil."** The sporophyte of *Psilotum nudum* (division Psilophyta) closely resembles 400-million-year-old fossils of psilophytes. Round, yellow sporangia are produced on the sides of leafless photosynthetic stems.

Lepidodendron

**Figure 33-9**
**The lush Carboniferous forests** were dominated by giant psilophytes, sphenophytes, and lycophytes. *Lepidodendron* was a giant lycopod that produced leaves and sporangia at the ends of its branches. All treelike lower vascular plants, including *Lepidodendron*, are now extinct, the victims of changing environmental conditions.

cies. Most species are tropical, and the majority are *epiphytes,* plants that grow on other plants. The largest genus, *Selaginella,* contains 700 species (Figure 33-3*b*). The most common genus in the United States is *Lycopodium,* commonly called club moss or ground pine because the sporophytes look like miniature pine trees (Figure 33-10). The spores of lycopods were used in firecrackers and flash bars (before flashbulbs were developed) because they ignite explosively, emitting light. Curiously, lycopod spores have also been used as baby powder.

Most lycopod sporophytes produce spores in a conelike **strobilus**, a terminal cluster of specialized leaves that produce sporangia. These specialized leaves are called **sporophylls**. The majority of lycopods are homosporous. In the few heterosporous lycopods, the megaspore grows into a **megagametophyte** with archegonia; microspores either germinate into a **microgametophyte** with antheridia or simply release flagellated sperm. (The prefixes "mega" [mega = large] and "micro" [micro = small] do not refer to the size of the spore or the size of the gametophyte, but rather to the relative sizes of the gametes they produce; a sperm is always smaller than an egg.) Water is required for fertilization in both homosporous and heterosporous lycopods.

## SPHENOPHYTA: THE HORSETAILS

Growing alongside the giant lycopods 350 million years ago were woody sphenophytes that grew 15 meters (50 feet) tall. But like the lycopods, all remaining sphenophytes are herbaceous. Today, the Sphenophyta includes 15 species, all of which are in one genus, *Equisetum,* the horsetails (Figure 33-11).

The upright sporophytes of horsetails have jointed, hollow stems arising from a horizontal stem, or rhizome. The vertical stems have distinctive longitudinal ⁻groves and whorls of leaves or branches at the nodes. The leaves of horsetails fuse at the base to form a sheath around each node. Strobili are produced at the ends of stems or from side branches, and homospory is more common than heterospory.

Each *Equisetum* spore is sur-

**Figure 33-10**
*Lycopodium annotinum* sporophytes with strobili. Each strobilus is a tight cluster of spore-producing leaves.

**Figure 33-11**
**Horsetails**. *Equisetum arvensis* has a horizontal rhizome that supports two types of hollow, upright stems—one topped with a strobilus (on the right in the photo) and the other producing orderly whorls of radiating branches (left). Although the stems appear leafless, they have small leaves that fuse at the base to form sheaths around each node. *Equisetum* is the only genus in the Sphenophyta, a once widely distributed, very diverse division.

rounded by coiled elaters, elongated flaps that unfurl as they dry. The movement of elaters helps to dislodge spores from the sporangium when the air is dry. Dry spores are lighter and remain windborne longer than wet spores, dispersing them over greater distances. The spores remain dormant until moistened and then grow into a tiny photosynthetic gametophyte about the size of the period at the end of this sentence. It isn't surprising that most gametophytes go completely unnoticed.

## PTEROPHYTA: FERNS

Like the other lower vascular plants, ferns are an ancient plant group, arising during the Devonian period 375 to 420 million years ago. Two-thirds of the 12,000 fern species living today are tropical. All ferns are restricted to habitats that are occasionally wet since they produce flagellated sperm that must swim to the egg during sexual reproduction.

The life cycle of a typical fern is illustrated in Figure 33-12. The familiar sporophyte of a fern produces *fronds* —leaflike structures that, unlike "true" leaves, have an apical meristem and clusters of sporangia called **sori**.

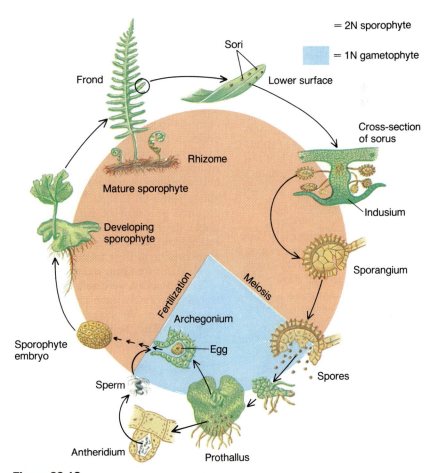

**Figure 33-12**
**Fern life cycle**. The fern sporophyte and gametophyte live independently, although the young sporophyte extracts food and water during its early development. The heart-shaped fern gametophyte (the prothallus) produces both anteridia and archegonia.

Fern fronds uncoil from tight spirals arising from rhizomes or upright stems.

Most ferns are homosporous. The spores are released from sporangia produced either along the margins or in clusters on the lower surface of the fronds. Clusters of sporangia are covered by an umbrella-shaped **indusium**, whereas the sporangia formed along frond margins are often sheathed by a flap called a **false indusium**. Both types of indusia shrivel and fold back as the air dries, releasing spores during dry periods when dispersal is favored.

Each fern sporangium is encircled by an **annulus**, a row of specialized cells with thick cell walls on five of its six sides. As water is lost through the thinnest, outward-facing wall, the bonds between remaining water molecules become strained and tension builds within each annulus cell. Increasing tension causes the sporangium to slowly bend. When the tension finally exceeds the cohesive force between water molecules, the hydrogen bonds snap and the thick, elastic cells of the annulus spring back to their original shape, jettisoning the spores into the dry air. (Figure 33-13).

Fern spores grow into a small, heart-shaped gametophyte called a **prothallus**. The photosynthetic, independently growing prothallus lacks vascular tissues and, like many gametophytes of lower vascular plants, is reminiscent of primitive ground-hugging plants.

## Seed Plants

The simple vascular tissues of the lower vascular plants were more than adequate for living in lowland swamps that were so widespread during the Devonian and Permian periods of the Paleozoic Era. The perpetually saturated soils also provided ample water for sperm and more than enough water for their ground-hugging, nonvascular gametophytes. But environmental conditions changed at the end of the Permian period. These changes were so dramatic and so abrupt that the course of plant (and animal) evolution shifted in a new direction.

The new environmental conditions developed when the earth's continents shifted positions (see Bioline: Conti-

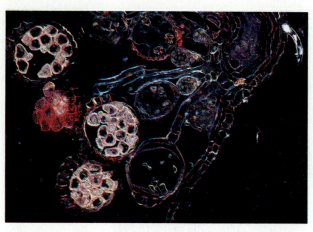

**Figure 33-13**
**Engineered for dispersing spores**, the sporangium of ferns is encircled by an annulus. As annulus cells lose water, the sporangium is gradually pulled backward. When water bonds snap, the sporangium shoots forward, flinging spores into the air.

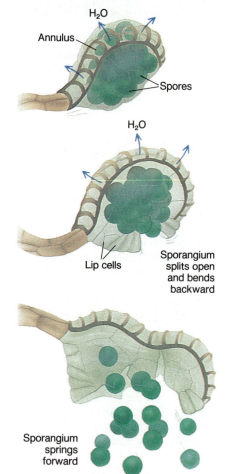

nental Drift Chapter 30), draining swampy lowlands and raising mountain ranges. The land became drier and deserts were formed. The climate began to fluctuate, and yearly seasonal cycles became established on earth. Most lower vascular plants were ill equipped to adapt to these changes and vanished. Among the surviving plants were those that had already acquired a means of surviving less stable, drier climates. These were the earliest seed plants, the primitive **gymnosperms** (gymno = naked, sperm = seed), plants that bear naked seeds.

Gymnosperms were able to quickly colonize scores of new habitats that opened up at the beginning of the Mesozoic Era because they were armed with the following adaptations:

- Extensive root systems for reaching deeper underground water supplies.

- Efficient conducting tissues for delivering water and nutrients to all parts of the plant.

- Secondary growth for replacing and adding new vascular tissues, and for increasing support (Chapter 13).

- Enclosure of gametes in protective structures that also prevent drying (sperm in pollen grains and eggs embedded in sporophyte tissues).

- "Dry" fertilization (sperm do not require water for dispersal to an egg).

- Encasement of the plant embryo in a protective seed that withstands harsh, dry conditions during dispersal, and delays germination until conditions become favorable for growth.

Although a great variety of gymnosperms evolved during the Mesozoic era, today some gymnosperms survive only as relics of the past—"living fossils"—with restricted distributions because in most habitats further environmental changes gave plants with other adaptations the competitive edge. Other gymnosperms still dominate many landscapes as they did throughout most of the Mesozoic Era.

**Figure 33-14**
Because of their appearance, cycads are sometimes mistaken for palms. But unlike flowering palms, cycads are gymnosperms that produce either male or female cones. Like the dinosaurs, cycads were abundant at the same time, between 100 and 200 million years ago.

## THE GYMNOSPERMS

All modern seed-producing plants are divided into two groups; the gymnosperms (pines, firs, cycads, etc.) and the angiosperms (angio = covered), or flowering plants. As noted earlier, gymnosperms have "naked seeds"; that is, their seeds are borne on the upper surface of scales and are not surrounded by additional fruit tissues as they are in angiosperms. The reproductive structures of gymnosperms are produced in cones, tight spirals of sporophylls.

The more than 700 species of living gymnosperms are grouped into four divisions: the Cycadophyta, Ginkgophyta, Gnetophyta, and Coniferophyta.

### Cycadophyta.

Cycads were most abundant and widespread during Mesozoic times, the Age of Dinosaurs. Now the 100 or so species of cycads grow only in tropical and subtropical regions, although they are used as ornamental plants in virtually all regions of the world. Some cycads superficially resemble palm trees (Figure 1-5), but like all other gymnosperms, cycads produce cones and bear naked seeds, not flowers and fruits as do palms (Figure 33-14).

Cycads are **dioecious**; that is, individual plants produce either male or female reproductive structures, but never both. Pollen is produced in male cones and is dispersed by wind. When a pollen grain lands on a female cone, it grows a pollen tube to within a few millimeters of the egg. Flagellated sperm are released from the pollen tube and swim a short distance through a miniature water chamber provided by the female gametophyte, a remnant of water fertilization.

### Ginkgophyta.

Like the cycads, a number of ginkgo species grew during the Mesozoic Era. Today only one species remains in the Ginkgophyta, *Ginkgo biloba* (the maidenhair tree). *Ginkgo biloba* escaped extinction because it has been cultivated for thousands of years as an ornamental plant in Asian temples and gardens (Figure 33-15). No "wild" or natural populations of ginkgos occur anywhere in the world today; they survive only where tended by humans.

### Gnetophyta.

The gnetophyta contains 70 species, in three genera — *Welwitschia, Ephedra,* and *Gnetum. Welwitschia* is the oldest genus, having fossils dating back more than 280 million years. The only living species, *Welwitschia mirabilis,* occurs in southwestern African deserts. As shown in Figure 33-16, *Welwitschia mirabilis* resembles a pile of tattered leaves growing from a wide, blunt stem. Yet *Welwitschia* produces only two leaves during its 100-year lifetime. Each leaf shreds as it grows, giving the appearance of multiple leaves. *Welwitschia*'s stubby, succulent stem and deep tap root store water, helping it to survive long periods of drought.

All 40 species of *Ephedra* are short shrubs, with jointed stems and scale-like leaves. Native Americans made flour and tea from *Ephedra* plants, and today drugs for treating asthma, emphysema, and hay fever come from extracts of *E. sinica* and *E. equisetina.*

The genus *Gnetum* includes 30 species of vines, woody shrubs, and trees, all of which are found in Asia, Africa, and Central and South America.

**Figure 33-15**
**Saved from extinction**, the maidenhair tree (*Ginkgo biloba*) has been cultivated for thousands of years in gardens, parks, and other tended sanctuaries. There are no natural populations anywhere in the world today. The maidenhair tree is the only living species in the division Ginkgophyta.

**Figure 33-16**
**Less than meets the eye.** Although *Welwitschia mirabilis* appears to have many leaves, each plant produces only two. The broad leaves shred and curl as they grow. *Welwitschia* grows flat against the hot, sandy soil in the Kalahari Desert of Africa. Seeds are produced in cones along leaf edges.

Coniferophyta.

The most familiar gymnosperms are the conifers that dominate many Northern Hemisphere forests. The more than 700 species of conifers are classified into 50 genera, including *Pinus* (pines), *Juniperus* (junipers), *Abies* (firs), and *Picea* (spruces). The largest plants, the giant sequoias (*Sequoiadendron gigantea*), and the tallest plants, the redwoods (*Sequoia sempervirens*), are conifers. The leaves of conifers are either long needles or short scales, both of which are covered with a thick cuticle to retard water vapor loss. Since many conifers retain green leaves throughout the year, they are frequently called evergreens.

Most conifers are **monoecious**; that is, male and female reproductive structures are produced on the same sporophyte individual. During the pine life cycle, illustrated in Figure 33-17, microspore mother cells in the male cones divide by meiosis to produce microspores. The microspores then divide mitotically to produce the male gametophyte, a pollen grain that contains sperm. The pollen of pines develop balloonlike "wings" that help them remain airborne during dispersal. To improve the chances of fertilization, a single pine tree releases billions of pollen grains, sometimes producing a cloud of yellow pollen.

Within the female cones, megaspore mother cells divide meiotically inside megasporangia to produce megaspores. Four megaspores are formed, three of which degenerate. The remaining megaspore divides repeatedly by mitosis to produce a multicellular female gametophyte containing two archegonia, each with a single egg. The combination of the female gametophyte, megasporangium, and surrounding sporophyte tissues forms the pine **ovule**.

The scales of female cones secrete a sticky substance that captures windborne pollen. Pollen grains form a pollen tube, which grows through the sporophyte and ovule tissues to deliver a sperm to the egg. Following fertilization, the zygote grows and develops into a pine embryo. The embryo, female gametophyte, and remaining ovule tissues combine to form the seed, or pine nut. Many seeds are

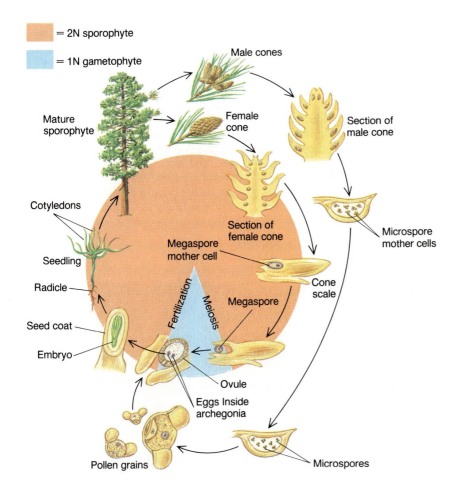

= 2N sporophyte

= 1N gametophyte

Male cones

Mature sporophyte

Female cone

Section of male cone

Cotyledons

Section of female cone

Megaspore mother cell

Microspore mother cells

Seedling

Radicle

Cone scale

Fertilization

Meiosis

Megaspore

Seed coat

Embryo

Ovule

Eggs Inside archegonia

Pollen grains

Microspores

**Figure 33-17**
**Pine life cycle**, a representative life cycle of the division Coniferophyta.

"winged" to aid dispersal by the wind.

The process described above normally takes more than two years to complete. Fertilization doesn't occur until 13 months after pollination; it takes that long for the pine ovule to develop. Another 12 months or so is required for the pine embryo and seed to mature.

The Coniferophyta is the most advanced division of gymnosperms. But even with an array of adaptations well suited for life on land, conifers are not the most diverse group of plants. The angiosperms far outnumber conifers in numbers of species and, except in northern and high altitude habitats, in numbers of individuals.

## THE ANGIOSPERMS— FLOWERING PLANTS

All plants that produce flowers belong to the division Anthophyta. This is the most recently evolved plant division, appearing about 100 million years ago. The Anthophyta evolved from a now extinct group of gymnosperms during the middle of the Mesozoic Era. The success of flowering plants was swift. Equipped with a more efficient vascular system and two unique key characteristics—the flower and fruit—angiosperms rapidly diversified; thousands of angiosperms appear in the fossil record by the end of the Mesozoic Era. Today the Anthophyta is the most diverse division of plants;

nearly 75 percent of all plant species are flowering plants.

Members of the Anthophyta are called angiosperms (meaning "enclosed seed") because their seeds are surrounded by fruit tissues formed from the mature ovary of the flower (Chapter 21). The more than 350,000 species of angiosperms that have been described range in size from tiny, stemless duckweeds to towering eucalyptus trees. Flowering plants grow in virtually every terrestrial and most aquatic habitats.

Flowers and fruits are devices for recruiting animals for pollination and seed dispersal. Gymnosperms rely on wind to indiscrimantly disperse pollen and seeds. A better means of dispersal is provided by insects and other animals that transfer pollen between individuals of the same species with greater precision. This method not only reduces the need to manufacture large numbers of pollen grains, but also increases the frequency of cross-pollination, which in turn, increases the genetic variability in a gene pool. Increased variability improves the chances of a species adapting to environmental changes. Seed dispersal by animals also improves the chances that offspring will be deposited in hospitable areas, since animals can only live where plants live. In contrast, the wind blows seeds everywhere, to the middle of the ocean or the top of a glacier, places where no plants can grow.

The Anthophyta is divided into two classes: the monocotyledons (or simply, monocots) and dicotyledons (dicots) (Chapter 21). There are over twice as many dicot as monocot species (250,000 versus 100,000) (Figure 33-18). Dicots include flowering trees and shrubs with wood secondary growth, as well as many kinds of herbs. Most fruits and vegetables available at supermarkets are dicots: these include tomatoes, potatoes, various melons, stone fruits (peaches, cherries, apricots, for example), grapes, zucchini, and broccoli. Many important agricultural staples, however, are monocot grasses—corn, wheat, rye, oats, barley, and rice. Monocots also include orchids, palms, lilies, and grasses. The sexual life cycle of flowering plants was discussed fully in Chapter 21.

a                    b                    c                    d

**Figure 33-18**
**A sampler of success.** Flowering plants (division Anthophyta) are now the most diverse and abundant type of plant on earth. The Anthrophyta is subdivided into Dicots (a + b) and Monocots (c + d).

# Synopsis

## MAIN CONCEPTS

- **Plants evolved from an aquatic filamentous green alga** that colonized the land about 400 million years ago.

- **The first land plants grew as ground-hugging mats,** remaining close to continuous water sources.

- **The evolution of vascular tissues and adaptations to control water loss enabled early plants to disperse into drier habitats.** Land plants quickly diversified, becoming the predominant form of terrestrial life. Most modern plants are terrestrial.

- **The colonization of the land by animals** became possible only after plants became established on land, thereby creating the terrestrial habitats and food sources needed by animals.

- **The roughly 400,000 species of living plants** include one division of nonvascular bryophytes and nine divisions of vascular tracheophytes.

- **Bryophytes and lower vascular plants still retain characteristics of semiaquatic ancestors** including the requirement for water as a medium for dispersing sperm to female sex organs that contain an egg.

- **During sexual reproduction, plants alternate between two multicellular generations**—a diploid sporophyte and a haploid gametophyte. In sporophytes, meiosis produces spores, which grow into sperm and egg producing gametophytes.

- **One evolutionary trend in plants is from a prominant gametophyte to a prominant sporophyte,** the diploid condition being genetically more advantageous for terrestrial life.

- **Between 350 and 150 million years ago, lower vascular plants were the predominant plants on earth.** Extensive mountain building and shifting continents during the past 150 million years created drier and more variable environmental conditions, causing most lower vascular plants to become extinct. With efficient vascular tissues, protected gametes (eggs in ovules and sperm in pollen grains), and covered embryos, the early seed plants (the gymnosperms) thrived under these new conditions.

- **The most recently evolved group of plants—the Anthophyta—is now the most diverse.** All produce flowers and develop a fruit from a ripened ovary. Flowers and fruits provide protection for gametes and embryos, and enable pollen and seeds to be dispersed with precision by insects and other animals.

## KEY TERM INTEGRATOR

Sexual Life Cycle

| | |
|---|---|
| **alternation of generations** | The sequential changes between a diploid sporophyte and haploid gametophyte during the plant's sexual life cycle. |
| **homosporous** | Production of identical spores that grow into gametophytes that contain sperm-producing and egg-producing **gametangia**. |
| **heterosporous** | Production of two types of spores, megaspores and microspores. Megaspores grow into **megagametophytes** that produce eggs; microspores grow into **microgametophytes** that produce sperm. |

Plant Classification

| | |
|---|---|
| **bryophytes** | Plants in the division *Bryophyta* (all nonvascular liverworts, hornworts, and mosses). Bryophytes grow a **thallus** with **rhizoids**. |
| **rhyniophytes** | Extinct group of plants; oldest plant fossils. |
| **tracheophytes** | Plants that contain vascular tissues; include nine divisions. Lower vascular plants include the *Psilophyta Sphenophyta, Lycophyta, and Pterophyta.* Most lycophytes produce a **strobilus** with clusters of spore-producing **sporophylls.** The fronds of fern sporophytes produce clusters of sporangia called **sori.** Each sorus is covered by an umbrella-shaped **indusium** or a flap of cells that forms a **false indusium.** The sporangium splits open at the **annulus** to disperse spores. The fern gametophyte is called a **prothallus.** |
| **gymnosperms** | Plants that produce naked seeds and belong to the divisions *Cycadophyta, Ginkgophyta, and Coniferophyta.* **Monoecious** plants produce both male and female productive structures, whereas **dioecious** plants produce one or the other, but never both. The eggs of gymnosperms and angiosperms are contained in **ovules.** |
| **angiosperms** | Plants that produce covered seeds; belong to the division *Anthophyta.* |

# Review and Synthesis

1. Complete the classification scheme for the Plant kingdom using the following terms:

| | | |
|---|---|---|
| liverworts | bryophytes | gymnosperms |
| hornworts | tracheophytes | angiosperms |
| Sphenophyta | Lycophyta | Psilophyta |
| Cycadophyta | Ginkgophyta | Gnetophyta |
| higher vascular plants | | |

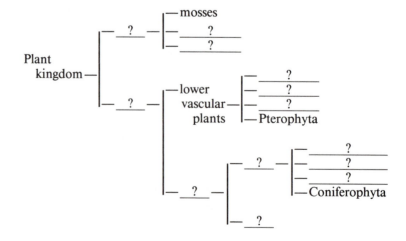

2. Explain why there are fewer plant species than animal species?

3. Match the structures with the appropriate plant group.

    ___ a. gymnosperms   1. prothallus
    ___ b. angiosperms   2. strobilus
    ___ c. ferns           3. rhizoids
    ___ d. mosses       4. fruits and flowers
    ___ e. club mosses   5. naked seeds

4. List the divisions of plants that produce the following: pollen; ovules; flowers.

5. How has the need for liquid water to promote fertilization affected the evolution of plants?

6. The perfect plant for a desert environment would include many of the adaptations discussed in this chapter. List those adaptations plus any additional ones you feel would be beneficial to survival and reproduction in periodically dry, very hot environments. Which adaptations are better suited to very wet habitats, such as a rainforest?

## Additional Readings

Banks, H. 1975 "Early land plants: proof and conjecture." *Bioscience* 25:730–37. (Intermediate.)

Barbour, M., J. Burk, and W. Pitts. 1987. *Terrestrial Plant Ecology.* 2nd ed. Benjamin/Cummings, Menlo Park, Calif. (Advanced.)

Briggs, D., and S. Walters. 1984. *Plant Variation and Evolution.* 2nd ed. Cambridge University Press, Cambridge. (Intermediate.)

Pickersgill, B. 1977. "Taxonomy and the origin and evolution of cultivated plants in the New World." *Nature* 268:591–95. (Intermediate.)

Chapter 34

# The Animal Kingdom

As individuals, animals are greatly outnumbered—by plants, by bacteria, and even by fungi. Yet there are more kinds of animals than any other type of organism. Almost 75 percent of all species (more than 1.3 million) are included in the Animal kingdom, and the tally is far from complete. Many taxonomists estimate that from 4 to 30 million animal species remain unnamed and unclassified, most of them insects.

Animals range in size from the minute *Trichoplax adhaerens,* a simple animal composed of a few thousand cells, to complex, giant blue whales (*Balaenoptera musculus*) that reach lengths up to 33 meters (110 feet) and weigh more than 160 tons (Figure 34-1). Between these extremes is an immense assortment of animals that includes sponges, jellyfish, mollusks, various worms (roundworms, flatworms, ribbon worms, and segmented worms), insects, fishes, amphibians, reptiles, birds, and mammals. Almost 97 percent of animal species are **invertebrates**, animals without a backbone; only 3 percent are the more familiar **vertebrates**, the backboned animals.

Despite their astounding diversity, members of the Animal kingdom all share three basic characteristics.

- *Animals are multicellular.* They have cells grouped into tissues (except *Trichoplax adhaerens* and sponges).

- *Animals are heterotrophic.* Most ingest their food, taking food particles into a digestive cavity or tube for digestion and absorption.

- *Animals are motile.* Animals are capable of locomotion at some stage of their development.

## Evolutionary Trends

Like plants, the first animals left no fossils, making it impossible to trace the origin of the Animal kingdom. However, because the cells that line the internal cavity of one of the most primitive type of animals (sponges) are virtually identical to some flagellated protozoa, most zoologists believe that animals evolved from a flagellated protozoan.

The first animals were most likely colonies of identical cells, formed when daughter cells failed to separate during cell division. Eventually, different cells in the colony specialized, some for the acquisition or digestion of food and others for reproduction, locomotion, and coordination. Later, the specialized cells became organized into tissues, and tissues complexed to form organs. These levels of organization improved efficiency in multicellular organisms and allowed animals to grow larger.

The animal fossil record goes back 600 million years. Animal evolution was already well advanced by then—fossils unearthed in 600-million-year-old deposits represent all major groups of aquatic invertebrates. The oldest vertebrate fossils are fishes that date back 500 million years. The first land animals evolved from lobe-finned lung fishes around 400 million years ago. These terrestrial pioneers were the ancestors of amphibians, reptiles, birds, and eventually warm-blooded mammals.

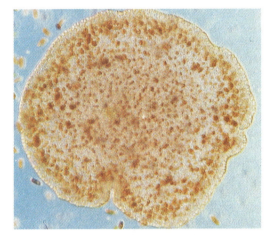

a

b

**Figure 34-1**
**From one extreme to the other.** Animal size ranges from the microscopic *Trichoplax adhaerens* (a) to enormous blue whales (b). Both live in the ocean, but *Trichoplax* is a tiny speck of life without tissues, whereas blue whales are the largest animals, with complex tissues and organs. (Blue whales are six times larger than the biggest dinosaur, Brachiosaurus.)

# Criteria for Classifying Animals

Animal taxonomists rely on similarities in structure and physiology to classify animals and to determine evolutionary relationships. Animal groups are distinguished on the basis of four criteria: body symmetry, body design, body cavity, and differences in embryonic development.

## Body Symmetry

A few animals have bodies with no particular symmetry; that is, there is no way the animal can be divided (sectioned) to produce similar-appearing halves, or mirror images. Animals that lack symmetry are **asymmetrical** like the sponge shown in this photo.

In animals with symmetrical bodies, a median plane through the center of the animal produces two sides that are roughly mirror images of each other. If more than one median plane can divide the animal into similar halves, the animal is said to have **radial symmetry**. Radially symmetrical animals often have similar body parts repeated in a circle, like the sea star ("starfish") shown here.

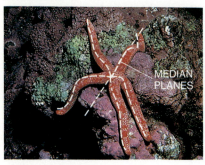

Radial symmetry enables animals to feed, and receive and respond to stimuli from any direction.

Most animals have **bilateral symmetry**, that is, only one median plane will produce mirror images, as in this arrow crab.

All bilaterally symmetrical animals, such as yourself, have right and left sides. For descriptive purposes, zoologists further subdivide bilaterally symmetrical animals using a number of median planes (Figure 34-2).

The evolution of bilateral symmetry led to the specialization of animal parts. For example, nervous tissues became concentrated at one end of the body to form a "head" end, a phenomenon called **cephalization**. (Without bilateral symmetry, there would be no "head" end, or any ends at all.) Cephalization produces a control center of sensory and "brain" tissues that improve environmental monitoring and coordination of responses.

## Body Design

Animals have two basic body designs for ingesting and digesting food: a **sac body** or a **tube-within-a-tube body**. A sac body has a single opening leading to a digestion chamber; this opening is used for both taking in food and expelling wastes. Animals with a tube-within-a-tube body have two openings; food enters through the mouth, and digestive wastes are released through the anus. These fundamental body designs are typified in the figure below.

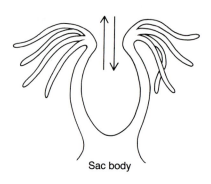

Sac body

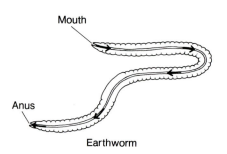

Mouth

Anus

Earthworm

## Body Cavity

As shown in Figure 34-3*a*, all animal tissues are grouped into two or three layers according to which germ layer produced them (see Chapter 24). In animals with three germ layers (ectoderm, mesoderm, and endoderm), taxonomists use characteristics of the mesoderm to further distinguish animals (Figure 34-3*b*).

If the mesoderm-derived region is packed solid with cells, the animals are called **acoelomates** (a = without, coelom = cavity with a cellular lining). Animals with a cavity in the mesoderm are either pseudocoelomates or coelomates, depending on whether the cavity is partially or fully lined with mesoderm cells. **Pseudocoelomates** do not have a body cavity fully lined with mesoderm cells, creating a partially lined pseudocoelom or "false coelom." Water pressure in the pseudocoelom helps support body parts by

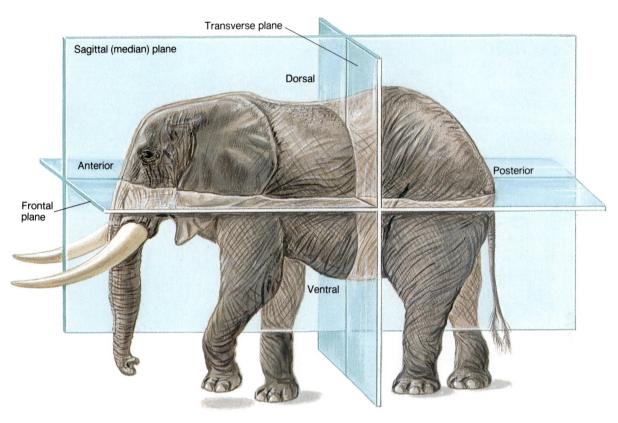

**Figure 34-2**
**Sectional planes** divide bilaterally symmetrical animals into different halves. A saggital plane delineates right and left sides; the transverse plane divides anterior (front) and posterior (rear) halves; and the frontal plane distinguishes dorsal (upper) and ventral (lower) halves.

creating a "hydrostatic skeleton" (Chapter 17). It also allows muscular contractions in the digestive system to operate independently of those used for locomotion. **Coelomates** are animals with a fully lined body cavity. That is, the cavity is lined with cells derived from the mesoderm, forming a true **coelom**. The cavity's lining is called the **peritoneum**. In complex animals, many organs and organ systems are suspended by **mesentery** extensions from the peritoneum (Figure 34-3*b*).

### Embryonic Development

The embryos of bilaterally symmetrical coelomate animals develop along one of two courses, producing two groups of animals: the **protostomes** (which includes mollusks, annelids, and arthropods) and **deuterostomes** (which includes echinoderms and chordates). Such distinctive developmental sequences suggest that coelomates evolved along two lines, as illustrated in Figure 34-4. Table 34-1 summarizes the differences in development of protostomes and deuterostomes.

## Overview of the Animal Kingdom

The 1.3 million animal species are grouped into two subkingdoms: (1) the **Parazoa**, simple organisms without tissues; and (2) the **Eumetazoa**, tissue-forming animals.

Our exploration of the Animal kingdom includes 9 of the 35 animal phyla, summarized in Table 34-2. We chose these phyla because they contain the greatest numbers of species and because they illustrate the principal evolutionary innovations that occurred during animal evolution. An evolutionary tree depicting the relationships between the major animal phyla is presented in Figure 34-4.

## Major Animal Phyla

### Phylum Porifera: Sponges

All 10,000 species of sponges are aquatic, but only 150 species are ocean-dwellers. The fossil record reveals that sponges have changed little

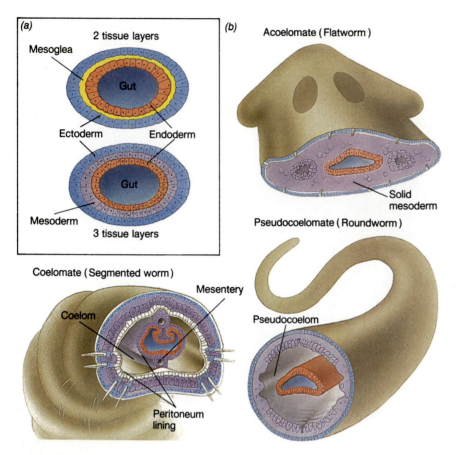

**(a)**

2 tissue layers

Mesoglea

Gut

Ectoderm — Endoderm

Mesoderm

3 tissue layers

Coelomate (Segmented worm)

Mesentery

Coelom

Peritoneum lining

**(b)**

Acoelomate (Flatworm)

Solid mesoderm

Pseudocoelomate (Roundworm)

Pseudocoelom

**Figure 34-3**

**Organization of animal tissues.** *(a)* Cross sections show how animal bodies are organized into two or three layers of embryonic or germ tissues. In animals with two tissue layers, a noncellular, jellylike mesoglea separates the ectoderm and endoderm. *(b)* Animals with three tissue layers either have a solid mesoderm (the acoelomates) or a body cavity within the mesoderm (psuedocoelomates and coelomates). In psuedocoelomates, the cavity is not fully lined with mesoderm-derived cells, whereas in coelomates (segmented worms and all more advanced animals) the cavity is fully lined with mesoderm-derived cells. In addition to this peritoneum, the mesoderm forms muscles, blood, an internal skeleton (if present), and other organs that are suspended by mesentery extensions.

over more than 550 million years. The **Porifera** forms a separate evolutionary line that has not acted as an ancestral group for any other known animal group.

Adult sponges are sessile, remaining attached to rocks, shells, coral reefs, pilings, and other hard surfaces. The simplest sponges have bodies shaped like a flower vase; that is, an internal cavity, the **spongocoel**, has a single opening at the top called an **osculum** (Figure 34-5a). More complicated sponges have folded bodies that form a complex system of channels leading to one or more spongocoels (Figure 34-5b).

Sponges have three body layers, only two of which (the outer and inner lining) developed from true germ layers. The third, middle layer is an aggregate of **mesenchyme** cells. The outer layer is constructed of flattened cells and thousands of **porocytes**, specialized cells equipped with a pore through which water is drawn into the spongocoel. (The name "Porifera" means "having pores.") The mesen-chyme layer contains migrating **amoebocytes**—cells that circulate nutrients and produce gametes for sexual reproduction. In many sponges, the amoebocytes manufacture sharp-pointed **spicules** made of glass (silicon dioxide) or calcium carbonate. Spicules form the sponge's "skeleton." The sponges we use for bathing or washing dishes manufacture a "skeleton" of interlocking protein fibers called **spongin**, not piercing spicules.

The inner layer of a sponge is constructed of **choanocytes**, or collar cells.

Protostomes

Deuterostomes

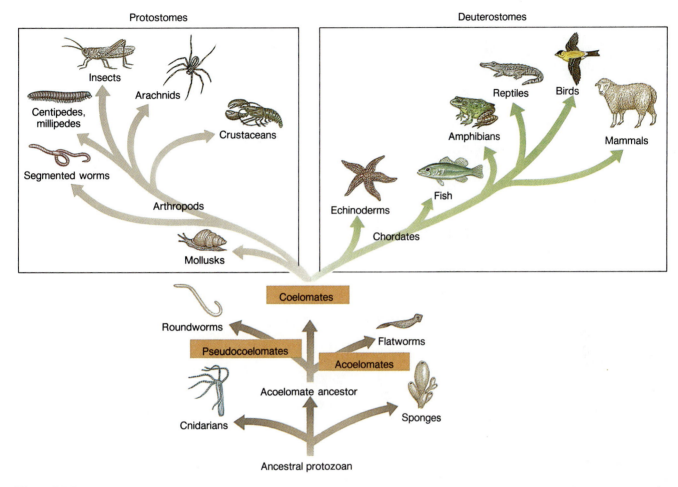

**Figure 34-4**
**Evolutionary tree of the animal kingdom.**

Each choanocyte has a filtering collar that encircles a flagellum. Movement of the flagellum produces an inward current that draws water and suspended food in through the porocytes. The collar filters out small organisms and bits of food as water is drawn in. Trapped food is then digested by the choanocyte or by an adjacent amoebocyte, and the filtered water exits through the osculum.

Sponges reproduce both sexually and asexually. Many sponges are hermaphrodites, each individual producing both sperm and eggs. During sexual reproduction, one sponge releases clouds of sperm through its osculum. Sperm become trapped in the choano-

cytes of another sponge, where an amoebocyte transports the trapped sperm to an egg in the mesenchyme. The resulting zygotes grow and develop into free-swimming larvae which eventually attach to a surface and grow into a new generation of adults.

Sponges reproduce asexually by developing either buds that pinch off from an adult or fragments released from a parent sponge. When conditions become unfavorable for growth, many sponges produce *gemmules,* clusters of amoebocytes surrounded by a hard, protective cell layer. Gemmules can survive drying and very cold temperatures, then regenerate a

new sponge when favorable conditions return. They are especially common in freshwater sponges.

## Phylum Cnidaria: Hydras, Jellyfish, Sea Anemones, and Corals

Members of the phylum **Cnidaria** are mostly marine animals that live in warm shallow water. All are radially symmetrical and have a sac body. Many taxonomists group cnidarians into three classes (Figure 34-6).

- The **Hydrozoa** (2700 species)—the Portuguese man o' war and freshwater species of *Hydra.*

TABLE 34-1

## Comparison of Protostome and Deuterostome Embryo Development

| Stage | Protostomes | Deuterostomes |
|---|---|---|
| Cleavage pattern | Spiral | Radial |
| Cell migration | None | Migration of cell groups |
| Organ formation from mesoderm | | |
| Larvae type | Trochophore* | Dipleurula* |
| Blastopore forms the: | | |

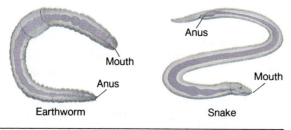

* Both types of larvae are free swimming, using cilia for propulsion. The larvae differ in shape and arrangement of cilia. The pear-shaped trochophore larva has a characteristic girdle of cilia around the middle, whereas the roundish dipleurula larva has winding ciliated bands.

TABLE 34-2

| Phylum | Cellular Organization |
|---|---|
| Porifera (sponges) 10,000 spp. | No tissues |
| Cnidaria (Hydra, corals, jellyfish) 9,000 spp. | Tissues |
| Platyhelmenthes (flatworms) 15,000 spp. | Organs and organ systems |
| Nematoda (roundworms) 10,000 spp. | Organ systems |
| Mollusca (squids, snails, clams) 100,000 spp. | Organ systems |
| Annelida (segmented worms) 9,000 spp. | Organ systems |
| Arthropoda (spiders, crabs, insects) 800,000 spp. | Organ systems |
| Echinodermata (sea urchins, sea stars, crinoids) 5,300 spp. | Organ systems |
| Chordata (tunicates, lancelets, vertebrates) 44,300 spp. | Organ systems |

- The **Scyphozoa** (200 species)— jellyfish.
- The **Anthozoa** (6200 species)—sea anemones and corals.

Taxonomists originally grouped cnidarians with all other radially symmetrical, sac-bodied animals into a single phylum named "Coelenterata," a term that means hollow sac. Today, however, coelenterates are split into two phyla: the Cnidaria, named for their specialized stinging cells called **cnidocytes**; and the Ctenophora (the

**Major Animal Phyla**

| Coelom | Circulatory System | Nervous System | Reproductive System | Distinguishing Characteristics |
|---|---|---|---|---|
| None | None (diffusion) | None | Male, female, or hermaphroditic | Aquatic filter feeders, porocytes, choanocytes |
| none | None (diffusion) | Nerve net, sensors over body surface | Male and female medusae, asexual budding | Stinging cnidocytes on tentacles |
| None | None (diffusion) | Ladder-type paired nerve cords, few ganglia | Hermaphroditic | Flatworms with a definite head and tail ends |
| Pseudocoelom | None (diffusion) | Simple "brain" dorsal and ventral nerve cords | Male or female, internal fertilization | Tapered body, cuticle covering, tube-within-a-tube digestion |
| Coelom | Open, 1 heart | Cerebral ganglia, nerve cords | Male, female, or hermaphroditic | Shells in most, free-swimming larvae in some |
| Coelom | Partly closed, 2 pumping vessels | Simple brain, paired ventral nerve cords | Male or female, external or internal fertilization, or hermaphroditic | Aquatic and terrestrial Repeating segments, but continuous digestive, nervous, and circulatory systems |
| Coelom | Open, 1 heart | Simple brain, ventral nerve cord | Male or female, mostly internal ferilization | Jointed appendages, chitinous exoskeleton, specialized segments, and sensory system |
| Coelom | Open, no heart | Nerve ring, radial nerves to each arm | Male or female, external fertilization | Radially symmetrical adult; bilaterally symmetrical, free-swimming larvae; spines, water vascular system |
| Coelom | Closed, 1 or no heart | Anterior brain, dorsal hollow nerve cord | Male, female, or hermaphroditic, internal or external fertilization | Notochord, gill slits, tail |

comb jellies), which lack cnidocytes and are structurally more complex than the cnidarians. We will limit our discussion to the cnidarians.

As a group, cnidarians have two body forms, a polyp and a medusa. **Polyps** are cylindrical, with the mouth and tentacles at one end facing upward (Figure 34-7a). **Medusas** are umbrella-shaped, with the mouth and tentacles on the lower, concave surface facing downward (Figure 34-7b).

Some cnidarians alternate between the two body types during their lifetimes. Others form either a polyp or a medusa, but not both. In species that alternate form, the polyp produces male or female medusae for sexual reproduction. Following fertilization, the zygote grows into a **planula**, which is a flattened, swimming larva that aids in dispersal. The planula eventually attaches to an object and develops into a new polyp.

Polyps reproduce asexually by forming small polyp buds. Although the buds usually separate from the parent to form new individuals, in some species the polyp buds remain attached, forming a colony of polyps. The Portuguese man o' war, for example, is not a single organism, but a community of polyps attached to one gas-filled bag.

The sac body of cnidarians has three layers: an ectodermis, endodermis, and intermediate **mesoglea** layer (see Figures 34-3a and 34-7). In some cnidarians the mesoglea lacks cells entirely, whereas in others, as in jellyfish, the mesoglea is a mass of cells that forms a thick jelly. The endodermis surrounds a single digestive cavity called the **gastrovascular cavity** (Chapter 18). The mouth leading to the gas-

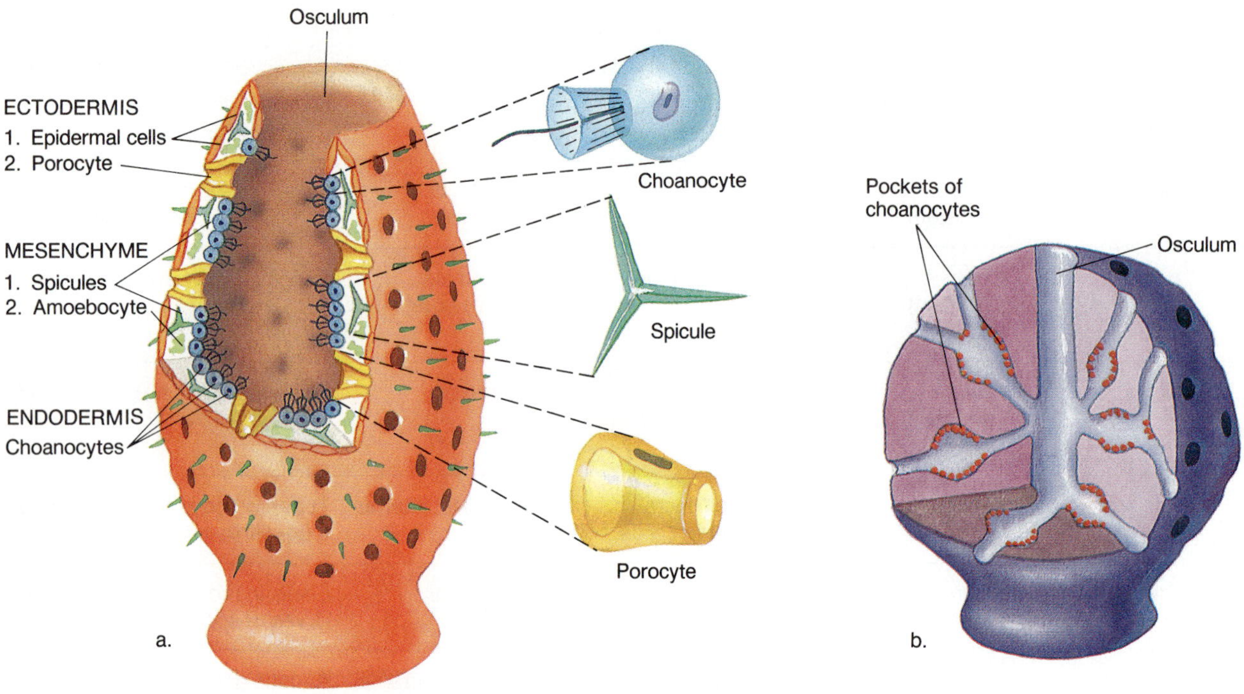

ECTODERMIS
1. Epidermal cells
2. Porocyte

MESENCHYME
1. Spicules
2. Amoebocyte

ENDODERMIS
Choanocytes

Osculum

Choanocyte

Spicule

Porocyte

Pockets of choanocytes

Osculum

a.

b.

**Figure 34-5**
**Sponge anatomy.** *(a)* Simple sponges have a single, central spongocoel chamber. *(b)* More complicated sponges have many smaller chambers created by repeated foldings of the sponge body. The blue arrows indicate the pathway water follows during feeding.

a

b

c

**Figure 34-6**
**Classes of cnidarians:** the Hydrozoa, hydra *(a),* Scyphozoa, jellyfish *(b),* and Anthozoa, sea anemones and corals *(c).* The tuft of tentacles at the top of anthozoans makes them appear so flowerlike that some of these animals are commonly mistaken for plants!

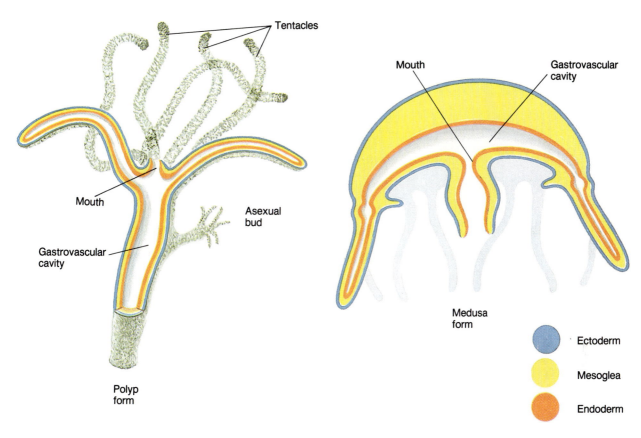

**Figure 34-7**
**Two cnidarian body types:** a polyp *(a)*, and a free-swimming medusa *(b)*.

trovascular cavity is surrounded by tentacles armed with cnidocytes which help cnidarians capture their prey.

All cnidarians are carnivores that use stinging cnidocytes to defend themselves, and to stun or kill their prey. Each cnidocyte contains a capsule called a **nematocyst** (Figure 34-8). Inside the nematocyst is a coiled thread attached to a spiked spear laden with powerful toxins. (The Portuguese man o' war produces a venom as potent as that of a cobra; toxins produced by the sea wasp *[Chironex fleckeri]* are powerful enough to kill a human within 30 seconds.) When stimulated, water rushes into the nematocyst. Within 3 milliseconds, hydrostatic pressure builds high enough to eject the tiny spear with a force capable of penetrating even the hard shells of some sea animals. When the harpoon-like spear hits its target, the barbs in-

ject immobilizing toxins and the trailing threads help to ensnare the prey. The cnidarian then grabs the incapacitated prey with its tentacles and stuffs it into its mouth.

## Phylum Platyhelminthes: Flatworms

The simplest animals with bilateral symmetry are members of the phylum **Platyhelminthes**, the flatworms. All 15,000 to 20,000 flatworm species are acoelomates; the only internal space is a digestive cavity. Flatworms have wide, thin bodies that are flattened dorsoventrally, from top (dorsal surface) to bottom (ventral surface) (see Figure 34-2). The phylum contains three classes (Figure 34-9): the **Turbellaria** (free-living flatworms), the **Trematoda** (parasitic flukes), and the **Cestoda** (parasitic tapeworms).

Flatworms have a sac body with a definite head and tail end. Unlike sponges and cnidarians, flatworms have organs and organ systems, including a nervous system, digestive system, and two muscle systems. One of the muscle systems enables flatworms to ingest and move food through the digestive system, and the other is used for locomotion.

Flatworms reproduce both sexually and asexually. Hermaphroditic flatworms mutually copulate; that is, the penis of each flatworm penetrates the female system of the other. During asexual reproduction, the flatworm simply divides in two along a median plane.

Some flatworms, such as the free-living planaria, *Dugesia* (Figure 34-10), have a simple central nervous system with nerve cords running the length of the animal. Clusters of

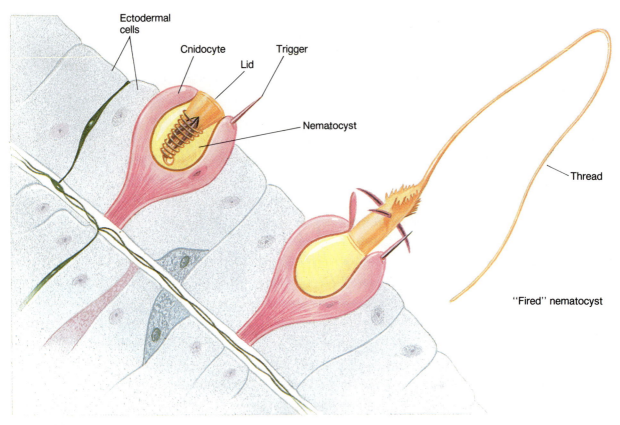

**Figure 34-8**
The tentacles of cnidarians are armed with numerous cnidocytes, each containing a nematocyst. When "fired", stinging barbs inject toxins, immobilizing prey. Prey also become entangled in the unraveled threads. Cnidocytes fire only once; they are then reabsorbed and new ones are formed.

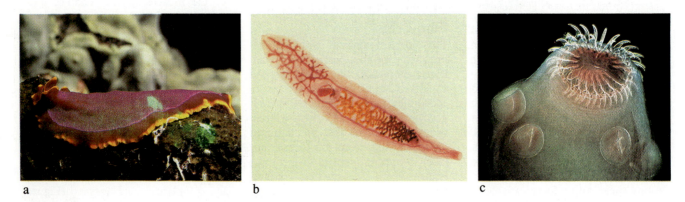

**Figure 34-9**
**Classes of flatworms:** *(a)* Turbellaria, free-living flatworms; *(b)* Trematoda, flukes; and *(c)* Cestoda, tapeworms.

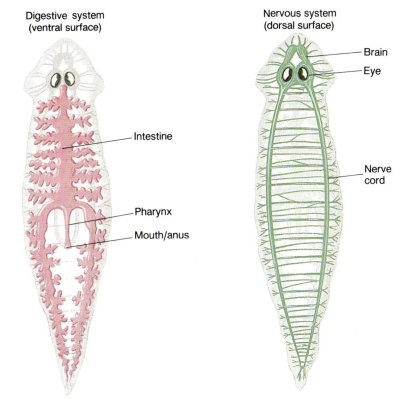

Digestive system
(ventral surface)

Intestine

Pharynx

Mouth/anus

Nervous system
(dorsal surface)

Brain

Eye

Nerve
cord

**Figure 34-10**
**Extensive branching of the digestive system** of a flatworm distributes food to body cells. The ladderlike nervous system has two nerve cords, terminating at a primitive brain. The eyes respond to light intensity; flatworms avoid bright light, behavior that helps them hide from predators.

ticle that is periodically shed and replaced as the nematode grows. In addition to a one-way digestive system, roundworms have a pseudocoelom.

Nematodes are either (1) decomposers, scavenging through moist soil for bits of organic material; (2) predators feeding on protozoa, earthworms, and each other; or (3) parasites. Parasitic nematodes attack other animals and many plants. They can quickly devastate an agricultural crop or kill a large vertebrate. A few species cause dangerous diseases in humans. Hookworms, for example, anchor to a person's intestinal wall by hooks, rasp an open sore, and then feed on the host's blood. Some nematodes invade human blood or the lymphatic system. In large numbers, they block lymph vessels, causing fluids to accumulate. Fluid accumulation in extremities can enlarge them to enormous propor-

neurons form a primitive brain at the head end. Flatworms lack a circulatory system to transport food and oxygen. Instead, they have flat thin bodies and a branched digestive system which enables all cells to be within diffusion distance from the surface of the worm for exchanging gases, as well as from the digestive system for absorbing food. Parasitic tapeworms lack a digestive system altogether. Since they live within a host's intestines, they absorb nutrients directly through their body walls.

The highly branched digestive system of flatworms not only acts as a chamber for digestion, but also transports nutrients throughout the organism. With a saclike body, food enters and wastes escape through the same opening. This system prevents flatworms from feeding and eliminating wastes at the same time; otherwise

they would ingest their own wastes. But the tube-within-a-tube digestive system of more complex animals, starting with the roundworms, forms a one-way tract with a mouth and an anus. This system improves feeding, digestion, and elimination of wastes.

## Phylum Nematoda: Roundworms

**Nematoda** is not only a diverse phylum (10,000 species), but possibly the most widespread and abundant animal group as well. Roundworms are established in virtually all habitats and occur in enormous numbers. In one study, 4.5 billion nematodes were found in one square meter of mud.

All nematodes have cylindrical, bilaterally symmetrical bodies that taper at both ends (Figure 34-11). The body is covered with a thick, yet flexible, cu-

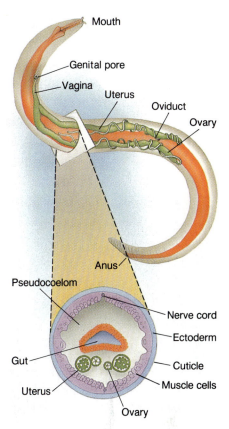

Mouth

Genital pore

Vagina

Uterus

Oviduct

Ovary

Anus

Pseudocoelom

Nerve cord

Ectoderm

Gut

Cuticle

Uterus

Muscle cells

Ovary

**Figure 34-11**
**The anatomy of a female nematode (Ascaris)** The transverse section shows the pseudocoelom and internal organs

tions, creating a condition named *elephantiasis*.

## Phylum Mollusca: Mollusks

With 100,000 plus species, the phylum **Mollusca** is second only to the Arthropoda in diversity. Nearly half of them are marine; the others live in fresh water (15,000 species) or on land (35,000 species). The three largest *classes* of mollusks are (see Figure 34-12):

- **Gastropods** (50,000 species—snails, abalones, nudibranchs, and slugs.

- **Bivalvia** (7500 species)—mollusks with hinged shells, such as clams, oysters, scallops, and mussels.

- **Cephalopoda** (200 species)—mollusks with tentacles or arms, such as squid, cuttlefish, and octopus. Cephalopods either have very reduced shells or lack them entirely.

Most mollusks have three body zones (Figure 34-13): a muscular **head-foot**, a **visceral mass** containing the internal organs, and the **mantle**, folds of tissue that cover the visceral mass. Openings to the digestive, reproductive, and excretory systems of gastropods empty into the **mantle cavity**, the area between the mantle and visceral mass.

The mantle of bivalves and most gastropods secretes a hard shell of calcium carbonate and protein. The shell helps protect against physical damage from pounding waves, and prevents drying out. (Abalones and limpets can clamp onto rocks tight enough to avoid desiccation while exposed to air during low tide.) Mollusk shells also create a refuge from predators. The lustrous inner layer of some mollusk shells is called the **nacre**, or "mother of pearl." In some bivalves, concentric layers of nacra are deposited around any foreign particle that lodges between the mantle and shell. After years, a pearl is formed.

Bivalves and gastropods both have an open circulatory system powered by a three-chambered heart. Two heart chambers (the atria) collect oxygenated blood from gills, while the third (the ventricle) pumps blood to the body tissues. Cephalopods have a closed circulatory system with auxiliary hearts to boost circulation.

The excretory system of mollusks consists of one or two *nephridia*, tubular organs that remove nitrogen-containing wastes and regulate the water and chemical balance of body fluids (Chapter 18). Each nephridium leads from the coelom to a bladder. When the heart contracts, body fluids are forced into the coelomic end of the nephridial tube. As the fluids flow through a nephridium, nutrients are absorbed back into the body, while nitrogen and other wastes are emptied into the bladder and expelled through an excretory pore.

Most mollusks are male or female, although some bivalves and many gastropods are hermaphroditic. A few sea slugs and oysters can even change their sex, not just once, but several times. Following fertilization, the zygotes of marine mollusks develop into free-swimming larvae that help disperse offspring.

Having a true coelom, free-swimming larvae, and nephridia are three evolutionary advances which the mollusks share with another large phylum of protostomes, the Annelida. However, unlike mollusks, annelids are segmented.

## Phylum Annelida: Segmented Worms

The phylum **Annelida** includes 9,000 species of segmented worms, such as earthworms, leeches, and marine bristle worms called polychaetes (Figure 34-14). Annelids are segmented, that is, their bodies are made up of a series

a

b

c

**Figure 34-12**
**Classes of mollusks.** *(a)* Gastropods (Mexican dancer nudibranch), *(b)* Bivalvia (bay scallop), and *(c)* Cephalopoda (squid).

of nearly identical units, each delineated by a visible groove that encircles the body (Figure 34-15). Segmentation increases flexibility, allowing the body to bend in a different direction at each segment. Increased flexibility improves locomotion. Some segments have become specialized for particular tasks. The first few units, for example, form the head with sensory appendages, brain, and feeding structures.

Although the body is segmented, the digestive system, main blood vessels, and ventral nerve cords pass through successive segments and are continuous throughout the annelid's body. On the other hand, muscles, nephridia, and other body parts are repeated in each segment. The coelom is compartmentalized by a partitioning septa between adjacent segments. Fluid pressure in the coelom creates a hydrostatic skeleton that helps support the body.

The annelid's circulatory system has muscularized blood vessels that contract, directing blood flow to the head through a dorsal vessel and to the posterior end through a ventral vessel.

All segmented worms, except leeches, have **setae**—short bristles from the sides of their body that aid in locomotion. Segmented worms lodge their setae into the substrate to brace their body against backward slippage as they contract their body muscles. In aquatic polychaete worms, setae are housed in fleshy paddles called **parapodia**. The large surface area of parapodia improves swimming and crawling on soft mud, and permits gas exchange. They are also used to generate feeding currents that direct food to the mouth.

Leeches rely on anterior and posterior suckers for locomotion. The anterior sucker anchors the body while muscles pull the body forward. The leech then attaches its posterior sucker, releases the anterior sucker, and then extends its body. The sequence is repeated, and the leech inches along the surface. Leeches also use their anterior sucker, which is equipped with a mouth, to suck blood from their hosts or to gulp down small prey.

Sexual reproduction in annelids is varied. Marine polychaetes have separate sexes. Gametes are usually ex-

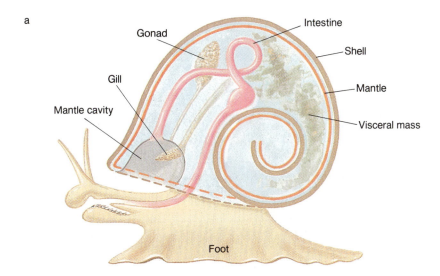

a

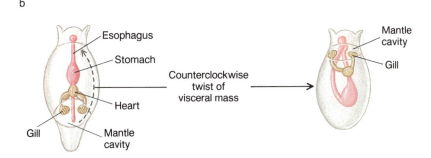

b

**Figure 34-13**
**Cut-away view of a snail.** *(a)* Like all gastropods, the digestive, respiratory, reproductive, and excretory systems empty into the mantle cavity. *(b)* During larval development, the visceral mass twists, positioning the mouth and anus at the anterior end of the body.

a

b

**Figure 34-14**
**Some representative annelids:** *(a)* leech and *(b)* feather duster, a polychaete worm.

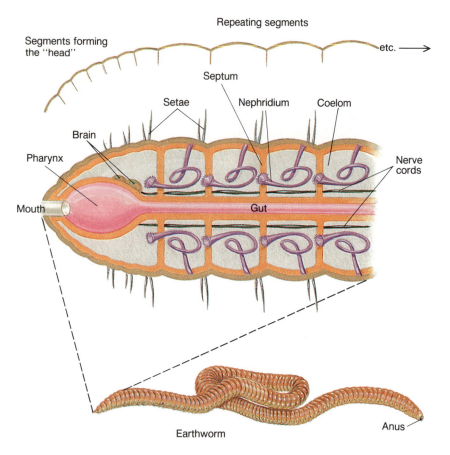

**Figure 34-15**
**Like other annelids,** earthworms are constructed of repeating segments, each with its own set of internal organs. The nervous, circulatory, and digestive systems, however, run the length of the worm.

pelled into the sea through nephridia openings. Following external fertilization, zygotes develop into motile trochophore larvae (see Table 34-1).

In hermaphroditic annelids, a few segments in the front third of the body contain male and female gonads. During mutual copulation, sperm are simultaneously passed between worms. Each worm then deposits its eggs and the transferred sperm into a cocoon. Fertilization occurs in the cocoon, and the zygotes develop into small worms.

## Phylum Arthropoda: Arthropods

Zoologists estimate there are a billion billion ($1 \times 10^{18}$) arthropods living at any given moment, making them close rivals to nematodes as the most abundant type of animal. But it is not only their enormous abundance that makes the arthropods an important animal group. Arthropods are the most diverse kind of organism on earth. More than 900,000 species are included in the phylum **Arthropoda**, a term that refers to their jointed limbs (arthros = jointed, poda = feet).

## CHARACTERISTICS OF ARTHROPODS

The enormous success of arthropods is the result of a combination of unique characteristics (Figure 34-16), including

1. *A hard exoskeleton made of chitin and protein.* The exoskeleton of arthropods helps protect them against predators and, in terrestrial species, helps prevent water loss by evaporation. This rigid cover also supports soft internal tissues and organs. Exoskeletons do not enlarge as the animal grows, however, so they must be repeatedly shed and replaced with a larger version. This shedding process is called *molting,* and the stages between molts are called *instars.*

2. *Jointed limbs.* Arthropods have jointed appendages capable of complex movements.

3. *Specialization of segments and appendages.* Different body segments are specialized for particular tasks. A grasshopper, for example, has segments adapted for sensing the environment, feeding, walking, and flying (Figure 34-16b). Some segments are fused to form the head, thorax, and abdomen.

4. *A three-region digestive system.* Arthropods have a foregut for ingesting, pulverizing, and storing food; a midgut for digesting and absorbing food; and a hindgut for absorbing water and eliminating feces.

5. *Specialized respiratory systems for different habitats.* Aquatic arthropods absorb oxygen and release $CO_2$ through gills. Terrestrial arthropods exchange gases either through *book lungs* or, more commonly, through a system of air-conducting tubes called *tracheae* (Chapter 19).

6. *Specialized sensory system.* An exoskeleton is a barrier to receiving external stimuli. Therefore, arthropod sensors either extend through the exoskeleton (hairs or bristles), or lie in fissures in the exoskeleton. Some arthropods have antennae with sensors for sound, smell, and taste. Arthropods also have eyes, ranging from simple versions with only a few photoreceptors, to large compound eyes with thousands of receptors.

## ARTHROPOD CLASSIFICATION

Arthropods are grouped into four subphyla (see Figure 34-17).

- Subphylum **Trilobita** — 10,000 species, all extinct.

- Subphylum **Chelicerata** — 80,000 species in three classes:

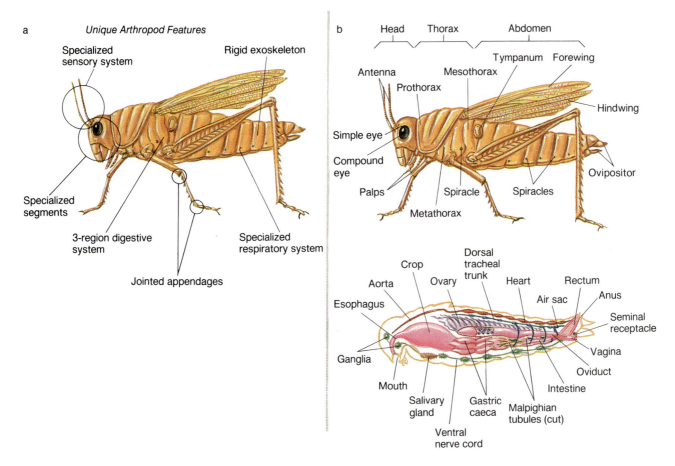

a    *Unique Arthropod Features*

Specialized sensory system

Rigid exoskeleton

Specialized segments

3-region digestive system

Jointed appendages

Specialized respiratory system

b

Head    Thorax    Abdomen

Antenna

Prothorax

Mesothorax

Tympanum    Forewing

Simple eye

Compound eye

Hindwing

Palps

Metathorax

Spiracle

Spiracles

Ovipositor

Crop

Dorsal tracheal trunk

Aorta

Ovary

Heart

Rectum

Esophagus

Air sac

Anus

Seminal receptacle

Ganglia

Vagina

Oviduct

Mouth

Intestine

Salivary gland

Gastric caeca

Malpighian tubules (cut)

Ventral nerve cord

**Figure 34-16**
**Unique characteristics of arthropods** as illustrated in a grasshopper *(a)*. Detail of grasshopper anatomy *(b)*.

a

b

c

**Figure 34-17**
**Subphyla of arthropods**
1. Trilobita. Although there were more than 10,000 species of marine trilobites during the Paleozoic Era, today all are extinct (Chapter 30). (Photo *(a)* is a fossilized trilobite.)
2. Chelicerata. Chelicerates have two body sections: a cephalothorax (head-thorax) covered with a single skeletal plate (carapace) and an abdomen. (Shown in *(b)* is a jumping spider.)
3. Crustacea. Crustaceans have specialized feeding appendages called mandibles and two pairs of antennae (Refer to arrow crab, page 606)
4. Uniramia. Uniramians have a pair of mandibles and one pair of antennae. (Shown in *(c)* is a centipede.)

1. *Merostomata* (5 species)—horseshoe crabs.

2. *Arachnida* (60,000 species)—scorpions, spiders, mites, ticks, and harvestmen (daddy long-legs).

3. *Pycnogonida* (1000 species)—sea spiders.

• Subphylum **Crustacea**—35,000 species. Crustaceans include crabs, shrimp, lobsters, crayfish, barnacles, water fleas, sand fleas, copepods, fairy shrimp, isopods, and amphipods.

• Subphylum **Uniramia**—more than 800,000 species in three classes:

1. *Diplopoda* (10,000 species)—millipedes.

2. *Chilopoda* (2500 species)—centipedes.

3. *Insecta* (800,000 species)—insects.

### THE INSECTS

Insects are the most diverse class of organisms on earth (Figure 34-18). Some live in complex societies where different individuals carry out specific functions that benefit the group as a whole (see Animal Behavior: Cooperative Groups).

Insects have a unique combination of features: three pairs of legs; a body divided into a head, thorax, and abdomen; a single pair of antenna; and simple or compound eyes (Chapter 15). Flying insects have wings that were produced from outgrowths of the thorax, not from modified limbs as those of birds and bats.

Most insects (about 90 percent) undergo dramatic morphological changes when they molt. This phenomenon is called metamorphosis (shown in Bioline: A Gallery of Remarkable Adaptations, Chapter 29). These changes produce distinct developmental stages: an egg, larva, pupa, and, finally, an adult. Although the insect does not feed during the egg and pupa stages, these stages can withstand long cold or dry periods, enabling the insect to survive unfavorable conditions. The larval stage is specialized for feeding, the winged adult stage for reproduction and dispersal. In general, each developmental stage has its own habitat, environmental requirements, and adaptations that differ from those of other stages.

Like other arthropods, mollusks, and annelids, insects are protostome coelomates. Among the phyla of deuterostome coelomates, the two largest are the Echinodermata and the Chordata.

a

c

b

**Figure 34-18**
**This gallery of insects** provides a glimpse of the most diverse, and perhaps most successful, group of animals on earth. Some biologists estimate there may be as many as 30 million kinds of living insects still to be named and classified by taxonomists. *(a)* Red-banded leaf hopper. *(b)* a beetle, *Neptunides stanleyi*. *(c)* New Zealand walking stick.

# ANIMAL BEHAVIOR IV

## COOPERATIVE GROUPS

Many animals live in flocks, herds, and schools. Some form complex societies with rigidly defined roles for members. These animals live together for the mutual benefit of the group. Here are a few examples of social behavior among animals.

- Young caterpillars crowd together, warming one another and speeding up each other's development.

- Bees beat their wings to warm their hive during cold spells; they gather water (a natural coolant) and fan the hive to keep it cool during hot periods.

- A flock of birds finds more food per individual than each could find on its own.

- Lions, hyenas, and wild dogs hunt in cooperative groups that can corner and bring down prey more surely than predators hunting as individuals.

- Quail huddle in a circle at night to keep warm. Their heads face outward, maintaining a 360° lookout for predators.

- A large school of water fleas confuses predators. With so many targets, a predator has difficulty focusing on one individual and often comes away empty-stomached.

- Juveniles, females, and dominant male baboons are encircled by low-ranking males for protection against lion attacks.

- A few adult emperor penguins tend and protect the young while the majority of adults fish for food.

- Several female wasps share the same burrow and cooperate in the task of defending the burrow against predators.

These examples illustrate that group living often improves conditions, increases food supply, aids in detection and avoidance of predators, and increases reproduction rates.

Insects form some of the most complex societies in the Animal kingdom. Many ant colonies, for example, have a rigid caste system with a queen that lays rounds of eggs, workers of varying sizes that tend the queen, maintain the nest, and gather food. Menacing soldiers aggressively defend the colony. Most ant colonies go through cycles, usually controlled by the queen's egg-producing cycle. For instance, army ants alternate between a three-week stationary phase, during which the queen lays thousands of eggs a day, and a nomadic phase in which the entire colony of a million or so ants goes out on massive and devastating raids. The raiding ant front moves at about 20 meters/hour (almost 80 feet per hour), killing everything that strays in their path. Insect nests (including those of other ants) are ransacked, the adults are slain and dismembered and the eggs, larvae, and pupae consumed on the spot. Should the trail give out, the ants automatically cooperate to form a living bridge of linked bodies that allows the migrating horde to cross virtually any gap, even watery ones such as rivers.

Some social animals demonstrate various degrees of *altruistic behavior,* self-sacrifice for the benefit of other individuals. In bumble bee societies, for example, worker bees sacrifice their own reproductive opportunity for the sake of the queen bee. Sterile female workers do all the work of building cells for the queen's eggs, tending the young, and gathering food. In these societies, the hive, not the individual, benefits from the altruistic behavior because workers are able to raise the young of a single queen in greater numbers and with more efficiency than each would be able to do alone. Other examples of altruism are soldier ants that lay down their lives to protect the colony, a chimpanzee that shares food with other individuals in its group, and the African monkey who warns others of an approaching predator by issuing a noisy alarm that puts him at personal risk of detection.

Animal societies rely on elaborate communication systems to cordinate activities. Ants use touch and smell more than vision to communicate. Olfactory communication in ants is based on the production of a wide variety of chemical messagers (pheromones) used for marking food trails, sounding defense alarms, and identifying colony members. Honeybees inform others in the hive of the location of food by using dance language. The length of a dance and the number of waggles (vibrations of the body) correspond to the distance to the food source, and the direction of the waggle correlates with the direction to the food source.

## Phylum Echinodermata: Echinoderms

The phylum **Echinodermata** includes 5300 species of sea urchins, sea stars, brittle stars, crinoids (feather stars and sea lilies), and sea cucumbers (Figure 34-19). All echinoderms have an internal skeleton made of many small, calcium carbonate plates. These plates have jutting spines—inspiring the name echinoderm (echino = spiny, derm = skin).

Echinoderms are marine animals. Their free-swimming, bilaterally symmetrical larvae develop into radially symmetrical adults with body extensions arranged around a central mouth and anus (Figure 34-20).

Echinoderms have a unique **water vascular system** which is used for loco-

**Figure 34-19**
**Echinoderms** range from familiar sea stars (see next figure) and sand dollars *(a)*, to exotic feather stars *(Melithaea)* that capture prey with its featherlike arms *(b)*.

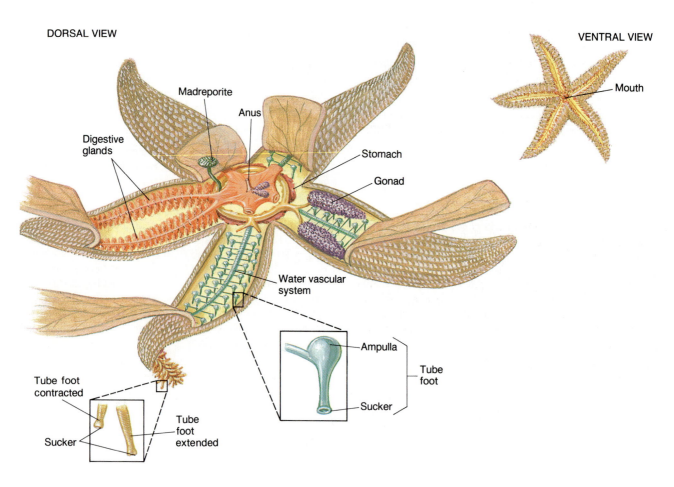

**Figure 34-20**
**The anatomy of a sea star**, detailing the vascular and digestive systems. (See text for a description of both systems.)

motion, respiration, sensory perception, clinging to surfaces, and attacking prey. Radial canals branch out into each arm from a central ring-shaped channel. Water is drawn in through a sievelike opening called the **madreporite** attached to the central channel. Each radiating canal has hundreds of short side branches that terminate with a hollow tube foot. In many echinoderms, the tube foot has a muscularized water bladder called an **ampulla** at one end and a sucker at the other. As muscles surrounding the ampulla contract, water is forced into the tube foot, causing it to elongate. When muscles running the length of the tube foot contract, water is forced back into the bladder and the foot shortens.

The movement of hundreds of tube feet is neurologically coordinated; the echinoderm can either move slowly or cling tightly to one spot. Once suckers are in place, by contracting muscles along the tube foot a sea star can create a sufficiently strong force to pry open the two shells of a mussel, oyster, or clam. Pried open the sea star then extrudes its stomach out through its mouth. The everted (inside out) stomach releases digestive enzymes and begins digesting the prey in its own shell. Partially digested food is taken into the gut, and digestion is completed within digestive glands that extend into each arm.

## Phylum Chordata: Chordates

Each member of the phylum **Chordata** possesses a stiff, yet flexible, rod called a **notochord** that runs down their dorsal surface. The combination of a flexible back support to which muscles are attached provides enormous versatility in movement, including walking, running, flying, slithering, crawling, hopping, and an ability in some chordates to move their extremities with enough precision to do extraordinarily fine manipulations.

Chordates are distinguished from other animal groups by four characteristics.

1. A stiff, yet flexible notochord.

2. A **dorsal hollow nerve cord** running the length of the animal.

3. **Gill slits** in the pharynx wall.

4. A **tail**.

In most chordates (especially the ones people are familiar with) the notochord and gill slits appear briefly in embryos, then disappear. The notochord develops into a vertebral column in adults. In a few chordates (no-tably humans) the tail also disappears.

About 45,000 species of chordates are alive today, including sea squirts, lampreys, sharks, trout, frogs, lizards, birds, elephants, and humans. Zoologists classify this diverse group of animals into three subphyla: the Urochordata (1250 species), the Cephalochordata (23 species), and the Vertebrata (more than 43,000 species).

### SUBPHYLUM UROCHORDATA: TUNICATES

Tunicates (subphylum **Urochordata**) are marine animals with an outer wall, or tunic, covering the adult body. In most, the surrounding tunic is made of cellulose-like carbohydrate; a few have a gelatinous tunic.

Tunicates undergo a remarkable transformation from the larva to adult stage (Figure 34-21): larvae are tadpole-shaped and free-swimming, whereas most adults are sac-shaped and sessile. Equipped with a notochord, the larvae are excellent swimmers. But with a poorly developed digestive system, the larvae are unable to feed at all, so they cannot swim very far. They attach themselves to a solid surface soon after being released, before they exhaust their stored energy supplies. Once attached, larvae de-

a

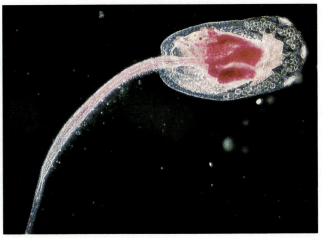
b

**Figure 34-21**
**Representing an evolutionary turning point,** adult tunicates *(a)* have only one chordate characteristic—gill slits. On the other hand, the larvae *(b)* possess all chordate characteristics—a dorsal nerve chord, notochord, gill slits, and tail.

velop into a flask-shaped adult; the tail, notochord, and nerve cord are absorbed, but the gill slits remain to function in gas exchange and feeding.

Adult tunicates are filter feeders. The sac-shaped body has two openings: an incurrent and excurrent siphon (Figure 34-22). Water is drawn in through the incurrent siphon by cilia that line the pharynx. As water passes through the gill slits, small food particles are strained out. The filtered seawater is then expelled through the excurrent siphon.

Because larvae of tunicates possess all four basic chordate characters, evolutionary zoologists speculate that the larvae of an ancient tunicate-like organism acquired the ability to produce gametes for sexual reproduction and acted as the ancestor for all other chordate groups. This primitive ancestor appeared more than 500 million years ago, the time when the fossil record reveals the emergence of a new group of chordates, the cephalochordates.

## SUBPHYLUM CEPHALOCHORDATA: LANCELETS

Members of the **Cephalochordata** subphylum (the lancelets) have changed very little over their 500-million-year history. Cephalochordates are small, fishlike animals that live in shallow oceans throughout the world. The body is flattened and pointed at both ends, giving it the appearance of a surgical lancet (Figure 34-23). Adults have gill slits, a dorsal, hollow nerve cord, and a prominent notochord that extends beyond the brain. (The name cephalochordate refers to this prominent head extension.)

Like tunicates, adult lancelets are filter-feeders. They wriggle into sand or mud, posterior end first. The protruding anterior end contains the mouth. Cilia draw in water and food, and the tentacles surrounding the mouth help separate out food particles. As water passes through the gill slits, food and oxygen are removed, and carbon dioxide is released. The water then passes through the *atrium* and exits through the *atriopore* near the tail.

## SUBPHYLUM VERTEBRATA: VERTEBRATES

All vertebrates have a vertebral column, a backbone composed of separate vertebrae constructed of bone or cartilage (Chapter 17). In virtually all adults, vertebrae either surround or replace the notochord and form the main axis of the internal skeleton. All members of the subphyllum **Vertebrata** have an elaborate nervous system with an encased brain and spinal cord, a closed circulatory system, a pair of kidneys for excreting nitrogenous wastes and maintaining homeostasis, and one pair of eyes. Vertebrate embryos have a series of **pharyngeal pouches** — outgrowths from the walls of the pharynx that during embryonic development break through the body surface to form gill slits.

Vertebrates are grouped into seven classes.

Class Agnatha (63 species) — jawless hagfish and lampreys.

Class Chondrichthyes (850 species) — fishes with cartilaginous skeletons such as sharks, skates, and rays.

Class Osteichthyes (18,000 species) — bony fishes, such as trout, salmon, and bass.

Class Amphibia (3200 species) — salamanders, frogs, and toads.

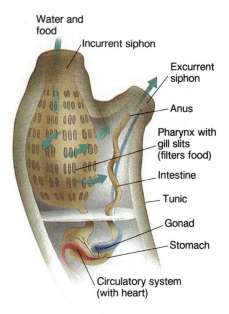

**Figure 34-22**
**Filter-feeding adult tunicates never move.** Drawing seawater in through an incurrent siphon, they filter out suspended food particles as the seawater passes through its gill slits. Filtered water, digestive wastes, and gametes exit through an excurrent siphon.

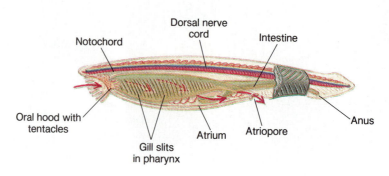

**Figure 34-23**
**Knifelike adult lancelets** wriggle into the seabed, heads protruding for feeding. Sea water enters through the oral hood, passes through the atrium and exists the atriopore.

Class Reptilia (6000 species)—scaly air-breathing vertebrates such as lizards, snakes, turtles, and crocodiles.

Class Aves (9000 species)—feathered vertebrates, such as woodpeckers, doves, and hummingbirds.

Class Mammalia (4500 species)—vertebrates that nurse their young with mammary glands, such as kangaroos, leopards, whales, and humans.

Because all members of the Agnatha, Chondrichthyes, and Osteichthyes classes are fishes, they are sometimes grouped together into the superclass Pisces.

## Pisces: The Fishes

A collection of adaptations enable fishes to successfully live in water. Blocks of muscles along the length of their bodies are attached to a stiff backbone or a notochord (Agnatha). Rhythmic muscle contractions move the tail from side to side, pushing the tail fin against the water and driving the fish forward. Fins on the sides, top, and bottom are used to turn or change depth. In addition to adaptations for swimming, most fishes have overlapping scales that produce a double-layered protective armor, while allowing the body to remain flexible.

Sight, taste, and smell are well developed in fishes. Another sense that is especially important in most fishes is the ability to detect changes in water pressure. A chain of pressure-sensitive cells are embedded along each side of the fish. These cells form part of the *lateral line* system that not only helps fish maintain a certain depth, but also allows detection of oncoming predators, helps prevent collisions with obstacles in the dark, and may participate in the schooling behavior.

The most primitive fishes belong to the class Agnatha. The Agnatha contains jawless hagfish and lampreys, fishes that have changed little in 400 million years (Figure 34-24). Although lampreys and hagfish look like eels, they do not have jaws, scales, a head, paired fins, or vertebrae as eels do. Instead of a vertebral column, the muscles of agnathans are attached directly to the notochord.

Many adult lampreys are parasites that use their suckerlike mouths to attach themselves to the outside of another fish. Spiny teeth on their tongues rasp a hole in the prey's flesh. The lamprey then feeds on the prey's body fluids. Hagfish and nonparasitic lampreys are scavengers, feeding on the remains of dead animals.

Fishes with a skeleton made of rubbery cartilage (class Chondrichthyes) and bony fishes (class Osteichthyes) share two evolutionary advances not found in agnathans:

• *Paired fins* for improved locomotion.

• *Jaws* for grasping, biting, and chewing food. The jaws are developed from modifications of skeletal arches between the gill slits (Figure 34-25).

Members of the Chondrichthyes include the somewhat rare, deep-water ratfish *(Chimaera),* and the more familiar sharks, skates, and rays (Figure 34-26). The bodies of skates, rays, and sharks are covered with small, sharp denticles, structures that are similar to teeth.

Skates and rays "fly" through the water, propelling their dorsoventrally flattened bodies by undulating enormous, winglike pectoral fins. Most are scavengers, spending most of their

a

b

**Figure 34-24**
**Jawless fishes,** class Agnatha, include parasitic lampreys *(a)* and scavenging hagfish *(b).*

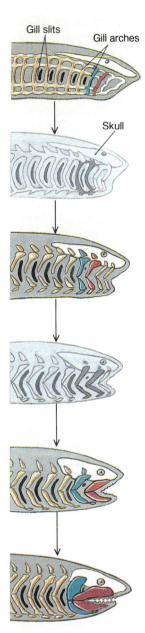

**Figure 34-25**
**Jaws evolved from gill arches.** This sequence traces the evolutionary changes from an ancient jawless ancestor to modern shark.

**Figure 34-26**
**The class Chondricthyes** includes the manta ray *(photo)* and sharks.

time skimming the ocean floor. The long tails of stingrays are armed with poisonous spines for protection.

Many sharks are ferocious predators with protruding jaws that enable them to feed on a variety of other fishes, crustaceans, mollusks, and an occasional human swimmer. The whale shark *(Rhincodon typus)* is the largest of all fishes, weighing up to 20 tons and growing to lengths of 18 meters (60 feet). Ironically, this giant doesn't consume large prey. A sieve in its pharynx filters out only minute crustaceans and drifting plankton.

The class Osteichthyes includes fishes with skeletons that are at least partly composed of true bone. With 18,000 species, the Osteichthyes is the most diverse class of fishes (Figure 34-27). Although a bony skeleton is heavier than a cartilaginous one, bony jaws aid feeding. The weight of a bony skeleton is usually counterbalanced by a gas-filled *swim bladder.* In addition to maintaining buoyancy, the swim bladder serves as an oxygen reserve.

A few species of bony fishes have lungs, which enable some of them to survive for long periods in dry mud.

The African lungfish, for example, secretes a mucous cocoon around itself when rivers begin to dry. In the dried mud, it uses its lungs to breathe air through a small opening in the cocoon.

### Amphibians

The class Amphibia contains 3150 species, grouped into three orders (Figure 34-28).

- *Caudata,* tailed amphibians (salamanders and newts).

- *Anura,* tail-less amphibians (frogs and toads).

- *Apoda,* wormlike, burrowing amphibians (caecilians).

As the name suggests, virtually all amphibians lead a double life — the larva or tadpole lives in water, and the adult on land. However, the adults must remain in damp habitats or stay close to water to prevent their most skin from drying out. Adult amphibians exchange $CO_2$ and $O_2$ through their skin, supplementing respiration in their simple saclike lungs (Chapter 19).

a

b

**Figure 34-27**
**The second most diverse group** of animals are the bony fishes (class Osteichthyes). *(a)* The blenny *(Ecsenius yaeyamaensis)* has eyes on top of its head to scan for danger. *(b)* A juvenile cowfish, *Lactoria diaphanus.*

All amphibians except caecilians return to water to breed or to lay eggs. Adult frogs and toads do not copulate but simultaneously release sperm and eggs directly into the water. The larvae tadpoles swim and "breathe" through gills. Eventually, they grow legs, absorb their gills and tail, develop lungs, and emerge from the water as miniature adults.

Amphibians were the first land vertebrates. They were abundant during the Carboniferous period 300 million years ago and remained the dominant land animal for over 100 million years. But because amphibians are tied to water for reproduction, they are not fully adapted for living on dry land.

About 250 million years ago, a new type of animal appeared with well-developed lungs and scaly, water-resistant skin. This new animal laid "land eggs" that resisted drying and were equipped with an internal supply of food and water for the developing embryo. Freed of the need to be near water or to return to water for reproduction, these animals were better adapted to living on land than amphibians. These were the first reptiles.

a

b

**Figure 34-28**
**A representative amphibian.** Amphibians include salamanders and newts (Order Caudata), frogs (Order Anura shown in photo *(a)*), and the less familiar caecilians (Order Apoda) shown in photo *(b)*.

## Reptiles

The 6000 species of living reptiles are divided into four orders (Figure 34-29).

- Squamata (5600 species)—reptiles with two openings in the temporal region of the skull (lizards and snakes).

- Chelonia (220 species)—reptiles without openings in the temporal region of the skull (turtles and tortoises).

- Crocodilia (22 species)—amphibious quadrapeds (crocodiles and alligators).

- Rhynochocephalia (1 species)—a "living fossil", the lizardlike tautara *(Sphenodon punctatum)*.

During the Permian and most of the Triassic periods, reptiles were the dominant land animal. Among these ancient reptiles was a group called the Archosauria, which included the pterosaurs (flying reptiles), crocodilians, and the most varied and fascinating group of reptiles, the dinosaurs. Hundreds of dinosaur species have been uncovered by scientists, from almost every part of the world. They ruled the land for more than 150 million years, then vanished by the end of the Mesozoic Era (see Bioline: Fall of the mighty, Chapter 30).

All animal groups discussed so far are *ectothermic,* that is, their body temperature varies with the temperature of their surroundings. Ectotherms employ many forms of behavior to maintain satisfactory body temperatures. For example, turtles bask in the morning sun and lizards do "pushups" to raise their body temperature. Snakes hide under bushes or rocks to lower their body temperature during dangerously hot periods. *Endotherms,* on the other hand, maintain a particular body temperature regardless of ambient temperature. This adaptation gives them a competitive edge in many environments. Endothermia is restricted to two classes of chordates: Aves (birds) and Mammalia (mammals).

## Birds

The class Aves contains 9000 species of birds, making this group of vertebrates second only to bony fish in diversity (Figure 34-30). Birds are classified into 27 orders and 160 families mainly on the basis of differences in beaks and feet. They are found in almost every terrestrial habitat. African ostriches are the largest birds, growing to 2 meters (7 feet) in height and weighing 136 kilograms (300 pounds). The smallest bird is the Helena's hummingbird of Cuba, shorter than 6 centimeters (2.3 inches) and weighing only 4 grams (0.1 ounces). Some hummingbirds are so small, they are prey for spiders.

Birds have retained some of the characteristics of their reptilian ancestors, such as scaly skin, amniote eggs, and a *cloaca,* a common chamber that receives discharges from the intestinal, excretory, and reproductive systems. But birds were equipped with two unique adaptations that enabled them to become excellent flyers: (1) the forearms became modified into wings, and (2) their scales became modified into two types of feathers, one for insulating against heat loss (fluffy down) and the other for flight. Flight feathers are constructed of parallel barbs with crisscrossing barbules hooked together to form an interlocking, lightweight mesh that supports the animal in air (Figure 34-31). Hollow, lightweight bones, internal airsacs, streamlined bodies, and powerful flight muscles attached to the breastbone also contribute to a bird's flying ability. This combination of adaptations works so well that even huge birds can fly. The great condors, for example, have wing spans greater than 3 meters (10 feet).

Many birds migrate from their springtime breeding grounds to spend winter in places were temperatures are warmer and food is more plentiful. Birds use their keen powers of navigation to migrate distances as far as 18,000 kilometers (10,800 miles) (see Animal Behavior: Migration). This ex-

a        b

**Figure 34-29**
**Representatives of living reptiles.** Reptiles include lizards, snakes, turtles, tortoises, and crocodiles and alligators. *(a)* In murky waters, this flattened tortoise (matamata) attracts little attention from passing fish. Any unwary fish quickly disappears into the matamata's wide mouth. *(b)* A flick of its head and this crocodile will soon be dining on a young frog.

# ANIMAL BEHAVIOR

## Migration

After swimming hundreds of miles out to sea, how do salmon find their way back to the stream in which they were born? How do long-distance migratory birds reach winter habitats and then return to the same breeding ground thousands of miles away in the spring? Animals rely on a variety of mechanisms to accomplish impressive navigation feats. Some birds are simply programmed to fly in a certain direction, for a certain period of time. European warblers, for example, have an internal compass and clock that triggers them to fly southwest for 40 days and then southeast for 20 to 30 days, taking them from Central Europe through Spain, across Gibraltar, then into southern Africa. But many birds and other vertebrates travel longer distances and to relatively smaller targets (Africa is a big target), requiring them to judge their position relative to home. In other words, they have a "map" sense.

Navigation is sometimes as simple as following a temperature or odor gradient; getting warmer or sensing increasingly stronger odor means the animal is on the right path. For example, evidence suggests that salmon may rely on their sense of smell to navigate to their home stream from the sea.

For animals that migrate long distances, the situation is much more complex. Long-distance navigation apparently involves more than one system, including the ability to

- determine compass direction from the earth's magnetic field or from the position of the sun (at any time of day) or stars at night;
- locate familiar landmarks near their destination; and
- possibly respond to tiny changes in barometric pressure and odor.

But a complete answer to how birds and wide-ranging vertebrates develop a "map" sense for navigation is still a mystery. Studies suggest that the answer may include other information-gathering systems, learning, and a sensitive internal clock. A biological clock is clearly needed by "sun navigators" for them to maintain the same direction of flight even as the sun changes position during the day.

**Figure 34-30**
**Class Aves — Birds.** The multitude of hummingbird species in the tropics have brilliant-colored body feathers, tails, and heads. The diversity of shapes and sizes of bills enables different hummingbirds to get nectar from various kinds of flowers.

traordinarily long journey is made by the arctic tern *(Sterna paradisaea)* which breeds during the summer in the Arctic Circle and winters on the other side of the globe in Antarctica.

## MAMMALS
The class Mammalia includes 4500 species. Two characteristics distinguish mammals from other vertebrates: (1) mammalian skin is endowed with hair and (2) mammals nourish their young with milk produced by mammary glands. Brain development is greatest among the mammals, enabling them to depend less on instinctive behavior and more on information learned from experience.

Taxonomists divide mammals into three groups (Figure 34-32).

- Monotremes, egg-laying mammals (duckbilled platypus and spiny ant-teater).

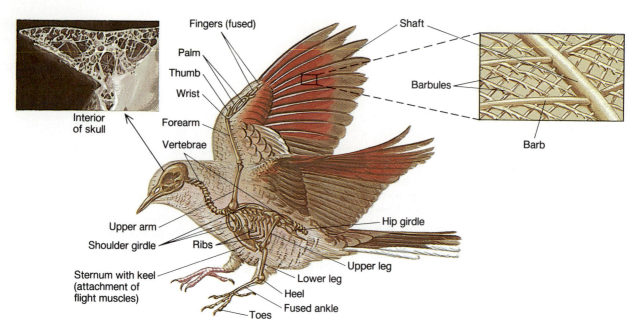

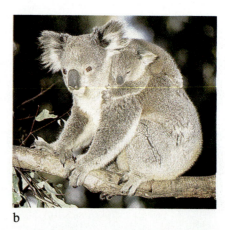

**Figure 34-31**
**Birds of a feather.** Although insects and bats can fly, birds have unparalleled powers of flight because of their feathers. Interconnecting barbules along with a lightweight skeleton make birds excellent flyers.

- Marsupials, mammals with a cloaca whose young are born immature and complete their development in an external pouch in the mother's skin (opossum, koala bears, and kangaroos).

- Placentals, mammals that nourish developing offspring internally, through a placenta (see Chapter 23). Placentals include a wide variety of animals such as insectivores (shrews, moles, and hedgehogs), primates (monkeys, apes, and humans), rodents (squirrels, rats, mice, and marmots), carnivores (bears, wolves, foxes, cats, dogs, raccoons), elephants, and cetaceans (whales and dophins).

Marsupials and placentals give birth to live young and care for offspring until maturity. Parental care improves the offspring's chances of survival, which is perhaps one of the reasons for the great success of mammals today.

**Figure 34-32**
**Class Mammalia** *(a)* Monotremes include strange looking spiny anteaters and this duckbilled platypus, the only mammals that lay eggs. *(b)* Young marsupials are born immature and complete development in their mother's pouch. A baby koala is born after only 30 days of development. Blind and with no back legs, the underdeveloped koala uses miniature front legs to crawl into its mother's pouch, where it stays for six months. When a young koala becomes too large for the pouch, it clings to its mother's back, returning to the pouch for an occasional drink of milk. *(c)* As with the mountain gorilla, most mammals nourish their embryos and fetuses through a placenta, then continue to nourish their newborn with maternal milk from mammary glands.

# Synopsis

## MAIN CONCEPTS

- **Animals are multicellular, generally motile organisms.** Most form tissues and organs, and most ingest food and digest it in an internal cavity.

- **All animals are heterotrophs,** dependent on autotrophs for their nourishment.

- **With more than 1.3 million described species, the Animal kingdom is the most diverse kingdom of organisms.** Most animals (800,000 species) are insects.

- **Animals probably evolved from flagellated protozoa.**

- **Taxonomists distinguish different animal groups on the basis of body symmetry, body design, presence of a body cavity, and characteristics of embryo development.**

- **The Animal kingdom is divided into 2 subkingdoms and 35 phyla.** Most phyla include marine animals.

- **Although some animals reproduce asexually, most produce gametes (nonmotile eggs and flagellated sperm) for sexual reproduction.**

## KEY TERM INTEGRATOR

### Characteristics for Classifying Animals

| | |
|---|---|
| **invertebrates** | Animals without a backbone. |
| **vertebrates** | Animals with a vertebral column or backbone. |
| body symmetry | Either **asymmetrical**, or **radial** or **bilateral symmetry**. **Cephalization** results in a definite head end in bilaterally symmetrical animals. |
| body design | Either a **sac body** plan or a **tube-within-a-tube body** with a mouth and anus. |
| body cavity | **Acoelomates**, which lack an internal body cavity; and **pseudocoelomates**, which have a partially lined body cavity. The body cavity of **coelomates** is fully lined with a **peritoneum** that forms **mesentery** extensions for organs. |
| embryo development | In bilateral coelomates, either **protostomes** or **deuterostomes**. |

### Animal Classification

| | |
|---|---|
| **Parazoa** | Animals without tissues. |
| | Phylum **Porifera**. Sponges have a **spongocoel** cavity with a single **osculum** opening at the top. **Porocytes** form many pores in the outer layer. The **mesenchyme** contains **amoebocytes** for circulating nutrients and producing gametes, and, in some, sharp **spicules** or **spongin** forms the sponge skeleton. Flagellated **choanocytes** lining the spongecoel filter food from water. |
| **Eumetazoa** | Animals that form tissues. |
| | Phylum **Cnidaria** (classes **Hydrozoa**, **Scyphozoa**, and **Anthozoa**). Hydras, jellyfish, sea anemones, and corals have stinging **cnidocytes** with a **nematocyst** capsule. They either have a **polyp** or **medusa** body form. Some produce a ciliated **planula** larva. The **mesoglea** layer separates the ectodermis and endodermis which surrounds a **gastrovascular cavity**. |
| | Phylum **Platyhelminthes** (classes **Turbellaria**—flatworms, **Trematoda**—flukes, and **Cestoda**—tapeworms). |
| | Phylum **Nematoda**—roundworms. |

Phylum **Mollusca** (classes **Gastropoda**, **Bivalvia**, and **Cephalopoda**). Most mollusks have a **head-foot**, **visceral mass**, and **mantle**. Organs empty into a **mantle cavity**. A lustrous **nacre** layer (mother-of-pearl) is sometimes produced.

Phylum **Annelida**. Nonparasitic segmented worms have **setae** or **parapodia** for locomotion.

Phylum **Arthropoda** (subphyla **Trilobita**, **Chelicerata**, **Crustacea**, **Uniramia**). Arthropods, the most diverse group of organisms, have an exoskeleton which is shed during **molting**.

Phylum **Echinodermata**. The **water vascular system** of echinoderms has a **madreporite** sieve opening. Water movement in each tube foot is controlled by a muscular **ampulla**.

Phylum **Chordata** (subphyla **Urochordata**, **Cephalochordata**, and **Vertebrata**). Chordates have a stiff **notochord**, a dorsal **hollow nerve cord**, gill slits, and a **tail** during some stage of their development. Vertebrate embryos have **pharyngeal pouches**.

# Review and Synopsis

1. Complete a classification scheme for the Animal kingdom by inserting the following in the appropriate place

| | | |
|---|---|---|
| Eumetazoa | radial symmetry | Platyhelminthes |
| coelomates | Nematoda | bilateral symmetry |
| Porifera | deuterostomes | Annelida |
| Arthropoda | Echinodermata | |

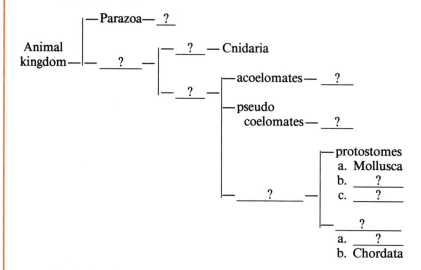

2. Match the following structures with the appropriate phylum:

____ vertebral column    A. Cnidaria
____ cnidocytes    B. Chordata
____ visceral mass    C. Porifera
____ choanocytes    D. Mollusca
____ exoskeleton    E. Annelida
____ setae or parapodia    F. Arthropoda

3. What characteristics distinguish mammals from other chordates?

4. Of the 35 phyla of animals, only 2, the Arthropoda and Chordata, are fully adapted to living on land. What characteristics do these phyla share that enable them to be truly terrestrial?

5. Almost 97 percent of animal species are invertebrates. What are the advantages and disadvantages of not having a backbone? List the habitats in which you would expect to find invertebrates. What makes these habitats suitable for invertebrates?

6. List the animal phyla that . . .
   have a sac body;
   have external fertilization;
   show cephalization.

7. Rank the animal phyla discussed in this chapter from the most diverse to the least.

8. Compare the advantages of radial symmetry and bilateral symmetry. For an active animal, which type of symmetry is more advantageous? For a sessile animal? Why?

## Additional Readings

Alcock, J. 1975. *Animal Behavior, An Evolutionary Approach.* Sinauer Associates, Inc., Sunderland, Mass. (Classic text.)

Barnes, R. 1980. *Invertebrate Zoology.* 4th ed. Saunders College Publishing, Philadelphia. (Intermediate.)

Eisner, T., and E. O. Wilson, eds. 1977. "The insects." Readings from *Scientific American.* W. H. Freeman, San Francisco. (Introductory.)

Hadley, N. 1986. "The arthropod cuticle." *Scientific American* 255:104–12. (Intermediate.)

Little, C. 1983. *The Colonisation of Land: Origins and Adaptations of Terrestrial Animals.* Cambridge University Press, New York. (Intermediate.)

Schneirla, R. 1971. *Army Ants.* W. H. Freeman, San Francisco. (Introductory.)

Tinbergen, N. 1965. *Social Behavior of Animals.* 2nd ed. John Wiley and Sons, New York. (Intermediate.)

Webb, P. 1984. "Form and function in fish swimming." *Scientific American* 251:72–82. (Intermediate.)

# 10 *Ecology*

Usually not the most obvious or complex organisms in a community, bacteria and fungi, like this beautiful coral fungus on the left, are critical to perpetuating life on earth. By decaying dead organisms, even simple decomposers form a critical link in the recycling of all chemical nutrients in all ecosystems.

Chapter 35

# The Biosphere

The Apollo astronauts watched the earth recede in the distance as they sped toward the moon. The artificial life support systems that sustained the astronauts' lives on the spacecraft temporarily replaced their natural life support system, that luminous cloud-streaked sphere suspended in the blackness of space, the earth (Figure 35-1).

Without artificial support systems, life as we know it cannot exist beyond the earth's atmosphere or deep beneath its solid surface. All life is restricted to a relatively narrow zone of air, water, and land called the **biosphere**, the thin envelope where living organisms are found (see Chapter 1). The biosphere is no thicker than 22.4 kilometers (14 miles), from the upper limits of life in the atmosphere to the depths of a few dark ocean trenches. Compared to the size of the earth, the thickness of the biosphere is equivalent to that of the skin on an apple. All life as we know it exists within this thin layer that envelops the earth.

## Boundaries of the Biosphere

The biosphere includes portions of the earth's hydrosphere, lithosphere, and atmosphere. Of the three, the *hydrosphere* of oceans, seas, lakes, ponds, rivers, and streams harbors the greatest quantity of life. Most aquatic organisms are found in shallow waters along shorelines where sunlight penetrates, providing the energy for photosynthesis. Similarly, most terrestrial organisms live near the illuminated surfaces of the *lithosphere,* the rocky crust that forms the earth's rigid plates and terrestrial habitats. In the gaseous *atmosphere,* organisms are found at altitudes below 7 kilometers. (4.3 miles).

There are a few exceptions to these general boundaries, however. Pollen, dormant bacteria, and fungal spores, for example, have been collected in the atmosphere at altitudes above 62 kilometers (100 miles). In the hydrosphere, gravity pulls dead organisms and their wastes downward, providing food for life in the darkened zones of the earth's waters. Thus, the biosphere is not uniform in width, but a region whose thickness varies from one location to another.

## Arenas of Life

In Chapter 1 you learned that the biosphere is the highest level of biological organization (see Figure 1-4, page 8). It consists of many distinct ecosystems, dynamic units composed of organisms and their physical environment.

### Aquatic Ecosystems

Almost three-quarters of the earth's surface is covered with water, of which nearly 71 percent is salty, forming the earth's marine habitats of seas and oceans. Other aquatic habitats include freshwater lakes, ponds, rivers, streams and estuaries of mixed fresh and salt water.

MARINE HABITATS
Open Oceans
Marine ecosystems are found in shallow waters around continents, islands and reefs, in intertidal zones, and in the **pelagic zone**, or open oceans (Figure 35-2).Biologists subdivide the vast pelagic zone into three layers: the sunlit **photic** or **epipelagic zone**, the dimly lighted **mesopelagic zone**, and the continually dark **aphotic** or **bathypelagic zone**.

The photic zone does not usually

**Figure 35-1**
**The earth from space.** Portions of Africa, Madagascar, and Antarctica can be seen beneath the swirling cloud layer. Seemingly calm at this distance, the earth teems with millions of species of organisms.

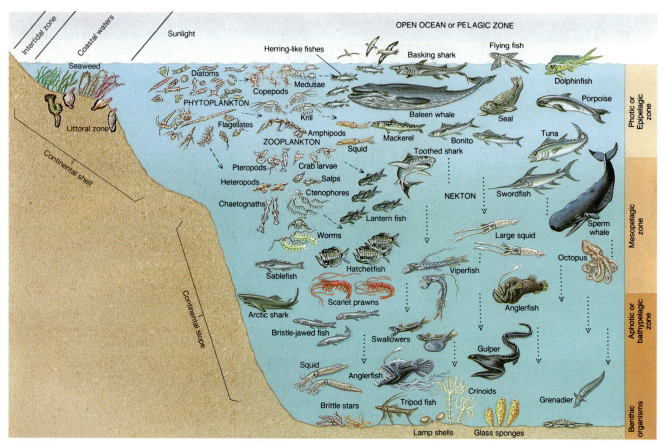

**Figure 35-2**
**Marine habitats and representative marine organisms.** (Sizes of organisms are not drawn to same scale.)

exceed depths of 100 meters (320 feet). Considering the average depth of open ocean is 3.9 kilometers (12,500 feet), the photic zone is very thin indeed. It contains enormous populations of **phytoplankton**, microscopic photosynthetic bacteria and algae that drift with the ocean currents. Phytoplankton provide the energy and nutrients for the open ocean's animal species. Most phytoplankton are eaten by **zooplankton**, tiny crustaceans (mostly copepods and shrimplike krill), larvae of invertebrates, and fish that are small enough to be swept along by ocean currents. Larger fishes and other animals feed on the tiny zooplankton, or on both phytoplankton and zooplankton (or simply, plankton). An adult blue whale, for example, guzzles some three tons of plankton in a single day.

Because the dim midwaters of the mesopelagic zone do not receive enough light to power photosynthesis, there are no phytoplankton. Inhabitants therefore make daily migrations up to the epipelagic zone or down to the bathypelagic zone to feed. Large fishes, whales, and squid are the principal animal predators of the mesopelagic zone. They eat smaller fishes that feed on the wastes or carcasses of organisms from the photic zone.

With the exception of a few unusual deep-vent communities (see Discovery: Living on the fringe of the Biosphere, Chapter 9), the aphotic zone has no energy-capturing plants or bacteria. The pitch-black bathypelagic zone is populated primarily by heterotrophic bacteria, together with benthic (bottom-dwelling) scavengers that feed on a constant rain of organic debris, wastes, and corpses that settle to the bottom, and predators that eat the scavengers and each other. In addition to sponges, sea anemones, sea cucumbers, worms, sea stars, and crustaceans, benthic animals include a collecion of odd-looking fish. Some of these have lanterns that light up to attract a meal or a mate (Figure 35-3).

## Coastal Waters

The greatest concentration of marine life is found in shallow **coastal waters** along the edges of continents (Figure 35-4). In addition to abundant light, coastal waters are rich in nutrients that continually drain from the surrounding land. Waves, winds, and tides constantly stir coastal waters, distributing the nutrients. Continually bathed in light and nutrients, photosynthetic organisms grow at fantastic rates and in

**Figure 35-3**
**Generating its own light,** the suspended lantern on this female angler fish lures prey in the black of the ocean's aphotic zone. Backward-pointing teeth prevent even large prey from escaping.

**Figure 35-4**
**The ocean's bounty.** Abundant light and constant circulation of nutrients make coastal waters the ocean's richest habitats.

great profusion, providing ample food and habitats for a multitude of fishes, arthropods, mollusks, worms, and mammals.

But coastal waters also present some unique problems for organisms. Rough, surging waters can shred or bash organisms against rocks. This is one reason why many coastal animals remain in burrows or have tough protective shells. Since surging waters can also quickly wash organisms onto the shore or pull them out to sea, many coastal organisms have adaptations for clinging to stationary objects. Algae attach to rocks by holdfasts, mussels anchor themselves with powerful cords, and abalone hold tight with their large, muscular foot.

The most fertile coastal waters lie along the edges of Peru, Portugal, Africa, and California. In these areas, persistent winds blow surface waters toward the sea. As surface water is blown out to sea, an upwelling of nutrient-laden water from below replaces the displaced surface water, creating an enriched environment for growth.

Shallow coastal waters are also found along coral reefs. Depending on their location, coral reefs are classified as one of three types (Figure 35-5):

- *Fringe reefs,* which extend out from the shores of islands and continents.

- *Barrier reefs,* which are separated from shores by channels or lagoons.

- *Atolls,* which are islands of coral with a shallow lagoon in the center.

Coral reefs develop only in tropical areas (between 30° north and 30° south latitudes) where water temperatures never fall below 16° C (61° F). Persistent cold currents prevent coral reefs from developing along the western edges of continents.

All coral reefs are built in the same way. Only the outer layer has living coral animals. Beneath this thin layer of tiny polyps are layer upon layer of skeletons of past generations. Each polyp has algae living inside it. Together, the animal and algae form a relationship beneficial to both: the photosynthesizing alga provides food for the coral, and the coral provides protection and nutrients for itself as well as its resident alga by capturing

a

b

c

**Figure 35-5**
**Coral reefs** *(a)* A barrier reef surrounding the island of Taiatea in French Polynesia. *(b)* A fringe reef around Society Island, Tahiti. *(c)* A circular atoll is a ring of coral built on a submerging volcanic crater, creating a central lagoon.

zooplankton.

Free-living algae grow in crevices of the coral and provide food for sea urchins, sea cucumbers, brittle stars, many kinds of mollusks, and the most colorful, fishes on earth. Moray eels, crabs, and other animal predators also lurk in the coral's recesses.

### The Intertidal Zone

At the interface between the ocean and the land is the **intertidal zone**. Recurring tides and the geology of the shoreline create a variety of intertidal habitats — tide pools, mud flats, sandy beaches, and rocky shores (Figure 35-6).

Organisms inhabiting the intertidal zones survive in both water and air as the flow of tides rhythmically submerge and expose their habitats. Four strata of organisms form in the interti-dal zone (Figure 35-7). The uppermost stratum is the (1) **splash zone** where organisms receive only sprays of water at high tides. Below the splash zone is the (2) **high intertidal zone**, followed by the (3) **middle intertidal** and (4) **low intertidal zones.** Organisms in the high, middle, and low intertidal zones are subjected to progressively shorter duration in air.

### ESTUARIES

**Estuaries** form where rivers and streams empty into oceans, mixing freshwater with salty water (Figure 35-8). An organism's location in an estuary depends on its ability to tolerate different concentrations of salt. But salinity changes as tides bring in new influxes of salty seawater or as storms increase the supply of freshwater. At any one spot in an estuary, salinity may change within minutes, sometimes from zero (freshwater) to the salinity of seawater, creating an ever-changing environment for the organisms that live there.

Despite such fluctuations in salinity, estuaries are tremendously fertile habitats for organisms, continually enriched with nutrients from rivers, and from debris washed in by tides. Plankton flourishes in the estuary's warm, murky waters. Phytoplankton and rooted plants along the edges provide food for crustaceans, fishes, shellfishes, and the young of many open-ocean animals that use rich estuaries for spawning.

### FRESHWATER HABITATS

Oceans contain 97.2 percent of the earth's total water. Of the remaining 2.8 percent that is freshwater, 2 per-

b

c

d

**Figure 35-6**
**Part-time oceans: intertidal habitats.** *(a) Tide pools* are reservoirs of seawater trapped in depressions during low tide. The pools are filled with flowerlike sea anemones, sea stars, crabs, small fishes, and a variety of seaweeds and other algae. *(b) A mud flat* in Magdalena Bay, Baja California, with red mangroves, cordgrass, and bamboo worms (Maldanidae). *(c) A sandy beach. (d) Rocky shores.* Large numbers of animals live in the moist tangle of brown seaweeds that drape the shore's rocks.

cent is permanently frozen in ice caps and glaciers, 0.6 percent is groundwater, 0.017 percent is in lakes and rivers, and 0.001 percent is in the atmosphere as ice and water vapor. Because the quantity of water on earth is so enormous, even a small percentage like 0.017 percent produces more than 52,000 cubic miles of freshwater habitats on earth. The principal freshwater habitats are flowing rivers and streams, and standing lakes and ponds.

## Rivers and Streams

Depending on its size, freshwater flowing over the land forms either a **stream** or a **river**; smaller streams usually converge into larger rivers. The constant flow of water erodes the land, sculpting dramatic landscapes and creating unique habitats that support particular groups of organisms.

The velocity of a current affects not only erosion, but also the deposition of sediments, and the supply of oxygen, carbon dioxide, and nutrients—factors that determine which organisms inhabit a stream or river. Even in one spot, the current varies. Friction along the edges and bottom slows water velocity, allowing algae to cling to rocky surfaces, plants to take root, and animals to reside without being swept away. Calmer water is also found behind rocks, pebbles, and mounds, creating stable habitats for smaller organisms.

In a river's open waters, the never-

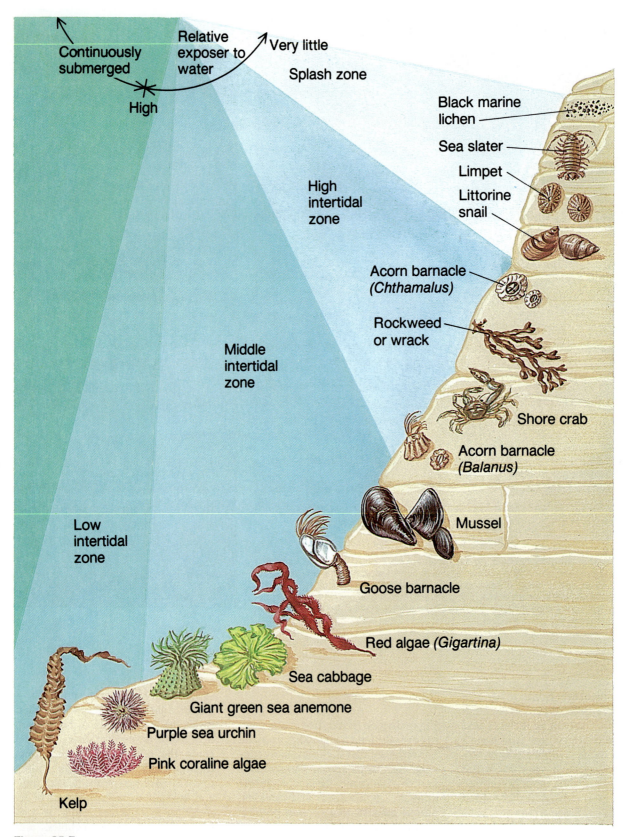

**Figure 35-7**
**Intertidal zones.** Organisms that live in the *splash zone* must survive with only periodic sprays of moisture, whereas organisms are submerged about 10 percent of the time in the *high intertidal zone,* 50 percent of the time in the *middle intertidal zone,* and 90 percent of the time in the *low intertidal zone.*

**Figure 35-8**
**Estuaries** like this are shallow inlets where freshwater mixes with salty seawater. Nearly one-half of the ocean's living matter is found in such nutrient-rich estuaries.

shallow **littoral** zone along the water's edge; a **limnetic** zone encompassing the lake's open, lighted waters; and a dark **profundal** zone at depths below which light is able to penetrate. Rooted plants and floating algae are the characteristic photosynthesizers of the littoral zone. In the limnetic zone, phytoplankton and photosynthetic plants provide food for zooplankton, fishes, and other animals. As in the aphotic zone of the oceans, heterotrophic predators and scavengers inhabit the profundal zone. Because oxygen can become scarce near the bottom of a lake, anaerobic bacteria are usually very abundant there as well.

In many lakes, bursts of biological activity occur in the spring and autumn when the lake's waters naturally circulate (Figure 35-9). In spring, the sun warms surface waters and winds stir the lake, driving heat and oxygen down to deep waters, and circulating nutrients up to the surface. This *spring overturn* stimulates a sudden increase in biological activity. As spring passes to summer, heated surface waters become more buoyant and resist mixing. Water circulation stops altogether, producing three temperature layers: a layer of warm surface water; a relatively thin middle layer in which water temperature declines rapidly (this

ending tug of the current presents unique challenges for animals. Some fishes use their fins as hydrofoils to force themselves down to less turbulent bottom waters. On the other hand, powerful trout and salmon swim fast enough to oppose the current, or even to swim upstream.

## Lakes and Ponds

**Lakes** and **ponds** are standing bodies of water that form in depressions of the earth's crust. As with streams and rivers, the difference between a lake or pond depends on its size: lakes are larger and usually deeper than ponds.

Generally, a lake has three zones: a

Spring and autumn overturns

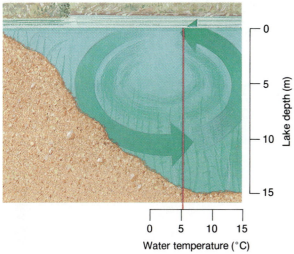

Summer thermal strata

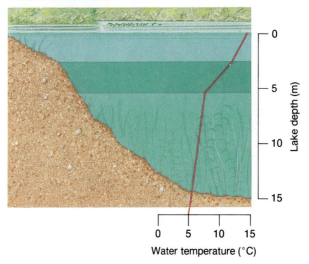

**Figure 35-9**
**Seasonal changes in a temperate lake.** Water temperature is nearly uniform during spring and autumn overturns. In summer, warmer surface waters resist mixing, forming distinct temperature zones.

layer is called a *thermocline*); and a bottom layer of cold, relatively still water. Without mixing, oxygen is quickly depleted on the lake's bottom, and nutrients begin to accumulate. With little oxygen at the bottom and no nutrient circulation, biological activity in the lake slows down throughout summer.

In autumn, the lake's thermal layers are disrupted as heat is dissipated from surface waters to cold air. The colder surface waters then sink. Combined with increasing autumn winds, circulation in the lake is restored, producing a more uniform range of temperature, nutrients, and oxygen. Rising warmer water from the bottom brings nutrients upward and warms the surface waters, producing another burst of biological activity during this *autumn overturn*.

Unlike rivers and streams, lakes and ponds eventually fill in with sediment and organic matter to become dry land. This natural aging process is called **eutrophication** (Figure 35-10). Eutrophication may take months, or it may take thousands of years, depending on the size of the lake or pond, the amount of biological activity, and the rate of filling and draining. Thus, biologists speak of a lake's "life span" or a lake's "natural aging process."

A recently formed young lake is termed *oligotrophic* (meaning little nourished) because it contains few nutrients. Oligotrophic lakes support very little life and, as a result, are usually crystal clear. Middle-aged lakes are *mesotrophic* (meaning moderately nourished). Mesotrophic lakes support large populations of organisms. The nutrient-rich waters of old, *eutrophic* (meaning fully nourished) lakes support the largest populations and the greatest diversity of species. As the organisms of eutrophic lakes add ever-increasing supplies of organic matter, the filling rate is accelerated. Outside sources of nutrients, such as human sewage or runoff from nutrient-soaked agricultural lands, also promote eutrophication.

## Biomes: Patterns of Life on the Land

Large terrestrial ecosystems are called **biomes**. Plants form the bulk of the living mass in a biome. Each biome is

b

a

**Figure 35-10**
**Three phases of eutrophication.** Newly formed *oligotrophic* lakes *(a)* are perfectly clear, virtually devoid of life. Oligotrophic lakes eventually begin to fill with sediment, providing nutrients for organisms. As sediment and life increases, oligotrophic lakes become *mesotrophic* lakes *(b)* which then become *eutrophic* lakes *(c)*. Eventually, the lake becomes completely filled with sediment and organic material.

c

therefore characterized by the predominant type of plant. Dense, tall trees form forest biomes, short woody plants form shrublands, and grasses and herbs form grasslands. Since the prevailing climate (especially temperature and moisture) is the primary factor in determining the types of plants that grow in an area, the earth's terrestrial biomes tend to follow global climate patterns (Figure 35-11). Climate also changes with elevation, producing successive layers of biomes on a single mountainside (Figure 35-12).

## FORESTS

Over 30 percent of the earth's land surfaces are covered by forests, dense patches of trees. Biologists recognize three forest biomes:

- **Tropical rainforests**, lush forests that grow in a broad belt around the equator.
- **Deciduous forests** with trees that drop their leaves during unfavorable seasons.
- **Coniferous forests** dominated by evergreen conifers.

## Tropical Rainforests

Warm temperature, abundant rainfall (over 250 mm — 100 inches — a year), and roughly constant daylength throughout the year produce rich tropical rainforests in Central and South America, Africa, India, Asia, and Australia (Figure 35-11). Over half of the earth's forests are tropical rainforests. They contain more species of plants

and animals than all other biomes combined. One hectare (10,000 square meters, or 2.5 acres) can have more than 100 species of trees, 300 species of orchids, and thousands of species of animals, most of them insects.

Trees towering over 50 meters (160 feet) tall form the upper story of the tropical rainforest (Figure 35-13). Below these giants is a layer of tall trees so densely packed that very little sunlight penetrates through their canopy. Since much of the light is blocked, plant growth is greatly suppressed below these two layers. Where a glimmer of sunlight does penetrate, ferns, shrubs, and mosses crowd the forest floor.

Of all biomes, the rate of decompo-

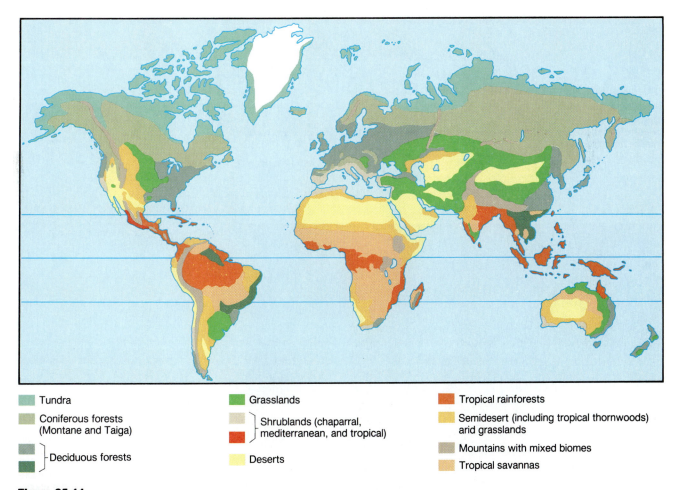

Tundra

Coniferous forests (Montane and Taiga)

Deciduous forests

Grasslands

Shrublands (chaparral, mediterranean, and tropical)

Deserts

Tropical rainforests

Semidesert (including tropical thornwoods) arid grasslands

Mountains with mixed biomes

Tropical savannas

**Figure 35-11**
**Climate and the earth's biomes.** Climate is the principal factor in determining where plants grow, which is the principal factor in determining where animals live. Not unexpectedly then, the locations of terrestrial biomes closely follow the earth's 11 main climate types.

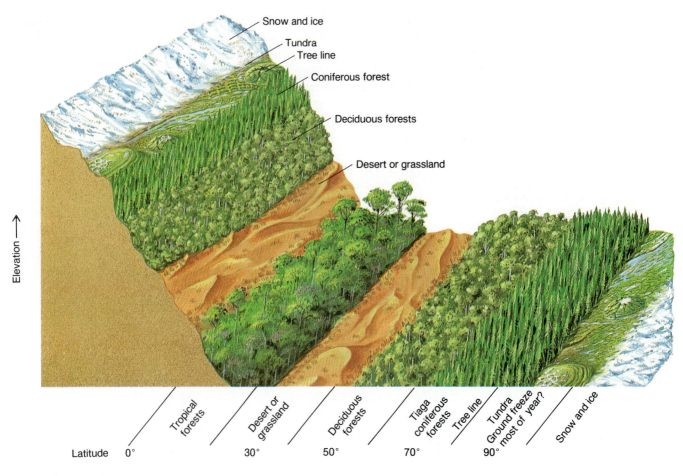

Snow and ice
Tundra
Tree line
Coniferous forest
Deciduous forests
Desert or grassland

Elevation →

Tropical forests

Desert or grassland

Deciduous forests

Tiaga coniferous forests

Tree line

Tundra Ground freeze most of year?

Snow and ice

Latitude    0°            30°         50°         70°      90°

**Figure 35-12**
**Elevation and the distribution of biomes.** Terrestrial biomes change according to elevation and distance from the equator.

**Figure 35-13**
**A picture of unrivaled diversity,** life in the tropical rainforest is more varied and more abundant than any other biome on earth. Immense, broad-leafed evergreen trees drip moisture on masses of clinging epiphytes and enormously long vines, some over 200 meters long. These giant trees have wide-spreading buttresses that supply the necessary support to keep them upright even in shallow tropical soil.

sition is fastest in the tropical rainforest. A dead animal or fallen tree is swiftly cleared from the the forest floor as hordes of fungi and bacteria promptly carry out decomposition. But despite such rapid decomposition, virtually no nutrients accumulate in the soil; they are either absorbed immediately by plants or washed away by steady rains. Surprisingly, the earth's most verdant forests have poor soils, an irony that has contributed to the ever-increasing destruction of the earth's tropical rainforests by humans. Here's why.

Because tropical rainforest soils are nutrient-poor, clearing and burning rainforests for agriculture (a technique called *slash-and-burn*) has only short-term benefits. Nutrients released in the ashes are either quickly washed away by regular downpours, or are used up by the first one or two generations of crops, leaving barren habitats that, at best, are very slow to recover.

Meanwhile the farmers slash and burn new areas to replace the "played out" land.

The practice of clearing, burning, and then abandoning one area after another has destroyed large tracts of tropical forests—over 50 acres each minute. The practice also adds huge doses of $CO_2$ to the atmosphere which traps heat, increasing the greenhouse effect of the atmosphere. If destruction continues at its present rate, nearly all tropical rainforests will disappear by the year 2000. Since tropical rainforests harbor the greatest diversity of life in the biosphere, the earth's largest treasury of organisms could soon be lost forever. But that's not all. The destruction of rainforests will permanently change the earth's climate by increasing the greenhouse effect and by eliminating large tracts of plants that will no longer absorb incoming solar energy or give off huge quantities of water vapor through transpiration.

The effects of destroying tropical rainforests will be felt by all organisms in the biosphere including humans. One out of four medicines now come from tropical plants, and an estimated 1400 tropical plants (plus countless undiscovered species) show promise for providing cures for cancer.

### Deciduous Forests

Away from the equator, distinct seasons develop during each year. Unfavorable growing conditions during one or more seasons produce deciduous forests with trees that leaf out during warm, wet seasons, and lose their leaves at the onset of the dry season (in tropical deciduous forests) or the cold season (in temperate deciduous forests) (Figure 35-14).

Deciduous forests are less dense than tropical rainforests, enabling sunlight to reach the forest floor. With adequate light, many layers of plants develop; the top layer contains decidu-

a

b

**Figure 35-14**
**The yearly rhythm of a deciduous forest.** Deciduous trees in temperate areas leaf out in the spring when temperatures warm. During summer *(a)*, the overstory trees have fully expanded leaves that shade the plants below. Shortening photoperiod and the cool temperatures of autumn trigger the breakdown of chlorophyll, revealing pigments that turn leaves brilliant shades of orange, red, and purple. With protected terminal buds, dormant deciduous trees withstand the bite of freezing winds and snows during the winter *(b)*.

ous trees. Plants growing below the upper layer receive varying amounts of sunlight throughout the year as deciduous trees gradually leaf out, bear fully expanded leaves, and then drop their leaves at the beginning of the unfavorable season. The growth and reproductive cycles of understory plants coincide with brief periods of maximum sunlight and favorable conditions.

### Coniferous Forests

Some of the most extensive forests in the world are populated by evergreen conifers such as pines, firs, and spruces (Figure 35-15). Belts of coniferous forests girdle the huge continental land masses in the Northern Hemisphere (Figure 35-11) and blanket higher elevations of mountain ranges in North, South, and Central America, and Europe.

The transition from deciduous forests to coniferous forests is a result of colder winters. During the spring and summer, melting snows fill lakes and form watery bogs and marshes within many coniferous forests. (This is why the coniferous forests of northern latitudes are called **taiga**, which is Russian for "swamp forest.") Because decomposition by fungi and bacteria is limited to brief warm periods, the forest floor accumulates a thick litter layer of needles, producing acidic and relatively infertile soils.

Coniferous forests typically have two stories of plants: a dense overstory of trees and an understory of shrubs, ferns, and mosses. Unlike the tropical rainforests, coniferous forests are usually not mixtures of many tree species, but rather expanses of many individuals of a few species.

### TUNDRA

Timberline, a zone where trees thin and eventually disappear, marks the boundary between the coniferous forest and the tundra biome (Figure 35-11 and 35-12). **Tundra**, another term adopted from Russian, means a marshy, unforested area. There are two types of tundra: *alpine tundra* at high elevations of mountain ranges, and *arctic tundra* at high latitudes (Figure 35-16). Both look similar, featur-

**Figure 35-15**
**A coniferous forest in the icy grip of winter.** Many animals hibernate or migrate to less sever habitats. Only a few, like this moose, remain active during harsh winter months, forced to travel over large areas in search of scarce food.

**Figure 35-16**
**Life is fleeting in the tundra.** During the brief summer growing period, ground-hugging tundra plants erupt with new growth, produce flowers, and set seeds before severe weather returns. In the arctic tundra shown here, permafrost prevents drainage of water from melting snow, producing multitudes of shallow ponds.

ing low-growing plants that often belong to the same species.

The climate that produces tundra is brutal. The growing season lasts only 50 to 90 days, with an average temperature in the warmest month less than 10° C (50° F).Thick snow covers the ground, and icy winds blow most of the time during winter, while melting snows create frigid marshes and ponds during the brief summer period. In the arctic tundra, only the top 0.5 meters (1.5 feet) thaws, leaving permanently frozen soil, or *permafrost,* that halts root growth. With such harsh conditions, it is not surprising that the diversity of plant and animal life in the tundra is low. The entire arctic tundra region of North America, which covers thousands of square miles, has a total of about 600 plant species, fewer than you would find in a single square mile of tropical rainforest.

## GRASSLANDS

Not too long ago, over 30 percent of the earth's lands were covered by **grasslands** of densely packed grasses and herbaceous plants (Figure 35-11 and 35-17). In North America, grasslands were once more widespread than any other biome. But the combination of rich soil and favorable growing (and living) conditions made grassland habitats prime targets for agriculture, livestock grazing, and building cities. Today, most of the earth's grasslands have been cleared for farming or have become damaged because of overgrazing.

Grasslands naturally develop in regions with cold winters, hot summers, and seasonal rainfall—more annual rainfall than in the desert, but not enough to support a forest. Natural fires periodically clear grasslands,

opening up space and releasing nutrients for new growth.

## SAVANNAS

A combination of grassland with scattered or clumped trees forms a **savanna** biome. *Tropical savannas* in South America, Africa, Southeast Asia, and Australia cover nearly 8 percent of the earth's land (Figure 35-18). In many temperate areas, pockets of savannas are found sandwiched between grasslands and forests.

Like pure grasslands, savannas have seasonal rainfall punctuated with a dry season. During the dry season, the above-ground stems of the grasses and herbs die, providing fuel for fast-moving surface fires. The grasses and herbs recover quickly from fires by resprouting from underground roots and stems, or, if killed, are replaced by fast-growing seedlings. Because ground

**Figure 35-17**
**Grassland** in New South Wales, Australia.

**Figure 35-18**
**The African savanna** has flat-topped acacias and dry grassland. Tropical savannas in central Africa have grasslands studded with palm trees.

**Figure 35-19**
**Chaparral** is a shrubland formed in regions of California with a mediterranean type of climate. In addition to being drought-tolerant, most chaparral plants are fire-adapted and are able to regrow rapidly after a fire. Layers of charcoal testify that fire is a natural component of the chaparral, as well as of forests, grasslands, savannas, and other shrublands.

fires move rapidly through the savanna, the trees are not usually damaged.

The tropical savannas of Africa support large populations of herbivores, including wildebeest, gazelles, impalas, zebras, and giraffes. Like many grasslands, tropical savannas are now being used for grazing. Overgrazing reduces grass cover and allows trees to invade the area. Declining grass cover and increasing numbers of trees reduce the number of animals that can inhabit overgrazed savannas.

## SHRUBLANDS

**Shrublands**, with scrub vegetation, are areas where woody shrubs rather than trees predominate. In regions of the earth with a mediterranean type of climate (hot, dry summers and cool, wet winters), shrubs grow so close together that there are usually no understory plants. The shrubs typically have small, leathery leaves with few stomates and thick cuticles to retard water vapor loss (Chapter 14). Remarkably similar shrublands develop in all areas with a mediterranean climate—around the Mediterranean Sea, on coastal mountains in California and Chile, at the tip of South Africa, and in southwestern Australia. In California, this type of shrubland is called **chaparral** (Figure 35-19).

Organisms living in mediterranean-type shrublands must survive many stressful periods. For one, water from rainfall is available only during the cool winters, whereas over the long, hot summers when water is most needed, there is little or no rainfall. Because the supply and need for water are directly out of phase, most plant and animal activity is restricted to the spring when temperatures are warm and the soils still moist from winter rains.

Fires are a common feature of mediterranean-type shrublands. The accumulation of dry, woody stems and highly flammable litter greatly increases the chances of fire during the hot summer. But most plants have evolved adaptations that allow them to cope with recurring fires. In the chaparral of California, for example, a burned area often recovers within just a few years. Even one month after an intense fire, many chaparral plants resprout from protected underground stems, or their seeds germinate, stimulated by the fire itself.

In tropical regions with a short wet season, another type of shrubland develops called **tropical thornwood**. Most thornwood plants lose their small leaves during the dry season, reducing transpiration and exposing sharp thorns that discourage even the hungriest browser. One type of thornwood, the *Acacia* plant, and certain ants have coevolved a close partnership that help both survive in these habitats. The plant provides food and shelter for the ants, and in turn, the ants patrol the ground and stems of the plant, aggressively warding off hungry herbivores and removing flammable debris from under the plant (Figure 35-20).

## DESERTS

Intense solar radiation, lashing winds, and little moisture (annual rainfall of less than 25 centimeters—10 inches —a year) create some of the harshest living conditions in the biosphere. These conditions characterize the **deserts.** Deserts cover about 14 percent of the earth's land and occur mainly near 30° north and south latitude where global air currents create belts of descending dry air. Some deserts are produced in the rainshadows of high mountain ranges, leeward slopes that face away from incoming storms and thereby receive little rainfall. With generally cloudless skies, the sun quickly heats the desert by day, producing the highest air temperatures in the biosphere. (The record is 57.8° C (138° F) in Death Valley, California.) High daytime temperatures and persistent winds accelerate water evaporation and transpiration of water vapor from plants. High evapotranspiration and low yearly rainfall characterize all deserts, producing sparse perennial vegetation of widely spaced shrubs (Figure 35-21*a*). But despite such harsh living conditions, only portions of the Sahara Desert in Africa and the Gobi Desert of Asia are totally devoid

**Figure 35-20**
**Mutual aid.** Swollen thorns of this acacia tree provide a home and brood place for ants while nectar and nutrient-rich plant structures produced at the tips of leaves provide the ants with a balanced diet. Ants protect the acacia by attacking, stinging and biting herbivores (or scientists). Ants also remove flammable debris from around the base of the tree, forming a natural fire break. Any plant seedling growing within this "bare zone" is quickly clipped and discarded, protecting the acacia from competing plants, including its own offspring.

of organisms, an affirmation of the tenacity of life and the power of evolution.

Plants have evolved many adaptations for surviving the rigors of the desert (see Discovery: Survival in the Desert, Chapter 14). For example, following brief spring and summer rains, the desert floor often becomes carpeted with masses of small, colorful annual plants (Figure 35-21*b*). These annuals germinate, grow, flower, and release seeds—all within the brief period when water is available and temperatures are warm. They avoid the stresses of the desert during the remainder of the year as dormant seeds. The evolution of succulent tissues

allows cacti and other succulents to avoid dehydration by storing water in specialized roots, stems, or leaves. During dry periods, succulents frugally tap their internal water reserves. Other adaptations of desert plants include some that grow exceptionally long tap roots to siphon deep groundwater supplies. Other plants simply endure extreme conditions by being able to survive various degrees of dehydration.

Many desert animals rely on learned and instinctive behavior to avoid the desert heat and dryness. They remain in cool, humid underground burrows during the day and search for food only at night when temperatures are

a

b

**Figure 35-21**
**The desert.** *(a)* During dry periods, a few hardy perennial bushes, cacti, and trees are scattered over the scorching desert floor. Many animals remain underground, in moist, cool burrows. *(b)* Following a spring rain, the desert floor blooms with colorful annual growth.

lower. If they do venture out during the day, their underground burrows act as heat sinks, quickly removing heat acquired while they scurry across the desert in search of food. Many birds avoid stressful desert conditions by simply flying to less hostile areas, while other animals *estivate,* that is, they sleep through the driest part of the year. Finally, the desert toad *(Bufo punctatus)* uses a survival strategy similar to that employed by succulent plants: it stores water in its urinary bladder, carryng its own version of an internal well.

# Synopsis

## MAIN CONCEPTS

- **The biosphere is the region of the earth's lands, waters, and air that supports all life as we know it.**

- **The earth's watery hydrosphere supports the greatest quantity of life.** Most aquatic life is concentrated in shallow coastal waters along the fringes of land and coral reefs.

- **Biomes cover vast expanses of the earth's land and are characterized by certain types of plants.** Since prevailing climate is the primary factor in determining plant distribution, terrestrial biomes follow global climatic patterns and elevational gradients.

- **Forests cover more than 30 percent of the earth's land.** The major forest biomes include (1) tropical rainforests, (2) deciduous forests, and (3) coniferous forests. Tropical rainforests contain more species than all other biomes combined. These rich forests are rapidly being destroyed by humans.

- **At one time, dense grasslands covered more than 30 percent of the earth's land.** Today most of the world's grasslands are used by people for agriculture, grazing, and other types of development.

- **Savannas of grass and scattered trees cover nearly 8 percent of the earth's land surface.** Overgrazing is reducing their ability to support animal life.

- **Dense shrublands develop in areas with a relatively long dry season.** Because fire is a frequent feature, many plants have evolved adaptations for quick recovery.

- **Nearly 14 percent of the land is covered with hot, dry deserts.** The harsh living conditions of the desert have produced many anatomical, physiological, and behavioral adaptations in desert organisms.

## KEY TERM INTEGRATOR

### Boundaries of Life

| | |
|---|---|
| **biosphere** | The region of the earth's water, land, and air in which living organisms are found. |

### Aquatic Habitats: Life in Water

| | |
|---|---|
| marine habitats | The **pelagic zone** (open ocean) is divided into the **epipelagic** or **photic**, **mesopelagic**, and **bathypelagic** or **aphotic** zones. Photosynthetic **phytoplankton** and small **zooplankton** form the foundation of the ocean's food chains.<br>Shallow **coastal waters** along continents, islands, and coral reefs contain most marine life.<br>At the intersection of the land and sea is the **intertidal zone** with four strata: the **splash zone**, and **high**, **middle**, and **low intertidal zones**. |
| **estuaries** | Form where rivers and streams empty into the sea, mixing salt and freshwater. |
| freshwater habitats | Include running **rivers** and **streams**, and standing **lakes** and **ponds**.<br>Lakes have three zones of life: the **littoral zone** in shallow water, the **limnetic zone** in open water, and the dark **profundal zone**. |
| **eutrophication** | The process by which lakes and ponds accumulate nutrients and organic debris, eventually filling in completely. |

---

Biomes: Life on Land

---

| | |
|---|---|
| **forests** | Include the **taiga** or **coniferous forests**, **deciduous forests**, and **tropical rainforests**. |
| **tundra** | Marshy, unforested areas. Low-growing arctic and alpine tundra occurs at high latitudes and high elevations. |
| **grasslands** | Expanses of dense grasses and other herbaceous plants that occur in regions with a long growing season and rich soil. Many grasslands are now used for agriculture, grazing, and developing cities. |
| **savannas** | Combinations of grasses and scattered trees. |
| **shrublands** | Dense **shrubs** formed in areas with a prolonged dry season. They include the **chaparral** in California and **tropical thorn-woods**. |
| **deserts** | Biomes characterized by low annual rainfall and high evapo-transpiration potential. |

---

## Review and Synthesis

1. Complete the chart on aquatic ecosystems by placing the following terms in their appropriate space:

| | | |
|---|---|---|
| pelagic zone | littoral | limnetic |
| standing | freshwater | aphotic |
| mesopelagic | ponds | profundal |
| moving | streams | epipelagic |

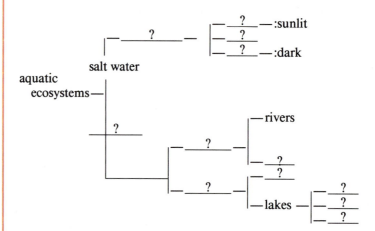

2. Rank the earth's terrestrial biomes starting with the one that covers the greatest land surface and ending with the one that covers the smallest area. Why do some biomes cover more land than others?

3. Species have evolved adaptations for particular types of habitats. For each of the following, list the major adaptations needed to survive such conditions.
   a. Prolonged hot and dry season.
   b. Daily rainfall and continually warm temperatures.
   c. Surging waters against a rocky coastline.
   d. Calm, warm freshwater with abundant nutrients and many species of living organisms.
   e. Scalding hot deep-sea vents.

4. Which habitat(s) in the biosphere is (are) best suited for human existence? Which biomes are the most affected by human activity and why?

5. What would happen if all tropical rainforests were indeed destroyed by the year 2000? How would this catastrophe affect life throughout the biosphere?

6. For each continuum below, identify the position of each terrestrial biome:

df = deciduous forest    tf = tropical rainforest
g = grassland    cf = coniferous forest
d = desert    t = tundra

$\Bigg($ Example: Species Diversity   tf     df   g   cf   t & d $\Bigg)$
                                    high                  low

Growing season
       long                        short

Decomposition rate
       fast                         slow

Yearly rainfall
       high                        low

Fires
       frequent                    rare

## Additional Readings

Boorer, M. 1975. *Forest Life. The Living Earth.* Danbury Press, Aldus Books Limited, London. (One of twenty volumes in The Living Earth series.) (Introductory.)

Newell, N. 1972. "The evolution of reefs." *Scientific American* 226:54–65. (Introductory.)

Parks, P. 1976. *The World You Never See. Underwater Life.* Rand McNally, Chicago. (Introductory.)

Sutton, A., and M. Sutton. 1966. *The Life of the Desert.* McGraw-Hill, New York. (Introductory.)

Sutton, A., and M. Sutton. 1979. *Wildlife of the Forests.* Harry N. Abrams, New York. (Introductory.)

Chapter 36

# Ecosystems and Communities

- A long drought destroys vast stretches of grasses and trees in the African savanna, causing thousands of animals to starve and leaving others frail and malnourished.

- Runoff from exceptionally heavy winter snows keeps streams cold throughout the summer, slowing hatching and larval development of blackflies, mayflies, and caddisflies, which in turn reduce the number and size of fishes that inhabit the streams.

- Acrid Los Angeles smog blows into mountain forests where it cuts photosynthesis by as much as 80 percent in pines. Unable to manufacture enough resin (sap) to protect themselves against burrowing insect larvae, many thousands of pines are dying from bark beetle infestations.

Each of these examples illustrates how organisms are affected by both other organisms and by the surrounding physical environment. Organisms and the physical, nonliving environment are inseparably linked, each continually affecting the other. Because of this continual interplay, the community of organisms and the physical environment function as an organized unit called an **ecosystem**.

## From Biomes to Microcosms

The variety of ecosystems in the biosphere is enormous. Ecosystems range in size from expansive biomes (Chapter 35) to countless small ecosystems within each biome (Figure 36-1). However, the boundaries that separate ecosystems are not clear delineations, for all ecosystems are linked with other ecosystems to some degree. For example, although a lake ecosystem and a cave ecosystem are separated by many miles, the rate at which the lake fills in with sediment affects the number of beetles living in the cave. Bats are the connecting link between the two "separate" ecosystems. Bats rest in caves during the day and forage for insects at night. As a lake fills with sediment and organic material, more plants can grow along its margins, providing more food for greater numbers of plant-eating insects, the dietary mainstay for the bats. When the well-fed bats return to the cave, they defecate large amounts of feces on the cave floor. The feces nourish the growth of a particular mold, on which the cave beetles dine. More sediment means more plants; more plants means more

**Figure 36-1**
**Ecosystems within ecosystems.** A South American tropical rainforest contains many ecosystems, each with a unique community of organisms and unique set of physical factors. The ecosystem hugging the Amazon River *(a)* supports colonies of aggressive army and leaf cutter ants that harvests leaves to cultivate a fungus garden ecosystem for food *(b)*. Perched high on a tree branch, the top of a bromeliad collects water, forming a tiny pond ecosystem *(c)*.

insects; more insects means more bat feces; more feces means more mold; and more mold means more beetles. In addition to beetles, crickets, flies, springtails, moths, spiders, centipedes, indeed almost the entire cave community, depend on the supply of bat droppings.

The lake is linked not only to a cave ecosystem, but also to other ecosystems, such as a nearby forest ecosystem and the ecosystem in the river that drains the lake. The fact that ecosystems are interconnected has led many biologists to view the entire biosphere as one giant global ecosystem of tremendous complexity and order.

## The Structure of Ecosystems

An ecosystem is a functional unit in which energy and nutrients flow between the physical or **abiotic environment**, and a community of organisms, the **biotic environment** (see "Levels of Biological Organization," Chapter 1). The abiotic and biotic environments continually affect each other, producing a complex of interdependent connections that often lead to a "balance in nature." Even a slight modification in one factor can disrupt this balance, causing the entire ecosystem to change (see Bioline: Reverberations Throughout an Ecosystem).

It is easy to visualize how the abiotic environment can change the biotic community. A long period of freezing temperatures, for example, usually kills many organisms, as would a devastating fire. Strong winds can uproot plants or blow flying insects and birds far away from their natural habitat. But the relationship is two-way—organisms change the abiotic environment as well as being changed by it. For instance, recall from Chapter 30 that the earth's original atmosphere was devoid of $O_2$. By releasing $O_2$ during photosynthesis, ancient photosynthetic organisms gradually added $O_2$ to the atmosphere. In other words, the earth's biotic component dramatically altered the abiotic environment, changing the atmosphere to its current 21 percent $O_2$ level. Similarly, plants contribute to the formation and fertility of soils. Coral reefs change the flow

and temperature of oceans. Dense forests modify humidity, temperature, and the amount of light and rain that reaches the forest floor.

## The Biotic and Abiotic Environments

Biologists often divide ecosystems into five subcomponents—two for abiotic factors and three for the biotic community (Figure 36-2). The two subcomponents of the abiotic environment are abiotic resources and abiotic conditions. *Abiotic resources* are the supplies needed by organisms, including (1) a source of energy from outside the ecosystem (usually sunlight); and (2) inorganic substances (nitrogen, $CO_2$, $H_2O$, phosphorus, potassium,

and so on) that are needed for the construction of organic compounds. *Abiotic conditions* are nonresource components and include (1) the substrate and/or medium (air, water, and soil) in which the organism lives, and (2) conditions such as temperature and water currents.

The three subcomponents of the biotic environment are (1) **primary producers**, all autotrophs (algae, bacteria, and plants), which use sunlight or chemical energy to manufacture food from inorganic substances; (2) **consumers**, heterotrophs that feed on other organisms or organic wastes; and (3) **decomposers** or **saprophytes**, mainly fungi and bacteria, that get their nutrition by breaking down organic compounds in wastes and dead organisms.

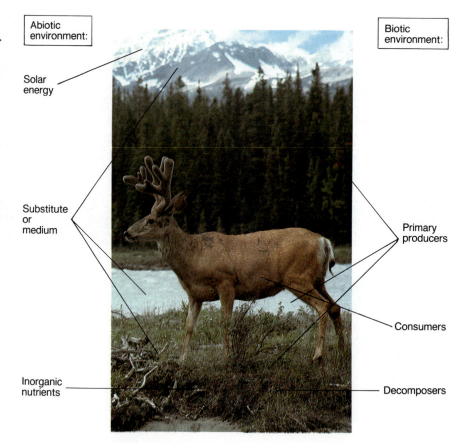

**Figure 36-2**
**The basic components of an ecosystem** are the abiotic environment and the biotic community. The abiotic environment includes energy, inorganic substances (essential elements), and the substrate and/or medium in which the community lives. Within the biotic community, primary producers convert energy and inorganic nutrients into organic food molecules, which are passed to consumers and decomposers. Respiration, excretion, and decomposition return essential elements back to the abiotic environment.

# B I O L I N E

## REVERBERATIONS THROUGHOUT AN ECOSYSTEM

Some great lessons in ecology are stumbled on by accident. For example, in the 1950s raising ducks on Long Island, New York, produced some unexpected and economically devastating consequences.

No one foresaw that allowing duck farms near the Great South Bay of Long Island would totally destroy the lucrative oyster industry in the bay, but it did. Before the ducks arrived, the famous blue-point oysters thrived in the bay, eating a normal mixture of diatoms, unicellular green algae, and dinoflagellates. Shortly after the duck farms were established along the tributaries to the bay, oysters and some other shellfish were found starving to death. Eventually, all oysters disappeared.

How could such a disaster happen? The large duck farms produced enormous quantities of duck droppings that washed into the tributaries and eventually into the bay. Because of the bay's slow circulation rate, the duck droppings quickly changed the nutrient balance, drastically altering the prevailing mixture of algae. One species of green algae that was ordinarily present in very small numbers prospered under these new nutrient-rich conditions, while the populations of other phytoplankton species plummeted. Unable to digest this species of green algae, the oysters quickly starved to death.

Since this disaster, several attempts have been made to reestablish a normal nutrient balance and reintroduce oysters to the Great South Bay, but all have failed. The ecological story of the "ducks and oysters" points out how a change in the biotic environment (increasing the number of ducks) can alter the abiotic environment (the levels of inorganic nutrients in the bay), which in turn changes the biotic environment (the proportion of algae species), causing another change in biotic environment (the death of oysters), which produces yet another change in the biotic environment (the loss of food for humans and the bay's other oyster eaters).

The story also illustrates how one seemingly minor alteration can send reverberations throughout an ecosystem—usually with unexpected and destructive results.

## TOLERANCE RANGE

Organisms are able to survive only within certain maximum and minimum limits for each environmental factor. (Lethal temperature limits for many animals, for example, are a minimum of just above 0° C (32° F) and a maximum of about 42° C—slightly higher than 107° F). The range between maximum and minimum limits is called an organism's **tolerance range** (Figure 36-3). Tolerance ranges can also be determined for an entire species by combining the tolerance ranges of all members of the species.

Tolerance ranges

• Differ for each abiotic factor; a marine organism may have a broad range of tolerance for temperature and a narrow range for salinity.

• Have an optimum point within the range where conditions are best for the organism (see the peaks of the graphs in Figure 36-3).

• One factor may affect the tolerance ranges of other factors, especially when conditions approach the maximum or minimum limits for one factor. (For example, grasses grown in nitrogen-deficient soils need more water than those grown in nitrogen-rich soils.)

• Change as an organism passes through different phases of its life cycle. (Eggs, gametes, embryos, and young organisms usually have narrower ranges than adults.)

• May differ slightly from population to population within a species. Pop- ulations with different, genetically fixed tolerance ranges are called **ecotypes** of a species (see Connections: The species concept, Chapter 38).

## LIMITING FACTORS

When the limits of tolerance for one or a few factors are approached, those factors usually take on greater importance than all others in determining where or how well an organism survives. These more critical factors become **limiting factors** because they impose restraints on the distribution, health, or activities of the organism (Figure 36-4). Because each species has a multitude of tolerance ranges (one for each abiotic factor), the ability to identify a few, key limiting factors helps ecologists understand otherwise

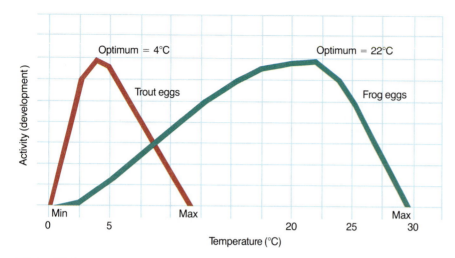

**Figure 36-3**
**The limits of life.** For each physical factor, organisms have minimum and maximum tolerance limits. Within most tolerance ranges (such as the temperature ranges for development of trout and frogs) lies an optimum at which growth is greatest.

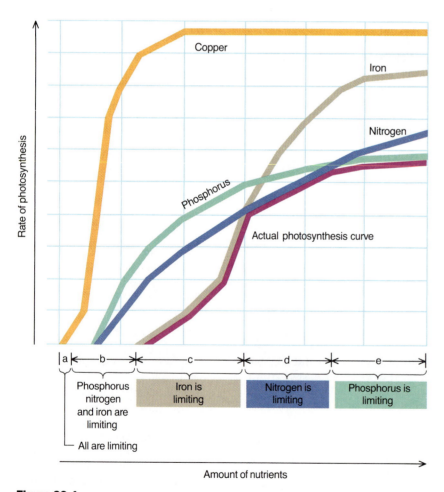

**Figure 36-4**
**Nutrient availability and photosynthesis.** These graphs illustrate the Law of the Minimum. The rate of photosynthesis in phytoplankton (black line) is limited by the nutrient in the shortest supply. In areas *(a)* and *(b)* on the graph, one or more essential nutrients are lacking, so there is no photosynthesis. Low supplies of iron limit photosynthesis in area *(c)*, nitrogen in area *(d)*, and phosphorus in area *(e)*.

incomprehensibly complex ecosystems.

This concept of limiting factors is summarized by the **Law of the Minimum**. Examples of this law are very common. Low amounts of phosphate in a lake drastically reduce algae growth, which in turn limits the growth of all consumers. The amount of zinc in soil is usually so scarce that it alone limits the yield of many agricultural crops.

Exceptions to the "Law of the Minimum" occur when one factor changes the tolerance for another. For example, when the oxygen content of water is low, American lobsters tolerate temperatures only to 29° C (84° F), whereas when the oxygen content is high, the temperature tolerance rises to 32° C (89° F). Another exception occurs when one factor substitutes for another, such as mollusks using strontium to build their shells when calcium is limiting.

## Ecological Niche and Guilds

Each organism in an ecosystem occupies a specific **habitat**, the physical location in which the organism lives—the bottom of a lake, under a rock, inside another organism, and so on. But in addition to needing space to live and reproduce, organisms also need energy and nutrients. The process by which they acquire them determines their "roles" in an ecosystem; they may be producers, consumers, decomposers, predators, prey, parasites, or competitors. All together, the habitat, role(s), requirements for environmental resources, and tolerance ranges for each abiotic condition form an organism's **ecological niche**. In other words, an ecological niche includes every aspect of an organism's existence—where it lives, plus all its activities, requirements, and effects. An ecological niche can be described for a specific organism or for an entire species.

When considering only two or three factors, it is possible to plot an ecological niche on a graph, with each axis representing a different factor (Figure 36-5). It is impossible, however, to draw a graph with *all* components of the ecological niche (all tolerance ranges, the total range of habitats, and all functional roles), for it would cir-

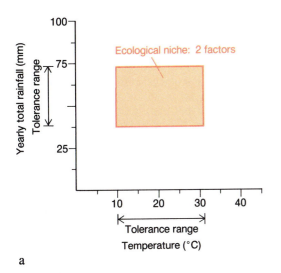

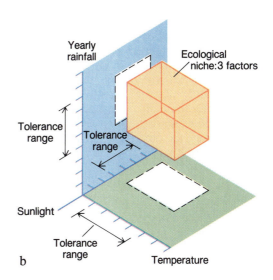

**Figure 36-5**
**An ecological niche** based on the tolerance range of two *(a)* and three *(b)* factors.

cumscribe a multidimensional area, what ecologists refer to as a **hypervolume**. The hypervolume represents the potential ecological niche, or what is called the **fundamental niche**. Most organisms never realize their full fundamental niche. For example, interactions with other organisms (such as competition) often reduce the range of available habitats and possible functional roles. The remaining portion of the hypervolume in which an organism actually exists and functions is called its **realized niche**. Put another way, the fundamental niche represents all possible ranges under which an organism or species can exist, but the realized niche is the range of conditions in which the organism actually lives and reproduces.

**Niche breadth** is a measure of the relative dimensions of a niche. Some species have broad niches. Hawks, for example, have a broad niche because they visit many habitats over large areas and feed on several kinds of organisms. On the other hand, species with narrow niches have very specific habitat and environmental requirements. The cotton boll weevil has a narrow niche, living, feeding, and reproducing only on cotton plants.

**Niche overlap** occurs when organisms share a habitat, have the same functional roles, or in some other way

have one or more identical environmental requirements. Niche overlap leads to competition for the same needed resource; the greater the overlap, the more intense the competition. Competition becomes so great that two species with *identical* niches cannot coexist for very long in the same ecosystem. One eventually excludes the other. (This is called the Competitive Exclusion Principle and is discussed in the next chapter.)

Although species with identical niches cannot coexist in the same ecosystem for long, species with similar niches can. In a forest, for example, all species of insect-eating birds have similar habitats, roles, and nutrient requirements, and yet are able to coexist. Such groups of species with similar ecological niches form a **guild**. Not surprisingly, however, members of a guild interact frequently, often with intense competition when resources become scarce.

Species that have similar ecological niches but live in different ecosystems are called **ecological equivalents**. Ecological equivalents usually have very similar characteristics, which are often the product of convergent evolution. For example, placental mammals and their marsupial counterparts in Australia are ecological equivalents (see Figure 29-20).

## Energy Flow Through Ecosystems

All organisms must secure energy and nutrients from their environment in order to remain alive. Biologists categorize organisms according to their energy-acquiring strategy; primary producers harvest energy through photosynthesis or chemosynthesis, whereas consumers and decomposers obtain energy and nutrients from other organisms. By harnessing light or the energy in inorganic chemicals, primary producers form a direct link between the abiotic and biotic environments.

### Productivity

The earth's primary producers manufacture more than 165 billion tons of organic material each year. The energy (sunlight) stored in this material is equivalent to the output from 2 billion nuclear power plants. The rate at which the primary producers convert radiant energy and inorganic chemicals into organic compounds is defined as the **gross primary productivity (GPP)** of an ecosystem and is usually measured in kcal/m²/year.

Autotrophs must use some of the gross primary productivity to keep themselves alive and to reproduce.

When the amount of organic compounds consumed by autotrophs during respiration (abbreviated $R_A$) is subtracted from the GPP, the result is called the **net primary productivity (NPP)** of an ecosystem (GPP − $R_A$ = NPP). The heterotrophs use some of the NPP. When heterotrophic respiration ($R_H$) is subtracted from the net primary productivity, the result is the **net community productivity NCP** (NPP − $R_H$ = NCP). Net community productivity is a measure of how much organic matter remains and accumulates in an ecosystem, which contributes to the **biomass**, the weight of organic material present at any one time. Productivity measurements in a stream and in a tropical rainforest ecosystem are compared in Figure 36-6.

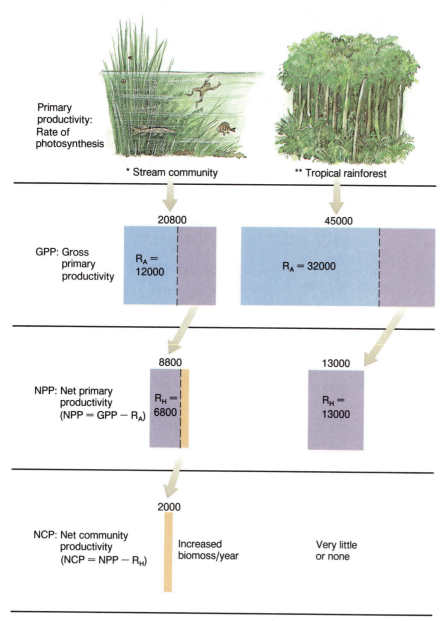

Primary productivity: Rate of photosynthesis

\* Stream community  \*\* Tropical rainforest

GPP: Gross primary productivity

20800  45000

$R_A = 12000$  $R_A = 32000$

NPP: Net primary productivity (NPP = GPP − $R_A$)

8800  13000

$R_H = 6800$  $R_H = 13000$

NCP: Net community productivity (NCP = NPP − $R_H$)

2000

Increased biomoss/year  Very little or none

\* *After H.T. Odum. 1957. Silver Springs, Florida.*

\*\* *After H.T. Odum and Pigeon. 1970. Puerto Rico.*

**Figure 36-6**
**Productivity** in a Florida stream and a tropical rainforest in Puerto Rico. In both ecosystems, over half of the *gross primary productivity* (GPP) is used by the autotrophs for their own respiration ($R_A$). Subtracting respiration by heterotrophs ($R_H$) gives *net community productivity*. Low or no net community productivity is characteristic of stable ecosystems. (All productivity measurements are given in kcal/m²/yr.)

## Trophic Levels

As you might guess from the above discussion, productivity is often very hard to measure. But tracing the transfer of food energy between the organisms in an ecosystem is not; it is relatively easy to count how many leaves a caterpillar eats, how much phytoplankton goes into a copepod, or how many prairie dogs a hawk consumes. Energy and nutrients are transferred step by step between organisms along a linear feeding path. For example, in a streamside community a hawk eats a snake that ate a frog that ate a butterfly that sipped nectar from the flower of a periwinkle plant that converted the sun's energy into chemical energy during photosynthesis.

This feeding pathway has the same organization as those of all ecosystems: it begins with a primary producer (the periwinkles in the streamside community); the primary producer provides food for a **primary consumer** (the butterfly); the primary consumer is eaten by a **secondary consumer** (the frog), which is eaten by a **tertiary consumer** (the snake), and so on until the final consumer dies and is disassembled by decomposers. Each step along a feeding pathway is called a **trophic level** (trophic = feeding).

The sequence of trophic levels maps out the course of energy (food) flow between functional groups of organisms (primary producers, primary consumers, secondary consumers, and so on). As shown in Figure 36-7, trophic levels are consecutively numbered to indicate the order of energy flow. Trophic Level 1 is always populated by primary producers, and trophic Level 2 is always populated by primary consumers. Levels of consumers above the primary consumer are then numbered sequentially. The final carnivore, called the **ultimate** or **top carnivore**, as well as organisms that escape being eaten, eventually die and are then consumed by decomposers.

## Food Chains, Multichannel Food Chains, and Food Webs

Transfers of food energy from organism to organism form a **food chain** (also shown in Figure 36-7). Like a trophic level diagram, a food chain is a flow chart that follows the course of food energy through an ecosystem. But unlike trophic level diagrams, food chains name the organisms (rather than the group) that occupy each link. The example given earlier of periwinkle plants, butterflies, frogs, snakes, and hawks is a food chain because the organism that occupies each step is identified.

Energy flows one way through the organisms of a food chain—it does not recycle. All the energy that enters a food chain through photosynthesis is eventually dissipated as unusable heat energy. (That is, the system's entropy is increased.) To keep life fueled and running, an extraterrestrial source of energy is needed to continually renew the supply; this source is radiant energy from the sun.

Most ecosystems contain many food chains. When more than one food chain originates from the same primary producer, a **multichannel food chain** is formed. In the example illustrated in Figure 36-8, different parts of a single manzanita shrub provide energy and nutrients for at least five independent food chains.

Organisms other than the primary producers often form linkages between food chains. In the streamside ecosystem, for example, frogs may sometimes eat flies instead of butterflies, and snakes may eat small rodents instead of frogs. When all interconnections between food chains are mapped out for an ecosystem, they form a **food web** (Figure 36-9). A food web illustrates all possible transfers of energy and nutrients among the organisms in an ecosystem, whereas a food chain traces only one pathway in the food web.

## Ecological Pyramids

Feeding relationships are graphically represented by plotting the energy content, numbers of organisms, or biomass at each trophic level. Such graphs are called **ecological pyramids** because of their shape. Each trophic level of a food chain forms a tier on the pyramid, with successive trophic levels stacked on top of the level that represents its food source.

A **pyramid of energy** illustrates the rate at which food energy moves through each trophic level of an ecosystem (measured in kcal/m$^2$/year). A pyramid of energy for a food chain in a

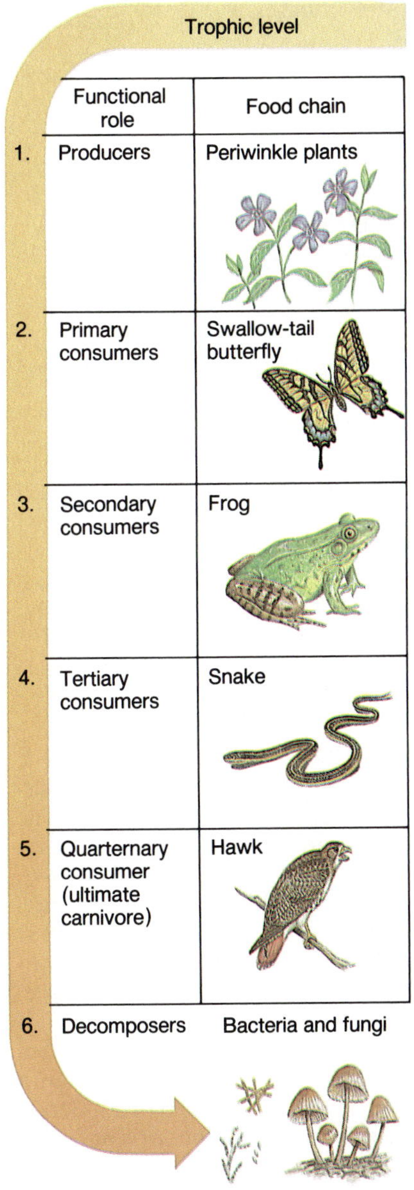

Trophic level
Functional role
Food chain
1. Producers
Periwinkle plants
2. Primary consumers
Swallow-tail butterfly
3. Secondary consumers
Frog
4. Tertiary consumers
Snake
5. Quarternary consumer (ultimate carnivore)
Hawk
6. Decomposers
Bacteria and fungi


**Figure 36-7**
**A feeding path in a streamside ecosystem.** Energy and nutrients flow through the biotic community as food passes from trophic level to trophic level. Trophic level numbers indicate the order of flow.

Primary producer
Primary consumers
Secondary consumers
Nectar
Fruit
Leaves
Food in phloem
Food stored in roots
Hummingbird
Cat
Other consumers
Brown bear
Parasite
Other consumers
Caterpillar
Bird
Other consumers
Aphid
Lady-bug beetle
Other consumers
Mycorrhizal fungi
Earthworm
Other consumers


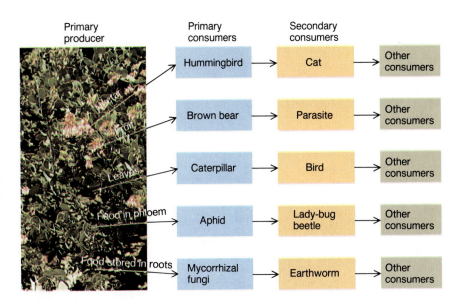

**Figure 36-8**
**Multichannel food chains.** This manzanita (*Arctostaphylos pringlei* var. *drupacea*) is the primary producer for many food chains, including the five drawn here.

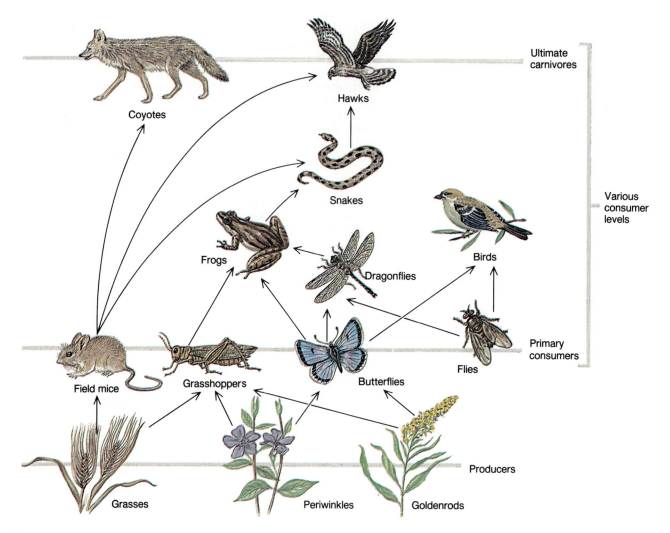

**Figure 36-9**
**A very simplified food web.** A complete food web charts all possible feeding transfers within an ecosystem.

grassland ecosystem during the spring (excluding the decomposers) looks like this:

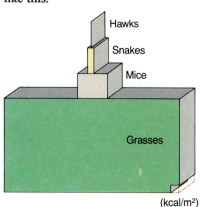

As you can see, there is a dramatic reduction in usable energy as food is transferred from one trophic level to the next. On an average, only about 10 percent of the energy in any trophic level is converted into biomass in the next level. The least efficient transfer is between trophic levels 1 and 2, where as much as 99 percent of the energy in producers does not get transferred to the primary consumer level.

What happens to this huge amount of lost energy? Some of it is lost as heat during energy conversions (the second law of thermodynamics). The organisms in each trophic level also use some of the energy for their own me-

tabolism and biological process. Furthermore, not all food in one trophic level is eaten by organisms in the next trophic level; some of the energy goes to nonbiological phenomena (fire, etc.). Finally, not all the food that is eaten is usable; some energy, for example, is excreted or defecated as waste. Because such a small fraction of energy gets passed on to successive trophic levels, the number of links in a food chain is limited and rarely exceeds more than four or five.

The rate of energy flow through a trophic level determines how many individuals can be supported, as well as their overall biomass. Because energy decreases with each successive trophic level, the number of organisms and their biomass will also decrease. The following **pyramid of numbers** illustrates this principle for a grassland ecosystem:

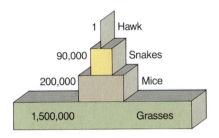

In any grassland, all primary producers are relatively small, so it takes a large number of plants to provide enough food for big primary consumers such as wildebeests or bison. However, in ecosystems where *larger* producers support *smaller* consumers, the pyramid of numbers becomes inverted. In a forest ecosystem, for example, a few large trees provide enough food to support many insects and insect-eating birds. The pyramid of numbers for such a forest looks like this:

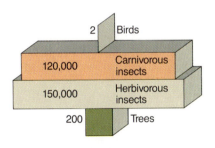

A **pyramid of biomass** represents the total dry weight (in gms/m²) of the organisms in each trophic level at a particular time. Although most pyramids of biomass are "upright," they too may become inverted depending on when samples are taken. For example, in the open oceans where producers are microscopic phytoplankton and consumers range all the way up to massive blue whales, the biomass of consumers may temporarily exceed that of the primary producers if sampling is done when the number of phytoplankton is low. During such sam-

pling periods, the pyramid of biomass would look like this:

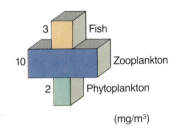

(mg/m³)

However, if samples are taken during the spring when phytoplankton populations are immensely large, or if multiple generations of phytoplankton are included, the pyramid of biomass assumes the upright pyramid shape:

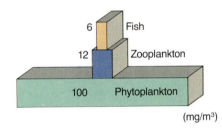

(mg/m³)

Although it is possible for pyramids of biomass and numbers to be inverted, an energy pyramid can never be inverted. There must always be more energy at lower levels to maintain life at higher trophic levels.

# Biogeochemical Cycles: Recycling Nutrients in Ecosystems

All organisms are composed of chemical elements (Chapter 3). Of the more than 90 naturally occurring elements in the biosphere, only some 30 are used by organisms. Some elements, such as carbon, hydrogen, oxygen, phosphorus, sulfur, and nitrogen, are needed in large supplies, whereas others, including sodium, potassium, manganese, calcium, iron, magnesium, iodine, cobalt, and boron, are needed in small, or even minute, amounts.

Unlike energy, there is no extraterrestrial source for the elements essen-

tial for life. For life to continue through time, then, essential elements must be recycled (reused) over and over again. Organisms today are using the same atoms that were present in the primitive earth. Some of the atoms that comprise your body may have been part of an ancient bacterium, a long-extinct tree fern, or a dinosaur.

During recycling, elements pass back and forth between the biotic and abiotic environments to form **biogeochemical cycles**. Primary producers introduce elements into the biotic environment by incorporating them into organic compounds; consumers and decomposers release the elements back into the abiotic environment by breaking down complex organic molecules into simple inorganic forms.

The rate of nutrient recycling depends primarily on where the element is found in the abiotic environment. For example, elements that cycle through the atmosphere or hydrosphere, forming *gaseous nutrient cycles*, recycle much faster than those of *sedimentary nutrient cycles* that cycle through the earth's soil and rocks.

## Gaseous Nutrient Cycles
### THE CARBON CYCLE

Because large amounts of carbon are continually exchanged between the atmosphere and the community of organisms, the recycling rate of carbon is especially rapid (Figure 36-10). Primary producers constantly extract $CO_2$ to build organic molecules during photosynthesis; organic molecules are then disassembled during cellular respiration, releasing carbon as $CO_2$.

On land, producers extract $CO_2$ gas directly from the atmosphere. In water, gaseous $CO_2$ must first dissolve in water before it can be incorporated into organic compounds by aquatic autotrophs. However, not all of the $CO_2$ dissolved in water is used in photosynthesis. As a result, carbon-containing bicarbonates and carbonates, compounds with low solubility, settle to the bottom of oceans, streams, and lakes. Additional carbon sediments are formed as the skeletons and shells

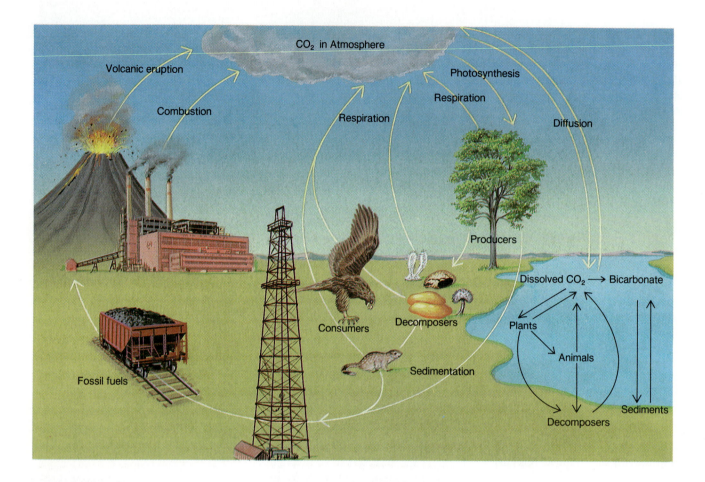

**Figure 36-10**
**The carbon cycle.** Atmospheric and dissolved carbon dioxide is used by primary producers to make energy-rich organic compounds during photosynthesis. When producers are eaten, carbon-containing organic compounds are passed to consumers and eventually to the decomposers. All organisms (producers, consumers, and decomposers) release carbon as $CO_2$ during respiration, most of which returns to the atmosphere. Organisms that are not decomposed may eventually form fossil fuels. Combustion of fossil fuels releases this carbon as $CO_2$ back into the atmosphere.

of organisms accumulate. The settling of carbonate, shells, and skeletons can tie up carbon for long periods in sediment, limestone, and several other forms of rock.

Under certain conditions, dead organisms and organic matter are not decayed by the decomposers. Nondecayed organic material may eventually form deposits that turn into fossil fuels, producing a carbon reservoir within the earth. Most of the world's fossil fuels were formed during the Carboniferous period, between 285 and 375 million years ago, when shallow seas repeatedly covered vast forests, thereby preventing decomposi-

tion. As we burn fossil fuels, we return this carbon to the atmosphere.

## THE NITROGEN CYCLE
Although the earth's atmosphere is 79 percent nitrogen gas ($N_2$), only a few microorganisms are able to tap this huge reservoir and begin a series of conversions and transfers that produces the nitrogen cycle (Figure 36-11). These "nitrogen-fixing" microorganisms include some bacteria and cyanobacteria. Nitrogen-fixers convert atmospheric nitrogen gas to ammonia ($NH_3$) in a process called **nitrogen fixation**. Once nitrogen is converted, other groups of soil bacteria,

collectively called *nitrifying bacteria,* convert ammonia into nitrites ($NO_2$) and then nitrates ($NO_3$). Plants absorb nitrates and incorporate the inorganic nitrogen into organic molecules—nucleotides and amino acids, the fundamental building blocks of DNA, RNA, and proteins. When the producers are eaten or die, nitrogen is passed to the consumers and decomposers in a food chain. The decomposers convert nitrogen-containing organic molecules into inorganic ammonia. Ammonia is also released directly from the consumers; many excrete ammonia (in urea) as a means of eliminating excess nitrogen that would

## Synopsis

### MAIN CONCEPTS

- **Ecosystems are dynamic, self-sustaining units of organisms and their physical environment.**

- **All ecosystems are interconnected to varying degrees.**

- **Within an ecosystem, organisms comprise the biotic environment (community), and the physical factors make up the abiotic environment.**

- **Each organism (or species) has a tolerance range for the various physical factors.** When the minimum or maximum tolerance is approached or exceeded, the factor limits the distribution, health, or activities of the organism.

- **Each organism (or species) has an ecological niche** composed of its habitat, functional role, and total environmental requirements and tolerances.

- **Energy flows through an ecosystem.** Trophic levels, food chains, and food webs track the flow of energy.

- **Essential inorganic nutrients cycle between the abiotic and biotic environment in biogeochemical cycles.** Gaseous nutrient cycles proceed faster than those with sedimentary nutrient cycles.

- **Nutrient cycles require producers and decomposers to cycle elements between the biotic and abiotic environments.** All organisms in the biosphere rely on other organisms for sources of essential nutrients.

- **Ecosystems inevitably change.** Permanent changes in the abiotic environment cause permanent changes in the community of organisms.

- **A climax community is one in which the composition and organization of organisms remains stable over long periods.**

### KEY TERM INTEGRATOR

Ecosystem Structure

| | |
|---|---|
| **ecosystems** | Functional units composed of the **abiotic environment** of physical factors and the **biotic environment** or community. |

The Biotic Environment

| | |
|---|---|
| biotic community | Includes **primary producers, consumers,** and **decomposers** or **saprophytes.** |
| tolerance range | Range of conditions in which an organism or species can survive. **Ecotypes** are populations of the same species with different tolerance ranges for the same factor. **Limiting factors** restrain the growth, health, distribution, and activity of organisms. This principle is called the **Law of the Minimum.** |
| ecological niche | Includes the **habitat,** functional roles, ranges of tolerance, and effects of an organism or species on the ecosystem. The **fundamental niche,** represented by a **hypervolume,** is the potential niche, whereas the **realized niche** is that in which an organism actually lives. |
| niche breadth | The relative dimension of an ecological niche. |
| niche overlap | Two or more species having identical requirements, habitats, or roles in an ecosystem. |

| | |
|---|---|
| **guild** | Groups of species in the same ecosystem with similar ecological niches. **Ecological equivalents** are species with the identical or very similar ecological niches that live in different ecosystems. |

## Energy Flow In Ecosystems

| | |
|---|---|
| **gross primary productivity** | The rate at which energy is converted by autotrophs. Subtracting the amount of energy consumed by autotrophs reveals the **net primary productivity**. The **net community productivity** is the net primary productivity minus the amount of energy consumed by the heterotrophs. |
| **biomass** | The weight of organic matter in an ecosystem at any given time. |
| **trophic levels** | Functional groups of organisms (**primary producers, primary consumers, secondary consumers, tertiary consumers,** and **ultimate carnivores**) in the transfer of energy and nutrients between organisms in an ecosystem. |
| **food chains** | Trace the path of food transfers between organisms. When more than one food chain originates from the same primary producer, a **multichannel food chain** is formed. A **food web** illustrates all of the food chains in an ecosystem. |
| **ecological pyramids** | Include **pyramids of energy, pyramids of numbers,** and **pyramids of biomass.** |

## Biogeochemical Cycles

| | |
|---|---|
| **biogeochemical cycles** | Formed when essential elements cycle between the biotic and abiotic environments. In one such cycle (the nitrogen cycle) **nitrogen fixation** converts nitrogen gas into ammonia. **Denitrifying bacteria** convert nitrites and nitrates into nitrogen gas during **denitrification.** |

## Succession

| | |
|---|---|
| **succession** | The sequence of communities that leads to the formation of a **climax community. Primary succession** occurs in areas where no community existed before, whereas **secondary succession** occurs in disturbed areas. |

# Review and Synthesis

1. Arrange the following subcomponents of an ecosystem:

biotic community     consumers
decomposers         abiotic environment
energy               inorganic elements

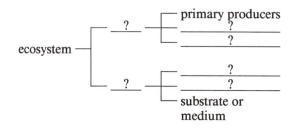

2. Why would niche overlap be greatest between members of the same species or between closely related species?

3. Give an example that illustrates each of the following:
   a. guild
   b. primary succession
   c. food chain
   d. niche overlap
   e. secondary succession

4. List instances in which the biotic community causes changes in the abiotic environment. (Do not use the examples given in the text.)

5. How might ecotypes evolve into a new species?

6. List as many components as you can think of for the fundamental niche and the realized niche of a whale (a relatively broad niche) and a tapeworm (a narrow niche). Now try it for humans.

7. Using your experience with house and garden plants, list some effects of limiting factors that you have observed for yourself.

## Additional Readings

*The Biosphere.* 1970. A Scientific American Book, Scientific American, Inc. W. H. Freeman, San Francisco. (Introductory.)

Brewer, R. 1979. *Principles of Ecology.* W. B. Saunders, Philadelphia. (Intermediate)

Oats, J. 1975. *Web of Life.* Danbury Press, Aldus Books Limited, London. (The Living Earth Series.) (Introductory.)

Odum, E. P. 1983. *Basic Ecology.* Saunders College Publishing, Philadelphia. (Intermediate.)

Thoreau, H. D. 1937. *Walden.* The Modern Library, New York. (Classic, Introductory.)

Chapter 37

# Community Ecology: Interactions Between Organisms

A lthough seemingly calm, ecosystems can be as busy as a city during rush hour—but usually without all the noise. During warmer months, the soil of a forest ecosystem teems with bacteria, fungi, nematodes, springtails, amoebas, mites, slugs, worms, beetles, spiders, and scores of other organisms that churn the ground as they move about, grow, and reproduce. Erupting through the soil surface, delicate plant seedlings absorb nutrients recycled by microorganisms and fungi. The seedlings eventually grow into herbs, shrubs, and trees that create habitats and manufacture food for countless herbivores. An assortment of carnivores devour the herbivores, as well as other carnivores.

Although the participants vary from one ecosystem to the next, the forest illustrates how some organisms interact with each other. Some interactions benefit one or both participants. Some have neutral consequences, whereas others harm either or both participants. In this chapter we discuss the kinds of interactions that develop between the organisms that make up a *community,* the collection of populations that coexist in a particular place at the same time. The general categories of interactions and the eventual outcome for the participants are previewed in Table 37-1.

## Symbiosis

Some organisms must interact because they live in close association with each other. A close, long-term relationship between two individuals of different species is called **symbiosis**, a term that means "to live together" (Figure 37-1). Symbiotic interactions can include forms of parasitism, commensalism, protocooperation, and mutualism. (The term "symbiosis" is sometimes incorrectly used as a synonym for mutualism, an obligate interaction that benefits both participants.)

## Competition: Interactions That Harm Both Participants

Species can coexist in a community as long as they have slightly different ecological niches (Chapter 36), even though their niches may overlap. When a shared resource is abundant, such as oxygen in terrestrial habitats or water in aquatic habitats, there are sufficient supplies for all. But most resources are in limited supply, and organisms with overlapping niches enter into **competition**. This type of interaction harms both participants because each reduces the other's supply of a needed resource. The more similar the requirements of organisms, the greater their niches overlap; the greater the niches overlap, the more intense the competition (Figure 37-2). Because members of the same species require many of the same resources, **intraspecific competition** between members of one species is often more intense than **interspecific competition** between members of different species.

### Competitive Exclusion Principle

Although species with small niche differences are often able to live in the same community, those with identical ecological niches cannot do so, even if only one shared resource is in limited supply. Competition becomes so intense that one species eventually eliminates the other from the community, either by taking over its habitat and displacing it from the community, or by causing its extinction. The "winner" species is successful because it possesses some characteristic(s) that gives it a slight advantage over its competitor. This advantage enables members of the "winner" species to capture a greater share of resources, which in turn increases their survival and reproduction. Greater numbers of offspring with better suited traits gradually displace members of the less efficient species. This "winner takes all" outcome is named the **competitive exclusion principle**. It was first demonstrated in mixed cultures of closely related paramecium species (Figure 37-3). The competitive exclusion principle has also been referred to as the "one niche, one species" rule.

a         b

**Figure 37-1**
**Symbiosis: living together.** *(a)* **Aphids and ants.** Some ants live with groups of aphids. Constantly refueling, aphids feed on the sugary juices of plants, often getting more than their bodies can hold. The excess passes through the aphid's digestive system and out its anus, forming a honeydew drop that is lapped up by the ants. The ants protect the aphids by aggressively keeping their predators away (such as ladybird beetles and syrphid fly larvae). *(b)* **Anemones and algae.** Some anemones have resident algae living inside their bodies. The algae conduct photosynthesis, producing food and oxygen for the anemone and, in turn, receive $CO_2$ and a safe habitat inside the animal's tissues.

TABLE 37-1

| Possible Interactions between Organisms in a Community | | |
|---|---|---|
| **Kind of Interaction** | **Organism 1** | **Organism 2** |
| Competition | harmed | harmed |
| Predation (including herbivory) | benefited | harmed |
| Parasitism* | benefited | harmed |
| Allelopathy | benefited | harmed |
| Commensalism* | benefited | unaffected |
| Protocooperation* | benefited | benefited |
| Mutualism* | benefited | benefited |

*\* May include symbiotic interactions.*

Although competitive exclusion has been demonstrated repeatedly in the laboratory, it is not always apparent in natural communities. For instance, six species of leafhoppers *(Erythoneura)* can live on the same sycamore tree and feed side by side on the same leaves. Not only are their habitats and food source the same, but also the phases of their life cycles are identical. Researchers could not find any niche differences between the six species—a perfect setup for competitive exclusion. Yet, researchers found no evidence that these species even harm each other, much less cause exclusion or extinction. Apparently, competitive exclusion is avoided because of the abundance of critical resources.

Competitive exclusion is not usually noticed in ecosystems where environmental conditions change frequently. There simply is not enough time for one species to displace another during the short periods when resources become limited. For example, in oceans and temperate lakes, changes sometimes occur too quickly for one species of phytoplankton to increase in numbers large enough to exclude another, despite intense competition between them for limited nutrients. Because steady changes occur during primary and secondary succession (Chapter 36), competitive exclusion has not been recorded in communities undergoing succession.

Another reason why the process of competitive exclusion may go unnoticed in natural communities is that it takes time for one species to exclude another. If researchers are unable to continuously observe the community, they may miss the process entirely. For example, goats were introduced on the island of Abingdon in the Galapagos Archipelago in 1957. The goats browsed on the same low-growing plants as the island's native tortoises, as well as on leaves on higher stems and branches. Without any predators, the goats reproduced rapidly and consumed all the low-growing food that

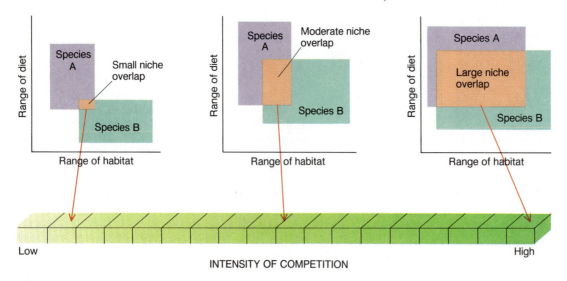

**Figure 37-2**
**Niche overlap and competition.** Each graph compares habitat and diet requirements for two species. Since resources are usually limited, the species compete. The greater the similarity in requirements, the greater the niche overlap and the more fierce the competition.

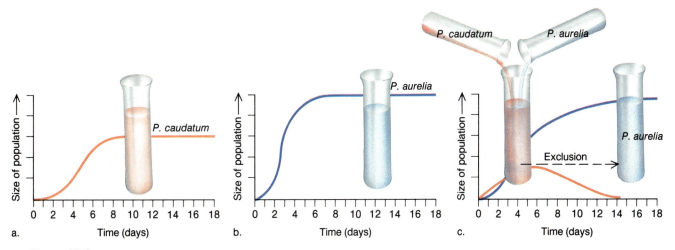

**Figure 37-3**
**Competitive exclusion: winner takes all.** Two closely related species of paramecium (*P. cauda-tum* and *P. aurelia*), with exactly the same niche cannot coexist for very long (unless resources are unlimited). Grown separately (*a* and *b*), the number of individuals of both paramecia increases rapidly, and then levels off and remains constant. But when cultured together (*c*), competition for limited food results in the death of *P. caudatum*, outcompeted by the more rapidly reproducing *P. aurelia*.

could be reached by the tortoises. By the time a research team revisited the island in 1962, all of the tortoises were gone. Competitive exclusion had caused the extinction of the Abingdon tortoise.

## The Results of Competition: Resource Partitioning and Character Displacement

Competitive exclusion is not the only outcome of competition. Sometimes a shared resource can be partitioned in a way that will allow potential competitors to use different portions of the same resource. Five species of North American warblers, for instance, feed in slightly different zones on the same spruce tree, enabling these very similar birds to coexist without competing (Figure 37-4). In addition to such spatial partitioning, a shared resource may be exploited at different times, producing temporal partitioning. Temporal partitioning occurs in a grassland ecosystem where a species of buttercup *(Ranunculus)* grows only in early spring, before competing perennial grasses begin development. Temporal and spatial partitioning of a resource are examples of **resource partitioning**.

Another alternative to competitive exclusion is **character displacement** where intense competition affects evolution, leading to the displacement of a characteristic rather than that of a species. An example of character displacement is seen in the forests of middle Europe, where six species of small birds called titmice *(Parus)* coexist because each species has a slightly different-sized bill. Differences in bill size prevents any two species from seeking the same food source in the same feeding area; titmice with longer beaks usually capture larger insects, for example. Biologists have documented that all six species evolved from an identical ancestor and that the variation in beak size seen today is the result of character displacement.

## Exploitative and Interference Competition

Organisms compete either *directly* or *indirectly* for a limited resource. Indirect competition occurs when competitors have equal access to a limited resource, but one species manages to get more of the resource, thereby reducing supplies to the competitor. This form of competition is called **exploitative competition**. An example of exploitative competition is taking place in the California deserts today between deep-rooted native plants and newly introduced tamarisk trees. Tamarisk trees were brought from the Middle East to act as windbreaks along freeways and railroad tracks. The rapidly spreading tamarisk trees are better able to tap groundwater supplies, reducing water availability to native trees, such as mesquite and desert willows.

In contrast to exploitative competition which is indirect, **interference competition** is direct: one species directly interferes with the ability of a competitor to gain access to a limited resource. Interference competition includes aggressive behavior in animals, such as is displayed when hyenas drive away vultures from the remains of a zebra, and **territoriality** as when animals establish and defend an area against other members of its species (see Animal Behavior: Territoriality).

## Interactions That Harm One and Benefit the Other

### Predation and Herbivory

All organisms need a source of energy and nutrients to survive and reproduce. In **predation**, one organism (the

**Figure 37-4**
**Resource partitioning: dividing the ecological pie.** Five species of North American warblers feed in the same spruce tree; however, each feeds in a slightly different zone. Foraging in different areas of a common resource at the same time is one form of resource partitioning. Resource partitioning reduces competition, enabling species with similar ecological niches to coexist in a community.

**predator**) acquires the needed resources by eating another organism (the **prey**). If the "victim" is a primary producer, the interaction is called **herbivory** and plant-eating animals are called **herbivores**. Organisms that eat other animals are **carnivores**, and those with a mixed diet of plants and animals are **omnivores**.

## PREDATOR AND PREY DYNAMICS

Although some predators limit their diets to one type of prey, most rely on more than one species for nourishment. The choice often depends on the abundance and accessibility of prey. During the summer, for example, a red fox mainly eats meadow mice. But as the availability of meadow mice dwindles, the fox shifts to eating more abundant white-footed mice. During winter when both kinds of mice become scarce, the fox turns to a more readily available food supply—

apples, berries, and other fruits.

As the availability of prey increases, the number of predators also increases. But as the number of predators increases, they consume greater numbers of prey, reducing the availability of prey, which in turn reduces the number of predators. This reciprocal interaction produces recurring cycles of increases and decreases in predator and prey abundance. We discuss this subject further in the next chapter.

## PREDATOR AND PREY ADAPTATIONS

Coevolution between predators and prey has produced an array of remarkable and effective adaptations. Some adaptations improve the skills of predators to capture prey, and some improve the prey's chances of escaping predators. Adaptations that aid prey survival can be grouped into two gen-

eral categories: escape adaptations and defense adaptations (Table 37-2).

### Camouflage

Some prey go unnoticed by predators because they blend in with their surroundings or because they look inanimate (like a dried twig, for example) or inedible (such as bird droppings). Such adaptations are called **camouflage** because the color, shape, and behavior of an organism make it difficult to detect, even when in plain sight. But camouflage is not reserved exclusively for prey; predators use camouflage to help conceal themselves while waiting to ambush prey. Examples of camouflage in both predators and prey are shown in Figure 37-5.

**Cryptic coloration** is a form of camouflage whereby the organism's color or patterning helps it resemble its background. The long-tailed weasel, for example, changes color to match seasonal changes in its surroundings.

# ANIMAL BEHAVIOR VI

## TERRITORIALITY

a

b

c

a. On the jagged cliffs of the Galapagos Islands, two marine iguanas charge at each other, heads lowered and mouths wide open. The two males carry out an instinctive duel for a few square meters of rock on which to live and breed with several females.

b. These two male cichlid fish are not kissing, but are grasping each other by the jaw, pushing and pulling until one gives up and swims away.

c. Two hippos wrestle head to head.

Ceremonial tournaments such as these are common among animals. A large number of animal species engage in contests to establish and defend territories that are suitable for survival and reproduction. **Territoriality** is a form of behavior in which individuals forcibly exclude intruders from a defined area.

Territoriality has a number of benefits.

- For some species, the size and quality of a territory attract females for reproduction.

- Territories provide access to adequate food supplies.

- Territoriality may reduce competition for food and breeding sites among members of the same species.

- Only the more vigorous individuals succeed in capturing the greatest share of environmental resources, the largest territories. These ''more fit'' individuals live longer and produce more offspring, passing on their favorable characteristics.

- Territoriality limits population size, providing a check on overpopulation.

- By becoming familiar with a particular territory, animals become more efficient in finding resources and utilizing their habitat.

Sometimes territoriality includes aggressive behavior. For many animals, aggressiveness is a basic biological phenomenon; that is, aggressive behavior can be elicited by stimulating specific areas of an animal's brain. Some biologists and behavioral psychologists believe that territoriality and aggressive behavior are inborn in humans as well, but expression of this genetic endowment is subject to modification by experience and learning.

During winter, its pure white coat helps it to blend in with the snow (and perhaps helps it conserve heat), whereas during summer, its brown coat helps it blend with the forest floor (Figure 37-6). The plumage color of several species of grouse and hares also changes seasonally. Laboratory experiments with willow grouse *(Lagopus lagopus)* reveal that the length of daylight (photoperiod) triggers hormonal changes that coordinate color change.

**Disruptive coloration** disguises the shape of an organism, as in the coloration of the moth shown in Figure 37-7. The color pattern breaks up the outline of the moth when it is resting on a dark tree trunk, concealing its shape.

Disruptive coloration sometimes includes camouflaging vital parts—frequently, the organism's eyes since many predators use eyes as an attack target. With its real eyes camouflaged, the predator's attention is often diverted away from a vital part (its head). Following an attack, a prey with false

TABLE 37-2

| How Some Organisms Avoid Becoming Prey | |
|---|---|
| **Escape Adaptations** | **Effect** |
| Camouflage | |
|   Cryptic coloration | Hides from predator |
|   Disruptive coloration | Distorts shape and confuses predator |
| Individual responses | |
|   Startle behavior | Confuses predator |
|   Playing dead | Confuses predator |
|   Shedding body parts | Escapes capture |
|   Outdistancing predators | Escapes capture |
| Group responses | Warn, protect, and confuse |
| | |
| **Defense Adaptations** | |
| Physical defenses | |
|   Armor | Deters an attack |
|   Aposematic coloration | Advertises noxious trait |
| Mimicry | |
|   Mullerian mimicry | Noxious species avoided |
|   Batesian mimicry | Harmless or palatable species avoided |
| Chemical defenses | |
|   Poisons | Kills predators |
|   Hormones | Disrupts predator development |
|   Allelochemicals | Repels predator |

a

c

b

d

**Figure 37-5**

**Camouflage: nature's masqueraders.** *(a)* The "vine" "growing" on this branch is really a grass green whip snake *(Dryophis)*. *(b)* This woodcock blends in with its surroundings, helping it escape without being noticed by predators. *(c)* A remarkable example of camouflage, the Malaysian horned frog has adaptations that help it blend in with dead leaves lying on the forest floor, including shading (which conceals the frog's eyes) and curly horns that resemble drying leaf tips. These characteristics help make the horned frog virtually indistinguishable from its surroundings and invisible to its prey. *(d)* Seeing through the deception, the Costa Rican clearwing butterfly *Ithomia* virtually disappears because of its transparent wings.

**Figure 37-6**
**Keeping a "low profile."** The long-tailed weasel (Mustela frenata) has fur that changes from white in winter to brown in summer. Both cryptic colors help the weasel blend with its surroundings and escape being noticed by predators.

eyes in a less vital area of the body (a wing, for example) usually escapes with only minor damage (Figure 37-8*a*). False eyes may not only divert a predator's attack, but also sometimes threaten or startle a predator by making the prey appear to be larger than it really is (Figure 37-8*b*).

Cryptic and disruptive color adaptations are not used solely by animals. Cryptic coloration in some plants helps them resemble less palatable plants, and cryptic and disruptive coloration helps to camouflage some plants from herbivores (Figure 37-9).

In addition to coloration, shape and behavior of an organism also contribute to a successful disguise. The crab spider, African thorn spider, and potoo bird in Figure 37-10 remain motionless, lessening the chance of their being detected.

### Individual Responses

When confronted with a predator, some prey rely on sudden escape responses. Generally, predators are momentarily stopped by the unexpected, especially if the escape response seems dangerous. A moment's hesitation often gives the prey a chance to escape. Examples of such "last ditch" responses include the following.

**Figure 37-7**
**Disruptive coloration** disguises the shape of this moth as it rests on a dark tree trunk, creating an image that goes unrecognized by its sharp-eyed bird predators.

a                                        b

**Figure 37-8**
**An eye-full.** *(a)* In addition to false eyes, the Malaysian back-to-front butterfly *(Zeltus amasa)* deflects predator attacks with false legs and fake antennae on its hind wingtips. An attack on this end does not damage vital organs, so the butterfly can dart to safety, minus only a wing fragment. *(b)* When confronted with a predator, this South American frog bends over and puffs up its body, revealing large, false eyes. The startled predator is usually frightened away by the intimidating display. However, if the predator proceeds with the attack, the frog releases an unpleasant secretion from glands near each false eye.

**Figure 37-9**
**Looking like stones,** pebble plants *(Dinteranthus)* usually go unnoticed by passing herbivores.

- An owl fluffs its feathers and spreads out its wings, a last-minute bluff that usually startles an attacking predator.

- A mosquito fish frantically splashes on the surface of a pond when approached by a voracious pickerel *(Esox)*, making it difficult for the pickerel to launch a pinpoint attack.

- A tiny bombadier beetle sprays hot chemical irritants at its predator, thwarting the attack.

- An opossum "rolls over and plays dead" when confronted by a predator; discouraged predators search elsewhere for a fresh meal.

A few organisms escape a predator's clutches by releasing the seized part of their body. A lizard, for example, quickly detaches its tail. Often, the severed tail continues to move for several minutes, keeping the predator occupied while the lizard scurries to safety to begin regenerating a new one.

Finally, some animals escape becoming a prey by simply outrunning their predator; a healthy antelope or impala can usually outdistance a lion. Like many predators, lions generally capture the young or weak.

### Group Responses

Schools, packs, colonies, and herds defend themselves more effectively than can a single organism. For example, the first smelt fish to notice an approaching predator releases chemicals into the water that immediately send the school fleeing in various directions. The confused predator does not know which way to turn. In grazing herds, stronger individuals generally surround the young and weak, protecting them from an advancing predator. There is indeed safety in numbers, for a predator is less able to pick out a single target among a swarming group.

### DEFENSE ADAPTATIONS
#### Physical Defenses

Organisms have evolved an arsenal of anatomical features to help protect themselves against direct attacks by predators. Many have protective shells (mollusks and turtles), barbed quills (porcupines), needlelike spines (sea urchins), and piercing thorns, spines, and stinging hairs on plants that can discourage even the hungriest predator. Your skin, in fact, is a protective armor against the daily invasions of millions of microbes. Imagine how effective the 9-inch thick hide and blubber of a whale is as a barrier; it sometimes even prevents the penetration of a high-velocity harpoon.

Many foul-tasting, poisonous, stinging, smelly, or biting animals have striking colors, and bold stripes and spots. The opposite of camouflage, **aposematic coloring,** or warning coloration, makes an organism stand out from its surroundings. (Aposematic refers to anything that serves to warn off potential attackers.) The distinc-

a

b

c

**Figure 37-10**
**Nature's impostors.** *(a)* The Borneo crab spider resembles bird droppings, a disguise that lures butterflies and other insect prey that eat genuine droppings. Birds, the predators of the spider, stay clear of what appears to be their own wastes. *(b)* The "twin thorns" on this acacia branch are really parts of an African thorn spider, resting motionless during the day to avoid predators. At night it spins a web to catch insect prey. *(c)* Unwary animals fall easy prey to this dead tree trunk (really a motionless potoo bird), with its eyes half open and neck outstretched. Not only is the adult potoo perfectly disguised, but its single spotted egg also blends with the broken tree stump, camouflaging it from predators.

tive aposematic stripes of a skunk advertise to potential predators that this animal can yield an obnoxiously smelly counterattack; it usually takes only one encounter for a potential predator to avoid any future entanglements with a skunk. In plants, warning coloration includes the red and black fruits of poisonous nightshades.

Having a bad taste is often of little help for the individual animal being attacked, for at least part of it must be eaten before the predator notices its foul taste. But the species as a whole profits, because individual predators learn to recognize the characteristics of a vile-tasting species and avoid them, sparing other members of the species.

Color is not the only aposematic defense. Recently, researchers learned that some foul-tasting moth species make a clicking sound when they are being pursued by bats. Bats learn to associate the clicking sound with bad-tasting moths, pursuing instead quiet, tasty species. However, some tasty moths avoid becoming prey for the bats by making the same clicking sound as the foul-tasting moths. The clicking of tasty moths is an example not only of an aposematic sound defense, but also of **mimicry**, in which one species resembles another in color, shape, behavior, or, in this instance, sound as a mechanism for defense or disguise.

## Mimicry

If similar-appearing species are equally obnoxious, the resemblance is called **Mullerian mimicry**. In the tropics, many species of beetles have bright orange wing-cases with bold black tips. When attacked, the beetles release drops of their own foul-tasting blood; just a taste deters a predator, sparing the beetle. Predators learn to stay away from other similarly colored beetle species. Examples of Mullerian mimicry abound in many tropical butterfly species, such as between monarch and *Acraea* butterflies (Figure 37-11).

In some cases of mimicry, a harmless or palatable species gets a "free ride" by resembling a vile-tasting or stinging species. When a good-tasting or harmless species (called the *mimic*) resembles a species with unpleasant predator-deterring traits (the *model*), the similarity is called **Batesian mimicry** (Figure 37-12).

**Figure 37-11**
**Equally distasteful,** the *Acraea* butterfly *(top)* and the African monarch *(below)* are Mullerian mimics. These butterflies are not even closely related, yet resemble each other almost exactly in color, pattern, behavior, and foul flavor.

**Figure 37-12**
**The deadly moray eel (top) and the harmless plesiopid fish (below)** are an example of Batesian mimicry. When pursued by a predator, the plesiopid fish swims head first into a rock crevice. The shape, color, pattern, and false eye of its exposed tail strongly resemble the head of the dangerous Moray eel, frightening predators away.

## Chemical Defenses

Some plants and animals release **allelochemicals**, chemicals that deter, kill, or in some other way discourage predators. Such defenses must be swift and effective in animals; otherwise, the predator may inflict fatal damage before the allelochemical takes affect. To accomplish this defense, many animals rely on noxious propellants, or painful, sometimes deadly bites and stings to blunt a predator's attack.

Some tropical toads and frogs secrete extremely poisonous chemicals. South American tribesmen, for example, simply touch the tip of their arrows to the skin of poisonous toads to produce a lethal missile that can kill an animal (including a human) within minutes. Other swift-killing poisons are manufactured by the Japanese puffer fish, Asian Goby fish, and the American newt.

Poisonous animals frequently have aposematic coloration, which serves as a blatant signal to experienced predators of the consequences of an attack. The conspicuous stripes on poisonous monarch butterfly larvae make it easy for its predators to recognize them (Figure 37-13). Interestingly, although both the larva and adult monarch butterfly are poisonous, neither manufactures the toxic chemicals themselves. Instead, larvae are poisonous because they eat milkweed plants that synthesize the toxic chemicals; adults are poisonous because the toxic milkweed chemicals are passed on from larvae to adults during metamorphosis. Like monarch butterflies, some nudibranchs also derive their defenses second hand. In an ironic twist, these nudibranchs arm themselves with their prey's stinging cnidocytes (see Figure 34-8) and use them to defend themselves against their own predators.

Plant chemical defenses against herbivores and competitors are discussed in the section on allelopathy.

## Parasitism

**Parasitism** is another type of interaction that benefits one organism and harms the other. A **parasite** secures its nourishment by living on or inside another organism called the **host**. Although parasitism is sometimes considered a form of predation, the victim of parasitism usually survives the interaction. The victim of predation is almost always killed.

Most parasites are **host-specific**, that is, their anatomy, morphology, metabolism, and life history are adapted specifically to those of their host. Such specificity is evidence of a long history of coevolution.

Apparently, no organism escapes being parasitized. Even parasites have parasites (Figure 37-14). Animals are hosts to a huge battery of parasites, including viruses, bacteria, fungi, protozoa, and other animals (flatworms, flukes, tapeworms, nematodes, mites, fleas, and lice, for example). The range of plant parasites is equally broad, and includes viruses, bacteria, fungi, nematodes, and even other plants, such as mistletoe and dodder.

Some organisms are host to a number of different kinds of parasites at the same time. A single bird may have 20 different parasites (Figure 37-15), and of each type, there may be hundreds of individuals. Researchers counted more than 1000 feather lice in the plumage of a single curlew.

Although most parasites use their host only as a source of nutrients, some parasites use their host as a haven for protection from predators. For example, the pearl fish *(Carapus)*

**Figure 37-13**
**Second-hand poison** obtained from eating milkweed plants makes the monarch butterfly and its larva toxic to predators. The bold stripes on the larva, and distinctive color and pattern of the adult broadcast danger to bird predators. After just a few tastes and subsequent episodes of vomiting, predators quickly learn to avoid these bold patterns.

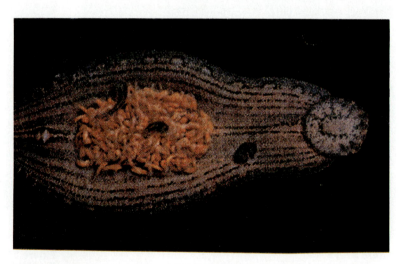

**Figure 37-14**
**Leeches have leeches,** illustrating that virtually all organisms, even parasites, have parasites.

**Figure 37-16**
**One very unusual habitat for a fish** is the anus of a sea cucumber. With poor eyesight and barely able to swim, a young pearl fish is quite vulnerable to predators. Soon after hatching, the pearl fish locates a sea cucumber. When the sea cucumber opens its anus to draw in water, the parasitic pearl fish enters.

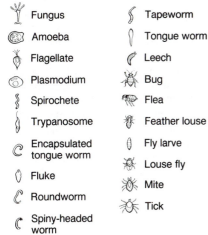

| | | | |
|---|---|---|---|
| Fungus | | Tapeworm | |
| Amoeba | | Tongue worm | |
| Flagellate | | Leech | |
| Plasmodium | | Bug | |
| Spirochete | | Flea | |
| Trypanosome | | Feather louse | |
| Encapsulated tongue worm | | Fly larve | |
| Fluke | | Louse fly | |
| Roundworm | | Mite | |
| Spiny-headed worm | | Tick | |

**Figure 37-15**
**The invasion begins immediately after hatching.** As many as 20 kinds of parasites can infest a single bird.

develops and lives in a safe, but very unusual place—the anus of a sea cucumber *(Actinopygia)* (Figure 37-16). As a sea cucumber draws in water for gas exchange through its anus, a newly hatched pearl fish swims in. The pearl fish remains there, feeding on the host's tissues, and taking periodic excursions outside its host to supplement its diet and to reproduce. It always returns to the sea cucumber for protection from predators.

In addition to using a host for nutrients or protection, some parasites exploit the behavior of their host, an interaction called **social parasitism**. Examples of social parasitism are provided by European cuckoos, American cowbirds, and African honey guides. After a host bird builds a nest and lays eggs, the parasite bird destroys one of the host's eggs and replaces it with her own. The egg is usually so similar in size and coloration that the host bird fails to recognize it as an alien egg. The host incubates the parasite's egg as its own. After hatching, the parasitic baby bird instinctively shoves all solid objects out of the nest, including the host's babies and any unhatched eggs. After clearing the nest of its rivals, the parasite snatches up all of the food brought to the nest by its duped "foster parents."

Many internal parasites are unable to live independently of their host. The tapeworm that feeds on humans, for example, lacks eyes, a digestive tract, and muscular systems. However, it has a combination of adaptations that are suited for living inside human intestines, including

- An outer cover that protects the tapeworm from powerful digestive enzymes, yet allows nutrients to be absorbed.

- A long, flat shape that creates a maximum absorptive surface area, yet prevents obstruction of the host's intestine that could endanger the host's life.

- Hooks on its "head" that anchor it to the host's intestinal lining.

- A reproductive system with male *and* female parts that allows for self-fertilization. This is an important reproductive strategy in a location where contact with another tapeworm is highly unlikely. (Internal parasites are often little more than reproduction "machines," producing millions and millions of offspring, increasing the chances of infecting a new host.)

Some internal parasites require more than one host to complete their reproductive cycle. The fox tapeworm, for example, requires not only a fox, as its name indicates, but also a rabbit. Inside the intestines of a fox, the tapeworm produces hundreds of eggs that are released with the fox's feces. When a rabbit eats a plant contaminated with fox feces, the tapeworm eggs enter the rabbit's digestive system. Inside the rabbit, the eggs hatch and the larvae bore their way out of the intestine and into the rabbit's muscles. The larvae then form cysts, a resting stage of a parasite. When an infected rabbit is eaten by a fox, the cysts become activated and develop into young tapeworms that attach themselves to the intestines of the new fox host. Within a short time, the fox tapeworm, like its parent, is producing hundreds of eggs a day, and the life cycle begins again.

Although most parasites do not kill their host, the larvae of some insect parasites do. Such larvae are called **parasitoids**. For example, after a female tarantula wasp captures and paralyzes a tarantula with her sting, she lays eggs in the spider's flesh. When the eggs hatch, the larvae gorge themselves on fresh tarantula tissues, killing the helpless spider.

## Allelopathy

Some organisms wage chemical warfare on other members of the community. **Allelopathy** is a type of interaction in which one organism releases allelochemicals that harm another organism. Although some of the chemical defenses described earlier for animals may also be examples of allelopathy, we will confine our discussion here to harmful chemical defenses of plants.

Some plants manufacture allelochemicals that kill herbivores or competing plants. For instance, chemicals released by some chaparral plants accumulate in the soil beneath them and block the germination and growth of other plants, thereby reducing competition for scarce water and nutrients.

Sometimes allelochemicals percolate deep into the soil and reach high enough concentrations to kill the very plant that produced them. This phenomenon is called *autotoxicity*. Although killing oneself goes against a basic "goal" of life—survival—some biologists argue that autotoxicity has adaptive value for the species as a whole. Because it takes many years for

allelochemicals to accumulate to toxic levels, only older plants with low reproductive ability die, reopening space for new, reproductively more vigorous individuals.

Some plants defend themselves against herbivores by fatally poisoning them. Members of the crucifer family (cabbages, broccoli, brussel sprouts, mustards, radishes, and so on) produce mustard oils, chemicals that are lethal to many herbivores, and disease-causing fungi and bacteria. But some herbivores have evolved *counteradaptations* that detoxify poisonous allelochemicals. Cabbage white butterfly caterpillars *(Pieris brassicae)* have been successfully reared on cultures containing more than ten times the concentration of mustard oils found in cabbage plants. Apparently, what started out as a means of protection against herbivores has backfired, for cabbage white butterflies now use the scent of mustard oil to locate plants on which to lay more eggs.

Recently, investigators discovered that some ferns, conifers, and flowering plants produce allelochemicals that disrupt the normal growth and development of insect herbivores. These allelochemicals are virtually identical to the insect's hormones that coordinate development during metamorphosis. When larvae consume these plants, the allelochemicals cause premature metamorphosis or produce sterile adults. Either way, herbivore reproduction is disrupted. Some of these hormone-mimicking allelochemicals are being considered for use as natural pesticides, chemicals that would cause considerably less environmental damage than synthetic insecticides.

## Commensalism: Interactions That Benefit One Participant and Have No Effect on the Other

The benefits of **commensalism** are one-sided: only one of the participants (called the commensal) profits while

the other is virtually unaffected. Nature exhibits many examples of commensalism. For instance, remoras attach themselves by suckers to the undersides of sharks, positioning themselves for gathering food scraps as sharks feed. Remoras benefit, but their presence apparently has little or no impact on the shark.

Some commensals simply live in a habitat created by another organism. The burrows of large "innkeeper" sea worms, for instance, house an array of "guests" that use the burrow for shelter but do not hinder or benefit the innkeeper worm (Figure 37-17). Epiphytes are commensals that grow on the branches of taller plants. Being higher up in the forest canopy, the epiphyte captures more light than it could if it occupied a position lower in the canopy. Barnacles encrusted on a humpback whale are also commensals, gaining a habitat and being carried to new sources of food. However, when the number of barnacles gets large, they can reduce a whale's buoyancy and increase resistance as it swims, hampering the whale's ability to move. Large numbers of barnacles (or epiphytes, for that matter) illustrate how commensalism can sometimes turn into parasitism.

## Interactions That Benefit Both Participants

### Protocooperation

Many mutually beneficial interactions are compulsory; that is, both organisms are completely dependent on each other. However, noncompulsory **protocooperation** interactions can also benefit both participants. For example, protocooperation between a fungus and algae forms a lichen (page 579). The fungus uses food produced by the algae, while the algae gain a habitat, plus water and minerals absorbed by the fungus. Both benefit from this form of symbiosis, though both could live on their own.

Another example of protocooperation is the relationship between an oxtail bird and a rhinoceros. The bird

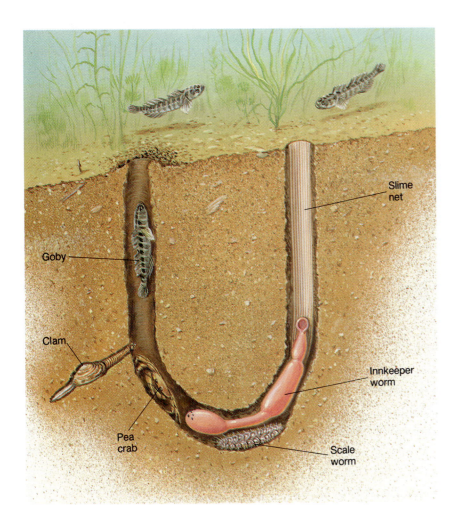

**Figure 37-17**
**Commensalism: one benefits, the other is unaffected.** The innkeeper worm bores a tunnel in the mud of shallow coastal waters. The worm spins a slime net that filters minute organisms as the worm pumps water through the net. When full, it gulps down the whole thing, net, trapped food, and all. But the innkeeper worm is not the only occupant of its tunnel. The Goby uses the burrow for protection, while the pea crab, clam, and scale worm feast on the innkeeper's leftovers. Although the guests benefit from the association, the innkeeper worm apparently neither profits nor suffers from their presence.

perches on the back of a rhinoceros and removes pests (blood-sucking ticks and flies). The oxtails benefit by getting food, warmth, and protection from predators; the rhino benefits by being kept relatively free of parasites. (The sharp eyed bird also alerts the dim visioned rhino of an approaching intruder.) Again, the relationship is facultative because both are capable of surviving on their own (Figure 37-18).

## Mutualism

**Mutualism** is a form of interaction in which both participants benefit. However, unlike protocooperation, the mutualistic interaction is essential to the survival or reproduction of both participants. Because many mutualistic interactions are symbiotic, it is sometimes called **obligate symbiosis**.

**Figure 37-18**
**Protocooperation between oxtail birds and rhinos.** Although neither requires the other for its survival, both benefit from their interaction.

The pollination of some flowers by specific insects, birds, or bats (Chapter 21), and the interaction between ants and the *Acacia* plant in the tropics (Chapter 35) are examples of mutualism that have been discussed earlier in the book.

Through coevolution, the adaptations of many mutualistic partners have become functionally interlocked. The partnership between many species of termites and their intestinal protozoa, for instance, goes beyond the termite simply providing food and housing, while the protozoa digest the cellulose in wood for the termite (Figure 37-19). Coevolution has led to synchronized life cycles between these organisms. The synchrony is so precise that the internal protozoa are transmitted from one developmental stage of the termite to the next during molting. The same hormones that trigger the termite to molt also trigger the protozoa to encyst. When the termite reingests its gut lining after molting, it "reinfects" itself with its mutually beneficial partner.

**Figure 37-19**
**Mutualism: obligate partnerships with mutual benefits.** Without internal protozoa, termites would starve to death, unable to digest the wood they consume. Linked through coevolution, the protozoa obtain a habitat and food supply inside the termite's digestive system, while the termites receive a supply of usable nutrients from the digestion of wood by the protozoa.

## Synopsis

### MAIN CONCEPTS

- **The organisms of a community interact in a variety of ways.**
  Competition harms both participants.
  During predation, one organism gains nutrients by eating another.
  Parasitism is a form of interaction in which one organism lives in or on a host, damaging it in the process.
  During commensalism, one organism benefits, while the other is unaffected.
  Protocooperation benefits both participants; however, each participant is able to survive independently.
  Mutualism benefits both participants; however, neither can survive or reproduce without its partner.

- **Species can coexist in the same community as long as they have slightly different ecological niches.**

- **When the niches of two species overlap, members of those species compete for the limited resources they both require.**

- **When the ecological niches of two species are identical, competition may result in one rival eliminating or excluding the other from the community.** Intense competition may lead to partitioning a resource, or it may cause the characteristics of competing organisms to eventually change.

- **Organisms have evolved a number of adaptations that enhance their predatory skills or help them escape predators.**

### KEY TERM INTEGRATOR

Close Associations between Species

| | |
|---|---|
| **symbiosis** | A close association between different organisms, such as when organisms live together as a single unit. |

Competition

| | |
|---|---|
| **competition** | Harms participants as they try to obtain supplies of the same limited resource. **Intraspecific competition** between members of the same species is often more intense than **interspecific competition** between individuals of different species. |
| **competitive exclusion principle** | States that when two species have identical ecological niches, one species will eventually exclude the other from the community. |
| **resource partitioning** | Occurs when a limited resource is shared in some way. |
| **character displacement** | Occurs when the characteristics of direct competitors diverge over time. |
| **interference competition** | Occurs when a competitor prevents another from gaining access to a limited resource, such as in **territoriality**. |
| **exploitative competition** | Results when one competitor secures a greater share of a limited resource than the other, reducing the supply of the resource to its competitor. |

Predation and Herbivory

| | |
|---|---|
| **predation** | Occurs when a **predator** eats a **prey**. |
| **herbivory** | A form of predation where an **herbivore** eats plants. |
| **carnivores** | Organisms that eat animals. |

| | |
|---|---|
| omnivores | Organisms that eat both plants and animals. |
| camouflage | Includes **cryptic coloration**, **disruptive coloration**, and **aposematic coloring**. |
| mimicry | One species resembling another for disguise or protection. Batesian mimicry is the resemblance of a non-noxious organism (the mimic) resembling a harmful or obnoxious organism (the model), whereas Mullerian mimicry is the resemblance between two equally harmful or obnoxious organisms. |
| allelopathy | The use of harmful **allelochemicals** in defense against predators or competitors. |

**Parasitism**

| | |
|---|---|
| parasitism | Interaction in which a **host-specific parasite** secures nourishment from a host. Social parasitism occurs when a parasite exploits the behavior of a host. **Parasitoids** are larval parasites that kill their hosts. |

**Cooperative Interactions**

| | |
|---|---|
| commensalism | A type of symbiosis in which one organism benefits and the other is unaffected. |
| protocooperation | A noncompulsory (facultative) interaction that benefits both participants. |
| mutualism | Often an **obligate symbiosis** that benefits both participants. |

# Review and Synthesis

1. Organize the following community interactions.
   mutualism  protocooperation  neutralism
   predation  commensalism  competition
   parasitism  allelopathy

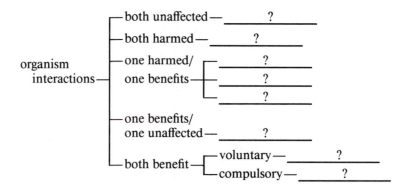

2. Consider two seedlings of different plant species growing right next to each other in a community. Both grow at about the same rate and grow roots to the same depth. List the resources for which both seedlings will compete as they grow. What is the probable outcome of this situation? What will happen if one suddenly outgrows the other?

3. Is there greater opportunity for resource partitioning in a tropical rainforest, a deciduous forest, or a desert? Why? How does each of these terrestrial biomes compare to potential resource partitioning in the Pelagic Zone of oceans?

4. Match the term with the example.
   ___A. One bee species chases away other bee species from flowers.
   ___B. Similar monkey species with different-sized teeth.
   ___C. Similar species of whales visit a feeding bay at different times of the year.
   ___D. One type of caterpillar consumes leaves faster than another.
   ___E. A plant releases chemicals that prevent other plants from growing beneath it.
   ___F. Contrasting colors distort the apparent shape of a fish as it swims through a coral reef.
   ___G. A harmless fly that looks like a stinging wasp.

   1. Mullerian mimicry
   2. Batesian mimicry
   3. disruptive coloration
   4. Allelopathy
   5. exploitative competition
   6. interference competition
   7. character displacement
   8. resource partitioning

5. Use examples to distinguish between cryptic coloration and aposematic coloring. In what ways do these adaptations help prey escape predation?

6. Monarch butterflies and their larvae do not manufacture toxic chemicals, yet both are poisonous to birds. How is this possible?

7. Under what conditions would competitive exclusion not take place in an ecosystem?

8. Why is competition rarely observed in natural ecosystems?

## Additional Readings

Barbour, M., J. Burk, and W. Pitts. 1987. *Terrestrial Plant Ecology.* Benjamin/Cummings, Menlo Park, Calif. (Intermediate.)

Brewer, R. 1979. *Principles of Ecology.* W. B. Saunders Co., Philadelphia. (Intermediate.)

Fogden, M., and P. Fogden. 1974. *Animals and Their Colors.* Crown Publishers, New York. (Introductory.)

Owen, D. 1980. *Survival in the Wild. Camouflage and Mimicry.* University of Chicago Press, Chicago. (Introductory.)

Putman, R., and S. Wratten. 1984. *Principles of Ecology.* University of California Press, Berkeley. (Intermediate.)

Tanner, O. 1978. *Animal Defenses. Wild, Wild World of Animals.* A Time-Life Television Book. Time-Life Films. (Introductory.)

**Chapter 38**

# Population Ecology

*E*lephants are among the slowest reproducers on earth. During her entire lifetime, a female elephant can give birth to only six babies. But even at that, the number of possible descendants from just one pair of mating elephants could total 5 billion ($5 \times 10^9$) after just 1000 years. After a brief 100,000 years, the descendants of one mating pair could fill the visible universe.

If such growth is possible with a slow reproducer, imagine what could happen with organisms that have faster reproduction rates—house flies, for example. In less than *one year,* the number of possible descendants from a single pair of house flies would exceed 5.5 trillion ($5.5 \times 10^{12}$)! (If you're having a hard time envisioning this number, consider this: 1 trillion seconds amounts to about 31,700 years. That means humans were in the Stone Age only 1 trillion seconds ago.)

Clearly, animals have a tremendous capacity to reproduce, and the reproductive potential of many plants is even greater. Yet the world is not tightly packed with elephants, flies, or any other kind of organism. Limited food and space, plus disease, parasitism, and predation, curb the potential number of individuals, often creating a balance between the number of individuals living in an area and the availability of resources, including space.

Predator and prey relationships illustrate such a balance. The number of prey in an area is controlled not only by the availability of its food, but also by the number of predators; more predators eat more prey. In turn, the number of prey determines how many predators can survive; more prey means more predators; fewer prey means fewer predators. Because of this reciprocal effect, a dynamic balance is often produced between the number of predators and the number of prey in an area, as shown in Figure 38-1.

## Population Structure

In most ecosystems, the community is made up of many populations, each of which consists of the individuals of the same species living in the same area at the same time. A yellow pine forest community, for example, contains populations of yellow pine trees, sugar pine trees, white fir trees, brown bears, Anna's hummingbirds, and so on.

The existence of ecotypes, genetically distinct populations of the same species, creates a dilemma in defining a species using morphology alone. This is discussed in Connections: The Species Concept.

To understand the structure and dynamics of each population, ecologists examine three fundamental properties of populations:

- *Population density* (the size of the population expressed as number of individuals in a given area at a particular time).

- *Distribution* of individuals throughout the habitat.

- *Growth rate* (increases or decreases in population density per unit of time).

## Population Density

Population density is the number of individuals of a species that live in a particular area. For example, the population density of people in Manhattan is 100,000 per square mile; that of

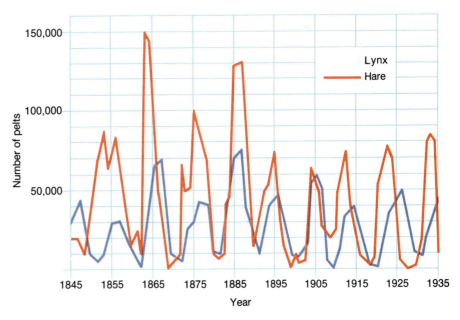

**Figure 38-1**
**The ups and downs of predator and prey populations.** The graph shows the number of lynx and snowshoe hare pelts sold by trappers to the Hudson Bay Company over a 90-year period. The lynx is a natural predator of snowshoe hares. When the number of hares increased, the number of lynx increased. As the number of lynxs grew, they ate more hares, lowering the hare population. A reduction in available prey caused a reduction in the number of predators, and so on over the 90-year period of records.

Other factors may also have affected the lynx and hare populations. Overbrowsing and fluctuations in plant growth may have altered the availability of food for the hares, while outbreaks of disease and climate changes could have affected the size of the lynx population.

sugar maple trees in Michigan is 300 per hectare (10,000 square meters); and that of dinoflagellates during a red tide is 8 million per liter of ocean water.

Sometimes ecologists simply count every individual in an area to determine a population's density. But more often the population density is estimated by counting in small, representative areas, and then extrapolating to the total area being studied. This technique is called *sampling* (Figure 38-2). To estimate the number of creosote bushes in California's Mojave Desert, for example, the number of individuals in ten randomly placed 100 m² plots were counted and averaged. The average was 30 bushes per sample plot. If we project to a 1-hectare area, we see that the population density is 3000 bushes per hectare.

## Distribution of Individuals

Although population density reveals the number of individuals in an area, it provides no information about how the individuals are arranged in space. The distribution of individuals is categorized into one of three general patterns—clumped, uniform, or random (Figure 38-3).

**Clumped** and **uniform** patterns are nonrandom distributions that result when members of a population have an effect on each other (which is almost always the case) or when environmental conditions favor growth in suitable patches. Uniform spacing re-

CONNECTIONS — Morphological Species (page 11)
Animal Versus Plant Species (page 380)
Defining Moneran Species (page 559)

## THE SPECIES CONCEPT: ECOLOGICAL SPECIES

Wide-ranging species present a problem for taxonomists because they are frequently made up of many separate populations. The genetically determined characteristics of the individuals in each population may differ slightly from those of individuals in other populations of the same species. These differences between ecotypes develop as natural selection favors different traits in different environments. The problem for the taxonomist is deciding whether the differences between ecotypes are great enough to justify establishing a new species, or whether the variants should simply be considered *subspecies* or *varieties* of the same species.

Differences in characteristics between individuals of the same species arise from two sources. First, the surrounding environment may affect the growth and development of each individual differently. Alternatively, the individuals differ genetically. Biologists agree that only genetically determined characteristics should form the foundation for defining a species and that variations due to environmental conditions do not distinguish one species from another. Therefore, scientists must determine if the different traits are genetically determined or environmentally induced.

To make this distinction, botanists perform *reciprocal transplant studies* on wide-ranging plant species. In these studies, individuals from each population are transplanted among each of the other populations that make up the species. For example, if a species is made up of populations growing along a beach, on a mountain peak, in a wet meadow, and on a dry site, individuals growing along the beach would be transplanted into the mountain, meadow, and dry site populations, and individuals from these populations would be transplanted to the beach. After the plants grow under the new set of environmental conditions, the characteristics that do not change or change in degree only (for instance, leaves may get longer, but they are still the same basic type of leaf) are identified as genetic traits that can be used to define the species.

Transplant studies and other research techniques that help scientists distinguish between genetic and environmental variability all come to the same conclusion—most species are composed of an assemblage of genetically unique ecotypes, and species with wider ranges have more ecotypes than

**Figure 38-2**
**Sampling populations.** Most often population density is estimated by random sampling. The number of individuals inside these sample plots is used to estimate the density and distribution of individuals in an entire population. In this aerial photograph, two equal-sized sample areas of different shapes show virtually the same number of each species.

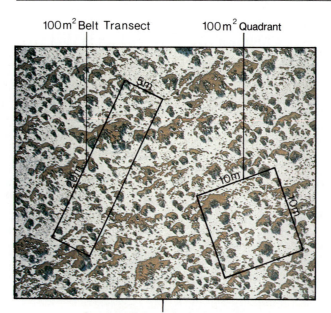

100 m² Belt Transect     100 m² Quadrant

Screened aerial photograph

those with narrower ranges. So we come back to the question raised earlier: when should ecotypes be considered variants of a single species, and when does an ecotype differ "significantly enough" to represent a new species? The answer often depends on the individual taxonomist, and taxonomists do not always agree. One taxonomist may define 20 species in a genus, while another equally qualified taxonomist identifies 40 species in the same genus. And both are right. Defining a species is a judgment call. It depends on what characteristics a taxonomist chooses to analyze, as well as where a taxonomist draws the dividing line between species. Some taxonomists, like the one who defined 20 species in our example, are called "lumpers" because they tend to view variants as subspecies or varieties. Other taxonomists, like the one who defined 40 species, are called "splitters" because they tend to view variants as separate species. This is one reason why in this textbook we sometimes included a range in numbers of species; for example, the Fungus kingdom contains between 75,000 and 200,000 species.

We have seen in this series on "The Species Concept" that the process of defining a species is not always the same; it varies according to the

- *Kind ot traits* considered—morphological, anatomical, biochemical, paleontological (fossil traits), developmental, and so on

- *Number of traits.*

- *Evolutionary relationships.*

- *Breeding success.*

- *Kingdom to which the organism belongs.*

- *Degree of variation.*

- *Judgment of the individual taxonomist.*

But, despite differences between taxonomists and approaches, a species is a natural distinction that identifies one of the millions of different kinds of organisms that populate the earth.

a

b

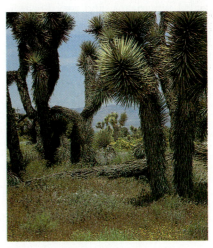

c

**Figure 38-3**
**Distribution patterns.** *(a) Clumped:* A Grove of clumped palms. A school of fish or a herd of elephants are also examples of dumped distributions. *(b) Uniform:* Oaks secrete chemicals that prevent growth of nearby oaks, creating more or less equal distances between trees. When animals defend territories, such as this hawk, individuals remain separated, producing a uniform distribution. *(c) Random:* Joshua trees have random distributions.

sults when members of a species repel one another, such as when the roots of some plants release chemicals that inhibit the growth of other members of its species, or when animals establish and defend territories. Both chemical inhibition and territoriality maintain maximum distance between members of a species, creating a more or less uniform pattern of individuals.

The most common distribution pattern is clumped, wherein individuals aggregate into groups to form groves, schools, flocks, herds, and so on. Some animals have a clumped distribution because they cooperate in societies or gain protection from predators by remaining in herds (Chapter 37). At least three factors can contribute to clumping in plants.

- Favorable conditions for germination and survival are usually not uniformly distributed throughout an ecosystem, so plants grow only in suitable patches.

- Many plant seeds are dispersed in groups.

- Asexual reproduction from runners (stolons), bulbs, branches, or rhizomes concentrates offspring near the parent plant.

**Random** spacing is the least common pattern in natural populations. For individuals to be randomly distributed, two requirements must be met: (1) the presence of one individual can in no way affect the location of another, and (2) environmental conditions must be more or less the same throughout an area. Both of these prerequisites are very rare in ecosystems.

Not all distribution patterns remain constant. In some animals, for example, seasonal changes trigger migrations, causing cyclic changes in distribution. Falling night-time temperatures and strong winds at the end of summer, for example, initiate migration behavior in several alpine tundra animals (marmots, mountain goats, mountain sheep, pikas, white-tailed ptmarigans, and rosy finches), causing entire populations to move to warmer regions. At the end of winter when conditions become less severe, the tundra animals migrate back (see Animal Behavior: Migration, Chapter 34).

## Aspects of Population Growth

### Factors Changing Population Size

Like ecosystems, populations inevitably change. Four factors cause the density of a population either to increase or decrease.

- **Natality** (births) increases density by adding new members to a population.

- **Immigration** increases density as new individuals permanently move into the area.

- **Mortality** (deaths) decreases density by removing individuals from a population.

- **Emigration** decreases density as individuals permanently move out of an area.

If the combination of natality and immigration exceeds that of mortality and emigration, the population grows and density increases. Conversely, population density decreases when the combination of mortality and emigration exceeds that of natality and immigration. **Zero population growth** occurs when the combined positive and negative factors are equal. When this happens, population density remains the same.

### Age – Sex Structure

The age of the members of a population can be used to predict natality and mortality. Red alder trees, for example, live to be about 100 years old. If the majority of red alders are older than 95 years, you can predict that the mortality rate will be high and that the population of trees will likely decline over the next five years.

Because each individual reproduces only during part of its lifetime, the ages of members of a population can also be used to predict natality. The number of individuals of a certain age and sex is used to produce an **age – sex structure** for a population. When a large share of individuals are at reproductive age (or younger), the age – sex structure tends to be shaped as a pyramid.

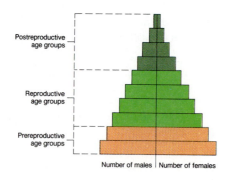

Number of males | Number of females

A broad-based population such as this one will increase in size; the broader the base, the more rapidly the population grows. On the other hand, a population with an inverted pyramid will decline because most individuals are past reproductive age.

A pyramid for a population with

zero population growth would resemble the following:

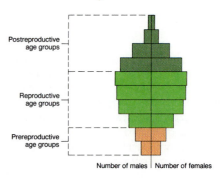

Number of males | Number of females

In this stable population, the number of prereproductive individuals balances the number of older individuals, so natality equals mortality. In this stable population either there is no migration at all, or emigration and immigration are equal.

### Mortality and Survivorship Curves

To predict population changes, ecologists must know the organism's life expectancy. When life expectancy is plotted on a graph, a **survivorship curve** is produced. Three general types of survivorship curves are shown in Figure 38-4. In a Type I curve, mortality remains low for much of the organisms' lifetimes and then increases sharply as individuals reach old age. Humans and other animals that provide long-term care for their young often have a Type I survivorship curve. A Type II survivorship curve is a straight, diagonal line, indicating that the chances of survival remain about the same throughout an individual's lifetime. Many birds and small aquatic animals have a Type II curve. A Type III curve is the opposite of a Type I: most individuals die when very young, and only a few adults survive to old age. Species that produce enormous numbers of offspring and provide no parental care (insects, frogs, and many plants, for instance) have a Type III curve.

### Biotic Potential

As we learned earlier in this chapter, all species have the capacity to eventually produce tremendously large num-

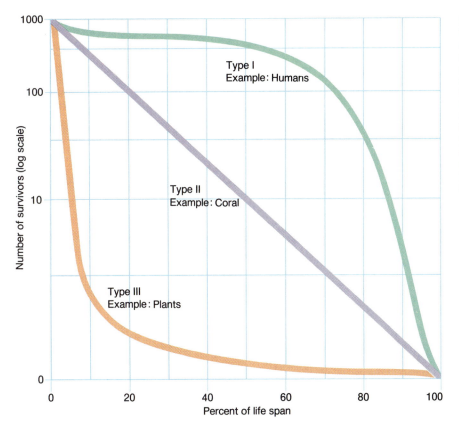

**Figure 38-4**
**Three types of survivorship curves** are produced by plotting the number of survivors (on a log scale) against the percent of the life span of a species. In the Type I curve, the mortality rate is low in the first years of life and then becomes higher at old age. Humans have a Type I curve. In the Type II curve, the chances of surviving or dying are virtually the same throughout an organism's entire life. In a Type III curve, nearly all the young die quickly, but the mortality rate is quite low for the few survivors until they reach old age. Oysters, some insects, and weedy plants have a Type III curve.

bers of descendants as long as there are no restrictions to curb population growth. This innate capacity to increase in numbers is called the **biotic potential** of a population.

When conditions are optimal and there are no limitations, the population grows at its **intrinsic rate of increase**, abbreviated $r_m$ (r = rate and m = maximum), the maximum increase in numbers of individuals per unit of time. Only occasionally do organisms find themselves in such conditions. Such rare opportunities occur when organisms colonize a new favorable habitat, when environmental conditions suddenly change for the better, or when people introduce organisms into new habitats in which there are no natural competitors or predators (Figure 38-5).

**Figure 38-5**
**Although it may have seemed like a great idea at the time,** importing plants into new habitats without natural population controls has produced some unexpected overpopulation disasters. For example, water hyacinth plants, with their orchidlike flowers and attractive stems and leaves, were brought from Venezuela to display at the 1884 New Orleans Cotton Exposition. Many exposition visitors were given clippings of the charming plant to place in ponds and streams near their homes. With no competitors, few predators, and plenty of available space and nutrients, water hyacinth spread rapidly. Today, as this photo dramatically shows, water hyacinths are choking streams, rivers, irrigation systems, and hydroelectric installations.

## Exponential Growth: The J-Shaped Growth Curve

When populations grow at their intrinsic rate of increase, the number of individuals increases exponentially. That is, the number increases by a fixed proportion, such as when a population doubles in size with each new generation—2, 4, 8, 16, 32, 64, 128 and so on. This **exponential growth** can be demonstrated by placing a single *E. coli* bacterium into a nutrient culture (Figure 38-6). The bacterium divides into two after 20 minutes; into four in another 20 minutes; and into eight cells 20 minutes later. The population continues to double every 20 minutes. After five hours, 32,768 bacteria have been produced from the original cell. After seven hours, there are more than 2 million bacteria. And after 36 hours, there would be enough bacteria to blanket the entire earth's surface with 1 foot (28.8 centimeters) of bacteria.

The growth of this bacterial culture illustrates an important point about population growth: the number of individuals in a population affects the number of offspring produced. In other words, the more reproducing individuals in a population, the greater the number of offspring. (During a 20-minute period, a population of *E. coli* increases to 2 when the population contains only the original bacterium, but during the same 20-minute period, the population can jump to 2 million when the population starts with 1 million bacteria.) If we express this principle mathematically, we see that the rate of exponential growth (G) equals the intrinsic rate of increase ($r_m$) multiplied by the number of individuals in the population (N):

$$G = r_m N.$$

The pattern of exponential growth is always the same: the size of a population increases gradually at first and then grows larger and larger in progressively shorter periods of time as more and more individuals are added to the population. Plotting exponential growth on a graph produces a curve that resembles the letter J, and so, not surprisingly, it is called a **J-shaped curve**. As Figure 38-6 illustrates, once a population "rounds the bend" of a J-shaped curve, the number of individuals added to a population begins to skyrocket.

## Environmental Resistance and Carrying Capacity

No natural ecosystem can support continuous exponential growth for any species; no ecosystem has unlimited resources, and environmental conditions never remain constant. Eventually, some environmental limitation will restrict population growth. The factor(s) that eventually limit the size of a population create an **environmental resistance** to population growth. Environmental resistances include such factors as competition, predation, hostile weather, limited food or water supply, restricted space, depleted soil nutrients, and the buildup of toxic byproducts from the organisms themselves.

The combined limitations of the environment form a ceiling for the number of individuals that can be supported in an area. The size of a population that can be supported indefinitely is called the **carrying capacity** (abbreviated **K**) of the environment.

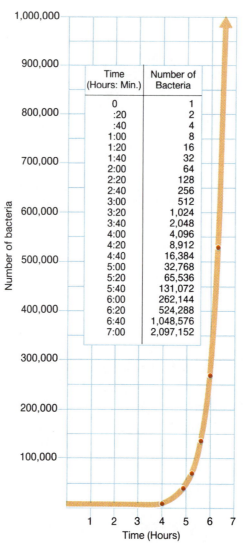

| Time (Hours: Min.) | Number of Bacteria |
|---|---|
| 0 | 1 |
| :20 | 2 |
| :40 | 4 |
| 1:00 | 8 |
| 1:20 | 16 |
| 1:40 | 32 |
| 2:00 | 64 |
| 2:20 | 128 |
| 2:40 | 256 |
| 3:00 | 512 |
| 3:20 | 1,024 |
| 3:40 | 2,048 |
| 4:00 | 4,096 |
| 4:20 | 8,912 |
| 4:40 | 16,384 |
| 5:00 | 32,768 |
| 5:20 | 65,536 |
| 5:40 | 131,072 |
| 6:00 | 262,144 |
| 6:20 | 524,288 |
| 6:40 | 1,048,576 |
| 7:00 | 2,097,152 |

**Figure 38-6**
**Exponential growth** of an *E. coli* bacterial culture produces a J-shaped growth curve. After rounding the "bend" of the J, the curve gets steeper and steeper until some limitation, such as depletion of food or the buildup of contaminating waste products, curbs further growth.

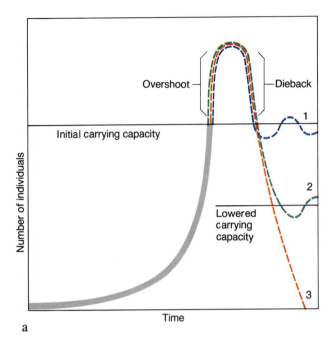

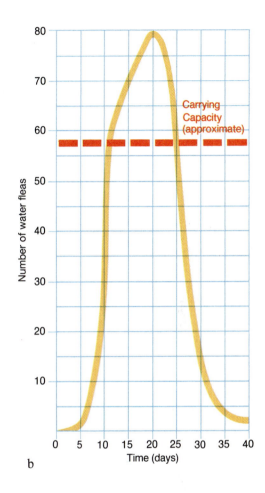

**Figure 38-7**
**The risks of overpopulation.** *(a)* A population undergoing exponential growth may overshoot the carrying capacity of the environment. When it does, the population experiences a dieback to one of three levels: (1) to the original carrying capacity (the least common outcome); (2) to a population density at a lower carrying capacity; or (3) to extinction or near extinction levels. *(b)* A natural population of water fleas *(Daphnia)* illustrates fate #3.

Populations undergoing exponential growth sometimes overshoot the carrying capacity before environmental resistances curb their growth. The greater the reproductive momentum, the more likely the population will exceed the environment's carrying capacity.

Both laboratory experiments and field observations identify three possible fates for populations that overshoot the carrying capacity of their environment (Figure 38-7). All begin with a *dieback*—the death of a proportion of the population. The least dramatic of the three fates is also the least common: the population simply dies back to the level of the original carrying capacity (curve 1 on the graph). However, in most cases the excess population damages the environment in some way, reducing its carrying capacity. For example, an excessively large population of caterpillars might consume all of the leaves on a plant, weakening or possibly killing the plant. When the carrying capacity is lowered, the population plunges. If the damage caused by overpopulation is not too severe, the population eventually comes into balance with a lower carrying capacity (curve 2). But when damage is extensive or when a vital resource is drastically depleted, the population crashes to a very low level or disappears altogether (curve 3).

Occasionally, a population exceeds the carrying capacity because a limiting factor does not take effect until a threshold level is reached. The population continues to grow exponentially until it reaches the threshold level, and then the factor causes a sharp decrease in population size. For example, the gradual buildup of toxic waste products may have no effect on a population until these chemicals reach some critical level, at which point large numbers of the individuals die. Some biologists warn that this might be the case for the human population, as air, soil, and water become more and more polluted.

## Logistic Growth: The Sigmoid Growth Curve

An alternative to unbridled exponential growth occurs when environmental resistance increases as a population approaches the carrying capacity of the environment, causing growth to slow down. When this happens, a **sigmoid growth curve** is produced (Figure 38-8). The S-shaped, sigmoid curve begins the same as the J-shaped curve;

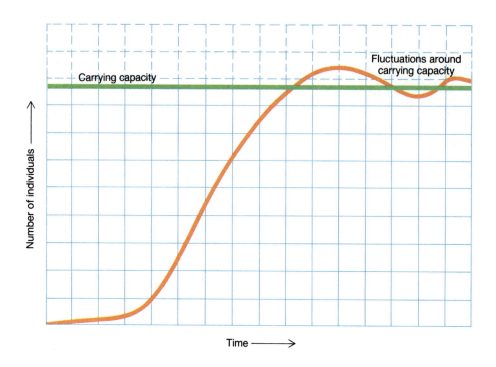

**Figure 38-8**
**The sigmoid growth curve.** *(a)* In a new habitat, a population increases slowly at first and then undergoes exponential growth until environmental resistances begin to reduce growth rates and stabilize population density around the carrying capacity. This common logistic growth pattern produces a sigmoid growth curve. *(b)* Sheep were introduced on Tasmania in the early 1800s. The population growth pattern produced a sigmoid growth curve, with the population oscillating between 1.5 and 2.0 million.

that is, numbers of individuals increase slowly at first, and then as the reproductive base builds, the population goes into a period of exponential growth. But unlike the result during logistic growth, the rate of growth gradually slows down as the population approaches the carrying capacity, forming the top of the S-shaped curve. Population density eventually oscillates around the carrying capacity: when the population slightly exceeds the carrying capacity, environmental resistance intensifies and causes a slight dieback; when the population falls below the carrying capacity, environmental resistance relaxes and the population gradually increases. In this way, the size of a population remains fairly stable, and the growth curve flattens out at the carrying capacity level. Populations that produce a sigmoid growth curve have **logistic growth**.

## Reproductive Strategies: r and K Selection

Just as natural selection favors the selection of physical traits that improve survival (camouflage, for example), so too are reproductive traits naturally selected. Over the course of evolution, species have evolved a range of reproductive strategies that offer advantages in different types of habitats. For example, species with the ability to reproduce large numbers of offspring will most likely dominate a disturbed or uninhabited area before slower reproducing species can do so. But in stable, climax communities (see Chapter 36) where competition for resources is intense, species that provide long-term care to a few offspring often have a greater chance of success than species that produce many offspring and provide no care.

Species that produce many offspring at one time are said to be **r-selected**, "r" referring to a high intrinsic rate of increase ($r_m$). In other words, r-selected species have adaptations that maximize $r_m$. Species that produce one or a few well-cared-for individuals at a time are said to be **K-selected**, "K" referring to strategies more favorable for populations near the carrying capacity of the environment. These opposite strategies are the extremes in a continuum of reproductive strategies found in organisms. Reproductive strategies have a number of components. The main components of reproductive strategy and their relation to r-selected and K-selected species are presented in Table 38-1.

Differences between r- and K-selection strategies become very important when we consider the fate of rare, threatened, or endangered species, the three categories of species that are the most probable candidates for extinction. (A rare species is one with only a few individuals, whereas endangered and threatened species are near extinction because of human activities.) For instance, it took less than 100 years for the passenger pigeon to decline to extinction, even though there were many millions of pigeons in the early 1800s. But because the passenger pigeon was K-selected—females laid only one egg each year—the production of offspring could not keep pace with the enormous number of pigeons killed by sportsmen. Thus, the species quickly died out.

With an innate ability to reproduce quickly, r-selected species may not become extinct as quickly as K-selected species. However, no amount of reproduction can save a species if its environment is destroyed or severely contaminated.

## Factors Controlling Population Growth

### Density-Dependent Factors

All interactions in an ecosystem—between the organisms themselves and between organisms and their physical environment—can influence population size. Some factors that affect population growth increase or decrease in intensity as the size of the population changes. For example, in locusts, when population density is low, young locusts develop normal-length wings. But when locust density is high, hormonal changes trigger the development of longer wings in offspring. Long wings increase emigration, which in turn reduces the population density in that area. Such factors that are influenced by the number of individuals in the population are called **density-dependent factors**.

TABLE 38-1

| **Ranges of Reproductive Strategies*** | | |
|---|---|---|
| | **r-Selected** | **K-Selected** |
| Number of offspring | many | few |
| Number of times an individual reproduces | once | many times |
| Size of young | small | large |
| Rate of development | fast | slow |
| Parental care | minimal | intensive |
| Life span | short | long |
| Survivorship curve | Type III | Type II | Type I |
| Energy to reproduction | high | low |
| Energy to increasing body size | low | high |
| **Examples** | | |
| **Animals** | oysters | insects | birds | elephants | humans |
| **Plants** | weeds | saguaro | oaks | pears | mangroves |

* r-selected and K-selected strategies are at the extremes of each continuum.

There is a direct relationship between the intensity of density-dependent factors and population size: as population density rises, the intensity of density-dependent regulatory mechanisms increases, dampening population growth by reducing natality or by boosting mortality or emigration (as in locust populations). Conversely, when population density falls, density-dependent mechanisms decrease in intensity, allowing population growth to accelerate. (This explains why there are small fluctuations around the carrying capacity of the environment during logistic growth.)

### Density-Independent Factors

But not all regulatory factors are affected by the size of a population. Factors that are not influenced by population size are called **density-independent factors**. An earthquake is an example of a density-independent factor because the population density neither caused the earthquake nor affected the magnitude of the quake or the percentage of individuals killed. Many catastrophic events, including fires, floods, hurricanes, tornadoes, volcanic eruptions, and avalanches, are density-independent factors.

Density-dependent and density-independent factors often combine to regulate the size of a population. For example, the number of aphids feeding on a sycamore leaf is affected not only by competition (a density-dependent factor that determines how many aphids may feed on the same leaf vein), but also by wind velocity (a density-independent factor). As winds increase, sycamore leaves brush up against one another, scraping off the aphids.

Laboratory studies on overcrowding in rats show that crowding increases aggressive behavior, delays sexual maturation, reduces sperm production in male rats, and causes irregular menstrual cycles in females. Overcrowding also reduces sexual contacts between males and females while increasing homosexual contact between males. Such density-dependent factors quickly curb population growth by reducing natality. Whether such dramatic effects occur in populations outside the laboratory or in other

species is not yet known. However, many demographers (scientists who study human population dynamics) believe that increased crime, drug abuse, and suicide are the result of overcrowding—a situation that worsens as the world human population swells by more than 250,000 people each day (a *net* increase of 3 persons per second).

## The Human Population Explosion

In 1987, the world human population reached 5.0 billion people. (A billion is a very large number; to travel 1 billion inches you would have to walk across the United States five times.) More than 380,000 people are born each day (over 4 babies every second), and about 130,000 people die each day. This means the world human population is growing by some 250,000 people every day (birth rate − death rate = growth rate in populations where immigration and emigration are absent). At this rate, almost 100 million people are added to the human population every year. What will the

future be like for those babies born into an exploding human population?

The answer has a great deal to do with where these babies are born. If these children are born in the United States in 1989, many will begin kindergarten in 1994, enter college around 2007, and be eligible to retire in the year 2054. By the time they enter college, the world human population will have swelled to more than 8 billion. By the time they retire, the world population will have climbed to 17 billion people, more than three times the number of people living on earth today—that is, unless human endeavor or density-dependent controls change the human population growth rate.

### Historical Overview

The first humans appeared sometime between 1 and 2 million years ago. The size of early human populations remained in balance with available food supply, increasing slowly as our early ancestors spread out from Africa and discovered new lands.

When the size of the human population is plotted against time on a log scale, as in Figure 38-9, three surges of

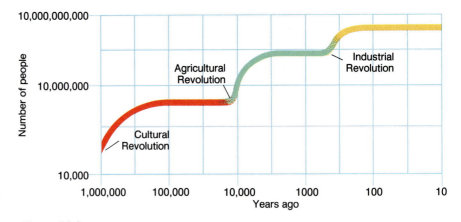

**Figure 38-9**
**Surges of human population growth** become clear when the size of the human population is plotted against time on a log-log scale. (In log-log graphs, both population size and time are plotted on log scales, allowing huge numbers of individuals and enormous time spans to be concentrated on a single sheet of graph paper.) Bursts of exponential growth occurred on three occasions corresponding to the cultural, agricultural, and industrial revolutions. What revolution do you predict will be next, and how will it affect human population growth?

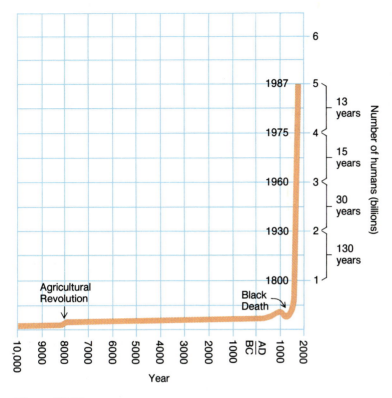

**Figure 38-10**
**The human population growth curve.** The rapid exponential growth of the human population is obvious in this graph. By the year 2000, there will be 8 to 14 billion people living on earth if our population continues to grow at its present rate.

TABLE 38-2

| Percent Growth Rate for Various Countries | |
|---|---|
| **Country** | **Percent Annual Increase** |
| Developing countries | |
| India | 2.3 |
| Brazil | 2.9 |
| Uganda | 3.3 |
| Mexico | 3.5 |
| Kenya | 4.1 |
| Developed countries | |
| United Kingdom | 0.0 |
| United States | 0.6 |
| Japan | 1.1 |

population growth become evident. The development of culture and the use of tools were responsible for the first population surge about 600,000 years ago. The development of agriculture (cultivating food in a village-farming society) produced the second population surge about 10,000 years ago. The most recent growth spurt occurred 200 years ago as humans entered the industrial revolution. Along with industry, humans continued to make advances in agriculture, medicine, and hygiene, the combination of which decreased the death rate and increased the average life span. These advances lessened former controls on human population growth, mainly food shortages and disease.

The overall growth curve for the world human population, shown in Figure 38-10, is clearly a J-shaped curve. It took some 2 million years for the world human population to reach 1 billion people. However, after a reproductive base of 1 billion people was established in 1800, it took progressively less time to add another 1 billion people to the human population—130 years to reach 2 billion, 30 years to reach 3 billion, 15 years to reach 4 billion, and then only 13 years to reach 5 billion in 1987.

GROWTH RATES AND DOUBLING TIMES

The current human birth rate is 27.7 babies per 1000 people per year. The death rate is 9.5 people per 1000 per year, making the current human population growth rate equal to 18.2 humans per 1000 people per year:

$$27.7 - 9.5 = 18.2$$
(birth   (death   (growth
rate)    rate)    rate)

Population increases are also expressed as the number of people added to the population per 100 individuals. This gives the **percent annual increase** in population. The average annual increase for all nations is currently 1.8 percent (see Table 38-2).

An annual increase of 1.8 percent means it will take less than 56 years for the world's population of people to double. (If you add 1.8 people to a population of 100 each year, it will take 55.6 years for the population to

TABLE 38-3

| | Estimated Human Population (bil.) | Doubling Time (mil.yrs.) | Percent Annual Increase |
|---|---|---|---|
| **Human Population Size Estimates, Doubling Times, and Percent Growth Rates** | | | |
| **Date** | | | |
| Pre 8000 B.C. | | | 0.0007 |
| 8000 B.C. | 0.005 | 1500 | 0.0015 |
| 1650 A.D. | 0.500 | 200 | 0.1 |
| 1850 A.D. | 1.000 | 80 | 0.8 |
| 1930 A.D. | 2.000 | 45 | 1.9 |
| 1975 A.D. | 4.000 | 35 | 2.0 |
| 1987 A.D. | 5.000 | 39 | 1.8 |

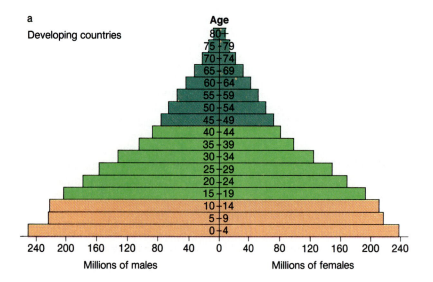

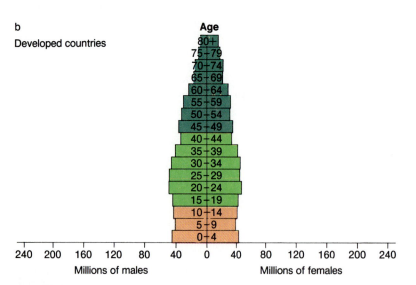

**Figure 38-11**
**Age–sex structure diagrams** for developing *(a)* and developed *(b)* countries differ radically. With a tremendous reproductive base of prereproductive and reproductive aged groups, developing countries will continue to grow rapidly while developed countries will decline in population. With three-quarters of humans living in developing countries, the world human population will continue its perilous rise.

reach 200 [55.6 × 1.8 = 100]). But as we learned in an earlier section in this chapter, population size affects the rate of increase; thus, a growth rate of 1.8 actually has a doubling time of only 39 years instead of 55.6 because of the large reproductive base. Doubling times for the world human population are given in Table 38-3.

## CURRENT AGE–SEX STRUCTURE

Figure 38-11 illustrates that there is a big difference between the age–sex structure of developing nations and that of developed countries. Developing nations have rapidly growing populations because they have large numbers of individuals moving into their reproductive years. Because developing countries are heavily populated, their effect on the world human population growth is great. As a result, the world human population will continue to increase at a rapid rate.

## FERTILITY RATES AND POPULATION GROWTH

Although age–sex structure diagrams help predict population growth, they do not take into consideration another factor that affects human population growth—**fertility rate**, the average number of children born to each woman between 15 and 44 years of age (the reproductive years). The average fertility rate for all nations in the world is now slightly below 2.1 births per woman. In developed countries, fertility rates average 1.9, compared to 4.5 in developing nations.

Given current mortality rates, fertility rates of between 2.1 and 2.5 are required to maintain zero population growth. With an average fertility rate of 1.9 in developed countries, populations in these countries will decline, while developing countries with an average fertility rate of 4.5 will rapidly increase. (In 1987, 92 percent of the human population growth occurred in developing nations.) As a result, the world human population will continue to increase by greater and greater numbers each year. Some scientists believe the world human population will begin leveling off in 30 to 40 years to between 8 and 14 billion people, while other scientists believe that human population will not reach these

# B I O L I N E

## THE HUMAN POPULATION EXPLOSION AND THE FUTURE OF LIFE ON EARTH

The ever-burgeoning human population is reducing the ability of the earth to support life.

Lowering the earth's carrying capacity affects not only millions of other organisms, but ultimately humans as well, for like all other organisms, humans are part of the biosphere.

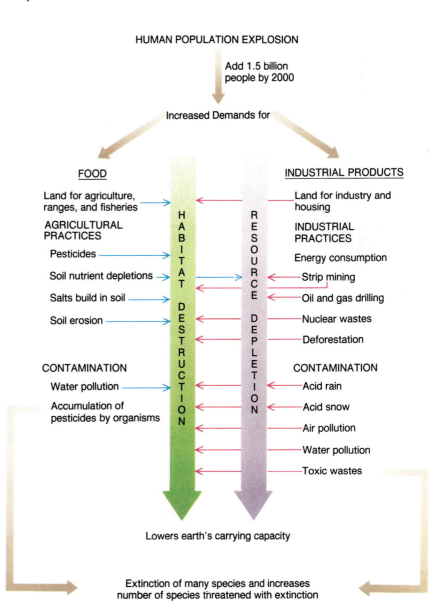

HUMAN POPULATION EXPLOSION

Add 1.5 billion people by 2000

Increased Demands for

FOOD

Land for agriculture, ranges, and fisheries

AGRICULTURAL PRACTICES

Pesticides

Soil nutrient depletions

Salts build in soil

Soil erosion

CONTAMINATION

Water pollution

Accumulation of pesticides by organisms

HABITAT DESTRUCTION

RESOURCE DEPLETION

INDUSTRIAL PRODUCTS

Land for industry and housing

INDUSTRIAL PRACTICES

Energy consumption

Strip mining

Oil and gas drilling

Nuclear wastes

Deforestation

CONTAMINATION

Acid rain

Acid snow

Air pollution

Water pollution

Toxic wastes

Lowers earth's carrying capacity

Extinction of many species and increases number of species threatened with extinction

We have learned how important photosynthetic plants are to life on earth; they supply chemical energy and nutrients to virtually all heterotrophs, humans included. With only 15 staple plant species standing between humankind and starvation, it is imperative that we minimize the damage to our environment. If just one of these staple crops were to become extinct, millions, or possibly billions, of people might perish.

Even simple, microscopic organisms like bacteria, fungi, and nematodes are important to human life (and life in general), for they contribute to the recycling of nutrients. Many organisms provide medical and agricultural remedies. The extinction of species not only reduces ecological diversity, but also cuts into the genetic bank, a bank of genes we have tapped a number of times to solve medical, industrial, and agricultural problems.

---

levels because we are very close or have already exceeded the earth's carrying capacity for humans.

## The Earth's Carrying Capacity and Future Population Trends

Density-dependent controls are already beginning to curb human population growth (see Bioline: Human Population Explosion). Between 10 and 20 million people—mostly children—die each year from starvation or malnutrition-related diseases. Even in affluent countries, overpopulation has accelerated the rate of environmental deterioration, lowering the quality of life and most likely reducing the environment's carrying capacity. Is it possible that the world human population is already in overshoot? If so, what lies ahead: a dieback to the original carrying capacity, a dieback to a lower carrying capacity, or a dieback to extinction (review Figure 38-7)?

Many biologists believe we can still avoid these undesirable consequences through efforts to curb population growth and protect the environment. These are the challenges now facing us. Biotechnological advances for increasing food production and methods for disposing of wastes that otherwise contaminate the biosphere (or other advances that would help control growth rates) may someday help us avert the disasters of overpopulation.

# Synopsis

## MAIN CONCEPTS

- **Communities are made up of populations.** Each population contains all of the individuals of the same species that occupy an area at the same time.

- **Every population has the potential to reproduce large numbers of offspring.**

- **A population growing at its maximum rate increases in size exponentially and produces a J-shaped growth curve.** Eventually, some factor or combination of factors (limited food or the buildup of toxic wastes) halts continued exponential growth, causing a population dieback, sometimes to extinction.

- **For most populations, growth is slow at first, increases rapidly, and then slows and levels off** at the carrying capacity of the environment, where it remains in dynamic balance with available nutrients and conditions.

- **Populations grow when natality and immigration exceed mortality and emigration.** In populations where there is no immigration or emigration, the rate of growth equals the difference between the birth rate and death rate.

- **Organisms have evolved a broad range of reproductive strategies,** from species that produce enormous numbers of offspring at one time to those that reproduce one or just a few offspring and provide them with extended care. Each strategy has advantages in certain environments.

- **The world human population reached 5 billion in 1987** and is continuing to grow exponentially, mainly from growth in developing countries.

## KEY TERM INTEGRATOR

### Population Structure

| | |
|---|---|
| **population** | All individuals of the same species that live in an area at the same time. Within a population, the distribution of individuals may be **clumped**, **uniform**, or **random**. |

### Population Growth

| | |
|---|---|
| population factors | **Natality** and **immigration** add new individuals to a population, and **mortality** and **emigration** subtract individuals from a population. All four factors contribute to changes in **population density**. **Zero population growth** results when the combination of natality and immigration equals that of mortality and emigration. |
| **age–sex structure** | Together with **survivorship curve** enable ecologists to predict natality and mortality. |
| **biotic potential** | The innate capacity of a population to produce large numbers of offspring. Under ideal conditions, this biotic potential results in a population to grow at its **intrinsic rate of increase ($r_m$)**. |
| **exponential growth** | Growth of populations at their intrinsic rate of increase. Plotting exponential growth on a graph produces a **J-shaped** growth curve. |
| **environmental resistances** | Limit population growth and establish a **carrying capacity**, the maximum population density that can be maintained indefinitely. |
| **sigmoid growth curve** | Produced by populations with **logistic growth**. Population density gradually levels off at the carrying capacity of the environment. |
| **r-selected, K-selected** | Reproductive strategies that are opposite extremes in the way species reproduce. |

---

Population Controls

---

| control factors | Factors that control population growth are either **density-dependent factors** if their intensity varies with population density, or they are **density-independent factors** if their intensity is not influenced by the size of the population. |

---

Human Population Growth

---

| **percent annual increase** | Equals the number of people added per 100 individuals per year. |
| **fertility rate** | Equals the average number of children each woman gives birth to during her reproductive years. |

---

# Review and Synthesis

1. Match the process (numbered column) with the natural outcome (lettered column).

   ——1. logistic growth     A. J-shape growth curve
   ——2. intrinsic rate of    B. Sigmoid growth curve
        increase
   ——3. exponential      C. a population dieback
        growth

2. Put a check next to the conditions that would cause a population to *decrease* in density.

   ——1. Natality that greatly exceeds mortality, immigration, and emigration.
   ——2. Pyramid-shaped age–sex structure where most individuals are at reproductive age.
   ——3. A Type III survivorship curve, high emigration, and high mortality.
   ——4. A population that has exceeded the carrying capacity of the environment.

3. In the following habitats, which reproductive strategy would be of greater advantage—K-selected species or r-selected species? Give a brief explanation for your answer.

   | Environment | r or K Strategy | Explanation |
   |---|---|---|
   | cool lava | | |
   | burn area | | |
   | climax rainforest | | |
   | abandoned farm | | |
   | desert | | |
   | streamside | | |

4. Listen to the evening news tonight and make a list of density-dependent and density-independent factors that have affected the world human population. (For example, a plane collision with another plane because of crowded airways would be a density-dependent factor, whereas a plane crash into a mountain because of faulty equipment would be a density-independent factor.)

5. Has the world human population exceeded the carrying capacity of the earth? Present an argument in favor of or against this question based on your own experience and knowledge. Imagine you can look 10, 100, and 1000 years into the future. How many humans do you predict there will be on earth, and what will life be like for them? Propose a strategy, other than increasing the earth's carrying capacity, that might prevent a massive and catastrophic dieback.

# Additional Readings

Ayensu, E., V. Heywood, G. Lucas, and R. Defilipps. 1984. *Our Green and Living World. The Wisdom to Save It.* Cambridge University Press, Cambridge, London. (Introductory.)

Council on Environmental Quality, 1981. *Global Future: Time to Act. Report to the President on Global Resources, Environment and Population.* U.S. Government Printing Office, Washington, D.C. (Introductory.)

Ehrlich, P., and A. Ehrlich. 1972. *Population, Resources, Environment. Issues in Human Ecology.* W. H. Freeman, San Francisco. (Introductory.)

Ehrlich, P., and A. Ehrlich. 1981. *Extinction.* Random House, New York. (Intermediate.)

Haupt, A., and T. Kane. 1978. *Population Handbook.* Population Reference Bureau Inc., Washington, D.C. (Introductory.)

Population Reference Bureau, Inc. 1972. *The World Population Dilemma.* Columbia Books, Washington, D.C. (Introductory.)

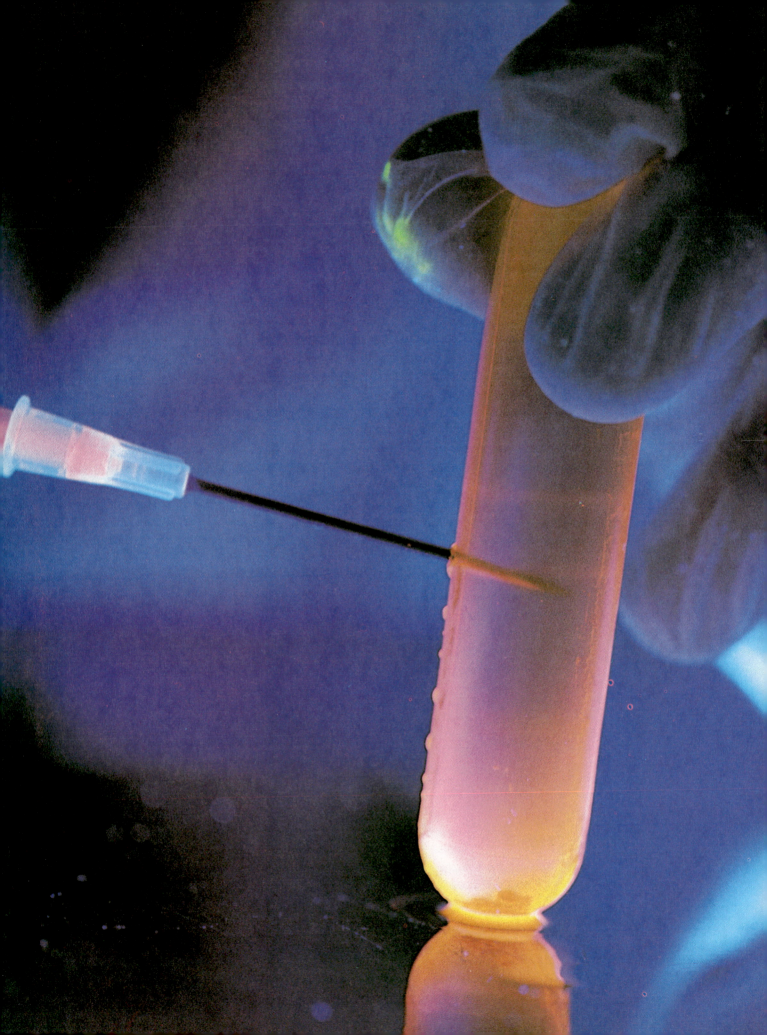

# 11

# *Biotechnology*

We are entering a new age of biotechnology, in which, biologists are able to genetically engineer microorganisms to produce valuable chemicals. In this photo, a biologist carefully extracts the DNA that will be used to genetically engineer organisms.

Chapter 39

# The Biotechnology Revolution

Watchers of Wall Street were stunned. The upstart company's stock began to sell the moment the opening bell launched the market day. Regarded by many as a novelty stock, it was the first company of its kind to be listed on the stock exchange, and its only products were still unmarketable medicines years away from federal approval. Yet the stock caused a frenzy of buying and selling, and by the end of its first day was worth $80 million.

The company was Genentech, the world's first stockholder-backed *biotechnology* organization, a company devoted to applying some of the modern lessons of biology. Its very existence is an offshoot of the scientific breakthroughs of a group of geneticists who are creating new ways of putting organisms to work for people and the environment. In doing so they have shocked industrial leaders, members of the news media, and even representatives of the legal community. (The U.S. Supreme Court ruled in 1980 that individuals can patent "life." They can literally own the rights to the new life forms "invented" in laboratories by genetic engineers.) These scientists are working at the leading edge of the "biotechnology revolution."

**Biotechnology** is a term that defies simple definition. Some people equate it with the new field of genetic engineering, while others take a broader viewpoint, defining it as any application of biological knowledge. This broad definition would encompass an enormous number of endeavors, from agriculture to modern genetic engineering (Figure 39-1). To reduce confusion, we will limit our interpretation to the two areas most often equated with biotechnology. One of these, the genetic engineering of organisms, is the endeavor that inspired the coinage of the term "biotechnology" in the 1970s. The other area consists of recent developments in the fields of tissue (and cell) culture, most notably those that have enabled us to fuse two different eukaryotic cells into a single cell that possesses the combined properties of both.

Although modern biotechnological innovations may herald the dawn of a "new age" in scientific and technological achievement, the roots of the field

a

b

c

**Figure 39-1**
**Practical applications of biology** range from agriculture to modern biotechnology. Here are a few. *(a) Agriculture* is one of the original uses of organisms by people. It includes cultivation of crops (including ornamental plants, such as the flowers in this picture) and raising of livestock animals. *(b) Biopesticides* use organisms to control pests. Although the most common approach is to release microorganisms that infect and kill the pest, the unusual technique shown here exploits the tendency of mosquito larvae to stick to mustard seeds and drown. *(c) Genetic engineering* is often equated with modern biotechnology. The golden flask in this photo contains a pure preparation of a nucleotide, providing one of the building blocks for making synthetic genes used for genetic engineering.

are thousands of years old, dating back to the beginning of practical biology.

## Practical Biology— A Brief Perspective

**Applied biology** is the application of biological knowledge to achieve practical objectives. Again, this creates conflicting opinions. Did applied biology begin with our first use of organisms or when we first applied our knowledge of biology to enhance our success at using organisms? The production of alcoholic beverages by fermentation typifies this conflict. Fermentation may have been the first endeavor in practical biology (Figure 39-2). Alcoholic fermentation was discovered by Egyptians thousands of

years before yeasts were shown (in the mid-1800s) to be responsible for the process. Modern biological research, however, is revealing methods for improving the performance of these microbes. Consequently, fermentation of alcoholic beverages is now indisputably a branch of applied biology.

Advances in applied biology are occurring at too rapid a rate to list them all here. Instead, this chapter provides a "snapshot" of the field, with emphasis on the modern methods of biotechnology. This glimpse will help you gain some insight into biotechnology —its future, its benefits, and its challenges. Biotechnologists are ushering in a new age. Like the space age, its promise is still much in the future, so we must discuss "what may be" as well as "what is". No one will remain untouched by the effects of this new bio-

logical technology. Modern biotechnology has been compared to the discovery of fire in its anticipated magnitude of importance to our species.

Before we discuss the potential of biotechnology, however, a look at the more conventional applications of biology will reveal "what already is" and how this existing technology influences your life.

## Conventional Applications of Organisms and Their Activities

Ancient uses of organisms to provide practical services were not limited to brewing, or even to bread baking, which uses alcoholic fermentation by yeasts to leaven bread dough. (The gas byproduct of fermentation, $CO_2$, forms bubbles that literally inflate the dough prior to cooking.) Like these two, other ancient techniques are still in use today. In pre-refrigeration days, microorganisms were unknowingly used to help slow decomposition of food. As noted in Chapter 31, yogurt, cheeses, sauerkraut, soysauce, and vinegar are all products of microbial growth that are more resistant to spoilage than are the unprocessed foods from which they are made. The microbes also impart pleasing tastes, smells, and textures to the preserved food. The blue stripes running through roquefort cheese, for example, are ribbons of the mold *Penicillium roqueforti,* and the "holes" in a slice of swiss cheese are trapped bubbles of $CO_2$ that are generated by the bacterium *Propionibacterium* as it transforms milk curd into cheese. Even chocolate depends on processing by microorganisms before becoming what "chocoholics" consider *the* most essential product of applied biology. Algae are used in growing our food, either indirectly (dried kelp—*Macrocystis*— used as fertilizer) or directly by growing them as a dietary supplement. The most commercially successful of these supplements is not really an alga at all, but the edible cyanobacterium *Spirolina.* An acre of this "blue-green alga"

**Figure 39-2**
**The dawn of practical biology?** Some biotechnologists believe that their field began as an accident some 6000 to 8000 years ago. The dusty whiteness that coats these grapes is a natural layer of the yeast, *Saccharomyces cerevisiae,* which ferments grape juice into wine. Two thousand years later, the same yeast was found to cause bread dough to rise. Consequently, the yeast (see inset) goes by the twin nicknames "brewer's yeast" and "baker's yeast." Brewers and bakers, however, had no idea that the agent was a living organism until the mid-1800s.

produces 40 times more protein than does an acre of soybeans. These and some other conventional applications of biological knowledge are listed in Table 39-1.

Although the newer "revolutionary" applications are presented in the upcoming section on biotechnology, many of the conventional techniques may seem surprisingly modern. For example, using bacteria to extract otherwise unobtainable copper from low grade ore sounds like a recent biological application. Instead of discarding this poor ore, it is piled into enormous mountains and soaked with a "leaching solution" containing rock-dissolving bacteria. The microbial attack releases copper ions from the rock, and the dissolved ions are carried in the solution to recovery facilities that precipitate the ions into pure copper. The technique is very successful, providing 10 percent of the copper mined in the United States. Yet this "modern" technique is ten centuries old, invented by Egyptians who lacked any knowledge of the existence of bacteria. Other metals, such as uranium, are also mined using bacteria.

Some of the more recently developed biological applications are helping protect the environment from serious disturbances, for example, those that result from spraying our crop fields with toxic chemical pesticides. **Biological pesticides**, organisms that reduce pest populations, are beginning to replace chemicals in the war against crop-eating insects that annually destroy one-third of the earth's human food supplies. (This loss continues even with the use of chemical pesticides.) Promising strategies include introducing predators that eat the pests or spraying a field with microorganisms that specifically kill the pest. These microbes spare the "friendly" insects and spiders that also help control pest populations. Another approach interferes with reproduction of a particular pest. Spraying a field with insect pheromones creates "chemical static" that interferes with the ability of the insects to locate partners for mating. Unable to reproduce, the pest population dies after one generation. These strategies may someday eliminate the practice of showering the environment with insect-killing chemicals that are toxic to many organisms other than the pest.

Another way to avoid pests is to grow the "crop" in closed containers. Although this is impractical for crop plants, some strains of bacteria may someday replace certain crop plants. One species of the bacterium *Acetobacter,* for example, produces cellulose fibers that can be used to spin real cotton, providing a new source of the material (Figure 39-3). Growing "cotton" on culture media in factories could stop the infamous boll weevil's devastating influence.

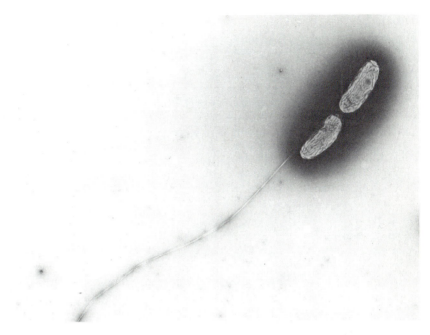

**Figure 39-3**
**Microscopic cotton spinner.** The bacterium *Acetobacter xylinum* naturally produces long fibers of cellulose, forming threads that can be woven into cotton fabric.

## Biotechnology — The Modern Revolution

In 1973, a team of scientists transferred a gene from a mammal into a bacterium. The bacteria followed the "foreign" gene's instructions just as if it were its own native gene. Equally important, all descendants of the bacterium received the "foreign" gene and expressed the mammalian trait. Scientists had "cloned a gene."

This innocuous-sounding achievement launched the modern biotechnological revolution. Gene cloning is the cornerstone of a field popularly called "genetic engineering." Most genetic engineering techniques require that the gene for the desired trait first be isolated (or synthesized in the laboratory), and then introduced into the organism being engineered. DNA recombination stabilizes the new gene so that it becomes a permanent part of the recipient's genetic library. In other words, geneticists can now (to a very limited extent) engineer the genotype of an organism, sometimes producing a creature with a combination of traits that has never occurred naturally in any organism.

But biotechnology refers to more than just genetic engineering. Another breakthrough that spurred the biotechnology revolution was the creation of **hybridoma technology**. Hy-

TABLE 39-1

| A Sampler of Biology's "Classical" Applications* | |
| --- | --- |
| **Application/Product** | **Role of Organism** |
| Cheese | Fungi or bacteria ripen milk curd; microbe provides characteristic flavor of cheese. |
| Fermentation | Yeasts ferment sugar to alcohol, producing wine or beer; bacteria ferment fruit juice to acetic acid (vinegar). |
| Yogurt | Fermentation byproducts produced by bacteria growing in skim milk provide characteristic texture and flavor. |
| Soysauce | Bacteria and fungi ferment wheat and soybean curd. The process takes about one year. |
| Food supplements | Microbes produce single-cell protein, vitamins, amino acids, and flavor enhancers. |
| Medicines | Microbes produce infection-fighting antibiotics and cancer-fighting chemotherapeutic agents; chemicals from fungi (cyclosporin) suppress rejection of transplanted organs; plants and lichens supply raw materials for many drugs. |
| Dyes | Dyes are made from pigments extracted from plants and lichens. |
| Hydroponics | Crop plants that are grown in nutrient-supplemented water rather than soil. |
| Sewage treatment | Harmless bacteria render sewage safe by replacing or directly killing pathogens in sewage; microbes also digest most organic pollutants in sewage so it can be safely returned to the environment. |
| Perfumes and other cosmetics | Bases for these cosmetics are supplied by plants, animal pheromones, fungi, and lichens. |
| Mining copper | Bacteria extract metal from poor quality ore (see text). |
| Solidifiers | These are hardeners used to solidify microbiological culture media (agar obtained from red algae). |
| Thickeners | Chemicals called alginates (acquired from algae) maintain the thickness of puddings, toothpastes, paints, soaps, creams, and other oil- and water-based products that would otherwise separate into a liquid mess. |
| Diatomaceous earth | The enormous surface area of diatom shells (about 1000 square feet in one gram) makes this material an excellent filter. It is also a superb polishing agent. |
| Prospect for gold | The bacterium *Bacillus cereus* grows to unusually high concentrations over gold deposits. |
| Enzyme production for food and detergents | Enzymes from microbes solidify milk for cheese production; protein-digesting enzymes tenderize meat; lactase added to dairy products reduces allergic reactions to milk; enzymes in laundry detergents disassemble protein stains in fabrics. |
| Fuel | The alga *Chlorella* converts garbage to combustible oil (see text). |
| Nonagricultural cotton | Bacteria produce cellulose fibers when grown in culture (see text). |
| Shellac | Extracts of the tiny lac insects of Asia are used to prepare varnish and insulation. |
| Vaccines | All disease-specific immunizing agents are produced from microorganisms. Traditionally, the pathogen is grown and inactivated or attenuated. Modern use of genetically engineered microbes to prepare vaccines is discussed in the biotechnology section of this chapter. |

*\* Traditional agriculture and modern biotechnological endeavors are not included in this table.*

bridomas are cells produced by the fusion of two different types of cells into a single entity that contains the genes of both original cells. A eukaryotic cell that produces small amounts of a desirable product can become a prolific producer of the product when fused with a cancer cell, which has lost its ability to turn off its metabolism. The remarkable potential of hybridoma technology is discussed in the section following genetic engineering.

## Genetic Engineering

For thousands of years, people have been selectively breeding organisms to increase their value to humans, mating individuals with desirable characteristics to produce improved progeny. Modern *genetic engineering,* however, can modify an organism's genotype by introducing genes that have never been in that particular species before. But genes and other short pieces of DNA are, by themselves, somewhat unstable in a cell and are quickly disassembled. They can be stabilized by splicing them into a larger strand of DNA, such as a cell's chromosome or small DNA circles called plasmids (see Connections: Plasmid Technology). This "gene splicing" technology has successfully produced new genetically stable types of microorganisms that act as microscopic "factories," churning out the product specified by the newly acquired gene. Genetically engineered microbes are now manufacturing products from drain cleaner to fuels to medicines. (One biotechnologist referred to these microbes as "the largest nonunion workforce in the world.")

### PRODUCTS FROM GENETICALLY ENGINEERED MICROBES

Some of the most notable successes in genetic engineering have been made in creating strains of microbes that either manufacture products more cheaply or provide a source of products that were simply unavailable by conventional means. One of the first products of this technology was *interferon,* a group of antiviral compounds that may also have value in treating a few forms of cancer. Before genetic engi-

neering, human cells were the only source of human-specific interferon. Not only are human cells relatively difficult to cultivate, but what little interferon they produce was deluged in a sea of unwanted protein. Separating interferon from the chemical debris was very expensive, and it was impossible to achieve adequate purity. In 1978, a single dose of impure interferon cost about $50,000. Now genetically engineered bacteria containing the interferon gene secrete large amounts of the drug into the culture medium where it is easily harvested and purified. This has reduced the per dose production price to around $1.00.

Prior to genetic engineering, each of the compounds in the following list was either inaccessible or harvested from mammals (or mammalian laboratory cultured cells) in very low quantities. They are now being produced by genetically engineered microbes (although many are still either in the experimental stage or are awaiting government approval of safety and effectiveness).

| | |
|---|---|
| *Interferons* | Fight viral infection; boost the immune system; possibly effective against melanoma (skin cancer) and some forms of leukemia; may help relieve rheumatoid arthritis. |
| *Interleukin 2* | Activates the immune system and so may help in treating cancer and immune system disorders. |
| *Insulin* | Controls diabetes mellitus symptoms. |
| *Growth hormone* | Counters pituitary dwarfism (lack of growth due to pituitary defect); also promotes healing. |
| *Plasminogen activator* | Dissolves blood clots, reducing probability of strokes and heart attacks. |
| *Tumor necrosis factor* | Attacks and kills tumors (cancer treatment). |
| *Factor VIII* | Stimulates blood clotting, even in persons with hemophilia. |
| *Erythropoietin* | Stimulates red blood cell production and so may be used to combat anemia. |
| *Beta endorphins* | Relieve pain (these are the body's "natural morphine"). |
| *Enzymes* | Perform a multitude of services, from driving chemical reactions for industry to supplementing human dietary enzymes. |
| *Protein vaccines* | Stimulate the body's immunity to one or two of a pathogen's antigens without the risk associated with conventional vaccines (Figure 39-4). |

Virtually any protein that is manufactured by an organism is a candidate for production by genetically engineered microbes.

One product that is not on this list may save the lives of millions of people dying of protein deficiencies in their diet. It is **single cell protein (SCP),** produced by easy-to-grow microorganisms such as yeasts and algae. These organisms also provide the most efficient means of protein production. Whereas a 1000-pound steer produces only one pound of new protein each day, and 1000 pounds of soybeans manufacture 80 pounds of protein daily, a 1000-pound inoculum of yeast generates 50,000 pounds of high-quality protein per day, protein that provides all amino acids essential to the human diet. Genetic engineers are taking this one step further—employ genetically engineered microbes that use our garbage as nutrients, turning it into high-quality edible protein. Although many people will find such foods aesthetically objectionable, they are still valuable as feed supplements for livestock. More than 150,000 tons of yeast are fed to livestock each year, serving as indirect sources of protein in human diets.

## PLASMID TECHNOLOGY: THE CORNERSTONE OF GENETIC ENGINEERING

Most genetic engineers earn their livelihood by coaxing bacteria to manufacture substances they are genetically incapable of producing. They do this by introducing new genes that direct the cell to produce the desired product—a human hormone, for example. But introducing genes into bacteria is far more difficult to achieve than it sounds. The biotechnologist must first obtain the "desired" gene and then stably incorporate it into the DNA of the bacterium in a way that allows the introduced gene to be replicated and passed on to new generations of bacteria as the culture grows.

The key to doing this kind of genetic engineering is *gene splicing,* the joining of a DNA segment from one organism to the DNA of another organism. The desired gene is usually spliced into a small circle of bacterial DNA, called a **plasmid** (see figure 1). The plasmid readily enters a bacterial cell, and is replicated along with the cell's own DNA. Furthermore, all genes on the plasmid are expressed by the cell as though they were native genes.

The plasmid has another valuable property: it can be passed from one cell to another, for example, by transformation (Chapter 27). Once a "foreign" gene (usually extracted from a eukaryotic chromosome) has been spliced into a plasmid, it acts like a "gene taxi," transporting the passenger gene into a bacterium. The gene-carrying plasmid is readily absorbed by the bacterium and replicates much like the cell's chromosome does, so that each daughter cell inherits the new gene. Every bacterium in the culture follows the new gene's instructions (for example, "produce the mammalian hormone insulin").

Splicing foreign genes into plasmids requires genetic recombination, the breakage and reunion of DNA. (Gene splicing is also known as **recombinant DNA technology**.) Until the early 1970s, scientists could not control recombination. Then a special group of DNA-cutting enzymes was discovered in bacteria. These newly discovered enzymes provided scientists with the ability to splice genes from different sources. These enzymes, called **restriction endonucleases**, are normally used by bacteria to protect themselves against viral infection. The enzymes carve viral DNA into harmless pieces without doing similar damage to the bacterium's own DNA. The enzymes are "sequence-specific"; they only cut DNA after a particular nucleotide sequence that is not found in the DNA of the enzyme's owner. This ability to recognize nucleotide sequences provides the genetic "scissors" for cutting genes out of human chromosomes and, as you will see, the "glue" needed to splice them into bacterial plasmids.

The enzyme generates an "offset cut" in the DNA because the paired complementary strands, which run in opposite directions, are nicked in different places, as shown in figure 2. This leaves short, single-stranded regions that attract and bind with a complementary single-stranded sequence. In other words, the ends are "sticky" by virtue of their complementary sequences of base pairs. A particular endonuclease always cuts DNA after the same nucleotide sequence, so it creates the same sticky ends regardless of whether the DNA is from a human, plant, bacterium, or virus. The sticky ends of a genetic segment cut from a human chromosome readily adhere to the open ends of a plasmid that has been cut with the same endonuclease (see figure 2). When these two DNA preparations are mixed together, the complementary sticky ends join the isolated human genetic segment with the plasmid, much like taping an extra paragraph into a sheet of instructions. As long as both papers are cut with scissors that make

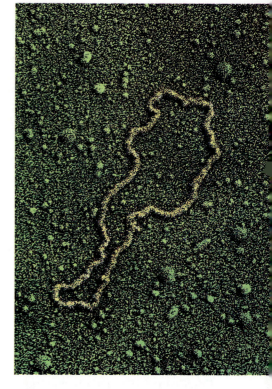

**Figure 1**
**Electron micrograph of a bacterial plasmid.**

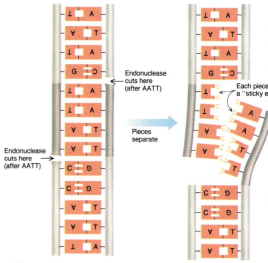

**Figure 2**

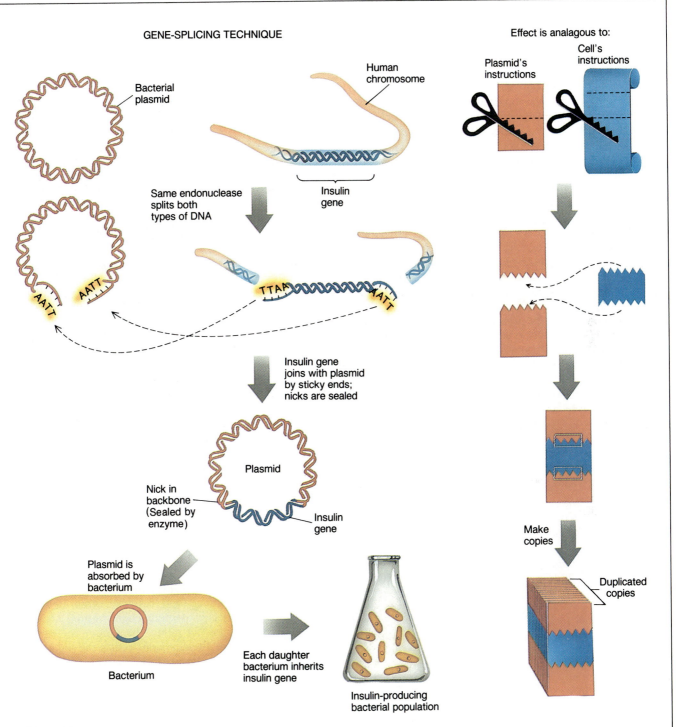

GENE-SPLICING TECHNIQUE

Effect is analogous to:

Plasmid's instructions

Cell's instructions

Bacterial plasmid

Human chromosome

Insulin gene

Same endonuclease splits both types of DNA

TTAA

AATT

AATT

AATT

Insulin gene joins with plasmid by sticky ends; nicks are sealed

Plasmid

Nick in backbone (Sealed by enzyme)

Insulin gene

Make copies

Plasmid is absorbed by bacterium

Bacterium

Each daughter bacterium inherits insulin gene

Insulin-producing bacterial population

Duplicated copies

**Figure 3**
**Gene splicing techniques.** Steps described in the text.

*Continued on next page*

the same pattern, the cuts fit and the pieces can be joined. (Restriction endonucleases are analogous to the scissors.)

Following splicing, each recombinant plasmid containing the new gene (for insulin in this example) is added to a bacterial culture and absorbed by a cell. These cells can be isolated and grown on culture medium, each daughter cell perpetuating the plasmid and expressing the incorporated gene. Such genetically engineered bacteria produce large amounts of high-quality insulin that is free of the contaminating animal protein that causes allergic reactions in some diabetics. (Insulin is conventionally obtained from the pancreas of animals.) Even though it is produced by bacteria, the product is identical to human insulin. It is human insulin.

A major difficulty in genetic engineering is obtaining the gene for the desired product. The probability is about a million to one that a particular fragment from the endonuclease-cleaved human chromosome will contain the insulin gene. Thus, to produce an insulin-producing bacterium, the above procedures might have to be repeated a million times or more, and each recombinant examined for the presence of the active insulin gene. One way scientists can avoid this tedious and costly approach is to use "synthetic genes." These are real genes that are abiotically synthesized in the laboratory rather than extracted from an organism. Once the amino acid sequence of the desired protein has been identified, the universal genetic code (page 464) makes it easy to deduce a nucleotide sequence that would code for the production of that protein. Scientists can then synthesize DNA with that sequence. The success of this technique has been enormously boosted by the invention of the so-called "gene machine", an automatic nucleotide assembler that rapidly constructs the exact DNA sequence requested, reducing the cost of making a gene by a hundredfold. In one day the gene machine does what used to take a whole year to accomplish.

## "Fingerprinting" the Chromosome

Gene-synthesizing machines also assemble nucleotides into sequences that are used as *DNA probes*, single-stranded stretches of DNA that base-pair with complementary regions of a cell's chromosome. DNA probes made with radioactive nucleotides can be easily tracked when mixed with chromosomes. If the chromosome in question contains the same sequence as the gene probe, the probe will base-pair with it, and a radioactive "hot spot" will be detected on the chromosome. This pinpoints the exact position of the corresponding sequence on a chromosome. In this way chromosomes can be mapped, taxonomic relationships can be determined, and hundreds of other questions in genetic research answered. But there is also a practical side to gene probes. They can be used to detect oncogenes (cancer-causing regions) in human chromosomes. Finding the genetic "fingerprint" of an oncogene in a person's chromosomes flags that individual as a probable candidate for cancer. The subsequent close monitoring can lead to earlier detection and treatment if the disease does develop, an advantage that can save the patient's life.

DNA "fingerprints" can even help solve crimes. A single human cell left at the crime scene can be identified as belonging to a particular suspect by characterizing its nucleotide sequence. DNA matching has already been accepted in many courts as conclusive evidence of the suspect's presence at the crime scene. Gone may be the days when a lawbreaker can wear a pair of gloves to prevent leaving incriminating fingerprints; he will have to cover his whole body in a sealed glove to prevent shedding a single cell that could be identified by DNA fingerprinting.

Fuel shortages may be relieved by genetically engineered microbes that inexpensively store solar energy in the chemical bonds of combustible organic compounds, providing a virtually inexhaustible fuel source. Genetic engineers are attempting to improve on the efficiency of natural systems by which some organisms can already do this (Figure 39-5a). In addition, genetic engineers have created "oil-producing" microorganisms that generate highly combustible organic compounds equivalent to petroleum in its potential energy. Another supply of energy is the earth's most abundant renewable resource, cellulose. Each year, well over 3.5 billion tons of paper, cardboard, and other cellulose-based products are manufactured in the United States (15 tons for every person in the country). About 900 million tons of this material ends up as refuse. Microbes can convert this huge resource to one of two valuable fuels: ethanol (the combustible alcohol in "gasohol") and methane, which is natural gas (Figure 39-5b). This approach has the triple advantage of energy production, waste disposal, and food production (SCP could be harvested as a byproduct.)

## RECOMBINANT ORGANISMS THAT WORK IN THE FIELD

Some genetically engineered bacteria and fungi must "get out into the world" before they can perform the task for which they were designed. One such organism has been constructed to reduce food loss when crops are subjected to freezing temperatures. Called the "ice minus" bacterium, this strain of *Pseudomonas* resists the tendency to trigger ice formation during freezing conditions, unlike the naturally occurring bacteria that live on the surfaces of plants (Figure 39-6). Bacteria normally act as "seed crystals" around which water freezes. The freezing water expands and forms sharp crystals that crush cells and slash membranes, killing cells and possibly the entire plant. The bacterial molecule that induces water solidification is the *ice nucleating protein*. Through recombinant DNA technology, a strain of bacteria has been developed that lacks the gene for producing the ice

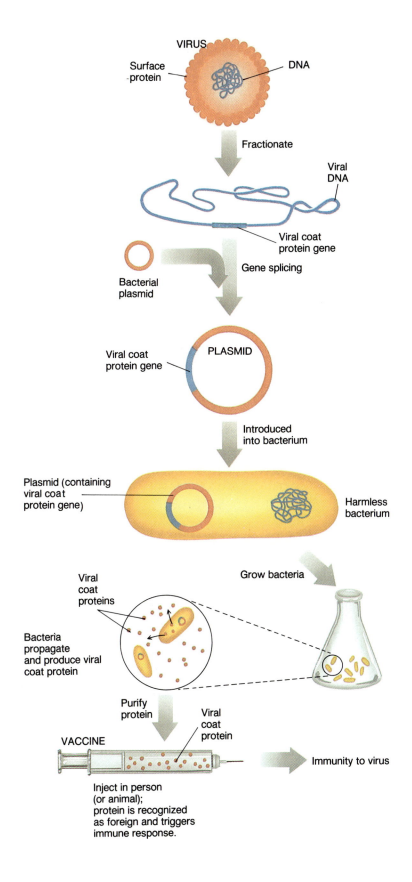

VIRUS

Surface-protein

DNA

Fractionate

Viral DNA

Viral coat protein gene

Bacterial plasmid

Gene splicing

Viral coat protein gene

PLASMID

Introduced into bacterium

Plasmid (containing viral coat protein gene)

Harmless bacterium

Grow bacteria

Viral coat proteins

Bacteria propagate and produce viral coat protein

Purify protein

VACCINE

Viral coat protein

Immunity to virus

Inject in person (or animal); protein is recognized as foreign and triggers immune response.

**Figure 39-4**
**Genetically engineered vaccines** are safer because they contain none of the antigens that contaminate traditionally prepared vaccines (killed or "disarmed" pathogens). Thus, the likelihood of allergic reactions is reduced and the danger of the vaccine organisms reactivating to their former pathogenicity is eliminated, since the vaccine contains no living organisms. This technique uses a restriction endonuclease to splice into a plasmid the gene(s) specifying a particular antigen of the pathogen (in this example, the coat protein of a virus). The gene-carrying plasmid is then introduced into harmless bacteria, which manufacture and secrete the antigen into the medium, where it can be easily harvested and purified for use as a safe immunizing agent in people or animals. The immune response that the antigen stimulates protects the individual against subsequent attacks by the pathogen. Such a vaccine against hoof and mouth disease is making the destruction of entire herds of infected livestock a thing of the past. A similar type of vaccine against the viruses that cause AIDS is one of the promising weapons being developed against this disease.

a

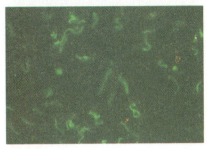

b

**Figure 39-5**
**Two natural fuel producers.** *(a)* Generally considered a nuisance that chokes waterways, this aquatic fern *(Salvina auriculata)* is being used to generate combustible fuel, methane. The drops are condensed water, not fuel. *(b)* Bacteria like these green fluorescent stained cells are especially prolific methane producers. Most biological fuel generators depend either on these kinds of cells or on different microbes that produce flammable alcohol.

**Figure 39-6**
**The "ice minus bacterium"** would have prevented formation of the ice crystals seen on these leaves. It is already being sprayed on crops and may soon be saving billions of pounds of food lost each year to frost damage.

nucleating protein. As a result, this bacterium fails to induce ice formation, even when temperatures drop to 25° F (−4° C). When sprayed on plants, the engineered "ice minus" bacteria compete with normal ice-forming bacteria, displacing them and reducing crop loss. In 1987, ice minus bacteria were sprayed on a plot of strawberry plants, becoming the first genetically engineered microorganism to be legally released into the environment.

Genetically engineered microbes can increase crop yields in other ways as well, such as improving the nitrogen-fixing capacity of *Rhizobium* (the bacterium in the root nodules of leguminous plants described in Chapters 14 and 31). Improved bacterial strains have increased soybean yields by 50 percent. Other genetic engineers are trying to develop a strain of *Azoto-*

*bacter* (a free-living nitrogen-fixing bacterium) that adheres to the roots of nonleguminous plants, such as corn, and proliferates, freeing corn of its need for applied nitrogenous fertilizer.

Genetic engineers are also releasing their microbial work force against a serious threat to the biosphere—that of pollution and toxic wastes. Many toxic pollutants are such new synthetic compounds that organisms capable of decomposing these chemicals have not yet evolved. Consequently, the compounds accumulate to dangerous levels in the environment. Genetic engineers are trying to accelerate evolution in the laboratory to develop bacteria that can quickly degrade several types of poisonous compounds. Already, a strain of bacteria that attacks petroleum has been developed and patented. (An oil spill may be one big meal for these bacteria.)

## GENETIC ENGINEERING OF MULTICELLULAR ORGANISMS

Biotechnologists are also attempting to modify the genetics of multicellular organisms. They are trying to improve plants, for example, by producing crop species that are more resistant to drought, disease, poor soil conditions, and chemical pesticides and herbicides. Another goal is to produce self-fertilizing plants that provide their own usable nitrogen. Plants rely on a few types of bacteria and cyanobacteria to fix the nitrogen into a biologically usable form (Chapter 36). Genetically engineered plants that contain the bacterial genes for nitrogen fixation could grow well even in nitrogen-poor soil.

Genetic engineers are also developing plants that produce their own pesticides. (One bite and the pest dies.) In 1986, the first such genetically engineered plant successfully passed a field test, growing undamaged by pests that ravaged the normal "unprotected" plants close by.

In another effort, the value of plant protein to the human diet is being improved by creating corn or beans that manufacture a "complete" protein, one with all the amino acids essential to the human diet. This would eliminate the need to mix vegetables to acquire all the essential amino acids from a strictly vegetarian diet. (Plants are usually deficient in one essential amino acid, usually methionine or lysine.)

Genetic engineers are also investigating ways to cure genetic deficiency diseases by "gene therapy"—implanting an effective gene to replace a defective or missing one. Diabetes, for example, may someday be treated by introducing functionally capable insulin genes into enough pancreatic cells to compensate for deficiencies in this metabolism-regulating hormone.

## Hybridoma Technology

This branch of biotechnology is an outgrowth of our ability to take two cells from different tissues of the same organism, or even from different organisms, and join them together into a single cell. This "hybrid" cell can then be conveniently cultured to form trillions of cells, each containing a complete set of genes from the two original cells. For example, one of the two original cells might be a human cell that specializes in the secretion of a potentially valuable product, such as a hormone or antibody. Because regulatory mechanisms in this normal cell restrain production of the product, it is secreted in very low quantities. However, if the cell is fused with a cancer cell that lacks these normal restraints on growth and protein synthesis, production increases dramatically. The resulting hybrid cell is called a **hybridoma** (hybrid = mixed origins, oma = cancer). Cultures of hybridoma cells produce huge quantities of the desired product.

Hybridomas are most often used to obtain antibodies, which are then harvested for diagnostic or therapeutic purposes, but virtually any kind of protein is amenable to the technique. (Hybridoma formation also provides a way to genetically cross species barriers in eukaryotes, something that cannot be accomplished by sexual fusion of gametes).

Because cells do not automatically fuse into one, scientists have devised laboratory techniques to coax them to merge (Figure 39-7). Once fused, however, hybridoma cells produce some remarkable proteins, the most exciting and versatile of which are **monoclonal antibodies**. These antibodies are produced by a *clone* of hybridoma cells, all of which descended from one cell. All the cells in the culture therefore make antibodies of the same specificity, that is, against one particular antigen (Chapter 20). This is quite unlike conventional antibody preparations obtained from the blood of immunized animals, which are heavily contaminated with "multiclonal"

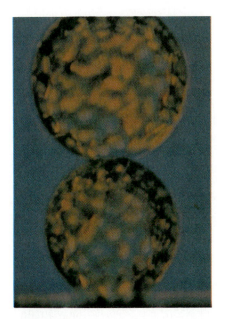 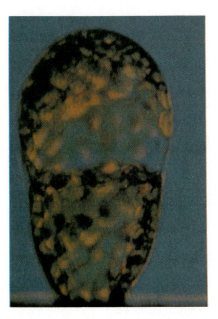 

**Figure 39-7**
**Electrofusion** joins two cells in a high-frequency electrical field, which causes cells to attract one another and fuse (as shown here). Although these two cells are of the same kind, electrofusion shows its greatest promise in joining cells of different types to form hybridomas. In many laboratories, it has replaced the more inefficient and time-consuming use of chemicals or viruses to promote cell fusion.

antibodies of unwanted specificities (multiclonal because they arise from many different types of antibody-producing cells). Hybridoma technology eliminates this problem by isolating one antibody-producing cell. This cell (and its descendants) manufactures only one type of antibody, which reacts with only one type of antigen. When this cell is fused with a cancer cell, the resultant hybridoma performs with the combined advantages of both cells — rapid growth and unrestrained production of the specific antibody. The antibody quickly accumulates in the medium and is easily harvested.

Monoclonal antibodies were initially used to diagnose disease and test for pregnancy. For example, soon after the technology was introduced, a diagnostic test for cancer was approved. This test uses tumor-specific antibodies to detect cancer long before symptoms or any other clinical indications develop. Monoclonal antibodies are also being used to treat cancer. Antibodies specific for tumor cells are armed with a lethally radioactive compound or a cytotoxin (a cell-killing chemical) and injected into the patient. The antibodies then act like "guided missiles" that carry their lethal cargo to the cancer cells, selectively destroying them without injuring the patient's healthy cells. The antibody specifically recognizes the corresponding antigen on the surface of the target cell and combines with it, concentrating the cell-killing compound at the target site. This dramatically reduces the side effects associated with conventional radiation and chemical therapy against cancer. This approach is 1000 times more lethal to tumors than the use of antitumor drugs or antitumor antibodies alone. The same strategy may also be a breakthrough in the treatment of allergy, autoimmune disease, and transplant rejection. Each of these conditions is caused by certain host cells that have "gone astray." If the monoclonal antibody-toxin complex can selectively eliminate wayward cells without harming the rest of the host's tissues, these often incurable conditions may cease to be the medical dilemmas they are today.

Hybridomas can be used to make other proteins too, not just antibodies. Computer engineers believe that biology is going to help create the next generation of computers. The molecular scale of these "biocomputers" will allow an enormous jump in information storage capacity with no increase in size. It also means that some day you may have to remember to feed your computer.

The biotechnological advances discussed in this chapter and throughout this book are not just ways to harness biology and put it to work for us. Virtually every advance provides new information that adds to our understanding of life. This is partially evident from examining Table 39-2. Even with this abbreviated list of advances, the information in this chapter is still a long way from a comprehensive panorama of this blossoming field.

This book has provided you a wide-angled view of modern biology's sweeping scope. If we have succeeded, your interest in biology has been kindled sufficiently for you to pay closer attention to the changes in the biosphere and in the field of biology, to strive to improve your understanding, and to invent ways to use your new knowledge.

Exploring life enriches life.

TABLE 39-2

| A Sampler of Additional Biotechnological Applications | | |
|---|---|---|
| **Method** | **Description** | **Value** |
| Liposome delivery systems | Liposomes are artificial membrane vesicles that carry substances inside or on their surface. | Vehicles for delivering DNA to cells; also safely deliver otherwise toxic drugs to target cells for treating infection or cancer. Substances stay entrapped in liposome until the artificial membrane fuses with the target cell's membrane and releases its cargo. Also used to boost immune responses against antigens in vaccines. |
| Protoplast regeneration | Each protoplast produced by enzymatic removal of plant cell walls can resynthesize its cell wall and regenerate an entire plant. | An excellent technique for cloning commercial plants. Saves valuable land space—100 million potential plants can be evaluated in a single flask. This facilitates selection of plants that do well in poor growing conditions. Plants with increased resistance to cold, heat, drought, and disease can be selected by growing millions of protoplasts (in culture media) while exposed to these conditions. Only resistant variants will survive. Plants grown from these surviving protoplasts may be similarly resistant. |
| Protoplast fusion | Two or more protoplasts are joined together, forming a single cell. | An application that brings together, in a single plant, genes of different species too unrelated to allow mixing by standard breeding techniques (pollination), thereby allowing creation of hardier hybrid plants. |
| In vitro fertilization | Mature ova removed from a female animal are fertilized in a petri dish and implanted into the uterus of the same animal or another of the same species. | The mixing of sperm and eggs from a single pair of desirable animals, with the fertilized ova implanted into hundreds of females for development. The method also allows infertile human couples to have children of their own. |
| Automated DNA sequencing | A machine is utilized that directly determines the base sequence of the DNA of a particular gene or a whole chromosome. | An application that allows rapid and more accurate testing for transplant compatibility, determining paternity, diagnosing genetic disorders, even in a fetus months before birth; may speed the development of gene therapy to replace defective genes with functional substitutes, allows "DNA fingerprints", identifying a person suspected of a crime by determining the DNA sequence of cells left at the crime scene. |

# Synopsis

## MAIN CONCEPTS

- **Traditional applications.** Agriculture notwithstanding, fermentation may have been the first successful attempt of humans to make use of other species. Microorganisms were also used to retard spoilage of food and to mine copper even before anyone knew of the existence of microbes. Our modern knowledge of biology has created new applications and improved on ancient ones.

- **Genetic engineering and plasmid technology.** The biotechnology "movement" was launched by the cloning of a gene. Genes for desired products can be cloned by recombinant DNA techniques that splice a gene for some desired product into a plasmid that carries it safely into a bacterial cell. The resulting bacterial culture produces the foreign gene's product. Multicellular organisms may be improved by using similar techniques for introducing foreign genes for desired characteristics.

- **Gene cloning depends on restriction endonucleases,** which sever DNA in a way that leaves "sticky ends" that join with DNA showing complementary sticky ends. The result is recombination between foreign and native genes.

- **Obtaining the desired gene for transfer.** Specific genes and DNA sequences needed for genetic engineering can now be synthesized artificially.

- **Hybridomas can produce valuable substances in enormous amounts.** Joining an antibody-producing cell and a tumor cell yields a hybridoma that, when grown in culture, generates huge quantities of monoclonal antibody, uncontaminated by antibodies that react with other types of antigens.

## KEY TERM INTEGRATOR

### Classical Applications of Biology

| | |
|---|---|
| **applied biology** | Application of biological knowledge to achieve practical objectives. It includes classical techniques such as fermentation and food production, as well as more modern advances, such as the use of **biological pesticides** and improved cultivation of **single cell protein**. Classical applied biology is being vastly improved by **biotechnology**. |

### Biotechnology

| | |
|---|---|
| **recombinant DNA technology** | The genetic engineering of organisms by introducing new genes. Most techniques rely on **restriction endonucleases** to splice the desired gene into **plasmids**, which carry the foreign gene into a cell. |
| **hybridoma technology** | Joining together two cells, such as a malignant cell and an antibody-producing cell. Hybridoma cell cultures produce large quantities of cell products, including **monoclonal antibodies**. |

# Review and Synthesis

1. Name three advantages that biological pesticides have over chemical pesticides.

2. How do bacteria help us recover copper from poor quality ore?

3. Select any two items from Table 39-1 and describe how they might be improved by genetic engineering.

4. Distinguish between the following: biotechnology and applied biology; plasmid and chromosome; conventionally produced interferon and interferon produced by genetically engineered *E. coli.;* restriction endonuclease and ligase.

5. Why is the production of "sticky ends" so essential to recombinant DNA techniques?

6. It was mentioned in this chapter that restriction endonucleases act as both "scissors" and "glue". Justify this analogy.

7. Describe the procedure you would design for genetically engineering *E. coli* to produce the human hormone insulin (a protein).

8. In what three ways might crop plants be improved by genetically engineering them?

9. To counter the possible danger from accidental release of genetically engineered bacteria, recombinant cells have been intentionally "crippled" so that they cannot survive in the absence of laboratory-induced supportive conditions. Which of the following precautions intentionally built into bacteria would likely be an effective safeguard? Sensitivity to normal ultraviolet rays in sunlight; defective cell wall requiring the protection of hypertonic media; inability to replicate DNA unless thymine is supplied in the medium; inactivated flagella, so bacteria cannot swim.

10. Why are monoclonal antibodies such powerful tools (compared to conventional antibody preparations) for diagnosing and treating disease?

## Additional Readings

Blanch, H., S. Drew, and I. Wang. 1985. *Comprehensive Biotechnology: The Principles, Applications, and Regulations of Biotechnology in Industry, Agriculture, and Medicine.* (4 vols.) Pergamon Press, New York. (Advanced.)

McKane, L., and J. Kandel. 1985. *Microbiology: Essentials and Applications.* (Chapter 25, "Microbial Biotechnology.") McGraw-Hill, New York. (Intermediate.)

Olson, S. 1986. *Biotechnology—An Industry Comes of Age.* National Academy Press. (Introductory to intermediate.)

Primrose, S. 1987. *Modern Biotechnology.* (Advanced.)

Watson, J., J. Tooze, and D. Kurtz. 1983. *Recombinant DNA: A Short Course.* W. H. Freeman, San Francisco. (Intermediate.)

# *Metric and Temperature Conversion Charts*

| | Metric Unit (symbol) | | Metric to English | English to Metric |
|---|---|---|---|---|
| **L E N G T H** | kilometer (km)<br>meter (m)<br><br>centimeter (cm)<br>millimeter (mm)<br>micrometer ($\mu$m)<br>nanometer (nm)<br>angstrom (Å) | = 1,000 ($10^3$) meters<br>= 100 centimeters<br><br>= 0.01 ($10^{-2}$) meter<br>= 0.001 ($10^{-3}$) meter<br>= 0.000001 ($10^{-6}$) meter<br>= 0.000000001 ($10^{-9}$) meter<br>= 0.0000000001 ($10^{-10}$) meter | 1 km = 0.62 mile<br>1 m = 1.09 yards<br>    = 3.28 feet<br>1 cm = 0.394 inch<br>1 mm = 0.039 inch | 1 mile = 1.609 km<br>1 yard = 0.914 m<br>1 foot = 0.305 m<br>1 inch = 2.54 cm<br>1 inch = 25.4 mm |
| **A R E A** | square kilometer ($km^2$)<br>hectare (ha)<br>square meter ($m^2$)<br><br>square centimeter ($cm^2$) | = 100 hectares<br>= 10,000 square meters<br>= 10,000 square centimeters<br><br>= 100 square millimeters | 1 $km^2$ = 0.386 square mile<br>1 ha = 2.471 acres<br>1 $m^2$ = 1.196 square yards<br>    = 10.764 square feet<br>1 $cm^2$ = 0.155 square inch | 1 square mile = 2.590 $km^2$<br>1 acre = 0.405 ha<br>1 square yard = 0.836 $m^2$<br>1 square foot = 0.093 $m^2$<br>1 square inch = 6.452 $cm^2$ |
| **M A S S** | metric ton (t)<br><br>kilogram (kg)<br>gram (g)<br>milligram (mg)<br>microgram ($\mu$g) | = 1,000 kilograms<br>= 1,000,000 grams<br>= 1,000 grams<br>= 1,000 milligrams<br>= 0.001 gram<br>= 0.000001 gram | 1 t = 1.103 tons<br><br>1 kg = 2.205 pounds<br>1 g = 0.035 ounce | 1 ton = 0.907 t<br><br>1 pound = 0.454 kg<br>1 ounce = 28.35 g |
| **V O L U M E** / **S O L I D S** | 1 cubic meter ($m^3$)<br><br>1 cubic centimeter ($cm^3$) | = 1,000,000 cubic centimeters<br><br>= 1,000 cubic millimeters | 1 $m^3$ = 1.308 cubic yards<br>    = 35.315 cubic feet<br>1 $cm^3$ = 0.061 cubic inch | 1 cubic yard = 0.765 $m^3$<br>1 cubic foot = 0.028 $m^3$<br>1 cubic inch = 16.387 $cm^3$ |
| **V O L U M E** / **L I Q U I D S** | kiloliter (kl)<br>liter (l)<br><br><br>milliliter (ml)<br>microliter ($\mu$l) | = 1,000 liters<br>= 1,000 milliliters<br><br><br>= 0.001 liter<br>= 0.000001 liter | 1 kl = 264.17 gallons<br>1 l = 1.06 quarts<br><br><br>1 ml = 0.034 fluid ounce | 1 gal = 3.785 l<br>1 qt = 0.94 l<br>1 pt = 0.47 l<br><br>1 fluid ounce = 29.57 ml |

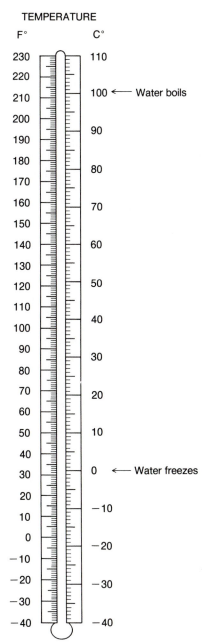

TEMPERATURE

Fahrenheit to Centigrade: °C = ⅝ (°F − 32)

Centigrade to Fahrenheit: °F = ⅘ (°C + 32)

# The Hardy-Weinberg Principle

If the allele for brown hair is dominant over that for blond hair, and curly hair is dominant over straight hair, then why don't all people by now have brown, curly hair? The **Hardy-Weinberg Principle** (developed independently by English mathematician G. H. Hardy and German physician W. Weinberg) demonstrates that the frequency of alleles remains the same from generation to generation unless influenced by outside factors. The outside factors that would cause allele frequencies to change are mutation, immigration and emigration (movement of individuals into and out of a breeding population, respectively), natural selection of particular traits, and breeding between members of a small population. In other words, unless one or more of these four forces influence hair color and curl, the relative number of people with brown and curly hair will not increase over those with blond and straight hair.

To illustrate the Hardy-Weinberg Principle, consider a single gene locus with two alleles, $A$ and $a$, in a breeding population. (If you wish, consider $A$ to be the allele for brown hair and $a$ to be the allele for blond hair.) Because there are only two alleles for the gene, the sum of the frequencies of $A$ and $a$ will equal 1.0. (By convention, allele frequencies are given in decimals instead of percentages.) Translating this into mathematical terms, if

$p$ = the frequency of allele $A$, and
$q$ = the frequency of allele $a$,

then $p + q = 1$.

If $A$ represented 80 percent of the alleles in the breeding population

($p = 0.8$), then according to this formula the frequency of $a$ must be 0.2 ($p + q = 0.8 + 0.2 = 1.0$).

After determining the allele frequency in a starting population, the predicted frequencies of alleles and genotypes in the next generation can be calculated. Setting up a Punnett square with starting allele frequencies of $p = 0.8$ and $q = 0.2$:

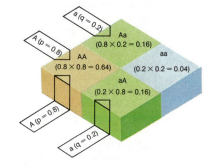

The chances of each offspring receiving any combination of the two alleles is the product of the probability of receiving one of the two alleles alone. In this example, the chances of an offspring receiving two $A$ alleles is $p \times p = p^2$, or $0.8 \times 0.8 = 0.64$. A frequency of 0.64 means that 64 percent of the next generation will be homozygous dominant ($AA$). The chances of an offspring receiving two $a$ alleles is $q^2 = 0.2 \times 0.2 = 0.04$, meaning 4 percent of the next generation is predicted to be $aa$. The predicted frequency of heterozygotes ($Aa$ or $aA$) is 0.32 or 2 $pq$, the sum the probability of an individual being $Aa$ ($p \times q = 0.8 \times 0.2 = 0.16$) plus the probability of an individual being $aA$ ($q \times p = 0.2 \times 0.8 = 0.16$). Just as all of the allele frequencies for a particular gene must add up to 1, so must all of the possible genotypes for a particular gene locus add up to 1. Thus, the Hardy-Weinberg Principle is

$$p^2 + 2pq + q^2 = 1$$
$$(0.64 + 0.32 + 0.04 = 1)$$

So after one generation, the frequency of possible genotypes is

$$AA = p^2 = 0.64$$
$$Aa = 2pq = 0.32$$
$$aa = q^2 = 0.04$$

Now let's determine the actual allele frequencies for $A$ and $a$ in the new generation. (Remember the original allele frequencies were 0.8 for allele $A$ and 0.2 for allele $a$. If the Hardy-Weinberg Principle is right, there will be no change in the frequency of either allele.) To do this we sum the frequencies for each genotype containing the allele. Since heterozygotes carry both alleles, the genotype frequency must be divided in half to determine the frequency of each allele. (In our example, heterozygote $Aa$ has a frequency of 0.32, 0.16 for allele $A$, plus 0.16 for allele $a$.) Summarizing then:

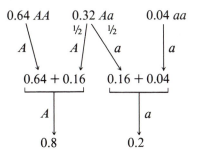

As predicted by the Hardy-Weinberg Principle, the frequency of allele $A$ remained 0.8 and the frequency of allele $a$ remained 0.2 in the new generation. Future generations can be calculated in exactly the same way, over and over again. As long as there are no mutations, no gene flow between populations, completely random mating (no natural selection), and no genetic drift, there will be no change in allele frequency, and therefore no evolution.

Population geneticists use the Hardy-Weinberg Principle to calculate a starting point allele frequency, a reference that can be compared to frequencies measured at some future time. The amount of deviation between observed allele frequencies and those predicted by the Hardy-Weinberg Principle indicates the degree of evolutionary change. Thus, this principle enables population geneticists to measure the rate of evolutionary change and identify the forces that cause changes in allele frequency.

# Glossary

## A

**ABDOMEN** The part of a higher vertebrate's body between the diaphragm and the pelvis containing the stomach and intestines. The segmented portion of an insect's body.

**ABSORPTION SPECTRUM** Region of the electromagnetic spectrum absorbed by a specific atom or molecule

**ABIOTIC ENVIRONMENT** Components of ecosystems that include all nonliving factors.

**ABSCISSION** Separation of leaves, fruit, and flowers from the stem.

**ABSCISIC ACID (ABA)** A plant hormone that inhibits growth and causes stomata to close. ABA may not be commonly involved in leaf drop.

**ACETYL CoA** Acetyl coenzyme A. A complex formed when acetic acid binds to a carrier coenzyme forming a bridge between the end products of glycolysis and the Krebs cycle in respiration.

**ACCESSORY PIGMENT** (see antenna pigments)

**ACIDS** Substances that release hydrogen ions (H+) when dissolved in water. Acids have a pH below 7.0.

**ACETYLCHOLINE** Neurotransmitter released by motor neurons at neuromuscular junctions.

**ACOELOMATES** Animals that lack a body cavity. The mesoderm of the embryo remains packed with cells.

**ACTION POTENTIAL** A change in polarity that travels along the membrane of the nerve or muscle cell. In neurons, it is associated with the nerve impulse, and in muscle cells with contraction.

**ACTIVATION ENERGY** Energy required to initiate a chemical reaction.

**ACTIVE SITE** Region on an enzyme that binds its specific substrates, making them more reactive.

**ACTIVE TRANSPORT** Movement of substances into or out of cells against a concentration gradient, i.e., from a region of lower concentration to a region of higher concentration. The process requires an expenditure of energy by the cell. (Compare with passive transport.)

**ACTOMYOSIN** Contractile protein complex of actin and myosin; generates a contractile force in muscle and other cells.

**ADAPTATION** A hereditary trait that improves an organism's chances of survival and/or reproduction.

**ADAPTIVE RADIATION** The divergence of many species from a single ancestral line.

**ADENOSINE TRIPHOSPHATE (ATP)** The molecule present in all living organisms that provides energy for cellular reactions in its phosphate bonds. ATP Is the universal energy currancy of cells.

**ADENYL CYCLASE** An enzyme activated by hormones that converts ATP to cyclic AMP, a molecule that activates resting enzymes or helps turn on suppressed genes.

**ADHESION** The force of attraction between contacting surfaces of unlike substances, such as water and glass.

**ADRENAL CORTEX** Outer layer of the adrenal glands. It secretes steroid hormones in response to ACTH.

**ADRENAL MEDULLA** An endocrine gland that controls metabolism, cardiovascular function, and stress responses.

**ADRENOCORTICOTROPIC HORMONE (ACTH)** An anterior pituitary hormone that stimulates the cortex of the adrenal glands to secrete cortisol and other steroid hormones.

**ADVENTITIOUS ROOT SYSTEM** Secondary roots that develop from stem or leaf tissues.

**AEROBE** An organism that requires oxygen to release energy from food molecules.

**AEROBIC RESPIRATION** Pathway by which glucose is completely oxidized to $CO_2$ and $H_2O$, requiring oxygen and an electron transport system.

**AFFERENT NEURONS** Neurons that conduct impulses from the sense organs to the central nervous system.

**AGE-SEX STRUCTURE** The number of individuals of a certain age and sex within a population.

**AGGREGATE FRUITS** Fruits that develop from many pistils in a single flower.

**ALBUMINOUS CELL** The specialized parenchyma that supports a sieve cell.

**ALDOSTERONE** A hormone secreted by the adrenal cortex that stimulates reabsorption of sodium and potassium from the distal tubules and conducting ducts of the kidneys.

**ALGAE** A group consisting mainly of unicellular or multicellular photosynthetic eukaryotes in the kingdom Protista.

**ALLANTOIS** Extraembryonic membrane that serves as a repository for nitrogenous wastes. In placental mammals, it helps form the vascular connections between mother and fetus.

**ALLELE** Alternative form of a gene at a particular site, or locus, on the chromosome.

**ALLELE FREQUENCY** The relative occurrence of a certain allele in individuals of a population.

**ALLELOCHEMICALS** Chemicals released by some plants and animals that deter or kill a predator or competitor.

**ALLELOPATHY** A type of interaction in which one organism releases allelochemicals that harm another organism.

**ALEURONE** Tissue layer between the seed coat and the endosperm in the seeds of monocots. It secretes alpha amylase, which digests the starch reserved in the endosperm.

**ALLOPATRIC SPECIATION** Formation of new species when gene flow between parts of a population is stopped by geographic isolation.

**ALLOSTERIC ENZYMES** Enzymes that change shape (and thus activity) in response to products of the metabolic pathway. They possess a site additional to the active site that enables metabolites to regulate enzyme activity.

**ALTERNATION OF GENERATIONS** Sequential change during the life cycle of a plant in which a haploid (1N) multicellular stage (gametophyte) alternates with a diploid (2N) multicellular stage (sporophyte).

**ALTRUISM** Self sacrificing behavior for the benefit of other individuals exhibited by various animal species (bees, ants, monkeys, humans, for example).

**ALVEOLUS** A tiny pouch in the lung where gas is exchanged between the blood and the air; the functional unit of the lung where $CO_2$ and $O_2$ are exchanged.

**AMINO ACIDS** Molecules containing an amino group ($NH_2$) and a carboxyl group (—COOH) attached to a central carbon atom. Amino acids are the subunits from which proteins are constructed.

**AMNION** Extraembryonic membrane that envelops the young embryo and encloses the amniotic fluid that suspends and cushions it.

**AMOEBA** A protozoan that employs pseudopods for motility.

**AMOEBOCYTE** Motile cell of sponges, which circulate nutrients and produce gametes for sexual reproduction.

**AMPHIBIA** A vertebrate class grouped into three orders: Caudata (tailed amphibians); Anura (tail-less amphibians); Apoda (rare wormlike, burrowing amphibians).

**AMPULLA** A muscularized water bladder in many echinoderms; found at one end of a tube foot, with a sucker at the other end. It is part of the water-vascular system, which functions in locomotion and grasping.

**ANABOLISM** Biosynthesis of complex molecules from simpler compounds. Anabolic reactions are endergonic, i.e., require energy.

**ANAEROBE** Organism that does not require oxygen to release energy from food molecules.

**ANAEROBIC RESPIRATION** Pathway by which glucose is completely oxidized, using an electron transport system but requiring a terminal electron acceptor other than oxygen. (Compare with fermentation.)

**ANALOGOUS STRUCTURES** Structures that perform the same function, such as wings in birds, bats, and insects, but did not originate from a common ancestor.

**ANAPHASE** Stage of mitosis when the centromeres split and the sister chromatids (now termed chromosomes) move to opposite poles of the spindle.

**ANASTRAL POLES** Spindle poles of cells without centrioles.

**ANATOMY** The structural chracteristics of an organism.

**ANGIOSPERM (ANTHOPHYTA)** Any plant having its seeds surrounded by fruit tissue formed from the mature ovary of the flowers.

**ANIMAL** A mobile, multicellular organism lacking the ability to manufacture nutrients required for nutrition, classified in the animal kingdom.

**ANION** A negatively charged ion.

**ANNELIDA** The phylum which contains segmented worms (earthworms, leeches, and marine worms).

**ANNUALS** Plants that live for one year or less.

**ANNULUS** A row of specialized cells encircling each sporangium on the fern frond; facilitates rupture of the sporangium and dispersal of spores.

**ANTENNA PIGMENTS** Components of photosystems that gather light energy of different wavelengths and then channel the absorbed to a reaction center.

**ANTERIOR** In anatomy, at or near the front of an animal; the opposite of posterior.

**ANTERIOR PITUITARY** A true endocrine gland manufacturing and releasing six hormones when stimulated by releasing factors from the hypothalamus.

**ANTHER** The swollen end of the stamen (male reproductive organ) of a flowering plant. Pollen grains are produced inside the anther lobes in pollen sacs.

**ANTIBODIES** Proteins produced by plasma cells. They react specifically with the antigen that stimulated their formation.

**ANTICODON** Triplet of nucleotides in tRNA that recognizes and base pairs with a particular codon in mRNA.

**ANTIDIURETIC HORMONE (ADH)** One of the two hormones released by the posterior pituitary. ADH increases water reabsorption in the kidney, which then produces a more concentrated urine.

**ANTIGEN** Specific foreign agents that trigger an immune response.

**ANTIPODAL CELLS** In the mature embryo sac of a flowering plant, the three cells farthest from the egg. Their function is unknown.

**ARCHENTERON** In gastrulation, the hollow core of the gastrula that becomes an animal's digestive tract.

**ATOM** The fundamental unit of matter that can enter into chemical reactions; the smallest unit of matter that possesses the qualities of an element.

**ATOMIC MASS** Combined number of protons and neutrons in the nucleus of an atom, also called atomic weight.

**ATOMIC NUMBER** The number of protons in the nucleus of an atom.

**ATOMIC WEIGHT** The mass roughly equivalent to the total number of protons and neutrons in the atomic nucleus.

**AORTA** Largest blood vessel in the body through which blood leaves the heart and enters the systemic circulation.

**APICAL MERISTEMS** Centers of growth located at the tips of shoots, axillary buds, and roots. Their cells divide by mitosis to produce new cells for primary growth in plants.

**APOSEMATIC COLORING** Warning coloration which makes an organism stand out from its surroundings.

**APPENDICULAR SKELETON** The bones of the appendages and of the pectoral and pelvic girdles.

**ARTERIES** Large, thick-walled vessels that carry blood away from the heart.

**ARTERIOLES** The smallest arteries, which carry blood toward capillary beds.

**ARTHROPODA** The most diverse phylum on earth, so called from the presence of jointed limbs. Includes insects, crabs, spiders, centipedes.

**ARTIFACTS** Alterations introduced during preparation of a specimen which can mislead or confuse proper understanding.

**ASCOSPORES** Sexual spores of the sac fungi, Ascomycota.

**ASEXUAL REPRODUCTION** Reproduction without the union of male and female gametes.

**ASSIMILATE STREAM** The moving phloem sap, which contains sugars and other organic molecules.

**ASTRAL SPINDLE POLE** Spindle fibers which radiate out from pairs of centrioles at each pole of a dividing cell.

**ASYMMETRICAL** Referring to a body form that cannot be divided to produce mirror images.

**ATRIOVENTRICULAR (AV) NODE** A neurological center of the heart, located at the top of the ventricles.

**ATRIUM** A contracting chamber of the heart which forces blood into the ventricle. There are two atria in the hearts of all vertebrates, except fish which have one atrium.

**AUSTRALOPITHECUS** An early homonoid with a humanlike body and an apelike head believed to be the first true homonoid.

**AUTONOMIC NERVOUS SYSTEM** The nerves that control the involuntary activities of the internal organs. It is composed of the parasympathetic system, which functions during normal activity, and the sympathetic system, which operates in times of emergency or prolonged exertion.

**AUTOSOMES** Chromosomes present in the same number in both sexes.

**AUTOTROPH** Organisms that satisfy their own nutritional needs by building organic molecules photosynthetically or chemosynthetically from inorganic substances.

**AUXINS** Plant growth hormones that promote cell elongation by softening cell walls.

**AXIAL SKELETON** The bones aligned along the long axis of the body, including the skull, vertebral column, and ribcage.

**AXIAL TRANSLOCATION SYSTEM** Cells produced by the vascular cambium that move substances up and down the stem of a plant.

**AXILLARY BUD** A bud that is directly above each leaf on the stem. It can develop into a new stem or a flower.

**AXON** The long, sometimes branch extension of a neuron which conducts impulses from the cell body to the synaptic knobs.

# B

**BACTERIOPHAGE** A virus attacking specific bacteria that multiplies in the bacterial host cell and usually destroys the bacterium as it reproduces.

**BALANCED POLYMORPHISM** The maintenance of two or more alleles for a single trait at fairly high frequencies.

**BARK** Common term for the periderm. A collective term for all plant tissues outside the secondary xylem.

**BARR BODY** In mammals, a darkly staining mass of chromatin normally found in the interphase cell nucleus of females, but not in males. A small percentage of cells in normal males have a Barr body.

**BASE** Substance that removes hydrogen ions (H+) from solutions. Bases have a pH above 7.0.

**BATESIAN MIMICRY** The resemblance of a good-tasting or harmless species to a species with unpleasant traits.

**B-CELL** A lymphocyte that is a part of the immune system; it produces antibodies when stimulated by an antigen.

**BICARBONATE ION** $HCO_3-$.

**BIENNIALS** Plants that live for two years.

**BILATERAL SYMMETRY** The quality possessed by organisms whose body can be divided into mirror images by only one median plane.

**BINARY FISSION** Process of cell division in simple prokaryotes in which replicated DNA and cytoplasm of a mother cell are split equally between two daughter cells.

**BINOMIAL** A term meaning "two names" or "two words". Applied to the system of nomenclature for categorizing living things with a genus and species name that is unique for each type of organism.

**BIOGEOCHEMICAL CYCLES** The exchanging of chemical elements between organisms and the abiotic environment.

**BIOMASS** The weight of organic material present in an ecosystem at any one time.

**BIOME** Broad geographic region with a characteristic array of organisms.

**BIOSPHERE** Zone of the earth's soil, water, and air in which living organisms are found.

**BIOSYNTHESIS** Construction of molecular components in the growing cell and the replacement of these compounds as they deteriorate.

**BIOTECHNOLOGY** A new field of genetic engineering; more generally, any practical application of biological knowledge.

**BIOTIC ENVIRONMENT** Living components of the environment.

**BIOTIC POTENTIAL** The innate capacity of a population to increase tremendously in size were it not for curbs on growth; maximum population growth rate.

**BLADE** Large, flattened area of a leaf; effective in collecting sunlight for photosynthesis.

**BLASTOCOEL** The hollow space in a blastula.

**BLASTOCYST** Early stage of a mammalian embryo, consisting of a mass of cells enclosed in a hollow ball of cells called the trophoblast.

**BLASTODISK** In bird and reptile development, the stage equivalent to a blastula. Because of the large amount of yolk, cleavage produces two flattened layers of cells with a thin blastocoel between them.

**BLASTOPORE** The opening of the archenteron that is the

embryonic predecessor of the anus in vertebrates and some other animals.

**BLASTULA** An early developmental stage in many animals. It is a hollow ball formed by single layer cells that encloses a cavity, the blastocoel.

**BLOOD** A type of connective tissue consisting of red blood cells, white blood cells, platelets, and plasma.

**BLOOD PRESSURE** Positive pressure within the cardiovascular system that propels blood through the vessels.

**BONE** A tissue composed of collagen fibers, calcium, and phosphorous that serves as a means of support, a reserve of calcium and phosphorous, and an attachment site for muscles.

**BOTANY** Branch of biology that studies the life cycles, structure, growth, and classification of plants.

**BOTTLENECK** The stage in growth/decline cycles of a population when there are relatively few individuals, reducing the gene pool.

**BOWMAN'S CAPSULE** A double-membraned container that is an invagination of the proximal end of the renal tubule that collects molecules and wastes from the blood.

**BRAIN** Mass of nerve tissue composing the main part of the central nervous system made up of gray and white matter.

**BRAINSTEM** The part of the brain made up of the mid- and hindbrain, which coordinate the automatic, involuntary body processes.

**BRONCHI** The two divisions of the trachea through which air enters each of the two lungs.

**BRYOPHYTA** Division of non-vascular terrestrial plants that include liverworts, mosses, and hornworts.

**BUDDING** Asexual process by which offspring develop as an outgrowth of a parent.

**BUFFERS** Chemicals that couple with free hydrogen and hydroxide ions thereby resisting changes in pH.

**BUNDLE SHEATH** Parenchyma cells that surround a leaf vein which regulate the uptake and release of materials between the vascular tissue and the mesophyll cells.

# C

**C₃ SYNTHESIS** The usual pathway for fixing $CO_2$ in the synthesis reactions of photosynthesis. It is so named because the first detectable organic molecule into which $CO_2$ is incorporated is a 3-carbon molecule, phosphoglycerate (PGA).

**CALCITONIN** A thyroid hormone which regulates blood calcium levels by inhibiting its release from bone.

**CALORIE** Energy (heat) necessary to elevate the temperature of one gram of water by one degree Centigrade (1°C).

**CALVIN CYCLE** The cyclical pathway in which $CO_2$ is incorporated into carbohydrate. See C₃ synthesis.

**CALYX** The outermost whorl of a flower, formed by the sepals.

**CAM** Crassulacean acid metabolism. A variation of the photosynthetic reactions in plants, biochemically identical to C₄ synthesis except that all reactions occur in the same cell and are separated by time. Because CAM plants open their stomates at night, they have a competitive advantage in hot, dry climates.

**CAMBIUM** A ring or cluster of meristematic cells that increase the width of stems and roots when they divide.

**CAMOUFLAGE** Adaptations of color, shape and behavior that make an organism more difficult to detect.

**CANCER** A disease resulting from uncontrolled cell divisions.

**CAPILLARIES** The tiniest blood vessels consisting of a single layer of flattened cells.

**CAPILLARY ACTION** Tendency of water to be pulled into a small-diameter tube.

**CARBAMINOHEMOGLOBIN** Loosely-bound complex of $CO_2$ and hemoglobin, formed when hemoglobin picks up $CO_2$ from body tissues.

**CARBOHYDRATES** A group of compounds that includes simple sugars and all larger molecules constructed of sugar subunits, e.g., polysaccharides.

**CARBON DIOXIDE ($CO_2$)** An inorganic molecule formed from the elements carbon and oxygen. The carbon source for photosynthesis and chemosynthesis.

**CARBON DIOXIDE FIXATION** In photosynthesis, the combination of $CO_2$ with carbon-accepting molecules to form organic compounds.

**CARBONIC ANHYDRASE** An enzyme in erythrocytes which catalyzes the breakdown of carbonic acid ($H_2CO_3$) to H+ and $HCO_3$−.

**CARCINOGEN** A cancer-causing agent.

**CARDIAC CYCLE** The sequence in which the various chambers of the heart contract. Specifically, the atria contract while the ventricles rest and vice versa, resulting in efficient pumping of blood.

**CARDIAC MUSCLE** One of the three types of muscle tissue; it pumps blood through the circulatory system.

**CARDIOVASCULAR SYSTEM** The organ system consisting of the heart and the vessels through which blood flows.

**CARNIVORE** A flesh-eating animal.

**CAROTENE** A red, yellow, or orange plant pigment that absorbs light in 400-500 nm wavelengths.

**CARPELS** Central whorl of a flower containing the female reproductive organs. Each separate carpel, or each unit of fused carpels, is called a pistil.

**CARRYING CAPACITY** The size of a population that can be supported indefinitely in a given environment.

**CARTILAGE** A firm but flexible connective tissue. In the human, most cartilage originally present in the embryo is transformed into bones.

**CAROTENOID PIGMENTS** Pigments that absorb light, mainly from 400-500 nm wavelengths. They give carrots, oranges, tomatoes, and autumn foliage their characteristic colors.

**CASPARIAN STRIP** The band of waxy suberin that surrounds each endodermal cell of a plant's root tissue.

**CATABOLISM** Reaction that degrades complex compounds into simpler molecules, usually with the release of the chemical energy that held the atoms of the larger molecule together.

**CATALYST**  A chemical substance that accelerates a reaction or causes a reaction to occur but remains unchanged by the reaction. Enzymes are biological catalysts.

**CATION**  A positively charged ion.

**CECUM**  A closed-ended sac extending from the intestine in grazing animals lacking a rumen (e.g., horses) that enables them to digest cellulose.

**CELL**  The basic structural unit of all organisms.

**CELL BODY**  Region of a neuron that contains most of the cytoplasm, the nucleus, and other organelles. It relays impulses from the dendrites to the axon.

**CELL CYCLE**  Total sequence of stages from one cell division to the next. The stages are denoted G, S, G2, and mitosis.

**CELL MEMBRANE**  (see plasma membrane)

**CELL PLATE**  In plants, the cell wall material deposited midway between the daughter cells during cytokinesis. Plate material (pectin) is deposited by small Golgi vesicles.

**CELL SAP**  Solution that fills the plant vacuole. In addition to water, it may contain pigments, salts, and even toxic chemicals.

**CELL THEORY**  The fundamental theory of biology that states: 1) all organisms are composed of one or more cells, 2) the cell is the basic organizational unit of life, 3) all cells arise from pre-existing cells.

**CELL WALL**  Rigid outer-casing of cells in plants and other organisms which gives support, slows dehydration, and prevents a cell from bursting when internal pressure builds due to an influx of water.

**CELLULAR RESPIRATION**  (See aerobic respiration)

**CELLULOSE**  The structural polysaccharide comprising the bulk of the plant cell wall. It is the most abundant polysaccharide in nature.

**CENTRAL NERVOUS SYSTEM**  The brain and spinal cord.

**CENTRIOLE**  A structure at each pole of a dividing animal cell that serves as an attachment point for the spindle fibers.

**CENTROMERE**  The attachment point of two duplicated chromatids. During cell division, the centromere splits and spindle fibers attached to the centromeres separate, causing the chromatids to move to opposite poles of the mother cell.

**CEPHALIZATION**  The clustering of neural tissues at the anterior (leading) end of the animal.

**CEPHALOCHORDATES**  A subphylum of Chordata which includes the lancelet, a small, fishlike animal that lives in shallow oceans.

**CEREBRAL CORTEX**  The outer, highly convoluted layer of the cerebrum. In the human, this is the center of higher brain functions, such as speech and reasoning.

**CEREBRUM**  The most dominant part of the human forebrain, composed of two cerebral hemispheres, generally associated with higher brain functions.

**CERVIX**  The lower tip of the uterus.

**CESTODA**  A class of platyhelminthes, consisting of parasitic tapeworms.

**CHARACTER DISPLACEMENT**  Divergence of a physical trait in closely related species in response to competition.

**CHEMICAL BONDS**  Linkage between atoms as a result of electrons being shared or donated.

**CHEMICAL EVOLUTION**  Spontaneous synthesis of increasingly complex organic compounds from simpler molecules.

**CHEMICAL REACTION**  Interaction between chemical reactants.

**CHEMIOSMOSIS**  The process by which a pH gradient drives the formation of ATP.

**CHEMOSYNTHESIS**  An energy conversion process in which inorganic substances (H, N, Fe, or S) provide energized electrons and hydrogen for carbohydrate formation.

**CHIASMATA**  Cross-shaped regions within a tetrad, occurring at points of crossing over or genetic exchange.

**CHITIN**  Structural polysaccharide that forms the hard, strong external skeleton of many arthropods and the cell walls of fungi.

**CHLOROPHYLL PIGMENTS**  Major light-absorbing pigments of photosynthesis.

**CHLOROPLASTS**  A plastid containing chlorophyll found in plant cells in which photosynthesis occurs.

**CHOANOCYTES**  Collar cells, which are the major component of the sponge endodermis.

**CHORDATE**  A member of the phylum Chordata possessing a skeletal rod of tissue called a notochord, a dorsal hollow nerve cord, gill slits, and a tail at some stage of its development.

**CHORION**  The outermost of the four extraembryonic membranes. In placental mammals, it forms the embryonic portion of the placenta.

**CHROMATID**  Each of the two identical subunits of a replicated chromosome.

**CHROMATIN**  Collection of DNA-protein fibers which, during prophase, condense to form the chromosomes.

**CHROMOSOMES**  Dark-staining structures in which the organism's genetic material (DNA) is organized. Each species has a characteristic number of chromosomes.

**CHROMOSOMAL PUFFS**  In the chromosomes of certain insect larvae, sites where chromatin unravels from its condensed state so that its DNA can be transcribed.

**CILIA**  Short, hairlike structures projecting from the surfaces of some cells. They beat in coordinated ways, are usually found in large numbers, and are densely packed.

**CILIATED MUCOSA**  Layer of ciliated epithelial cells lining the respiratory tract. The beating of cilia propels an associated mucous layer and trapped foreign particles.

**CIRCADIAN RHYTHM**  Behavioral patterns that cycle during approximately 24 hour intervals.

**CIRCULATORY SYSTEM**  The system that circulates internal fluids throughout an organism to deliver oxygen and nutrients to cells and to remove metabolic wastes.

**CISTERNAE** Flattened membrane sacs of the endoplasmic reticulum and Golgi complex.

**CLASS (TAXONOMIC)** A level of the taxonomic hierarchy that groups together members of related orders.

**CLEAVAGE** Successive mitotic divisions in the early embryo. There is no cell growth between divisions.

**CLEAVAGE FUROW** Constriction around the middle of a dividing cell caused by constriction of microfilaments.

**CLIMAX COMMUNITY** Community that remains essentially the same over long periods of time; final stage of ecological succession.

**CLITORIS** A protrusion at the point where the labia minora merge; rich in sensory neurons and erectile tissue.

**CLONES** Offspring identical to the parent, produced by asexual processes.

**CLOSED CIRCULATORY SYSTEM** Circulatory system in which blood travels throughout the body in a continuous network of closed tubes. (Compare with open circulatory system).

**CNIDARIA** A phylum that consists of radial symmetrical animals that have two cell layers. There are three classes: 1) Hydrazoa (Hydra), 2) Scyphozoa (jellyfish), 3) Anthozoa (sea anemones, corals). Most are marine forms that live in warm, shallow water.

**CNIDOCYTES** Specialized stinging cells found in the members of the phylum Cnidaria.

**COACERVATES** Polymer-rich droplets formed by combining proteins and nucleic acids under controlled laboratory conditions.

**CODOMINANCE** The simultaneous expression of both alleles at a genetic locus in a heterozygous individual.

**CODON** Linear array of three nucleotides in mRNA. Each triplet specifies a particular amino acid during the process of translation.

**COELOMATES** Animals in which the body cavity is completely lined by mesodermally-derived tissues.

**COENZYME** A small organic molecule associated with an enzyme, enabling the enzyme to perform its catalytic function.

**COEVOLUTION** Evolutionary changes that result from reciprocal interactions between two species, e.g., flowering plants and their insect pollinators.

**COFACTOR** A metallic ion which temporarily associates with an enzyme, enabling the enzyme to perform its catalytic function.

**COHESION** The tendency of different parts of a substance to hold together because of forces acting between its molecules.

**COITUS** Sexual union in mammals.

**COLEOPTILE** Sheath surrounding the tip of the monocot seedling, protecting the young stem and leaves as they emerge from the soil.

**COLLOIDAL SUSPENSION (COLLOID)** Suspension in a fluid of very small particles which neither dissolve nor settle out.

**COMBINED IMMUNE RESPONSE** An immune response in which both the T-cell and B-cell systems are activated.

**COMMENSALISM** A form of symbiosis in which one organism benefits from the union while the other member neither gains nor loses.

**COMMUNITY** The populations of all species living in a given area.

**COMPACT BONE** The solid, hard outer regions of a bone surrounding the honey-combed mass of spongy bone.

**COMPANION CELL** Specialized parenchyma cell associated with a sieve-tube member in phloem.

**COMPETITION** Interaction among organisms that require the same resource. It is of two types: 1) intraspecific (between members of the same species); 2) interspecific (between members of different species).

**COMPETITIVE EXCLUSION PRINCIPLE (GAUSE'S PRINCIPLE)** Competition in which a winner species captures a greater share of resources, increasing its survival and reproductive capacity. The other species is gradually displaced.

**COMPETITIVE INHIBITION** Prevention of normal binding of a substrate to its enzyme by the presence of an inhibitory compound that competes with the substrate for the active site on the enzyme.

**COMPLEMENT** Blood proteins with which some antibodies combine following attachment to antigen (the surface of microorganisms). The bound complement punches the tiny holes in the plasma membrane of the foreign cell, causing it to burst.

**COMPLETE DIGESTIVE SYSTEMS** Systems that have a digestive tract with openings at both ends—a mouth for entry and an anus for exit.

**COMPLETE FLOWER** A flower containing all four whorls of modified leaves—sepels, petals, stamen, and carpels.

**COMPOUND** Chemical substances composed of atoms of more than one element.

**CONCENTRATION GRADIENT** Regions in a system of differing concentration representing potential energy, such as exist in a cell and its environment, that cause molecules to move from areas of higher concentration to lower concentration.

**CONJUGATION** A method of reproduction in single-celled organisms in which two cells link and exchange nuclear material.

**CONNECTIVE TISSUES** Tissues that protect, support, and hold together the internal organs and other structures of animals. Includes bone, cartilage, tendons, and other tissues, all of which have large amounts of extracellular material.

**CONSUMERS** Heterotrophs in a biotic environment that feed on other organisms or organic waste.

**CONTACT INHIBITION** A regulatory mechanism that prevents cellular overgrowth when tissues make direct contact.

**CONTINENTAL DRIFT** The continuous shifting of the earth's land masses explained by the theory of plate tectonics.

**CONTINUOUS VARIATION** An inheritance pattern in which there is graded change between the two extremes in a phenotype (compare with discontinuous variation).

**CONTRACTILE PROTEINS** Actin and myosin, the protein filaments that comprise the bulk of the muscle mass. During contraction of skeletal muscle, these filaments form a temporary association, actomyosin, and slide past each other, generating the contractile force.

**CONTROL (EXPERIMENTAL)** A duplicate of the experiment identical in every way except for the one fact being tested. Use of a control is necessary to demonstrate cause and effect.

**CONVERGENT EVOLUTION** The evolution of similar structures in distantly related organisms in response to similar environments.

**CORK CAMBIUM** In the stem and roots of perennials, a secondary meristem that produces the outer protective layer of the bark.

**CORONARY ARTERIES** Two large arteries that branch immediately from the aorta, providing oxygen-rich blood to the cardiac muscle.

**CORPUS LUTEUM** In the mammalian ovary, the structure that develops from the follicle after release of the egg. It secretes hormones that prepare the uterine endometrium to receive the developing embryo.

**CORPUS SPONGIOSUM** The lower erectile body of the penis lying outside the penis sheath.

**CORTEX** In the stem or root of plants, the region between the epidermis and the vascular tissues. Composed of ground tissue. In animals, the outermost portion of some organs.

**CORTICOSTEROIDS** A family of steroid hormones secreted by the adrenal cortex.

**COTYLEDON** The seed leaf of a dicot embryo containing stored nutrients required for the germinated seed to grow and develop, or a food digesting seed leaf in a monocot embryo.

**COVALENT BONDS** Linkage between two atoms which share the same electrons in their outermost shells.

**CRISTAE** the convolutions of the inner membrane of the mitochondrion. Embedded within them are the components of the electron transport system and proton channels for chemiosmosis.

**CROSSING OVER** During synapsis, the process by which homologues exchange segments with each other.

**CRYPTIC COLORATION** A form of camouflage wherein an organism's color or patterning helps it resemble its background.

**CULTURAL EVOLUTION** Process of passing knowledge from generation to generation.

**CUTICLE** Waxy layer covering the outer cell walls of plant epidermal cells. It retards water vapor loss and helps prevent dehydration.

**CYCLIC AMP (cAMP (Cyclic adenosine monophosphate))** A ring-shaped molecular version of an ATP minus two phosphates. A regulatory molecule formed by the enzyme adenyl cyclase which converts ATP to cAMP.

**CYCLIC PATHWAYS** Metabolic pathways in which the intermediates of the reaction are regenerated while assisting the conversion of the substrate to product.

**CYCLIC PHOTOPHOSPHORYLATION** A pathway that produces ATP, but not NADPH, in the light reactions of photosynthesis. Energized electrons are shuttled from a reaction center, along a molecular pathway, back to the original reaction center, generating ATP en route.

**CYCLOSIS (CYTOPLASMIC STREAMING)** The one-way streaming of a cell's contents, produced by contractile protein filaments.

**CYTOCHROMES** Iron-containing proteins in the cell membrane that form a chain of electron carriers.

**CYTOKINESIS** Final event in eukaryotic cell division in which the cell's cytoplasm and the new nuclei are partitioned into separate daughter cells.

**CYTOKININS** Growth-producing plant hormones which stimulate rapid cell division.

**CYTOPLASM** General term that includes all parts of the cell, except the plasma membrane and the nucleus.

**CYTOPLASMIC LATTICE (CYTOSKELETON)** Framework of microfilaments, intermediate filaments, and microtubules, extending through the cytoplasm. It gives the cell its shape and supports various organelles.

# D

**DAUGTER CELL** Cell formed from the division of a mother cell that contains a copy of the mother cell's DNA.

**DAY NEUTRAL PLANTS** Plants that flower at any time of the year, independent of the relative lengths of daylight and darkness.

**DECIDUOUS** Trees or shrubs that shed their leaves in a particular season, usually autumn, before entering a period of dormancy.

**DECOMPOSERS (SAPROPHYTES)** Organisms that obtain nutrients by breaking down organic compounds in wastes and dead organisms. Includes fungi, bacteria, and some insects.

**DEHYDRATION SYNTHESIS** Formation of a covalent bond between two molecules (e.g., simple sugars, amino acids) by the removal of a molecule of water.

**DELETION** Loss of a portion of a chromosome, following breakage of DNA.

**DENATURATION** Change in the normal folding of a protein as a result of heat, acidity, or alkalinity. Such changes result in a loss of enzyme functioning.

**DENDRITES** Cytoplasmic extensions of the cell body as a neuron. They carry impulses from the area of stimulation to the cell body.

**DENSITY-DEPENDENT FACTORS** Factors that control population growth which are influenced by population size.

**DENSITY-INDEPENDENT FACTORS** Factors that control population growth which are not affected by population size.

**DENITRIFICATION** The conversion by denitrifying bacteria of nitrites and nitrates into nitrogen gas.

**DEOXYRIBONUCLEIC ACID (DNA)** Double-stranded polynucleotide comprised of deoxyribose (a sugar), phosphate, and four bases (adenine, guanine, cytosine, and thymine). Encoded in the sequence of nucleotides are the instructions for making proteins or ribosomal RNA. DNA is the genetic material in all organisms except certain viruses.

**DERMIS**  Layer of cells below the epidermis in which connective tissue predominates. Embedded within it are vessels, various glands, smooth muscle, nerves, and follicles.

**DESERT**  Biome characterized by intense solar radiation, very little rainfall, and high winds.

**DEUTEROSTOME**  One path of development exhibited by coelomate animals (e.g., echinoderms and chordates) characterized by radial cleavage.

**DIABETES MELLITUS**  A disease caused by insufficient insulin production, preventing glucose from being absorbed by the cells.

**DIAPHRAGM**  A sheet of muscle that separates the thoracic cavity from the abdominal wall.

**DICOTYLEDONAE (DICOTS)**  One of the two classes of flowering plants, characterized by having seeds with two cotyledons, flower parts in 4s or 5s, net-veined leaves, one main root, and vascular bundles in a circular array within the stem. (Compare with Monocotylenodonae).

**DIFFERENTIALLY PERMEABLE**  Term applied to the plasma membrane because it allows some material to pass through freely, yet it completely restricts the passage of others.

**DIFFERENTIATION**  The development of distinct cell types with different forms and specializations, though all possess the same genes.

**DIFFUSION**  Tendency of molecules to move from a region of higher concentration to a region of lower concentration, until they are uniformly dispersed.

**DIGESTION**  The process by which food particles are disassembled into molecules small enough to be absorbed into the organism's cells and tissues.

**DIGESTIVE SYSTEM**  System of specialized organs that ingests food, converts nutrients to a form that can be distributed throughout the animal's body, and eliminates undigested residues.

**DIMORPHISM**  Presence of two forms of a trait within a population, resulting from diversifying selection.

**DINOFLAGELLATES**  Photosynthetic, single-celled algae that have two flagella. Important component of the food chain in marine ecosystems.

**DIOECIOUS**  Having male and female flowers on separate plants.

**DIPLOID**  Having two sets of chromosomes. Often written 2n.

**DIRECTIONAL SELECTION**  The steady shift of phenotypes toward one extreme.

**DISCONTINUOUS VARIATION**  An inheritance pattern in which the phenomenon of all possible phenotypes fall into distinct categories. (Compare with continuous variation).

**DISRUPTIVE COLORATION**  Coloration that disguises the shape of an organism by breaking up its outline.

**DIVERGENT EVOLUTION**  The emergence of new species as branches from a single ancestral lineage.

**DIVERSIFYING SELECTION**  The increasing frequency of extreme phenotypes because individuals with average phenotypes die off.

**DIVISION (OR PHYLUM)**  A level of the taxonomic hierarchy that groups together members or related classes.

**DNA POLYMERASE**  Enzyme responsible for replication of DNA. It assembles free nucleotides, aligning them with the complementary ones in the unpaired region of each single strand.

**DNA PROBES**  Single-stranded stretches that base-pair with complementary regions of a cell's chromosome. When radioactively tagged, such probes can pinpoint the location of genes on the chromosome.

**DNA RECOMBINATION**  The breakage and rejoining of DNA molecules during the process of chromosomal crossing over.

**DOMINANT**  The form of an allele that masks the presence of other alleles for the same trait.

**DORMANCY**  A resting period, such as seed dormancy in plants or hibernation in animals, in which organisms maintain reduced metabolic rates.

**DORSAL**  In anatomy, the back of an animal.

**DUPLICATION**  The repetition of a segment of a chromosome following the breakage of two chromosomes and their misaligned reunion during repair.

**DYNAMIC EQUILIBRIUM**  In a chemical reaction, the points at which the rates of conversion of both products and reactants are equal resulting in no further net change.

# E

**ECOLOGY**  The branch of biology that studies interactions among organisms as well as the interactions of organisms and their environment.

**ECYDYSIS**  Molting process by which an arthropod periodically discards its exoskeleton and replaces it with a larger version. The process is controlled by the hormone ecdysone.

**ECHINODERMATA**  A phylum composed of animals having an internal skeleton made of many small calcium carbonate plates which have jutting spines. Includes sea stars, sea urchins, etc.

**ECOLOGICAL NICHE**  The habitat, functional role(s), requirements for environmental resources and tolerance ranges for each abiotic condition in relation to an organism.

**ECOLOGICAL PYRAMID**  Illustration showing the energy content, numbers of organisms, or biomass at each trophic level.

**ECOSYSTEM**  Unit comprised of organisms interacting among themselves and with their physical environment.

**ECOTYPES**  Populations of a single species with different, genetically fixed tolerance ranges.

**ECTOTHERMS**  Animals that lack an internal mechanism for regulating body temperature. "Cold-blooded" animals.

**EFFECTORS**  Muscle fibers and glands that are activated by neural stimulation.

**EFFECTOR MOLECULE**  A cellular molecule that determines whether the structural genes of the operon are turned on or switched off.

**EFFERENT NERVES**   The nerves that carry messages from the central nervous system to the effectors, the muscles, and glands. They are divided into two systems: somatic and automatic.

**EGG**   Female gamete, also called an ovum. A fertilized egg is the product of the union of female and male gametes (egg and sperm cells).

**EGG APPARATUS**   In the mature embryo sac of a flowering plant, the three cells near the micropyle. One of these three is the egg gamete.

**EJACULATORY DUCTS**   The pathways through which sperm from the vas deferens enters the urethra.

**ELECTRONS**   Negatively charged particles that orbit the atomic nucleus.

**ELECTRON CARRIER**   Substances (such as $NAD^+$ and FAD) that transport electrons from one step of a metabolic pathway to the next or from metabolic reactions to biosynthetic reactions.

**ELECTRON TRANSPORT SYSTEM**   Highly organized assembly of cytochromes and other proteins which transfer electrons. During transport, which occurs within the inner membranes of mitochondria and chloroplasts, the energy extracted from the electrons is used to make ATP.

**ELEMENT**   Substance composed of only one type of atom.

**EMBRYO**   An organism in the early stages of development, beginning with the first division of the zygote.

**EMBRYO SAC**   The sum of the egg apparatus, antipodal cells, and the endosperm mother cell, forming a fully developed female gametophyte within the ovule of the flower.

**EMBRYONIC INDUCTION**   The phenomenon by which one region or tissue layer of an embryo induces the differentiation of an adjacent part.

**ENDERGONIC REACTIONS**   Chemical reactions that require energy input from another source in order to occur.

**ENDOCRINE GLANDS**   Ductless glands, which secrete hormones directly into surrounding tissue fluids and blood vessels for distribution to the rest of the body by the circulatory system.

**ENDOCYTOSIS**   A type of active transport that imports particles or small cells into a cell. There are two types of endocytic processes: phagocytosis, where large particles are ingested by the cell, and pinocytosis, where small droplets are taken in.

**ENDODERMIS**   The innermost cylindrical layer of cortex surrounding the vascular stele of the root. The closely pressed cells of the endodermis have a waxy covering, forming a waterproof layer, the Casparian strip.

**ENDOSKELETON**   The internal support structure found in all vertebrates and a few invertebrates (sponges and sea stars).

**ENDOSPERM**   Nutritive tissue in seeds.

**ENDOSYMBIOSIS THEORY**   A theory to explain the development of complex eukaryotic cells by proposing that some organelles once were free-living prokaryotic cells that then moved into another larger such cell, forming a beneficial union with it.

**ENDOPLASMIC RETICULUM (ER)**   An elaborate system of folded, stacked and tubular membranes contained in the cytoplasm.

**ENERGY**   The ability to do work.

**ENTROPY**   Energy that is not available for doing work; measure of disorganization or randomness.

**ENVIRONMENTAL RESISTANCE**   The factors that eventually limit the size of a population.

**ENZYME**   Biological catalyst; a protein molecule that accelerates the rate of a chemical reaction.

**EOSINIPHIL**   A type of nonphagocytic white blood cell.

**EPICOTYL**   The portion of the embryo of a dicot plant above the cotyledons. The epicotyl gives rise to the shoot.

**EPIDERMIS**   In animals, the outer layer of the skin, consisting of superficial layers of dead cells produced by the underlying living epithelial cells. In plants, the outer layer of cells covering leaves, primary stem, and primary root.

**EPIDIDYMIS**   Mass of convoluted tubules attached to each testis in mammals. After leaving the testis, sperm enter the tubules where they finish maturing and acquire motility.

**EPIGLOTTIS**   A flap of tissue that covers the glottis during swallowing to prevent food and liquids from entering the lower respiratory tract.

**EPIPHYSEAL PLATES**   The action centers for ossification (bone formation).

**EPISTASIS**   A type of gene interaction in which a particular gene blocks the expression of another gene at another locus.

**EPITHELIAL TISSUE**   Continuous sheets of tightly packed cells that cover the body and line the digestive and respiratory tracts. Epithelium is a fundamental tissue type in animals.

**EQUILIBRIUM**   A state in which a system will remain if it is undisturbed.

**ERYTHROCYTES**   Red blood cells.

**ESSENTIAL AMINO ACIDS**   Eight amino acids that must be acquired from dietary protein. If even one is missing from the human diet, the synthesis of proteins is prevented.

**ESSENTIAL FATTY ACIDS**   Linolenic and linoleic acids, which are required for phospholipid construction and must be acquired from a dietary source.

**ESSENTIAL NUTRIENTS**   The 16 minerals essential for plant growth, divided into two groups: macronutrients, which are required in large quantities, and micronutrients, which are needed in small amounts.

**ESTROGEN**   A female sex hormone secreted by the ovaries when stimulated by pituitary gonadotrophins.

**ESTUARIES**   Areas found where rivers and streams empty into oceans, mixing fresh water with salt water.

**ETHYLENE GAS**   A plant hormone that stimulates fruit ripening.

**ETIOLATION**   The condition of rapid shoot elongation, small underdeveloped leaves, bent shoot-hook, and lack of chlorophyll, all due to lack of light.

**EUKARYOTIC**   Referring to organisms whose cellular anatomy includes a true nucleus with a nuclear envelope, as well as other membrane-bound organelles.

**EUTROPHICATION**  The natural aging process of lakes and ponds, whereby they become marshes and, eventually, terrestrial environments.

**EVOLUTION**  A process whereby the characteristics of a species change over time, eventually leading to the formation of new species that go about life in new ways.

**EXCITATORY NEURONS**  Neurons that stimulate their target cells into activity.

**EXCRETION**  Removal of metabolic wastes from an organism.

**EXCRETORY SYSTEM**  The organ system that eliminates metabolic wastes from the body.

**EXERGONIC REACTIONS**  Chemical reactions that release energy.

**EXCRETION**  The process used by organisms to remove metabolic wastes.

**EXOCYTOSIS**  A form of active transport used by cells to move molecules, particles, or other cells contained in vesicles across the plasma membrane to the cell's environment.

**EXOCRINE GLANDS**  Glands which secrete their products through ducts directly to their sites of action, e.g., tear glands.

**EXONS**  Structural gene segments that are transcribed and whose genetic information is subsequently translated into protein.

**EXOSKELETONS**  Hard external coverings found in some animals (e.g., lobsters, insects) for protection, support, or both. Such organisms grow by the process of molting.

**EXPLOITATIVE COMPETITION**  A competition in which one species manages to get more of a resource, thereby reducing supplies for a competitor.

**EXTERNAL FERTILIZATION**  The union of sperm and egg outside the body. This is the usual mode of production in aquatic animals.

**EXTINCTION**  The loss of a species.

**EXTRACELLULAR DIGESTION**  Digestion occurring outside the cell; occurs in bacteria, fungi, and multicellular animals.

# F

**F1**  First filial generation. The first generation of offspring in a genetic cross.

**F2**  Second filial generation. The offspring of an F1 cross.

**FACILITATED DIFFUSION**  The transport of molecules into cells with the aid of "carrier" proteins embedded in the plasma membrane. This carrier-assisted transport does not require the expenditure of energy by the cell.

**FAD**  Flavin adenine dinucleotide. A coenzyme that functions as an electron carrier in metabolic reactions. When it is reduced to $FADH_2$, this molecule becomes a cellular energy source.

**FAMILY**  A level of the taxonomic hierarchy that groups together members of related genera.

**FAT**  A triglyceride that is solid at room temperature (25°C).

**FAUNA**  The animals in a particular region.

**FEEDBACK INHIBITION (NEGATIVE FEEDBACK)**  A mechanism for regulating enzyme activity by temporarily inactivating a key enzyme in a biosynthetic pathway when the concentration of the end product is elevated.

**FERMENTATION**  The direct donation of the electrons of NADH to an organic compound without their passing through an electron transport system.

**FERTILITY RATE**  In humans, the average number of children born to each woman between 15 and 44 years of age.

**FERTILIZATION**  The process in which two haploid nuclei fuse to form a zygote.

**FETUS**  The term used for the human embryo during the last seven months in the uterus. During the fetal stage, organ refinement accompanies overall growth.

**FIBRINOGEN**  A rod-shaped plasma protein that, converted to fibrin, generates a tangled net of fibers that binds a wound and stops blood loss until new cells replace the damaged tissue.

**FILAMENTOUS FUNGI**  Multicellular fungi comprised mostly of living threads, called hyphae, that grow by division at their tips. These fungi are commonly referred to as "molds."

**FLAGELLA**  Cellular extensions that are longer than cilia but fewer in number. Their undulations propel cells like sperm and many protozoans, through their aqueous environment.

**FLORA**  The plants in a particular region.

**FLORIGEN**  A chemical hormone that is produced in the leaves and stimulates flowering.

**FLUID MOSAIC MODEL**  The model proposes that the phospholipid bilayer has a viscosity similar to that of light household oil and that globular proteins float like icebergs within this bilayer. The now favored explanation for the architecture of the plasma membrane.

**FOLLICLE (OVARIAN)**  A chamber of cells housing the developing oocytes.

**FOOD CHAIN**  Transfers of food energy from organism to organism, in a linear fashion.

**FOOD WEB**  The map of all interconnections between food chains for an ecosystem.

**FOREBRAIN**  Most prominent part of the human brain, containing the cerebrum, which is responsible for the most cognitive processes. In lower vertebrates, the forebrain, and especially the cerebrum, is much smaller.

**FOREST BIOMES**  Three types of broad geographic regions, each with characteristic vegetation: 1) tropical rain forests (lush forests in a broad band around the equator), 2) deciduous forests (trees and shrubs drop their leaves during unfavorable seasons), 3) coniferous forest (evergreen conifers).

**FOSSILS**  The preserved remains of organisms from a former geologic age.

**FOUNDER EFFECT**  The potentially dramatic difference in allele frequency of a small founding population as compared to the original population.

**FOUNDER POPULATION** The individuals, usually few, that colonize a new habitat.

**FRAMESHIFT MUTATION** The insertion or deletion of nucleotides in a gene that throws off the reading frame.

**FRONDS** The large leaf-like structures of ferns. Unlike true leaves, fronds have an apical meristem and clusters of sporangia called sori.

**FRUIT** A mature plant ovary (flower) containing seeds with plant embryos. Fruits protect seeds and aid in their dispersal.

**FRUITING BODY** Spore-producing structures in certain fungi (e.g., the mushroom).

**FSH** Follicle stimulating hormone. A hormone secreted by the anterior pituitary that prepares a female for ovulation by stimulating the primary follicle to ripen or stimulates spermatogenesis in males.

**FUNCTIONAL GROUPS** Accessory chemical entities (e.g., —OH, —NH$_2$, —CH$_3$), which help determine the identity and chemical properties of a compound.

# G

**G1 STAGE** The first of three consecutive stages of interphase. During G1, cell growth and normal functions occur. The duration of this stage is most variable.

**G2 STAGE** The final stage of interphase in which the final preparations for mitosis occur.

**GAS EXCHANGE MEMBRANES** Surface through which gases must pass in order to enter or leave the body of an animal. It may be the plasma membrane of a protistan or the complex tissues of the gills or the lungs in multicellular animals.

**GASTROINTESTINAL TRACT** The stomach and the intestines.

**GASTROVASCULAR CAVITY** In cnidarians and flatworms, the branched cavity with only one opening. It functions in both digestion and transport of nutrients.

**GASTRULA** The embryonic stage formed by the inward migration of cells in the blastula. It consists of: an outer ectoderm, an inner ectoderm, a hollow core called the archenteron, and a blastospore, the opening into the archenteron.

**GENE FLOW** Exchange of genes among members of a single population or from one population to another.

**GENE POOL** All the genes in all the individuals of a population.

**GENES** Discrete units of inheritance which determine hereditary traits.

**GENETIC DRIFT** Random changes in allele frequency that occur by chance alone.

**GENETIC MARKERS** Mutations that show up as observable traits which enable the tracking of an allele from generation to generation.

**GENETIC RECOMBINATION** The reshuffling of gene combinations caused by breakage of DNA and its reunion at a different chromosomal point.

**GENOTYPE** An individual's genetic makeup.

**GENUS** Taxonomic group containing related species.

**GERMINATION** The sprouting of a seed, beginning with the radicle of the embryo breaking through the seed coat.

**GERM LAYERS** Collective name for the endoderm, ectoderm, and mesoderm, from which all the structures of the mature animal develop.

**GERMINAL TISSUES** Animal tissues that produce germ cells or gametes. These are found in the ovaries of the female and in the testes of the male.

**GIBBERELLINS** More than 50 compounds that promote growth by stimulating both cell elongation and cell division.

**GLANS** The most sensitive part of the penis which contains the urethral opening through which semen, the sperm-containing fluid, is expelled during ejaculation.

**GLIAL CELLS** Nonconducting support cells of neurons in the central nervous system.

**GLOMERULUS** A capillary bundle embedded in a double-membraned container, Bowman's capsule, through which blood for the kidney first passes.

**GLOTTIS** Opening leading to the larynx and lower respiratory tract.

**GLUCAGON** A hormone secreted by the Islets of Langerhans that promotes glycogen breakdown to glucose.

**GLUCOCORTICOIDS** Steroid hormones which regulate sugar and protein metabolism. They are secreted by the adrenal cortex.

**GLYCOSIDIC BOND** The covalent bond between individual molecules in carbohydrates.

**GLYCOLYSIS** Cleavage, releasing energy, of the six-carbon glucose molecule into two molecules of pyruvic acid, each containing three carbons.

**GOBLET CELLS** Mucous secreting cells in the epithelium of respiratory tract. The mucous humidifies the air and also traps airborn particles.

**GOLGI BODY** A single stack of cisternae within the Golgi complex. In plant cells it is sometimes called a dictysome.

**GOLGI COMPLEX** All the Golgi bodies in a cell. A system of flattened membranous sacs, or cisternae, which package substances for secretion from the cell.

**GONADOTROPINS** Two anterior pituitary hormones which act on the gonads. Both FSH (follicle-stimulating hormone) and LH (leutinizing hormone) promote gamete development and stimulate the gonads to produce sex hormones.

**GONADOTROPIN RELEASING HORMONE (GNRH)** A hormone secreted by the hypothalamus which, by its quantity, dictates the amount of LH and FSH to be produced.

**GONADS** Gamete-producing structures in animals: ovaries in females, testes in males.

**GRANA (sing. GRANUM)** Stacks of thylakoids within a chloroplast.

**GRASSLANDS** Areas of densely packed grasses and herbaceous plants.

**GRAVITROPISMS (GEOTROPISMS)** Changes in plant growth caused by gravity. Growth away from gravitational force is called negative gravitropism; growth toward it is positive.

**GRAY MATTER** Gray-colored neural tissue in the cerebral cortex of the brain and in the butterfly-shaped interior of the spinal cord. Composed of nonmyelinated cell bodies and dendrites of neurons.

**GROSS PRIMARY PRODUCTIVITY (GPP)** The rate at which primary producers convert energy and inorganic chemicals into organic compounds.

**GROUND MERISTEM** One of the three primary meristems of plants; produces the cells and tissues of the ground tissue system.

**GROUND TISSUE SYSTEM** All plant tissues except those in the dermal and vascular tissues.

**GROWTH** An increase in size, resulting from cell division and/or an increase in the volume of individual cells.

**GROWTH HORMONE (GH)** Hormone produced by the anterior pituitary; stimulates protein synthesis and bone elongation.

**GUARD CELLS** Specialized epidermal plant cells that flank each stomated pore of a leaf. They regulate the rate of gas diffusion and transpiration.

**GUILD** Group of species with similar ecological niches.

**GUTTATION** The forcing of water and mineral completely out to the tips of leaves as a result of positive root pressure.

**GYMNOSPERMS** The earliest seed plants, bearing naked seeds. Includes the pines, hemlocks, and firs.

# H

**HALF-LIFE** The time required for half the mass of a radioactive element to decay into its stable, non-radioactive form.

**HAVERSIAN CANALS** A system of microscopic canals in compact bone that transport bone that transport nutrients to and remove wastes from osteocytes.

**HERBACEOUS** Plants having only primary growth and thus composed entirely of primary tissue.

**HEMIZYGOUS** Referring to the single does of X-linked genes in an XY male.

**HEMOGLOBIN** The iron-containing blood protein that temporarily binds $O_2$ and releases it into the tissues.

**HEREDITY** The passage of genetic traits to offspring which consequently are similar or identical to the parent(s).

**HERMAPHRODITES** Animals that possess gonads of both the male and the female.

**HETEROSPOROUS** Higher vascular plants producing two types of spores, a megaspore which grows into a female gametophyte and a microspore which grows into a male gametophyte.

**HETEROZYGOUS** A term applied to organisms that possess two different alleles for a trait. Often, one allelle (A) is dominant, masking the presence of the other (a), the recessive.

**HISTONES** Small alkaline proteins that are complexed with DNA to form nucleosomes, the basic structural components of the chromatin fiber.

**HOMEOSTASIS** Maintenance of fairly constant internal conditions (e.g., blood glucose level, pH, body temperature, etc.)

**HOMOLOGOUS STRUCTURES** Anatomical structures that may have different functions but develop from the same embryonic tissues, suggesting a common evolutionary origin.

**HOMOLOGUES** Members of a chromosome pair, which have a similar shape and the same sequence of genes along their length.

**HOMOSPOROUS** Plants that manufacture only one type of spore, which develops into a gametophyte containing both male and female reproductive structures.

**HOMOZYGOUS** A term applied to an organism that has two identical alles for a particular trait.

**HORMONES** Chemical messengers that direct tissues to change their activities and correct imbalances in body chemistry.

**HUMAN CHORIONIC GONADOTROPIN (HCG)** A hormone that prevents the corpus luteum from degenerating, thereby maintaining an adequate level of progesterone during pregnancy. It is produced by cells of the early embryo.

**HUMAN IMMUNODEFICIENCY VIRUS (HIV)** The infectious agent that causes AIDS, a disease in which the immune system is seriously disabled.

**HYDROGEN BONDS** Relatively weak chemical bonds formed when two molecules share an atom of hydrogen.

**HYDROLYSIS** Splitting of a covalent bond by donating the $H^+$ or $OH^-$ of a water molecule to the two components.

**HYDROPHILIC MOLECULES** Polar molecules that are attracted to water molecules and readily dissolve in water.

**HYDROPHOBIC MOLECULES** Nonpolar substances, insoluble in water, which form aggregates to minimize exposure to their polar surroundings.

**HYDROSTATIC SKELETONS** Body support systems found usually in underwater animals (e.g., marine worms). Body shape is protected against gravity and other physical forces by internal hydrostatic pressure produced by contracting muscles encircling their closed, fluid-filled chambers.

**HYPERTONIC SOLUTIONS** Solutions with higher solute concentrations than found inside the cell. These cause a cell to lose water and shrink.

**HYPOCOTYL** Portion of the embryo below the cotyledons. The hypocotyl gives rise to the root and, very often, to the lower part of the stem.

**HYPOTHALAMUS** The area of the brain below the thalamus that regulates body temperature, blood pressure, etc.

**HYPOTHESIS** A tentative explanation for an observation or a phenomenon, phrased so that it can be tested by experimentation.

**HYPOTONIC SOLUTIONS** Solutions with lower solute concentrations than found inside the cell. These cause a cell to accumulate water and swell.

# I

**IMBIBITION** Entry of water into the seed, causing swelling and subsequent eruption of the seed coat.

**IMMUNE SYSTEM** A system in invertebrates for the surveillance and destruction of disease-causing microorganisms and cancer cells. Composed of lymphocytes, particularly B-cells and T-cells, and triggered by the introduction of antigens into the body which makes the body, upon their destruction, resistant to a recurrence of the same disease.

**IMPERFECT FLOWERS** Flowers that contain either stamens or carpels, making them male or female flowers, respectively.

**INCOMPLETE (PARTIAL) DOMINANCE** A phenomenon in which heterozygous individuals are phenotypically distinguishable from either homozygous type.

**INCOMPLETE FLOWER** Flowers lacking one or more whorls.

**INDEPENDENT ASSORTMENT** The shuffling of members of homologous chromosome pairs in meiosis I. As a result, there are new chromosome combinations in the daughter cells, which later produce offspring with random mixtures of traits from both parents.

**INDOLEATIC ACID (IAA)** An auxin responsible for many plant growth responses including apical dominance, a growth pattern in which shoot tips prevent axillary buds from sprouting.

**INFLAMMATION** A body strategy initiated by the release of chemicals following injury or infection which brings additional blood with its protective cells to the injured area.

**INHIBITORY NEURONS** Neurons that oppose a response in the target cells.

**INHIBITORY NEUROTRANSMITTERS** Substances released from inhibitory neurons where they synapse with the target cell.

**INSULIN** One of the two hormones secreted by several endocrine centers called Islets of Langerhans; promotes glucose absorption, utilization, and storage. Insulin is secreted by them when the concentration of glucose in the blood begins to exceed the normal level.

**INTEGUMENTARY SYSTEM** The body's protective external covering, consisting of skin and subcutaneous tissue.

**INTEGUMENTS** Protective covering of the ovule.

**INTERCALATED DISKS** Folded plasma membranes between adjacent cells of cardian muscle fibers. They facilitate conduction of nerve impulses from cell to cell.

**INTERFERENCE COMPETITION** Competition in which one species directly interferes with the ability of a competitor to gain access to a limited resource.

**INTEGRANAL THYLAKOIDS** Membranes connecting adjacent grana in plant chloroplasts.

**INTERNEURONS** Neurons of the central nervous system that carry impulses from sensory to motor neurons.

**INTERNODE** The region between two nodes along a stem.

**INTERPHASE** Usually the longest stage of the cell cycle during which the cell grows, carries out normal metabolic functions, and replicates its DNA in preparation for cell division.

**INTERSTITIAL CELLS** Cells in the testes that produce testosterone, the major male sex hormone.

**INTERNAL FERTILIZATION** The union of sperm and egg inside a chamber in the body. This is the mode of reproduction in terrestrial animals.

**INTERTIDAL ZONE** The region of beach exposed to air between low and high tides.

**INTRINSIC RATE OF INCREASE ($r_m$)** the maximum growth rate of a population under conditions of maximum birth rate and minimum death rate.

**INTRONS** Intervening sequences of DNA in the middle of structural genes, separating axons.

**INVERTEBRATES** Animals that lack a vertebral column, or backbone.

**ION** An electrically charged atom created by the gain or loss of electrons.

**IONIC BOND** The chemical linkage formed by the attraction of oppositely charged ions. Technically, the association does not produce a molecule.

**ISOLATING MECHANISMS** Barriers that prevent gene flow between populations or among segments of a single population.

**ISOTONIC SOLUTIONS** Solutions in which the solute concentration outside the cell is the same as that inside the cell.

**ISOTOPES** Different forms of the same element, results from different numbers of neutrons.

# J

**J-SHAPED CURVE** A curve which represents the exponential growth phase of a population.

# K

**K-SELECTED SPECIES** Species that produce one or a few well-cared for individuals at a time.

**KARYOKINESIS** The first stage in eukaryotic cell division in which the nucleus and its contents divide to form new nuclei.

**KARYOTYPE** A visual display of an individual's chromosomes.

**KIDNEYS** Paired excretory organs which, in humans, are fist-sized and attached to the lower spine. In vertebrates, the kidneys remove nitrogenous wastes from the blood and regulate ion and water levels in the body.

**KILLER T-CELLS** A type of lymphocyte that functions in the destruction of virus-infected cells and cancer cells.

**KINETIC ENERGY** Energy in motion.

**KINGDOM** A level of the taxonomic hierarchy that groups together members of related phyla or divisions. Modern taxonomy divides all organisms into five Kingdoms: Monera, Protista, Fungi, Plantae, and Animalia.

**KRANZ ANATOMY** The characteristic leaf anatomy of all $C_4$ plants which contain photosynthetic mesophyll cells, but also bundle sheath cells with many chloroplasts. The special arrangements of these cell types and the compartmentalization of different synthesis pathways allows glucose to be synthesized even when the $CO_2$ concentration in the leaf is very low.

**KREBS CYCLE** A pathway in aerobic respiration that completely oxidizes the two pyruvic acids from glycolysis.

# L

**LABIA** The lips of the vulva.

**LAMELLA** In bone, concentric cylinders of calcified collagen deposited by the osteocytes. The laminated layers produce a greatly strengthened structure.

**LAMPBRUSH CHROMOSOMES** In oocytes, a synapsed meiotic chromosome from whose main axis loops of DNA extend. The loops of DNA contain genes that direct the synthesis of proteins needed by the zygote and early embryo.

**LARGE INTESTINE** Portion of the intestine in which water and salts are reabsorbed. It is so named because of its large diameter. The large intestine, except for the rectum, is called the colon.

**LARVA** A self-feeding, sexually, and developmentally immature form of an animal.

**LARYNX** The short passageway connecting the pharynx with the lower airways.

**LAW OF INDEPENDENT ASSORTMENT** Alleles on nonhomologous chromosomes segregate independently of one another.

**LAW OF SEGREGATION** During gamete formation, pairs of alleles separate so that each sperm or egg cell has one gene for a trait.

**LAW OF THE MINIMUM** The ecological principle that a species' distribution will be limited by whichever abiotic factor is most deficient in the environment.

**LAWS OF THERMODYNAMICS** Physical laws that describe the relationship of heat and mechanical energy. The first law states that energy cannot be created or destroyed, but one form can change into another. The second law states that the total energy of a system decreases as energy conversions occur and some energy is lost as heat.

**LENTICELS** Loosely packed cells in the periderm of the stem that create air channels for transferring $CO_2$, $H_2O$, and $O_2$.

**LEUKOCYTES** White blood cells.

**LH** Leuteinizing hormone. A hormone secreted by the anterior pituitary that stimulates testosterone production in males and triggers ovulation and the transformation of the follicle into the corpus luteum in females.

**LICHEN** Symbiotic associations between certain fungi and algae.

**LIFE CYCLE** The sequence of events during the lifetime of an organism from zygote to reproduction.

**LIGHT REACTIONS** First stage of photosynthesis in which light energy is converted to chemical energy in the form of energy-rich ATP and NADPH.

**LIGAMENTS** Strong straps of connective tissue that hold together the bones in articulating joints or support an organ in place.

**LIMBIC SYSTEM** A series of an interconnected group of brain structures, including the thalamus and hypothalamus, controlling memory and emotions.

**LIMITING FACTORS** The critical factors which impose restraints of the distribution, health, or activities of an organism.

**LINKAGE** The tendency of genes of the same chromosome to stay together rather than to assort independently.

**LINKAGE GROUPS** Groups of genes located on the same chromosome. The genes of each linkage group assort independently of the genes of other linkage groups. In all eukaryotic organisms, the number of linkage groups is equal to the haploid number of chromosomes.

**LINKAGE MAPPING** A technique used to construct genetic maps that show relative positions and distances of the genes on a chromosome.

**LIPIDS** Organic compounds which contain fatty acids of the four-ringed steroid structure.

**LIPOPROTEINS** A molecular hybrid of lipid and protein. In membrane lipoproteins, the lipid portion of helps integrate the functional protein into the hydrophobic interior of the membrane.

**LOCUS** The chromosomal location of a gene.

**LOGISTIC GROWTH** Population growth producing a sigmoid, or S-shaped, growth curve.

**LONG-DAY PLANTS** Plants that flower when the length of daylight exceeds some critical period.

**LONGITUDINAL FISSION** The division pattern in flagellated protozoans, where division is along the length of the cell.

**LOOP OF HENLE** An elongated section of the renal tubule that dips down into the kidney's medulla and then ascends back up to the cortex, separating the proximal and distal convoluted tubules.

**LUMEN** A space within an hollow organ or tube.

**LYMPH** The colorless fluid in lymphatic vessels.

**LYMPHATIC SYSTEM** Network of fluid-carrying vessels and associated organs that participate in immunity and in the return of tissue fluid to the main circulation.

**LYMPHOCYTES** A group of non-phagocytic white blood cells which combat microbial invasion, fight cancer, and neutralize toxic chemicals. The two classes of lymphocytes, B-cells and T-cells, are the heart of the immune system.

**LYMPHOID ORGANS** Organs associated with production of blood cells and the lymphatic system, including the thymus, spleen, appendix, bone marrow, and lymph nodes.

**LYMPHOKINES** Chemicals that enhance the phagocytic activities of macrophages and stimulate division of certain cells of the immune system. Lymphokines are produced by helper T-cells.

**LYSIS** To split or dissolve.

**LYSOMES** A type of storage vesicle produced by the Golgi complex, containing hydrolytic (digestive) enzymes capable of digesting many kinds of macromolecules in the cell. The membrane around them keeps them sequestered.

# M

**MACROMOLECULES** Large polymers, such as proteins, nucleic acids, and polysaccharides.

**MACROPHAGES** Phagocytic cells that develop from monocytes.

**MACROSCOPIC** Referring to biological observations made with the naked eye or a hand lens.

**MADREPORITE** Sieve-like openings in echinoderms, attached to the central channel to draw water.

**MAMMALS** A class of vertebrates that possesses skin covered with hair and that nourishes their young with milk from mammary glands.

**MAMMARY GLANDS** Glands contained in the breasts of mammalian mothers that produce breast milk.

**MANTLE** In a mollusk, the cavity between the mantle flap and the visceral mass.

**MARSUPIALS** Mammals with a cloaca whose young are born immature and complete their development in an external pouch in the mother's skin.

**MEDULLA** The center-most portion of some organs.

**MEDULLAR CAVITY** The hollow core of long bones in which fat is stored.

**MEDUSA** The motile, umbrella-shaped body form of some members of the phylum enidaria, with mouth and tentacles on the lower, concave service. (Compare with polyp.)

**MEGASPORE MOTHER CELL** Large diploid cell in the ovule of a flower. It gives rise to the megaspores by meiosis.

**MEGASPORES** Spores that divide by mitosis to produce female gametophytes that produce the eggs.

**MEIOSIS** The division process that produces cells with one-half the number of chromosomes in each somatic cell. Each resulting daughter cell is haploid (1n).

**MEIOSIS I** A process of reductional division in which homologous chromosomes pair and then segregate. Homologues are partitioned into separate daughter cells.

**MEIOSIS II** Second meiotic division. A division process resembling mitosis, except that the haploid number of chromosomes is present. After the chromosomes line up at the metaphase plate, the centromeres split and the two sister chromatids separate.

**MELANIN** A brown pigment that gives skin and hair its color.

**MEMBRANE INVAGINATION THEORY** A theory which attempts to explain the development of complex eukaryotic cells. It proposes that the plasma membrane of the ancestral prokaryotic cell gradually folded in on itself, forming pockets (organelles) that isolated specific metabolic reactions.

**MEMORY CELLS** Lymphocytes responsible for active immunity. They recall a previous exposure to an antigen and, on subsequent exposure to the same antigen, proliferate rapidly into plasma cells and produce large quantities of antibodies in a short time. This protection typically lasts for many years.

**MENINGES** A water-tight membrane encasing the entire surface of the central nervous system, preventing the cerebrospinal fluid from seeping into surrounding tissues.

**MENSTRUAL CYCLE** The repetitive monthly changes in the uterus that prepare the endometrium for receiving and supporting an embryo.

**MERISTEMS** In plants, clusters of cells that retain their ability to divide, thereby producing new cells. One of the four basic tissues in plants.

**MESENCHYME** The middle body layer of sponges.

**MESENTERIES** Extensions from the peritonium on which are suspended the organs and organ systems of complex animals.

**MESODERM** In animals, the middle germ cell layer of the gastrula. It gives rise to muscle, bone, connective tissue, gonads, and kidney.

**MESOGLEA** The middle body layer of cnidarians.

**MESOPHYLL** Layers of cells in a leaf between the upper and lower epidermis; produced by the ground meristem.

**MESSENGER RNA (mRNA)** The RNA that carries genetic information from the DNA in the nucleus to the ribosomes in the cytoplasm, where the sequence of bases in the mRNA is translated into a sequence of amino acids.

**METABOLIC INTERMEDIATES** Compounds produced as a substrate are converted to end product in a series of enzymatic reactions.

**METABOLIC PATHWAYS** Set of enzymatic reactions involved in either building or dismantling complex molecules.

**METABOLISM** The sum of all the chemical reactions in an organism; includes all anabolic and catabolic reactions.

**METAPHASE** The stage of mitosis when the chromosomes line-up along the mataphase plate, a plate that lies midway between the spindle poles.

**METAMORPHOSIS** Transformation from one form into another form during develoment.

**METAZOAN** A member of the Eumetazoan subkingdom; one of 33 million tissue forming phyla of the animal kingdom.

**MICROBES** Microscopic organisms.

**MICROBIOLOGY** The branch of biology that studies microorganisms.

**MICROBODIES** Membrane-enclosed organelle that contains enzymes that catalyze the conversion of fats to carbohydrates and the disposal of toxic hydrogen peroxide.

**MICROFIBRILS** Bundles formed from the intertwining of cellulose molecules, i.e., long chains of glucose molecules in the cell walls of plants.

**MICROFILAMENTS** Thin protein fibers that are responsible for maintenance of cell shape, muscle contraction and cyclosis.

**MICROMETER** One millionth (1/1,000,000) of a meter.

**MICROPYLE** A small opening in the integuments of the ovule through which the pollen tube grows to deliver sperm.

**MICROSPORE MOTHER CELL** Diploid cell in the anther of a flower. It gives rise to microspores by meiosis.

**MICROSPORES** Spores within anthers of flowers. They divide by mitosis to form pollen grains, the male gametophytes that produce the plant's sperm.

**MICROTUBULES** Thin, hollow tubes in cells; built from repeating protein units of tubulin. Microtubules are components of cilia, flagella, and the cytoskeleton.

**MICROVILLI** The small projections on the cells that comprise each villus of the intestinal wall, further increasing the absorption surface area of the small intestine.

**MIDDLE LAMELLA** Intermediate layer that forms when adjacent cell walls of multicellular plants bond together.

**MIGRATION** Movements of a population into or out of an area.

**MIMICRY** A defense mechanism where one species resembles another in color, shape, behavior, or sound.

**MINERALOCORTICOIDS** Steroid hormones which regulate the level of sodium and potassium in the blood.

**MITOCHONDRIA** Organelles that contain the biochemical machinery for the Krebs cycle and the electron transport system of aerobic respiration. They are composed of two membranes, the inner one forming folds, or cristae.

**MITOSIS** The process of cell division producing daughter cells with exactly the same number of chromosomes as in the mother cell.

**MOLECULAR HYBRIDS** Molecules that contain two or more biochemical types, e.g., glycoproteins.

**MOLECULE** Chemical substance formed when two or more atoms bond together; the smallest unit of matter that possesses the qualities of a compound.

**MOLLUSCA** A phylum, second only to Arthropoda in diversity. Composed of three classes: 1) Gastropoda (spiral-shelled), 2) Bivalvia (hinged shells), 3) Aphalopoda (with tentacles or arms and no, or very reduced shells).

**MOLTING (ECDYSIS)** Shedding process by which certain arthropods lose their exoskeletons as their bodies grow larger.

**MONERA** The taxonomic kingdom comprised of single-celled prokaryotes such as bacteria, cyanobacteria, and archebacteria.

**MONOCLONAL ANTIBODIES** Antibodies produced by a clone of hybridoma cells, all of which descended from one cell.

**MONOCOTYLEDAE (MONOCOTS)** One of the two divisions of flowering plants, characterized by seeds with a single cotyledon, flower parts in 3s, parallel veins in leaves, many roots of approximately equal size, scattered vascular bundles in its stem anatomy, pith in its root anatomy, and no secondary growth capacity.

**MONOECIOUS** Having both male and female flowers on the same plant.

**MONOMERS** Small molecular subunits which are the building blocks of macromolecules. The macromolecules in living systems are constructed of some 40 different monomers.

**MORPHOGENESIS** The development of forms in the animal embryo, beginning with the inward migration of cells during gastrulation.

**MORPHOLOGY** The branch of biology that studies form and structure of organisms.

**MORULA** Stage in animal development where the embryo is a solid ball of cells, the result of cleavage.

**MOTILE** Capable of independent movement.

**MOTOR NEURONS** Nerve cells which carry outgoing impulses to their efforts, either glands or muscles.

**MOTOR UNIT** A muscle fiber together with the motor neuron that innervates it.

**MUCOSA** The cell layer that lines the digestive tract and secretes a lubricating layer of mucus.

**MULLERIAN MIMICRY** Resemblance of different species, each of which is equally obnoxious to predators.

**MULTICELLULAR** Consisting of many cells.

**MULTICHANNEL FOOD CHAIN** Where the same primary producer supplies the energy for more than one food chain.

**MULTIPLE ALLELE SYSTEM** Three or more possible alleles for a given trait, such as ABO blood groups in humans.

**MULTIPLE FISSION** Division of the cell's nucleus without a corresponding division of cytoplasm.

**MULTIPLE FRUITS** Fruits that develop from pistils of separate flowers.

**MULTIUNIT SMOOTH MUSCLE** Muscle found at non-visceral locations, such as around blood vessels, in the eyes, and at the base of hair shafts.

**MUSCLE FIBERS** The multicellular fiber that results from the fusion of several pre-muscle cells during embryonic development.

**MUSCLE TISSUE** Bundles and sheets of contractile cells that shorten when stimulated, providing force for controlled movement.

**MUTAGENS** Chemical or physical agents that induce genetic change.

**MUTATION** Random heritable changes in DNA that introduce new alleles into the gene pool.

**MUTUALISM** The symbiotic interaction in which both participants benefit.

**MYCOLOGY** The branch of biology that studies fungi.

**MYCORRHIZAE** Soil fungi associated with the roots of vascular plants, increasing the plant's ability to extract water and minerals from the soil.

**MYELIN SHEATH** In vertebrates, a jacket which covers the axons of high-velocity neurons, thereby increasing the speed of a neurological impulse.

**MYOFIBRILS** In striated muscle, the banded fibers that lie parallel to each other, constituting the bulk of the muscle fiber's interior and powering contraction.

# N

**NAD⁺** Nicotinamide adenine dinucleotide. A coenzyme that functions as an electron carrier in metabolic reactions. When reduced to NADH, the molecule becomes a cellular energy source.

**NATURAL SELECTION** Differential survival and reproduction of organisms with a resultant increase in the frequency of those best adapted to the environment.

**NEGATIVE FEEDBACK** Any regulatory mechanism in which the increased level of a substance inhibits further production of that substance, thereby preventing harmful accumulation. A type of homeostatic mechanism.

**NEGATIVE PRESSURE BREATHING** Intake of air into the lungs caused by decreased pressure inside the lungs. The pressure drop results from contraction of the diaphragm which increased the volume of the thoratic cavity.

**NEMATOCYST** Within the stinging cell (cnidocyte) of cnidarians, a capsule that contains a coiled thread which, when triggered, harpoons prey and injects powerful toxins.

**NEMATODA** The widespread and abundant animal phylum containing the roundworms.

**NEPHRIDIUM** A tube surrounded by capillaries found in an organism's excretory organs that removes nitrogenous wastes and regulates the water and chemical balance of body fluids.

**NEPHRON** The functional unit of the kidney, consisting of the glomerulus, Bowman's capsule, proximal and distal conjointed tabules, and loop of Henle.

**NERVE** Parallel bundles of neurons and their supporting cells.

**NERVE IMPULSE** An action potential.

**NERVOUS TISSUE** Excitable cells that receive stimuli and, in response, transmit an impulse to another part of the animal.

**NET COMMUNITY PRODUCTIVITY (NCP)** Net primary productivity (NPP) minus the energy usage during respiration by heterotrophs ($R_H$). Formula: $NCP = NPP - R_H$.

**NET PRIMARY PRODUCTIVITY (NPP)** The rate at which organic matter accumulated by autotrophs in an ecosystem (Gross Primary Productivity, or GPP) minus the energy usage by autototrophs during respiration ($R_A$). Formula: $NPP = GPP - R_A$.

**NEURON** A nerve cell

**NEUROTRANSMITTERS** Chemicals released by neurons into the synaptic cleft, stimulating or inhibiting the post-synaptic cell.

**NEURULATION** Formation by embryonic induction of the neural tube in a developing embryo.

**NEUTRONS** Electrically neutral (uncharged) particles contained within the nucleus of the atom.

**NEUTROPHIL** Phagocytic leukocyte, most numerous in the human body.

**NICHE** An organism's habitat, role, resource requirements, and tolerance ranges for each abiotic condition.

**NITROGEN FIXATION** the conversion of atmospheric nitrogen gas $N_2$ into ammonia ($NH_3$) by certain bacteria and cyanobacteria.

**NODES** The attachment points of leaves to a stem.

**NODES OF RANVIER** Uninsulated (nonmyelinated) gaps along the axon of a neuron.

**NON-CYCLIC PHOTOPHOSPHORYLATION** The pathway in the light reactions of photosynthesis in which electrons pass from water, through two photosystems, and then ultimately to NADP. During the process, both ATP and NADPH are produced. It is so named because the electrons do not return to their reaction center.

**NONDISJUNCTION** Failure of synapsed chromosomes to separate properly at meiosis I. The result is that one daughter will receive an extra chromosome and the other gets one less.

**NONPOLAR MOLECULES** Molecules which have an equal charge distribution throughout their structure and thus lack regions with a localized positive or negative charge.

**NORMAL FLORA** Microbes indigenous to the surface of an animal's body.

**NOTOCHORD** A flexible rod that is below the dorsal surface of the chordate embryo, beneath the nerve chord. In most chordates, it is replaced by the vertebral column.

**NUCLEAR ENVELOPE** A double membrane that separates the contents of the nucleus from the rest of the eukaryotic cell.

**NUCLEIC ACIDS** DNA and RNA; linear polymers of nucleotides, responsible for the storage and expression of genetic information.

**NUCLEOUS (pl. nucleoli)** One or more darker regions of a nucleus where each ribosomal subunit is assembled from RNA and protein.

**NUCLEOSOMES** Nuclear protein complex consisting of a length of DNA wrapped around a central cluster of 8 histones.

**NUCLEOTIDES** Monomers of which DNA and RNA are built. Each consists of a 5-carbon sugar, phosphate, and a nitrogenous base.

**NUCLEUS** The large membrane-enclosed organelle that contains the DNA of eukaryotic cells.

**NUCLEUS, ATOMIC** The center of an atom containing protons and neutrons.

**NYMPH** Immature stage in those insects where metamorphosis does not include a pupal stage. Unlike a larva or pupa, the nymphs resemble the adult; they go through several molts in which they shed their hard outer skin and secrete a new cuticle over a larger body.

# O

**OLFACTION** The sense of smell.

**OLIGOTROPHIC** Little nourished, as a young lake that has few nutrients and supports little life.

**OMNIVORE** An animal that obtains its nutritional needs by consuming plants and other animals.

**ONCOGENE** A gene that causes cancer, perhaps activated by mutation or a change in its chromosomal location.

**OOCYTE** An egg cell. Primary oocytes are precursors of mature egg cells; the mature oocyte and its cell layer are the primary follicles.

**OOGENESIS** The process of egg production.

**OPEN CIRCULATORY SYSTEM** Circulatory system in which blood travels from vessels to tissue spaces, through which it percolates prior to returning to the main vessel (compare with closed circulatory system).

**OPERATOR** A regulatory gene in the operon of bacteria. It is the short DNA segment to which the repressor binds, thus preventing RNA polymerase from attaching to the promoter.

**OPERON** A regulatory unit in prokaryotic cells that controls the expression of structural genes. The operon consists of structural genes that produce enzymes for a particular metabolic pathway, a regulator region composed of a promoter and an operator, and R (regulator) gene.

**ORDER** A level of the taxonomic hierarchy that groups together members of related families.

**ORGANELLE** A specialized part of a cell having some particular function.

**ORGAN** Body part composed of several tissues that performs specialized functions.

**ORGANISM** A living entity able to maintain its organization, obtain and use energy, reproduce, grow, respond to stimuli, and display homeostatis.

**ORGAN SYSTEM** Group of functionally related organs.

**ORGANIC COMPOUNDS** Chemical compounds that contain carbon.

**ORGANOGENESIS** Organ formation in which two or more specialized tissue types develop in a precise temporal and spatial relationship to each other.

**OSMOREGULATION** The maintenance of a stable fluid environment using hormones to regulate osmotic gradients that adjust fluid concentration, as in the nephrons of the kidneys.

**OSMOSIS** The diffusion of water through a differentially permeable membrane.

**OSSIFICATION** Synthesis of a new bone.

**OSTEOCLAST** A type of bone cell which breaks down the bone, thereby releasing calcium into the bloodstream for use by the body. Osteoclasts are activated by hormones released by the parathyroid glands.

**OSTEOCYTES** Living bone cells embedded within the calcified matrix they manufacture.

**OVARIAN CYCLE** The cycle of egg production within the mammalian ovary.

**OVARIAN FOLLICLE** In a mammalian ovary, a chamber of cells in which the oocyte develops.

**OVARY** In animals, the egg-producing gonad of the female. In flowering plants, the enlarged base of the pistil, in which seeds develop.

**OVIDUCT (FALLOPIAN TUBE)** The tube in the female reproductive organ that connects the ovaries and uterus and where fertilization takes place.

**OVIPAROUS** Referring to an animal that produces eggs that hatch outside the mother's body.

**OVULATION** The release of an egg (ovum) from the ovarian follicle.

**OVULE** In seed plants, the structure containing the female gametophyte, nucellus, and integuments. After fertilization, the ovule develops into a seed.

**OVUM** An unfertilized egg cell; a female gamete.

**OXIDATION** The loss of electrons by an atom or molecule.

**OXIDATIVE PHOSPHORYLATION** The formation of ATP from ADP and inorganic phosphoate that occurs in the electron-transport chain of cellular respiration.

**OXYHEMAGLOBIN** A complex of oxygen and hemoglobin, formed when blood passes through the lungs and is dissociated in body tissues, where oxygen is released.

**OXYTOCIN** A female hormone released by the posterior pituitary which triggers uterine contractions during childbirth and the release of milk during nursing.

## P

**P680 REACTION CENTER (P = PIGMENT)** Special chlorophyll molecule in Photosystem II that traps the energy absorbed by the other pigment molecules. It absorbs light energy maximally at 680 nm.

**PANCREAS** In vertebrates, a large gland that produces digestive enzymes and two hormones.

**PARASITE** An organism that secures its nourishment by living on or inside another called a host.

**PARASYMPATHETIC NERVOUS SYSTEM** Part of the autonomic nervous system active during relaxed activity.

**PARATHYROID GLANDS** Four glands attached to the thyroid gland which secrete parathyroid hormone (PTH). When blood calcium levels are low, PTH is secreted, causing calcium to be released from bone.

**PARENCHYMA** The most prevalent cell type in herbaceous plants. These thin-walled, polygonal-shaped cells function in photosynthesis and storage.

**PARTHENOCARPY** The development of fruits without fertilization.

**PARTHENOGENESIS** Process by which offspring are produced without egg fertilization.

**PASSIVE TRANSPORT** Movement of substances into or out of cells along a concentration gradient, i.e., from a region of higher concentration to a region of lower concentration. The process does not require an expenditure of energy by the cell. (Compare with active transport).

**PATHOGEN** A disease-causing microorganism.

**PECTORAL GIRDLE** The two scapulae (shoulder blades) and two clavicles (collarbones) which support and articulate with the bones of the upper arm.

**PEDICEL** A shortened stem carrying a flower.

**PELAGIC ZONE** The open oceans, divided into three layers: 1) photo- or epipelagic (sunlit), 2) mesopelagic (dim light), 3) aphotic or bathypelagic (always dark).

**PENIS** An intrusive structure in the male animal which releases male gametes into the female's sex receptacle.

**PEPTIDE BOND** The covalent bond between the amino group of one amino acid and the carboxyl group of another.

**PEPTIDOGLYCAN** A chemical component of the prokaryotic cell wall.

**PERENNIALS** Plants that live longer than two years.

**PERFECT FLOWER** Flowers that contain both stamens and carpels.

**PERIDERM** Secondary tissue that replaces the epidermis of stems and roots. Consists of cork, cork cambium, and a layer of parenchyma cells.

**PERIPHERAL NERVOUS SYSTEM** Neurons, excluding those of the brain and spinal cord, that permeate the rest of the body.

**PERISTALSIS** Sequential waves of muscle contractions that propel a substance through a tube.

**PERITONEUM** The smooth connective tissue that lines the abdominal and pleural cavities.

**PERMEABILITY** The ability to be penetrable, such as a membrane allowing molecules to pass freely across it.

**pH** A scale that measures the concentration of hydrogen ions in a solution. The pH scale extends from 0 to 14. Acidic solutions have a pH of less than 7; alkaline solutions have a pH above 7; neutral solutions have a pH equal to 7.

**PHAGOCYTOSIS** Engulfing of food particles and foreign cells by amoebae and white blood cells. A type of endocytosis.

**PHARYNGEAL POUCHES** In the vertebrate embryo, outgrowths from the walls of the pharynx that break through the body surface to form gill slits.

**PHENOTYPE** An individual's observable characteristics that are the expression of its genotype.

**PHEROMONES** Chemicals that, when released by an animal, elicit a particular behavior in other animals of the same species.

**PHLOEM** The vascular tissue that transports sugars and other organic molecules from sites of photosynthesis and storage to the rest of the plant.

**PHYLETIC EVOLUTION** The gradual evolution of one species into another.

**PHOSPHOLIPIDS** Lipids that contain a phosphate and a variable organic group that form polar, hydrophilic regions on an otherwise nonpolar, hydrophobic molecule. They are the major structural components of membranes.

**PHOSPHORYLATION** A chemical reaction in which a phosphate group is added to a molecule or atom; the reaction within cells required to store energy in the form of ATP.

**PHOTOEXCITATION** Absorption of light energy by photosynthesis pigments, causing their electrons to be raised to a higher energy level.

**PHOTON** A theoretical particle of light energy.

**PHOTOPERIODISM** Changes in the behavior and physiology of an organism in response to the relative lengths of daylight and darkness, i.e., the photoperiod.

**PHOTOSYNTHESIS** The conversion by plants of light energy into chemicals. The overall reaction is: $6CO_2 + 12H_2O$ light $C_6H_{12}O_6 + 6O_2 + 6H_2O$ energy

**PHOTOSYSTEMS** Highly organized clusters of photosynthetic pigments and electron/hydrogen carriers embedded in the thylakoid membranes of chloroplasts. There are two photosystems, which together carry out the light reactions of photosynthesis.

**PHOTOSYSTEM I** Photosystem with a P700 reaction center; participates in cyclic photophosphorylation as well as in noncyclic photophosphorylation.

**PHOTOSYSTEM II** Photosystem activated by a P680 reaction center; participates only in noncyclic photophosphorylation and is associated with photolysis of water.

**PHOTOTROPISM** The growth responses of a plant to light.

**PHYCOBILINS** Light-trapping photosynthetic pigments absorbing wavelengths of light between 450 and 650 nm. Found only in cyanobacteria and red algae.

**PHYLOGENY** Evolutionary history of a species.

**PHYSIOLOGY** The branch of biology that studies how living things function.

**PHYTOCHROME** A light-absorbing pigment in plants which controls many plant responses, including photoperiodism.

**PHYTOPLANKTON** Microscopic, photosynthetic microorganisms that drift with ocean currents in the photic zone of the oceans.

**PINOCYTOSIS** Uptake of small droplets by cells. A type of endocytosis.

**PLACENTA** In mammals (exclusive of marsupials and monotremes), the structure through which nutrients and wastes are exchanged between the mother and embryo/fetus. Develops from both embryonic and uterine tissues.

**PLANT** Multicellular, autotrophic organism able to manufacture food through photosynthesis.

**PLANULA** Free-swimming, ciliated larva of cnidarians.

**PLASMA** In veretrates, the liquid portion of the blood, containing water, proteins (including fibrinogen), salts, and nutrients.

**PLASMA MEMBRANE** The selectively permeable, molecular boundary that separates the cytoplasm of a cell from the external environment.

**PLASMID** A small circle of DNA in bacteria in addition to its own chromosome.

**PLASMOIDAL SLIME MOLDS** A large, multinucleated amoeboid form, often classified as a protist. It reproduces by forming fruiting bodies which release spores.

**PLASMOLYSIS** The shrinking of a plant cell away from its cell wall when the cell is placed in a hypertonic solution.

**PLASTIDS** Organelles found only in plant cells specialized for photosynthesis (chloroplasts) and other functions.

**PLATELETS** Small, cell-like fragments derived from the special white blood cells. They function in clotting.

**PLATYHELMINTHES** The phylum containing simple, bilaterally symmetrical animals, the flatworms.

**PLEIOTROPY** Where a single mutant gene produces two or more phenotypic effects.

**POINT MUTATIONS** Changes that occur at one point within a gene, often involving one nucleotide in the DNA.

**POLAR MOLECULE** A molecule with an unequal charge distribution that creates distinct positive and negative regions or poles.

**POLLEN** The male gametophyte of seed plants, comprised of a generative nucleus and a tube nucleus surrounded by a tough wall.

**POLLINATION** The transfer of pollen grains from the anther of one flower to the stigma of another. The transfer is mediated by wind, insects, water, and animals.

**POLYGENIC INHERITANCE** An inheritance pattern in which a phenotype is determined by two or more genes at different loci. In humans, examples include height and pigmentation.

**POLYMER** A macromolecule formed of monomers joined by covalent bonds. Includes proteins, polysaccharides, and nucleic acids.

**POLYP** Stationary body form of some members of the phylum Cnidaria, with mouth and tentacles facing upward. (Compare with medusa).

**POLYPLOIDY** An organism or cell containing three or more complete sets of chromosomes. Polyploidy is rare in animals but common in plants.

**POLYSACCHARIDE** A carbohydrate molecule consisting of two or more monosaccharides.

**POLYSOME** A complex of ribosomes found in chains, linked by mRNA. Polysomes contain the ribosomes that are actively assembling proteins.

**POPULATION** Individuals of the same species inhabiting the same area.

**POSITIVE PRESSURE BREATHING** The intake of air into the lungs, forced in by increasing external pressures.

**POSTERIOR PITUITARY** A gland which manufactures no hormones but receives and later releases hormones produced by the cell bodies of neurons in the hypothalamus.

**POTENTIAL ENERGY** Stored energy, such as occurs in chemical bonds.

**PRECAPILLARY SPHINCTORS** Muscules whose contractions regulate the amount of blood entering the capillary bed of a particular tissue.

**PRECELLS** Simple forerunners of cells that, presumably, were able to concentrate organic molecules, allowing for more frequent molecular reactions.

**PREDATION** Ingestion of prey by a predator for energy and nutrients.

**PREPUCE** The foreskin covering of the glans of the flaccid penis.

**PRIMARY ENDOSPERM CELL** In angiosperms, the triploid (3n) product that results when a sperm nucleus fuses with the diploid endosperm mother cell; gives rise to endosperm, the food storage tissue in the seed.

**PRIMARY FOLLICLE** In the mammalian ovary, a structure composed of the mature oocyte and its surrounding cell layer.

**PRIMARY IMMUNE RESPONSE** Process of antibody production following the first exposure to an antigen. There is a lag time from exposure until the appearance in the blood of protective levels of antibodies.

**PRIMARY GROWTH** Growth from apical meristems, resulting in an increase in the lengths of shoots and roots in plants.

**PRIMARY PRODUCERS** All autotrophs in a biotic environment that use sunlight or chemical energy to manufacture food from inorganic substances.

**PRIMARY SEXUAL CHARACTERISTICS** Gonads and external genitals.

**PRIMATES** Order of mammals that includes humans, apes, monkeys, and lemurs.

**PRIMITIVE (CHARACTERISTICS)** Characteristics that arose early in evolution.

**PROCAMBIUM** One of three primary meristems; produces the vascular tissue systems of plants with cells to transport food, minerals and water throughout the plant.

**PRODUCTS** In a chemical reaction, the compounds into which the reactants are transformed.

**PROGENY** Offspring

**PROGESTERONE** A hormone produced by the corpus luteum within the ovary. It prepares and maintains the uterus for pregnancy, participates in milk production, and prevents the ovary from releasing additional eggs late in the cycle or during pregnancy.

**PROKARYOTIC** Referring to single-celled organisms that have no membrane separating the DNA from the cytoplasm and lack membrane-enclosed organelles. Prokaryotes are confined to the kingdom Monera.

**PROLACTIN** A hormone produced by the anterior pituitary, stimulating milk production by mammary glands.

**PROMOTER** A short segment of DNA to which RNA polymerase attaches at the start of trancription.

**PROPHASE** Longest phase of mitosis, involving the formation of a spindle, coiling of chromatin fibers into condensed chromosomes, and movement of the chromosomes to the center of the cell.

**PROSTAGLANDINS** Hormones secreted by endocrine cells scattered throughout the body responsible for such diverse functions as contraction of uterine muscles, triggering the inflammatory response, and blood clotting.

**PROSTATE GLAND** A muscular gland which produces and releases fluids that make up a substantial portion of the semen.

**PROTEINS** Long chains of amino acids, linked together by peptide bonds. They are folded into specific shapes essential to their functions.

**PROTENOIDS** Artificially-synthesized vesicles, made up of hundreds of amino acids. The vesicles enlarge and bud, suggesting they may be similar to precells.

**PROTHALLUS** The small, heart-shaped gametophyte of a fern.

**PROTISTS** A member of the kingdom Protista; simple eukaryotic organisms that share broad taxonomic similarities.

**PROTOCOOPERATION** Non-compulsory interactions that benefit two organisms, e.g., lichens.

**PROTODERM** one of three primary meristems. It forms the dermal tissue system or epidermis of the plant.

**PROTOSTOMES** One path of development exhibited by coelomate animals (e.g., mollusks, annelids, and arthropods).

**PROTONS** Positively charged particles within in the nucleus of an atom.

**PROTOPLASM** The living portion of a plant cell, i.e., the plasma membrane and cytoplasm, but not the cell wall.

**PROTOZOA** Heterotrophic, unicellular protists, most of which lack cell walls, ingest food particles, and are motile.

**PROVIRUS** DNA copy of a virus' nucleic acid that becomes integrated into the host cell's chromosome.

**PSEUDOCOELAMATES** Animals in which the body cavity is lined with mesodermally-derived cells on only one surface, adjacent to the ectodermis.

**PSEUDOPOD** An organelle for locomotion in certain protozoa (e.g., amoeba); fingerlike projections of cytoplasm that flow forward enabling protozoan to move and to gather food.

**PUBERTY** Development of reproductive capacity, often accompanied by the appearance of secondary sexual characteristics.

**PULMONARY CIRCULATION** The loop of the circulatory system that channels blood to the lungs for oxygenation.

**PUNCTUATED EQUILIBRIUM THEORY** A theory to explain the phenomenon of the relatively sudden appearance of new species, followed by long periods of little or no change.

**PUNNETT SQUARE METHOD** A visual method for predicting the possible genotypes and their expected ratios from a cross.

**PUPA** In insects, the stage in metamorphosis between the larva and the adult. Within the pupal case, there is dramatic transformation in body form as some larval tissues die and others differentiate into those of the adult.

**PURINE** A nitrogenous base found in DNA and RNA having a double ring structure. Adenine and guanine are purines.

**PYRIMIDINE** A nitrogenous base found in DNA and RNA having a single ring structure. Cytosine, thymine, and uracil are pyrimidines.

## Q

**QUIESCENT CENTER** The region in the apical meristem of a root containing relatively inactive cells.

## R

**R (REGULATOR) GENE** A gene that directs the synthesis of a repressor protein which prevents RNA polymerase from attaching to the promoter of the operon, thereby preventing transcription.

**R-GROUP** The variable portion of a molecule; R stands for radical.

**RADICAL SYMMETRY** The quality possessed by animals whose bodies can be divided into mirror images by more than one median plane.

**RADIANT ENERGY** Energy issued from a source as rays, such as heat and light from the sun.

**RADIATION** The emission from atoms of high-speed atomic particles or electromagnetic rays.

**RADICLE** In the plant embryo, the tip of the hypocotyl that eventually develops into the root system.

**RADIOACTIVE** Emission of detectable radiation.

**RADIOLARIAN** A prozoan member of the protitan group Sarcodina that secretes silicon shells through which it captures food.

**RADIOISOTOPE** An unstable isotope of an element that emits high-energy radiation as it decays to a more stable element.

**REACTANTS** Molecules or atoms that are changed to products during a chemical reaction.

**REACTION CENTER** A special chlorophyll molecule in a photosystem ($P_{700}$ in Photosystem I, $P_{680}$ in Photosystem II).

**REACTION** A chemical change in which starting molecules (reactants) are transformed into new molecules (products).

**RECEPTOR SITE** A site on a cell's plasma membrane to which a chemical such as a hormone binds. Each surface site permits the attachment of only one kind of hormone.

**RECESSIVE** An allele whose expression is masked by the dominant allele for the same trait.

**RECOMBINANT DNA** DNA formed by the insertion of foreign genes or foreign sections of DNA from one organism into the chromosomes of a host cell.

**RECOMBINATION** The rejoining of DNA pieces with those of a different strand or with the same strand at a point different from where the break occurred.

**RED MARROW** The soft tissue in the interior of bones

**REDOX REACTION** An electron-releasing oxidation reaction coupled to an electron-accepting reduction reaction.

**REDUCTION** The gain of electrons by an atom or molecule.

**REFLEX ARC** The simplest example of central nervous system control, involving a sensory neuron, interneuron, and motor neuron.

**REGENERATION** Ability of certain animals to replace injured or lost limbs parts by growth and differentiation of undifferentiated stems' cells.

**REGULATORY GENES** Genes whose sole function is to control the expression of structural genes.

**RENAL** Referring to the kidney.

**REPLICATION** Duplication of DNA, usually prior to cell division.

**REPRESSION** Inhibition of transcription of a gene which, in an operon, occurs when repressor protein binds to the operator.

**REPRODUCTION** The process by which an organism produces offspring.

**REPTILES** Members of class Reptilia, scaly, air-breathing, egg-laying vertebrates such as lizards, snakes, turtles, and crocodiles.

**RESOLVING POWER** The ability of an optical instrument (eye, microscopes) to discern whether two very close objects are separate from each other.

**RESOURCE PARTITIONING** Temporal or spatial sharing of a resource by different species.

**RESPIRATION** Process used by organisms to exchange gases with the environment; the source of oxygen required for metabolism. The process organisms use to oxidize glucose to $CO_2$ and $H_2O$ using an electron transport system to extract energy from electrons and store it in the high-energy bonds of ATP.

**RESTING POTENTIAL** The electrical charge across the plasma membrane of a neuron when the cell is not carrying an impulse.

**RESTRICTION ENDONUCLEASE** A DNA-cutting enzyme found in bacteria.

**RETROVIRUSES** RNA viruses that reverse the typical flow of genetic information; within the infected cell, the viral DNA serves as a template for synthesis of a DNA copy. Examples include HIV, which causes AIDS, and certain cancer viruses.

**REVERSE TRANSCRIPTASE** An enzyme present in retroviruses that transcribes a strand of DNA, using viral RNA as the template.

**RHIZOIDS** In ferns and bryophytes, slender cells that anchor the gametophyte but do not absorb water and minerals.

**RHYNIOPHYTES** Ancient plants having vascular tissue which thrived in marshy areas during the Silurian period.

**RIBONUCLEIC ACID (RNA)** Single-stranded polynucleotide comprised of ribose (a sugar), phosphate, and one of four bases (adenine, guanine, cytosine, and uracil). The sequence of nucleotides in RNA is dictated by DNA, from which it is transcribed. There are three classes of RNA: mRNA, tRNA, and rRNA, all required for protein synthesis.

**RIBOSOMES** Organelles involved in protein synthesis in the cytoplasm of the cell.

**RICKETTSIAS** A group of obligate intracellular parasites, smaller than the typical prokaryote. They cause serious diseases such as typhus.

**RNA POLYMERASE** The enzyme that directs transcription and assembling RNA nucleotides in the growing chain.

**ROOT CAP** A protective cellular helmet at the tip of a root that surrounds delicate meristematic cells and shields them from abrasion and directs the growth downward.

**ROOT HAIRS** Elongated surface cells near the tip of each root for the absorption of water and minerals.

**ROOT NODULES** Knobby structures on the roots of certain plants. They house nitrogen-fixing bacteria which supply nitrogen in a form that can be used by the plant.

**ROUGH ER (RER)** Endoplasmatic reticulum with many ribosomes attached. As a result, they appear rough in electron micrographs.

**RUMINANT** Grazing mammals that possess an additional stomach chamber called rumen which is heavily fortified with cellulose-digesting microorganisms.

# S

**S STAGE** The second stage of interphase in which the materials needed for cell division are synthesized and an exact copy of cell's DNA is made by DNA replication.

**SAC BODY** The body plan of simple animals, like cnidarians, where there is a single opening leading to and from a digestion chamber.

**SAP** Fluid found in xylem or sieve of phloem.

**SAPHROPHYTE** Organisms, mainly fungi and bacteria, that get their nutrition by breaking down organic wastes and dead organisms, also called decomposers.

**SARCOLEMMA** The plasma membrane of a muscle fiber.

**SARCOMEREA** The contractile unit of a myofibril in skeletal muscle.

**SARCOPLASMIC RETICULUM (SR)** In skeletal muscle, modified version of the endoplasmic reticulum that stores calcium ions.

**SAVANNA** A grassland biome with alternating dry and rainy seasons. The grasses and scattered trees support large numbers of grazing animals.

**SCANNING ELECTRON MICROSCOPE (SEM)** A microscope which operates by showering electrons back and forth across the surface of a specimen prepared with a thin metal coating. The resultant image shows three-dimensional features of the specimen's surface.

**SCIENTIFIC METHOD** The step-by-step approach to scientific inquiry; begins with observations, proceeds to formulation of an hypothesis, experimentation, and eventually of a theory.

**SCLEREIDS** Irregularly-shaped sclerenchyma cells, all having thick cell walls; a component of seed coats and nuts.

**SCLERENCHYMA** Component of the ground tissue system of plants. They are thick walled cells of various shapes and sizes, providing support or protection. They continue to function after the cell dies.

**SCHIZOPHYTA** One of the two divisions of monerans. The common name for the division is bacteria.

**SCHWANN CELLS** The cells that are wrapped around the axon of neurons, forming the myelin sheath.

**SCROTUM** Sac containing the testes of the male animal.

**SCUTELLUM** The single cotyledon of a monocot embryo. It absorbs nutrients from the surrounding endosperm.

**SECONDARY CELL WALL** An additional cell wall that improves the strength and resiliency of specialized plant cells, particularly those cells found in stems that support leaves, flowers, and fruit.

**SECONDARY IMMUNE RESPONSE** The rapid release

of high levels of antibody when the individual is challenged with an antigen that was previously encountered.

**SECONDARY SEX CHARACTERISTICS**   Physical attributes that promote sexual recognition and arousal of reproductive behavior in a potential mate of the same species, but of opposite sex.

**SECONDARY GROWTH**   Growth from cambia in perennials; results in an increase in the diameter of stems and roots.

**SECONDARY MERISTEMS (vascular cambium, cork cambrium)**   Rings or clusters of meristematic cells that increase the width of stems and roots when they divide.

**SECRETION**   the process of exporting materials produced by the cell.

**SEED**   A mature ovule consisting of the embryo, endosperm, and seed coat.

**SELECTIVELY PERMEABLE**   A term applied to the plasma membrane because membrane proteins control which molecules are transported. Enables a cell to import and accumulate the molecules essential for normal metabolism.

**SEMEN**   The fluid discharged during a male orgasm.

**SEMICONSEVATIVE REPLICATION**   The manner in which DNA replicates; half of the original DNA strand is conserved in each new double helix.

**SEMINAL VESICLES**   The organs which produce most of the ejaculatory fluid.

**SEMINIFEROUS TUBULES**   Within the testes, highly coiled and compacted tubules, lined with a self-perpetuating layer of spermatogonia, which develop into sperm.

**SENESCENCE**   Aging and eventual death of an organism, organ or tissue.

**SENSORY NEURONS**   Neurons which detect changes in the external and internal environments and relay impulses to the central nervous system.

**SETAE**   Short bristles from the sides of the bodies of all segmented worms, except leeches, that aid in locomotion.

**SEX CHROMOSOMES**   The one chromosomal pair that is not identical in the karyotypes of males and females of the same animal species.

**SEX HORMONES**   Steroid hormones which influence the production of gametes and the development of male or female sex characteristics.

**SEX-INFLUENCED DOMINANCE**   An inheritable trait that is expressed as a dominant in one gender and as a recessive in the other, as a result of sex hormone differences, e.g., baldness in humans.

**SEX-LIMITED GENE EXPRESSION**   An inheritable trait that is expressed in only one sex, as a result of sex hormone differences, e.g., beards in men.

**SEX LINKED TRAITS**   Traits controlled by genes located on the X-chromosome.

**SEXUAL DIMORPHISM**   Differences in the appearance of males and females in the same species.

**SEXUAL REPRODUCTION**   The process by which haploid gametes are formed and fuse during fertilization to form a zygote.

**SEXUAL SELECTION**   The natural selection of adaptations that improve the chances for mating and reproducing.

**SHOOT**   In angiosperms, the system consisting of stems, leaves, flowers and fruits.

**SHORT-DAY PLANTS**   Plants that flower in late summer or fall when the length of daylight becomes shorter than some critical period.

**SIEVE ELEMENTS**   Phloem-conducting cells that have clusters of pores in their cell walls that resemble a sieve.

**SIGMOID GROWTH CURVE**   An S-shaped curve illustrating the lag phase, exponential growth, and eventual approach of a population to its carrying capacity.

**SIMPLE EPITHELIUM**   Single layer of cells that acts primarily as a filter.

**SIMPLE FRUITS**   Fruits that develop from the ovary of one pistil.

**SINGLE CELL PROTEIN (SCP)**   A product of easy-to-grow microorganisms such as yeasts and algae, providing a most efficient means of protein production.

**SINOATRIAL (SA) NODE**   A collection of cells that generates an action potential regulating heart beat; the heart's pacemaker.

**SKELETAL MUSCLES**   Separate bundles of parallel, striated muscle fibers anchored to the bone, which they can move in a coordinated fashion. They are under voluntary control.

**SKELETON**   A rigid form of support found in most animals either surrounding the body with a protective encasement or providing a living girder system within the animal.

**SKULL**   A vault of unyielding bones which surrounds and protects the neural tissue of the brain.

**SMALL INTESTINE**   Portion of the intestine in which most of the digestion and absorption of nutrients takes place. It is so named because of its narrow diameter. There are three sections: duodenum, jejunum, and ilium.

**SMOOTH ER (SER)**   Membranes of the endoplasmic reticulum that have no ribosomes on their surface. SER is generally more tubular than the RER.

**SMOOTH MUSCLE**   The muscles of the internal organs (digestive tract, glands, etc.). Composed of uninneleats, spindle-shaped cells that interlace to form sheets of visceral muscle.

**SOLUTE**   A substance dissolved in a solvent.

**SOLUTION**   The resulting mixture of a solvent and a solute.

**SOLVENT**   A substance in which another material dissolves by separating into individual molecules or ions.

**SOMATIC CELLS**   All cells that make up the body of an animal or plant, except germinal cells.

**SOMATIC NERVOUS SYSTEM**   The nerves that carry messages to the muscles that move the skeleton either voluntarily or by reflex.

**SOMITES**   In the vertebrate embryo, blocks of mesodermon either side of the netochord that give rise to the musculature of the animal.

**SPECIATION**   Divergence of gene pools in a population.

**SPECIES** Taxonomic subdivisions of a genus. Each species has recognizable features that distinguish it from every other species. Members of one species generally will not interbreed with members of other species.

**SPECIFIC EPITHET** In taxonomy, the second term in an organism's scientific name identifying its species within a particular genus.

**SPERMATOGENESIS** The production of sperm.

**SPERMATOZOA** Male gametes.

**SPHINCTORS** Circularly arranged muscles that close off the various tubes in the body.

**SPINAL CORD** A centralized mass of neurons for processing neurological messages and linking the brain with that part of peripheral nervous system not reached by the cranial nerves.

**SPINDLE APPARATUS** In dividing eukaryotic cells, the complex rigging, made of microtubules, that aligns and separates duplicated chromosomes.

**SPINDLE FIBERS** Groups of microtubules that extend from pole to pole during cell division. They move the chromosomes poleward during anaphase.

**SPONGOCOEL** The internal cavity of the simplest sponges with a single opening at the top—the osculum.

**SPORANGIOSPORES** Black, asexual spores of the zygomycete fungi.

**SPORANGIUM** A hollow structure in which spores are formed.

**SPORES** In plants, haploid cells that develop into the gametophyte generation. In fungi, an asexual reproductive cell that gives rise to a new mycelium.

**SPOROPHYTE** The diploid spore producing generation in plants.

**STABILIZING SELECTION** Natural selection favoring an intermediate phenotype over the extremes.

**STARCH** Polysaccharides used by plants to store energy.

**STAMEN** The flower's male reproductive organ, consisting of the pollen-producing anther supported by a slender stalk, the filament.

**STEM** In plants, the organ that supports the leaves, flowers, and fruits.

**STEROIDS** Compounds classified as lipids which have the basic four-ringed molecular skeleton as represented by cholesterol. Two examples of steroid hormones are the sex hormones; testosterone in males and estrogen in females.

**STIGMA** The sticky area at the top of each pistil to which pollen adheres.

**STOMACH** A muscular sac that is part of the digestive system where food received from the esophagus is stored and mixed, some breakdown of food occurs, and the chemical degradation of nutrients begins.

**STOMATES (Pl. STOMATA)** Microscopic pores in the epidermis of the leaves and stems which allow gases to be exchanged between the plant and the external environment.

**STRATIFIED EPITHELIA** Multicellular layers of protective epithelium.

**STRETCH RECEPTORS** Sensory receptors embedded in muscle tissue enabling muscles to respond reflexively when stretched.

**STRIATED** Referring to the striped appearance of skeletal and cardiac muscle fibers.

**STROBILUS** In lycopids, terminal, cone-like clusters of specialized leaves that produce sporangia.

**STROMA** The fluid interior of plastids.

**STRUCTURAL GENES** DNA segments that direct the formation of enzymes or structural proteins.

**STYLE** The portion of a pistil which joins the stigma to the ovary.

**SUBSTRATES** The reactants which bind to enzymes and are subsequently converted to products.

**SUCCESSION** The orderly progression of communities leading to a climax community. It is one of two types: primary, which occurs in areas where no community existed before; and secondary, which occurs in disturbed habitats where some soil and perhaps some organisms remain after the disturbance.

**SUCCULENTS** Plants having fleshy, water-storing stems or leaves.

**SURFACE AREA-TO-VOLUME RATIO** The ratio of the surface area of the plasma membrane to the volume of the cell, which determines the rate of exchange of materials between the cell and its environment.

**SURFACE TENSION** The resistance of a liquid's surface to being disrupted. In aqueous solutions, it is caused by the attraction between water molecules.

**SURVIVORSHIP CURVE** Graph of life expectancy, plotted as the number of survivors versus age.

**SYMBIOSIS** A close, long-term relationship between two individuals of different species.

**SYMPATRIC SPECIATION** Speciation that occurs in populations with overlapping distributions. It is common in plants when polyploidy arises within a population.

**SYNAPSIS** The pairing of homologous chromosomes during prophase of meiosis I.

**SYNAPSE** Juncture of a neuron and its target cell (another neuron, muscle fiber, gland cell).

**SYNAPTIC CLEFT** Small space between the synaptic knobs of a neuron and its target cell.

**SYNAPTIC KNOBS** The swellings that branch from the end of the axon. They deliver the neurological impulse to the target cell.

**SYNOVIAL CAVITIES** Fluid-filled sacs around joints, the function of which is to lubricate and separate articulating bone surfaces.

**SYNTHESIS REACTION (DARK REACTIONS)** The second stage of photosynthesis in which $CO_2$ is added to an inorganic molecule, which is subsequently reduced to carbohydrate. Both ATP and NADPH are required for the process.

**SYSTEMIC CIRCULATION** Part of the circulatory system that delivers oxygenated blood to the tissues and routes deoxygenated blood back to the heart.

# T

**TAP ROOT SYSTEM**   Root system of plants having one main root and many smaller lateral roots. Typical of conifers and dicots.

**TAXONOMY**   The science of classifying and grouping organisms based on their morphology and evolution.

**T-CELL**   Lymphocytes that are part of the immune system. They respond to antigen stimulation by becoming lymphokin-producing cells, killer cells, and memory cells.

**TELEOSTS**   Modern bony fish possessing a swim bladder.

**TELOPHASE**   The final stage of mitosis which begins when the chromosomes reach their spindle poles and ends when cytokinesis is completed and two daughter cells are produced.

**TERMINAL ELECTRON ACCEPTOR**   In aerobic respiration, the molecule of O which removes the electron pair from the final cytochrome of the respiratory chain.

**TERRESTRIAL**   Living on land.

**TERRITORIALITY**   An animal's defense of an area against other members of the same or different species.

**TESTCROSS**   An experimental procedure in which an individual exhibiting a dominant trait is crossed to a homozygous recessive to determine whether the first individual is homozygous or heterozygous.

**TESTIS**   In animals, the sperm-producing gonad of the male.

**TESTOSTERONE**   The male sex hormone secreted by the testes when stimulated by pituitary gonadotropins.

**TETRAD**   A unit of four chromatids formed by a synapsed pair of homologous chromosomes, each of which has two chromatids.

**THALLUS**   The flat, ground-hugging body of liverworts. It lacks roots, stems, leaves, and vascular tissues.

**THEORY**   A well accepted principle explaining a natural phenomenon, verified by observation and reported.

**THERMODYNAMICS**   The branch of physics that deals with the relationship of heat and mechanical energy and the conversion of one into the other.

**THERMOREGULATION**   The process of maintaining a constant internal body temperature in spite of fluctuations in external temperatures.

**THIGMOTROPISM**   Changes in plant growth stimulated by contact with another object, e.g., vines climbing on cement walls.

**THORACIC CAVITY**   The anterial portion of the body cavity in which the lungs are suspended.

**THYLAKOIDS**   Flattened membrane sacs within the chloroplast. Embedded in these membranes are the light-capturing pigments and other components that carry out the light reactions of photosynthesis.

**THYMUS**   An endocrine gland located in the chest.

**THYROID GLAND**   A butterfly-shaped gland that lies just in front of the human windpipe, producing two metabolism-regulating hormones, thyroxin and triodothyronine.

**THYROID STIMULATING HORMONE (TSH)**   An anterior pituitary hormone which stimulates secretion by the thyroid gland.

**TISSUE**   A group of cells that carry out a specialized function.

**TOLERANCE RANGE**   The range between the maximum and minimum limits for an environmental factor that is necessary for an organism's survival.

**TONOPLAST**   The membrane surrounding the large vacuole in a plant cell. It controls which materials are exchanged between the cytoplasm and the vacuole interior.

**TOTIPOTENT**   The genetic potential for one type of cell from a multicellular organism to give rise to any of the organism's cell types, even to generate a whole new organism.

**TRACHEID**   A type of conducting cell found in xylem functioning when a cell is dead to transport water and dissolved minerals through its hollow interior.

**TRANSCRIPTION**   The process by which a strand of RNA assembles along one of the DNA strands.

**TRANSDUCTION**   A type of genetic recombination resulting from transfer of genes from one organism to another by a virus.

**TRANSFER RNA (tRNA)**   A type of RNA that decodes mRNA's codon message and translates it into amino acids.

**TRANSFORMATION**   A genetic transfer mechanism that produces new DNA in bacteria when DNA from a new organism is combined with the DNA of the host cell.

**TRANSLATION**   The cell process that converts a sequence of nucleotides in mRNA into a sequence of amino acids.

**TRANSLOCATION**   The joining of segments of two non-homologous chromosomes.

**TRANSMISSION ELECTRON MICROSCOPE (TEM)**   A microscope that works by shooting electrons through very thinly sliced specimens. The result is an enormously magnified image, two-dimensional, of remarkable detail.

**TRANSPIRATION**   Water vapor loss from plant surfaces.

**TRANSPIRATION PULL**   The principle means of water and mineral transport in plants, initiated by transpiration.

**TRANSPOSABLE ELEMENTS**   Nomadic segments of DNA that are moved to new chromosome locations; "jumping genes".

**TRANSVERSE FISSION**   The division pattern in ciliated protozoans where the plane of division is perpendicular to the cell's length.

**TRANSVERSE TUBULES (T-TUBULES)**   Tubular branchings of the sarcolemma that reach into the core of the cell and carry the action potential to myofibrils in the interior of the muscle fiber.

**TRIPLOID**   Having three sets of chromosomes, abbreviated 3n.

**TRIMESTER**   Each of the three stages comprising the 266-day period between conception and birth in humans.

**TROPHIC LEVEL**   Each step along a feeding pathway.

**TROPISMS** Changes in the direction of plant growth in response to environmental stimuli, e.g., light, gravity, touch.

**TRUE-BREEDER** Organisms that, when bred with themselves, always produce offspring identical to the parent for a given trait.

**TUMOR** A mass of uncontrolled cells within an otherwise normal tissue.

**TURGOR PRESSURE** The internal pressure in a plant cell caused by the diffusion of water into the cell. Because of the rigid cell wall, pressure can increase to where it eventually stops the influx of more water.

**TUNDRA** The marshy, unforested biome in the arctic and at high elevations. Frigid temperatures for most of the year prevent the subsoil from thawing, which produces marshes and ponds. Dominant vegetation includes low growing plants, lichens, and mosses.

# U

**UNICELLULAR** The description of an organism where the cell is the organism.

**URETHRA** A tube that extends from the uniary bladder through the length of the penis, conveying both sperm and urine, though not simultaneously.

**URINARY TRACT** The structures that form and export urine: kidneys, ureters, urinary bladder, and urethra.

**URINE** The excretory fluid consisting of urea, other nitrogenous substances, and salts dissolved in water. It is formed by the kidneys.

**UROCHORDATA** A subphylum of chordata, containing the tunicates, marine animals with an outerwall, or tunic, covering the adult body.

**UTERINE (MENSTRUAL) CYCLE** the repetitive monthly changes in the uterus that prepare the endometrium for receiving and sustaining an embryo.

**UTERUS** An organ in the female reproductive system in which an embryo implants and is maintained during development.

**VACCINES** Modified forms of disease-causing microbes which cannot cause disease but retain the same antigens of it. They permit the immune system to build memory cells without diseases developing during the primary immune response.

**VACUOLE** A large organelle found in mature plant cells, occupying most of the cell's volume, sometimes more than 90% of it.

**VAGINA** The female mammal's copulatory organ and birth canal.

**VASCULAR BUNDLES** Groups of vascular tissues (xylem and phloem) in the shoot of a plant.

**VASCULAR CAMBRIUM** In perennials, a secondary meristem that produces new vascular tissues.

**VASCULAR CYLINDER** Groups of vascular tissues in the central region of the root.

**VASCULAR PLANTS** Plants having a specialized conducting system of vessels and tubes for transporting water, minerals, food, etc., from one region to another.

**VAS DEFERENS** The tube that conducts sperm from the epididymis toward the urethra.

**VASOCONSTRICTION** Reduction in the diameter of blood vessels.

**VEINS** In plants, vascular bundles in leaves. In animals, blood vessels that return blood to the heart.

**VENA CAVA** Each of the two large veins which return deoxygenated blood from body tissues to the heart.

**VENTRICLE** Lower chamber of the heart which pumps blood through arteries. There is one ventricle in the heart of lower vertebrates and two ventricles in the four-chambered heart of birds and mammals.

**VENULES** Small veins that collect blood from the capillaries. They empty into larger veins for return to the heart.

**VERNALIZATION** The promotion of flowering or germination after a plant is exposed to low temperatures.

**VERTEBRAE** The bones that form the backbone. In the human there are 33 bones arranged in a gracefully curved line along the bone, cushioned from one another by disks of cartilage.

**VERTEBRAL COLUMN** the backbone, which encases and protects the spinal cord.

**VERTEBRATES** Animals with a backbone.

**VESICLES** Small membrane-enclosed sacs which form from cisternae of the ER and Golgi Complex. Some store chemicals in the cells; others move to the surface and fuse with the plasma membrane to secrete their contents to the outside.

**VESSEL** A tube or connecting duct containing or circulating body fluids.

**VESSEL MEMBER** A type of conducting cell in xylem functioning when the cell is dead to transport water and dissolved minerals through its hollow interior; also called a vessel element.

**VESTIGIAL STRUCTURE** Remains of ancestral structures or organs which were, at one time, useful.

**VILLI** Finger-like projections of the intestinal wall that increase the absorption surface of the intestine.

**VIRUS** Minute structures composed of only heredity information (DNA or RNA), surrounded by a protein or protein/lipid coat. After infection, the viral mucleic acid subverts the metabolism of the host cell, which then manufactures new virus particles.

**VISIBLE LIGHT** The portion of the electromagnetic spectrum producing radiation from 380 nm to 750 nm detectable by the human eye.

**VITAMINS** Any of a group of organic compounds essential in small quantities for normal metabolism.

**VIVIPAROUS** Referring to an animal whose offspring are born rather than hatched through an egg.

**VULVA** The collective name for the external features of the female's genitals.

# W

**WATER VASCULAR SYSTEM**   A system for locomotion, respiration, etc., unique to echinoderms.

**WAVELENGTH**   The distance separating successive crests of a wave.

**WHITE MATTER**   Regions of the brain and spinal cord containing myelinated axons, which confer the white color.

**WILTING**   Drooping of stems or leaves of a plant caused by water loss.

**WOOD**   Secondary xylem.

# X

**XEROPHYTE**   A plant adapted to arid habitats.

**X CHROMOSOME**   The sex chromosome present in two doses in cells of a female, and in one dose in the cells of a male.

**X-RAY**   High-frequency electromagnetic radiation.

**XYLEM**   The vascular tissue that transports water and minerals from the roots to the rest of the plant. Composed of tracheids and vessel members.

# Y

**Y CHROMOSOME**   The sex chromosome found in the cells of a male. When Y-carrying sperm unite with an egg, all of which carry a single X chromosome, a male is produced.

**Y-LINKED INHERITANCE**   Genes carried only on Y chromosomes. There are relatively few Y-linked traits; maleness being the most important such trait in mammals.

**YEASTS**   Unicellular fungi that reproduce by budding. They also form colonies similar to those of bacteria when growing on the surface of solid nutrient medium.

**YOLK**   A deposit of lipids, proteins, and other nutrients that nourishes a developing embryo.

**YOLK SAC**   A sac formed by an extraembryoid membrane. In humans, it manufactures blood cells for the early embryo and later helps to form the umbilical cord.

# Z

**Z-DNA**   A form of DNA in which the spirals coil to the left, as distinct from classical B-DNA, in which they coiled to the right.

**Z LINE**   The darkest line of the striations in the myofibrils of a muscle; a membrane disk that separates adjacent sacromeres.

**Z-SCHEME**   The representation of the complex electron course of noncyclic photophosphorylation.

**ZERO POPULATION GROWTH**   In a population, the result when the combined positive growth factors (births and immigration).

**ZOOPLANKTON**   Protozoa, small crustaceans and other tiny animals that drift with ocean currents and feed on phytoplankton.

**ZYGOSPORE**   The diploid spores of the zygomycete fungi, which include Rhizopus, a common bread mold. After a period of dormancy, the zygospore undergoes meiosis and germinates.

**ZYGOTE**   A fertilized egg. The diploid cell that results from the union of a sperm and egg.

# Index

# *Photo Credits*

**Section I Opener:** J. Carmichael/The Image Bank.

**Chapter 1**

Figures 1.1 and 1.2*a*: Frans Lanting. Figure 1.2*b*: The Image Bank. Figure 1.2*c*: Michel Viard/Peter Arnold. Figure 1.3*a*: Tim Rock/ Animals, Animals. Figure 1.3*b*: Hans Pfletschinger/Peter Arnold. Figure 1.3*c*: Charles G. Summers, Jr./Colorado Nature Photographic. Figure 1.3*d* © 1988 Bianca Lavies. Figure 1.3*e*: Steve Allen/Peter Arnold. Figure 1.3*f*: Georg Gerster/Photo Researchers. Figure 1.3*g*: Akira Uchiyama/ Photo Researchers. Figure 1.5*a*: (top left) Leonard Lee Rue III/Earth Scenes/Animals, Animals; (top right) Bradley Smith/Earth Scenes/Animals, Animals. Figure 1.5*b*: (bottom left) R. Andrew Odum/Peter Arnold; (bottom right) Michael Fogden/Animals, Animals. Figure 1.8*a*: CNRI/Science Photo Library/Photo Researchers. Figure 1.8*b*: Eric Gravé/Photo Researchers. Figure 1.8*c*: John Stern/Earth Scenes/Animals, Animals. Figure 1.8*d*: R.H. Armstrong/Earth Scenes/Animals, Animals. Figure 1.8*e*: Zig Leszczynski/ Animals, Animals. Figure 1.9*a*: Geoff Tompkinson/Aspect Picture Library, Ltd. Figure 1.9*b*: Runk/Schoenberger/Grant Heilman.

**Chapter 2**

Figure 2.3: Buzz Soard/Concept West. *Discover*–Page 27: (left) Time McCabe/ Taurus Photos; (center) Laurence Gould/Earth Scenes/Animals, Animals; (right) Peter Veit/ DRK Photo. Figure 2.5: John Macgregor/Peter Arnold.

**Section II Opener:** Michel Tcherevkoff/The Image Bank.

**Chapter 3**

Figure 3.1: Four By Five. *Bioline*–Page 37: J. Y. Kerban/The Image Bank. Figure 3.7: Marty Stouffer/Animals, Animals. Figure 3.9: Alex Kerstitch. Connection–Page 45 Figure 1.*b*: Courtesy Computer Graphics Laboratory, Regents of The University of California, San Francisco. Figure 3.10: Hermann Eisenbeiss/ Photo Researchers.

**Chapter 4**

Figure 4.3: Zig Leszczynski/Animals, Animals. Figure 4.8*a*: Jim Frazier/Mantis Wildlife Films/Animals, Animals. Figure

4.10: Hemoglobin illustration copyright by Irving Geis.

**Chapter 5**

Figure 5.1*a*: A. Kerstitch/Taurus Photos. Figure 5. l*b*: S.R. Morris/Oxford Scientific Films/Animals, Animals. Figure 5.3*b*: Courtesy Stan Koszelek, Ph.D., University of California, Riverside.

**Section III Opener:** Reproduced from the Journal of Cell Biology, 1987, Volume 104, No. 1, by copyright permission of The Rockefeller University Press. Photo courtesy Gary J. Gorbsky, Ph.D., University of Virginia.

**Chapter 6**

Figure 6.1: (left) CNRI/Science Photo Library/Photo Researchers; (right) Oxford Scientific Films/Animals, Animals. *Bioline*–Page 79: (Figure *d*) Courtesy Hale L. Wedberg. *Bioline*–Page 79: (Figure *c*) Courtesy Naval Research Laboratory Photo. Figure 6.5*b*: Courtesy Dr. R. Malcolm Brown, Jr., Department of Botany, University of Texas at Austin. Figure 6.5*d*: Courtesy R. D. Preston, University of Leeds, England. Figure 6.6: Courtesy C.F. Kettering Research Laboratory. Figure 6.8: Courtesy Richard Chao, California State University, Northridge. Figure 6.9: Lennart Nilsson. From "The Incredible Machine" © 1987 National Geographic Society. Figure 6.10: Courtesy D. Branton, 1966. Proc. Nat. Acad. Sci. U.S. 55: 1048. Figure 6.11*a*: Courtesy D.W. Fawcett, Harvard Medical School. Figure 6.11*b*: D.W. Fawcett/Photo Researchers. Figure 6.12*a*: K.R. Porter/Photo Researchers. Figure 6.13*a*: Courtesy G. E. Palade. Figures 6.14*a* and 6.16: Courtesy USDA. Figure 6.17: From S.E. Frederick & E.H. Newcomb, *Journal of Cellular Biology*, 43, 343 (1969), by permission of The Rockefeller University Press, NY. Figure 6.18*a*: Grant Heilman. Figure 6.20*a*: Gary W. Grimes/Taurus Photos. Figure 6.20*b*: David Phillips/Taurus Photos.

**Chapter 7**

Figure 7.6: Ed Reschke. Figure 7.9: Courtesy Dr. Birgit H. Stair, Albert Einstein College of Medicine.

**Chapter 9**

Figure 9.1*a*: Robert Archibald/Earth Scenes/ Animals, Animals. Figure 9.1*b*: Time

McCabe/Taurus Photos. Figure 9.5: Courtesy T. Elliot Weier. *Bioline*–Page 133: (left) J. Frederick Grassle/Woods Hole Oceanographic Institution; (right) Jack Donnelly/Woods Hole Oceanographic Institution.

**Chapter 10**

Figure 10.8*a:* Kjell B. Sandved/Photo Researchers. Figure 10.8*b*: Dick Rowan/Photo Researchers.

**Chapter 11**

Figure 11.1: CNRI/Science Photo Library/ Photo Researchers. Figure 11.3*a*: Dr. R. Vernon/Phototake. Figure 11.3*b*: Omikron/ Photo Researchers. Figure 11.3*c*: Biophoto Associates/Science Source/Photo Researchers. Figure 11.4: Dr. G. Schatten/Science Photo Library/Photo Researchers. Figure 11.5: Courtesy Ripon Microslides Laboratory. Figure 11.7: Dr. Andrew S. Bajer, University of Oregon. Figure 11.8: Courtesy Professor Richard G. Kessel. Figure 11.9*b*: David M. Phillips/ Taurus Photos.

**Chapter 12**

Figure 12.3: A. M. Winchester, Ph.D. Figure 12.4: Courtesy Jaroslav Cervenka, M.D., University of Minnesota.

**Section IV Opener:** Don Landwehrle/The Image Bank.

**Chapter 13**

Figure 13.3: R. F. Evert, University of Wisconsin, Madison. *Bioline*–Page 183: (Figure *a*) Michel Viard/Peter Arnold. *Bioline*–Page 183: (Figure *b*) Gil Brum. Figure 13.4: (left) Carolina Biological Supply Company; (right) Richard Burda/Taurus Photos. Figure 13.6*a*: DSIR Library Centre. Figure 13.6*b*: S. Rannels/Grant Heilman. Figure 13.7: Dr. Jeremy Burgess/Science Photo Library/Photo Researchers. Figure 13.8*a*: DSIR Library Centre. Figure 13.8*b*: David Scharf/Peter Arnold. Figure 13.8*c*: Jon M. Holy from *Botany Principles and Applications* by Roy H. Saigo and Barbara Woodworth Saigo, p. 131. © 1983 by Prentice-Hall, Inc. Used by permission. Figure 13.9: Ken W. Davis/Tom Stack & Associates. Figure 13.10: (left) Runk/Schoenberger/Grant Heilman; (right) From *Botany Principles and Applications* by Ron H. Saigo and Barbara Woodworth Saigo, p. 131. © 1983 by Prentice-Hall, Inc. Used by permission. Figure

13.11*a*: R. F. Evert, University of Wisconsin, Madison. Figure 13.11*b*: Ed Reschke. Figure 13.12: Courtesy Thomas A. Kuster, Forest Products Laboratory/USDA. Figure 13.14: Carolina Biological Supply Company. Figure 13.15: DSIR Library Centre. Figure 13.16: R. F. Evert, University of Wisconsin, Madison. Figure 13.17: Runk/Schoenberger/Grant Heilman. *Bioline*–Page 193: Spencer Swanger/Tom Stack & Associates. Figure 13.22: R. F. Evert, University of Wisconsin, Madison. Figure 13.24: Biophoto Associates/Science Source/Photo Researchers.

**Chapter 14**

Figure 14.1*a*: Courtesy C.P.P. Redi, Department of Forestry, University of Florida. Figure 14.1*b*: Dr. J. Burgess/Photo Researchers. Figure 14.3: Courtesy Frank Salisbury, Plant Science Department, Utah State University. Figure 14.6*a*: Biophoto Associates/Photo Researchers. Figure 14.6*b*: A. Pasicka/Taurus Photos. Figure 14.6*c*: DSIR Library Centre. *Discovery*–Page 211: (Figures *a* and *b*) Martin Zimmerman.

**Section V Opener:** Stephen Dalton/Oxford Scientific Films/Animals, Animals.

**Chapter 15**

Figure 15.1: Doug Wedsler/Animals, Animals. Figures 15.6 and 15.10: Lennart Nilsson, *Behold Man*, Little Brown & Company, Boston. Figure 15.15: Courtesy Robert B. Livingston, M.D., Laboratory for Quantitative Morphology, University of California, San Diego. Figure 15.19: Dr. Tuttle/Photo Researchers. Figure 15.20*b*: Kjell B. Sandved. Figure 15.21: G. Matthew Brady/Tom Stack & Associates. Figure 15.23*a*: Oxford Scientific Films/Animals, Animals. Figure 15.23*b*: Kjell B. Sandved/Photo Researchers. Figure 15.23*c*: George Shelley, (inset) Raymond A. Mendez/Animals, Animals. Figure 15.24*b*: Anthony Bannister/NHPA.

**Chapter 16**

Figure 16.1*a*: Danny Brass/Photo Researchers. Figure 16.1*b*: (left) Fred McConnaughey/Phototake; (right) Al Grotell.

**Chapter 17**

Figure 17.1: Wallin/Taurus Photos. Figure 17.2*a*: E. S. Ross/Phototake. Figures 17.3*a* and 17.3*b*: Al Grotell. Figure 17.3*c*: Mia & Klaus/Four by Five. Figure 17.3*d*: Christopher Newbert/Four by Five. Figure 17.5: Lennart Nilsson, *Behold Man*, Little Brown & Company, Boston. Figure 17.7 insets: (top) Lennart Nilsson, *Behold Man*, Little Brown & Company, Boston; (bottom) Michael Abbey/Science Source/Photo Researchers. Figure 17.10: (inset) A. M. Siegelman/FPG International. Figure 17.11: Franzini Armstrong/Photo Researchers.

**Chapter 18**

Figure 18.1*a*: Courtesy Gregory Antipa, San Francisco State University. From *Journal of Protozoology*, (1970) 17:250. Used by permission. Figure 18.1*b*: Runk/Schoenberger/Grant Heilman. Figure 18.2*a*: David Hall/Photo Researchers. Figure 18.2*b*: Brian Parker/Tom Stack & Associates. *Bioline*–Page 285 and Figure 18.5: Lennart Nilsson, *Behold Man*, Little Brown & Company, Boston. Figure 18.6: (inset) Lennart Nilsson, *The Incredible Machine*. © 1987 National Geographic Society. Figure 18.10: (inset) © Richard Kessel and Gene Shih.

**Chapter 19**

Figure 19.1: David Doubilet. Figure 19.3: Robert F. Sisson © 1972 National Geographic Society. Figure 19.4: CNRI/Science Photo Library/Photo Researchers. Figure 19.6: Lennart Nilsson, *Behold Man*, Little Brown & Company, Boston.

**Chapter 20**

Figure 20.1*a*: Runk/Schoenberger/Grant Heilman. Figure 20.2: (inset) Biophoto Associates/Photo Researchers. Figure 20.4: Lennart Nilsson, *Behold Man*, Little Brown & Company, Boston. Figure 20.9: CNRI/Science Photo Library/Photo Researchers. Figure 20.10: (inset) Lennart Nilsson, *Behold Man*, Little Brown & Company, Boston. Figure 20.14: Dr. Tony Brain/Science Photo Library/Photo Researchers. Figure 20.15: (inset) Manfred Kage/Peter Arnold. Figure 20.17: Lennart Nilsson, Boehringer Ingelheim International GmbH. Figure 20.18: (inset) Courtesy Dr. Arthur J. Olson, Scripps Clinic and Research Foundation. Figure 20.19: Francis Leroy, Biocosmos/Science Photo Library/Photo Researchers.

**Section VI Opener:** J. Carmichael/The Image Bank.

**Chapter 21**

Figure 21.1: G.I. Bernard/Oxford Scientific Films/Earth Scenes/Animals, Animals. Figures 21.3: Dr. E. R. Degginger. Figure 21.4*a*: Gil Brum. Figure 21.4*b*: Phil Degginger. Figure 21.4*c*: Dr. E. R. Degginger. Figure 21.5*a*: Runk/Schoenberger/Grant Heilman. Figure 21.5*b*: Stephen J. Krasemann/Valan Photos. Figure 21.6*a*: Robert Harding Picture Library. Figure 21.6*b*: Richard Parker/Photo Researchers. Figure 21.6*c*: Jack Wilburn/Earth Scenes/Animals, Animals. Figure 21.9*a*: Jonathan D. Eisenback/Biological Photo Service. Figure 21.9*b*: Runk/Schoenberger/Grant Heilman. *Bioline*–Page 351: (Case 1) Biological Photo Service; (Case 2) P.H. & S. L. Ward/Natural Science Photos; (Case 3) Biological Photo Service; (Case 4, top) G. I. Bernard/Animals, Animals; (Case 4, bottom) Oxford Scientific Films/Animals, Animals. Figure 21.12: E. R. Degginger/Bruce Coleman. Figure 21.15: Courtesy Media Resources, California State Polytechnic University. Figure 21.16*a*: John Fowler/Valan Photos. Figure 21.16*b*: Richard Kolar/Earth Scenes/Animals, Animals. Figure 21.17: Russ Kinne/Comstock.

**Chapter 22**

Figure 22.1*a*: Gary Milburn/Tom Stack & Associates. Figure 22.1*b*: Kevin Schafer/Tom Stack & Associates. Figure 22.5: Courtesy Professor T. Brian, Chair of Botany at Cambridge. Figure 22.7: Courtesy Dr. Harlan K. Pratt, University of California, Davis. Figure 22.9*a*: Dr. E. R. Degginger. Figure 22.9*b*: Fritz Prenze/Earth Scenes/Animals, Animals. Figure 22.12*a*: Pam Hickman/Valan Photos. Figure 22.12*b*: R. F. Head/Earth Scenes/Animals, Animals.

**Chapter 23**

Figure 23.1*a*: R. La Salle/Valan Photos. Figure 23.1*b*: Oxford Scientific Films/Animals, Animals. Figure 23.1*c*: N. A. Callow/Robert Harding Picture Library. Figure 23.2: Charles Nickin/Westlight. Figure 23.3: Peter Ward/Bruce Coleman. Figure 23.4*a*: © 1988 Bianca Lavies. Figure 23.4*b*: Densey Clyne/Oxford Scientific Films Animals, Animals. Figure 23.4*c*: Biological Photo Service. *Animal Behavior*–Page 383: © 1988 Bianca Lavies. Figure 23.5: Courtesy Mia Tegner, Scripps Institution of Oceanography. Figure 23.7: (inset) © Richard Kessel and Randy Kardon. From *Tissues and Organs*, W. H. Freeman and Company. Used by permission.

**Chapter 24**

Figure 24.2*a*: © Richard Kessel and Gene Shih. Figure 24.9: (left) Oxford Scintific Films/Animals, Animals; (right) M.P.L. Fogden/Bruce Coleman. Figure 24.10: Natural Science Photos. Figure 24.12*a*–*f*: Lennart Nilsson/Bonnier Fakta.

**Section VII Opener:** Lennart Nilsson. From "The Incredible Machine" © 1987 The National Geographic Society.

**Chapter 25**

Figure 25.1: Erich Hartmann/Magnum Photos. *Discovery*–Page 431: Courtesy Bildarchive d. Ost National Biblioteque. Figure 25.8: Bruce Davidson/Animals, Animals.

**Chapter 26**

Figure 26.1: Biological Photo Service. Figure 26.2: Robert Noonan. Figure 26.6: Courtesy Richmond Products. Figure 26.8*a*: Courtesy Dr. Murray L. Barr. Figure 26.8*b*: Courtesy Jackson Laboratories.

Brown/Animals, Animals. Figure 34.27*b*: Christopher Newbert/Four by Five. Figure 34.28*a*: Zig Leszczynski/ Animals, Animals. Figure 34.28*b*: Chris Mattison/Natural Science Photos. Figure 34.29*a*: Michael Fogden/Animals, Animals. Figure 34.29*b*: Johnthan Blair/Woodfin Camp. Figure 34.30: Bob & Clara Calhoun/Bruce Coleman. Figure 34.32*a*: Ferrero/NHPA. Figure 34.32*b*: G. R. Roberts. Figure 34.32*c*: Nancy Adams/Tom Stack & Associates.

**Section X Opener:** Patti Murry/Earth Scenes/ Animals, Animals.

## Chapter 35

Figure 35.1: Courtesy NASA. Figure 35.3: Peter David/Seaphot Limited: Planet Earth Pictures. Figure 35.4: Monterey Bay Aquarium/Biological Photo Service. Figure 35.5*a* and 35.5*b*: Nicholas Devore III/Bruce Coleman. Figure 35.5*c*: Juergen Schmitt/The Image Bank. Figure 35.6*a*: Alexander Lowry/ Photo Researchers. Figure 35.6*b:* Tom McHugh/Photo Researchers. Figure 35.6*c*: Gregory G. Dimijian/Photo Researchers. Figure 35.6*d*: Fred Bavendam/Peter Arnold. Figure 35.8: Tony Morrison/South American Pictures. Figure 35.10*a:* K. Gunnar/Bruce Coleman. Figure 35.10*b*: Claude Rives/ Agence C.E.D.R.I. Figure 35.10*c*: Maurice Mauritius/Serraillier-Rapho. Figure 35.13: Robert W. Hernandez/Photo Researchers. Figure 35.1: James P. Jackson/Photo Researchers. Figure 35.15: Steve McCutcheon/Alaska Pictorial Service. Figure 35.16: Stephen J. Krasemann/Photo Researchers. Figure 35.17: Bill Bachman/Phot Researchers. Figure 35.18: Comstock. Figure 35.19: Gil Brum. Figure 35.20: B. G. Murray, Jr./Animals, Animals. Figure 35.21: Gil Brum.

## Chapter 36

Figure 36.1: (inset) Laura Riley/Bruce Coleman. Figure 36.1*a*: Courtesy Edward S. Ross, California Academy of Sciences. Figure 36.1*b*: Ulrike Welsch/Photo Researchers. Figure 36.1*c*: Michael Fogden. Figure 36.2: Calvin Larsen/Photo Researchers. Figure 36.8: Gil Brum. Bigure 36.13: Charlie Ott/Photo Researchers. Figure 36.14: (left) Krafft Explorer/Phot Researchers; (right) Pat & Tom Leeson/Photo Researchers.

## Chapter 37

Figure 37.1*a*: Treat Davidson/Photo Researchers. Figure 37.1*b*: Al Grotell, *Animal Behavior*–Page 679: (Figure *a*) Thomas A. Wiewandt, Ph.D.,; (Figure *b*) Liz & Tony Bomford/Ardea London, Ltd.; (Figure *c*) Courtesy Dr. Richard M. Laws, University of Cambridge. Figure 37.5*a*: Cosmos Blank/ Photo Researchers. Figure 37.5*b*: Karl & Steve Maslowski/Photo Researchers. Figure 37.5*c*: Steinhart Aquarium/Photo Researchers.

Figure 37.5*d*: Gregory G. Dimijian, M. D. / Photo Researchers. Figure 37.6: B. & C. Calhoun/Bruce Coleman. Figure 37.7: Kjell B. Sandved/Photo Researchers. Figure 37.8*a*: J. Carmichael/The Image Bank. Figure 37.8*b*: Zig Lesczynski/Animals, Animals. Figure 37.9: Anthony Bannister/NHPA. Figure 37.10*a*: Adrian Warren/Ardea London, Ltd. Figure 37.10*b*: Peter Ward/Bruce Coleman. Figure 37.10*c*: Michael Fogden. Figure 37.11: P.H. & S.L. Ward/Natural Science Photos. Figure 37.12: Tom McHugh/Photo Researchers. Figure 37.13: Dr. E. R. Degginger. Figure 37.14: William H. Amos. Figure 37.16: Dr. J.A.L. Cooke/Oxford Scientific Films/Animals, Animals. Figure 37.18: Peter J. Bryant/Biological Photo Service. Figure 37.19: Raymond A. Mendez/ Animals, Animals.

## Chapter 38

Figure 38.2: William Garnett. Figure 38.3*a*: Robert J. Ashworth/Phot Researchers. Figure 38.3*b:* M. P. Kahl/Photo Researchers. Figure 38.3*c*: C. G. Maxwell/Photo Researchers. Figure 38.5: Robert E. Pelham/Bruce Coleman.

**Section XI Opener:** Ted Spiegel/Black Star.

## Chapter 39

Figure 39.1*a*: Craig Aurness/Westlight. Figure 39.1*b*: Courtesy John T. Barber and Lawrence Y. Yatsu, Tulane University. Figure 39.1*c*: Erich Hartmann/Magnum Photos. Figure 39.2: R. Smith/FPG International. Figure 39.2: (inset) Jeremy Burgess/Science Photo Library/ Photo Researchers. Figure 39.3: Courtesy Dr. R. Malcolm Brown, Jr., Department of Bontany, University of Texas at Austin. *Connections*–Page 718: (Figure 1) Professor Stanley N. Cohen/Photo Researchers. Figure 39.5*a*: Kjell B. Sandved. Figure 39.5*b*: Courtesy Lakshmi Bhatnagar, Ph.D., Michigan Biotechnology Institute. Figure 39.6: John Shaw/Tom Stack & Associates.